CHEMISTRY OF HAZARDOUS MATERIALS

6th Edition

Eugene Meyer

PEARSON

Boston   Columbus   Indianapolis   New York   San Francisco   Upper Saddle River
Amsterdam   Cape Town   Dubai   London   Madrid   Milan   Munich   Paris   Montreal   Toronto
Delhi   Mexico City   São Paulo   Sydney   Hong Kong   Seoul   Singapore   Taipei   Tokyo

Publisher: Julie Levin Alexander
Publisher's Assistant: Regina Bruno
Editor-in-Chief: Marlene McHugh Pratt
Senior Acquisitions Editor: Stephen Smith
Program Manager: Monica Moosang
Development Editor: Eileen Clawson
Editorial Assistant: Mary Ellen Ruitenberg
Director of Marketing: David Gesell
Executive Marketing Manager: Brian Hoehl
Marketing Specialist: Michael Sirinides
Marketing Assistant: Crystal Gonzalez
Team Lead for Brady and Health Professions:
    Cynthia Zonneveld

Project Manager: Julie Boddorf
Full-Service Project Manager: Kelly Ricci, Aptara®, Inc.
Manufacturing Manager: Vincent Scelta
Manufacturing Buyer: Nancy Maneri
Editorial Media Manager: Amy Peltier
Media Project Manager: Ellen Martino
Creative Director: Jayne Conte
Cover Designer: Karen Noferi
Cover Image: Shutterstock/TFoxFoto
Composition: Aptara®, Inc.
Printer/Binder: R.R. Donnelley/Willard
Cover Printer: R. R. Donnelley

Credits and acknowledgments borrowed from other sources and reproduced, with permission, in this textbook appear on appropriate pages with text.

Library of Congress Cataloging-in-Publication Data

Meyer, Eugene
    Chemistry of hazardous materials / Eugene Meyer. — 6th ed.
        pages   cm.
    Includes index.
    ISBN-13: 978-0-13-314688-2
    1. Hazardous substances—Fires and fire prevention—Textbooks.
    2. Hazardous substances—Textbooks.   I. Title.
    TH9446.H38M48 2014
    628.9'2—dc22

                                                    2013012430

5 16

ISBN-13: 978-0-13-314688-2
ISBN-10:   0-13-314688-X

DEDICATION

It was with love and heartfelt appreciation that I dedicated the first five editions of this book to my wife of 47 years, Phyllis Annette Meyer. In January 2010, Phyllis passed away following a three-year-long battle with breast cancer and non-Hodgkin's lymphoma. Prior to her death, she encouraged me to write this edition of *Chemistry of Hazardous Materials* despite the colossal amount of work she knew it would entail. It was due to her special encouragement that this edition ultimately came to exist. It is therefore proper and fitting that I dedicate it posthumously to her memory.

CONTENTS

Chapter 2: Some Features of Matter and Energy   34

Chapter 5: Principles of Chemical Reactions 135

Chapter 7: Chemistry of Some Common Elements 221

Chapter 8: Chemistry of Some Corrosive Materials   270

Chapter 7: Chemistry of Some Common Elements   221

Chapter 10: Chemistry of Some Toxic Substances   345

Chapter 11: Chemistry of Some Oxidizers 421

Chapter 13: Chemistry of Some Hazardous Organic Compounds: Part II   533

Chapter 15: Chemistry of Some Explosives   645

Emergency responders use at least three means to rapidly identify the dangers, if any, associated with encountering chemical substances. The first is to acknowledge the numbers entered in the quadrants of hazard diamonds affixed to permanent structures. This procedure, devised by the National Fire Protection Agency (NFPA), has been used in the United States by professional firefighters for decades. In fact, most firefighters are familiar with the significance of the numbers posted in hazard diamonds before taking a course on hazardous materials.

Another way that emergency responders identify the dangers associated with chemical substances is to recognize the manner in which shippers and carriers affix labels, markings, and placards on shipping containers and transport vehicles. In the United States, these procedures initially were required by the U.S. Department of Transportation (DOT) to provide useful information to transportation personnel and emergency responders at the scenes of transportation mishaps. Today, a significant component of these procedures consists of similar, modified methods developed by the United Nations. The labels, markings, and placards provide emergency responders with chemical and health information about the commodities being shipped: Are they explosive, flammable, corrosive, radioactive, poisonous, or potentially dangerous in some other way?

A third means for identifying the dangers associated with chemical substances is to acknowledge the signal words, hazard and precautionary statements, and pictograms displayed on container labels. The Occupational Safety and Health Administration (OSHA), a division of the U.S. Department of Labor, adopted procedures devised by the United Nations that address the posting of hazard-warning messages on labels affixed to containers of hazardous chemicals by their manufacturers, distributors, and importers. These procedures are elements of the Globally Harmonized System (GHS) initially devised by the United Nations. Although they were intended primarily to advise employees to exercise caution when using these substances in their workplace, the messages also serve as a valuable resource when emergency responders encounter chemical substances at disaster scenes.

In this sixth edition, considerable emphasis has been placed on the use of all three systems of hazard identification. In particular, the NFPA hazard diamonds and GHS pictograms are components of the marginal art displayed throughout this book. In addition, Chapter 6 is devoted solely to the DOT labeling, marking, and placarding regulations that shippers and carriers use when transporting hazardous materials from place to place. The DOT shipping names of hazardous materials are provided in other chapters as the features of individual hazardous materials are noted, as well as in Appendix C.

Not surprisingly, the systems used to identify physical hazards are technically based on chemical phenomena. Although some first responders have been introduced to chemistry in high school or college courses, this book is written at a level that assumes little or no previous exposure to the subject. As students proceed from one chapter to the next, their comprehension of chemistry should expand so that they can grasp the technical basis for the myriad of occupational, transportation, and environmental regulations that affect their daily work.

I have also written this book as an introduction to the subject; thus, despite its size, *Chemistry of Hazardous Materials* is intentionally noncomprehensive in scope. I have reorganized the material presented in earlier editions to facilitate understanding and show responders how to grasp the essential principles more easily and apply them to actual situations. This is accomplished in part by the inclusion within each chapter of solved exercises and end-of-the chapter review exercises. Many exercises were intentionally created to demonstrate how responders may use the material introduced in the chapters to predict the behavior of hazardous materials encountered at disaster scenes. The exercises also provide the opportunity for students to determine just how effectively they comprehend the subject matter.

Although some students find that learning chemistry was an unpleasant experience, it doesn't need to be. Emergency responders who plan to directly use the subject matter to save lives, property, and the environment often find that learning chemistry was enjoyable, albeit challenging at times. It has been personally rewarding to me to learn that some students were so affected by the exposure they received from reading this book that they subsequently enrolled in more advanced chemistry courses.

Writing this textbook has also provided me with a higher level of appreciation for the work accomplished by firefighters and other emergency responders. I hereby salute all those who have chosen this career path or are actively volunteering to accomplish its needs. By so doing, I join the ranks of all those who are profoundly grateful for your service.

WHAT'S NEW

Changes to this edition include:

- Updated correlation guide to FESHE curriculum learning objectives
- Updated color photos and illustrations throughout the text
- Chapter objectives now grouped together at the beginning of the chapter
- Information on the nature of contingency plans and emergency-action plans for implementation by emergency responders in the event of a release of hazardous substances to the environment
- Aspects of the Globally Harmonized System for Labeling and Characterizing Hazardous Chemicals (GHS), and the presentation of GHS pictograms and NFPA hazard diamonds in the marginal art throughout the text
- Labeling Information Based on Canada's Workplace Hazardous Materials Information System (WHMIS)
- Contemporary issues including the Fukushima Dai-ichi disaster, fracking natural gas from shale, dust explosions, biofuels, etc.
- OSHA's flammability criteria and its consistency with GHS definitions
- References to the original technical sources upon which the author relied to prepare specific subject matters
- Adverse health effects resulting from exposure to specific hazardous materials
- Updated information regarding the global-warming potential of greenhouse gases
- Updated DOT regulations pertaining to the labeling, marking, and placarding of hazardous materials for shipment
- Updated information on acceptable practices for handling and storage of hazardous materials
- Updated information on the contemporary use of alternative motor fuels (hydrogen, natural gas, liquefied petroleum gas, methanol, ethanol, coal-derived liquid fuels, and biofuels)
- Information concerning exposure to carcinogenic hazardous materials
- New solved exercises and end-of-chapter review exercises
- Expanded and updated coverage of warning labels and safety data sheets used by emergency responders
- New and updated lists of hazardous materials

ACKNOWLEDGMENTS

It is a pleasure to acknowledge the help of the following people and to thank them for their indispensable roles in reviewing selected chapters for this sixth edition:

Anthony R. DeAngelo, MEng, CHP
Adjunct Professor
Excelsior College
Albany, New York

Christopher L. Gilbert, Ph.D.
Senior Environmental Specialist. Hazardous Materials Enforcement Officer /
*   Emergency Spill Response - On site Hazmat Coordinator.*
Alachua County Environmental Protection Department—Hazardous Materials
Gainesville, Florida

Terry L. Heyns, AB Saint Louis University, MA University of Kansas, PhD Kansas State University
Professor of Fire Science and Education
Department of Fire Science
School of Arts, Letters, and Emergency Services
Lake Superior State University
Sault Ste Marie, MI 49783

Kurt Larson, BS, MS, EFO, CFO, MiFirE
Associate Professor, Fire Chief (ret.)
Pensacola, FL

Jeffery Laskowske (A.A.S. Fire Technology)
Fire Captain & Fire Technology Program Director
Grand Forks Fire Department & Northland Community and Technical College
Grand Forks, ND & East Grand Forks, MN

Robert Massicotte, M.S. Fire Science, B.S. Fire Science Technology—University of New Haven; Executive Fire Officer.
Instructor; Fire Chief (Retired) Waterbury, CT.
University of New Haven
West Haven, Connecticut

Lawrence A. Mauerman, MAS (Masters of Administrative Science, Johns Hopkins University), PE (Professional Engineer, Safety—Massachusetts), CSP (Certified Safety Professional, Board of Certified Safety Professionals); Instructor, HazMat Chemistry, FEMA
Coordinator, OSH&E Degree Programs
Southeastern Louisiana University
Hammond, Louisiana

Michael McKenna / BS Economics, AA Fire Technology
Instructor Fire Technology
American River College
Sacramento Regional Public Safety Training Center
City/State: McClellan CA

Tom Sitz, Lt
Painesville Twp FD
Painesville Ohio 44077

Michael Teague
Fire Captain/Hazardous Materials Specialist
Sacramento Metropolitan Fire District
Mather, CA

I am also deeply grateful to all those persons whose individual talents guided the production of this edition from a manuscript into a book, especially Eileen Clawson, Developmental Editor; Monica Moosang, Program Manager, Brady Publishing/Pearson Education; Kelly Ricci, Full-Service Project Manager, Aptara Corporation; Stephen Smith, Editor, Brady Publishing/Pearson Education; Bret Workman, Copyeditor; and Project Manager, Julie Boddorf.

Eugene Meyer

ABOUT THE AUTHOR

Eugene Meyer earned a Ph.D. in chemistry from Florida State University, Tallahassee, Florida, in 1964. He completed a postdoctoral fellowship at the Instituut voor Kernphysisch Onderzoek, Amsterdam, The Netherlands, in 1965. He served as Professor of Chemistry with tenure at Lewis University, Lockport, Illinois, from 1965 to 1979. He joined the technical staff of the U.S. Environmental Protection Agency, Region 5, in 1979, serving as Regional Expert in the Chemistry of Hazardous Waste and Chief of the Technical Programs Section of the Hazardous Waste Division. In 1982, he became President of Meyer Environmental Consultants, Inc., where during the course of his work he consulted with attorneys at the U.S. Department of Justice and private law firms and served in the capacity of an expert witness on more than 200 legal matters concerned with the transportation, treatment, storage, and disposal of hazardous substances. He is now retired and lives in Las Vegas, Nevada.

FIRE AND EMERGENCY SERVICES HIGHER EDUCATION (FESHE) GRID

The following grid outlines the course requirements of the Hazardous Materials Chemistry course developed as part of the FESHE model curriculum. For your convenience, we have indicated specific chapters where these requirements are located in the text.

COURSE REQUIREMENTS	1	2	3	4	5	6	7	8	9	10	11	12	13	14	15	16
Identify the common elements by their atomic symbols on the periodic table and demonstrate an understanding of why the table is organized into columns and groups.				X												
Differentiate among elements, compounds, and mixtures, and give examples of each.				X												
Explain the difference between ionic and covalent bonding and be able to predict when each will occur.				X												
Identify, name, and understand the basic chemistry involved with common hydrocarbon derivatives.												X	X	X		
Comprehend the basic chemical and physical properties of gases, liquids, and solids, and predict the behavior of a substance under adverse conditions.		X	X													
Identify, name, and understand the basic chemistry and hazards involved with the nine U.S. Department of Transportation hazard classes and their divisions.						X	X	X	X	X	X	X	X		X	X
Analyze facility occupancy, transportation documents, shape and size of containers, and material safety data sheets (MSDS) to recognize the physical state and potential hazards of reactivity related to firefighter health and safety.	X		X			X										
Demonstrate the ability to utilize guidebooks to determine an initial course of action for emergency responders.						X				X						

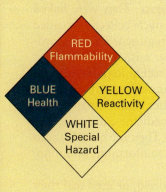

RED
Flammability

BLUE
Health

YELLOW
Reactivity

WHITE
Special
Hazard

KEY TERMS

OBJECTIVES

- Describe the general nature of the information that the Consumer Product Safety Commission (CPSC) requires on the labels affixed to containers of consumer products.
- Name the 16 categories of information about a hazardous chemical that OSHA requires employers to evaluate and provide to their employees on a safety data sheet.
- Identify the physical or health hazard associated with each GHS pictogram.
- Provide the meanings of the numbers entered in the top three quadrants of hazard diamonds.

Emergency incidents have always been caused by natural phenomena that firefighters, law enforcement officials, and other emergency responders have attempted to mitigate. These phenomena include wildfires, flash floods, earthquakes, erupting volcanoes, droughts, and typhoons. They often trigger **natural disasters** that were not wholly predictable at the time of their occurrence nor entirely controllable or preventable.

In the 1940s, new types of emergency incidents began to surface. They were linked with the behavior of certain chemical, petroleum, and nuclear products collectively called hazardous materials. These types of emergency incidents most often occur when hazardous materials are misused or involved in unintended fires and other mishaps.

Hazardous materials are needed for the operations of our technology-based society. They are used in connection with the production and manufacture of a host of products. Among the most common of them are plastics, rubber, leather, paper, textiles, paints, fertilizers, pesticides, solvents, detergents, fuels, medicines, automobiles, televisions, building materials, electronics, and sporting equipment. Their use has advanced society and assisted in making our lives longer, healthier, more comfortable, and more enjoyable.

It is when hazardous materials are improperly stored, transported, or used that their hazardous nature becomes evident. To eliminate them from our society would not only be impractical, it would be undesirable and, in some instances, altogether impossible. Instead, to reduce the loss of lives and property and protect the environment, we must learn to live safely in their presence and effectively respond to the emergency-related incidents they produce.

On September 11, 2001, the United States became the victim to a third type of emergency incident. On 9-11, as it is now called, members of al-Qaida, an international organization of Islamic extremists, hijacked and deliberately flew two commercial planes as missiles into the towers of the World Trade Center in New York. Other al-Qaida terrorists flew a third plane into the Pentagon. A fourth plane also was hijacked by al-Qaida terrorists but did not reach its intended target, instead crash-landing in rural Pennsylvania. In connection with these **terrorist events**, 2973 innocent people died together with 19 hijackers, and at least another 857 people were injured; 411 courageous emergency responders (firefighters, police officers, paramedics, and emergency medical technicians) gave their lives. To execute the evil acts, civilian aircraft carrying a hazardous material—aviation fuel—were used as a weapon of mass destruction.

Al-Qaida at that time was led by Osama bin Laden, a radical, wealthy Saudi who once vowed to kill Americans anywhere. During a covert raid in Pakistan in 2011, bin Laden was killed by Navy SEALs. Although the international reach of al-Qaida was struck a heavy blow by his killing, the death of bin Laden did not bring closure to America's war on terrorism. Furthermore, America has not been entirely successful in curbing the violence that the group has been able to unleash.

Even now, al-Qaida remains actively engaged in attempting to disrupt the American way of life. In 2012, on the anniversary of 9-11, al-Qaida operatives attacked the U.S. consulate in Libya and killed the American ambassador and three other Americans. Al-Qaida also continues to train future fighters in Algerian, Iranian, Iraqi, Malian, Somali, Tunisian, Yemeni, and other camps, for engaging in future evil acts, whether on domestic or international soils. The role of emergency responders is to remain vigilant and prepared to disrupt their activities.

Before the events of 9-11, Islamic extremists had chosen the World Trade Center as the site for a terrorist attack. In 1993, they detonated more than a half ton of explosives in the parking garage of the building, thereby killing six and injuring approximately 1,000 people. Hence, the 9-11 terrorist incidents were not the first actions undertaken by Islamic extremists against the United States. Nonetheless, the 9-11 incidents generally are cited as

natural disaster ■ A large-scale incident caused by extreme weather (hailstorm, blizzard, ice storm, etc.), tectonic movements, volcanism, or other natural phenomena and potentially associated with damaging property, disrupting transportation, or endangering human life

terrorist event ■ A premeditated, unlawful act of violence whose aim is to injure or kill people, damage property, generate social unrest, create economic turmoil, or promote panic and hysteria to achieve a political, ethnic, or religious objective

the events that initiated the war against al-Qaida and other Islamic terrorist groups in which the United States and its allies are now engaged.

During the aftermath of 9-11, emergency responders widely acknowledged the evolution of a new component of their workload: the necessity to conduct rescue and recovery work at incidents where hazardous materials were intentionally released to the environment. Although emergency responders continue to serve the nation by saving lives, property, and the environment, they now also must help secure the homeland and preserve our way of life.

The 9-11 incidents changed the course of world events and drew attention to the presence of a religious ideology that promotes terrorism internationally. Islamic extremists have now expanded into virtually every civilized country, where they sponsor suicide attacks against the local population. The more notorious incidents spearheaded by radical Islamic terrorists occurred in 2004, when bombings of Madrid's commuter trains killed 191 people; in 2005, when bombings in London's underground transit system killed 52 people; in 2008, when bombings in Mumbai's financial district killed 166 people; in 2010, when synchronized twin bombings in Kampala (Uganda) killed 76 people; and in 2011, when a suicide bomb at Moscow's Domodedovo Airport killed 37 people. In each instance, the extremists chose to produce injury and death by detonating explosives.

Americans have also been attacked by terrorists within our domestic borders. In 2013, for example, two terrorists, ethnic Chechens, detonated two shrapnel-packed pressure-cooker bombs within seconds of each other near the finish line of the Boston Marathon. Now considered the worst terrorist attack on American soil since the events of 9-11, the lawless deed killed three people and injured over 260.

The nature of hazardous materials did not change since the 9-11 incidents, nor did the manner in which we respond at hazardous materials scenes; but the 9-11 terrorist events did spawn a heightened awareness of the unorthodox ways in which hazardous materials may be used as weapons of mass destruction to intentionally kill and injure vast numbers of people, cause substantial property damage, and affect economic stability. They also alerted emergency responders to perform their jobs with an intensified sense of vigilance by anticipating that the acts may be executed within their communities, especially during large crowd gatherings.

We begin our study in this first chapter by broadly examining the general features of all hazardous materials. We also observe how federal statutes aim to eliminate or reduce the risks associated with the usage, storage, and transportation of hazardous materials, and how they may assist emergency responders at disaster scenes. Finally, we learn that allied professionals stand ready to help responders when hazardous materials are involved in emergencies.

1.1 WHY MUST EMERGENCY RESPONDERS STUDY CHEMISTRY?

Odds are you have never considered what life would be like without **chemistry**. If you had, it would soon be apparent that chemistry affects everything we do. There is not a single instant during which we are not affected by a chemical substance or a chemical process.

Chemistry is regarded as a natural science; that is, it is a subject concerned with studying natural phenomena. Specifically, the science of chemistry is the study of substances, their composition, properties, and the changes they undergo. Chemistry concerns itself not only with substances that occur naturally but also with synthetic substances.

By understanding the interplay of various substances, chemists have successfully contributed to combating the scourges of hunger, disease, and human deprivation throughout the world. By using established methods, chemists aspire to further improve the quality of

chemistry ■ The natural science concerned with the properties, composition, and reactions of substances

our lives in the future. For instance, during the twenty-first century, we anticipate that chemistry will radically alter the methods used to treat sicknesses and disease.

Chemistry is broadly divided into two main branches: **organic chemistry** and **inorganic chemistry**. At one time, chemists presumed that certain substances could exist in the world only if they had originated in living things—plants and animals. These substances included sugars, alcohols, waxes, fats, and oils, all of which were called organic substances, because their origin was believed to require the vital force of life itself.

In 1828, this hypothesis was disproved when Friedrich Wöhler prepared a substance called urea, a known constituent of urine. Wöhler successfully prepared urea from substances having no apparent connection to plants or animals. This achievement caused chemists to abandon the idea that a vital force was required for the production of certain substances. Despite abandonment of the hypothesis, however, the name of this branch of chemistry has been historically retained.

Organic chemistry is now recognized as the study of substances that contain carbon in their chemical composition. Although some organic substances are actually found in living systems, many others have been synthesized that have no known natural counterpart. The study of organic substances that affect the life process is now regarded as a subdivision of organic chemistry called biochemistry. By contrast, the study of those substances that do not contain carbon in their composition is known as inorganic chemistry. Inorganic substances include aluminum, iron, sulfur, oxygen, table salt, and many others.

To be an effective emergency responder, is it necessary to study organic and inorganic chemistry? It would be misleading to give the impression that the academic pursuit of chemistry is essential for achievement of successful careers as firefighters, police officers, and other emergency responders. Nonetheless, the study of hazardous materials by emergency responders achieves the major objective of enriching their sense of inquiry while investigating accident scenes at which hazardous materials are implicated. Without the knowledge acquired from the study of hazardous materials, emergency responders are severely limited when they must select a course of action during the performance of duty.

1.2 FEDERAL HAZARDOUS SUBSTANCES ACT

The **Federal Hazardous Substances Act,** or **FHSA,** was first enacted by Congress in 1960. Thereafter, it was amended to focus on the protection of users against unsuspecting exposure to hazardous substances contained within consumer products. The statute is administered by the Consumer Product Safety Commission (CPSC).

One way by which CPSC fulfills its congressional mandate is to require manufacturers to affix appropriate labeling on the containers of consumer products that contain hazardous substances. This labeling information is directed primarily at consumers, but it is potentially helpful to emergency responders and others. At 16 C.F.R. §1500.121,[1] CPSC requires the following information to be clearly and conspicuously provided on the labels affixed to containers of consumer products having hazardous substances as components:

- The name and place of business of the manufacturer, packer, distributor, or seller
- The common or usual name or the chemical name of each hazardous ingredient
- The signal word DANGER for products that are corrosive, extremely flammable, or highly toxic
- The signal word CAUTION or WARNING for all other hazardous products
- The word POISON in addition to the signal word DANGER for the highly toxic products listed at 16 C.F.R. §1500.129

---

[1]Relevant citations to federal regulations are noted throughout this text. The reference to 16 C.F.R. §1500.121 means Title 16 of the *Code of Federal Regulations*, Part 1500, Section 121. The parts are also divided into subparts, which are denoted with capital letters A, B, C, and so on. Readers may electronically access all regulations cited in this text on the Internet.

organic chemistry ■ The study of the properties of substances that contain carbon in their chemical composition

inorganic chemistry ■ The study of the properties of substances that do not contain carbon in their chemical composition

Federal Hazardous Substances Act ■ The federal statute that empowers the CPSC to protect the public against unsuspecting exposure to hazardous substances contained within household products

- An affirmative statement of the principal hazard or hazards represented by the product, such as FLAMMABLE, COMBUSTIBLE, VAPOR HARMFUL, CAUSES BURNS, POISON, SKIN AND EYE IRRITANT, or similar wording that is descriptive of the hazard
- Precautionary statements telling users what they must do or what actions they must avoid to protect themselves
- Instructions, when necessary or appropriate, for first-aid treatment to perform when the product is the cause of someone's injury
- Instructions for handling and storage of packages that require special care during handling or storage
- The statement KEEP OUT OF REACH OF CHILDREN

The signal word CAUTION appears on products that are least harmful to the user, WARNING appears when a product is more harmful than one bearing a CAUTION label, and DANGER appears on products that are poisonous or corrosive.

When products are stored within self-pressurized containers, CPSC requires the manufacturer at 16 C.F.R. §1500.130 to include the following statements or their equivalents on product labels:

> **WARNING–CONTENTS UNDER PRESSURE**
> **DO NOT PUNCTURE OR INCINERATE CONTAINER**
> **DO NOT EXPOSE TO HEAT OR STORE AT TEMPERATURES ABOVE**
> **120°F (49°C)**
> **KEEP OUT OF REACH OF CHILDREN**

When the misuse of a product is likely to adversely affect one's health, CPSC requires the manufacturer to include advisory warnings and first-aid actions on product labels. For example, the statements provided on the label in Figure 1.1 provide the user of a car battery with two follow-up actions to be implemented when the product is splashed into the eyes.

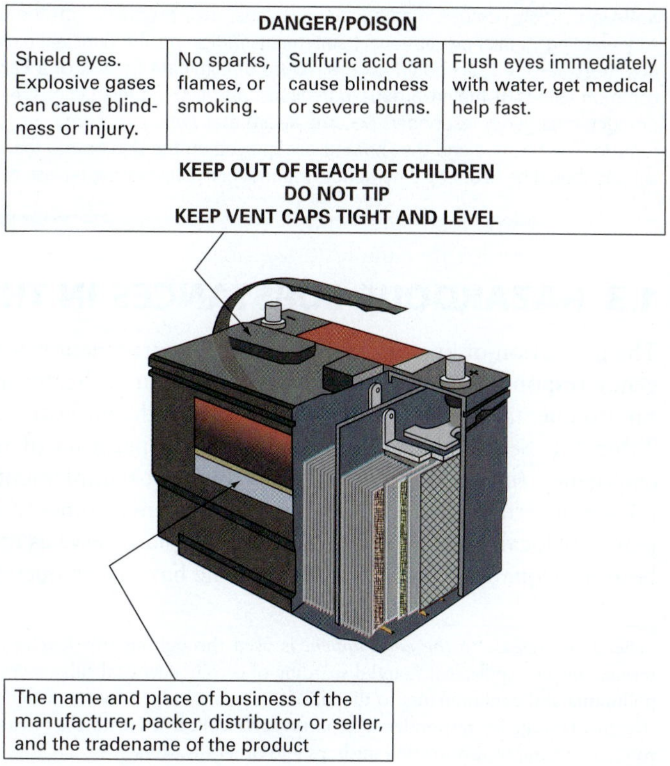

DANGER/POISON			
Shield eyes. Explosive gases can cause blindness or injury.	No sparks, flames, or smoking.	Sulfuric acid can cause blindness or severe burns.	Flush eyes immediately with water, get medical help fast.

KEEP OUT OF REACH OF CHILDREN
DO NOT TIP
KEEP VENT CAPS TIGHT AND LEVEL

The name and place of business of the manufacturer, packer, distributor, or seller, and the tradename of the product

FIGURE 1.1 The Federal Hazardous Substances Act provides CPSC with the authority to compel manufacturers to label their products with the specific information denoted at 16 C.F.R. §§1500.3 and 1500.130. The label on this car battery advises the consumer to beware of its potential flammable and health hazards. (To provide confidentiality, the name of the manufacturer, packer, distributor, or seller and the tradename of the battery have been intentionally omitted.)

H & L CLEANING SOLUTION

Directions For Use: For general cleaning, use ½ cup H & L Cleaning Solution to 1 gallon of hot water for refrigerators and other appliances, sinks, bathrooms, dishes, glassware, pots and pans, garbage pails, windows, mirrors, linoleum, ceramic tile floors and venetian blinds. For cleaning special woodwork and painted walls, add 1 cup H & L Cleaning Solution, ½ cup vinegar, and ¼ cup baking soda to 1 gallon hot water. For cleaning ovens and broilers, leave a dish of H & L Cleaning Solution in oven or broiler overnight and wipe surfaces in the morning.

CAUTION

EYE IRRITANT

KEEP OUT OF REACH OF CHILDREN

Do not mix with other household products such as chlorine-based bleaches, toilet bowl, wall, or tile cleaners. Avoid contact with eyes and prolonged contact with skin. Do not take internally. Avoid inhalation of vapor. Use in well-ventilated area.

Storage: Store bottle upright and tightly capped.

First Aid: As appropriate, flush eyes 10 to 15 minutes with water, *or* wash skin thoroughly with water, *or* immediately give large amounts of milk or water and do not induce vomiting. In all cases, IMMEDIATELY CONTACT A PHYSICIAN.

Ingredients: Ammonium hydroxide and surfactant.

A manufacturer has affixed the preceding label to 2-quart bottles containing H & L Cleaning Solution. How is this information useful to emergency responders who encounter 50 pallets of this product, each containing 200 bottles, during a warehouse fire?

Solution: CPSC compels manufacturers to label their products with the following minimum information: advisory warnings; initial precautionary statements indicating the principal hazards associated with use of the product; measures the user should take to reduce or eliminate the risk of a hazard during storage and handling; the common names of the hazardous ingredients; and first-aid instructions. By reading the label information of this product, emergency responders become aware that the contents may pose a health risk, especially if the bottles rupture when heated and the contents are sprayed on the skin or into the eyes. Logic dictates that precautionary actions should be taken by emergency responders to prevent the release of the contents into the environment.

1.3 HAZARDOUS SUBSTANCES IN THE ENVIRONMENT

The protection of lives, property, and the environment is a responsibility of today's emergency responders. To assure protection of human health and the environment, the U.S. Environmental Protection Agency (EPA) uses the authority of the federal statutes listed in Table 1.1. Several require the owners and operators of regulated facilities to prepare emergency response or contingency plans for implementation during an unintended release of certain chemical substances to the environment.[2] Among the recipients of these plans are local response organizations. The plans serve as one means by which firefighters become acquainted with the nature of the hazardous operations that occur in their areas

---

[2]The clause *release to the environment* is used throughout this text in its broadest sense. Examples of such releases include spills, leaks, and discarding of certain chemical substances in soil; emissions of gases containing pollutants and contaminants to the air; and the discharge of wastewaters and water runoff to rivers, lakes, or streams. Emergency responders are particularly concerned with the release of chemical substances to the environment during transportation mishaps and other disasters.

TABLE 1.1	Some Federal Environmental Statutes
STATUTE	**EPA'S RESPONSIBILITY**
Clean Air Act	Establishes national ambient air-quality standards for specified air contaminants; publishes technology-based emission-reduction standards for hazardous air pollutants; requires facilities to develop a risk management plan that addresses procedures for reducing risks associated with the mismanagement of certain regulated substances; directs the U.S. Chemical Safety and Hazard Investigation Board to independently investigate certain accidents involving chemical substances
Clean Water Act	Restores and maintains the chemical, physical, and biological integrity of the nation's waters by regulating the discharge of toxic pollutants and hazardous substances into waters of the United States
Resource Conservation and Recovery Act	Regulates the generation, transportation, treatment, storage, and disposal of hazardous wastes within a cradle-to-grave framework
Comprehensive Environmental Response, Compensation, and Liability Act	Leads efforts to clean up sites at which hazardous substances were improperly treated, stored, or disposed
Superfund Amendments and Reauthorization Act	Considers other environmental laws and regulations and includes citizen participation when selecting a remedial response action at a hazardous waste site; provides a means for increasing the public's access to information about the chemical substances within their communities and planning for emergencies associated with their potential release
Federal Insecticide, Fungicide, and Rodenticide Act	Regulates the sale, distribution, and use of pesticides in the United States; reviews and registers pesticides for specified uses
Toxic Substances Control Act	Collects data on toxic substances for risk evaluation, mitigation, and control; bans the manufacture and industrial use of substances that pose an unreasonable risk to public health and the environment
Emergency Planning and Community Right-To-Know Act	Requires the owners and operators of facilities that manufacture, use, or store certain chemical substances in excess of specified threshold quantities to report their presence to fire departments and state and local governments, prepare an emergency plan, and notify the local emergency planning committee in the event of a release of the substances to the environment

of jurisdiction. They also involve them in the process of emergency planning and preparedness procedures so a response action is more likely to be efficient and effective when it is executed during an actual disaster.

1.3-A CLEAN AIR ACT

In 1970, Congress first empowered EPA with regulating the quality of the air we breathe with the passage in 1970 of the **Clean Air Act**, or CAA. EPA accomplished its mandate by establishing national ambient air-quality standards for a number of air contaminants called **criteria air pollutants**. Standards for six criteria air pollutants have been established: carbon monoxide, ground-level ozone, lead, nitrogen dioxide, particulate matter, and sulfur dioxide. Typical urban concentrations of several of these pollutants are illustrated in Figure 1.2. Their properties are noted in appropriate sections of future chapters.

Clean Air Act (CAA) ■
The federal statute that empowers EPA to establish national ambient air-quality standards for specified air contaminants

criteria air pollutant ■
Any air pollutant for which EPA has developed emission standards based on specific health criteria

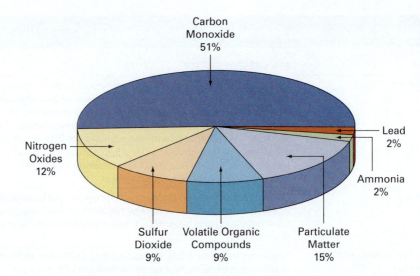

FIGURE 1.2 Typical concentrations of several air pollutants in an urban environment. The concentrations of volatile organic compounds (Section 7.1-J) and nitrogen oxides (Section 10.14) are an indirect measure of the ozone concentration, because ozone is produced in the lower atmosphere when these substances chemically interact. (*Courtesy of U. S. Environmental Protection Agency, Washington, DC.*)

Each national ambient air-quality standard is composed of two parts:

- *Primary standard.* This standard identifies a contaminant concentration in the ambient air that protects human health. When EPA sets the primary standard, it considers not only healthy individuals but the health of sensitive persons such as children, people with asthma, and the elderly.
- *Secondary standard.* This standard identifies a contaminant concentration that protects public welfare. In setting the standard, EPA accounts for such subjects as reduced air visibility and damage to animals and vegetation.

Daily and/or annual contaminant concentrations may be set for both standards.

The CAA regulations also provide EPA with the authority to regulate the prevention of accidental, potentially catastrophic releases of certain airborne hazardous substances. This program obligates the owners and operators of facilities producing, using, or storing certain regulated substances in amounts equal to or exceeding threshold amounts to identify the sources from which they could be potentially released. A list of these regulated substances and their threshold quantities for accidental release-prevention is published in four tables at 40 C.F.R. §68.130.

To assist in the prevention of releases of airborne hazardous substances, the CAA regulations obligate the owners and operators of facilities that produce, use, or store regulated amounts of the substances to develop risk management plans. The regulations require them to cite key information that may be used by emergency planners, firefighters, and local residents to reduce risks from the potential mismanagement of the regulated substances. By so doing, it places a responsibility on the owners and operators to identify the means to prevent and minimize the risks associated with potential catastrophes.

The owners and operators of facilities that release airborne hazardous substances are also required to incorporate certain subjects into their risk management plans, including coordination with local emergency planning and response agencies such as the fire department having responsibility within the appropriate jurisdiction. Emergency responders should be familiar with the content of these plans so that their execution is likely to be effective in the event that a chemical accident actually occurs within their jurisdictions.

An important outcome of the CAA was the establishment of the **U.S. Chemical Safety and Hazard Investigation Board**, or **CSB**. Modeled after the National Transportation Safety Board (Section 1.7-B), the CSB conducts investigations aimed at identifying the facts, conditions, and circumstances that impact the root causes of chemical accidents. CSB typically focuses on accidents at chemical plants and refineries, especially those that result in fatalities, serious injuries, or substantial property damage.

The findings of the CSB are made public on the Internet and elsewhere. They are used to recommend actions to improve the safety of operations involved in the production, transportation, industrial handling, use, and disposal of chemical substances. Although the CSB does not have regulatory powers, its recommendations have significantly impacted regulatory decision making, risk assessment, industry practices, and the codes of professional safety associations.

CSB has also produced videos in which chemical accidents are reenacted. These videos may be accessed on the Internet so that emergency responders and other viewers may readily observe how accidents involving hazardous materials happen and how chemical missteps contribute to their development. The videos have been praised for their inception and serve as an excellent resource for the training of emergency responders.

U.S. Chemical Safety and Hazard Investigation Board (CSB) ▪ The group that conducts chemical-accident investigations to determine their root causes or management-system failures that result in an accident

1.3-B CLEAN WATER ACT

In 1948, Congress passed the **Federal Water Pollution Prevention and Control Act**, or **FWPPCA**. Today, FWPPCA and its combination of subsequent amendments are commonly called the **Clean Water Act,** or **CWA**. It has two basic goals: to regulate the concentration of pollutants that are discharged from a point source into the nation's waterways and to provide water quality levels that make the waterways safe for recreational use and the survival of aquatic life.

One of the pollutants regulated by EPA is oil. The CWA authorizes EPA to address the hazards associated with a facility's storage of an oil of any type (petroleum, animal, or vegetable) above a threshold amount. EPA requires its owner to prepare for a worst-case scenario involving the discharge of the oil into navigable waters or upon adjoining shorelines or beaches. EPA requires the facility to prepare and submit a response plan whose basic requirements include immediate spill notification to the National Response Center (Section 1.13), timely deployment of spill-response equipment, and oil-spill monitoring and response. The regulations are published at 40 C.F.R. §§112.20 and 112.21.

Federal Water Pollution Prevention and Control Act (FWPPCA) ▪ The federal statute that empowers EPA to restore and maintain the chemical, physical, and biological integrity of the waters of the United States

Clean Water Act (CWA) ▪ The combination of amendments to FWPPCA

1.3-C RESOURCE CONSERVATION AND RECOVERY ACT

In 1976, Congress enacted the **Resource Conservation and Recovery Act**, or **RCRA**, whose aim is to ensure protection of public health and the environment from the activities of facilities that generate, transport, treat, store, or dispose of *hazardous wastes*. The locations at which hazardous wastes have been treated, stored, or disposed are called *hazardous waste sites*.

Hazardous wastes are distinguished from all other wastes in two ways. First, EPA *lists* a series of wastes at 40 C.F.R. §§261.31, 261.32, 261.33(e), and 261.33(f). Second, it provides criteria for *characterizing* a hazardous waste as ignitable, corrosive, reactive, or toxic. The nature of these defining criteria is identified at 40 C.F.R. §§261.21, 261.22, 261.23, and 261.24. We revisit them later, in Sections 3.2, 5.9, 8.15, and 10.1-B.

The RCRA regulations require the owners and operators of facilities that treat, store, or dispose of hazardous waste to prepare a contingency plan that identifies the emergency procedures to be implemented when responding to a release of a hazardous waste to the environment. A feature of the contingency plan is that it describes the individual roles that local police, fire departments, contractors, and other emergency response teams will assume during the response action.

Resource Conservation and Recovery Act (RCRA) ▪ The federal statute that empowers EPA to regulate the treatment, storage, and disposal of hazardous wastes

At 40 C.F.R. §264.53, RCRA requires the owners or operators of hazardous waste facilities to submit copies of their contingency plans to all local police departments, fire departments, hospitals, contractors, and other state and local emergency response teams that may be called upon to provide emergency services. Familiarity with the information cited in these plans is essential if their execution is to be successful.

1.3-D COMPREHENSIVE ENVIRONMENTAL RESPONSE, COMPENSATION, AND LIABILITY ACT

Comprehensive Environmental Response, Compensation, and Liability Act (CERCLA) ■ The federal statute that empowers EPA to identify and restore sites at which hazardous substances were released to the environment

In 1980, Congress enacted the **Comprehensive Environmental Response, Compensation, and Liability Act**, or **CERCLA**, whose aim is to correct for the past mistakes resulting from the improper treatment, storage, and disposal of hazardous substances. Outside the legal profession, the statute is also known colloquially as the *Superfund law*. The locations at which hazardous substances have come to be located are called *Superfund sites*, many of which are abandoned or uncontrolled.

CERCLA addresses the prevention and cleanup of hazardous substances released at superfund sites and the restoration and replacement of the natural resources that were damaged or lost due to the release. EPA promulgated regulations that include a listing of hazardous substances at 40 C.F.R. §302.4. Crude petroleum, petroleum fractions, and petroleum products are excluded from the definition.

EPA initially paid for correcting the problems associated with Superfund sites through the use of a corporate tax imposed on the chemical and petroleum industries. The tax monies were deposited into a federal trust fund called the *Superfund*. When those responsible could not be located, or when responsible parties were unwilling or unable to pay for the damages they had caused, EPA used Superfund monies to investigate and remediate the damages. The imposition of the tax expired in 1995, and congressional attempts to reimpose it have failed. Nonetheless, EPA used the authority of CERCLA to clean up over 1000 sites and recover the costs from the parties responsible for their contamination.

Superfund Amendments and Reauthorization Act (SARA) ■ The amendment to CERCLA whose directives include the provision of training for emergency responders, increasing the public's access to information about the chemical substances within their communities, and planning for emergencies associated with the release of these substances

In 1986, Congress amended CERCLA by enacting the **Superfund Amendments and Reauthorization Act**, or **SARA**. Among other directives, this statute required EPA to consider the standards identified in other environmental laws and include citizen participation in the decision-making process when it selected a remedial action at a site where hazardous substances are located.

SARA also mandated the development of a health and safety standard for protecting workers at hazardous waste sites. The standard is cited at 29 C.F.R. §1910.120 and 40 C.F.R. §311. Included is the establishment of an assistance program for training and educating workers engaged in **hazardous waste operations and emergency response**, or **HAZWOPER**, activities. The latter addresses clean-up operations and corrective actions at Superfund sites, and includes all emergency response operations that apply to the release of hazardous substances, regardless of where the release occurred.

hazardous waste operations and emergency response (HAZWOPER) ■ The regulatory program established through SARA for training and educating workers engaged in activities such as clean-up operations and emergency response efforts

In Section 1.6, we note that the authority of SARA was also used by EPA to provide a means for the public to learn about the presence of hazardous substances within the communities where they live.

1.4 HAZARDOUS CHEMICAL PRODUCTS

Some chemical substances are considered so hazardous that special action needs to be taken to reduce or eliminate the threat they can pose to public health and the environment. Two federal statutes that address certain of these substances are the Toxic Substances Control Act and the Federal Insecticide, Fungicide, and Rodenticide Act. Both are enforced by EPA. We briefly note them in the two sections that follow.

1.4-A TOXIC SUBSTANCES CONTROL ACT

In 1976, Congress passed the **Toxic Substances Control Act**, or **TSCA**. This law provides EPA with the broad authority to ensure that the potential effects resulting from exposure to certain chemical substances are identified and properly controlled before the substances are placed into commerce.

EPA uses the authority of TSCA to select a control action for preventing unreasonable risks to public health and the environment from exposure to these chemical products. The nature of the control action varies with the severity of the risks. In some instances, EPA requires the manufacturer to post a hazard warning on product labels; in other instances it bans outright the manufacture of a product or severely restricts its use.

We shall note the properties of hazardous substances that TSCA bans or severely restricts in future chapters in this text. The banned substances include asbestos (Section 10.19), polychlorinated biphenyls (PCBs) (Section 12.16), 2,3,7,8-tetrachlorodibenzo-p-dioxin (TCDD) (Section 13.4-B), nitrosating agents used in metalworking operations (Section 13.8-B), and acrylamide grout (Section 14.6-F). When these substances are encountered during emergency response actions, special precaution should be exercised to avoid exposure to them.

1.4-B FEDERAL INSECTICIDE, FUNGICIDE, AND RODENTICIDE ACT

In 1947, Congress passed the **Federal Insecticide, Fungicide, and Rodenticide Act**, or **FIFRA**, and in 1970, it provided EPA with the authority to administer the statute. EPA uses FIFRA to regulate the manufacture, use, and disposal of pesticides for agricultural, forestry, household, and other activities.

At 40 C.F.R. §152.3, EPA defines a **pesticide** in part as follows: any substance or mixture of substances intended for preventing, destroying, repelling, or mitigating any pest, or any substance or mixture of substances intended for use as a plant regulator, defoliant, or drying agent. FIFRA regulations require pesticide manufacturers to obtain an establishment number, to register their pesticides, and to classify them for general or restricted use. EPA assigns a number to each pesticide approved for use when the prevailing information indicates that its use will not pose an unreasonable adverse impact on public health and the environment. FIFRA also provides EPA with the authority to cancel a pesticide registration if unreasonable adverse effects to the environment and public health develop when a product is used according to widespread and commonly recognized practice.

At 40 C.F.R. §156.10, FIFRA provides EPA with the authority to require pesticide manufacturers and distributors to provide informative and accurate labeling on pesticide containers. The label information is primarily intended to be useful to pesticide users, but it is also potentially useful to emergency responders who encounter pesticides during transport mishaps, unintended fires, and other emergencies.

1.5 HAZARDOUS COMPONENTS OF PRODUCTS USED IN THE WORKPLACE

In 1970, Congress enacted the **Occupational Safety and Health Act**, whose aim is to protect employees in the workplace from occupational illnesses and injuries. By passing this Act, Congress empowered the Occupational Safety and Health Administration (OSHA), a division of the U.S. Department of Labor, to reduce or eliminate the incidence of chemically induced occupational illnesses and injuries by regulating certain conditions in the workplace.

An important feature of the Occupational Safety and Health Act is its **general duty clause**, which states the following requirements for employers and employees:

> Each employer (1) shall furnish to each of his employees employment and a place of employment which are free from recognized hazards that are causing or are likely to

Toxic Substances Control Act (TSCA) ■ The federal statute that empowers EPA to regulate all aspects of the manufacture of toxic substances other than pesticides and drugs

Federal Insecticide, Fungicide, and Rodenticide Act (FIFRA) ■ The federal statute that empowers EPA to regulate the manufacture, use, and disposal of pesticides

pesticide ■ For purposes of FIFRA regulations, any pesticide used to prevent, repel, or mitigate any pest, or any substance whose intended use is a plant regulator, defoliant, or desiccant

Occupational Safety and Health Act ■ The federal statute that empowers OSHA to protect employees from occupational illnesses and injuries, including those caused by exposure to hazardous chemicals in the workplace

general duty clause ■ The directive requiring employers to furnish a safe and healthful working environment for their employees, and requiring employees to abide by OSHA's rules, regulations, and orders

cause death or serious harm to his employees; (2) shall comply with occupational safety and health standards promulgated under this Act.

Each employee shall comply with occupational safety and health standards and all rules, regulations, and orders issued by this Act that are applicable to his own actions and conduct.

The OSHA regulation at 29 C.F.R. §1910.1200(b)(2) specifically directs employers to provide a safe and healthful working environment to their employees when they must be exposed to hazardous chemicals or other substances. The term *hazardous chemical* essentially refers to any substance or mixture of substances that can potentially pose a physical or health hazard. To accomplish its directive, OSHA establishes requirements for worker training, safe work practices, labeling, ventilation, proper storage procedures, worker-protective equipment, the posting of signs that warn of potential dangers, and the limiting of employee exposure to certain limits averaged over the 8-hour workday. OSHA publishes these limits in its regulations.

OSHA is assisted in accomplishing its mandate by the National Institute for Occupational Safety and Health (NIOSH). This federal agency is charged with the responsibility of developing recommendations for safety and health standards, establishing safe levels of exposure to toxic materials and harmful physical agents, and conducting its own research to establish new safety and health standards.

In 1983, OSHA first enacted a standard that sets minimum requirements to which employers must adhere for communicating information about hazardous chemicals to workers; it is often referred to as the **Hazard Communication Standard (HCS), HAZCOM Standard**, or the **employee right-to-know law**. Briefly, its intent is to assure workers of their right to know the identities and potential hazards of the hazardous chemicals to which they may be exposed in their places of employment, as well as the measures that may be taken to prevent jeopardizing their health and safety when their use is required.

The HCS requires chemical manufacturers and importers to assess the hazards of the chemical substances and products they produce or import and to transmit the hazard information to their users. This information must be sufficiently comprehensive to allow user-employers to develop appropriate employee-protection programs and give employees the information they require to protect themselves against the potential risks associated with exposure to them. This applies to all chemical substances known to be present within the workplace, substances to which employees could be exposed under normal working conditions, and substances to which employees could be exposed in a foreseeable emergency.

Hazard Communication Standard (HCS), (HAZCOM Standard, employee right-to-know law) ■ For purposes of OSHA regulations, the standard that addresses workers' right-to-know about the hazards associated with substances to which they are exposed in the workplace

1.5-A WARNING LABELS

In the HCS, OSHA requires each chemical manufacturer and importer of a hazardous substance to label, tag, or mark on its container the identity of the product, an appropriate hazard-warning statement, storage and handling information, and the name and address of the manufacturer, importer, or other responsible party. The labeling requirements at 29 C.F.R. §1910.1200(f) now comply with the requirements of the United Nations' Globally Harmonized System (GHS) (Section 1.9), which is now adopted for use in the United States. Hazard-warning messages are conveyed on labels through the posting of standardized signal words (DANGER or WARNING), hazard statements (e.g., FATAL IF SWALLOWED), precautionary statements (e.g., KEEP OUT OF REACH OF CHILDREN), and pictograms composed of symbols and pictures.

An example of the OSHA information required on a warning label is illustrated in Figure 1.3 for a 35% hydrogen peroxide solution in water. The OSHA/GHS label requirements are noted more fully in Section 1.9.

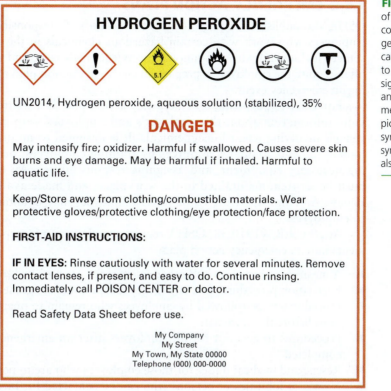

FIGURE 1.3 This portion of the label affixed to containers of 35% hydrogen peroxide communicates hazard information to the observer by using a signal word, GHS hazard and precautionary statements, GHS and transport pictograms, and WHMIS symbols. WHMIS hazard symbols (Section 1.10) are also provided.

1.5-B SAFETY DATA SHEETS

A **safety data sheet**, or **SDS**, is a technical bulletin containing detailed information about a hazardous chemical. Each SDS is now prepared to comply with the requirements of the GHS.[3] All employees who are exposed to a hazardous chemical in the workplace must be provided access to the appropriate SDS and appropriate training on its use.

At 29 C.F.R. §1910.1200(g), OSHA requires that an SDS address the following categories of information in distinct sections:

- Hazardous chemical identification
- Hazards identification
- Chemical composition/Information on ingredients
- First-aid measures
- Firefighting measures
- Accidental release measures
- Handling and storage information
- Physical and chemical properties
- Chemical stability and reactivity
- Toxicological properties
- Ecological information
- Disposal considerations
- Transport information
- Regulatory information
- Other information deemed important

safety data sheet (SDS) ■ A technical bulletin prepared by manufacturers and marketers in accordance with the provisions of 29 C.F.R. §1910.1200 to provide workers with detailed information about the properties of a commercial product that contains hazardous substances

An example of an SDS is provided in Appendix A at the back of this text. This example consists of seven pages of information about a 35% solution of hydrogen peroxide. We note the specific properties of hydrogen peroxide in Section 11.5.

---

[3]An SDS was formerly known as a "Material Safety Data Sheet," or "MSDS," but the word "Material" is not used in the GHS. In practice, the two expressions are still used interchangeably.

1.5-C EMERGENCY ACTION PLANS

OSHA also publishes unique standards that identify the responsibilities of employers and employees who work with certain hazardous chemicals in the workplace. These substances include anhydrous ammonia, acetylene, hydrogen, and oxygen. The standards often require the employer to prepare an emergency action plan that is implemented during an emergency event.

For OSHA's purposes, an emergency action plan is a written document that provides information about the employer's and employees' responsibilities during emergencies involving hazardous chemicals. It is designed to minimize injury and loss of human life and company resources by training employees, procuring and maintaining necessary equipment, and assigning responsibilities. An emergency action plan must be written, maintained in the workplace, and made available to employees for review. An employer with 10 or fewer employees may communicate the plan orally to employees.

At 29 C.F.R. §1910.38, OSHA requires employers to incorporate the following elements into an emergency action plan:

- A means by which fires and other emergencies are reported to employees
- Evacuation procedures and emergency escape-route assignments
- Procedures to be followed by employees who remain to operate critical plant operations before they evacuate
- Procedures to account for all employees after an emergency evacuation has been completed
- Rescue and medical duties for those employees who are to perform them
- Names or job titles of persons who can be contacted for further information or explanation of the duties stipulated in the plan

Employers generally report fires and other emergencies to employees by means of an alarm system that recognizes the distinctive nature of the emergency and signals employees to initiate the procedures provided in the emergency action plan. An individual and an alternative are often identified as the emergency plan manager, whose responsibilities include coordinating all emergency procedures with local public resources, such as fire department and emergency medical personnel. Emergency responders share responsibilities with the emergency plan manager to ensure that response procedures are performed in precisely the order and fashion as they are identified in the plan.

1.6 HAZARDOUS SUBSTANCES STORED AND USED IN COMMUNITIES

Emergency Planning and Community Right-to-Know Act (EPCRA) ■ The federal statute that empowers EPA to identify the chemical substances located within communities and plan for emergencies associated with their potential release

In 1986, EPA used the authority of SARA to enact the **Emergency Planning and Community Right-to-Know Act**, or **EPCRA**, to increase the public's access to information concerning the presence of chemical substances within their communities. The statute requires the owners and operators of all facilities handling certain chemical substances in amounts equal to or greater than specified threshold quantities to provide information concerning the substances to the following three groups:

- Local emergency planning committee (LEPC)
- State emergency response commission (SERC)
- Local fire department having jurisdiction over the facility

The *local emergency planning committee* is a body of individuals appointed by a SERC to assume responsibility for emergency planning for chemical incidents at the local level. At least one member of the LEPC is usually a representative of the local fire department. The *state emergency response* commission is the body of individuals that selects LEPCs, oversees emergency planning for chemical incidents on a statewide basis, and provides oversight and guidance to the LEPCs.

EPCRA requires the owners and operators of facilities that manufacture, store, or use certain hazardous substances to identify them to the LEPC, SERC, and the fire department in their jurisdictions by submission of a chemical inventory report. In addition, they must either submit copies of relevant SDSs or provide a list of the hazardous chemicals stored or used at their facilities. In the latter instance, they must also provide for each hazardous chemical its acute health hazard, chronic health hazard, fire hazard, sudden-release-of-pressure hazard, and reactivity hazard. EPCRA requires facility owners and operators to submit this information within 60 days of receipt of a hazardous chemical.

There are several purposes associated with this combination of information. First, compiling the information is directly helpful to the owners and operators of facilities during their preparation of an emergency plan. The submission also informs emergency responders that certain chemical substances are present at a facility within their community and exactly where they are present. The information also provides a means for emergency responders and other citizens to be prepared in the event of a natural disaster or other event that potentially could cause the release of these substances to the environment. When information concerning this submission is utilized effectively, it provides emergency responders and other citizens with sufficient knowledge to prepare for the likelihood that adverse incidents may actually occur within their community. It also serves to improve the quality of accident-planning exercises, preparedness, mitigation, response, and recovery capabilities.

The LEPC draws upon the combined expertise of the personnel associated with a facility, SERC, and the fire department to incorporate each of the following into the facility's emergency plan:

- Identification of the affected facilities and their transportation routes
- Description of emergency notification and response procedures
- Designation of community and facility emergency coordinators
- Description of methods to determine the occurrence and extent of a release
- Identification of available response equipment and personnel
- Outline of evacuation plans
- Description of training and practice programs and schedules
- Description of methods and schedules for exercising the plan

Some EPCRA requirements for preparing an emergency plan overlap with the Clean Air Act requirements for preparing a risk management plan.

A major purpose associated with providing information to the LEPC (including the fire department) and SERC concerns notification, when appropriate, by the owner or operator of a facility that an actual release of a hazardous substance or extremely hazardous substance to the environment has occurred. CERCLA requires facility owners and operators to immediately report to the National Response Center (Section 1.13) any release of a hazardous substance in an amount that equals or exceeds its reportable quantity (Section 6.2-B) within a 24-hour period. EPCRA requires the owners and operators to immediately report to the LEPC and SERC the release of a hazardous substance or extremely hazardous substance in an amount that exceeds a reportable quantity within a 24-hour period when the release possesses the potential to migrate off-site and affect off-site persons. The reportable quantities of extremely hazardous substances are also provided in Appendices A and B following 40 C.F.R. Part 355. The notification then triggers action by the LEPC to inform nearby residents to take immediate action to mitigate the harm that is linked with exposure to the substance. It also triggers action for first-on-the-scene responders to initiate the actions in the emergency plan to protect human health and the environment from fires, explosions, or other unplanned events.

The notification of a release of a hazardous substance or extremely hazardous substance must minimally include the following information:

- The chemical name or identity of the substance involved in the release
- An indication as to whether the substance is an extremely hazardous substance
- An estimate of the quantity released to the environment
- The time and duration of the release
- The medium or media into which the release occurred
- Any known or anticipated acute or chronic health risks associated with exposure to the substance, and when appropriate, advice regarding medical attention necessary to treat exposed individuals
- Proper precautions to be taken as a result of the release, including evacuation
- The names and telephone numbers of the person or persons to be contacted for further information

1.7 HAZARDOUS MATERIALS IN TRANSIT

Federal Hazardous Materials Transportation Law ■ The federal statute that empowers DOT to regulate certain activities of shippers and carriers who offer or accept hazardous materials for transportation

Using the authority of the **Federal Hazardous Materials Transportation Law**, as amended by the Homeland Security Act of 2002, the Pipeline and Hazardous Materials Safety Administration of the U.S. Department of Transportation (DOT) regulates shippers and carriers who offer or accept hazardous materials for intrastate, interstate, or international transportation through the use of certain marking, labeling, placarding, and packaging requirements. These regulations are now consistent with the procedures implemented in the GHS (Section 1.9). The information conveyed by the warning markings, labels, placards, and shipping papers is especially critical for emergency responders. We discuss the details of this information specifically in Chapter 6 and in appropriate sections of Chapters 7 through 16.

1.7-A MARINE ACTIVITIES REGULATED BY DOT

Oil and hazardous substances are often transferred from place to place by means of pipelines and transported in bulk as cargo onboard ships, tanks, and other watercraft. Off- and onshore facilities, such as watercraft and pipelines, respectively, potentially serve as a means of substantial environmental harm when oil or hazardous substances are discharged from them into navigable waters or upon adjoining shorelines or beaches.

To ensure protection of human health and the marine environment, DOT requires the owners and operators of vessels and facilities handling oil and hazardous substances to prepare for a worst-case scenario involving the discharge of the oil or hazardous substances to the environment. The preparation includes the preparation and submittal of a vessel response plan or a facility response plan to DOT. When oil is stored at an onshore facility, the responsibility for review of the facility-response plan is shared by DOT and EPA.

Vessel and facility response plans must address the potential threat to human health and the environment, including if appropriate, the release of oil and hazardous substances from vessels or facilities into navigable waters or upon adjoining shorelines and beaches. The specific requirements regarding the preparation of these plans are published at 49 C.F.R. Part 194 and 33 C.F.R. Part 154, Subpart F.

National Transportation Safety Board (NTSB) ■ The agency charged with determining the probable causes of transportation accidents promoting transportation safety, and assisting victims of transportation accidents and their families

1.7-B NATIONAL TRANSPORTATION SAFETY BOARD

In 1967, Congress established the **National Transportation Safety Board**, or **NTSB**, as an independent federal agency, but situated it within the DOT for administrative purposes. The mission of the NTSB is to determine the probable causes of the following:

- All U.S. civil aviation accidents and certain public-use aircraft accidents
- Selected highway accidents

- Railroad accidents involving passenger trains or any train accident that results in at least one fatality or major property damage
- Major marine accidents and marine accidents involving watercraft
- Pipeline accidents involving a fatality or substantial property damage
- Releases of hazardous materials in all forms of transportation
- Selected transportation accidents involving problems of a recurring nature

The NTSB has investigated numerous accidents and developed factual records and safety recommendations. Its findings are made public on the Internet and elsewhere.

1.8 INTEGRATED CONTINGENCY PLANS

Although the owners and operators of a facility may be subject to preparing and implementing one or more separate, stand-alone emergency response plans, they may also elect to consolidate them into a single contingency plan called an **integrated contingency plan, ICP,** or **one plan**.[4] This document sets forth the following goals:

- To provide a single integrated plan that consolidates the requirements of multiple facility emergency response plans
- To improve coordination of planning and response activities within a facility and with public and commercial emergency responders
- To minimize duplication

If a facility's owners or operators elect to prepare an ICP, the document must comply with the individual requirements of the plans concerned with emergency response activities mandated by several federal statutes and administered by multiple participating agencies. The following five plans noted in this chapter may be incorporated into an ICP, as they are needed:

- Risk management plan mandated by Clean Air Act regulations (40 C.F.R. Part 68)
- Facility response plan mandated by Clean Water Act regulations (40 C.F.R. §§112.20 and 112.21)
- Contingency plan mandated by RCRA regulations (40 C.F.R. §§264.52 and 265.52)
- Emergency action plan mandated by OSHA regulations (29 C.F.R. §§1910.38)
- Vessel or facility response plan mandated by DOT regulations (49 C.F.R. Part 194 and 33 C.F.R. Part 154, Subpart F)

Standard formats for the preparation of ICPs have been developed by EPA. This format may be adopted by facility owners and operators as they integrate multiple plans into a single plan.

integrated contingency plan (ICP, one plan) ■ A single emergency response plan that incorporates into one document the regulatory requirements of the emergency and contingency plans mandated by multiple federal statutes

1.9 GLOBAL HARMONIZATION CONCERNING THE CLASSIFICATION AND LABELING OF HAZARDOUS CHEMICALS

Over the years, most civilized countries have implemented methods for identifying hazards and classifying and labeling chemical products. In many instances, these methods vary significantly from country to country. Even within the United States, four federal authorities—OSHA, CPSC, EPA, and DOT—have the responsibility to warn consumers that certain products are potentially harmful. Although similarities exist, the agencies'

United Nations

---

[4]61 *Fed. Reg.* 28642 (June 5, 1966).

inconsistent and conflicting methods of classification and labeling have created unnecessary confusion for the public.

To circumvent this use of multiple systems for defining and classifying hazards and communicating hazard information, the United Nations designed the **GHS**, known formally as the Globally Harmonized System of Classification and Labeling of Chemical Substances.[5] The basis of the system is a set of rules with a common format and content for worldwide use when classifying the hazards of chemical products on warning labels and Safety Data Sheets.

In the United States, DOT and OSHA revised their regulations to align with the GHS, and EPA and CPSC are studying its adoption. Chemical manufacturers are now in the process of complying with these regulations.

The GHS provides a basis for a greater consistency in the classification and labeling of *hazardous chemicals*, thus enhancing their safe handling and storage in the workplace or a consumer-use setting. GHS broadly uses the term to include unique substances, products, and other preparations that could potentially cause harm. It affords emergency responders with another means of rapidly identifying the hazard(s) associated with exposure to substances that could cause an unreasonable risk of injury to public health or the environment. Its worldwide utilization also provides a means for rapidly identifying the nature of a hazardous material that was manufactured abroad but imported and distributed in the United States.

GHS communicates hazard information in the three major hazard groups listed in Table 1.2: physical hazards, health hazards, and environmental hazards. Most hazard classes are further subdivided into multiple categories, each of which is derived from criteria that are based on the results of prescribed testing procedures. These individual categories are not of interest here.

Inherent in the GHS is information that manufacturers and distributors of products affix on the labels to product containers. This information consists of the following:

- The product identifier (product name, UN/NA identification number, and DOT proper shipping name)[6]
- Either of the signal words WARNING or DANGER to denote the more and less severe hazard categories of the product, respectively
- One or more hazard statements
- One or more pictograms, of which there are two types: GHS pictograms that are displayed on the labels of chemical products, and transport pictograms that are displayed when the products are transported in commerce
- One or more precautionary statements including first-aid instructions
- Supplier identification (i.e., the name, address, and telephone number of the manufacturer or distributor)

The GHS pictograms associated with the hazards exhibited by chemical products are displayed in Table 1.2. They are black-and-white, red-bordered, diamond-shaped warning signs. Multiple pictograms may be displayed on the same container label to warn the observer of the potential hazards posed by the container's contents.

The transport pictograms are displayed in Table 1.3. They have background colors and symbols identical to those used on the DOT labels and placards noted later in Figures 6.5 and 6.12, respectively. When a transport pictogram is displayed on a label, the red-bordered pictogram for the same hazard is not shown.

---

[5]Globally Harmonized System of Classification and Labeling of Chemical Substances (GHS), (New York, NY and Geneva, Switzerland, 2011), ISBN-13: 978-92-1-117006-1.

[6]The nature of the UN/NA identification number and DOT proper shipping name is discussed in Section 6.1.

TABLE 1.2	**GHS Pictograms for Labeling the Containers of Hazardous Chemicals**	

NATURE OF GHS PICTOGRAM	TYPES OF HAZARDOUS MATERIALS REPRESENTED BY GHS PICTOGRAM	GHS PICTOGRAM
Exploding grenade	Explosives; certain self-reactive substances and mixtures; certain organic peroxides	
Flame	Flammable gases; flammable aerosols; flammable liquids; flammable solids; certain self-reactive substances and mixtures; pyrophoric liquids and solids; self-heating substances and mixtures; substances and mixtures, that, in contact with water, emit flammable gases; certain organic peroxides	
Flame over the letter "O"	Oxidizing gases, liquids, and solids	
Gas cylinder	Gases under pressure	
Corrosion	Substances that corrode skin and serious eye damage; substances that corrode metals	
Skull and crossbones	Certain acute toxicants	
Health hazard	Carcinogens (substances that cause cancer or are suspected of causing cancer); respiratory sensitizers; reproductive toxicants; specific target organ toxicants (single exposure); certain specific target organ toxicants (single exposure) and target organ toxicants (repeated exposure); germ-cell mutagens;[a] aspirants	
Exclamation point	Certain acute toxicants, skin irritants, eye irritants, skin sensitizers, and specific target organ toxicants (single exposure)	
Environment (aquatic toxicity)	Aquatic toxicants	

[a]Germ-cell mutagens are substances that cause a permanent change in the amount or structure of a cell's genetic material.

TABLE 1.3	GHS Pictograms for Transporting Hazardous Chemicals				
TRANSPORT PICTOGRAM					
NAME OF TRANSPORT PICTOGRAM	Explosive 1.1	Explosive 1.2	Explosive 1.3	Explosive 1.4	Explosive 1.5
TRANSPORT PICTOGRAM					
NAME OF TRANSPORT PICTOGRAM	Explosive 1.6	Flammable Gas	Non-Flammable Gas	Poison Gas	Flammable Liquid
TRANSPORT PICTOGRAM					
NAME OF TRANSPORT PICTOGRAM	Flammable Solid	Spontaneously Combustible Material	Dangerous When Wet Material	Oxidizer	Organic Peroxide
TRANSPORT PICTOGRAM					
NAME OF TRANSPORT PICTOGRAM	Poison Inhalation Hazard	Corrosive Material			

The hazard and precautionary statements are respectively coded with numbers preceded by an *H* or *P*, as relevant. Some representative examples are provided in Table 1.4. Aside from their use on the labels of chemical products, these statements are also components of the "Hazards Information" section of an SDS for a given chemical product.

OSHA's adoption of the GHS requires each chemical manufacturer, distributor, and importer to select the appropriate pictograms that describe their products. The use of the environmental pictogram is optional because environmental hazards are regulated by EPA, not OSHA. Emergency responders may rapidly identify the hazards associated with the chemical products by looking at their labels and acknowledging the significance of the pictograms.

When we study the properties of individual hazardous materials beginning in Chapter 7, GHS pictograms will be displayed with their hazard diamonds (Section 1.11) in the page margin near the point at which a discussion of each hazardous material first begins.

TABLE 1.4 — Some Representative Hazard and Precautionary Statements for the GHS Hazard Classes

HAZARD CLASS					
PHYSICAL HAZARD GROUP	**HAZARD STATEMENTS**			**PRECAUTIONARY STATEMENTS**	
	CODE	**EXAMPLE**	**CODE**	**EXAMPLE**	
Explosives (Division 1.1)	H201	Explosive; mass explosion hazard.	P201	Obtain special instructions before use.	
			P210	Keep away from heat/sparks/open flames/hot surfaces. No smoking.	
			P202	Do not fight fire when fire reaches explosives.	
Flammable gases	H220	Extremely flammable gas.	P210	Keep away from heat/sparks/open flames/hot surfaces. No smoking.	
			P377	Leaking gas fire: Do not extinguish, unless leak can be stopped safely.	
Flammable aerosols	H222	Extremely flammable aerosol.	P251	Pressurized container: Do not pierce or burn, even after use.	
			P211	Do not spray on an open flame or other ignition source.	
Oxidizing gases	H270	May cause or intensify fire; oxidizer.	P244	Keep reduction valves free from grease and oil.	
Gases under pressure	H280	Contains gas under pressure; may explode if heated.	P410	Protect from sunlight.	
			P403	Store in a well-ventilated place.	
Flammable liquids	H224	Extremely flammable liquid and vapor.	P240	Ground/bond container and receiving equipment.	
			P241	Use explosion-proof electrical/ventilating/lighting/…/equipment.	
			P242	Use only non-sparking tools.	
			P243	Take precautionary measures against static electricity.	
Flammable solids	H228	Flammable solid.	P210	Keep away from heat/sparks/open flames/hot surfaces. No smoking.	
Self-reactive substances and mixtures	H240	Heating may cause an explosion.	P220	Keep away from clothing/…/combustible materials.	
Pyrophoric liquids	H250	Catches fire spontaneously if exposed to air.	P222	Do not allow contact with air.	
Pyrophoric solids	H250	Catches fire spontaneously if exposed to air.	P335	Brush off loose particles from skin.	
			P334	Immerse in cool water/wrap with wet bandages.	
Self-heating substances and mixtures	H252	Self-heating in large quantities; may catch fire.	P407	Maintain air gap between stacks/pallets.	
			P403	Store in a well-ventilated place.	
Substances and mixtures which, in contact with water, emit flammable gases	H260	In contact with water releases flammable gases that may ignite spontaneously.	P223	Keep away from possible contact with water, because of violent reaction and possible flash fire.	
			P234	Keep only in original container.	
Oxidizing liquids	H271	May cause fire or explosion; strong oxidizer.	P283	Wear fire/flame resistant/retardant clothing.	

(Continued)

HAZARD CLASS		HAZARD STATEMENTS	PRECAUTIONARY STATEMENTS	
PHYSICAL HAZARD GROUP	CODE	EXAMPLE	CODE	EXAMPLE
			P220	Keep/Store away from clothing/combustible materials.
			P280	Wear protective gloves/protective clothing/ eye protection/face protection.
Oxidizing solids	H272	May intensify fire; oxidizer.	P378	Use _____ for fire extinction.
Organic peroxides	H240	Heating may cause an explosion.	P411	Store at temperatures not exceeding ___ °F (__ °C).
			P410	Protect from sunlight.
Substances corrosive to metals	H290	May be corrosive to metals.	P390	Absorb spillage to prevent material damage.
			P406	Store in corrosion-resistant/... container with a resistant inner liner.
Health Hazard Group				
Acute toxicity, oral	H301	Toxic if swallowed.	P310	Immediately call a POISON CENTER or doctor/ physician.
Acute toxicity, inhalation	H331	Toxic if inhaled.	P271	Use only outdoors or in a well-ventilated area.
			P340	Remove victim to fresh air and keep at rest in a position comfortable for breathing.
Acute toxicity, dermal	H311	Toxic in contact with skin.	P361	Remove/Take off immediately all contaminated clothing.
Skin corrosion/irritation	H314	Causes severe skin burns and eye damage.	P350	Gently wash with plenty of soap and water.
Serious eye damage/eye irritation	H320	Causes serious eye damage.	P305	Rinse cautiously with water for several minutes
			P351	Remove contact lenses, if present and if easy to do.
			P338	Continue rinsing.
Skin sensitization	H317	May cause an allergic skin reaction.	P350	Wash with plenty of soap and water.
Germ cell mutagenicity	H340	May cause genetic defects (Included in the statement is the route of exposure, if it is conclusively proven that no other routes of exposure cause the hazard.)	P281	Use personal protective equipment as required.
Carcinogenicity (cancer-causing hazard)	H350	May cause cancer (Included in the statement is the route of exposure, if known, when it has been conclusively proven that no other routes of exposure cause the hazard.)	P313	Get medical advice/attention.
Reproductive toxicity	H360	May damage fertility or the unborn child (Included in the statement are the specific effect, if known, and the route of exposure, when it has been conclusively proven that no other routes of exposure cause the hazard.)	P405	Store locked up.

(Continued)

HAZARD CLASS		HAZARD STATEMENTS		PRECAUTIONARY STATEMENTS	
PHYSICAL HAZARD GROUP	**CODE**	**EXAMPLE**	**CODE**	**EXAMPLE**	
Specific target organ toxicity, single exposure	H372	Causes damage to organ (Included in the statement are the organ affected, if known, and the route of exposure, when it has been conclusively proven that no other routes of exposure cause the hazard.)	P264	Wash thoroughly after handling	
Specific target organ toxicity, repeated exposures	H373	May cause damage to organs through prolonged or repeated exposure. (Included in the statement are the organ affected, if known, and the route of exposure, when it has been conclusively proven that no other routes of exposure cause the hazard.)	P260	Do not breathe dust/fume/gas/mist/-vapor/spray.	
Aspiration hazard	H304	May be fatal if swallowed and enters airways.	P331	Do not induce vomiting.	
Environmental Hazard Group					
Hazardous to the aquatic environment (acute hazard)	H400	Very toxic to aquatic life.	P273	Avoid release to the environment.	
Hazardous to the aquatic environment (chronic hazard)	H401	Toxic to aquatic life.	P391	Collect spillage.	
Hazardous to the ozone layer	H420	Harms public health and the environment by destroying ozone in the upper atmosphere.	P502	Refer to manufacturer/supplier for information on recovery/recycling.	

The table header: **TABLE 1.4** Some Representative Hazard and Precautionary Statements for the GHS Hazard Classes *(Continued)*

1.10 CANADA'S WORKPLACE HAZARDOUS MATERIALS INFORMATION SYSTEM

Chemical products are regularly transported across the common borders shared by the United States with Canada and Mexico. Although Mexico has adopted the GHS for voluntary use, Canada was still considering the potential implementation in its workplace regulations during 2013. Hence, it is relevant to note here how Canada communicates information concerning the hazards of chemical products to workers.

In Canada, when a hazardous material is encountered in the workplace, it is referred to as a controlled product. The Canadian Centre for Occupational Health and Safety, or CCOHS, regulates certain aspects of the controlled products used by employees through its Workplace Hazardous Materials Information System, or WHMIS. These regulations require suppliers to use the symbols for six hazard classes (denoted A through F) when labeling the containers of controlled products. They also require employers to ensure that their workers understand the meaning of these symbols and their use on MSDSs and labels.

The hazard classes of controlled products are depicted by the eight hazard symbols shown in Table 1.5. The symbols consist of black encircled pictograms on a white background.

TABLE 1.5	WHMIS Hazard Symbols	
CLASS OF CONTROLLED PRODUCT	**NATURE OF CONTROLLED PRODUCTS**	**HAZARD SYMBOL**
Class A: Compressed Gas	Compressed gases, dissolved gases, and gases liquefied by compression or refrigeration	
Class B: Flammable and Combustible Material	Solids, liquids, and gases capable of catching fire in the presence of a spark or open flame under normal working conditions	
Class C: Oxidizing Material	Materials that increase the risk of fire if they contact flammable or combustible materials	
Class D: Poisonous and Infectious Material Division 1: Materials causing immediate and serious toxic effects	Materials that cause death or immediate injury when a person is exposed to small amounts	
Class D: Poisonous and Infectious Material Division 2: Materials causing other toxic effects	Materials that can cause life-threatening and serious long-term health problems as well as less severe but immediate reactions in a person who is repeatedly exposed to small amounts	
Class D: Poisonous and Infectious Material Division 3: Biohazardous Infectious Material	Materials containing an organism that has been shown to cause disease or to be a probable cause of disease in persons or animals	
Class E: Corrosive Material	Includes caustic and acidic materials that can destroy the skin or "eat" through metals	
Class F: Dangerously Reactive Material	Materials that self-react dangerously (e.g., they may explode) upon standing or when exposed to physical shock or increased pressure or temperature; materials that decompose or polymerize vigorously; and materials that react with water to release a toxic gas.	

Although they are intended primarily to communicate hazard information to employees in the workplace, the hazard symbols are also useful to emergency responders who encounter controlled products during a transportation mishap or elsewhere.

Given the close proximity between the United States and Canada, GHS *and* WHMIS information is likely to be cited on the labels on chemical products imported from Canada into the United States and exported into Canada from the United States. The label illustrated in Figure 1.3 provides an example of the manner by which the GHS and WHMIS convey hazard information associated with a 35% solution of hydrogen peroxide. The GHS pictograms and WHMIS symbols inform the observer that the solution is an oxidizer and corrosive material that damages the skin and eyes upon exposure. GHS also includes the product identifier, signal word, and hazard and precautionary statements including first-aid instructions.

1.11 NFPA SYSTEM OF IDENTIFYING POTENTIAL HAZARDS

At the scene of an emergency, how is it possible to identify the potential hazards associated with the presence of a given hazardous material? The answer to this question is based on recognizing certain markings that are posted on stationary tanks, exterior building walls, pipelines, and other fixed facilities at which hazardous materials are stored or used.

The **National Fire Protection Association**, or **NFPA**[7] uses a procedure that provides a means for rapidly identifying the relative degree of three chemical properties associated with a given hazardous material: its health, flammability, and instability hazards.[8] NFPA implements this procedure by assigning one of five numbers, 0 through 4, to each property for a given hazardous material. The numbers in Table 1.6 identify the relative degree of hazard that corresponds to a relevant property. The number 0 signifies that the hazardous material at issue does not possess the relevant hazard, whereas the number 4 denotes that it possesses the highest degree of that hazard.

National Fire Protection Association (NFPA) ■
The professional organization that promotes and improves fire protection and prevention and establishes safeguards against loss of life and property by fire

SOLVED EXERCISE 1.2

When firefighters respond to a fire in a tank farm, what information is immediately needed to help them effectively fight the fire?

Solution: When they respond to a fire within a tank farm, firefighters require readily accessible and accurate information concerning the hazards of the tank contents. Accordingly, the information marked on the tanks should be easily visible and legible and should include the names of the liquids stored within them, their general hazards, and special firefighting precautions of which the firefighters should be aware.

To convey hazard information, NFPA recommends the use of number codes that are marked within the top three quadrants of a diamond-shaped symbol painted on, or otherwise affixed to, the exterior surfaces of storage tanks. Beginning with the left-hand quadrant and proceeding clockwise, the number codes denote the relative degrees of health, flammability, and instability hazards of the substances stored within each tank. Firefighters compare each number with the appropriate information listed in Table 1.6. Special codes are provided in the bottom quadrant. This combination of information helps firefighters determine which actions to take and not to take when they first encounter a fire within a storage tank.

---

[7]National Fire Protection Association, 1 Batterymarch Park, Quincy, MA 02269-9101.

[8]NFPA 704-2012, *System for the Identification of the Hazards for Emergency Response*, Copyright © 2011 (Quincy, MA: National Fire Protection Association).

IDENTIFICATION OF HEALTH		**IDENTIFICATION OF FLAMMABILITY**[a]		**IDENTIFICATION OF INSTABILITY**[b]	
HAZARD COLOR CODE: BLUE		**HAZARD COLOR CODE: RED**		**HAZARD COLOR CODE: YELLOW**	
SIGNAL	**TYPE OF POSSIBLE INJURY**	**SIGNAL**	**SUSCEPTIBILITY OF MATERIALS TO BURNING**	**SIGNAL**	**SUSCEPTIBILITY TO RELEASE OF ENERGY**
4	Materials that on very short exposure could cause death or serious residual injury even though prompt medical treatment was given	4	Materials that will rapidly or completely vaporize at normal pressure and temperature, or is readily dispersed in air, and will burn readily	4	Materials that in themselves are readily capable of detonation or of explosive decomposition or reaction at normal temperature and pressures
3	Materials that on short exposure could cause serious temporary or residual injury even though prompt medical treatment was given	3	Liquids and solids that can be ignited under almost all ambient conditions	3	Materials that in themselves are capable of detonation or explosive reaction but require a strong initiating source or that must be heated under confinement before initiation, or that react explosively with water
2	Materials that on intense or continued exposure could cause temporary incapacitation or possible residual injury unless prompt medical treatment is given	2	Materials that must be moderately heated or exposed to relatively high temperature before ignition can occur	2	Materials that in themselves are normally unstable and readily undergo violent decomposition but do not detonate; also, materials that may react violently with water or may form potentially explosive mixtures with water
1	Materials that on exposure would cause irritation but only minor residual injury even if no treatment is given	1	Materials that must be preheated before ignition can occur	1	Materials that in themselves are normally stable but which can become unstable at elevated temperatures and pressures or which may react with water with some release of energy but not violently
0	Materials that on exposure under fire conditions would offer no hazard beyond that of ordinary combustible materials	0	Materials that will not burn	0	Materials that in themselves are normally stable, even under fire exposure conditions, and which are not reactive with water

[a]See also Section 3.1-C.
[b]Before 1966, NFPA referred to this property as "chemical reactivity."

hazard diamond ■ A diamond-shaped figure divided into four quadrants, each of which is color-coded for each of a substance's three hazards—health, fire, and instability—and marked with a number designating the relative degree of the hazard

Each number is then displayed in the appropriate top three quadrants of the diamond-shaped diagram in Figure 1.4, called the **hazard diamond.** Each quadrant of the diamond is color-coded, beginning with the left-hand quadrant and proceeding clockwise: blue for the health hazard, red for the fire hazard, and yellow for the instability hazard. When warranted, the following symbols are displayed in the bottom white-colored quadrant of the diamond:

■ The letter W with a line drawn through its center (W̶) to caution firefighters against the application of water
■ The letters CRY to indicate the storage of a cryogen (Section 2.13)
■ The letters OX or OXY to indicate the storage of an oxidizer (Section 11.1)

FIGURE 1.4 Four potential hazards of a substance may be rapidly and simultaneously identified by the use of a color-coded numeral system on this hazard diamond. Beginning with the left-hand quadrant and advancing clockwise, the quadrants are color-coded as follows: blue for the health hazard; red for the fire hazard; and yellow for the instability hazard. A number from 0 to 4 is entered in each of the top three quadrants consistent with the information compiled in Table 1.6 to identify the severity of the health, fire, and chemical reactivity hazards. Each of several symbols, such as W and/or OXY, may also be entered in the bottom white quadrant to provide additional hazard information. (*Reprinted with permission from NFPA 704-2012, System for the Identification of the Hazards of Materials for Emergency Response, Copyright © 2012, National Fire Protection Association. This reprinted material is not the complete and official position of the NFPA on the referenced subject, which is represented solely by the standard in its entirety. The classification of any particular material within this system is the sole responsibility of the user and not the NFPA. NFPA bears no responsibility for any determinations of any values for any particular material classified or represented using this system.*)

- The letters *AS* to indicate the storage of nitrogen, helium, neon, argon, krypton, or xenon, each of which is a simple asphyxiant (Section 10.3-A)
- The radiation hazard symbol, which resembles a three-bladed propeller, or trefoil (Section 16.3), to indicate the storage of a radioactive material
- The word *ACID* or the letters *ALK* or *CORR* to indicate the storage of an acid, alkaline material, or corrosive material, respectively.

Only W and *OX* are cited by NFPA, but the other symbols have such widespread use that they are included here.

As noted earlier, beginning in Chapter 7, appropriate GHS pictograms and hazard diamonds will be displayed in the page margin near the point at which a discussion of each hazardous material first begins.

1.12 CHEMTREC

The Chemical Transportation Emergency Center (**CHEMTREC**) serves as a state-of-the-art communications center that deals with transportation mishaps by reinforcing the effectiveness of specialized emergency response groups and enhancing hazardous materials transportation security. Within the United States, CHEMTREC may be contacted by telephoning the following number, which is often posted on cargo tanks, rail tankcars, and other bulk packaging used to transport hazardous materials:[9]

CHEMTREC ■ Formally known as the Chemical Transportation Emergency Center, a public-service hotline for firefighters, law enforcement, and other emergency responders for obtaining information and assistance for incidents involving hazardous materials

CHEMTREC
(800) 424-9300

Companies that list CHEMTREC's emergency number on their packaging must be registered with CHEMTREC, and pay an annual fee.

---

[9]For calls originating outside the United States, telephone collect (703) 527-3887. Within Canada, telephone Le Centre canadien d'urgence transport du Ministère des transports (the Canadian Transport Emergency Centre of the Department of Transport), or CANUTEC, at (613) 996-6666. Within Mexico, telephone the Secretaría de Comunicaciones y Transportes (Secretariat of Communications and Transportation of Mexico), or SCT, at 52-5-684-1275. These numbers have been widely circulated in the professional literature distributed to emergency service personnel, shippers and carriers, and members of the chemical industry, and they have been further circulated in bulletins of governmental agencies, trade associations, and similar groups.

During an average year, CHEMTREC receives more than 10,000 calls relating to transportation incidents involving the release of hazardous materials. The notification of a transportation emergency involving hazardous materials to CHEMTREC does not constitute compliance with the RCRA, CERCLA, or DOT reporting requirements noted in Sections 1.6 and 6.8.

During each phone conversation, the specialist on duty at CHEMTREC attempts to determine the following information from the caller:

- The nature of the transportation incident including where and when it occurred
- The hazardous materials involved, their amounts, UN/NA identification numbers (Section 6.1-C), and other information concerning them
- The type and condition of the tanks or containers
- The name of the shipper or manufacturer and the shipping point
- The name of the carrier or other consignee
- The general nature and extent of the injuries, if any, to people, property, and the environment
- The prevailing local weather conditions
- The nature of the surrounding area
- The name of the caller and a means by which the caller can be located
- A means by which telephone contact can be reestablished with the caller or another responsible person at the scene

With the caller remaining on the line, the communicator draws on the best available information concerning the hazardous materials involved in the emergency response effort. As immediate first steps in controlling the emergency, specific hazard information and directions on what to do and what not to do are provided to the caller. CHEMTREC obtains this information by searching among the nearly 2.8 million SDSs in its database. To preclude unfounded personal speculation regarding the specific features of an emergency, the communicator is under instructions to abide strictly by information previously prepared for use.

Having advised the caller, the communicator proceeds immediately to notify the shipper by telephone. The known particulars of the emergency incident having thus been relayed, the responsibility for further guidance to the emergency response crew, including dispatching personnel to the scene, passes to the shipper.

Although proceeding to the second stage of assistance becomes more difficult when the shipper is unknown, communicators are armed with other resources on which they can rely. More than likely, the hazardous materials team from the local jurisdiction is immediately contacted. CHEMTREC may also request support from national resources like the Chemical Biological Incident Response Force, or CBIRF, when responding to incidents involving the release of chemical, biological, or explosive materials. CBIRF is a special unit in the United States Marine Corps that responds, upon request, to all types of emergency situations, not only transportation incidents. It was established to assist local, state, and federal agencies as well as international groups by offering capabilities for chemical detection and identification, casualty search and extraction, technical rescue, personnel decontamination, and emergency medical care and stabilization of contaminated victims.

1.13 NATIONAL RESPONSE CENTER

National Response Center ■ The national point of contact for reporting hazardous substance releases to the environment within the United States and its territories

The **National Response Center**, or **NRC**, first began operating when EPA required the services of a federal organization to whom responsible parties could report spills, discharges, fires, explosions, and other incidents in which oil and hazardous substances were released into waters of the United States. The existence of this center was first mandated by the CWA and has been operated by the U.S. Coast Guard since 1974.

The responsibilities of the NRC were broadened subsequent to enactment of federal transportation and environmental regulations. Today, the NRC serves as the sole federal point of contact for reporting oil, hazardous materials, hazardous substances, biological, etiological (infectious), and radiological spills throughout the United States and its territories. The NRC serves as the link to the **National Response Team,** or **NRT**. The latter is a group of individuals from 16 federal agencies having varying interests and expertise in emergency response matters. When the situation warrants, the NRT is dispatched to the scene of a release incident to provide assistance.

Federal law requires that the NRC must be notified *immediately* whenever anyone who releases to the environment a reportable quantity (Section 6.2-B) of a hazardous substance (including oil when water is or may be affected) or a material identified as a marine pollutant (Section 6.2-C). Reporting a release incident initially involves telephoning NRC at (800) 424-8802 or, in the District of Columbia, at (202) 426-2675; or through the use of their website at www.nrc.uscg.mil/pls/apex/f?p=201. The notification should minimally include such information as the date, time, nature of the release incident, the quantity and type of material involved in the release, and the extent of injuries, if any. A follow-up written report to the relevant federal agency is also required.

The NRC maintains a communication link with CHEMTREC personnel, who provide the caller with emergency response information. The NRC also informs the Chemical Safety and Hazard Investigation Board, which subsequently may choose to investigate the nature of the physical evidence found at the accident scene.

EPA dispatches an on-scene coordinator to the scene who ensures that the subsequent cleanup follows the relevant environmental laws. When the release occurs in navigable waterways, the United States Coast Guard is the on-scene coordinator.

Additional information regarding the reporting of a release of a hazardous substance is noted in Section 6.8.

National Response Team (NRT) ■ A federal interagency group responsible for distributing technical, financial, and operational information about hazardous substance releases and oil spills

REVIEW EXERCISES

Hazardous Components of Consumer Products

1. The unsafe practice known as *glue sniffing* involves the willful act of inhaling the toluene vapor emitted from certain glue products like hobby glue. Individuals who regularly sniff glue risk permanent damage to their liver, heart, and lungs. Why are consumer products containing toluene subject to CPSC's labeling requirement?

2. The following label is affixed to the exterior surface of an aerosol can containing an air deodorizer:

WARNING

FLAMMABLE, EYE IRRITANT, CONTENTS UNDER PRESSURE

DIRECTIONS FOR USE: To deodorize the air, spray whenever odors are a problem or occur. Allow a few seconds for superfine mist to bloom. To deodorize fabrics, hold can 14 to 15 in. from surface and spray evenly in a sweeping motion to obtain maximum penetration.

CAUTION: May cause severe eye irritation. Prolonged or repeated contact may dry or defat the skin and lead to inflammation (i.e., dermatitis). Inhalation may be irritating to the eyes, nose, and respiratory tract. Nausea, vomiting, and stomach upset may occur. Misuse or intentional misuse may be harmful or fatal. May affect the nervous system, eyes, or skin upon prolonged or repeated overexposure.

FIRST-AID INSTRUCTIONS: EYES—Immediately flush with plenty of water for at least 15 minutes. Get prompt medical attention. **SKIN**—Wash with soap and plenty of water. **INHALATION**—Remove to fresh air. If symptoms develop, seek immediate medical attention. If not breathing, give artificial respiration. **INGESTION**—In the unlikely event of swallowing, get immediate medical attention. Do NOT induce vomiting unless directed by medical personnel. Rinse mouth with water and give another cupful to drink. Never give anything by mouth to an unconscious person.

KEEP AWAY FROM HEAT, SPARKS, AND FLAMES. Do not puncture or incinerate container. Do not expose to heat or store at temperatures above 120°F (49°C), as container may burst. Avoid prolonged exposure to sunlight. Use with adequate ventilation.

KEEP OUT OF REACH OF CHILDREN.

INGREDIENTS: Acetone, fragrance, and liquefied petroleum gas.

(To provide confidentiality, the name and place of business of the manufacturer, packer, distributor, or seller and the trade name of the deodorizer have been intentionally omitted.) Which federal law provides the CPSC with the legal authority to require manufacturers to list the information on this label?

Hazardous Substances in the Environment and Certain Chemical Products

3. Small particles that form during the incomplete combustion of fuels are called *particulates*. Their inhalation has been linked with an increase in cardiovascular and pulmonary diseases in susceptible individuals. EPA regulates the concentration of particulates in the ambient air. Which federal statute provides EPA the legal authority to take this action?

4. EPA regulates the treatment, storage, and disposal of spent cyanide-plating solutions. Which federal statute provides EPA with the legal authority to take this action?

5. The chemical substances called *N*-nitrosamines are known to cause cancer in humans. To reduce or eliminate human exposure to them, EPA prohibits the mixing of substances with metalworking fluids used as lubricants or coolants when the fluids are likely to form *N*-nitrosamines. Which federal statute provides EPA with the legal authority to take this action?

6. In 2010, a natural gas pipeline in a residential area of San Bruno, California ruptured, resulting in the release of approximately 47.6 million cubic feet (1.4 million m³) of natural gas to the environment. Which federal statute provided the National Transportation Safety Board with the legal authority to inspect the scene of this disaster?

Hazardous Substances in the Workplace

7. An adhesive caulk used by carpenters is characterized as a thick white or tan paste with a mild, sweet odor. Canisters of the caulk bear the following warning statements:

WARNING

May cause eye, skin, nose, throat, and respiratory tract irritation. Harmful if swallowed or absorbed through the skin. Presents little or no hazard (if spilled) and/or no unusual hazard if involved in a fire.

POTENTIAL HEALTH EFFECTS

EYE CONTACT: May cause eye irritation.

SKIN CONTACT: May cause allergic skin reaction or sensitization. Prolonged or repeated contact with skin may cause irritation.

INHALATION: Harmful if swallowed.

CHRONIC HAZARDS: Repeated or prolonged exposure may cause skin, respiratory, kidney, cardiovascular, and liver damage.

How does the manufacturer of this adhesive use this warning information when designing the safety data sheet for this product?

Hazardous Substances in Communities

8. Why are the SDSs for chemical products manufactured and stored within communities often compiled for review at local fire departments?

Globally Harmonized System

9. Use the information on the label shown in Figure 1.5 to determine each of the following:
   (a) The product identifier
   (b) Signal word
   (c) Hazard statements
   (d) Precautionary statements
   (e) GHS and transport pictograms

10. Identify the potential hazards associated with exposure to a chemical product having the following GHS pictograms displayed on its label:

FIGURE 1.5 The portion of the label affixed to the containers holding the hypothetical chemical product, Offensive Acid No. 1. The GHS skull-and-crossbones, exclamation, and transport pictograms indicate that the product is a poisonous corrosive material. The exclamation pictogram indicates that the product may adversely impact health in other ways. Also provided here are the WHMIS pictograms that show the container holds a corrosive material, exposure to which causes an immediate and serious toxic effect.

OFFENSIVE ACID NO. 1

UNxxxx, Offensive Acid No. 1 (contains 95% acid)

DANGER

May be harmful if swallowed. Causes severe skin burns and eye damage. May be corrosive to metals. Fatal if inhaled. Toxic to aquatic life.

Keep out of reach of children. Do not breathe dust, fume, gas, mist, vapor, or spray. Absorb spillage to prevent material damage. Store in corrosion-resistant container or tank with a resistant inner liner. Wear protective gloves, protective clothing, eye protection, and face protection. Wear respiratory protection.

FIRST-AID INSTRUCTIONS:

IF IN EYES: Rinse cautiously with water for several minutes. Remove contact lenses, if present, and easy to do. Continue rinsing. Immediately call POISON CENTER or doctor/physician.

IF SWALLOWED: Immediately call POISON CENTER or doctor/physician.

IF ON SKIN: Remove immediately all contaminated clothing. Rinse skin with water/shower.

My Company
My Street
My Town
My State 00000
Telephone (000) 000-0000

Hazard Diamonds

11. Proceeding clockwise from the far left, first-on-the-scene responders observe the numbers 4, 2, and 2 displayed within the top three quadrants of a hazard diamond affixed to the front of a burning chemical storage shed. Describe how each number is useful to the emergency responders.

12. For each of the following pairs of hazard diamonds, identify the appropriate code that conveys the greater degree of the following hazards:

    (a) Fire hazard

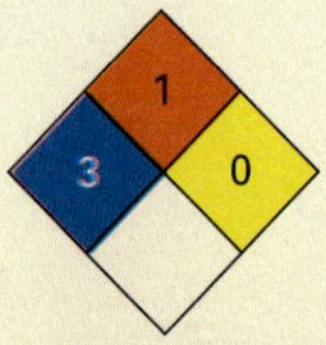

(*o*-Chlorophenol) 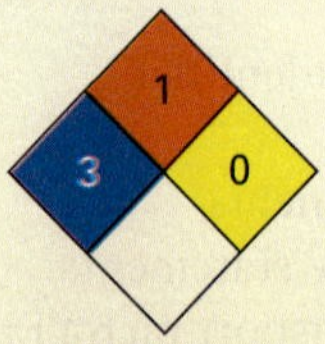　　　　(Beryllium)

    (b) Health hazard

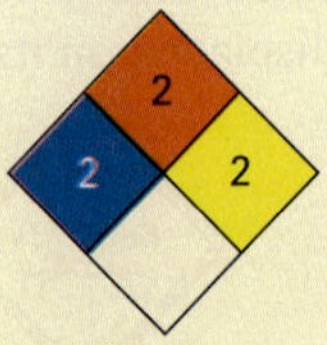

(Diketene) 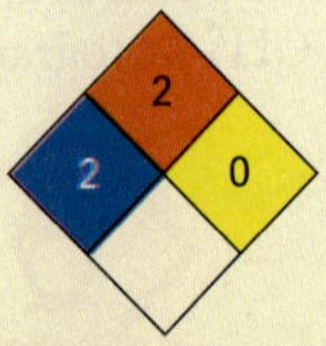　　　　(Benzaldehyde)

(c) Instability hazard

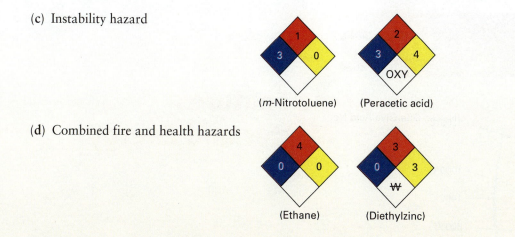

(*m*-Nitrotoluene)　(Peracetic acid)

(d) Combined fire and health hazards

(Ethane)　(Diethylzinc)

13. Using the following pair of hazard diamonds, determine whether the greater degree of fire hazard, health hazard, and instability is posed by the presence of epichlorohydrin or ethyl chloride within a workplace setting:

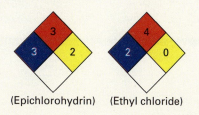

(Epichlorohydrin)　(Ethyl chloride)

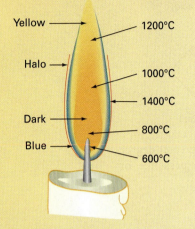

Yellow — 1200°C

Halo —

— 1000°C

— 1400°C

Dark — 800°C

Blue — 600°C

KEY TERMS

OBJECTIVES

- Identify the common properties that characterize solids, liquids, and gases.
- Use the factor-unit method to convert measurements expressed in customary units into their equivalents in appropriate metric units and vice versa.
- Describe the concepts of density, specific gravity, vapor density, and vapor pressure and cite examples of their usefulness to emergency responders.
- Convert temperature readings on one temperature scale into their equivalents on the other temperature scales noted in this chapter.
- Describe the mechanisms that contribute to the spread of fire from one location to another.
- Convert heat measurements expressed in Btus to measurements expressed in calories and vice versa.
- Use the gas laws to calculate the volume of gases subjected to different temperature and pressure conditions.
- Describe the potential danger associated with the expansion of a heated gas or vapor that is confined in a storage vessel.
- Identify the general hazards that emergency responders encounter when exposed to cryogens.

At first glance, the mere number of hazardous materials is likely to overwhelm the average nonscientist. How is it possible to learn the individual properties of so many substances and recall them under the stressful conditions that often prevail when lives and property are in jeopardy?

Fortunately, we can associate the hazardous properties of many substances with their state of matter. We learn, for instance, that all gases possess certain common properties; on studying the chemistry of gases, we learn to identify these common properties and then turn our attention later to the features that cause individual gases to be regarded as unique substances.

In this chapter, we learn about some of the general properties of matter and energy and how they influence certain phenomena such as the spread of fire. Also, as we review the properties of matter and energy, we learn how they relate to the issues in fire science. Specifically, in this chapter we learn how the modes of heat transfer contribute to the propagation of fire, the reason water often effectively extinguishes a fire, and why gas cylinders are likely to rupture when their contents are excessively heated.

2.1 MATTER DEFINED

Each day, we encounter air, water, metals, stone, dirt, animals, and plants. These are the materials of which our world is made. Scientists refer to them as the different kinds of **matter**. When we search for common features among its forms, we note that ordinary matter possesses mass and occupies space. In other words, matter is distinguishable from empty space by its presence in it.

matter ■ Anything that possesses mass and occupies space

mass ■ The quantity of matter possessed by an object regardless of its location in the universe

weight ■ The force with which a mass is pulled toward the center of Earth

gravity ■ The force of attraction between two bodies

physical state of matter ■ Any form in which matter is generally encountered: solid, liquid, and gas

volume ■ The capacity of matter confined within a tank or container

solid ■ Matter that possesses a definite volume and a definite shape

The concept of **mass** is closely related to the concept of **weight**. In our universe, every form of matter is attracted to all other forms by the force we call **gravity**. On Earth, the weight of matter is a measure of the force with which gravity pulls it toward Earth's center. As we leave Earth's surface, the gravitational pull decreases until it becomes virtually insignificant. The *weight* of matter accordingly reduces to zero, yet the matter still possesses the same *mass* as it did on the surface of Earth. For this reason, the expressions "has a mass of" and "weighs" are essentially equivalent on Earth's surface.

As usually experienced, matter exists in three different forms or states of aggregation: solid, liquid, and gas.[1] These are called the **physical states of matter**.

Because matter occupies space, a given form of matter is also associated with a definite volume or capacity. Space should not be confused with air, because air is itself a form of matter. **Volume** refers to the actual amount of space that a given form of matter occupies.

2.1-A SOLIDS

A **solid** is the form of matter existing in a rigid state independent of the size and shape of its container. Consider this book. It retains its shape regardless of its position in space and does not need to be placed into a container to retain that shape. Left to itself, it will never spontaneously assume a shape different from what it has now.

Solids also occupy a definite volume at a given temperature and pressure. We can squeeze most solids with all our might or heat or cool them, but their total volume change is relatively small.

Certain solid plastics can behave differently from most solids. Because of their inherent elasticity, they can assume different shapes and volumes when squeezed, stretched, or otherwise manipulated. Solid foams also can behave uncharacteristically, because they contain encapsulated air.

2.1-B LIQUIDS

liquid ■ Matter that possesses a definite volume but lacks a definite shape

A **liquid** is a form of matter that does not possess a characteristic shape; rather, its shape depends on the shape of the container it occupies. Consider water within a glass. The liquid takes the shape of the glass up to the level that it occupies. If we pour the water into a cup, the water takes the shape of the cup; or if we pour it into a bowl, the water takes the shape of the bowl. Of course, sufficient space must always be available for the water within the container; otherwise, the liquid overflows. Assuming that space is available, however, any liquid assumes whatever shape its container possesses.

Like solids, liquids occupy a specific volume at a given temperature and pressure. They tend to maintain a relatively fixed volume when they are exposed to a change in either of these conditions. Liquids and solids are often considered incompressible, because the application of pressure barely changes their volumes. When heated, liquids do expand, much more than solids do, but not nearly to the extent that gases expand. We revisit the expansion of heated liquids in Section 2.11.

2.1-C GASES

gas ■ Matter that possesses neither a definite volume nor a definite shape

A **gas**, or **vapor**, is another form of matter that does not possess a characteristic shape and assumes the shape of its container. If a gas or mixture of gases, such as air, is put into a balloon, it assumes the shape of the balloon; or if it is transferred into a tire, it assumes the shape of the tire.

vapor ■ The gaseous form of a substance that exists as a solid or liquid at normal ambient conditions

---

[1]Other physical states of matter are known besides solid, liquid, and gas. A fourth state, *plasma*, exists only at very high temperatures, whereas a fifth state, called the *Bose–Einstein condensate*, exists only at very low temperatures. Although we will not encounter materials in the fourth and fifth states during our study of hazardous materials, it is worthy to note that the sun, other stars, and other forms of intergalactic matter exist in the plasma state. It is the predominant state of ordinary matter within the universe.

Gases also lack a characteristic volume. When confined to a container with non-rigid, flexible walls, the volume that a confined gas occupies depends on its temperature and pressure. When confined to a balloon, for instance, the gas's volume expands and contracts depending on the prevailing temperature and pressure. When confined to a container with rigid walls, however, the volume of the gas is forced to remain constant. This property of gases can cause rigid containers to explode, a topic we note later in Section 2.12.

The properties of solids, liquids, and gases noted in this section can now be used to formally define the three states of matter:

- Matter in the solid state possesses a definite volume and a definite shape.
- Matter in the liquid state possesses a definite volume but lacks a definite shape.
- Matter in the gaseous state possesses neither a definite volume nor a definite shape.

2.2 UNITS OF MEASUREMENT

The necessity to measure, and to measure with accuracy, is essential to any kind of scientific or technological endeavor. To *measure* means to find the number of units in a sample of something. For instance, when we measure the distance from one point to another along a wall, we generally determine how many feet, yards, or meters are between the two points. The foot, yard, and meter are examples of the common units of length.

Two systems of units have survived the test of time: the United States **customary system of weights and measures**, and the SI or metric system. The United States is the only remaining industrialized country that does not exclusively use the SI system. The customary system is based on the use of an array of units that appear to have no obvious interrelationship, such as inches (in.), feet (ft), yards (yd), miles (mi), ounces (oz), pounds (lb), pints (pt), quarts (qt), and gallons (gal).

customary system of weights and measures ▪ Any unit of weight and measures based on the yard and pound and used in normal commerce within the United States

The majority of the world's population and the worldwide scientific community use a system that is historically called the metric system of measurement. This system has been modified so that today, we actually use what is called the **SI system of units**, where "SI" is an abbreviation of the system's official French name, *Le Système International d'Unités*. The SI system encourages the use of certain fundamental units from which all other measurements are constructed. These fundamental units are called the **SI base units**. Examples of SI base units are the meter and kilogram, for length and mass, respectively.

SI system of units ▪ Known historically as the metric system, the scientific standard of measurement that employs a set of units describing length, mass, time, and other characteristics of matter

Certain prefixes are used in the SI system to denote multiples and fractions of the units of measurement. Each prefix is a fraction or multiple of the number 10. For example, when we wish to refer to 1000 meters, we use the word *kilometer*. The prefix *kilo* means 1000 times the meter, the SI base unit.

The prefixes listed in Table 2.1 are commonly used in studying the chemistry of hazardous materials. These particular prefixes should be committed to memory. They are used to measure certain properties of matter, particularly its length, mass, and volume. It is appropriate to discuss each type of measurement separately.

SI base units ▪ The accepted units of measurement adopted internationally such as the meter for length, kilogram for mass, and kelvin for temperature

2.2-A LENGTH

Today, the **meter (m)** is defined as the length of the path traveled by light in a vacuum in $1/299,792,458$ of a second. In ordinary practice, we measure length in the metric system with a metric ruler. By so doing, we discover that one meter is slightly longer than a yard; specifically, one meter equals 39.37 inches (in.).

meter (m) ▪ The SI unit of length in the metric system of measurement

$$1 \text{ m} = 39.37 \text{ in.}$$

One meter is equivalent to 100 centimeters (cm) and to 1000 millimeters (mm).

$$1 \text{ m} = 100 \text{ cm} = 1000 \text{ mm}$$

TABLE 2.1	Common Prefixes Used in the Metric System	
PREFIX	**SYMBOL**	**MEANING**
hecta-	h	One hundred (10^2) times the SI base unit
kilo-	k	One thousand (10^3) times the SI base unit[a]
deci-	d	One-tenth (10^{-1}) of the SI base unit
centi-	c	One-hundredth (10^{-2}) of the SI base unit
milli-	m	One-thousandth (10^{-3}) of the SI base unit
micro-	μ	One-millionth (10^{-6}) of the SI base unit
nano-	n	One-billionth (10^{-9}) of the SI unit
pico-	p	One-trillionth (10^{-12}) of the SI unit

[a]The SI base unit of mass is an exception.

For very large lengths, we use the kilometer (km); once again, 1000 meters is equivalent to 1 kilometer.

$$1 \text{ km} = 1000 \text{ m}$$

For very small lengths, we use the micron (μm). One micron is one-millionth of a meter.

$$1 \text{ μm} = 0.000001 \text{ m}$$

The relationship between the inch and centimeter is shown in Figure 2.1. One inch equals 2.54 centimeters.

$$1 \text{ in.} = 2.54 \text{ cm}$$

2.2-B MASS

The SI unit of mass is the *kilogram*. A bit of attention needs to be given to constructing the multiples and fractions of mass measurements, because this unit of mass already contains a prefix (kilo-). The names of the various multiples and fractions of the unit of mass are constructed by attaching the appropriate prefix to the word **gram (g)**, not kilogram. In other words, the gram is used as though it were the SI unit of mass, even though it actually is not.

One kilogram is equivalent to 2.2 pounds. One gram is approximately the mass of a peanut.

Three common metric units of mass are the milligram, microgram, and kilogram. One one-thousandth of a gram is a *milligram* (mg); one one-millionth of a gram is a *microgram* (μg); and 1000 grams is a kilogram (kg).

$$1 \text{ mg} = 0.001 \text{ g}$$
$$1 \text{ μg} = 0.000001 \text{ g}$$
$$1 \text{ kg} = 1000 \text{ g}$$

gram (g) ■ One one-thousandth of the mass of 1 kilogram

FIGURE 2.1 The relationship between the inch and the centimeter. Note that 1 inch (in.) equals 2.54 centimeters (cm).

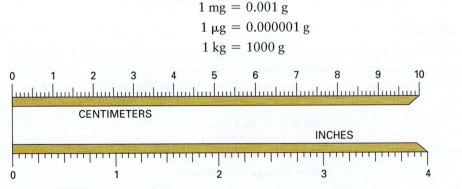

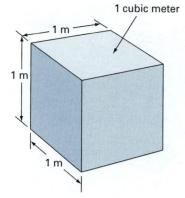

1 cubic meter

FIGURE 2.2 Because this cube measures 1 meter on each edge, its volume is 1 cubic meter (m^3), the approved SI unit of volume.

1 m

1 m

1 m

One milligram is approximately the mass of a grain of sand, whereas one microgram is approximately the mass of a fleck of dust.

For relatively large mass measurements, use is made of the metric ton, or **tonne (t)**.

$$1\ t = 1000\ kg$$

To verbally distinguish the ton and the tonne, the latter is pronounced "tunny."

tonne (t) ■ The metric unit of mass equivalent to 1000 kilograms

2.2-C VOLUME

The approved SI unit of volume is the *cubic meter* (m^3). This unit is derived directly from the SI unit of length and is not a SI base unit itself. We can easily derive the cubic meter by considering the cube in Figure 2.2, which measures 1 meter on each edge. The volume of this cube is determined by taking the product of its length, width, and height.

$$\text{Volume} = 1\ m \times 1\ m \times 1\ m = 1\ m^3$$

Using simple arithmetic, the volume of this cube is determined to be 1 cubic meter. Because it is possible to derive the cubic meter directly from a previously defined unit, it is unnecessary to define some other unit as the SI unit of volume.

Although the cubic meter is the approved unit of volume in the SI system, another unit has been used for many years to measure volume. This unit is the **liter (L)**. One quart is slightly less than one liter.

$$946\ mL = 1\ qt$$

liter (L) ■ The volume occupied by a cube measuring 10 centimeters to an edge

Chemists continue to use the liter for measuring volume because its size is so convenient for laboratory-scale measurements. The cubic meter is comparatively too large.

One liter is equivalent to one one-thousandth of a cubic meter.

$$1\ L = 0.001\ m^3$$

If we construct a cube measuring 1 meter on each edge and then divide the resulting volume into 1000 equal-size cubes, the volume of each small cube is one liter. Imagine further dividing each liter cube into another 1000 equal-size cubes. These subdivisions are illustrated in Figure 2.3. Because the prefix *milli-* means one one-thousandth of a unit, the volume of each of the new cubes is one *milliliter* (mL).

$$1\ L = 1000\ mL$$

Formerly, the milliliter was known as a cubic centimeter (cm^3). Although the use of the cubic centimeter has essentially been phased out in chemistry, it is still used in the medical field.

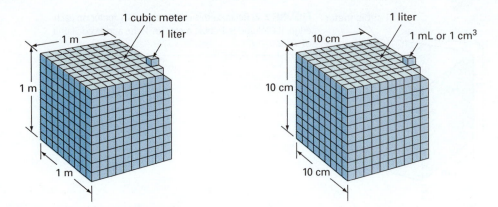

FIGURE 2.3 The cube on the left measures 1 meter on each edge and has been divided into 1000 equal-size cubes. The volume occupied by each of the smaller cubes is 0.001 cubic meters, or 1 liter (L). The cube on the right measures 10 centimeters on each edge and has been subdivided into 1000 equal-size cubes. The volume of the larger cube is 1 liter, and the volume of each smaller cube is 1 milliliter (mL), or 1 cubic centimeter (cm^3).

2.2-D GENERAL USE OF THE METRIC SYSTEM IN THE UNITED STATES

In the international community, the United States is the sole industrialized nation that has not officially adopted the metric system. The United States' nonuse of the metric system at one point was contributing to unsuccessful competition in some international markets. To address this concern, Congress passed the *Metric Conversion Act* in 1975. Notwithstanding its passage, the act called only for voluntary compliance and accomplished little change in the manner by which Americans measure the properties of matter.

Then, in 1988, Congress passed the **Omnibus Trade and Competitiveness Act**, which mandated the following:

■ Identified the metric system as the preferred system for U.S. trade and commerce
■ Required use of the metric system by all U.S. government agencies
■ Required use of metric units for all federally funded construction projects costing over $10 million
■ Required use of metric units for all federally assisted highway construction projects that began on or after October 1, 1996

This act obligated building and highway construction companies to begin using the metric system when seeking compensation for work on federally funded projects.

The use of metric units is also evident on warning labels for a variety of products whose manufacture or use is regulated by agencies of the federal government, especially when the relevant measurements have been obtained by scientists. For example, a high-salt (sodium chloride) diet has been linked with ailments like high blood pressure, heart attacks, and strokes. To warn about these potential ill effects, the FDA requires information concerning the dietary content of sodium to be printed on labels affixed to containers of prepared foods. Three labeling definitions are used to denote the sodium content per serving: sodium-free, less than 5 milligrams; very low sodium, less than 35 milligrams; and low sodium, less than 40 milligrams.

Even when the use of the metric system is not specifically mandated, consumer product manufacturers are printing metric units on labels together with their customary counterparts. Look at a beverage can, for instance. Its volume may be listed as 32 fluid ounces (1 qt) together with 946 milliliters. All in all, use of the metric system is slowly creeping into the United States and taking us one step closer toward common world-wide usage.

Omnibus Trade and Competitiveness Act ■ The federal statute that enhances the competitiveness of American industry throughout international markets based in part on the required use of the metric system

2.3 CONVERTING BETWEEN UNITS OF THE SAME KIND

Suppose we wish to convert between grams and kilograms, pounds and grams, or liters and gallons. How do we accomplish this task? Problems like these are best solved by using a simple procedure called the **factor-unit method**. Briefly, it consists of the following key steps:

- Identify the desired unit.
- Choose the proper **conversion factor(s)**.
- Multiply the given measurement by the conversion factor(s), being certain to multiply and divide numbers and cancel equal units.

Choosing the proper conversion factor is a crucial step to obtaining the correct answer to a problem. The conversion factor is a fraction numerically equal to 1 that equates the quantity in the numerator to the quantity in the denominator. For example, we know there are 1000 meters in 1 kilometer. Either of the following fractions serves as a correct conversion factor:

$$\frac{1000 \text{ m}}{1 \text{ km}} \quad \text{or} \quad \frac{1 \text{ km}}{1000 \text{ m}}$$

The factors are respectively read as follows: 1000 meters per kilometer; one kilometer per 1000 meters.

Suppose we know the height of Moscow's Mercury City Tower (at right), the tallest building in Europe. It measures 338.82 meters from ground level to the top. What is its height in kilometers?

Using the factor-unit method, we simply multiply the given measurement, 338.82 meters, by a conversion factor that relates meters to kilometers.

$$338.82 \cancel{\text{ m}} \times \frac{1 \text{ km}}{1000 \cancel{\text{ m}}} = 0.3388 \text{ km}$$

Note that the *m* in *338.82 m* cancels with the *m* in *1000 m*. Then, dividing 338.82 by 1000 gives 0.3388 kilometers (to four significant figures). Individuals who are experienced in using the metric system simply move the decimal point from left to right, or vice versa, as the need arises.

Some of the more frequently used metric units and their equivalent customary counterparts are given in Table 2.2. These relationships permit us to write conversion factors like the following:

$$\frac{1 \text{ m}}{39.37 \text{ in.}} \quad \frac{2.54 \text{ cm}}{1 \text{ in.}} \quad \frac{1 \text{ kg}}{2.2 \text{ lb}} \quad \frac{946 \text{ mL}}{1 \text{ qt}}$$

factor-unit method ■ A procedure for changing a quantity expressed in one unit to a quantity of another unit by multiplying, dividing, and arithmetically canceling numbers and units

conversion factor ■ A fraction equal to 1 in which the magnitude of one unit is related in the numerator to the magnitude of another unit of the same type in the denominator

TABLE 2.2	Common SI (Metric) Units and Their Customary Equivalents
METRIC UNIT	**CUSTOMARY UNIT**
1 m	39.37 in.
2.54 cm	1 in.
1 kg	2.2 lb
454 g	1 lb
946 mL	1 qt

A class I standpipe system contains a 2½-inch hose connection for use by fire departments and others trained in handling heavy fire streams. Express the equivalent length in centimeters.

Solution: We first need one or more conversion factors to convert 2½ inches into its equivalent length in centimeters. Because one inch is 2.54 centimeters, we can construct the following conversion factor:

$$\frac{2.54 \text{ cm}}{1 \text{ in.}}$$

Through use of the factor-unit method, we then convert 2½ inches (or 2.5 inches) into centimeters as follows:

$$2.5 \text{ in.} \times \frac{2.54 \text{ cm}}{1 \text{ in.}} = 6.4 \text{ cm}$$

A firefighter must possess the physical agility to easily carry a 60-pound bundle from the base to the top of a 148-foot ladder. Express the bundle mass in kilograms and the length of the ladder in meters.

Solution: According to the information in Table 2.2, 1 kg = 2.2 lb, and 1 m = 39.37 in. By the factor-unit method, a 60-pound bundle is equivalent to a 27-kilogram bundle.

$$60 \text{ lb} \times \frac{1 \text{ kg}}{2.2 \text{ lb}} = 27 \text{ kg}$$

Furthermore, a 148-foot ladder is equivalent to a 45-meter ladder.

$$148 \text{ ft} \times \frac{12 \text{ in.}}{1 \text{ ft}} \times \frac{1 \text{ m}}{39.37 \text{ in.}} = 45 \text{ m}$$

2.4 CONCENTRATION

concentration ■ The relative amount of one substance mixed or dissolved in a specified amount of a second substance

percentage (%) ■ The concentration of a mixture expressed as parts of a unit in 100 of the same units

Scientists often measure the amount of a substance as a stated unit in a given mass or volume of a mixture. This value is called the substance's **concentration**. A variety of units are used by professionals to measure concentration. Emergency responders often use concentrations provided by the federal government in DOT, OSHA, and EPA regulations. For instance, when noting an airborne concentration of hydrogen cyanide at a fire scene, you may wish to compare it to OSHA's 8-hour permissible exposure limit in the workplace. This concentration is 11 milligrams per cubic meter (11 mg/m^3).

Concentration is sometimes expressed as a percentage. The word **percentage** means parts per hundred. To calculate the percentage of an item in a total, we divide the number of items by the total number and multiply the product by 100. The percentage is expressed as a number followed by the percent sign (%). In chemistry, the concentration of a substance in a mixture is expressed as either a percentage by mass or a percentage by volume. In the former instance, we use mass units of the same type, and in the latter case, we use volume units of the same type. For example, suppose we have 100 grams of a solution in which 5 grams of table salt is dissolved. The concentration of table salt in the solution is expressed as 5% by mass. If we have 100 milliliters of a solution in which 5 milliliters of a substance is dissolved, the concentration of the substance in the solution is expressed as 5% by volume.

In this text, we shall also encounter hazardous materials in the air or water whose concentrations are expressed in **parts per million (ppm)** or **parts per billion (ppb)**. A concentration of 1 ppm means 1 part of a substance measured as an arbitrary unit in a million parts of the same unit. For example, depending on the units used, 1 ppm could be one molecule in a million molecules, one gram in a million grams, or one ounce in a million ounces. A concentration of 1 ppm is extremely minute. It corresponds approximately to 1 cent in $10,000.

Similarly, one part per billion means 1 part of a substance measured as a unit in a billion parts of the same unit. Parts per million are related to parts per billion as follows:

$$1 \text{ ppm} = 1000 \text{ ppb}$$

This means that 1 ppb is one one-thousandth of 1 ppm. One part per billion is equivalent to one cent in $10 million.

Concentrations in standard units like pounds per cubic foot (lb/ft^3) or kilograms per cubic meter (kg/m^3) may be converted to parts per million or parts per billion. The easiest procedure is to use the online service at http://www.unitconversion.org.

parts per million (ppm) ∎ The concentration of a mixture in which a certain number of units are contained within a million of the same units

parts per billion (ppb) ∎ The concentration of a mixture in which a certain number of units are contained within a billion of the same units

2.5 DENSITY OF MATTER

If the volume of a given mass of a substance is known, we can readily use division to compute its mass per unit volume. This ratio is a property of substances known as the **density**.

density ∎ The property of a substance measured by dividing its mass by the volume it occupies

$$\text{Density} = \frac{\text{mass}}{\text{volume}}$$

For instance, one milliliter of liquid mercury has a mass of 13.6 grams at 68°F (20°C). Hence, the density of mercury is 13.6 g/mL at 68°F (20°C).

In the customary system of units, the densities of solids and liquids are usually expressed in pounds per gallon (lb/gal) or pounds per cubic foot (lb/ft^3), whereas the densities of gases or vapors are expressed in pounds per cubic foot. In the metric system, the densities of solids and liquids are normally expressed in grams per milliliter (g/mL) or kilograms per cubic meter (kg/m^3), whereas the densities of gases and vapors are expressed in either grams per cubic meter (g/m^3) or kilograms per cubic meter (kg/m^3).

Scientists measure densities in the laboratory by determining the volume occupied by a given mass of a substance. Suppose we weigh exactly 50 pounds of water on a scale.

$$\text{Mass} = 50.00 \text{ lb}$$

Then, we transfer this water into containers of known capacity. When this simple exercise is performed, we discover that 50.00 pounds of water completely fills six 1-gallon containers to the brim at 68°F (20°C).

$$\text{Volume} = 6.00 \text{ gal}$$

We then compute the density of water to be 8.33 lb/gal as follows:

$$\text{Density of water} = \frac{50.00 \text{ lb}}{6.00 \text{ gal}} = 8.33 \text{ lb/gal}$$

Scientists establish the densities of other substances by conducting similar exercises. Some results are provided in Table 2.3.

We can also convert a density in one unit to its equivalent in other units. For example, we can determine that a density of 8.33 lb/gal is equivalent to 62.3 lb/ft^3. We calculate this latter value using the conversion factor 1 ft^3 = 7.48 gal, or 7.48 gal/ft^3.

$$8.33 \text{ lb/gal} \times 7.48 \text{ gal/ft}^3 = 62.3 \text{ lb/ft}^3$$

When the density of water is determined using metric units, the obvious result at 20°C is 1.00 g/mL, or 1000 kg/m³. Historically, water was once selected as the reference standard for establishing the units of mass and volume in the metric system. One gram was formerly defined as the mass of water that occupies a volume of one milliliter at 4°C, the temperature at which water has its maximum density.

SOLVED EXERCISE 2.3

The density of dry air in kilograms per cubic meter (kg/m³) varies with temperature as follows: 1.29 (0°C), 1.25 (10°C), 1.21 (20°C), and 1.16 (30°C). What do these data illustrate about the nature of dry air during building fires?

Solution: These data indicate that the density of dry air decreases as the temperature of the air increases. In other words, a given volume of dry air gets lighter in mass as it becomes hotter. This lighter, hotter air is buoyant; that is, it rises upward. During building fires, the lighter, hotter air concentrates near the undersides of ceilings and roofs, whereas the denser, colder air concentrates near the floors.

Each substance possesses a density that varies with the prevailing temperature, and possesses a maximum density at a unique temperature. Figure 2.4 illustrates that the maximum density of water is 1.0000 g/mL at 39.1°F (3.97°C). The density of water decreases at temperatures less than and greater than 39.1°F (3.97°C).

2.5-A SPECIFIC GRAVITY

Often, the mass of a liquid or solid substance is compared with the mass of an equal volume of water. This comparison yields a dimensionless number called **specific gravity**.

specific gravity ■ The mass of a given volume of matter compared with the mass of an equal volume of water

TABLE 2.3	Densities of Some Common Liquids and Solids	
	DENSITY AT 68°F (20°C)	
SUBSTANCE	**g/mL**	**lb/gal**
Acetone	0.790	6.58
Aluminum	2.7	22.5
Benzene	0.879	7.32
Carbon disulfide	1.274	10.6
Carbon tetrachloride	1.595	13.3
Chloroform	1.489	12.4
Diethyl ether	0.715	5.96
Ethanol, 180 proof	0.828	6.90
Gasoline	0.66–0.69	5.5–5.8
Kerosene	0.82	6.8
Lead	11.34	94.5
Mercury	13.6	113
Silver	10.5	87.5
Sulfur	2.07	17.3
Trichloroethylene	1.460	12.2
Water[a]	1.00	8.3

[a]At 39°F (4°C).

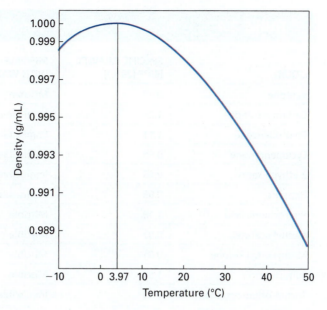

FIGURE 2.4 The density of water at atmospheric pressure near its freezing point as a function of temperature. The density of water is at its maximum at 39.1°F (3.97°C).

For example, when 1 gallon of sulfuric acid is weighed, we find that the scale reads 15.33 pounds. This means that the density of sulfuric acid is 15.33 lb/gal. We can compare the density of sulfuric acid with the density of water to compute the specific gravity of sulfuric acid as follows:

$$\text{Specific gravity} = \frac{15.33 \text{ lb/gal}}{8.33 \text{ lb/gal}} = 1.84$$

This computation informs us that any volume of sulfuric acid is 1.84 times heavier than an equal volume of water.

The specific gravities of liquids are directly determined through the use of a device called a *hydrometer*, a cylindrical glass stem containing a bulb weighted with shot so that the stem floats upright in a liquid. A graduated scale is printed on the hydrometer's stem. A sample of the liquid is generally poured into a tall cylinder or jar as in Figure 2.5, and the hydrometer is lowered into the sample until it floats. The specific gravity of the liquid is read from the graduated scale at the point where the surface of the liquid touches the stem.

FIGURE 2.5 A hydrometer is an instrument that is often used to determine the specific gravity of a liquid. It is a sealed graduated tube, weighted at one end with shot. When immersed into a liquid contained within a tall cylinder or jar, the hydrometer floats upright at a unique depth. The hydrometer operates on the principle that an object floats "high" in a liquid of relatively high specific gravity (greater than 1.0) and "low" in a liquid of relatively low specific gravity (less than 1.0). Commercial hydrometers are generally calibrated at 68°F (20°C). The specific gravity is read from the graduated scale at the point where the surface of the liquid touches the stem. The specific gravity of the liquid in this cylinder is 0.68.

TABLE 2.4	Properties of Some Common Liquids		
LIQUID	SPECIFIC GRAVITY [68°F (20°C)]	MISCIBLE/IMMISCIBLE WITH WATER	FLAMMABLE/ NONFLAMMABLE
Acetone	0.79	Miscible	Flammable
Carbon disulfide	1.26	Immiscible	Flammable
Chlorobenzene	1.11	Immiscible	Flammable
Cyclopentanone	0.95	Immiscible	Flammable
2-Ethylhexanol	0.83	Immiscible	Flammable
Heptane	0.68	Immiscible	Flammable
Hydrochloric acid	1.18	Miscible	Nonflammable
Methyl acetate	0.97	Miscible	Flammable
Methyl ethyl ketone	0.81	Miscible	Flammable
Sulfuric acid	1.84	Miscible	Nonflammable
Trichlorofluoromethane	1.49	Immiscible	Nonflammable

The specific gravities of some commercially important liquids are provided in Table 2.4. Because the specific gravity is determined by comparing the mass of a liquid or solid with the mass of an equal volume of water, the specific gravity of water is 1.0.

Many substances are soluble in water to some extent. They are said to be **miscible** with water. There are also substances that are relatively insoluble in water. When a liquid substance does not appreciably dissolve in water, the substance and water are regarded as mutually **immiscible**. Oil and water are examples of two mutually immiscible liquids. When combined, they coexist as two separate and distinct phases, one on top of the other.

When a liquid is immiscible with water, its specific gravity tells us how the liquid behaves when it is mixed with water as follows:

- An immiscible liquid *floats* on water when its specific gravity is less than 1.0.
- An immiscible liquid *sinks* below water when its specific gravity is greater than 1.0.

All grades of fuel oils have specific gravities of less than 1.0. Consequently, when an oil tanker ruptures at sea, the oil that spills from the tanker floats on the water and forms an oil slick.

Knowledge of the specific gravity of a flammable liquid tells us whether water functions effectively when it is applied to a fire as a fire extinguishing agent. The usefulness of this information is demonstrated in Figure 2.6 by the following two situations:

- In the first situation, a burning liquid immiscible with water and having a specific gravity of less than 1.0 is confined within a drum or tank. When water is added, it sinks below the surface of the liquid, where it is incapable of extinguishing the fire. If excess water is added, the burning liquid overflows the container or tank before the water does. Consequently, the application of water could potentially worsen the situation by spreading the fire to adjoining areas.
- In the second situation, a burning liquid immiscible with water and having a specific gravity of greater than 1.0 is confined within a drum or tank. When water is added, it floats on the heavier liquid and smothers it. In this instance, the fire is effectively extinguished.

miscible ■ Capable of combining, blending, or mixing in all proportions to form a single phase

immiscible ■ Incapable of combining, blending, or mixing to form a single phase

FIGURE 2.6 On the left, water is added from a hose to a drum of burning benzene. Water and benzene are immiscible liquids. Because the specific gravity of benzene is 0.879, the water settles *below* the benzene and does not ordinarily extinguish the fire. On the right, water is added from a hose to a drum of burning carbon disulfide. Water and carbon disulfide also are immiscible liquids. Because the specific gravity of carbon disulfide is 1.274, the water floats *on* the carbon disulfide and smothers the fire.

2.5-B VAPOR DENSITY

Air is a mixture of several gases. Its density is 0.08 lb/ft^3, or 1.29 g/L, at 32°F (0°C) at atmospheric pressure. Often, the mass of a given volume of a substance is compared with the mass of an equal volume of air. The comparison yields the **vapor density** of that substance. The vapor density is a dimensionless property of all gases.

Suppose we compare the masses of equal volumes of oxygen and air. A liter of oxygen weighs 1.43 g. This information can be used to compute the vapor density of oxygen relative to air as follows:

$$\text{Vapor density of oxygen} = \frac{1.43 \text{ g/L}}{1.29 \text{ g/L}} = 1.11$$

vapor density ■ The mass of a vapor or gas compared with the mass of an equal volume of another gas or vapor, generally air, at the same temperature and pressure

The value tells us that an arbitrary volume of oxygen is 1.11 times heavier in mass than an equal volume of air. The vapor densities of the gases and vapors in Table 2.5 were calculated from their densities in an identical fashion.

An awareness of the vapor density of a gas or vapor is useful for identifying the initial location of these substances when they are released into the air. The information can be summarized as follows:

■ When the vapor density of a gas or vapor is greater than 1.0, any arbitrary volume of the gas or vapor is heavier than the same volume of air. When the gas or vapor is released from its container, it tends to settle or concentrate in low spots before it dissipates in the air.

■ When the vapor density of a gas or vapor is less than 1.0, any arbitrary volume of the gas or vapor is lighter than the same volume of air. When the gas or vapor is released from its container, it tends to rise upwards before it dissipates in the air.

A leak of a gaseous substance whose vapor density is greater than 1.0 may pose a risk to health and safety. The gas displaces the air and concentrates in low spots before ultimately dissipating into the air. When the substance is poisonous, its presence may pose a risk of inhalation toxicity, and when the substance is flammable, its presence may pose a risk of fire and explosion. Consider gasoline vapor, for example. A given volume of gasoline vapor is heavier than an equal volume of air. When gasoline evaporates, its vapor

TABLE 2.5	Vapor Densities of Some Common Gases or Vapors
SUBSTANCE	**VAPOR DENSITY (AIR = 1)**
Acetylene	0.899
Ammonia	0.589
Carbon dioxide	1.52
Carbon monoxide	0.969
Chlorine	2.49
Fluorine	1.7
Gasoline	3.0–4.0
Hydrogen	0.069
Hydrogen chloride	1.26
Hydrogen cyanide	0.938
Hydrogen sulfide	1.18
Methane	0.553
Nitrogen	0.969
Oxygen	1.11
Ozone	1.66
Propane	1.52
Sulfur dioxide	2.22

accumulates in low spots such as the bilges in boats, where exposure to an ignition source could ignite the vapor.

In contrast, a gaseous substance whose vapor density is less than 1.0 rises in the atmosphere before it ultimately dissipates. In the open air, a gaseous substance whose vapor density is less than 1.0 dissipates rapidly. When this substance is released within a room, however, it first accumulates near its ceiling, where, if the substance is flammable, the localized accumulation poses a pronounced risk of fire and explosion.

All gases and vapors are totally miscible with air. This means that given adequate time, they become completely mixed with the components of the air. Gas and vapors are regarded as *infinitely miscible* with air.

SOLVED EXERCISE 2.4

Carbon monoxide is a poisonous gas emitted into the atmosphere from the exhaust pipes of operating motor vehicles. When first emitted from an exhaust pipe, does carbon monoxide tend to concentrate in the lower or upper regions of a confined garage?

Solution: Table 2.5 reveals that carbon monoxide has a vapor density of 0.969, which indicates that an arbitrary volume of carbon monoxide is only slightly lighter than an equal volume of air. In addition, the gases emitted from operating vehicular exhaust pipes are hotter than the air into which they are released. From these two facts, we can conclude that the carbon monoxide emitted from a vehicle's exhaust pipe is likely to rise toward the ceiling of the garage. Because the exhaust is hot and the densities of carbon monoxide and air are so similar, the carbon monoxide does not concentrate near the ceiling for long. Instead, it disperses rapidly within the confined space of the garage and mixes with the air.

2.6 ENERGY

Energy is defined as the capacity to do work; thus, energy is proportional to work. In a broad sense, when an effort is expanded to accomplish some act, the activity is referred to as work. Scientists describe work in terms of a force applied over a distance—like a push or pull of matter that results in moving it.

Energy is broadly classified as potential energy and kinetic energy. **Potential energy** is the energy of configuration or position. **Kinetic energy** is energy of motion. These two forms change into each other as chemical reactions occur.

Energy exists in a variety of forms, although these forms are often abstract: radiant energy (light, heat, and X rays), thermal energy (heat), acoustical energy (sound), mechanical energy, chemical energy, electrical energy, and atomic energy. All forms of matter possess chemical energy, which is the energy stored in its chemical makeup. Each hazardous material possesses chemical energy that can constitute the basis for its hazardous nature. Dynamite, for example, possesses a large amount of chemical energy. When dynamite detonates, some of its chemical energy is released into the environment as light, heat, and sound.

Energy is measured in various units. In the United States, energy units like the British thermal unit (Btu) and the calorie (cal) are commonly encountered. The SI unit of energy is the **joule**, or **J**. One joule is the kinetic energy possessed by a 2-kilogram mass moving at a velocity of 1 meter per second. We revisit these units in Section 2.10.

Mass and energy are closely interrelated. Neither mass nor energy can be created or destroyed. Certain phenomena require the conversion of matter to energy, and vice versa; but regardless of the nature of the transformation, the total amount of mass and energy in the universe remains constant. This observation is an important law of nature known as the **law of conservation of mass and energy**.

> **energy** ■ The property of matter that enables it to do work
>
> **potential energy** ■ The energy of configuration or position
>
> **kinetic energy** ■ The energy of motion
>
> **joule (J)** ■ The SI unit of energy, equal to the work done by a force of one newton moving one meter
>
> **law of conservation of mass and energy** ■ The observation that the total amount of mass and energy in the universe is constant

2.7 TEMPERATURE AND ITS MEASUREMENT

Temperature is the property of matter associated with its degree of hotness or coldness. Hot matter is associated with high temperatures, and cold matter is associated with low temperatures. To say that something is hot or cold, however, merely points out a relative condition of matter, whereas a temperature reading is an actual measure of its hotness or coldness. The temperature is an indication of a condition of matter, just as the measurement of its size is an indication of how large or small it is.

In the past, the temperature of many substances was most commonly determined using the simple mercury-in-glass column thermometer.[2] This device is a sealed glass capillary tube partially filled with mercury and then calibrated according to a prescribed procedure. When the thermometer is inserted into a substance, the temperature of the glass and mercury rises or falls—causing the mercury to expand or contract, respectively—until it reaches the temperature of the substance itself. The height of the calibrated column of mercury corresponds to the temperature of this substance.

Historically, temperature was measured on a number of scales, four of which have survived the test of time. They are illustrated in Figure 2.7 and discussed in the next two sections.

> **temperature** ■ The condition of a body that determines the transfer of heat to or from other bodies by comparison with established values for water, which is used as the standard

2.7-A THE FAHRENHEIT AND CELSIUS TEMPERATURE SCALES

Suppose an unmarked capillary tube containing mercury is placed in the steam that evolves above boiling water at sea level. The mercury level rises within the tube and then

---

[2]Because mercury vapor is toxic, mercury-in-glass thermometers have been phased out for most uses and replaced with digital and electronic thermometers.

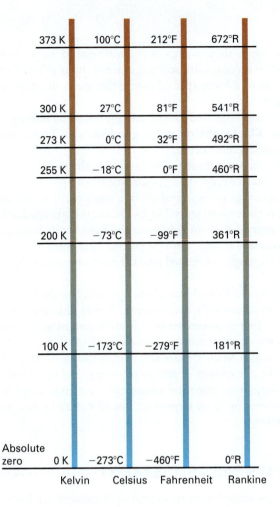

Kelvin	Celsius	Fahrenheit	Rankine
373 K	100°C	212°F	672°R
300 K	27°C	81°F	541°R
273 K	0°C	32°F	492°R
255 K	−18°C	0°F	460°R
200 K	−73°C	−99°F	361°R
100 K	−173°C	−279°F	181°R

Absolute zero

| 0 K | −273°C | −460°F | 0°R |

steam point ▪ The temperature at which liquid water and steam are in equilibrium at standard atmospheric pressure, assigned as 212°F (0°C)

ice point ▪ The temperature at which liquid water and ice are in equilibrium at standard atmospheric pressure, assigned as −32°F (0°C)

Fahrenheit scale ▪ The temperature scale on which water freezes at 32°F and boils at 212°F at 1 atm

Celsius scale ▪ The temperature scale on which water freezes at 0°C and boils at 100°C at 1 atm

remains stationary. This height can now be etched on the glass and serves as the first reference point, called the **steam point**. If the same tube is placed in ice water, the mercury level drops and ultimately becomes stationary. This height of the mercury column serves as the second reference point, called the **ice point**. These calibrations are assigned temperature units in at least two common ways:

▪ On the **Fahrenheit scale**, the ice and steam points are assigned the values of 32 degrees and 212 degrees, respectively. We recognize them as the freezing point (or melting point) and boiling point of water. There are 180 equally spaced divisions between these two reference points, and each division is called one degree Fahrenheit (°F). The capillary tube calibrated in this fashion is called a Fahrenheit thermometer.

▪ On the **Celsius scale**, the melting and boiling points are assigned the values of 0 degrees and 100 degrees, respectively. There are 100 equally spaced divisions between these two reference points and each division is called one degree Celsius (°C). This tube is called a Celsius thermometer.

Thus, the Fahrenheit and Celsius thermometers are alike insofar as the freezing and boiling points of water are used to define their reference points, but they differ in the numbers that are assigned to these points.

Because the same temperature interval is divided into 180 degrees on the Fahrenheit scale and 100 degrees on the Celsius scale, the temperature range corresponding to one Celsius degree is 180/100, or 9/5, as great as the temperature range corresponding to one

Fahrenheit degree. This factor can be used to relate these two temperature scales to each other using formulas such as the following:[3]

$$t(°F) = 9/5\ t(°C) + 32$$
$$t(°C) = 5/9[t(°F) - 32]$$

When using either formula, it is essential to perform the arithmetic exactly as the formula is written. To determine a Fahrenheit reading, nine-fifths of a Celsius reading is taken first, and 32 degrees is added to the product. Suppose we desire to convert 46°C to its equivalent temperature reading on the Fahrenheit scale. We take nine-fifths of 46, which equals 83, and add 32:

$$t(°F) = (9/5 \times 46) + 32 = 83 + 32 = 115°F$$

To determine a Celsius reading, 32 is subtracted from the Fahrenheit reading, and then five-ninths of this difference is taken. Suppose we desire to convert 115°F to its equivalent temperature on the Celsius scale. We subtract 32 from 115, which equals 83, and then take five-ninths of this number:

$$t(°C) = 5/9 \times (115 - 32) = 5/9 \times 83 = 46°C$$

Subdivision of the Fahrenheit and Celsius scales can be continued above and below the two reference points. When a thermometer is read below the zero mark, the divisions are read as minus degrees, or as degrees Fahrenheit or Celsius *below zero*. When a temperature reading below zero on either scale is converted to a reading on the other scale, it is important to account for the negative signs algebraically.

2.7-B THE KELVIN AND RANKINE TEMPERATURE SCALES

No upper temperature limit appears to exist. Conversely, scientific theory and experiment establish the limit below which it is impossible to cool matter. This coldest temperature is −273.15°C (−459.67°F), which is called **absolute zero**. Although this precise temperature cannot be attained experimentally, scientists have successfully reached a temperature of 450 ± 80 picokelvins.[4] (One picokelvin (pK) is one trillionth, or 10^{-12}, kelvin.)

Two additional temperature scales are identified by reference to this lowest attainable temperature as zero: the kelvin and Rankine temperature scales. The **Kelvin temperature** is defined as follows:

$$T(K) = t(°C) + 273.15°C$$

Thus, a Kelvin temperature is obtained by merely adding 273.15 to any Celsius temperature reading. The unit of temperature on the kelvin scale is called the **kelvin**, abbreviated **K** without a degree symbol.

The **Rankine temperature** is defined as follows:

$$T(°R) = t(°F) + 459.67°F$$

A Rankine temperature is thus obtained by adding 459.67 degrees to the Fahrenheit temperature reading. The unit of temperature on the Rankine scale is the degree Rankine (°R).

Because 0 K and 0°R represent the lowest attainable temperature, negative numbers are not encountered on the Kelvin or the Rankine temperature scales.

Astrophysicists have measured the temperature of the sun's core as 15.7 million degrees kelvin, the highest known natural temperature. Scientists have also attained extraordinarily high temperatures on Earth during fusion (Section 7.2-A) and laser-focusing experiments.

absolute zero ■ The lowest temperature theoretically attainable, equal to −459.67°F (−273.15°C)

Kelvin temperature ■ The absolute temperature obtained by adding 273.15°C to a Celsius temperature

kelvin (K) ■ The unit of temperature on the kelvin scale

Rankine temperature ■ The absolute temperature obtained by adding 459.67°F to a Fahrenheit temperature

---

[3]In this text, the lowercase *t* is used to denote a temperature on the Fahrenheit and Celsius scales, whereas the capital *T* is used to denote a temperature on the Kelvin and Rankine scales.

[4]A.E. Leanhardt et al., "Cooling Bose-Einstein condensates below 500 picokelvins," *Science*, Vol. 301 (2003), pp. 1513–1515.

2.8 PRESSURE AND ITS MEASUREMENT

pressure ■ A force applied to a unit area

pounds per square inch (psi) ■ A common unit of pressure in the English system of measurement (psia means pounds per square inch absolute; psig means pounds per square inch gauge)

newton (N) ■ The force required to accelerate a mass of 1 kilogram 1 m/s² per second

pascal (Pa) ■ The SI unit of pressure equal to 1 N/m²

atmospheric pressure ■ The force exerted on matter by the mass of the overlying air

atmosphere (atm) ■ The gaseous layer that surrounds Earth

standard atmospheric pressure ■ The average value of the atmospheric pressure at sea level

barometer ■ An instrument used to determine the atmospheric pressure, usually expressed as the height in millimeters or inches of a column of mercury

gauge pressure ■ The amount of pressure by which a confined gas exceeds atmospheric pressure

absolute pressure ■ The pressure exerted on a material; for confined gases, the sum of the gauge pressure and the atmospheric pressure

Pressure is the force exerted over a specified area; hence, pressure in customary units often is expressed as a unit of force per unit of area such as **pounds per square inch**, or **psi**. In the SI system, the unit of force is the **newton**, or **N**; it is the force that accelerates a one-kilogram body one meter per second each second. The SI unit of pressure is the newton per square meter, or N/m², which is also called a **pascal**, or **Pa**. A metric unit of pressure commonly encountered is the kilopascal (kPa).

The phenomenon called **atmospheric pressure** results from the exertion of Earth's gravity on the atmosphere. Earth's gravity is exerted in all directions and gives the atmosphere an average downward force of 14.7 psi at sea level; that is, on average, 14.7 pounds of force bears down on each square inch of Earth's surface. The average pressure exerted by the atmosphere at sea level supports a column of mercury 760 millimeters high (760 mmHg); this pressure is also called 1 **atmosphere**, or **atm**. For the purpose of converting a pressure that is expressed in one unit to its equivalent in another unit, each of the following applies:

$$1 \text{ atm} = 760 \text{ mmHg} = 760 \text{ torr} = 101.3 \text{ kPa} = 29.9 \text{ inHg} = 14.7 \text{ psi}$$

Each value is referred to as **standard atmospheric pressure**.

Atmospheric pressure readings are normally recorded on an instrument called a **barometer**. A simple barometer may easily be constructed by filling a glass tube, longer than 760 millimeters and closed at one end, with mercury. When the tube is inverted, open-side down, in a dish of mercury, the liquid flows out of the submerged open bottom until the level in the tube reaches an average height of 760 millimeters, or 1 atmosphere. When the atmospheric pressure decreases below the average value of 1 atmosphere, the mercury level on the barometer is said to "fall"; when it increases above 1 atmosphere, the mercury level is said to "rise." Such changes in the atmospheric pressure are closely monitored by meteorologists and used to forewarn of inclement weather conditions.

Modern barometers are usually constructed to record the atmospheric pressure in inches of mercury (inHg) or hectapascals (hPa). A measurement in inches of mercury may be converted to hectapascals by using the following factor:

$$29.5 \text{ inHg} = 1000.0 \text{ hPa}$$

A pressure reading is also used for monitoring the amount of a gas or vapor confined within a cylinder or tank. For such purposes, the affixed pressure gauge shown in Figure 2.8 is read. This reading provides the pressure exerted by the contents of the vessel. As portions of the contents are removed, the amount that remains in the vessel decreases as the gauge reading approaches zero. Because atmospheric pressure is continuously exerted on all forms of matter on Earth, a pressure gauge measures the amount of pressure by which the gas or vapor exceeds atmospheric pressure. The reading is called the **gauge pressure**. The units of gauge pressure are expressed as psig.

The true pressure of the contents within a containment vessel, called its **absolute pressure**, is the sum of the gauge pressure and the atmospheric pressure. The absolute pressure is measured with respect to a value of zero, whereas the gauge pressure is measured with respect to the prevailing pressure of the atmosphere; thus, in the English system of units, the absolute pressure on average is expressed as follows:

$$P(\text{psia}) = P(\text{psig}) + 14.7 \text{ psi}$$

The units of absolute pressure are sometimes noted by inclusion of an "a" with the unit, as shown in the previous expression. Thus, a gauge pressure of 19.4 psig corresponds to an average absolute pressure of 34.1 psia.

$$19.4 \text{ psig} + 14.7 \text{ psi} = 34.1 \text{ psia}$$

FIGURE 2.8 When the needle points to zero on a pressure gauge, the gauge pressure is zero. When the pressure gauge is connected to a gas cylinder, the cylinder is empty when the gauge reads zero. A gauge pressure of zero corresponds to an actual or absolute pressure of 14.7 psi, or 101.3 kPa.

We sometimes refer to the "normal" or "standard" temperature and pressure. For our purposes, the term refers to a temperature of 68°F (20°C) and a pressure of 14.7 psi (101.3 kPa).

When a liquid is confined within a closed container at a given temperature, it evaporates into the headspace above the liquid until equilibrium between the liquid and its vapor is attained. This equilibrium is characterized by two opposing changes that occur simultaneously: the rate at which the liquid evaporates, and the rate at which the vapor condenses. At equilibrium, these rates are equal. The pressure exerted by the vapor in equilibrium with its liquid is called the liquid's **vapor pressure**.

Although the vapor pressure can be expressed in any unit of pressure, chemists normally measure it in millimeters of mercury (mmHg) or kilopascals (kPa) at the prevailing temperature. In this text, the vapor pressure is exclusively expressed in mmHg at 68°F (20°C). When expressed in this fashion, the vapor pressure is the height of a column of mercury (measured on a meterstick) that the liquid's vapor supports at a given temperature.

The vapor pressure of liquids is a characteristic property. As their temperature increases, more and more of the substances vaporize into the headspace of the container in which each is confined. As Table 2.6 shows for three representative liquids, this causes the substances' vapor pressure to increase accordingly.

Liquids with low boiling points have relatively high vapor pressures, because they evaporate readily. Conversely, liquids with high boiling points have relatively low vapor pressures, because they evaporate more slowly. Liquids that evaporate readily at room temperature are said to be **volatile.**

The **evaporation rate** of a liquid is expressed as a unitless number that indicates the number of times it takes a liquid to evaporate at a given temperature when compared with the rate at which a benchmark substance like diethyl ether or *n*-butyl acetate (BuAc) evaporates. For this purpose, diethyl ether and *n*-butyl acetate are liquids assigned an evaporation rate of 1.0. For example, benzene is a liquid that has an evaporation rate of 2.8 (ether = 1). This number indicates that it takes 2.8 times as long for benzene to evaporate as the same quantity of diethyl ether, provided that both liquids are identically exposed to the atmosphere at the same temperature.

vapor pressure ■ The pressure exerted within a confinement vessel by the vapor of a liquid above its surface in a closed container; a measure of a substance's propensity to evaporate

volatile ■ The ability of a liquid to evaporate readily at room temperature

evaporation rate ■ The ratio of the time required for a measured volume of a liquid to vaporize at a given temperature relative to the time required for the same volume of a reference liquid to vaporize at the same temperature

TABLE 2.6	Vapor Pressures of Some Common Liquids as a Function of Increasing Temperature		
TEMPERATURE		**LIQUID**	
°F	°C	WATER (mmHg)	ETHANOL (mmHg)	BENZENE (mmHg)
14	−10	2.1	5.6	15
32	0	4.6	12.2	27
50	10	9.2	23.6	45
68	20	17.5	43.9	75
86	30	31.8	78.8	118
122	50	92.5	222.2	271
167	75	289.1	666.1	643
212	100	760	1693.3	1360

The evaporation rates of liquids are loosely classified as fast, medium, and slow when the numerical values are greater than 3.0, between 0.8 and 3.0, and less than 0.8, respectively. Using this subjective system, the evaporation rate of benzene is classified as medium.

Generally speaking, the vapor of a substance is its most hazardous physical form. It is the vapor of a flammable liquid—not the liquid itself— that burns, and it is a toxic liquid's vapor that causes adverse health effects when inhaled. Clearly, the vapor pressure of a flammable or toxic substance has a direct impact on its potential fire and health hazards, respectively.

When a flammable or toxic liquid has a relatively low vapor pressure, little vapor evolves at the prevailing temperature. It can neither pose a significant fire and explosion hazard nor an inhalation toxicity hazard. A flammable liquid with a comparatively high vapor pressure, however, poses a fire and explosion hazard, and a poisonous liquid with a comparatively high vapor pressure poses an inhalation toxicity hazard.

A flammable or poisonous liquid with a relatively high vapor pressure presents unique transportation problems. During the course of its transportation, a liquid confined within a tank or other vessel can increase in temperature by absorbing heat from its surroundings. As the temperature of the liquid increases, a considerable volume of vapor is produced within the tank. This vapor enters the headspace above the liquid and exerts pressure on the walls of the confining vessel. To retain its integrity, the vessel must be constructed to withstand this internal pressure during the time that it is used to transport the liquid.

2.9 HEAT AND ITS TRANSMISSION

heat ■ The form of energy transferred from one body to another because of their temperature difference; energy arising from atomic or molecular motion

Heat is the form of energy associated with the motion of atoms or molecules (Sections 4.4 and 4.6), small particles of which all matter is composed. Heat is manifested when changes in any of the following occur:

endothermic ■ The nature of any process that absorbs heat from its surroundings

- A material's temperature
- The physical state of a substance
- The chemical identity of a substance

exothermic ■ The nature of any process that emits heat into its surroundings

Heat is either absorbed or emitted as these processes occur. When heat is absorbed, the phenomenon is called an **endothermic** process; and when heat is released to the surroundings, the phenomenon is called an **exothermic** process.

In Section 2.6, it was noted that energy is often measured in British thermal units and calories. One **British thermal unit,** or **Btu,** represents the heat that must be supplied to raise the temperature of 1 pound of water 1 degree Fahrenheit, specified at the temperature of water's maximum density, 39.1°F (3.97°C). One **calorie,** or **cal,** is defined as the amount of heat required to raise the temperature of 1 gram of water 1 degree Celsius, from 14.5 to 15.5°C. Two hundred fifty-two (252) calories is equivalent to one British thermal unit, and 1 calorie is equivalent to 4.184 joules.

$$1 \text{ Btu} = 252.0 \text{ cal} = 1055 \text{ J}$$
$$1 \text{ cal} = 4.187 \text{ J}$$

When chemical reactions occur, the reactants change their chemical identities by transforming into one or more other substances. Thermal energy accompanies these chemical changes; it is called their **heat of reaction.**

Heat is evolved during a variety of combustion processes such as the burning of gasoline, wood, natural gas, and other materials. When these fuels burn, they provide energy and serve as sources of heat, power, and light. On the negative side, however, they can also pose a risk of fire and explosion. When these materials burn in bulk, they usually ignite secondary fires involving nearby materials.

The proper control of the heat emitted during combustion is an extremely important factor in fire control. Fires continue as self-sustained phenomena only when sufficient heat is released during combustion to substitute for the input energy initially provided from an ignition source. Conversely, many fires cannot be extinguished until some means is undertaken to reduce or eliminate this heat.

Emergency responders also encounter heat as a form of energy in situations having nothing to do with combustion phenomena. Paramedics, for example, often use instant heat packs and cold packs when administering first-aid to individuals with muscular injuries. One type of portable pack consists of a pouch of water and a solid chemical substance: magnesium sulfate for heat packs, and ammonium nitrate for cold packs. When the separate packs are squeezed, the water commingles with the solids and produces their solutions. As a magnesium sulfate solution forms, the heat that is evolved may be used to warm an injury. As an ammonium nitrate solution forms, the heat that is absorbed may be used to cool an injury.

Heat is always transferred from warm materials to cooler ones. If several materials near one another have different temperatures, those that are warm become cooler, and those that are cool become warmer, until they achieve a common temperature. Heat is transmitted from one material to another or from one spot to another spot by three independent modes: conduction, convection, and radiation. The nature of each mode of heat transmission is an important factor associated with understanding how fire spreads.

2.9-A CONDUCTION

The handle of a metal spoon that has been inserted into hot coffee becomes hot itself. This transfer of heat between two or more materials in contact—in this case, from the coffee to the spoon—is called **conduction.**

Every material conducts heat to some extent. Metals, such as silver, copper, iron, and aluminum, conduct heat most efficiently; nonmetals, such as glass and air, are not good conductors of heat. Materials that are not good conductors are good insulators, because they delay the transfer of heat.

2.9-B CONVECTION

Heat can also be transferred from spot to spot or even between two or more substances by the natural mixing of their component parts. This happens when cold milk or cream dis-

British thermal unit (Btu) ■ The amount of heat required to raise the temperature of 1 pound of water 1 degree Fahrenheit

calorie (cal) ■ The amount of heat required to raise the temperature of 1 gram of water 1 degree Celsius

heat of reaction ■ The energy that is absorbed or evolved when a chemical reaction occurs

conduction ■ The mechanism by which heat is transferred to the parts of a stationary material or from one material to another with which it is in contact

perses throughout hot coffee without the use of a spoon, or when the air in a room gets warmer when it is hot outside the room. This transmission of heat within a substance or between substances by means of natural mixing is known as **convection**.

convection ■ The mechanism by which heat is transferred by the movement of a heated material from spot to spot

The circulation of warm air that exits a heat vent into a room is caused by convection. The popular expression "heat rises" actually means "hot air rises." As warm air issues from a heat vent into a room, it rises toward the ceiling. The surrounding cool air descends to replace the warm air that rose, is heated at the heat vent, and then rises toward the ceiling. In this fashion, warm and cool air exchange to generate air currents.

The origin of the convection phenomenon is associated with the impact that gravity has on matter. In zero gravity, convection can occur only if it is artificially induced. When a substance burns, convection moves its combustion products away from the flame to dissipate into the surrounding atmosphere; but where zero gravity exists, as in a spaceship orbiting Earth, the combustion products remain in the immediate area of the combustion zone, where they quickly extinguish the flame.

2.9-C RADIATION

Imagine a 200-watt light bulb hanging from a ceiling. When the light is turned on, heat can be felt when we hold our hands *around* the bulb. This transfer of heat from the bulb to our hands cannot be caused by conduction, since the air between the bulb and our hands is a poor conductor of heat. It also cannot be caused by convection, because hot air currents rise upward. The heat from the bulb is transmitted by a third means called **radiation**.

radiation ■ The mechanism by which heat is transferred between two materials not in contact

Unlike conduction and convection, radiation occurs even when there is no material contact between two objects. Heat from the sun radiates through space and warms Earth and other celestial bodies.

All matter radiates heat at elevated temperatures; that is, it exhibits the phenomenon called *incandescence*. When the temperature of a heated object exceeds approximately 932°F (500°C), the radiation becomes visible. Burning flames, glowing coals, and molten metal are examples of matter hot enough to radiate visible light. Hot objects may also radiate energy at lower temperatures, but the energy emitted is generally in the infrared region of the electromagnetic spectrum, which cannot be observed with the naked eye.

2.9-D SPREAD OF FIRE

The conduction process does not significantly contribute to the sustenance and spread of a normal fire. Figure 2.9 illustrates that convection and radiation are the modes of heat transmission primarily responsible for sustaining and spreading freely burning fires in the following ways:

■ Convection affects the spread of fire by means of the natural movement of hot and cold air. As hot air rises, heat is transmitted to adjoining materials and initiates their ignition. Cool, fresh air simultaneously flows inward into the fire to replace the hot air, thereby providing the fire with a supply of oxygen.

■ Radiation affects the spread of fire by transmitting heat, primarily in a lateral direction, from the immediate fire scene to nearby materials. The transmission of radiant heat initiates the combustion of these materials.

2.10 CALCULATION OF HEAT

heat capacity ■ The amount of heat needed to raise the temperature of 1 pound of a substance 1 degree Fahrenheit or 1 gram of a substance 1 degree Celsius

Two or more substances differ from one another by the quantity of heat required to produce a given temperature change in the same mass of each substance. This quantity of heat is called the **heat capacity**. The heat capacities of some common liquids are provided in Table 2.7. For water, the magnitude of the heat capacity varies as a function of its temperature, as indicated. The units of heat capacity in the English and metric systems are the

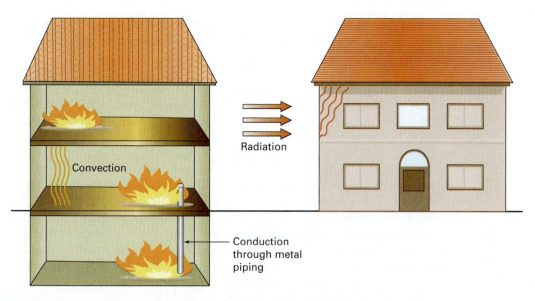

Radiation

Convection

Conduction
through metal
piping

FIGURE 2.9 Conduction, convection, and radiation contribute to the spread of fire. In the building on the left, metal piping first transmits heat by conduction from an unattended basement fire to the first floor. Heat is then transmitted by convention to the second floor and throughout the remainder of the building by convection. Heat is transmitted toward the nearby adjacent building on the right by radiation.

British thermal unit per pound per degree Fahrenheit (Btu/lb °F) and the calorie per gram per degree Celsius (cal/g °C), respectively. The magnitude of these units is equal:

$$1 \text{ Btu/lb °F} = 1 \text{ cal/g °C}$$

The ratio of the heat capacity of a substance to the heat capacity of water at the same temperature is a dimensionless number called the **specific heat** of the substance. The specific heats of most substances are less than 1.

When heat is transferred to a substance, it may cause a change in its physical state or its chemical identity, or it may merely cause the substance to assume a different temperature. The heat that is transferred to a substance causing a change in its temperature, but no change in its physical state or chemical identity, is calculated by the following formula,

specific heat ■ The ratio of the heat capacity of a substance to the heat capacity of water at the same temperature

TABLE 2.7	Heat Capacities of Some Common Liquids
LIQUID	**HEAT CAPACITY (cal/g °C or Btu/lb °F)**
Acetone	0.506
Benzene	0.406
Carbon tetrachloride	0.201
Chloroform	0.234
Diethyl ether	0.547
Ethanol	0.581
Methanol	0.600
Turpentine	0.411
Water [32 to 212°F (0 to 100°C)]	1.00
Water [< 32°F and > 212°F (< 0°C and > 100°C)]	0.5

in which Q is the heat, m is the mass of the substance, C is its heat capacity, and Δt (delta t) is the difference between the final and initial temperatures of the substance:

$$Q = m \times C \times \Delta t$$

For example, given that copper has a heat capacity of 0.093 cal/g°C, what amount of heat is transferred to 100 grams of copper when it is heated from 30 to 100°C? First, the difference in temperature, Δt, is 100°C − 30°C, or 70°C. Then, the heat transferred is calculated to be 651 calories by multiplication as follows:

$$Q = 100 \text{ g} \times 0.093 \text{ cal/g °C} \times 70°C = 651 \text{ cal}$$

As noted, a change in temperature can also result in a change in the physical state of matter. If water is exposed to the cold, its temperature falls until 32°F (0°C) is attained. Then, the temperature remains fixed until the entire liquid mass freezes. This latter temperature is called the **freezing point** of water. The quantity of heat per unit mass required to change a liquid to a solid at its freezing point is called the **latent heat of fusion** of the substance. For example, the conversion of water to ice liberates 144 Btu/lb, or 80 cal/g. This is the latent heat of fusion of water.

We can similarly heat a mass of water to 212°F (100°C). Then, the temperature remains fixed while the liquid changes into steam. This is the temperature at which the vapor pressure of water equals the atmospheric pressure of 1 atmosphere; it is called the **boiling point** of water. The amount of heat that must be supplied to a material at its boiling point to convert it from a liquid to a vapor is called its **latent heat of vaporization**. The heat of vaporization of water is 970 Btu/lb, or 540 cal/g. This is a relatively high value when it is compared to the heats of vaporization of other substances.

Two phenomena are closely associated with freezing and boiling, respectively:

- *Melting* is the reverse of freezing. As ice melts, 144 Btu/lb, or 80 cal/g, is absorbed from the surroundings.
- *Condensation* is the reverse of boiling. When steam condenses, 970 Btu/lb, or 540 cal/g, is liberated into the surroundings. This explains why exposure of the skin to steam at 212°F (100°C) causes a more severe burn than exposure to liquid water at the same temperature.

The amount of heat associated with the conversion of 1 pound of water into 1 pound of steam over the temperature range from −20 to 300°F (−29 to 149°C) is illustrated in Figure 2.10. The relevant quantities of heat in Btu can be calculated at each juncture as follows:

- Ice undergoes a change in temperature from −20 to 32°F (−29 to 0°C)

$$Q = 1 \text{ lb} \times 0.5 \text{ Btu/lb °F} \times 52°F = 26 \text{ Btu}$$

- Ice melts at 32°F (0°C)

$$Q = 1 \text{ lb} \times 144 \text{ Btu/lb} = 144 \text{ Btu}$$

- Water undergoes a change in temperature from 32 to 212°F (0 to 100°C)

$$Q = 1 \text{ lb} \times 1.0 \text{ Btu/lb °F} \times 180°F = 180 \text{ Btu}$$

- Water evaporates into steam at 212°F (100°C)

$$Q = 1 \text{ lb} \times 970 \text{ Btu/lb} = 970 \text{ Btu}$$

- Steam is superheated from 212 to 300°F (100 to 149°C)

$$Q = 1 \text{ lb} \times 0.5 \text{ Btu/lb °F} \times 88°F = 44 \text{ Btu}$$

These calculations clearly demonstrate that the largest quantity of heat is associated with the evaporation of water.

freezing point ■ The temperature at which the liquid and solid states of a substance coexist at 1 atmosphere (101.3 kPa)

latent heat of fusion ■ The amount of heat required to convert a unit mass of a solid substance into a liquid

boiling point ■ The temperature at which the vapor pressure of a substance equals the average atmospheric pressure

latent heat of vaporization ■ The amount of heat required to convert a unit mass of a liquid substance into a gas

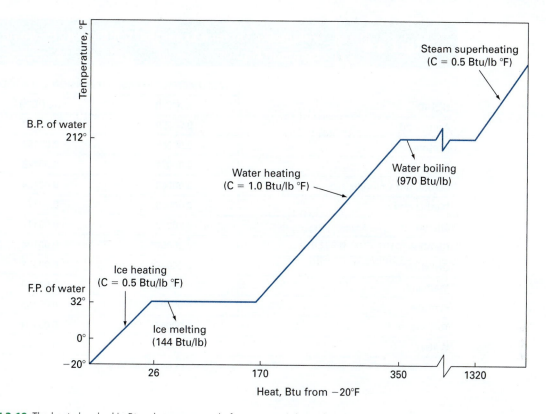

FIGURE 2.10 The heat absorbed in Btu when one pound of pure water is heated at atmospheric pressure. B.P. and F.P. refer to the boiling and freezing points of water, respectively.

Water is commonly employed as a fire extinguishing agent, because it is nonflammable, plentiful, and inexpensive; but the primary reason water effectively extinguishes fire is associated with its ability to remove heat from the burning material. As water is applied to a fire, it absorbs heat and ultimately vaporizes, each pound carrying 970 Btu of heat into the atmosphere and away from the burning material. Simultaneously, the temperature of the burning material decreases until it ultimately drops to the ambient temperature of the surroundings.

2.11 VOLUMETRIC EXPANSION OF LIQUIDS RESULTING FROM A CHANGE IN TEMPERATURE

With few exceptions, the dimensions of a material expand when it is heated and contract when it is cooled. As we note in more detail in Section 2.12, gases and vapors expand more than liquids and solids; liquids expand less; and solids even less. Water is the common exception to this general rule: Water begins to expand when it is cooled below 39.1°F (3.97°C).

The volume to which liquids and solids expand when heated is determined from the following formula:

$$V_2 = V_1 + (V_1 \times \alpha_v \times \Delta t)$$

We first determine the product of V_1, α_v, and Δt and then add the product to V_1. V_1 and V_2 are the initial and final volumes of an arbitrary liquid, respectively; Δt is the difference between its initial and final temperature; and the Greek letter alpha (α) subscripted with a V (α_v) is a measure of the change in volume per unit of original volume per

TABLE 2.8 — Volumetric Thermal-Expansion Coefficients for Some Common Liquids

LIQUID	VOLUMETRIC THERMAL-EXPANSION COEFFICIENT	
	$\alpha_V(°F^{-1})$	$\alpha_V(°C^{-1})$
Acetic acid	0.00059	0.00107
Acetone	0.00085	0.00153
Benzene	0.00071	0.00128
Carbon tetrachloride	0.00069	0.00124
Diethyl ether	0.00098	0.00176
Ethanol	0.00062	0.00112
Gasoline	0.00080[a]	0.00144
Glycerine	0.00028	0.00051
Methanol	0.00072	0.00130
Pentane	0.00093	0.00168
Toluene	0.00063	0.00113
Water	0.00012	0.00022

[a]The volumetric thermal-expansion coefficient for gasoline varies from 0.00055 to 0.00090°F^{-1}.

volumetric thermal-expansion coefficient ■ The change in volume of a liquid per change in its temperature on the Fahrenheit or Celsius temperature scale

degree change in temperature. The last term is called the **volumetric thermal-expansion coefficient**, the units of which are reciprocal degrees Fahrenheit (°F^{-1}) and reciprocal degrees Celsius (°C^{-1}). Numerical values of this coefficient are provided in Table 2.8 for some common liquids.

Imagine a tank or metal drum filled to its brim with a liquid. When the liquid expands, it must overflow the brim. If the tank or container is sealed so that the contents cannot overflow, it must either expand to compensate for the increase in liquid volume or rupture. To prevent its rupture, a tank or container should *never* be completely filled with a liquid product. The amount by which a tank or container of liquid falls short of being completely filled is called its **outage**, or *ullage*, which generally is expressed in percentage by volume.

outage ■ The percentage by volume by which a container of liquid falls short of being filled to the brim

The manufacturers of liquid chemical products normally acknowledge that an outage must be provided in containers. Nonetheless, the allowance could be inadequate to compensate for the volume increase experienced by liquid materials during a fire. Consider a 55-gallon steel drum filled to the brim with flammable benzene liquid at 68°F (20°C) that is subsequently heated during a fire to 170°F (77°C). As the liquid benzene is heated, it expands to occupy a volume of 59 gallons at the new temperature:

$$V_2 = 55\ \text{gal} + \left(55\ \text{gal} \times \frac{0.00071}{°F} \times 102°F \right) = 59\ \text{gal}$$

The coefficient listed in the table is a representative value.

This increase in volume is 4 gallons, or 7.3% by volume. To relieve the internal pressure accompanying this volume increase, the drum ruptures and releases the benzene into the environment. When the benzene vapor is exposed to an ignition source, it is likely to burst into flame.

When liquids are transferred into containers, rail tankcars, and other cargo tanks for transportation, DOT requires the presence of a headspace above the liquid so the vessels are not completely filled to the brim. DOT requires this space to prevent leakage from, or permanent distortion of, the transport vessel by accounting for the expansion of its contents due to a temperature rise likely to be encountered during transportation. Sufficient outage must also be provided within containers of 110 gallons or less so that they are not completely filled with liquid at 130°F (55°C). In rail tankcars, an adequate headspace must be provided within the shell when the *dome* of the tankcar does not itself provide sufficient outage. DOT's requirements for outage and filling limits for nonbulk and bulk containers are published at 49 C.F.R. §§173.24a(d) and 173.24b(a), respectively.

2.12 GENERAL PROPERTIES OF THE GASEOUS STATE OF MATTER

Of the three states of matter, only the gaseous state is capable of being described in comparatively simple terms. This description relates the volume of a gas to its temperature and pressure. We shall independently examine the laws of nature that apply to the gaseous state of matter: Boyle's law, Charles's law, and the combined gas law.

2.12-A BOYLE'S LAW

Robert Boyle, an Irish physicist, demonstrated experimentally that the volume of a confined gas varies inversely with its absolute pressure when the temperature remains fixed. The constant-temperature experiments performed by Boyle demonstrated that the mathematical product of the volume and absolute pressure of a gas is always constant. This observation is commonly called **Boyle's law**. It is expressed mathematically as follows:

$$V_1 \times P_1 = V_2 \times P_2$$

Boyle's law ■ The observation that at constant temperature the volume of a confined gas is inversely proportional to its absolute pressure

Here, V_1 and P_1 are the initial volume and initial absolute pressure, respectively; V_2 and P_2 are the final volume and final absolute pressure, respectively.

Boyle's law is used to determine the volume that a gas occupies when it remains at a fixed temperature but undergoes a change in pressure. Suppose the pressure of 40 cubic feet (40 ft³) of an arbitrary gas is changed from 14.7 psia to 450 psia at a fixed temperature. What is the new volume assumed by the gas? Common sense tells us that the new volume must be less than 40 cubic feet, because the application of any pressure squeezes the gas into a smaller volume. Using Boyle's law, we can readily compute the new volume as follows:

$$V_2 = 40 \text{ ft}^3 \times \frac{14.7 \text{ psia}}{450 \text{ psia}} = 1.3 \text{ ft}^3$$

2.12-B CHARLES'S LAW

Jacques Charles and Joseph Gay-Lussac, two French scientists, independently demonstrated experimentally that the volume of a confined gas increases proportionately to the increase in its absolute temperature when the pressure remains fixed. This statement is now known as **Charles's law**. It is expressed mathematically as follows:

$$V_1 \times T_2 = V_2 \times T_1$$

Charles's law ■ The observation that at constant pressure the volume of a confined gas is directly proportional to its absolute temperature

Here, V_1 and T_1 are the initial volume and initial absolute temperature, respectively; V_2 and T_2 are the final volume and final absolute temperature, respectively.

Suppose an arbitrary gas at atmospheric pressure occupies a volume of 300 milliliters at 0°C; what volume does it occupy at 100°C and the same pressure? Common sense tells us that a heated gas always occupies a larger volume. We must first convert the two temperatures in degrees Celsius into absolute temperatures in kelvins. On the kelvin scale, the temperature readings are as follows:

$$0°C = 273 \text{ K}$$
$$100°C = 373 \text{ K}$$

Now, we can use Charles's law to compute the new volume.

$$V_2 = 300 \text{ mL} \times \frac{373 \text{ K}}{273 \text{ K}} = 410 \text{ mL}$$

2.12-C COMBINED GAS LAW

Boyle's and Charles's laws can also be combined into the following mathematical expression:

$$V_1 \times P_1 \times T_2 = V_2 \times P_2 \times T_1$$

combined gas law ■ The observation that the volume of a confined gas is directly proportional to its absolute temperature and inversely proportional to its absolute pressure

This is called the **combined gas law**. It is used to calculate the volume of a gas at a new temperature and pressure.

The volume of a gas can be forced to remain fixed, as when a gas is confined within a steel cylinder or other storage vessel. Under this circumstance, $V_1 = V_2$, and the combined gas law assumes the following form:

$$P_1 \times T_2 = P_2 \times T_1$$

The heating of a gas that is initially enclosed within a sealed metal tube at 1000 psia at 70°F (21°C) is illustrated in Figure 2.11. When the gas is heated, its pressure increases accordingly. When the gas is heated to 570°F, the pressure of the gas nearly doubles; when it is heated to 1070°F, it nearly triples. These temperatures are akin to those routinely encountered during building fires. To relieve the strain on their walls, the second and third tubes are likely to rupture.

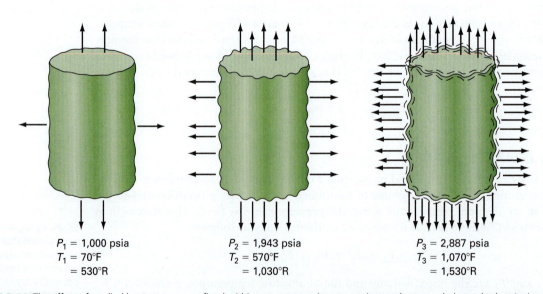

P_1 = 1,000 psia	P_2 = 1,943 psia	P_3 = 2,887 psia
T_1 = 70°F	T_2 = 570°F	T_3 = 1,070°F
= 530°R	= 1,030°R	= 1,530°R

FIGURE 2.11 The effect of applied heat on a gas confined within a constant-volume container such as a sealed metal tube. An increase in the temperature of the gas of 500°F and 1000°F causes the internal pressure to nearly double and triple, respectively, which usually causes the containers to rupture.

A welder purchases a gas cylinder of flammable hydrogen gas and chains it to a wall within a workshop. After periodic usage, the gauge pressure reads 235 psig when the temperature is 65°F. What is the gauge pressure reading when the temperature of the cylinder contents becomes 350°F during a fire?

Solution: Because a steel cylinder is a constant-volume container, $P_1 \times T_2 = P_2 \times T_1$, where these symbols refer to the *absolute* initial and final pressures and temperatures, respectively.

$$P_1 = 235 \text{ psig} + 14.7 \text{ psi} = 250 \text{ psia}$$

$$T_1 = 65°F + 459.67°F = 525°R$$

$$T_2 = 350°F + 459.67°F = 810°R$$

$$P_2 = 250 \text{ psia} \times \frac{810°R}{525°R} = 386 \text{ psia}$$

Then, we determine the gauge pressure by subtracting 14.7 psi from the absolute pressure and obtain 371 psig.

$$P_2 = 386 \text{ psia} - 14.7 \text{ psi} = 371 \text{ psig}$$

The absorption of heat causes the internal pressure of the cylinder contents to increase from 235 psig to 371 psig.

2.13 CRYOGENS

DOT defines **cryogenics** at 49 C.F.R. 173.115(g) as the study of matter at temperatures less than approximately –130°F (–90°C). Matter that has been cooled to these very low temperatures is called a **cryogen**, or **cryogenic liquid**.

A gas or vapor may eventually be reduced to a liquid by compressing it and/or lowering its temperature. This liquid may then be reduced to a solid by lowering its temperature and/or increasing its pressure further. Nonetheless, there is a temperature above which an increased pressure alone cannot cause a gas or vapor to condense; this is called the **critical temperature** of that substance.

Consider gaseous carbon dioxide. When it is cooled to a minimum temperature of 88°F (31°C) and compressed by sufficient pressure, carbon dioxide liquefies. But when its temperature is elevated above 88°F (31°C), applied pressure is unable to liquefy carbon dioxide. The pressure required to liquefy a gas or vapor maintained at its critical temperature is called the **critical pressure**. The critical pressure of carbon dioxide is 1070 psig (7373 kPa). Consequently, when carbon dioxide gas is confined within a vessel and maintained at a temperature equal to or less than 88°F (31°C), it liquefies only when compressed by a pressure equal to or greater than 1070 psig (7374 kPa).

The critical temperatures, critical pressures, and other physical properties of carbon dioxide and several other common cryogens are listed in Table 2.9.

To retain a gas or vapor as a cryogen, it must be stored in a specially designed and insulated vessel like the tank whose structural silhouette is illustrated in Figure 2.12. It is fitted with pressure-regulating valves to control its internal pressure.

When carriers transport cryogens, DOT requires them to use MC-338 cargo tanks and DOT-113/AAR-304 rail tankcars. DOT regulates the design specifications for these transport vehicles, including the nature of their temperature and pressure control systems, venting mechanisms, valves, and piping. These regulations are published at 49 C.F.R. §§173.319 and 178.338, respectively.

cryogenics ■ The study of the properties of matter at extremely cold temperatures

cryogen (cryogenic liquid) ■ A liquefied compressed gas that generally has a boiling point colder than −130°F (−90°C) at 14.7 psia (101.3 kPa)

critical temperature ■ The temperature above which the vapor of a liquid cannot be condensed by pressure alone

critical pressure ■ The minimum pressure that causes a gas to liquefy at its critical temperature

TABLE 2.9 — Physical Properties of Some Cryogens[a]

	CARBON DIOXIDE	HELIUM	HYDROGEN
Boiling point	−108°F (−78°C)	−452°F (−269°C)	−423°F (−252.8°C)
Critical temperature	88°F (31.1°C)	−450°F (−268°C)	−450°F (−268°C)
Critical pressure	1070 psig (7374 kPa)	34 psig (0.230 kPa)	183 psig (1256 kPa)
Liquid density	47.6 lb/ft³ (762 kg/m³)[b]	7.801 lb/ft³ (124.98 kg/m³)[c]	4.43 lb/ft³ (70.96 kg/m³)[c]
Gas density	0.1444 lb/ft³ (1.8333 kg/m³)[b]	0.0103 lb/ft³ (0.165 kg/m³)[b]	0.00521 lb/ft³ (0.08342 kg/m³)[b]
Liquid-to-gas expansion ratio	790	780	865
	METHANE	**NITROGEN**	**OXYGEN**
Boiling point	−258°F (−161°C)	−321°F (−196°C)	−297°F (−183°C)
Critical temperature	−116°F (−82°C)	−232°F (−147°C)	−181°F (−118°C)
Critical pressure	673 psig (4638 kPa)	492 psig (3390 kPa)	736 psig (5072 kPa)
Liquid density	26.57 lb/ft³ (425.61 kg/m³)[c]	50.7 lb/ft³ (808.5 kg/m³)[c]	71.23 lb/ft³ (1141 kg/m³)[c]
Gas density	0.04235 lb/ft³ (0.6784 kg/m³)[d]	0.072 lb/ft³ (1.153 kg/m³)[b]	0.083279 lb/ft³ (1.326kg/m)³)[b]
Liquid-to-gas expansion ratio	650	696	861

[a]The physical data quoted throughout this text have been selected from multiple authoritative sources, including the SDSs prepared by chemical manufacturers.
[b]At 70°F (21.1°C) and 1 atm.
[c]At boiling point and 1 atm.
[d]At 60°F (15.6°C) and 1 atm.

2.13-A EXPANSION OF CRYOGENS DURING VAPORIZATION

When a cryogenic liquid is accidentally released from confinement, it first forms a pool, or, on a flat terrain, it spreads into a thin layer in the immediate vicinity of its release. In both instances, heat transfer causes the liquid to boil or evaporate rapidly. Its vapors then disperse by atmospheric turbulence and may cause flammable or toxic hazards in areas far removed from the scene of the release.

Incidents also occur during which it becomes impossible to maintain the temperature and pressure conditions of a confined cryogenic liquid. Under these conditions, the evaporation of the cryogen causes the release of vapor to occur catastrophically because the difference in volume between the liquid and its gas or vapor is always substantial.

Consider cryogenic methane confined within a storage vessel whose pressure-reducing equipment malfunctions. When the entire bulk of the liquid vaporizes, it assumes a volume approximately 650 times its initial volume. This results in a buildup of internal pressure within the storage vessel. This increased pressure exerts force on the walls of the vessel and causes the vessel to rupture. As the vessel bursts, the entire mass of methane is released into the environment at once. Because methane is a flammable substance, this situation poses a pronounced risk of fire and explosion.

Fortunately, the storage vessels for cryogens are equipped with *pressure-relief valves* from which the vapor escapes. When the storage vessel has been

FIGURE 2.12 The structural silhouette of a tank used for the storage of a cryogenic liquid.

properly designed, the gas or vapor escapes so slowly that the vessel does not rupture. Nonetheless, when the pressure-reducing equipment fails, adequate ventilation must be provided so that the escaping gas fails to assume a concentration that can pose an unreasonable risk to health and safety.

2.13-B IMPACT OF CRYOGENS ON OTHER MATTER

Cryogens are so cold they can affect matter in a number of ways. In particular, when in contact with cryogenic liquids, many materials solidify, become brittle, and easily snap or break. For example, rubber items freeze when in contact with a cryogenic liquid and break into dozens of small pieces when they are dropped on a hard surface.

When bulk quantities of cryogens like hydrogen and helium vaporize into the atmosphere, they cause water vapor and other components of the air to solidify. When cryogens are transferred from one closed vessel into another, the solidification of an air component constitutes a major hazard, because the solid may block venting valves and the passageways within tubing. As noted in Section 2.13-A, this prevents the release of internal pressure and an increased likelihood of vessel failure.

When cryogenic oxygen vaporizes, it causes gases with a boiling point below −297°F (−183°C) to condense. These gases include atmospheric oxygen itself. A leak of cryogenic oxygen from its containment vessel generates an environment enriched in oxygen. Although oxygen does not burn, it supports combustion. The presence of an oxygen-enriched environment may cause explosions and other violent chemical reactions.

2.13-C ILL EFFECTS CAUSED BY EXPOSURE TO CRYOGENS

Exposure to cryogens may potentially cause several ill effects. In particular, the exposure may cause the development of serious skin burns. Depending on the length of exposure and the depth to which the skin tissue has been affected, these skin burns resemble first-, second-, and third-degree thermal burns in physical appearance. Dermatologists take advantage of this characteristic when they remove superficial skin growths such as warts. Liquid carbon dioxide or nitrogen is applied to freeze the growths. The skin surrounding them becomes swollen and red and may blister. The skin growths subsequently drop away, leaving healthy new skin.

To avoid the emergence of serious burns, individuals should exercise caution when they handle cryogenic liquids. A face shield or visor should be used to protect the eyes, and loosely fitted gloves and boots should be worn to protect the hands and feet. If a mishap occurs during which a cryogen accidentally flows inside gloves or boots, these items should be removed immediately to minimize the length of time during which the skin and liquid remain in contact.

Living tissue may solidify when it is exposed to cryogenic liquids for an extended period. The solidification causes a local arrest in the circulation of blood. Widespread cellular damage may occur within the solidified tissue, causing it to become vulnerable to bacterial infection. When solidification lasts for hours, physicians may decide to surgically amputate the affected tissue to protect against the onset of gangrene.

The extreme coldness of a cryogenic liquid or solid can also cause bonding between a body part and the uninsulated vessels or pipes containing the cryogen. The bond may be so firm that the flesh rips or tears when a separation attempt is made.

Body temperature decreases when an individual is exposed to a cryogen or is otherwise restricted to a cold environment for an extended period. When the body temperature becomes less than 95°F (35°C), we experience the signs and symptoms of **hypothermia** listed in Table 2.10. In particular, shivering occurs: Shivering is the involuntary contraction of nearly every muscle in the body. The heat generated during shivering serves as a

hypothermia ■ The condition experienced by an individual when exposed to a very cold environment for a length of time sufficient to cause a reduction in body temperature below its norm

TABLE 2.10		Signs and Symptoms of Hypothermia
BODY TEMPERATURE		
°F	°C	**SIGNS AND SYMPTOMS**
98	37	Normal body temperature
95–97	35–36	Uncontrollable shivering
91–95	33–35	Mental confusion, amnesia, drowsiness, and pallor
86–90	30–32	Muscular rigidity and stiffness
79–86	26–30	Unconsciousness, erratic heartbeat
<79	<26	Heart failure, death

natural mechanism to increase the body temperature. The elderly are more susceptible to becoming hypothermic than others because the body becomes progressively incapable of maintaining an even body temperature as we age.

Normal body temperature [37°F (98.6°C)] should be slowly restored to tissues that have been exposed to cryogens. This is best accomplished by flushing the affected area with large volumes of tepid—not hot—water. To avoid aggravating the injury, affected body parts should never be rubbed or massaged. The gradual restoration of normal body temperature minimizes the potential for further tissue damage.

Individuals generally experience discomfort when they breathe the cold gas or vapor of a cryogen. Prolonged breathing of a very cold gas or vapor causes serious lung damage and may trigger the onset of a heart attack.

Finally, caution must be exercised as a cryogenic liquid vaporizes, because its gas or vapor displaces a corresponding amount of air. In a confined or unventilated area, the resulting atmosphere may contain a reduced concentration of oxygen that is incapable of supporting life. Individuals who breathe this atmosphere can suffocate or experience other ill effects.

Conversions Between Units of Measurement

1. The members of a firefighting brigade in Cincinnati, Ohio, roast a 25-pound turkey in an aluminum roasting pan that measures 12 inches × 16 inches × 3.75 inches. A similar group in Montréal, Quebec, roasts a 11.33-kilogram turkey in an identical-size pan. What are the measurements in SI units of the roasting pan used by the Montréal brigade?
2. During World War II, American prisoners of war tunneled from Stalag Luft III by excavating 200 tons of sand under the noses of their captors. How many tonnes of sand did the prisoners excavate?
3. The circulatory system of an adult human contains approximately 1½ gallons of blood. What is the equivalent approximate volume when expressed in liters?
4. Aluminum foil is available commercially in several dimensions, including 200-square-foot sheets measuring 66½ yards × 12 inches. For use during cooking, pieces of aluminum foil are wrapped as covers over containers to prevent food from splattering and to promote even cooking. What equivalent metric units are used to describe the dimensions of this consumer product?

Density of Matter

5. At 68°F (20°C), 5 gallons of a flammable liquid weighs approximately 36 pounds.
   (a) Use Table 2.3 to identify this liquid.
   (b) What is the approximate specific gravity of the liquid?
6. A piece of metal weighs 13.21 grams. When inserted into a cylinder containing exactly 50.00 milliliters of water, the metal and water occupy a volume of 54.89 milliliters at 68°F (20°C). Use Table 2.3 to identify the metal.

Conversion Between Temperature Readings

7. The four reaction zones of color and temperature of a candle flame are illustrated in Figure 2.13. Incandescent soot particles burning in the *yellow zone*, sometimes called the *luminous zone*, provide most of the light from a burning candle.
   (a) Identify the coolest and hottest zones of the flame.
   (b) What is the temperature in degrees Fahrenheit of the blue zone?
   (c) What is the temperature in degrees Rankine of the yellow zone?
8. In 1989, after 22 years of drilling, Russian geologists ceased their efforts to drill a borehole into the Kola Peninsula near Finland to obtain scientific information about Earth's interior. They ceased their efforts because, at a depth of 40,230 feet (12,262 m), they discovered that the temperature of the surrounding rock was 356°F, much hotter than had been anticipated.
   (a) What is the equivalent Celsius temperature of the deepest point drilled within the Kola borehole?
   (b) What is the equivalent temperature when expressed in kelvins?

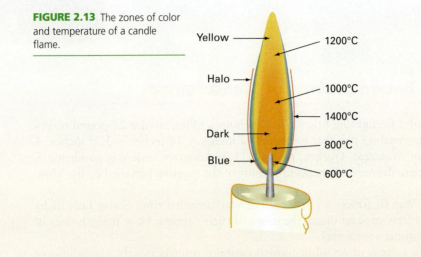

FIGURE 2.13 The zones of color and temperature of a candle flame.

Yellow — 1200°C

Halo — 1000°C

— 1400°C

Dark — 800°C

Blue — 600°C

Pressure Readings and Their Significance

9. The OSHA regulation at 29 C.F.R. §1910.158(a)(3)(iii) stipulates that employers must provide their personnel in fire brigades, industrial fire departments, and private or contractually arranged fire departments with fire hose of such a length that friction loss resulting from the flow of water through the hose will still provide a dynamic pressure at the nozzle within the approximate range of 210 kPa to 860 kPa. What pressure range at the nozzle does OSHA require in the fire hose when the pressure is expressed in pounds per square inch (psi)?

10. The OSHA regulation at 29 C.F.R. §1910.157(f)(6) stipulates that employers must hydrostatically test carbon dioxide hose assemblies having a shutoff nozzle at 300 psi. If an employer hydrostatically tests such a hose assembly at 2100 kPa, is the company in compliance with this OSHA regulation?

Energy

11. The U.S. Department of Transportation has estimated that the amount of energy needed to transport a single passenger one mile by high-speed rail ranges from 1200 to 1800 Btu. Based on this estimate, what range of energy in kilojoules is required to transport a passenger one kilometer?

Transmission of Heat

12. When fire begins on the fourth floor of a high-rise building, why are firefighters more concerned about its immediate spread to upper floors than to lower floors?

13. Why do firefighters often ventilate a burning building by opening a large hole at the highest point on its roof?

Significance of Heat

14. The human body is approximately 60% water by mass. How is it capable of maintaining a relatively steady internal temperature of 98.6°F (37°C) regardless of the surrounding temperature to which it is exposed?

General Properties of the Gaseous State of Matter

15. CPSC requires manufacturers to provide the following hazard-warning statements on aerosol cans whose contents have been pressurized:

> **CONTENTS UNDER PRESSURE**
>
> **DO NOT INCINERATE CONTAINER**
>
> **DO NOT EXPOSE TO HEAT OR STORE AT TEMPERATURES ABOVE 120°F**

   (a) What is the most likely reason CPSC requires manufacturers to provide these hazard-warning statements to consumers?

   (b) What is the scientific basis for these hazard-warning statements?

16. The air within a tire on a fire truck is compressed at 32.0 psig at 68°F (20°C). Following use of the fire truck, the new air pressure in the tire is measured as 36.0 psig. What is the new temperature in degrees Celsius of the air within the tire?

Cryogenic Liquids

17. Use the data in Table 2.9 to determine whether carbon dioxide exists as a compressed gas or a cryogenic liquid at 80.6°F (27°C) and 500 psig (3446 kPa).

18. Although nitrogen is neither flammable nor toxic, when emergency response crews respond to an incident involving a leak or spill of cryogenic nitrogen, what is the most likely reason they are encouraged to use self-contained breathing apparatus (SCBA) and wear total-encapsulating suits?

3

Flammable Gases and Liquids

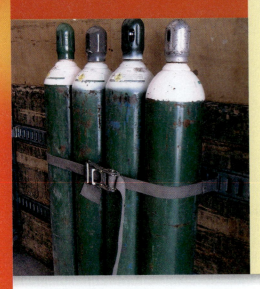

Courtesy of Eugene Meyer.

KEY TERMS

- Determine the OSHA category and NFPA class of flammable liquids from knowledge of their flashpoints and boiling points.
- Identify the information that DOT requires manufacturers to mark on gas cylinders and rail tankcars used to transport gases.
- Describe the general safety practices that are recommended when handling, storing, and transporting compressed gases.
- Identify the practices for safely storing flammable liquids in containers, portable tanks, and stationary tanks in the workplace.
- Describe the general features of pressure and nonpressure rail tankcars.
- Identify the conditions that trigger BLEVEs, and identify the actions emergency responders should implement to prevent the occurrence of BLEVEs.

Large inventories of flammable gases and flammable liquids are often stored within industrial settings in connection with their use during the manufacturing and processing of commercial products. Some of them are also located within residential environments, albeit in lesser amounts. Given that they are commonplace, thousands of emergency incidents involving them occur annually throughout the nation.

Although we note the properties of individual flammable gases and liquids in several chapters within this text, we note their common features here. We also examine the safety practices that OSHA mandates when employers store flammable gases and liquids in the workplace. Finally, we examine the procedures recommended for emergency responders when they encounter incidents involving their release into the environment.

3.1 FLAMMABILITY CRITERIA

Even when a flammable gas or the vapor of a flammable liquid is exposed to an ignition source in air, there are unique minimum and maximum gas or vapor concentrations below and above which it does not burn. These minimum and maximum concentrations are expressed in percent by volume, and are called the **lower flammable limit**, or **lower explosive limit**, and the **upper flammable limit**, or **upper explosive limit**, respectively. Flammable limits are always measured in ambient air, that is, the mixture of approximately 21% oxygen and 78% nitrogen by volume. These lower and upper flammable limits are different when this composition differs.

For example, the flammable liquid xylene[1] has lower and upper flammable limits in air of 1.1% and 7.0% by volume, respectively. The first limit tells us that a mixture having a concentration of less than 1.1% xylene vapor by volume in air will not burn. We say that it is too "lean" in xylene vapor. Similarly, the second limit tells us that a mixture containing more than 7.0% xylene by volume will not burn. It is said to be too "rich" in xylene vapor.

The **flammable range**, sometimes called the **explosive range**, of flammable gases and the vapors of flammable liquids is the set of their concentrations that lie between their upper and lower flammable limits. Thus, the flammable range of xylene vapor in air is 1.1% to 7.0% by volume. A flammable gas or vapor burns only when its concentration is exposed to an ignition source *and* when its concentration in the air lies within its flammable range. Thus, all concentrations of xylene vapor between 1.1% and 7.0% by volume ignite when exposed to an ignition source.

A gas or vapor often escapes from its confinement vessel at hatches, apertures, vents, and other openings. If the gas or vapor is flammable, it may ignite upon release into the

lower flammable limit (lower explosive limit)
■ The concentration of a gas or vapor in air below which a flame does not propagate when exposed to an ignition source

upper flammable limit (upper explosive limit)
■ The concentration of a gas or vapor in the air above which a flame does not propagate when exposed to an ignition source

flammable range (explosive range) ■ The percentage range of a gas or vapor in air, by volume, within which ignition can occur

---

[1]In Section 12.11-A, we note that xylene is a flammable liquid that has three different forms called *o*-, *m*-, and *p*-xylene. The properties quoted here are those of *m*-xylene.

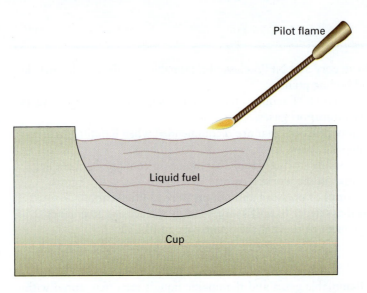

FIGURE 3.1 The test cup of a Pensky-Martens closed-cup flashpoint apparatus. In theory, an analyst determines the flashpoint of a liquid by heating a sample of it in the test cup while simultaneously applying a pilot flame within the vapor space above the liquid. When a flash of fire is momentarily observed, the analyst records the temperature of the liquid. This temperature is the flashpoint of the liquid.

Pilot flame

Liquid fuel

Cup

air. Immediate corrective action must be taken when the leaking gas or vapor is identified during an emergency response action. Generally, action must be taken to seal the point from which the leak is occurring.

Three temperatures are used to describe the ability with which a flammable liquid burns: the flashpoint, fire point, and autoignition point. They are also characteristic properties of certain flammable gases and flammable solids.

flashpoint ■ The lowest temperature of any liquid or solid at which it generates sufficient vapor to form an ignitable mixture with air near the surface of the liquid when exposed to an ignition source

fire point ■ The lowest temperature at which a flammable liquid ignites and achieves sustained burning when exposed to a test flame

autoignition point ■ The lowest temperature at which a liquid or solid ignites without exposure to an ignition source

■ *Flashpoint.* The **flashpoint** of a flammable liquid is the minimum temperature at which it gives off sufficient vapor to form an ignitable mixture with air across the surface of the liquid or within a test vessel. One procedure for measuring the flashpoint of a flammable liquid uses the specialized apparatus shown in Figure 3.1. Through its use, the flashpoint of xylene has been determined to be 84°F (29°C).

■ *Fire Point.* Above the flashpoint of a substance is a temperature at which self-sustained combustion occurs when the substance is exposed to a test flame. This temperature is called the **fire point**. At its fire point, a flammable liquid gives off sufficient vapor so that it continues to burn after initiation from exposure to an ignition source. The fire points of many liquids are approximately 30 to 50°F (17 to 27°C) higher than their flashpoints; for example, the fire point of xylene vapor is 111°F (44°C), almost 30°F (17°C) higher than its flashpoint.

■ *Autoignition Point.* As the temperature of the confined vapor of a flammable liquid further increases, a minimum temperature is attained at which self-sustained combustion occurs even in the absence of an ignition source. This is called the **autoignition point**. The autoignition point of xylene is 982°F (528°C).

SOLVED EXERCISE 3.1

Acetone and ethanol are flammable liquids having flashpoints of 0°F and 12°C, respectively. Other factors being equal, which liquid poses the greater risk of fire and explosion?

Solution: When two liquids have flashpoints that are widely different, the liquid having the lower flashpoint generally poses the greater risk of fire and explosion. To compare them, their flashpoints must be expressed on the same temperature scale. We calculate that a temperature of 0°F is equivalent to −18°C:

$$t(°C) = 5/9 \times [0 - 32] = -18\ °C$$

Because the flashpoint of acetone is substantially less than the flashpoint of ethanol, acetone poses the greater risk of fire and explosion.

The flashpoints, fire points, and autoignition points of the most commonly used liquids are compiled in the scientific literature. Emergency responders should identify an easily accessible means (e.g., the Internet or other source) to identify these values prior to their need. Emergency responders should always assume that the ignition of a flammable liquid is imminent when the temperature of the surroundings exceeds the liquid's flashpoint. In a comparison of the risk of fire and explosion for multiple flammable liquids, the greatest risk is potentially posed by the liquid that possesses the lowest flashpoint, ignoring all other factors.

3.1-A OSHA'S DEFINITION OF A FLAMMABLE LIQUID

OSHA defines a **flammable liquid** at 29 C.F.R. §1926.155 as any liquid having a vapor pressure not exceeding 40 psi at 100°F (37.8°C) and having a flashpoint at or below 199.4°F (≤ 93°C). OSHA further identifies four categories of flammable liquids:[2]

- A *category 1 flammable liquid* is a liquid with a flashpoint below 73.4°F (< 23°C) and an initial boiling point at or below 95°F (≤ 35°C).
- A *category 2 flammable liquid* is a liquid with a flashpoint below 73.4°F (< 23°C) and an initial boiling point at or above 95°F (≥ 35°C).
- A *category 3 flammable liquid* is a liquid with a flashpoint at or above 73.4°F (≥ 23°C) and at or below 140°F (≤ 60°C).
- A category 4 *flammable liquid* is a liquid with a flashpoint above 140°F (> 60°C) and at or below 199.4°F (≤ 93°C).

flammable liquid ■ For purposes of OSHA regulations, any liquid having a vapor pressure not exceeding 40 psi at 100°F (37.8°C) and having a flashpoint at or below 199.4°F (≤ 93°C)

OSHA requires the manufacturers, distributors, and importers of flammable liquids to display at least the flame pictogram on container labels of chemical products characterized as flammable liquids in categories 1, 2, and 3. However, OSHA does not require the display of a pictogram on the labels of chemical products that are category 4 flammable liquids.

3.1-B NFPA'S DEFINITIONS OF FLAMMABLE AND COMBUSTIBLE LIQUIDS

NFPA defines a **flammable liquid** as any liquid that has a flashpoint below 100°F (37.8°C) and a boiling point below 100°F (37.8°C). The organization recognizes the following three classes of flammable liquids:

- A *class IA flammable liquid* is a liquid with a flashpoint below 73°F (< 22.8°C) and a boiling point below 100°F (< 37.8°C).
- A *class IB flammable liquid* is a liquid with a flashpoint below 73°F (< 22.8°C) and a boiling point at or above 100°F (≥ 37.8°C).
- A *class IC flammable liquid* is a liquid with a flashpoint at or above 73°F (≥ 22.8°C) and below 100°F (< 37.8°C).

flammable liquid ■ For purposes of NFPA codes and standards, any liquid that has a flashpoint below 100°F (< 37.8°C) and a boiling point below 100°F (< 37.8°C)

NFPA also defines a **combustible liquid** as any liquid having a flashpoint at or above 100°F (≥ 37.8°C). There are three classes of combustible liquids:

- A *class II combustible liquid* is a liquid with a flashpoint at or above 100°F (≥ 37.8°C) but below 140°F (< 60°C), other than a liquid mixture having 99% or more of the volume of its components with flashpoints equal to or greater than 200°F (93.3°C). An example of a class II combustible liquid is acetic acid (Section 8.13), because its flashpoint is 109°F (43°C).

combustible liquid ■ For purposes of NFPA codes and standards, a liquid having a flashpoint at or above 100°F (≥ 37.8°C)

---

[2]Before 2012, OSHA and NFPA defined flammable and combustible liquids identically. However, in connection with implementing the GHS (Section 1.9), OSHA adopted the language noted in this section for the four categories of flammable liquids used in the GHS. OSHA no longer uses the term *combustible liquid*. The manner in which NFPA defines flammable and combustible liquids is noted in Section 3.1-B. Although NFPA still uses these terms in its codes and standards, we shall use the term *flammable liquid* throughout this text to jointly denote both classifications.

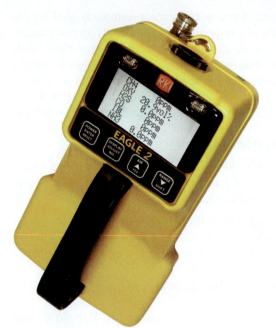

FIGURE 3.2 The use of this personal confined-space entry monitor, called the Eagle 2 Multi-Gas Meter, helps to ascertain whether it is safe to enter an enclosure in which a flammable gas has been released. Before an air–gas mixture reaches a concentration at which it will ignite, the meter triggers audible and visual alarms. The threshold concentration for the alarm is generally set at 10% of the lower flammable limit of the flammable gas. (*Photo by Eugene Meyer and courtesy of RKI Instruments, Inc., Union City, California.*)

■ A *class IIIA combustible liquid* is a liquid with a flashpoint at or above 140°F (≥ 60°C) but below 200°F (< 93.3°C). An example of a class IIIA liquid is creosote oil (Section 7.6-E), because its flashpoint ranges from 165 to 185°F (74 to 85°C).

■ A *class IIIB combustible liquid* is a liquid having a flashpoint at or above 200°F (≥ 93.3°C). An example of a class IIIB liquid is ethylene glycol (Section 13.2-J), a common antifreeze agent, because its flashpoint is 232°F (111°C).

Upon comparing the OSHA and NFPA definitions, it is apparent that the OSHA category 1, 2, and 4 flammable liquids are equal in definition to the NFPA class IA, IB, and III flammable liquids, respectively. The OSHA category 3 flammable liquid is subdivided into two groups having a flashpoint less than and greater than 100°F (37.8°C), respectively. These latter groups are the class IC and II flammable liquids, respectively, defined by NFPA.

Considering no other factors that impact liquid flammability, class IA flammable liquids are regarded as posing a high risk of fire and explosion; class IB and class IC flammable liquids pose a moderate risk; and combustible liquids pose a low risk.

When responding to an emergency involving the release of a flammable gas or vapor, experts advise first-on-the-scene responders to use a portable combustible gas monitor or sensor to determine whether the concentration of the flammable substance in the air is too hazardous for implementation of a response action. A **combustible gas monitor** is a handheld electronic device like that shown in Figure 3.2 that activates an alarm when the concentration of a flammable gas or vapor in the nearby atmosphere is 10% of the lower flammable limit or greater. Consider, for example, the use of a combustible gas monitor in an atmosphere containing methane, whose flammable range is 5% to 15% by volume. The combustible gas monitor first activates when a methane concentration of 0.5% is attained in air. The activation provides time for emergency responders to exit the area before the methane ignites.

A reading on a combustible gas monitor or sensor must always be used cautiously at scenes where a flammable gas or vapor is released. Because wind currents and other factors interplay, readings must be taken frequently, as the meter response may change from one moment to the next. These scenes are generally unsafe for emergency responders even under the best of weather conditions. When a combustible gas monitor or sensor is unavailable for use, they should respond to the emergency only from a distance.

combustible gas monitor ■ A handheld electronic instrument that activates when the concentration of a flammable gas or vapor in the nearby atmosphere is 10% of the lower flammable limit

3.1-C ASSIGNING NUMBERS TO THE RED QUADRANT OF THE HAZARD DIAGRAM

To forewarn emergency response personnel of a substance's potential to ignite and burn, five numbers are assigned to the uppermost red quadrants of a hazard diamond (Section 1.11). These numbers are assigned as follows:[3]

- For those gases and liquids that have flashpoints below 73°F (<23°C) and boiling points below 100°F (<37.8°C), the number "4" is entered in the quadrant.
- For gases and liquids that have flashpoints at or below 73°F (≤23°C) and boiling points at or above 100°F (≥37.8°C), or liquids that have flashpoints between 73°F (≥23°C) and 100°F (37.8°), the number "3" is entered in the quadrant.
- For liquids having flashpoints at or above 100°F (≥37.8°C) and below 200°F (<93.3°C) or solids that readily give off flammable vapor, the number "2" is entered in the quadrant.
- For liquids, solids, and semisolids with flashpoints at or above 200°F (≥93.3°C), the number "1" is entered in the quadrant.
- For materials that do not burn, including materials that do not burn in air when exposed to a temperature of 1500°F (816°C) for 5 minutes, the number "0" is entered in the quadrant.

3.1-D CPSC'S DEFINITION OF A FLAMMABLE LIQUID

At 16 C.F.R. §1500.3(c)(6), the CPSC regulations distinguish between liquid consumer products that are flammable, extremely flammable, and combustible as follows:

- **Flammable liquid** refers to a liquid having a flashpoint above 20°F (>−6.7°C) and below 100°F (<37.8°C)
- *Extremely flammable* liquid refers to a liquid having a flashpoint at or below 20°F (≤ −6.7°C)
- *Combustible* liquid refers to a liquid having a flashpoint at or above 100°F (≥ 37.8°C) to and including 150°F (65.6°C).

flammable liquid ■ For purposes of CPSC regulations, any liquid consumer product having a flashpoint above 20°F (> –6.7°C) and below 100°F (< 37.8°C)

3.1-E DOT'S DEFINITION OF A FLAMMABLE LIQUID

DOT defines a flammable liquid at 49 C.F.R. §173.120 as any liquid having a flashpoint of not more than 140°F (60°C), or any material in a liquid phase with a flashpoint at or above 100°F (37.8°C) that is intentionally heated and offered for transportation or transported at or above its flashpoint in bulk packaging. Several exceptions, not applicable here, apply.

flammable liquid ■ For purposes of DOT regulations, any liquid having a flashpoint of not more than 140°F (60°C), or any material in a liquid phase with a flashpoint at or above 100°F (37.8°C) that is intentionally heated and offered for transportation or transported at or above its flashpoint in bulk packaging

3.1-F FIRE AND TOXICITY HAZARDS IN "EMPTIED" CONTAINERS AND TANKS

When a flammable liquid is stored within a container or tank, its vapor evolves into the headspace above it and mixes with the confined air. When the liquid is removed from the container or tank, some residual vapor always remains. This "emptied" container or tank poses a risk to health and safety under the following circumstances:

- It is especially hazardous to conduct cutting or welding operations on a container or tank in which a flammable liquid was previously stored, unless specific actions to completely discharge the residual vapor from the tank have first been implemented.
- It is also hazardous to enter an "emptied" tank that had been used to store a flammable or toxic liquid. Although the inhalation of a nontoxic gas or vapor can cause dizziness, illness, or suffocation, the inhalation of a toxic gas can be fatal.

---

[3]NFPA 704-2012: *Standard System for the Identification of the Hazards of Materials for Emergency Response* (Copyright © 2012, National Fire Protection Association.)

For this combination of reasons, precautions should always be taken when handling or working in or around a container or tank in which a flammable or toxic liquid was once stored.

3.1-G ACCELERANTS AND THEIR USE BY ARSONISTS

accelerant ■ Any flammable or combustible material used to initiate fire or promote destruction through the use of fire

In the broadest sense, an **accelerant** is a flammable or combustible material whose intended use is to initiate and promote fire in either a beneficial or detrimental fashion. On the positive side, firefighters use liquid accelerants to affect a controlled burn; on the negative side, arsonists use them to intentionally initiate a fire or accelerate the rate at which they destroy buildings or other property for profit or perverse excitement.

Because they have low flashpoints and high vapor pressures, class IA, IB, and IC flammable liquids are popular accelerants used by arsonists. Their ignition can initiate massive fires that could subsequently destroy entire buildings.

Forensic investigators often inquire about the origin and cause of fires. When arson is suspected, they first search for an area where a liquid accelerant appears to have been used to initiate a fire. Then, by collecting samples of fire debris at the scene and submitting them to a chemical laboratory for analysis, they relate the cause of the fire with an arsonist's use of a specific flammable liquid. By linking the chemical nature of the accelerant with an arsonist's purchase or ownership of the same substance, investigators establish his or her guilt for instigating the crime.

Fire investigators examine the specific burn patterns that develop when an accelerant is used to initiate or promote a fire. A common burn pattern is the letter *V*, which outlines the upward movement of burning vapor as it rises and spreads away from a liquid spilled on a lower surface. Investigators easily trace the origin of the fire by locating the notch of the letter *V*.

Many professional arsonists have learned to avoid the use of flammable liquids for initiating fires. Instead, they use accelerants that fire investigators cannot easily identify at the crime scene. These are often solid combustible materials such as items made of Styrofoam, wicker, or cardboard. Massive fires can result when these materials are ignited; but, when samples of fire debris are subsequently collected and submitted to chemical analysis, the data are difficult to link easily with a specific accelerant.

3.2 RCRA CHARACTERISTIC OF IGNITABILITY

As previously noted in Section 1.3-C, the Resource Conservation and Recovery Act (RCRA) provides EPA with authority to regulate the treatment, storage, and disposal of hazardous wastes. Although EPA denotes some hazardous wastes by their chemical names, the nature of others is established by determining whether they exhibit certain characteristics, one of which is relevant here.

RCRA ignitability characteristic ■ For purposes of RCRA regulations, a characteristic of any waste having a property noted at 40 C.F.R. §261.21

A waste exhibits the **RCRA ignitability characteristic** if it is any one of the following:

■ A *flammable gas* (Section 3.3)
■ A liquid that possesses a flashpoint equal to or less than 140°F (≤ 60°C), other than an aqueous solution (with water as the solvent) containing less than 24% alcohol by volume, when tested through use of a specific type of closed-cup tester
■ A material other than a liquid that is capable "of causing fire by friction, absorption of moisture, or spontaneous chemical changes and, when ignited, burns so vigorously and persistently as to create a hazard"
■ A *flammable solid* (Section 9.1-A)
■ An *oxidizer* (Section 11.1)

A waste exhibiting the RCRA ignitability characteristic is a hazardous waste subject to EPA's treatment, storage, and disposal regulations. A waste exhibiting the characteristic of ignitability is assigned the hazardous waste number D001.

Three terms are routinely used to describe the potential for a liquid to burn: flammable, combustible, and ignitable. What are the differences among them in NFPA standards and codes and OSHA and EPA regulations?

Solution: Although all three terms are commonly used to denote the relative ease with which a liquid burns, these terms have specific meanings in NFPA codes and standards and OSHA and EPA regulations as follows:

- NFPA defines a *flammable liquid* as a liquid having a flashpoint below 100°F (< 37.8°C), unless it is a liquid mixture having 99% or more of the volume of its components with flashpoints at or above 100°F (≤ 37.8°C).
- NFPA define a *combustible liquid* as any liquid having a flashpoint at or above 100°F (≥ 37.8°C).
- OSHA defines a *flammable liquid* as any liquid having a vapor pressure not exceeding 40 psi at 100°F (37.8°C) and having a flashpoint at or below 199.4°F (≤ 93°C).
- EPA characterizes an *ignitable liquid* as a liquid waste, other than an aqueous alcoholic solution containing less than 24% alcohol by volume, possessing a flashpoint equal to or less than 140°F (≤ 60°C).

3.3 COMPRESSED GASES

During its storage and shipment, a gas confined under pressure is called a **compressed gas**. Examples are hydrogen, oxygen, and nitrogen. When the confined gas liquefies under the application of moderate pressure or refrigeration, it is called a **liquefied compressed gas**. Examples are methane and propane. A gas may also be dissolved in a solvent and confined under low to moderate pressure. An example is acetylene dissolved in acetone. In this text, the term *compressed gas* is used generally to represent any of the forms confined under pressure: a gas, dissolved gas, or liquid.

At 29 C.F.R. §1910.1200(c), OSHA defines a compressed gas as a gas confined in a container at an absolute pressure exceeding 40 psia (276 kPa) at 80°F (21.1°C); or, a gas or mixture of gases having in a container an absolute pressure exceeding 104 psia at 130°F (54.4°C) regardless of the pressure at 70°F (21.1°C); or, a liquid having a vapor pressure exceeding 40 psi (276 kPa) at 100°F (37.8°C) as determined by a specified test procedure.

A compressed gas may be either flammable or nonflammable. Hydrogen, methane, and propane are examples of flammable gases, whereas oxygen and nitrogen are examples of nonflammable gases.

A compressed gas is generally stored under pressure within a **cylinder** or a tank like those illustrated in Figure 3.3. When the gas is a liquefied compressed gas, it actually exists within the storage vessel as two phases: a liquid that settles to the bottom of the vessel and a gas that coexists in the headspace above it. During emergency response actions, this liquid may drip from valves, fittings, or openings. Once exposed to the environment, it immediately vaporizes and, if flammable, poses a fire and explosion risk; if toxic, the vapor poses a health risk.

compressed gas ■ For general purposes, any material that is a gas at normal temperature and pressure, but, when confined under pressure, is a gas, dissolved gas, or liquid; for purposes of OSHA regulations, a gas or mixture of gases having the temperature and pressure conditions noted at 29 C.F.R. §1910.1200(c); for purposes of the GHS, a gas that, when packaged under pressure, is entirely gaseous at –58°F (–50°C), including all gases with a critical temperature equal to or less than –58°F (–50°C)

liquefied compressed gas ■ For purposes of DOT regulations, a gas packaged under pressure for transportation that is partially liquid at temperatures above –58°F (> –50°C)

cylinder ■ For purposes of DOT regulations, a vessel having a circular cross-section and designed to safely store and transport compressed gases at pressures greater than 40 psia (275.6 kPa)

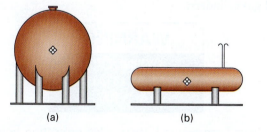

(a)   (b)

FIGURE 3.3 A high-pressure spherical tank and high-pressure horizontal tank used for the storage of bulk volumes of compressed gases, including flammable gases. The high-pressure horizontal tank is also used to store flammable liquids having relatively high vapor pressures.

flammable gas ■ For purposes of DOT regulations, a material that is a gas at 68°F or less (< 20°C) and 14.7 psia of pressure (101.3 kPa) *and* (a) is ignitable in a mixture of 13% or less by volume in air; *or* (b) has a flammable range with air of at least 12% regardless of the lower flammable limit.

nonflammable gas ■ For purposes of DOT regulations, any material that does one of the following: (a) exerts in its packaging an absolute pressure of 41 psi or greater (≥ 280 kPa) at 68°F (20°C); (b) is neither a flammable gas nor a poisonous gas.

poisonous gas (poison gas; gas poisonous-by-inhalation) ■ For purposes of DOT regulations, any material that is a gas at 68°F or less (≤ 20°C) and 14.7 psia of pressure (101.3 kPa) and is known to be so toxic to humans as to pose a health hazard during transportation; *or* in the absence of adequate data on human toxicity, is presumed to be toxic to humans because when tested on laboratory animals, it has an LC_{50} less than 5000 mL/m³.

In the United States, three classes of gases are identified for regulatory purposes. Their definitions in the DOT regulations are representative:

■ A **flammable gas** is a material that is a gas at 68°F (20°C) or less and 14.7 psia of pressure (101.3 kPa) *and* (a) is ignitable in a mixture of 13% or less by volume in air; *or* (b) has a flammable range with air of at least 12% regardless of the lower flammable limit.

■ A **nonflammable gas** is any material or mixture that does one of the following: (a) Exerts in its packaging an absolute pressure of 41 psi or greater (≥ 280 kPa) at 68°F (20°C); (b) is neither a flammable gas nor a poisonous gas.

■ A **poisonous gas** (also called a **poison gas** and **gas poisonous-by-inhalation**) is any material that is a gas at 68°F or less (<20°C) and 14.7 psia of pressure (101.3 kPa) and is known to be so toxic to humans as to pose a health hazard during transportation; *or* in the absence of adequate data on human toxicity, is presumed to be toxic to humans because when tested on laboratory animals, it has an LC_{50} (Section 10.6-B) less than 5000 mL/m³.

The GHS defines a compressed gas as any gas that, when packaged under pressure, is entirely gaseous at –58°F (−50°C), including all gases with a critical temperature equal to or less than −58°F (−50°C). The GHS gas-cylinder pictogram is affixed to the labels of cylinders and other containers in which compressed gases are confined. When the compressed gases are flammable, poisonous, or oxidizing gases, the flame, skull-and-crossbones, or flame-over-the-letter-"O" pictograms are also included on these labels, as relevant.

3.3-A STORAGE AND USE OF NONBULK VOLUMES OF COMPRESSED GASES

Nonbulk volumes of compressed gases are generally stored in the steel cylinders that were used to transport them from their place of purchase. When they are not in use, the gas cylinders are often stored within cabinets at one or more locations at a workplace. For example, at hospitals and clinics, cylinders of medical gases are generally encountered within specially identified cabinets, where they are safely segregated from other medical supplies. A gas cylinder storage area should be clearly marked as shown below:

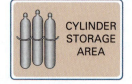

OSHA requires workers to exercise appropriate precaution so that gas cylinders are not damaged during their storage and use. Experts recommend maintaining them in an upright position and strapping, chaining, bracing, or otherwise securing them in a fixed position like that shown in Figure 3.4. OSHA also requires employers to post the following warning sign within the workplace to reduce or eliminate the physical hazards associated with the use of gas cylinders:

Once the cylinders have been secured, their valve-protection hoods should be kept in place and removed only when use of the gas is required.

When multiple gas cylinders are stored in near proximity, experts recommend that they be "nested." The practice of **nesting** refers to a method of storing gas cylinders by

nesting ■ The method of securing gas cylinders upright in a tight mass using a contiguous three-point contact system wherein the cylinders within a group have a minimum of three points of contact with other cylinders, walls, or bracing

FIGURE 3.4 When transported, compressed gas cylinders should always be strapped, chained, or otherwise secured to a trailer frame or other permanent structure. Their hoods should also be kept in place until the contents are to be used, and care should always be exercised to assure that they are not exposed to sources of heat or ignition. (*Courtesy of Eugene Meyer.*)

situating them upright in a tight-mass group using a contiguous three-point contact system. Each cylinder within the group has a minimum of three points of contact with other cylinders, walls, or bracing. This manner of storing cylinders helps to keep them upright.

When workers must move cylinders from place to place, they should first secure or restrain them by straps or chains to a cylinder cart having wheels. Cylinders should never be rolled or dragged across a floor or other surface, as this action could seriously damage them and cause the release of their contents. Dropping the cylinders could also result in the catastrophic release of their contents.

NFPA stresses that dedicated areas should be used to store gas cylinders to ensure proper management and protection. It also specifies the following limitations on the number of gas cylinders allowed in the same storage area: three 10 inches × 50 inches (25 cm × 127 cm) cylinders containing either flammable gases or oxygen; and three 4 inches × 15 inches (10 cm × 38 cm) cylinders containing toxic gases.

When cylinders containing flammable gases are stored, an area should be selected that is isolated from the area used for the storage of cylinders containing oxygen or other oxidizing gases. To reduce or eliminate the likelihood of fire and explosion, precaution should always be exercised to ensure the absence of ignition sources in the storage area. Smoking tobacco products, using open flames and hot surfaces, conducting cutting and welding operations, and performing all activities that potentially provide static, electrical, and mechanical sparks should be strictly avoided. Finally, workers should be trained to use at any given time only the smallest amount of a flammable gas that accomplishes a required task.

SOLVED EXERCISE 3.3

Why do experts recommend turning the valve on a gas cylinder to the closed position when the cylinder is essentially empty?

Solution: Residual gas remains within any cylinder that has been essentially emptied. If the valve is turned to the open position, air may enter the tank and mix with the contents. This situation increases the likelihood of an unwanted fire and explosion when the mixture is exposed to an ignition source.

A cylinder containing a flammable gas requires special attention during its storage and use. In particular, it should be properly grounded and bonded to prevent the build-up of static electricity. The practice of **grounding** refers to the use of wire cables, chains, or straps connecting the cylinder to an earth-ground such as a rod, water pipe, or other approved item. The cylinder is considered grounded when an electrical conductor connects it and an independent source such as the underlying earth. Grounding quickly drains away a static charge and prevents the formation of sparks that could otherwise ignite the flammable gas.

When a flammable gas is transferred from one cylinder or tank to another through a closed piping system, the cylinders or tanks should also be bonded. The practice of **bonding** refers to the act of connecting the cylinders or tanks involved in the transfer operation with a wire and clips to prevent the production of sparks and to ensure that they have the same electrical charge. A difference in charge on two cylinders or two tanks causes the generation of a spark that jumps from one to the other. Although bonding the containers or tanks does not eliminate the static charge, it prevents the formation of a spark between them and reduces the likelihood of a fire or explosion.

3.3-B TRANSPORTATION OF NONBULK VOLUMES OF COMPRESSED GASES

Nonbulk volumes of compressed gases are generally shipped in steel or aluminum cylinders. To ensure that they do not rupture during transit, DOT regulates their design, construction, and maintenance.

The design features of a typical gas cylinder are illustrated in Figure 3.5. Some features are readily recognizable. For example, a gas cylinder is equipped with a valve that is opened to release the desired quantity of the confined gas, after which the valve is closed.

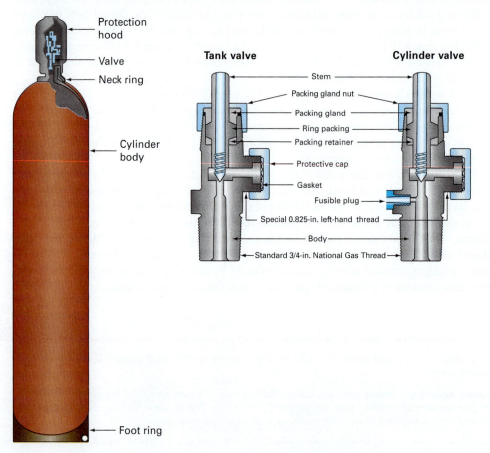

FIGURE 3.5 The design features of a typical DOT-approved steel cylinder used for the storage and transport of compressed gases. Located at the top of the cylinder is a single discharge valve in which a fusible plug is located within the valve body just below the seat. Two common types of fusible plugs are called CG-2 and CG-3, and have nominal melting temperatures of 165°F (73.9°C) and 212°F (100°C), respectively.

A major safety component of a gas cylinder is its **fusible plug**, a block of metal that melts when the cylinder is exposed to a nominal melting temperature. The fusible plug serves as a pressure-relief device that affords protection against the build-up of excessive pressure when the cylinder is exposed to heat. Two common types are the CG-2 and CG-3 fusible plugs, which have nominal melting temperatures of 165°F (73.9°C) and 212°F (100°C), respectively.[4] DOT requires nearly all cylinders to be equipped with fusible plugs. Their melting allows the contents to be slowly released into the environment, thereby preventing them from rupturing.

Gas cylinders may also be equipped with rupture discs or a combination of a rupture disc and a fusible plug. A **rupture disc** is also known as a **frangible disc** or **bursting disc**. It is generally made from metal and functions by means of bursting at a unique pressure, thereby releasing the cylinder's contents into the environment.

DOT regulates the design and construction of gas cylinders so they will maintain their integrity while their contents are under pressure and exposed to the range of temperatures likely to be encountered during transportation. DOT assigns a unique pressure at 70°F (21°C) to each type of gas cylinder. It is called the cylinder's **service pressure**. DOT requires the service pressure to be marked on the exterior surface of the cylinder in units of psig. For instance, when the service pressure of a DOT-3E gas cylinder is 1800 psig (1792 kPa), it is marked on the cylinder as DOT-3E1800. The DOT-3E indicates a type of cylinder that has been fabricated in compliance with DOT's specifications for a cylinder whose class is 3E.

DOT requires manufacturers to verify the integrity of each type of gas cylinder when the temperature of the contents is increased from 70°F (21°C) to 130°F (54°C). The cylinder is approved for use only when the test results reveal that the pressure of its contents will not exceed five-fourths of the cylinder's service pressure.

DOT also regulates the amount of gas that its manufacturer may fill into a cylinder. At 49 C.F.R. §§173.301 through 173.306, DOT identifies specific filling limits for compressed gases contained within cylinders. DOT also requires gas cylinders containing non-bulk volumes of compressed gases to be marked and labeled prior to transportation. We examine the nature of these marking and labeling requirements in Chapter 6.

Once they have been correctly marked and labeled, properly designed and constructed gas cylinders are organized for transportation only when they have been securely strapped or chained, sometimes in a **bundle**, to the body of a motor trailer or rail car. DOT also allows jumbo-sized cylinders to be packed horizontally and transported in a tube car or tube trailer.

fusible plug ▪ A device usually consisting of a low-melting-point metal inserted as a plug in certain compressed gas cylinders to function by melting in a narrow range of temperatures

rupture disc (frangible disc; bursting disc) ▪ A device usually consisting of a metal inserted in certain compressed gas cylinders to function by rupturing or bursting when a predetermined pressure is attained

service pressure ▪ For purposes of DOT regulations, the pressure designated in units of psig at 70°F (21°C) for an authorized type of compressed gas cylinder

bundle ▪ For purposes of DOT regulations, an assembly of cylinders fastened together, interconnected, and transported as a unit

SOLVED EXERCISE 3.4

DOT-3A2000 is marked on a steel cylinder whose gauge pressure reads 1750 psi at 70°F (21°C). Is this cylinder likely to rupture during a fire if the temperature of its contents rises to 300°F (149°C)?

Solution: To ensure the integrity of a gas cylinder at elevated temperatures, DOT specifies that the pressure of its contents cannot exceed five-fourths of the service pressure when the cylinder is subjected to a temperature of 130°F (54°C). The service pressure of a cylinder marked DOT-3A2000 is 2000 psi. If the gas manufacturer has complied with this DOT regulation, the integrity of the cylinder will be maintained as long as the pressure of the contents does not exceed 5/4 × 2000 psi, or 2500 psi, at a temperature of 130°F (54°C).

However, an internal pressure greater than 2500 psi is generated when the temperature of the gas cylinder exceeds 130°F (54°C). Consequently, a cylinder marked DOT-3A2000 is likely to rupture when heated to 300°F (149°C).

---

[4]Compressed Gas Association, *Handbook of Compressed Gases* (Dordrecht, The Netherlands: Kluwer Academic Publishers Group, 1999), p. 119.

3.4 ENCOUNTERING COMPRESSED GASES DURING EMERGENCY RESPONSE ACTIONS

Cylinders used for the storage of compressed gases have been designed, constructed, and tested to potentially provide them with the highest degree of safety. Hence, when heated slowly, they do not ordinarily rupture, even when their contents experience a moderate increase in temperature. As we previously noted, the inclusion of pressure-relief devices or fusible plugs allows the contents to slowly escape into the environment, thereby preventing the build-up of the internal pressure that could cause them to rupture.

Nonetheless, when they are *rapidly* heated to temperatures exceeding approximately 130°F (54°C), gas cylinders may readily burst and discharge their contents into the immediate environment. At such temperatures, the cylinders have clearly exceeded their authorized service pressures. The heated contents cannot discharge rapidly enough through pressure-relief valves or fusible plugs to adequately reduce the internal pressure. If unsecured, these damaged cylinders can travel like active airborne missiles from spot to spot, perhaps for hundreds of yards, until virtually all their contents have been released.

When a cylinder containing a flammable gas initially ruptures, the concentration of the released gas is usually within its flammable range. As it mixes with the air, the gas ignites explosively, typically engulfing the entire area in flames. Even when the contents are nonflammable, the immediate environment in which the rupture occurs can be a daunting scene. The immense force that is exerted on the surroundings is sufficient to shatter windows and cause other types of physical destruction within the immediate area. This force may also rupture eardrums, disembowel, or otherwise injure those who are struck by the flying shrapnel.

stationary tank
■ Packaging designed primarily to be installed in a stationary position and not intended for loading, unloading, or attachment to a transport vehicle

Stationary tanks containing compressed gases are equally dangerous. When they are situated inside a building, sensing devices electronically linked to exhaust fans are often installed to provide maximum protection for workers against potential exposure to a gas that may inadvertently be released into the working environment. Once activated, the fans evacuate escaping gases from the enclosure into the outside atmosphere at very high speeds, thereby minimizing the possibility that they will concentrate indoors. When a tank is used to store gases with vapor densities greater than 1.0, these sensing devices and fans are positioned near the floors of the buildings in which the tank is located, but when a tank is used to store gases with vapor densities less than 1.0, the sensing devices and fans are positioned near the ceilings.

3.5 RESPONDING TO INCIDENTS INVOLVING THE RELEASE OF FLAMMABLE GASES

Emergency responders are often called to scenes where cylinders, storage tanks, and transport vessels containing gases have already ruptured or are likely to rupture when they are exposed to intense heat. As these scenes are being assessed, certain actions should be quickly executed. The following list, although not intended to be exhaustive, is illustrative:

■ If the response action involves the storage of a compressed gas within a stationary tank, personnel should use the information posted on a hazard diamond to ascertain the degree of hazard relating to fire, health, and instability. It is particularly relevant to determine whether the tank contents are flammable or toxic.

■ When there is a fire in an area where cylinders or tanks containing compressed gases are located, unmanned monitors should be situated to cool them with direct streams of water—but only when the monitors may be placed without risk to personnel.

■ When a small leak from a cylinder containing a compressed gas is discovered, a water fog (Section 5.12-A) should be used to disperse the gas or vapor.

■ When personnel decide to move a leaking cylinder containing a flammable gas, a combustible gas monitor should be used to estimate the degree of fire hazard within the immediate area. No attempt to move a leaking cylinder should ever be made if the concentration of an escaping gas or vapor is within its flammable range. Furthermore, no attempt to stop the escape of the gas from its cylinder should be made unless the concentration can be reduced to less than its lower flammable limit. This may be accomplished by dispersing the gas or vapor with a water fog.

■ A slowly leaking cylinder containing a flammable gas may be moved into an open isolated area, but *only* if the gas has not ignited. When a leaking cylinder is moved, it is prudent to wear a total-encapsulating suit and use self-contained breathing apparatus. The latter action is warranted, because the exposure to flammable gases may cause responders to experience dizziness, illness, or suffocation.

■ When a cylinder is moved into an open area, it should be rotated so that the point of leakage becomes the uppermost part of the cylinder. Emergency responders may elect to allow the gas to slowly escape into the air, but sometimes the leak may easily be stopped by closing the valve or tightening the packing gland nut.

■ Experience indicates that vessels generally rupture at a tank's head or end, near the seam where the head or end was welded to the tank's body. For this reason, emergency responders should *never* approach the head or end of a cylinder or tank containing a flammable gas during an ongoing fire.

■ Because a compressed gas exerts pressure uniformly on all internal points of a cylinder or tank in which it is confined, an overturned, nonleaking rail tankcar may be safely uprighted, when practical, without first unloading its contents. This action should be undertaken *only* by experienced personnel. When a decision to lighten a load is made, the responders first transfer a portion of the contents into another tankcar. When a flammable gas is transferred from one tank to another, the transfer line must be electrically grounded and bonded.

■ Based on the combined experience of fire personnel, attempts to combat a major fire involving a flammable gas should not be undertaken *unless* the escape of the gas from its cylinder or tank can be stopped. Sealing a gas leak requires a special expertise and should be attempted only by experienced personnel.

■ Because the unintentional explosion of a flammable gas can cause structural failure of buildings and their ultimate collapse, emergency response personnel must acknowledge the presence of workers, nearby residents, and others to avoid multiple deaths and injuries.

3.6 STORING FLAMMABLE LIQUIDS

At manufacturing and production facilities, flammable liquids are routinely stored for future use in safety cans, bottles, metal cans, metal drums, barrels, portable tanks, or stationary tanks. When they are not in use, these containers and tanks should always be closed by means of a lid or other device that prevents the escape of vapor.

To protect against the possibility that static electricity will serve as an ignition source, a category I flammable liquid should be dispensed from or into metal containers or tanks only when they have been positively grounded and electrically bonded through wires attached to both containers or through a common ground system. An example of acceptable grounding and bonding is illustrated in Figure 3.6.

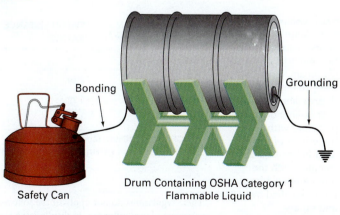

FIGURE 3.6 An OSHA category 1 (NFPA class IA) flammable liquid should only be transferred from a steel drum into a safety can by mutually bonding the drum and safety can and grounding the drum using metal wires as illustrated.

FIGURE 3.7 These safety cans have been constructed with leak-tight gasketed lids, dual-density flame arresters, and other requirements identified at 29 C.F.R. §1910.106 and NFPA Code 30, *Flammable and Combustible Liquids Code.* To prevent their rupture in the event of a fire, they slowly vent their confined vapors between 3 and 5 psig. The OSHA regulation at 29 C.F.R. §1910.144 requires safety cans and other portable containers of flammable liquids having a flashpoint at or below 80°F (27°C) to be painted red with some additional clearly visible identification either in the form of a yellow band around the can or the name of the contents conspicuously stenciled or painted on the can in yellow. *(Courtesy of Justrite Manufacturing Company, Des Plaines, Illinois.)*

safety can ■ For purposes of OSHA regulations, an approved, closed container of not more than 5-gallon (18.9-L) capacity, having a flash-arresting screen, spring-closing lid, and spout cover, and so designed that it will safely relieve internal pressure when subjected to fire exposure

fire area ■ The area of a building separated from its remainder by special construction and having a fire-resistance rating of at least 1 hour and all communicating openings properly protected by an assembly that also has a fire-resistance rating of at least 1 hour

fire-resistance rating ■ The length of time in hours that a construction material or assembly (beam, girder, or truss; column, floor, or floor-ceiling; roof or roof-ceiling; or wall or partition) withstands the effects of a standard fire exposure and complies with the requirements for acceptance specified by NFPA and others

As another precaution against its ignition, a flammable liquid should be transferred into a safety can from which it is then dispensed in the needed amount. A **safety can** is an approved container of not more than 5-gallon (19-L) capacity, having a spring-closing lid and spout cover and so designed that it will safely relieve internal pressure when exposed to fire. Two examples are shown in Figure 3.7.

To minimize or eliminate employee exposure to flammable liquids, OSHA regulates the manner in which they are stored within the workplace, and NFPA recommends similar storage practices at virtually all locations. Because the OSHA regulations and NFPA standards are incorporated into the fire codes of every major city, it is appropriate to examine certain of their aspects that are most relevant to fire-service personnel.[5]

When employers choose to store a flammable liquid in an inside room at a workplace, they must comply with the size and construction allowances listed in Table 3.1. These allowances specifically pertain to the **fire area**, that is, the area separated from the remainder of a building by special construction and having a fire-resistance rating of at least 1 hour, with all communicating openings properly protected by an assembly that also has a fire-resistance rating of at least 1 hour. The construction allowances for a flammable liquid storage room pertain to providing a 1- or 2-hour **fire-resistance rating**

TABLE 3.1	Storage of Flammable Liquids within Inside Rooms[a]				
		MAXIMUM SIZE		**TOTAL ALLOWABLE VOLUMES**	
FIRE-PROTECTION SYSTEM NEEDED[b]	**FIRE-RESISTANCE RATING**	ft²	m²	gal/ft²/floor area	L/m²/floor area
Yes	2 hr	500	46.5	10	410
No	2 hr	500	46.5	5	200
Yes	1 hr	150	13.9	4	160
No	1 hr	150	13.9	2	80

[a]29 C.F.R. §1910.106(d), Table H-13.
[b]The fire-protection system must be sprinkler, water spray, carbon dioxide, or a similarly approved system.

[5]These regulations do not apply to the following circumstances: liquids stored in bulk plants, service stations, processing plants, refineries, or distilleries; a liquid used in vehicle tanks, portable engines, or stationary engines; when the liquid is a paint, oil, or varnish used for maintenance; a liquid beverage stored in containers having a capacity of less than 1 gallon (3.8 L); and liquid medicines, beverages, foodstuffs, cosmetics, and other common consumer items.

FIGURE 3.8 This cabinet is specifically marked to identify its sole utilization for the storage of flammable liquids. It was constructed in compliance with OSHA's regulations at 29 C.F.R. §1910.106(d)(3)(ii)(a). *(Courtesy of A & A Sheet Metal Products, Inc., SE-Cur-All® Products, LaPorte, Indiana)*

based on whether a fire-protection system has or has not been provided. The fire-resistance rating is determined by exposing the construction material or assembly to standardized gas burners within a furnace. Materials having a fire-resistance rating of 1 hour and 2 hours have satisfactorily met the 1-hour and 2-hour furnace test, respectively. Also relevant in the construction of a flammable liquid storage room is compliance with relevant NFPA standards that include the use of liquid-tight sills or ramps at least 4 inches (10 cm) in height, approved self-closing doors, and a liquid-tight seal between the walls and floor.

When employers choose to store flammable liquids in the workplace, OSHA requires that they provide an inside room or storage cabinet used solely for storage purposes. Figure 3.8 illustrates a metal storage cabinet designed and constructed to prevent ignition of the contents by exposure to external fire or other sources of ignition. At 29 C.F.R. 1910.106(d)(3)(ii)(a), OSHA requires metal storage cabinets to be constructed in the following manner when the cabinet is intended for storage of flammable liquids:

- The bottom, top, and sides of the cabinet are constructed from No. 18-gauge sheet metal.
- The cabinet must be doubled-walled with a 1-inch airspace.
- The joints must be riveted, welded, or made tight by some equally effective means.
- The door shall have a three-point hatch.
- The door sill must be raised at least 2 inches above the cabinet bottom to retain spilled liquid within the cabinet.
- The cabinet must be conspicuously stenciled, imprinted, or otherwise marked with the following message:

> **FLAMMABLE**
> **KEEP FIRE AWAY**

When employers use a cabinet for the storage of flammable liquids, OSHA requires them at 29 C.F.R. §1910.106(d)(3)(i) to store no more than 60 gallons (230 L) of a category 1, 2, and/or 3 flammable liquid, and no more than 120 gallons (454 L) of a category 4 flammable liquid in a storage cabinet. Furthermore, NFPA requires employers to

TABLE 3.2	Maximum Allowable Sizes of Containers and Portable Tanks for Flammable Liquids[a]			
CONTAINER TYPE	CATEGORY 1	CATEGORY 2	CATEGORY 3	CATEGORY 4
Glass or approved plastic	1 pt (0.5 L)	1 qt (0.95 L)	1 gal (4 L)	1 gal (4 L)
Metal (other than DOT drums)	1 gal (4 L)	5 gal (19 L)	5 gal (19 L)	5 gal (19 L)
Safety cans	2 gal (8 L)	5 gal (19 L)	5 gal (19 L)	5 gal (19 L)
Metal drums (DOT specifications)	60 gal (230 L)	60 gal (230 L)	60 gal (230 L)	60 gal (230 L)
Approved portable tanks	660 gal (2500 L)	660 gal (2500 L)	660 (2500 L)	660 gal (2500 L)

[a]29 C.F.R. §1910.106(d), Table H-12.

use only three storage cabinets per fire area when they are located in a nonsprinklered fire area or when the cabinets are separated from one another by a distance of at least 100 feet (30 m), or six storage cabinets per fire area that is equipped with an automatic sprinkler system.

OSHA also allows the storage of specified limited amounts of flammable liquids outside a liquid storage room or storage cabinet. At 29 C.F.R. §1910.106(e)(2)(ii)(b), OSHA limits the volume of flammable liquids that are located *outside* of an inside storage room, storage cabinet in a building, or in any one fire area of a building to the following:

- 25 gallons (95 L) of category 1 flammable liquids within containers
- 120 gallons (454 L) of category 2, 3, or 4 flammable liquids within containers
- 660 gallons (2500 L) of category 2, 3, or 4 flammable liquids within a single portable tank

OSHA also restricts the volumes of flammable liquids that employers may store within different types of containers and portable tanks. These restrictions are listed in Table 3.2.

SOLVED EXERCISE 3.5

Methyl formate is a liquid that boils at 90°F (32°C) and has a flashpoint of −2°F (−19°C). Because it has a pleasant odor, methyl formate often is a component of commercial air fumigants. In a workplace, a manufacturer of air fumigants constructs an indoor room from fire-resistant materials and equips it with a sprinkling system. What is the maximum number of 55-gallon DOT-specification drums containing methyl formate that OSHA allows to be stored per stacked pile within the room?

Solution: From the boiling point and flashpoint data, we determine that methyl formate is an OSHA category 1 flammable liquid. We also determine from Table 3.3 that when the storage is protected, the maximum number of 55-gallon drums of methyl formate that OSHA allows to be stored per pile within an indoors storage room on either the ground or upper floors is 50. (When the storage is unprotected, the maximum number of 55-gallon drums of methyl formate that OSHA allows to be stored per pile in an indoor storage room is only 12.)

3.6-A STORAGE IN CONTAINERS

When employers store flammable liquids in the workplace, restrictions are imposed on the maximum volumes that OSHA permits. These allowable volumes vary with the category of the flammable liquid, the manner in which the liquid is stored, and whether the storage occurs indoors or outdoors.

		PROTECTED STORAGE, MAXIMUM PER PILE		UNPROTECTED STORAGE, MAXIMUM PER PILE	
TABLE 3.3	**Indoor Container Storage**[a]				
CATEGORY OF LIQUID	STORAGE LEVEL	gal	L	gal	L
1	Ground and upper floors	2750 (50)[b]	10,400	660 (12)	2500
	Basement	Not permitted	Not permitted	Not permitted	Not permitted
2	Ground and upper floors	5500 (100)	21,000	1375 (25)	5200
	Basement	Not permitted	Not permitted	Not permitted	Not permitted
3 [FP <100°F (37.8°C)][c]	Ground and upper floors	16,500 (300)	62,400	4125 (75)	15,600
	Basement	Not permitted	Not permitted	Not permitted	Not permitted
3 [FP >100°F (37.8°C)]	Ground and upper floors	16,500 (300)	62,400	4125 (75)	15,600
	Basement	5500 (100)	20,800	Not permitted	Not permitted
4	Ground and upper floors	55,000 (1000)	208,000	13,750 (250)	52,000
	Basement	8250 (450)	31,200	Not permitted	Not permitted

[a]29 C.F.R. §1910.106(d), Table H-14.
[b]The numbers in parentheses below the listed gallons refer to the number of 55-gallon drums that are equivalent to the indicated volumes.
[c]FP is flashpoint.

When employers choose to store flammable liquids indoors in glass or metal containers, they must comply with the restrictions in Table 3.3. When two or more categories of flammable materials are stored in containers that have been stacked into piles in an inside storage room, the maximum volume allowed in the pile is the smallest of their separate maximum volumes.

OSHA also requires employers to situate all containers more than 12 feet (4 m) from an aisle. The main aisles must be at least 3 feet (0.9 m) wide, and the side aisles must be at least 4 feet (1.2 m) wide. OSHA regulations at 29 C.F.R. §1910.106(d)(4)(v) prohibit the stacking of containers holding more than 30 gallons (113.5 L) of flammable liquids in inside rooms.

When employers choose to store flammable liquids outdoors within containers, they must comply with the restrictions in Table 3.4. When two or more classes of flammable liquids are stored within containers in piles, the maximum allowable volume in each pile is the smallest of their separate maximum volumes. OSHA also requires the availability of a 12-foot- (4-m-) wide access way for the approach of fire control apparatus.

The distances in Table 3.4 apply to a facility that is "protected for exposure," meaning that all adjacent structures on the facility are adequately provided with fire-protection systems. When these structures are not protected for exposure, the distances in the fourth column of Table 3.4 must be doubled. When the total volume of the flammable or combustible liquids does not exceed 50% of the maximum per pile, the distances in the fourth and fifth columns may be reduced by half, but in any event, the distance must never be less than 3 feet (0.9 m).

TABLE 3.4 | Outdoor Container Storage[a]

CATEGORY OF FLAMMABLE LIQUID	MAXIMUM PER PILE		DISTANCE BETWEEN PILES		DISTANCE TO THE PROPERTY LINE THAT MAY BE STORED UPON		DISTANCE TO STREET, ALLEY, OR OTHER PUBLIC WAY	
	gal	L	ft	m	ft	m	ft	m
1	1100	4200	5	1.5	20	6	10	3
2	2200	8300	5	1.5	20	6	10	3
3 [FP <100°F (<37.8°C)][b]	4400	16,600	5	1.5	20	6	10	3
3 [FP >100°F (>37.8°C)]	8800	33,300	5	1.5	10	3	5	1.5
4	22,000	83,000	5	1.5	10	3	5	1.5

[a]29 C.F.R. §1910.106(d), Table H-17.
[b]FP is the liquid's flashpoint.

SOLVED EXERCISE 3.6

Ethanol and acetone are flammable liquids that boil at 174°F (79°C) and 133°F (56°C) and have flashpoints of 54°F (12°C) and 0°F (−18°C), respectively. What is the maximum volume of these liquids that OSHA allows to be stored in containers outside a flammable liquid storage cabinet?

Solution: The boiling point and flashpoint data indicate that ethanol and acetone are OSHA category 2 flammable liquids. The OSHA regulation at 29 C.F.R. §1910.106(e)(2)(ii)(b) stipulates that no more than a total of 120 gallons (454 L) of category 2 flammable liquids may be stored outside a storage cabinet within a building. Thus, OSHA allows no more than 120 gallons (454 L) of ethanol and acetone to be stored outside a flammable liquid storage cabinet.

SOLVED EXERCISE 3.7

A major fire at a manufacturing plant within a densely populated industrial park threatens to engulf a stationary tank located on adjacent property. The captain of the responding firefighting team is informed that the tank contains Cellosolve acetate, an OSHA category 2 flammable liquid whose lower flammable limit is 1.75% by volume and whose flashpoint is 131°F (55°C). Aside from extinguishing the major fire, what action should the responding team undertake?

Solution: When emergency response personnel arrive at an emergency incident involving the potential release of a flammable liquid from a storage tank, it is critical that they undertake immediate action to assure that the structural integrity of the tank is maintained. This may be achieved by discharging cooling water from unmanned monitors upon the tank. The water should be directed at the uppermost area of the tank and the actual sites where flame contact with the shell of the tank occurs.

3.6-B STORAGE IN PORTABLE TANKS

The term **portable tank** is defined in OSHA and DOT regulations as follows:

■ For purposes of OSHA regulations, a portable tank is a closed vessel having a liquid capacity over 60 gallons (227 L) and not intended for fixed installation

■ For purposes of DOT regulations, a portable tank is a form of bulk packaging designed primarily to be loaded onto, or on, or temporarily attached to a **transport vehicle** or ship and equipped with skids, mounting, or accessories to facilitate handling of the tank by mechanical means

A structural silhouette of a common type of portable tank is illustrated in Figure 3.9. This type has been constructed with mountings to facilitate handling by mechanical means. The various types of portable tanks may be used for storing and transporting either liquids or solids. Their manufacturers must comply with standards established by the United Nation's globally harmonized system (Section 1.9) and adopted by DOT.

In the DOT regulations, portable tanks are designed and constructed to provide for the venting of vapor through top-mounted emergency and pressure-activated vents. The top-mounted vents are capable of limiting internal pressure under fire-exposure conditions to 10 psig (69 kPa) or 30% of the tank's bursting pressure, whichever is greater. The pressure-activated vents are set to open at not less than 5 psig (34.5 kPa).

When employers choose to store flammable liquids indoors in portable tanks, they must comply with the OSHA/NFPA restrictions noted in Table 3.5. When one or more classes of flammable liquids are stacked and stored within a single pile of portable tanks, the maximum volume allowed is the smallest of the separate maximum volumes. The requirements stipulate that all piles must be separated by at least 4 feet (1.2 m).

When employers choose to store flammable liquids outdoors within portable tanks, they must comply with the restrictions in Table 3.6. When one or more classes of a flammable or combustible liquid are stacked and stored in a single pile of portable tanks, the maximum volume allowed is the smallest of the separate maximum volumes.

FIGURE 3.9 The structural silhouette of a portable tank used to store and transport a bulk volume of a flammable liquid. This portable tank may be stacked into piles or moved from one location to another by means of a forklift.

portable tank ■ For purposes of OSHA regulations, a closed container having a liquid capacity over 60 gallons (227 L) and not intended for fixed installation; for purposes of DOT regulations, a bulk packaging other than a cargo tank, rail tankcar, or intermediate bulk container designed primarily to be loaded onto, or on, or temporarily attached to a transport vehicle or ship and equipped with skids, mounting, or accessories to facilitate handling of the tank by mechanical means

transport vehicle ■ A motor vehicle or rail car used for the transportation of cargo by any mode

TABLE 3.5	Indoor Portable Tank Storage[a]				
CATEGORY OF FLAMMABLE LIQUID	STORAGE LEVEL	PROTECTED STORAGE, MAXIMUM PER PILE		UNPROTECTED STORAGE, MAXIMUM PER PILE	
		gal	L	gal	L
1	Ground and upper floors	Not permitted	Not permitted	Not permitted	Not permitted
	Basement	Not permitted	Not permitted	Not permitted	Not permitted
2	Ground and upper floors	20,000	75,700	2000	7570
	Basement	Not permitted	Not permitted	Not permitted	Not permitted
3 [FP <100°F (<37.8°C)][b]	Ground and upper floors	40,000	151,000	5500	20,800
	Basement	Not permitted	Not permitted	Not permitted	Not permitted
3 [FP >100°F (>37.8°C)]	Ground and upper floors	40,000	151,000	5500	20,800
	Basement	20,000	75,700	Not permitted	Not permitted
4	Ground and upper floors	60,000	227,000	22,000	83,000
	Basement	20,000	75,700	Not permitted	Not permitted

[a]29 C.F.R. §1910.106(d), Table H-15.
[b]FP is the liquid's flashpoint.

TABLE 3.6	Outdoor Portable Tank Storage[a]							
CATEGORY OF FLAMMABLE LIQUID	MAXIMUM PER PILE		DISTANCE BETWEEN PILES		DISTANCE TO THE PROPERTY LINE THAT MAY BE BUILT UPON		DISTANCE TO THE STREET, ALLEY, OR OTHER PUBLIC WAY	
	gal	L	ft	m	ft	m	ft	m
1	2200	8300	5	1.5	20	6	10	3
2	4400	16,600	5	1.5	20	6	10	3
3 [FP < 100°F (< 37.8°C)][b]	8800	33,300	5	1.5	20	6	10	3
3 [FP >100°F (> 37.8°C)]	17,600	66,600	5	1.5	10	3	5	1.5
4	44,000	166,000	5	1.5	10	3	5	1.5

[a]29 C.F.R. §1910.106(d), Table H-17.
[b]FP is the liquid's flashpoint.

The distances in Table 3.6 apply to a property that is protected for exposure. When the property is not protected for exposure, the distances in the fourth column of Table 3.6 must be doubled. When the total volume of a flammable or combustible liquid does not exceed 50% of the maximum per pile, the distances in the fourth and fifth columns may be reduced by one half, but in any event, the distance must never be less than 3 feet (0.9 m).

Although some types of portable tanks may be stacked securely, OSHA regulations at 29 C.F.R. §1910.106(d)(4)(v) prohibit the stacking of portable tanks in inside rooms in a workplace when the tanks contain more than 30 gallons (113.5 L) of a flammable or combustible liquid.

3.6-C STORAGE IN STATIONARY TANKS

Flammable liquids are also stored within stationary tanks. There are three common types of storage tanks, each distinguished by its operating pressure as follows:

- An **atmospheric tank** is a storage tank designed to operate at pressures ranging from atmospheric pressure through 0.5 psig (3 kPa).
- A **low-pressure tank** is a storage tank designed to operate at pressures at or above 0.5 psig (3 kPa), but not more than 15 psig (100 kPa).
- A **pressure vessel**, or **pressure tank**, is a storage tank or other vessel designed to operate at pressures above 15 psig (100 kPa).

The structural silhouettes of the most common types of stationary tanks are illustrated in Figure 3.10.

All tanks used to store flammable gases and liquids are inherently hazardous, and the widespread practices for storing flammable gases and liquids in them are often faulty. For example, in 2011, the U.S. Chemical Safety & Hazard Investigation Board (Section 1.3-A) noted that since 1983, 44 people have been killed and another 25 injured from fires and explosions associated with the mismanagement of tanks in which flammable gases and oil were stored.[6] These incidents could have been prevented by implementing sound safety practices including the use of warning signs, full fencing, locked gates, locks on hatches, and design features such as flame arresters, pressure-vacuum vents, floating roofs, and vapor

atmospheric tank ■ A tank designed to store a flammable liquid at pressures ranging from atmospheric pressure through 0.5 psig (3 kPa)

low-pressure tank ■ A tank designed to store flammable liquids at pressures above 0.5 psig (3 kPa) but not more than 15 psig (100 kPa)

pressure vessel (pressure tank) ■ A closed vessel designed to store flammable liquids at pressures above 15 psig (100 kPa)

[6]U.S. Chemical Safety & Hazard Investigation Board, "Investigative Study: Public Safety at Oil and Gas Storage Facilities," Report No. 2011-H-1 (September 2011), p. 16.

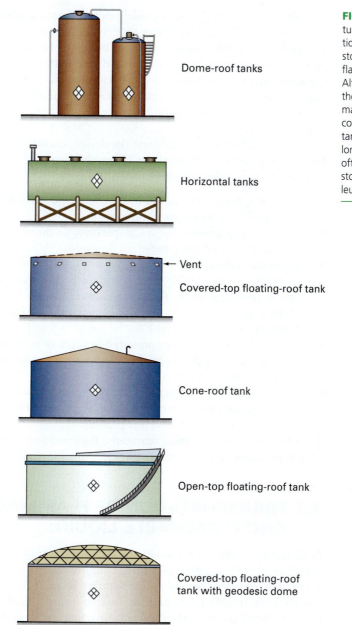

Dome-roof tanks

Horizontal tanks

Vent

Covered-top floating-roof tank

Cone-roof tank

Open-top floating-roof tank

Covered-top floating-roof tank with geodesic dome

FIGURE 3.10 The structural silhouettes of stationary tanks used to store bulk volumes of flammable liquids. Although the capacity of these tanks varies, the maximum capacity of the covered-top floating-roof tank is 20,000,000 gallons (75,680 m³). It is often used for the bulk storage of crude petroleum at oil refineries.

recovery systems. Firefighters should be aware of the existence of the storage-tank sites at which flammable gases and flammable liquids are stored in their areas of jurisdiction, and advise their owners to institute measures to prevent the ignition of their contents.

Multiple stationary tanks of varying sizes and shapes are often situated at adjacent positions on a parcel of property. This collection of storage tanks is called a **tank farm**. Water monitors, other cooling devices, and fire extinguishing agents are generally available within a tank farm for use in the event of their need.

When they are encountered, one or more stationary tanks are routinely situated within individual **secondary containment** areas designed to provide a degree of protection in the event of their failure. The liquids generated from overfilling, maintenance activities, and leaks from piping and other accoutrements collect within these areas and help to control the further release through sewers or drains to a public sewer system, surface waters, and groundwater.

tank farm ■ An installation in which liquids are stored in multiple stationary storage tanks

secondary containment ■ A safeguarding method (such as diking around a primary containment vessel) used for the purpose of preventing the unplanned release of a hazardous material into the environment

NFPA's Code 30 provides guidelines for situating stationary tanks used for the storage of flammable liquids. These guidelines include restrictions on the distances from property lines, public ways, and major buildings on the same property.[7] NFPA stipulates that the storage tanks be situated within secondary-containment areas having individual capacities equal to at least the capacity of the largest tank within the area. In certain instances, diversion curbs or grading must be provided to protect the adjoining property. The released liquids then drain from the area in which the tank is located into a remote impounding area.

OSHA and NFPA also address the design, construction, fabrication, and installation features that pertain to these tanks. When they are used to store flammable liquids, OSHA and NFPA require the following:

- An atmospheric tank should never be used to store a flammable liquid at a temperature equal to or above its boiling point.
- The operating pressure of a low-pressure or pressure vessel should never exceed the design pressure of the vessel.
- When located outside buildings, the distance between any two *adjacent* aboveground tanks storing flammable liquids cannot be less than 3 feet (0.9 m) or one-sixth the sum of their diameters. When the diameter of one tank is less than half the diameter of the adjacent tank, the distance between the two tanks cannot be less than half the diameter of the smaller tank.
- When multiple tanks are connected in three or more rows or in an irregular pattern, sufficient spacing between them must be provided so that firefighting crews may access the innermost tanks.
- Where crude petroleum tanks are located at production facilities in isolated areas and have capacities not exceeding 126,000 gallons (477 m^3), the distance between such tanks cannot be less than 3 feet (0.9 m).
- The minimum separation between a liquefied petroleum gas container and an aboveground atmospheric tank storing a flammable liquid must be 20 feet (6 m).

3.7 TRANSPORTATION OF FLAMMABLE GASES AND FLAMMABLE LIQUIDS

Both nonbulk and bulk packaging are used to transport flammable gases and flammable liquids by railway and public highway. The steel cylinder and 55-gallon steel drum, respectively, are the most popular forms of nonbulk packaging used to containerize flammable gases and flammable liquids. Both are loaded for transportation into rail boxcars, rail flatcars, cargo truck vans, and other motor vehicles. DOT authorizes their shipment only when their individual packaging and the transport vessels have been designed and constructed according to prescribed specifications, and the shipper and carrier have complied with testing, maintenance, marking, placarding, and other relevant requirements. Because DOT's marking and placarding requirements are especially relevant to the training of emergency responders, we note them in some detail in Chapter 6.

3.7-A TRANSPORTATION BY RAILWAY

Rail tankcars are often used to transport bulk volumes of compressed gases. Their design and construction specifications are regulated by DOT and the Association of American Railroads (AAR).

---

[7]NFPA No. 30, *Flammable and Combustible Liquids Code* (Quincy, Massachusetts: National Fire Protection Association).

The most common types of rail tankcar used to transport compressed gases (including flammable gases) are the DOT-105 and -112 models. Their structural features differ in part as noted below:

- The DOT-105 tankcars are equipped with hemispherical (dome-shaped) **headshields** and a **thermal-protection system** incorporated with steel jackets.
- The DOT-112 tankcars are equipped with hemispherical headshields only; headshields and a sprayed-on or rolled-on thermal-protection system (but no jacket); or headshields with a thermal-protection system enclosed in steel jackets.

The most common types of rail tankcar used to transport flammable liquids are the DOT-103 and -111 models, but the DOT-105 and -112 tankcars are also used to transport flammable liquids having relatively high vapor pressures.

Two types of marking requirements apply to the shipment by railway of *all* hazardous materials, including flammable gases and flammable liquids. One marking is provided by the tankcar manufacturer and the other is provided by its owner.

Marking Requirements of Tankcar Manufacturers

DOT requires rail tankcar manufacturers to legibly mark certain information on the exterior surfaces of their tankcars. This information includes a **tankcar specification marking**, an example of which is DOT-105J500NW5. As shown in Figure 3.11, DOT is the agency that issued the design and construction specification of a class of tankcar designated as 105. The 500 in the marking denotes the maximum pressure in psig (500 psi) that was used to test whether the tankcar can safely confine and transport its contents. The W denotes that the tank was built using fusion-welded construction methods. (When an F is listed instead of a W, the letter denotes that the tank was fabricated using forged welding; and when neither a W nor an F is provided, its absence signifies that the tank was constructed in a seamless manner.) When a number follows the W or F, it represents a code that denotes the nature of applicable fittings, materials, or linings; for example, the number 5 following the welding type indicates that the tankcar is rubber-lined.

The capital letter between the railcar class designation and the tank pressure is called the **separator character.** It is often designated as A, to which no specific meaning is assigned. Other separator characters are S, J, and T, each of which is used in a specification marking to designate the information provided at 49 C.F.R. §179.22 as follows:

- S indicates the tankcar is equipped with protective headshields.
- J indicates that the tankcar is equipped with headshields and a thermal-protection system incorporated with a steel jacket.
- T indicates that the tankcar is equipped with a sprayed-on or rolled-on thermal-protection system, but no jacket.

DOT requires manufacturers to test certain rail tankcars for leakage using a prescribed **tankhead puncture-resistance system.** At 49 C.F.R. §179.16, DOT requires their manufacturers to provide puncture-resistant protection that is capable of sustaining, without loss of lading, coupler-to-tankhead impacts at relative car speeds of 18 miles per hour (29 km/hr) when:

- The weight of the impact car is at least 263,000 pounds (119,295 kg);
- The impact car is coupled to one or more backup cars having a total weight of at least 480,000 pounds (217,724 kg) and the handbrake is applied on the last backup car; and
- The impact car is pressurized to at least 100 psig (689 kPa).

When tankcars satisfactorily comply with this standard but are not equipped with a thermal-protection system, DOT requires the manufacturer at 49 C.F.R. §173.31 to substitute the letter M for the A or S in their specification marking. When tankcars satisfactorily comply with this

headshield ▪ A supplemental steel plate affixed on the ends of certain rail tankcars to lessen the chance of tank-head puncture by the coupler of any adjacent car in the event of excessive end impact or derailment

thermal-protection system ▪ A rail tankcar's insulation system that prevents the release of a flammable gas, other than through the tankcar's safety relief valve, when subjected to a pool fire for 100 minutes and a torch fire for 30 minutes

tankcar specification marking ▪ A series of letters and numbers marked on the exterior surface of a rail tankcar to identify its design and construction features

separator character ▪ Any of the capital letters A, S, T, M, or N denoted in a tankcar specification marking

tankhead puncture-resistance system ▪ For purposes of DOT regulations, a performance standard set forth at 49 C.F.R. §179.16 to determine whether a rail tankcar is capable of sustaining, without loss of lading, coupler-to-tank-head impacts at relative car speeds of 18 miles per hour (29 km/hr) under test conditions

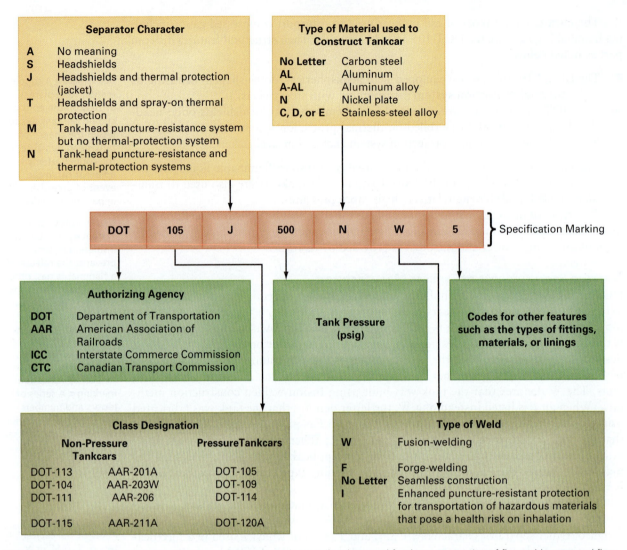

FIGURE 3.11 An example of the specification marking embossed on a rail tankcar used for the transportation of flammable gases and flammable liquids. Each marking component designates design and construction information pertaining to the tankcar.

standard and are equipped with a thermal-protection system, DOT requires substitution of the letter N for the A, J, M, S, or T in the specification marking.

Rail tankcars manufactured after March 16, 2009 are subject to the components of an enhanced regulation relating to puncture resistance when the tankcars are intended for use in transporting certain hazardous materials such as poisonous gases and liquids whose vapors pose a health hazard by inhalation. The performance standard for these tankcars is generally referred to as the **enhanced puncture-resistance system.** DOT requires their manufacturers at 49 C.F.R. §179.244 to provide better puncture resistance from side impacts through a combination of thicker inner shells and/or thicker outer jackets. At 49 C.F.R. §179.22, DOT requires manufacturers who construct tankcars with an enhanced puncture-resistance system to substitute the letter I for the letter W in their specification markings.

Railcar manufacturers have used carbon steel to build over 90% of the tankcars used in the United States, but other metals may also be used to construct them. These metals are sometimes cited in a specification marking by the use of the following capital letters: AL for aluminum; A-AL for an aluminum alloy; N for nickel plate; and C, D, or E for a

enhanced puncture-resistance system
■ DOT regulations applicable to rail tank-cars intended for use in transporting poisonous gases and liquids whose vapors pose a health hazard by inhalation

stainless-steel alloy. When capital letters are not cited between the tank test pressure and the welding type, their absence signifies that the rail tankcar was constructed of carbon steel. When they are cited, the letters appear between the tank pressure and welding type.

DOT also requires manufacturers to mark on the tankcar its maximum weight in pounds and kilograms to which it may be filled, its weight when empty, and the month and year of the tankcar's construction. DOT requires this information to be imprinted on the left side of the tankcar as an observer faces the tankcar, generally below the owner's reporting mark described below. The following is a representative example:

DOT-105A500W		
UTLX901794		
LD LMT	**181700 LB**	**82400 KG**
LT WT	**81300 LB**	**36900 KG**
NEW	**03 95**	

In this example, UTLX90174 is the reporting mark of Union Tank Car Company, a manufacturer that leases rail tankcars to shippers of hazardous materials. LD LMT refers to the load limit that the tankcar may safely carry and LT WT refers to its light weight (i.e., the weight of the tankcar without a load).

Marking Requirements of Tankcar Owners

DOT also obligates the owner of a tankcar to legibly imprint certain information on the exterior surface of both its sides and ends when the tankcar is used to ship a hazardous material. This includes a **reporting mark**, which is a sequence of several identification letters followed by several digits. These letters are assigned by the American Association of Railroads[8] to freight carriers who operate within North America, whereas the digits uniquely identify a specific tankcar.

When a compressed gas is shipped by means of a rail tankcar, DOT generally requires the name of the gas to be imprinted on its right side as an observer faces the tankcar. In Section 6.5-B, we also note that DOT requires the names of certain other substances aside from compressed gases to be imprinted on the surface of a tankcar.

Regardless of this limitation, when emergency responders encounter a reporting mark on the exterior surface of a tankcar during a transportation mishap, they are able to rapidly determine the tankcar's owner by consulting a computer compilation.[9] Then, in combination with the information on the waybill that accompanies the shipment, they may easily identify the tankcar's contents.

More commonly, however, emergency responders provide the reporting mark (letters and the digits) to CHEMTREC personnel, who contact the tankcar's owner directly to determine the tank's contents and an appropriate action. CHEMTREC provides this information to the emergency responders.

DOT also requires rail tankcar owners to maintain, inspect, and service their cargo tanks to ensure that they continue to qualify for service. At 49 C.F.R. §180.511, DOT requires the following periodic inspections and tests:

- Visual inspection to determine structural defects that may cause leakage or failure of the tank
- Structural integrity inspection and test to determine structural defects that may initiate cracks or propagate cracks and cause failure of the tank

reporting mark ■ A sequence of two to six identification letters assigned by the American Association of Railroads to rail carriers who operate in North America

---

[8]The American Association of Railroads (AAR) is an industry trade group whose members include representatives of the major freight railroads of Canada, United States, and Mexico.

[9]For example, the reporting marks of shippers can be obtained by accessing the following site on the Internet: en.wikipedia.org/wiki/List_of_AAR_reporting_marks:_A.

- Service-life shell-thickness inspection to ensure that the tank shell and heads show no thickness reduction
- Safety-system inspection to ensure that the thermal-protection system, tankhead puncture-resistance system, coupler vertical restraint system, and system used to protect discontinuities conform to DOT standards
- Lining and coating inspection to ensure conformation with the owner's acceptance criteria
- Leakage pressure test to ensure all piping, fittings, and closures do not leak
- Hydrostatic pressure test to ensure that the tank does not leak, distort, or excessively expand during service

A tankcar is marked after it has been tested, inspected, and determined to comply with DOT's regulations for qualification. An example of this marking follows:

DOT 112J340W	QUALIFIED	DUE
TANK QUALIFICATION	2014	2024
SHELL-THICKNESS INSPECTION	2014	2024
SAFETY SYSTEM INSPECTION	2019	2024
PRESSURE-RELIEF VALVE (75 PSI)	2014	2024
HEADS-THICKNESS INSPECTION	2014	2024
INTERIOR HEATING COILS	2014	2024
LINING INSPECTION	2019	2024
PRESSURE TEST FOR LEAKAGE OF PIPING, FITTINGS, AND CLOSURES	2014	2024
HYDROSTATIC PRESSURE TEST	2014	2024

When flammable liquids are transported by railway, they are encountered in nonpressure tankcars and pressure tankcars.

- *Nonpressure tankcars.* These are DOT-103, -104, -111, and -115 tankcars. Before use, their integrity is tested by applying an internal pressure ranging from 60 to 100 psi (410 to 690 kPa). Although the outlets and external heater lines are located on the bottom of these tankcars, all other fittings, valves and pressure-relief devices are located topside and externally within the dome. Approximately three-fourths of the rail tankcars that are used to transport flammable liquids are nonpressure tankcars.
- *Pressure tankcars.* These are DOT-105, -109, -112, -114, and -120 tankcars. Their integrity is tested by applying an internal pressure ranging from 100 to 600 psi (690 to 4100 kPa). Although the DOT-114 tankcar has a bottom unloading valve, the fittings, valves, and pressure-relief devices on the DOT-105, -109, -112, and -120 tankcars are located topside and externally within the dome. Approximately one-fourth of the rail tankcars that are used to transport flammable liquids are pressure tankcars.

The rail tankcars in use today are designed and constructed so that their contents are protected during transit in either of the following ways:

- *Thermally insulated tankcars* are constructed so that external heat is transferred to the contents very slowly under the *normal conditions* of transport.
- *Thermally protected tankcars* are constructed so that external heat cannot be transferred to the contents by either conduction or radiation under *abnormal conditions*, such as a transportation mishap.

3.7-B TRANSPORTATION BY PUBLIC HIGHWAY

Usually, bulk volumes of flammable gases are transported by public highway in **cargo tanks.** The term **tanker truck** is often used in everyday practice to distinguish a cargo tank from other types of transport vessels but the term does not uniquely identify tanks that transport cargo by highway; for example, the term oil tanker is used to identify a watercraft that transports oil in bulk. The design and construction specifications for cargo tanks manufactured in the United States are regulated by DOT's Motor Carrier Safety Administration. The title of a unique specification for a cargo tank is prefixed by the capital letters MC (motor carrier) followed by the class of cargo tank and occasionally a DOT specification.

The most common type of cargo tank used to transport flammable gases and other compressed gases by public highway is the MC-331/DOT-431 noninsulated, high-pressure tank. It is designed and constructed with hemispherical heads and ends from quenched and tempered steel or carbon steel to withstand an internal pressure of 300 psig (2067 kPa). The MC-331 cargo tank has a maximum capacity of 11,500 gallons (43.5 m^3). Its design and construction features are published at 49 C.F.R. §178.337.

Two cargo tanks are most commonly used for the transportation of flammable liquids by public highway. They are called the MC-306/DOT-406 and MC-307/DOT-407 cargo tanks whose design and construction features are published at 49 C.F.R. §§178.341 and 178.342, respectively. Their principal features are the following:

■ The MC-306 cargo tank is a single-shell, nonpressure [<3 psig (21 kPa)] tank constructed from an aluminum alloy with an oval cross-section and a typical capacity of 9000 gallons (34 m^3). Older models were constructed from steel. The MC-306 cargo tank is subject to more inspections than any other transport vehicle, because it is frequently used to transport gasoline.

■ The MC-307 tank is a double-shell, low-pressure [25 psig (172.4 kPa) to 40 psig (276 kPa)] tank constructed from stainless steel with a circular cross-section. It has one or two compartments with a total capacity of 6000 to 7000 gallons (23 to 27 m^3).

Specific information concerning the design and construction features of a cargo tank may be obtained from markings required by DOT on its **nameplate** and **specification plate**. These plates are attached, either individually or in combination, to the exterior surface of its left side by brazing, welding, or other means in a place suitable for inspection. The information required on each plate is provided in Table 3.7.

Information concerning the service and maintenance of cargo tanks may be obtained from other markings on the cargo tank. DOT requires the month and year of a test or inspection and the type of test or inspection performed to be marked on the tank. The abbreviations of the six required types of tests or inspections are provided in Table 3.8. For example, the following marking indicates that during November 2013, the relevant cargo tank was subjected to and complied with the requirements of the prescribed pressure test, lining inspection and test, and leakage test:

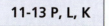

11-13 P, L, K

DOT also requires shippers to mark the proper shipping name of the compressed gas on each side and end of the tank and the type of steel (QT or NQT) used in its construction. QT means quenched and tempered steel, and NQT means other than quenched and tempered steel.

cargo tank (tanker truck) ■ For purposes of DOT regulations, a tank intended primarily for the carriage of liquids or gases that is not fabricated under a specification for cylinders, intermediate bulk containers, multiunit tank-car tanks, portable cars, or tankcars

nameplate ■ A corrosion-resistant metal plate containing information pertaining to the design and construction features of a cargo tank

specification plate ■ A corrosion-resistant metal plate containing information pertaining to the unique specification requirements for a class of cargo tank

TABLE 3.7	Marking Information on Cargo Tanks
NAMEPLATE[a]	**SPECIFICATION PLATE**[b]
DOT Specification Number	Cargo tank motor vehicle manufacture
Original test date (month and year)	Cargo tank motor vehicle certification date
Maximum allowable working pressure (psig)	Cargo tank manufacturer
Test pressure (psig)	Cargo tank date of manufacture (month and year)
Design temperature range (__°F to __°F)	Maximum weight of lading (lb)
Nominal water capacity (gal)	Maximum loading rate (gal/min)
Maximum design density of lading (lb/gal)	Maximum unloading rate (gal/min)
Material specification number for shell (alloy designation number and type)	Lining material, if applicable
Material specification number for heads (alloy designation number and type)	Heating system design pressure, if applicable (psig)
Weld material	Heating system design temperature, if applicable (__°F)
Minimum thickness for shell (in.)	
Minimum thickness for heads (in.)	
Manufactured thickness for shell (in.)	
Manufactured thickness for heads (in.)	
Exposed surface area (ft^2)	

[a]49 C.F.R. §§178.337 -17(b) and 178.345-14(b).
[b]49 C.F.R. §178.345-14(c).

TABLE 3.8	Periodic Test and Inspection Requirements on Cargo Tanks[a]	
TEST OR INSPECTION	**SYMBOL**	**MINIMUM TIME INTERVAL**
External Visual Inspection	V	Annual
Internal Visual Inspection	I	Every 5 y
Lining Inspection and Test	L	Annual
Leakage Test	K	Annual[b]
Pressure Test	P	Every 5 y[b]
Thickness Test	T	Biannual

[a]49 C.F.R. §§178.337-16, 180.407, and 180.415.
[b]When a cargo tank is used to transport chlorine, it must be tested for leakage and pressure biannually.

3.8 RESPONDING TO INCIDENTS INVOLVING THE RELEASE OF FLAMMABLE LIQUIDS

Countless mishaps involving the release of flammable liquids occur with an alarming degree of regularity. They occur not only while the liquids are transported but also when they are stored at manufacturing and processing industries. Most commonly, they originate when the liquids leak or spill from their containment vessels. The liquids then vaporize and their vapors attain a concentration within their flammable range as they mix with the air. Subsequently, the mixtures ignite when exposed to an ignition source.

3.8-A BOILING-LIQUID EXPANDING-VAPOR EXPLOSIONS

The presence of a flammable liquid in an intact transport vessel can give rise to a potentially dangerous incident when an ongoing fire impinges on the vessel. During a rail tank-car mishap, for example, the fire associated with a burning tankcar may impinge on the exterior metal surface of a second tankcar containing a flammable liquid. Then, either of the following events is likely to occur:

■ The fire can impinge on the metal of the second tankcar at an area associated with the vapor space above the liquid. Then, the heat produced by the fire is primarily absorbed by the metal, causing the metal to lose its tensile strength. The shell of the tankcar can become so severely weakened that the tankcar ruptures at the point of the impingement.

■ The fire can impinge on the metal of the second tankcar at an area associated with the presence of liquid. At first, the liquid absorbs the heat produced by the fire, but over an extended time, vapor is produced that accumulates within the vapor space. This increases the internal pressure within the vessel and ultimately causes the tankcar to rupture.

In both instances, a colossal mass of vapor is usually discharged into the atmosphere and ignites in a mushrooming fireball. The phenomenon is called a **boiling-liquid, expanding-vapor explosion**, or **BLEVE** (pronounced ble-vē or blē-vēe). BLEVEs produce fireballs having radii that vary with the amount of vapor instantaneously released to the environment. The relative size of the fireball in Figure 3.12 may be estimated by visually comparing it with the size of the standing water tower in the foreground.

When first-on-the-scene responders arrive at a mishap involving the potential release of a flammable liquid from a rupturing tankcar, it is critical to prevent the occurrence of a BLEVE. They may do this by situating unmanned monitors so that cooling water is

boiling-liquid, expanding-vapor explosion (BLEVE)
■ The phenomenon during which the rapid build-up of internal pressure within a failed container or tank is relieved by its rapid and violent rupture, accompanied by the release of a vapor or gas to the atmosphere and propulsion of the container or its pieces

FIGURE 3.12 A boiling-liquid, expanding-vapor explosion (BLEVE) results when a flammable liquid is rapidly heated to a temperature significantly above its boiling point. Under this condition, the venting mechanism of a tankcar is generally incapable of releasing the buildup of internal pressure that accompanies the expansion of the vapor. Structural failure of the tankcar occurs, resulting in its rupture. During the massive explosion that follows, the vapor is released as a fireball. (*Courtesy of Eugene Meyer.*)

directed at the vessel's main vulnerable points: the uppermost area of the tankcar where the vapor phase is present, and the actual site of flame contact on the shell of the vessel. As precautionary guidance only, DOT recommends the cooling of a 1544-gallon (4000-L) propane tank at a rate approximating 158 gallons per minute (598 L/min).[10] Unless the vessel is effectively cooled, failure in its metal structure is likely to occur within 20 minutes after initial contact of the fire.

The danger to firefighters, bystanders, and property from the occurrence of a BLEVE cannot be overemphasized. The fragments of the rupturing vessel often travel like missiles for hundreds of feet in all directions, potentially causing death or injury to anyone in their pathway. The secondary fires that originate from radiation and rocketing shells can destroy nearby structures. Following the occurrence of a BLEVE, the main assistance that firefighters may provide is often limited to extinguishing these secondary fires.

3.8-B GENERAL EMERGENCY RESPONSE PROCEDURES

The nature of an emergency response action involving flammable liquids does not always approach the magnitude of an operation concerned with preventing a BLEVE. As they assess the nature of the emergency scene, first-on-the-scene responders should proceed to undertake the following actions as deemed necessary and appropriate:

■ As quickly as possible, the chemical or product name of the flammable liquid should be determined, either by reviewing shipping papers, labels, markings, and placards, or by speaking with plant or transportation personnel.

■ When the emergency incident is associated with a transportation mishap, contact should be made with CHEMTREC and, if warranted, the National Response Center.

■ If the response action involves the storage of a flammable liquid within a stationary tank, personnel should use the information posted on the relevant hazard diamond to ascertain the degrees of hazard relating to fire, health, and instability.

■ To safeguard nearby residents, emergency responders should consider the issuing of an order to evacuate the area.

■ Once the chemical nature and approximate volume of the flammable liquid are established, emergency responders may select an appropriate fire extinguisher. The use of either water or carbon dioxide is typically effective when extinguishing a fire involving a relatively small volume of a flammable liquid, whereas the use of an aqueous-film-forming foam (Section 5.12-B) is warranted when extinguishing a class B fire involving a bulk volume of a flammable liquid.

■ When a small spill or leak of a flammable liquid from a container or tank is identified, a water fog should be used to disperse the vapor.

■ Water, fog, foam, or carbon dioxide should be applied for extinguishing or cooling purposes when a container or tank containing a flammable liquid is situated in an area near an ongoing fire, or when it is exposed directly to flames.

■ Quick attention should be given to cooling vessels used to store flammable liquids, as when they are heated, they are particularly susceptible to rupturing. To prevent their rupture, the vessels should be cooled with direct streams of water.

■ When a liquid's vapor has not yet ignited, or when its fire has been momentarily extinguished, the concentration of the flammable vapor in the air should be measured. If the reading establishes that the vapor concentration is within the flammable range, emergency responders should vacate the area until the use of a water fog within the area reduces the concentration to less than the lower flammable limit. Emergency response personnel may cautiously attempt to seal or otherwise stop a leak from a stationary tank

---

[10]Table titled BLEVE (USE WITH CAUTION), *Emergency Response Guidebook* (Washington, D.C.: U.S. Department of Transportation, 2012), p. 367.

or transport vessel but only when the vapor has not ignited and the vapor concentration in the air is less than the lower flammable limit. Sometimes, this task involves nothing more than tightening a valve. If it is determined that the source of the leak cannot be sealed, it may be necessary to transfer the liquid into a new storage vessel. When transfer of the liquid is impractical, the tank or transport vessel should be cooled with unmanned water monitors, but no further attempt to extinguish the fire should be made.

■ Special consideration is warranted when water is used to extinguish the fire of a burning liquid that is immiscible with, and heavier than, water. As noted in Section 2.5-A, the water sinks below the burning liquid. When too much water is applied as a fire extinguishing agent, it causes the liquid to overtop its containment vessel. Instead of extinguishing the fire, the water causes the spread of fire to other locations.

■ The use of water for cooling a leaking storage tank often generates a volume of water so large that the secondary containment system is unable to restrain its release. This water is contaminated with the flammable liquid that has leaked from the tank. When the water is likely to overtop the secondary containment system, action should be taken to preclude its entrance into nearby sewers or bodies of water or flow to areas where other flammable materials are stored. This may be accomplished by pumping the mixture into a portable tank or another temporary storage system.

■ When emergency responders encounter a massive leak of a flammable liquid, a foam blanket should be applied to confine the flammable vapor. To the extent feasible, pools of the liquid should be dammed or diked to prevent widespread exposure and contamination.

■ During mishaps associated with the transportation of a flammable liquid, upright cabs and railcars should be disconnected and moved from overturned tankcars whenever practical.

■ When the vapor of a flammable liquid has been released into the atmosphere, it is prudent to wear total-encapsulating suits, use self-contained breathing apparatus, and perform as many response actions as practical from an upwind location. Even when the vapor is nontoxic, the use of respiratory protection equipment is warranted, because its inhalation may cause dizziness, illness, or suffocation.

■ An overturned *pressure* transport tank may be safely righted without first unloading its liquid contents; however, an overturned *nonpressure* transport tank should not be uprighted until the majority of the tank's contents have first been transferred into a different tankcar. This transfer should be attempted only by experienced personnel.

Flammability

1. Given the boiling point and flashpoint of chlorobenzene as 270°F and 84°F, respectively:
   (a) Identify its OSHA category of flammable liquid.
   (b) Are manufacturers of products containing chlorobenzene required to display the GHS flame pictogram on the labels of this chemical product?

2. Use the following data to determine the OSHA category of flammable liquid to which each substance belongs:

Substance	Boiling point	Flashpoint
(a) cyclohexylamine	275°F (135°C)	90°F (32°C)
(b) vinyl ether	102°F (39°C)	−22°F (−30°C)
(c) *n*-pentane	97°F (36.1°C)	−40°F (−40°C)
(d) amyl nitrite	220°F (104°C)	50°F (10°C)

3. Cumene is a flammable liquid whose lower and upper flammable limits are 0.9% and 6.5% by volume, respectively. Can cumene vapor ignite at each of the following concentrations in air by volume when it is exposed to an ignition source?
   (a) 0.09%
   (b) 6%
   (c) 10%

4. Propylene oxide is a gas used to fumigate and sterilize packaged foods. Some of its physical properties are listed below:

Melting point	−170°F (−112°C)
Boiling point	93°F (34°C)
Specific gravity at 68°F (20°C)	0.830
Vapor density (air = 1)	2.0
Vapor pressure at 68°F (20°C)	445 mmHg
Flashpoint	−35°F (−37°C)
Autoignition point	840°F (449°C)
Lower flammable limit	2.3% by volume
Upper explosive limit	36% by volume
Solubility in water	Soluble [59% by weight at 77°F (25°C)]
Evaporation rate (BuAc = 1)	33.7

   (a) In which range of temperature does propylene oxide exist as a liquid?
   (b) In which category of flammable liquid is propylene oxide classified by OSHA?
   (c) Identify the number that is entered in the uppermost quadrant of a hazard diamond to describe the relative degree of fire hazard represented by propylene oxide.
   (d) If a 1-gallon metal drum containing propylene oxide ruptures, is its vapor likely to ignite in the absence of an ignition source when the surrounding temperature reaches 482°F (250°C)?
   (e) If a 1-gallon metal drum containing propylene oxide ruptures and produces a vapor concentration ranging from 10% to 30% by volume in air, is the vapor likely to ignite if exposed to an ignition source at 68°F (20°C)?

(f) If a 1-gallon metal drum containing propylene oxide ruptures, will its vapor initially accumulate in low areas or diffuse upward?

(g) Describe the extent to which water will effectively extinguish a fire involving propylene oxide confined in a metal drum.

(h) Are manufacturers of products containing propylene oxide required to display the GHS flame pictogram on the product labels?

5. *n*-Pentane and vinyl acetate are flammable liquids. Their evaporation rates (ether = 1) are 0.97 and 3.3, respectively. Based solely on this information, determine which liquid poses the greater risk of fire and explosion.

Storing and Transporting Compressed Gases

6. DOT-3E1800 is marked on a steel cylinder when its gauge pressure is 1610 psi at 70°F (21°C). Is this cylinder likely to rupture during a fire if the temperature of its contents rises to 300°F (149°C)?

General Hazards of Compressed Gases

7. Why is the rupture of a steel cylinder containing compressed helium gas gravely hazardous, despite the fact that helium is a nonflammable gas?

Responding to Incidents Involving the Release of Compressed Gases

8. When a leaking propane cylinder is remotely moved from a burning building into an open area, why do experts recommend that it be rotated so that the point of leakage becomes the uppermost part of the cylinder?

9. On arrival at a transportation mishap, the first-on-the-scene responders observe an ongoing fire and learn from the driver of an overturned cargo tank that the consignment is compressed liquefied butane. They also observe a ruptured discharge line from which releasing vapor has ignited. A pool of butane is not evident. What immediate action should the team take on arriving at this scene?

Storing Flammable Liquids

10. Methyl acetate is a liquid that boils at 135°F (57°C) and has a flashpoint of 15°F (−9°C). It is a constituent of paint removers. A company that manufactures paint removers has constructed an indoor storage room for flammable liquids from fire-resistant materials and equipped it with a sprinkling system. What is the maximum number of 55-gallon drums containing methyl acetate that may be stored per pile in the room in compliance with OSHA's requirements?

11. Use the data in Review Exercise 3.4 to determine the maximum volume in gallons, if any, of propylene oxide that OSHA permits to be stored as follows:
    (a) In containers in an indoors room located on the ground level of a facility that is protected for exposure
    (b) Outdoors in containers 20 feet (5 m) from the property line
    (c) In portable tanks in an indoors room located on the first floor of a facility that is not protected for exposure
    (d) Outdoors in portable tanks located 20 feet (5 m) from the property line

12. Adjoining property and waterways can be protected from the hazards posed by the presence of a flammable or combustible liquid in a stationary storage tank by constructing a means to drain a spilled or leaked liquid from the tank into a remote impounding area. For such purposes, NFPA No. 30, *Flammable and Combustible Liquids Code*, advocates each of the following for a 100,000-gallon (378-m^3) storage tank:
    - A slope of not less than 1% away from the tank for at least 50 feet (15 m) toward the impounding area
    - A capacity of the impounding area not less than the capacity of the tank
    - A route of drainage so that if ignited, the burning liquid within the drainage system will not seriously expose tanks or adjoining property
    - The location of the impounding area not closer than 50 feet (15 m) from any property line that is or may be built on, or from any tank

These conditions are illustrated below:

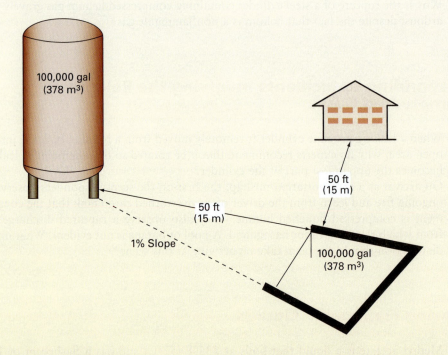

If a processing company stores 50,000 gallons of *n*-amyl alcohol, a class IC liquid, in a 100,000-gallon (378-m^3) tank located 50 feet (15 m) from the tank's remote 45,000-gallon impounding area, has the company complied with this NFPA recommendation?

Responding to Incidents Involving the Release of Flammable Liquids

13. *n*-Propyl bromide is a multipurpose colorless liquid. When it is encountered, *n*-propyl bromide is usually stored in bulk tanks to which the following hazard diamond has been affixed:

When first-on-the scene responders determine that *n*-propyl bromide is stored in two 1000-gallon (3785-L) adjacent tanks at a dry-cleaning facility engulfed in fire, what actions should they take to safeguard public health, safety, and the environment?

14. What information do emergency responders obtain upon observing each of the following markings on rail tankcars?
    (a) DOT105A100ALW
    (b) DOT112A200W

15. Upon arriving at the scene of a train derailment, the first-on-the-scene responders note that a fire in a damaged tankcar is impinging on the exterior surface of an adjacent, intact tankcar. No information is immediately available concerning the chemical nature of the contents held in either tankcar, but the following reporting marks are respectively embossed on each of their sides and ends: MDSB 124665 and MDSB 124666. How are these reporting marks useful to the emergency responders who observed them?

4
Chemical Forms of Matter

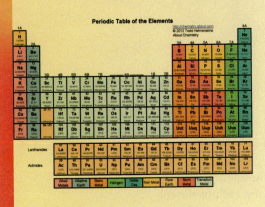

Periodic Table of the Elements

Courtesy of Todd Helmenstine, About Chemistry (2010).

KEY TERMS

alkali metal family, *p. 115*

alkaline earth family, *p. 115*

atom, *p. 112*

atomic nucleus, *p. 112*

atomic number, *p. 113*

atomic orbital, *p. 112*

atomic weight, *p. 113*

Avogadro number, *p. 131*

binary acid, *p. 130*

chemical bond, *p. 111*

chemical change (chemical reaction), *p. 111*

chemical formula, *p. 123*

chemical property, *p. 111*

chemical symbol, *p. 109*

compound (compound substance), *p. 107*

covalent bond, *p. 120*

covalent compound, *p. 122*

double bond, *p. 120*

electron, *p. 112*

element, *p. 108*

family of elements, *p. 114*

formula weight, *p. 130*

halogen family, *p. 116*

ion, *p. 117*

ionic bond, *p. 119*

ionic compound, *p. 122*

isotope, *p. 113*

Lewis structure, *p. 120*

Lewis symbol, *p. 118*

metalloid (semimetal), *p. 110*

metal, *p. 110*

mixture, *p. 107*

mole, *p. 131*

molecular weight, *p. 130*

molecule, *p. 116*

neutron, *p. 112*

noble gas family, *p. 116*

nonbonding electron, *p. 118*

nonmetal, *p. 110*

nonpolar solvent, *p. 123*

nonpolar substance, *p. 116*

octet rule, *p. 118*

oxyacid, *p. 130*

oxygen family, *p. 116*

period, *p. 115*

periodic law, *p. 114*

periodic table, *p. 114*

physical change, *p. 111*

physical property, *p. 111*

polar solvent, *p. 123*

polar substance, *p. 116*

polyatomic ion, *p. 118*

proton, *p. 112*

solute, *p. 122*

solution, *p. 122*

solvent, *p. 122*

substance, *p. 107*

triple bond, *p. 120*

valence electron, *p. 118*

- Distinguish between elements and compounds.
- Memorize the symbols of the elements listed in Table 4.2.
- Identify representative physical and chemical changes of a substance.
- Identify the names and common properties of the major particles of which all atoms are composed.
- Distinguish between the nature of ionic and covalent bonding.
- Use Lewis symbols of the elements to describe the covalent bonding in molecules.
- Memorize the symbols of the ions listed in Table 4.7.
- Write the chemical formulas of ionic compounds when their names are provided and vice versa.
- Memorize the chemical formulas of the covalent compounds named in Tables 4.8 and 4.10.

Commercial chemical products like paint, pesticide formulations, ammunition, gasoline, and other petroleum fuels are hazardous materials that consist of mixtures of distinctly different components, each separable from the others. Paint, for example, is a mixture of some resin or other film-forming compound, a solvent or thinner, and organic or inorganic pigments. Each chemical product is characterized in a similar fashion as a uniform or nonuniform blend of different components, each of which may be separated from one another.

When any mixture has been separated into its components, the individual components are unique forms of matter. In this chapter, we identify them as either elements or compounds possessing a characteristic set of physical and chemical properties. We also observe how chemists describe the structural nature of these substances in terms of atoms, molecules, and ions. Finally, we learn how chemists name these substances and write their chemical formulas.

4.1 ELEMENTS AND COMPOUNDS

A material that has been separated from all other materials is called a *pure substance* or, more simply, a **substance**. Examples include aluminum metal, copper metal, oxygen, distilled water, and table sugar. No matter how they originated or what procedures are used to purify them, samples of the same substance are indistinguishable from one another. For example, all samples of distilled water are alike and indistinguishable from all other samples; that is, water is the same substance whether it has been distilled from rainwater, well water, seawater, or any other source.

A substance is characterized as a material having a fixed composition, usually expressed in terms of percentage by mass. Distilled water is a pure substance consisting of 11.2% hydrogen and 88.8% oxygen by mass. By contrast, coal is not a pure substance; its carbon content alone may vary anywhere from 35% to 90% by mass. Such materials that are not pure substances are called **mixtures**.

Suppose we take calcium carbonate and heat it as shown in the laboratory exercise in Figure 4.1. The limestone ultimately crumbles into a white powder. Careful examination shows that carbon dioxide evolved from the calcium carbonate while it was heated. Substances like calcium carbonate that can be broken down into two or more simpler substances are called **compound substances** or, simply, **compounds**. There are more than 128 million different compounds.

substance ■ Any homogeneous material having a constant, fixed chemical composition; an element or compound

mixture ■ Matter consisting of two or more substances in varying proportions that are not chemically combined

compound substances (compounds) ■ Any substance composed of two or more elements in chemical combination

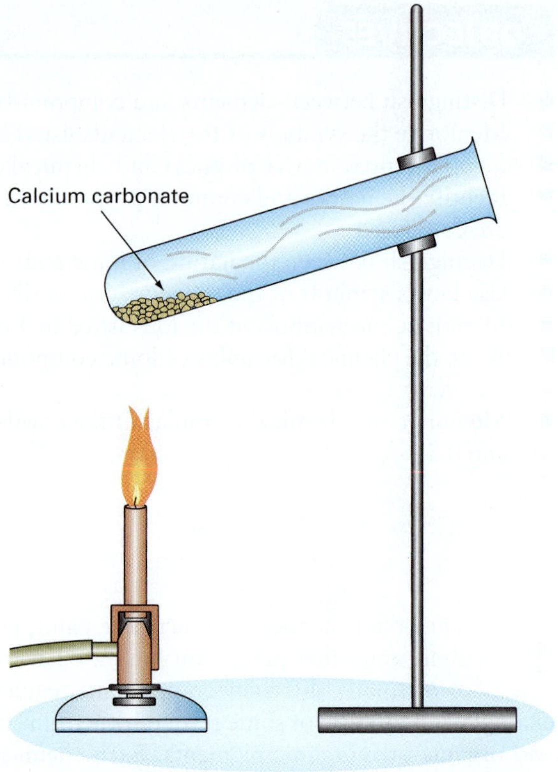

FIGURE 4.1 In this laboratory experiment, calcium carbonate is heated in a test tube. Carbon dioxide evolves as it thermally decomposes. This simple experiment establishes that calcium carbonate is a compound and not an element.

Calcium carbonate

element ▪ Any substance composed of only one kind of atom

The substances that cannot be decomposed into simpler forms of matter are called **elements**. The elements are the fundamental substances of which all compounds are composed. Although there are only 118 known elements, there are millions of known compounds. Of the 118 elements, only 88 are present on Earth in detectable amounts, and many are very rare. Table 4.1 shows that only 10 elements make up approximately 99% by mass of Earth, including its crust, the atmosphere, the surface layer, and the bodies of

TABLE 4.1	Natural Abundance of the Elements (Earth's Crust, Oceans, and the Atmosphere)		
Oxygen	49.5%	Chlorine	0.19%
Silicon	25.7%	Phosphorus	0.12%
Aluminum	7.5%	Manganese	0.09%
Iron	4.7%	Carbon	0.08%
Calcium	3.4%	Sulfur	0.06%
Sodium	2.6%	Barium	0.04%
Potassium	2.4%	Chromium	0.033%
Magnesium	1.9%	Nitrogen	0.030%
Hydrogen	0.9%	Fluorine	0.027%
Titanium	0.58%	Zirconium	0.023%
		All others	<0.1%

TABLE 4.2	Symbols of the Most Common Elements		
ELEMENT	**SYMBOL**	**ELEMENT**	**SYMBOL**
Aluminum	Al	Lithium	Li
Antimony	Sb	Magnesium	Mg
Argon	Ar	Manganese	Mn
Arsenic	As	Mercury	Hg
Barium	Ba	Neon	Ne
Bismuth	Bi	Nickel	Ni
Boron	B	Nitrogen	N
Bromine	Br	Oxygen	O
Cadmium	Cd	Phosphorus	P
Calcium	Ca	Platinum	Pt
Carbon	C	Potassium	K
Chlorine	Cl	Radium	Ra
Chromium	Cr	Silicon	Si
Cobalt	Co	Silver	Ag
Copper	Cu	Sodium	Na
Fluorine	F	Strontium	Sr
Gold	Au	Sulfur	S
Helium	He	Tin	Sn
Hydrogen	H	Titanium	Ti
Iodine	I	Tungsten	W
Iron	Fe	Uranium	U
Lead	Pb	Zinc	Zn

water. Oxygen is the most abundant element on Earth and is found in the free state in the atmosphere, as well as in combined form with other elements in numerous minerals and ores. Astatine and protactinium are among the rarest elements on the planet.

Each element has a specific *name* and **chemical symbol**. The names and symbols of some common elements are listed in Table 4.2. The symbols of the elements consist of either one or two letters with the first letter capitalized. The selection of the symbols of many elements was originally based on their Latin names: Pb for *plumbum* (lead), Ag for *argentum* (silver), and Sn for *stannum* (tin). It is best to memorize the symbols of the elements listed in Table 4.2, because they are frequently encountered in the study of hazardous materials.

chemical symbol ■ A notation using one or two letters to represent an element

As noted more fully in Section 4.4, each element is assigned an atomic number. For practical purposes, there are 118 atomic numbers, each of which uniquely identifies an element. The elements beyond atomic number 92 (uranium) do not exist naturally on Earth, but scientists have prepared some of them in nuclear laboratories. The following names for the elements having atomic numbers 104 through 112 have been accepted by the *International Union of Pure and Applied Chemistry* (IUPAC): rutherfordium (104), symbol Rf; dubnium (105), symbol Db; seaborgium (106), symbol Sg; bohrium (107), symbol Bh; hassium (108), symbol Hs; meitnerium (109), symbol Mt; darmstadtium (110), symbol Ds; roentgenium (111), symbol Rg; and copernicium (112), symbol Cn. The elements having atomic numbers 112 through 118 have been discovered but have not yet been named.[1,2] The element having an atomic number of 117 is the newest element that has been discovered.[3]

4.2 METALS, NONMETALS, AND METALLOIDS

metal ■ An element that conducts heat and electricity well, has a high physical strength, and is ductile and malleable

Each element can be classified as a metal, nonmetal, or metalloid. **Metals** are elements that usually have the following properties:

■ Conduct heat and electricity well
■ Melt and boil at relatively high temperatures
■ Have relatively high densities
■ Are normally malleable (able to be hammered into sheets) and ductile (able to be drawn into a wire)
■ Display a brilliant luster

Examples of metals are iron, platinum, and silver. Almost all metals are solids at room temperature [68°F (20°C)]. Mercury is the only common metal that is liquid at room temperature. No metal is gaseous at room temperature.

nonmetal ■ An element that does not conduct heat and electricity well, has a low physical strength, and is neither ductile nor malleable

Elements that do not have the physical properties of metals are called **nonmetals**. These elements have the following general properties:

■ Melt and boil at relatively low temperatures
■ Do not have a luster
■ Are less dense than metals
■ Weakly conduct heat and electricity

At room temperature, most nonmetals generally exist as either solids or gases. Bromine is the only liquid nonmetal at room temperature. Oxygen, fluorine, and nitrogen are examples of gaseous nonmetals, whereas carbon, sulfur, and phosphorus are examples of solid nonmetals.

Not all metals and nonmetals exhibit the general properties noted here. There are exceptions; for instance, carbon conducts heat and electricity very well, and one form of carbon (diamond) melts at about 6872°F (~3800°C), a relatively high temperature.

metalloid (semimetal) ■ A metal that exhibits the general physical properties of both metals and nonmetals

Several elements have properties resembling those of both metals and nonmetals. They are called **metalloids**, or **semimetals**. The metalloids are boron, silicon, germanium, arsenic, antimony, tellurium, and polonium. For example, silicon and germanium have the luster associated with metals, but they do not conduct heat and electricity well.

[1]In 2004, an international research team composed of scientists from the Joint Institute for Nuclear Research (JINR), Dubna, Russia, and the Lawrence Livermore National Laboratory (LLNL), Berkeley, California, announced the discovery of elements having atomic numbers 113, 114, 115, and 116. In 2004, 2005, and 2012, scientists at the RIKEN Nishina Center for Accelerator-Based Science in Japan reported the results of separate experiments that produced the element having an atomic number of 113. In 2006, JINR and LLNL announced that teams from both laboratories had produced the element having an atomic number 118. In 2009, researchers at LLNL reported that they had independently confirmed the synthesis of the element having an atomic number of 114.

[2]Although they are not yet officially named, the elements having atomic numbers 113 through 118 have respectively been given the following provisional names: ununtrium, ununquadium, ununpentium, ununhexium, ununseptium, and ununoctium. Names for the elements having atomic numbers 114 and 116 have been proposed as flerovium and livermorium, respectively, but these names have not been officially approved.

[3]Ununseptium was produced by scientists in April 2010 at Russia's U400 cyclotron.

4.3 CHEMICAL AND PHYSICAL CHANGES

The constant composition associated with a given substance is maintained by internal linkages among its units. These linkages are called **chemical bonds**. When a particular process occurs that makes or breaks these bonds, we say that a **chemical change**, or a **chemical reaction**, has occurred. Combustion and corrosion are common examples of chemical changes associated with some hazardous materials.

Let's briefly consider the nature of the combustion process. When something burns, it combines with oxygen. The resulting products of combustion are compounds containing oxygen, called *oxides*. For instance, many commercial gasolines are mixtures of several substances, including one called octane. When octane burns completely, it becomes carbon dioxide gas and water vapor. Carbon dioxide is an oxide of carbon, as its name implies, whereas water is an oxide of hydrogen. Carbon dioxide and water vapor are unlike octane or other gasoline components. They have different properties and different compositions. This conversion of octane to carbon dioxide and water is typical of a chemical change.

By contrast, substances can undergo changes during which their compositions remain the same. Such changes are called **physical changes**. Let's consider octane again. Some of its physical and chemical changes are illustrated in Figure 4.2. When exposed to the ambient environment, liquid octane evaporates, but its chemical composition remains unchanged. Such alterations in the physical state of a substance, such as from a liquid to a vapor, are considered physical changes. Other examples of physical changes are melting, freezing, boiling, crushing, and pulverizing.

The types of behavior that a substance exhibits when undergoing chemical changes are called its **chemical properties**. The characteristics that do not involve changes in the chemical identity of a substance are called its **physical properties**. All substances can be distinguished from one another by these properties, in much the same way as certain features—fingerprints or DNA, for example—distinguish one human being from another. The study of hazardous materials is concerned to a great extent with learning the chemical and physical properties of appropriate substances, some examples of which are listed in Table 4.3.

chemical bond ■ The force by which the atoms of one element become attached to, or associated with, other atoms in a compound

chemical change (chemical reaction) ■ Any modification or transformation that results in an alteration of the chemical identity of a substance

physical change ■ A modification or transformation that does not result in an alteration of the chemical identity of a substance

chemical property ■ Any type of behavior that a substance exhibits when it undergoes a chemical change

physical property ■ Any phenomenon that a substance exhibits when it undergoes a physical change

FIGURE 4.2 Examples of physical and chemical changes in the components of gasoline. On the left, the components evaporate; that is, they change their physical state from the liquid to the vapor. This phenomenon constitutes a physical change. On the right, a spill of gasoline ignites and burns, during which the components become carbon dioxide and water vapor. This phenomenon constitutes a chemical change.

TABLE 4.3	Characteristics of Some Substances	
SUBSTANCE	**PHYSICAL PROPERTIES**	**CHEMICAL PROPERTIES**
Oxygen, an element	Odorless, colorless gas; does not conduct heat or electricity; density = 1.43 g/L; becomes liquid at −297°F (−183°C)	Combines readily with many elements (a chemical reaction called oxidation)
Phosphorus, an element	White or red solid; does not conduct heat or electricity; density = 1.82 g/mL (white) and 2.34 g/mL (red)	Readily combines with oxygen, chlorine, and fluorine; white form spontaneously ignites in dry air
Carbon dioxide, a compound	Odorless and colorless gas; solidifies at −83°F (−67°C) under pressure, forming dry ice; soluble in water under pressure	Does not burn; reacts with water-soluble metal compounds, forming metallic carbonates
Hydrogen chloride, a compound	Strong-smelling, colorless gas; density = 1.20 g/mL; soluble in water, forming hydrochloric acid	Reacts with many minerals, forming water-soluble products; reacts with ammonia, forming ammonium chloride

4.4 SOME BASIC FEATURES OF ATOMS

atom ■ The smallest particle of an element that can be identified with that element

electron ■ An atomic particle that has an electric charge of −1

proton ■ A particle in the nucleus of an atom that has an electric charge of +1

neutron ■ A particle in the nucleus of an atom that has no electric charge

atomic nucleus ■ The region at the center of an atom occupied by protons and neutrons

atomic orbital ■ The region in the space around an atomic nucleus in which electrons are most likely to be found

If a small piece of an element, say aluminum, could be hypothetically divided and subdivided into smaller and smaller pieces until subdivision was no longer possible, the result would be one particle of aluminum. This smallest particle of the element that is still representative of the element is called an **atom**, from the Greek word *atomos*, meaning "indivisible."

Although an atom is infinitesimally small, it is also composed of even smaller particles known as electrons, protons, and neutrons. **Electrons** are negatively charged particles that are responsible for the chemical reactivity of a given element. **Protons** are positively charged particles, and **neutrons** are neutral particles. Electrons and protons bear the same magnitude of charge but are of opposite signs. For convenience, the electron has a charge of −1, the proton of +1, and the neutron of 0.

Protons are relatively heavy particles; they are 1836 times more massive than electrons. Neutrons are slightly more massive than protons. The fundamental characteristics of electrons, protons, and neutrons are summarized in Table 4.4.

The protons and neutrons of an atom reside in a central area called the **atomic nucleus**. Electrons reside primarily in designated regions of space surrounding the nucleus, called **atomic orbitals**. There are several types of atomic orbitals; some are close to the nucleus, whereas others are relatively remote from it. Scientists have learned

TABLE 4.4	Some Basic Atomic Particles		
PARTICLE	**PROTON**	**ELECTRON**	**NEUTRON**
Symbol	p	e	n
Relative charge	+1	−1	0
Relative mass	1	About 0[a]	1

[a]The mass of an electron is 1/1836 the mass of the proton.

that only a prescribed number of electrons reside in a given type of atomic orbital. Two electrons are always close to the nucleus, in an atom's innermost atomic orbital (with the exception of a hydrogen atom, which possesses only one electron). Most atoms have additional electrons in the atomic orbitals that are located some distance from the nucleus.

The number of protons in an atom is called the **atomic number**. The atomic number is often used in the study of chemistry to determine the number of electrons possessed by a neutral atom of an element. An atom of hydrogen has one electron, helium has two, lithium has three, and so forth. Carbon is an element composed only of carbon atoms, and all carbon atoms exhibit nearly the same physical and chemical properties. Some carbon atoms may have slightly different masses due to different numbers of neutrons in their nuclei, but they act the same when they undergo chemical changes. Similarly, oxygen is an element composed of oxygen atoms, and all oxygen atoms possess nearly the same properties. But carbon and oxygen atoms are not alike, because their atoms possess different numbers of electrons, protons, and neutrons.

Although neutral atoms of the same element have an identical number of protons and electrons, they may differ by the number of neutrons in their nuclei. Atoms of the same element having different numbers of neutrons are called the **isotopes** of that element. We note later, in Chapter 16, that some isotopes are radioactive. When the isotopes of an element are not radioactive, they are said to be stable. The elements exist in nature as mixtures of their stable isotopes. The majority have two or more. Forty-two (42) elements have no stable isotopes, and 18 have only one.

Over the past century, scientists have determined with considerable accuracy the absolute masses of the stable atoms of the known elements. Immediately apparent is the fact that the atomic masses are extremely small numbers. Because it is inconvenient to work with them, scientists selected a specific carbon isotope called carbon-12 as the atom whose mass now serves as the reference standard for assigning atomic masses to all the other stable isotopes. Each atom of carbon-12 has six electrons, six protons, and six neutrons. It is assigned a mass of exactly 12.

Scientists use modern analytical instruments to determine the natural abundance of an element's stable isotopes in a sample of a terrestrial material. Then, they can calculate their weighted average in the sample. A weighted average is determined by multiplying the abundance of each stable isotope by its atomic mass and summing the products. The weighted average mass of the stable isotopes of an element in a given source is called the element's **atomic weight.** Because atomic weights are relative parameters, they are unitless numbers. The atomic weights of the elements are numbers that tell us how the masses of their stable atoms from a given source compare with the mass of the carbon-12 atom.

When the natural abundances of an element's stable isotopes do not vary among the many sources from which they are obtained, the atomic weight of the element is calculated as a single number. However, when the abundances of the stable isotopes of an element vary with the locations around the world from which the element is obtained, the atomic weight of the element is dependent on its source. When an element's isotopic abundance varies from sample to sample, its atomic weight also varies, thereby providing a range of values. This latter situation is especially true for hydrogen, lithium, boron, carbon, nitrogen, oxygen, silicon, sulfur, chlorine, and thallium. The range of their atomic weights is often written as two bracketed numbers separated by a semicolon. For example, the atomic weight of hydrogen is written as [1.00784; 1.00811], meaning that hydrogen in a given source has an atomic weight that is a number equal to or greater than 1.00784 and equal to or less than 1.00811.

The atomic numbers and atomic weights of all the elements are listed in Appendix B at the back of this text.

atomic number ■ The number of protons in an atomic nucleus; the number of electrons in an electrically neutral atom

isotope ■ Any of a group of nuclei having the same number of protons but different numbers of neutrons

atomic weight ■ An abundance-weighted sum of the atomic masses of an element's naturally occurring isotopes from a specified source

The analysis of multiple samples of strontium collected from sources around the world shows that the element has four stable isotopes: strontium-84, strontium-86, strontium-87, and strontium-88. Their isotopic composition is essentially independent of the sources from which the strontium is collected, and is noted below:

Isotope	Atomic mass	Natural abundance (%)
Strontium-84	83.9134	0.50
Strontium-86	85.9094	9.90
Strontium-87	86.9089	7.00
Strontium-88	87.9056	82.60

Show that the atomic weight of strontium in these multiple samples is 87.62.

Solution: The word percentage means parts per hundred (Section 2.4); hence, 0.50%, 9.90%, 7.00%, and 82.60% equal 0.0050, 0.0990, 0.0700, and 0.8260, respectively. Because the atomic weight of an element is the weighted average of the atomic masses of its stable isotopes in a given sample, the atomic weight of strontium is calculated to be 87.62, as follows:

$$
\begin{aligned}
83.9134 \times 0.0050 &= 0.4196 \\
85.9094 \times 0.0990 &= 8.5050 \\
86.9089 \times 0.0700 &= 6.0836 \\
87.9056 \times 0.8260 &= \underline{72.6100} \\
& 87.62
\end{aligned}
$$

4.5 THE PERIODIC CLASSIFICATION OF THE ELEMENTS

During the last half of the nineteenth century, several scientists first noted that the chemical properties of any given element were similar to the chemical properties of certain other elements. For example, they noted that sodium metal reacts explosively with water and burns spontaneously in the air. When these two chemical properties of sodium were compared with the properties of other elements, they found that potassium also explodes on contact with water and burns spontaneously in air. These scientists summarized this observation in the **periodic law**: The properties of the elements vary periodically with their atomic numbers. The term *periodic* reflects this repetition of chemical properties.

Suppose we list each element in a square and then arrange the squares by order of increasing atomic number. This means that the total number of electrons possessed by each element increases in this arrangement, one at a time, as we move from one square to the next. Then, let's further arrange them into columns of elements that possess similar properties. Of course, we would need to know a great deal of chemistry to accomplish this feat. A similar exercise was first performed more than 130 years ago, when many elements known today had not yet been discovered.

Such an arrangement of the chemical elements into a chart designed to represent the periodic law is called a **periodic table**. Although a number of versions exist, the periodic table shown in Figure 4.3 provides significant information for emergency responders. In this version, the elements are grouped into columns numbered 1A through 8A and 1B through 8B. In another common version, the columns are numbered consecutively across the table from 1 to 18. To avoid confusion, we shall use the first version exclusively.

The elements positioned within the same column of the periodic table are called a **family of elements**. Each family is identified by a number and a capital letter at the top of the column, such as 1A and 2A. Thus, for example, helium, neon, argon, krypton, xenon, and radon belong to the same family, identified by 8A.

periodic law ▪ The observation that the chemical properties of the elements are periodic functions of their atomic numbers

periodic table ▪ A display of the known elements into periods and families, arranged by increasing atomic number so that elements with similar chemical properties are located in the same column (or family)

family of elements ▪ The group of elements listed in a vertical column of the periodic table

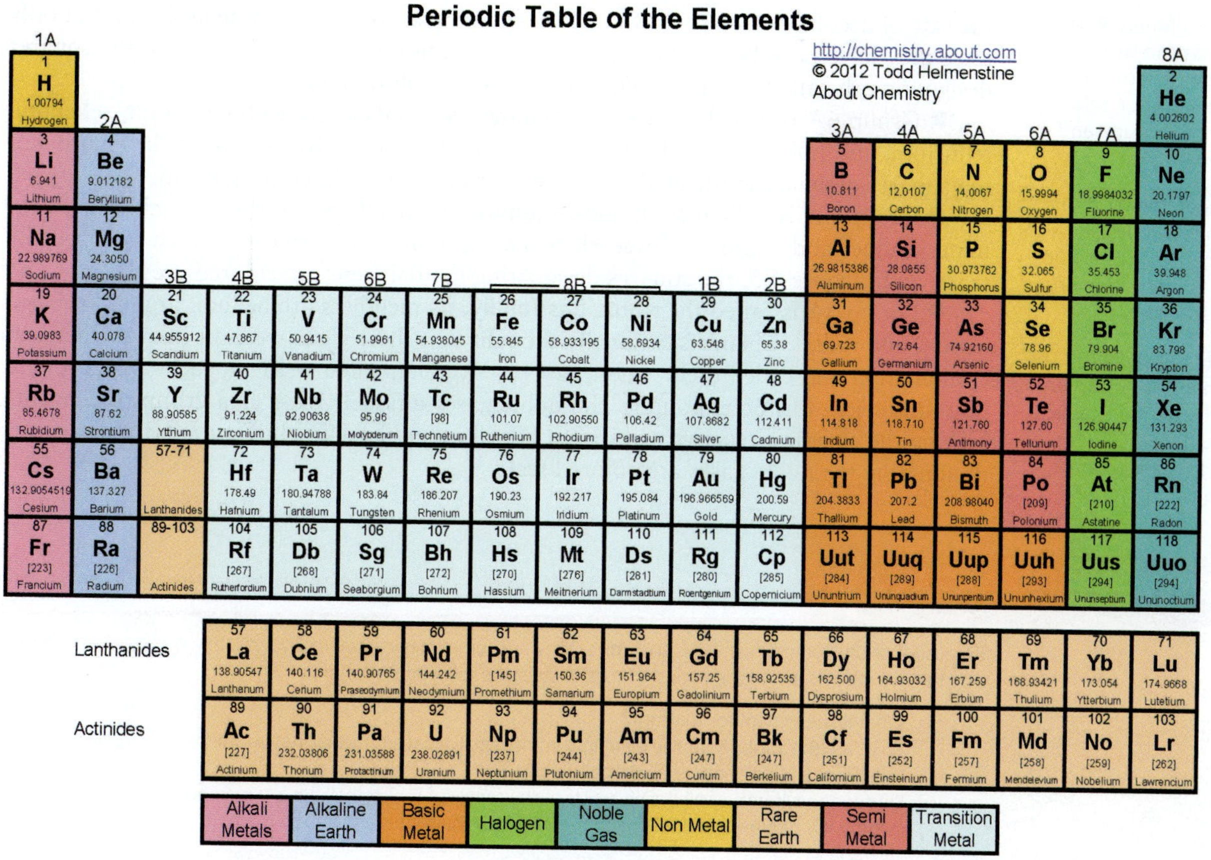

Periodic Table of the Elements

FIGURE 4.3 A modern version of the periodic table of the elements. [*Courtesy of Todd Helmenstine,* About Chemistry (2010).]

Elements in the same row of a periodic table are said to belong to the same **period**. The periods are numbered on the far left of the table from 1 to 7. There is one period of 2 elements, two periods of 8 elements, two of 18 elements, and two more periods of 32 elements.

The periodic table of the elements is one of the most powerful icons in science: a single table that consolidates an array of valuable information. Some version of the table hangs on the wall of nearly every chemistry laboratory throughout the world. Simply by glancing at it, we can observe immediately the atomic number of any element. We can also readily distinguish among those elements that are metals, nonmetals, and metalloids. In Figure 4.3, the salmon-colored squares contain the symbols of the elements that are metalloids. They separate the metals from the nonmetals. Generally, the metals are the elements that fall to the left of the group, and the nonmetals are the elements that fall to the right of it.

The usefulness of a periodic table consists in the manner by which it displays the periodicity, or repetition, in the properties of the elements at regular intervals. In particular, when we observe the elements as members of the same family, we know that they possess similar chemical properties. Five families deserve special recognition in this regard. They are identified by unique names:

■ Group 1A is called the **alkali metal family**; its members are lithium, sodium, potassium, rubidium, cesium, and francium. Each metal reacts with water, although lithium reacts slowly. Each metal also ignites in air, especially when exposed to a moist atmosphere. In Figure 4.3, the squares designating the alkali metals are colored hot pink.

■ Group 2A is called the **alkaline earth family**; its members are beryllium, magnesium, calcium, strontium, barium, and radium. These elements are also chemically reactive, but not nearly as reactive as the alkali metals. They cause water to decompose, but

period ■ A horizontal row on the periodic table

alkali metal family ■ The elements of Group 1A on the periodic table: lithium, sodium, potassium, rubidium, cesium, and francium

alkaline earth family ■ The elements in Group 2A on the periodic table: beryllium, magnesium, calcium, strontium, barium, and radium

Chapter 4 Chemical Forms of Matter **115**

oxygen family ■ The elements in Group 6A on the periodic table: oxygen, sulfur, selenium, tellurium, and polonium

halogen family ■ The elements in Group 7A on the periodic table: fluorine, chlorine, bromine, iodine, and astatine

noble gas family ■ The elements in Group 8A on the periodic table: helium, neon, argon, krypton, xenon, and radon

the rate of decomposition is slow at ambient temperatures. They ignite in the air, but only after they have been heated or exposed to an ignition source. In Figure 4.3, the squares designating the alkaline earth metals are colored medium blue.

■ Group 6A is called the **oxygen family**; its members are oxygen, sulfur, selenium, tellurium, and polonium, each of which is a moderately active substance. In Figure 4.3, the squares designating the chalcogens are not uniquely colored as a group.

■ Group 7A is called the **halogen family**; its members are fluorine, chlorine, bromine, iodine, and astatine. These elements are nonmetals, each of which is especially reactive. In Figure 4.3, the squares designating the halogens are colored green.

■ Group 8A is called the **noble gas family**; its members are helium, argon, krypton, xenon, and radon. Chemists originally thought that these gases were all inert to chemical combination and called them "inert gases." Although some of them, such as xenon, are now known to form compounds, the noble gases uniquely exist as a group of elements lacking the chemical reactivity observed for all other elements. In Figure 4.3, the squares designating the noble gases are colored teal.

SOLVED EXERCISE 4.2

Because thallium compounds are highly toxic, they are sometimes commercially available in rodenticides, products that kill rodents. Using the periodic table, answer the following questions:

(a) What is the symbol for thallium?
(b) What are the symbols for the elements immediately adjacent to thallium on the periodic table?
(c) Is thallium a metal, nonmetal, or metalloid?
(d) Provide the symbols of all the elements in the family of which thallium is a member.
(e) Identify the group number of the family of which thallium is a member.

Solution:

(a) Thallium is located in the sixth period. Its chemical symbol is Tl.
(b) The elements immediately adjacent to thallium on the periodic table are mercury and lead, whose symbols are Hg and Pb, respectively.
(c) Thallium is a metal because it is located to the left of the salmon-colored metalloids on the periodic table.
(d) Boron, aluminum, gallium, indium, and thallium are members of the same family of elements.
(e) Thallium is a member of the family of elements denoted as 3A.

4.6 MOLECULES AND IONS

molecule ■ The smallest neutral unit of some elements and compounds, composed of at least two atoms

nonpolar substance ■ Any element or compound having a symmetrical distribution of charges about its center

polar substance ■ Any element or compound having an asymmetrical distribution of charges about its center

Although the smallest representative particle of an element is the atom, not all uncombined elements exist as single atoms. In fact, only six elements actually exist as single atoms. They are the noble gases. They are said to be "monatomic."

Other gases or liquids at room conditions consist of units containing pairs of like atoms. They are called **molecules**. For example, hydrogen, oxygen, nitrogen, and chlorine are gaseous elements as we generally encounter them, each of which is composed of a molecule having two atoms. These molecules are said to be "diatomic" and are symbolized by the formulas H_2, O_2, N_2, and Cl_2, respectively. Their structures are illustrated in Figure 4.4. Because their charges are symmetrical about their molecular centers, they are called **nonpolar substances.**

By contrast, the charges in substances like hydrogen fluoride and water are nonsymmetrical about the centers of their molecules. These compounds are called **polar substances**, and are symbolized by the formulas HF and H_2O, respectively. Their structures are provided in Figure 4.5. Their nature is due to the congregation of positive and negative charges on opposite ends of the molecules.

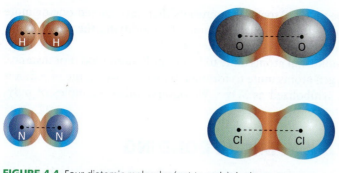

FIGURE 4.4 Four diatomic molecules (not to scale): hydrogen, oxygen, nitrogen, and chlorine. Chemists denote them as H_2, O_2, N_2, and Cl_2, respectively. There is no excess charge at their ends; hence, they are examples of the molecular forms of four nonpolar substances.

FIGURE 4.5 Two molecules (not to scale): hydrogen fluoride and water. Chemists denote them as HF and H_2O, respectively. They are examples of polar substances because their positive and negative charges are localized as shown.

The smallest particle of many compounds is the molecule. Molecules of compounds contain atoms of two or more elements. For example, the methane molecule consists of one carbon atom and four hydrogen atoms. Chemists denote it by the formula CH_4.

Not all compounds occur naturally as molecules. Many occur as aggregates of oppositely charged atoms or groups of atoms called **ions**. Atoms become charged by gaining or losing some of their electrons. In general, metal atoms *lose* electrons, whereas nonmetal atoms *gain* electrons. Atoms of metals that lose their electrons become positively charged; atoms of nonmetals that gain electrons become negatively charged.

> **ion** ■ An atom (or group of atoms bound together) with a net electric charge acquired from the loss or gain of electrons

Let's consider the difference between the sodium atom and the sodium ion. By examining Figure 4.3, we learn that the atomic number of sodium is 11. Thus, the neutral sodium atom has 11 electrons and 11 protons. If a sodium atom loses one electron, it then has only 10 left, although it still retains its 11 protons. By losing the electron, the sodium atom becomes a sodium ion. Its net charge is $+1$; that is, $+11 + (-10) = +1$.

Consider magnesium as a second example. The atomic number of magnesium is 12. Thus, the neutral magnesium atom has 12 electrons and 12 protons. When a magnesium atom loses two electrons, it becomes a magnesium ion. The magnesium ion possesses 10 electrons and 12 protons, and its net charge is $+2$.

These two ionization processes for metals can be represented as follows:

$$Na \longrightarrow Na^+ + e^-$$
$$Mg \longrightarrow Mg^{2+} + 2e^-$$

Here, e^- represents an electron; writing e^- to the right of the arrow means an electron is lost from the atom of the metal whose symbol appears to the left of the arrow. Na^+ and Mg^{2+} denote the sodium ion and the magnesium ion, respectively.

As noted earlier, the atoms of a nonmetal tend to *gain* electrons. Let's consider fluorine. The atomic number of fluorine is 9. The neutral fluorine atom possesses nine electrons and nine protons. If the fluorine atom somehow obtains another electron, it will then have 10 electrons, but still only 9 protons. Its net charge is -1, that is, $+9 + (-10) = -1$. Because fluorine exists in the form of diatomic molecules, we represent the ionization of fluorine using F_2 to symbolize the fluorine molecule on the left of the arrow. This process can be designated as follows:

$$F_2 + 2e^- \longrightarrow 2F^-$$

Here, a 2 is written in front of e^- and F^- so that an equal number of electrons exist on each side of the arrow.

Metal atoms that have lost one or more electrons to become positively charged ions retain the name of the metal. As noted earlier, Na^+ and Mg^{2+} are called the sodium ion and the magnesium ion, respectively. Atoms of nonmetals that have gained one or more electrons to become negatively charged ions are named by modifying the name of the nonmetal so that it ends in *-ide*; F^- thus is named the *fluoride ion*.

polyatomic ion ■ An ion having more than one constituent atom

Frequently, two or more nonmetal atoms unite to form a **polyatomic ion**. For instance, one sulfur atom and four oxygen atoms unite to form an ion with a net charge of -2; it is called the *sulfate ion* and is symbolized as SO_4^{2-}. We observe other examples of polyatomic ions throughout this text.

4.7 THE NATURE OF CHEMICAL BONDING

As we noted earlier, when compounds form, the atoms of one element become attached to, or associated with, atoms of other elements by forces called chemical bonds. But how do these chemical bonds form? Chemists have pondered the answer to this question for centuries.

A clue to understanding the manner in which chemical bonds form can be deduced from the observation that the noble gases are relatively inert substances. This implies that the electrons in the atoms of the noble gases are specially arranged. Today, we know that the electronic stability of the noble gases is associated with the number of electrons in their outmost atomic orbitals as follows:

■ An atom of helium has two electrons, both located in the innermost atomic orbital.
■ The atoms of the noble gases other than helium have eight electrons in their outermost atomic orbitals, those orbitals farthest from their nuclei.

Scientists discovered long ago that molecules and ions often form when their constituent atoms acquire the electronic stability of the noble gases. Because there are only a prescribed number of electrons in an atom, they must somehow interact with electrons from other atoms to acquire the electronic structure of the noble gases. For the elements in the first and second periods, this process occurs as follows:

■ Hydrogen, lithium, and beryllium atoms attain the electronic structure of the helium atom. This means that electronic stability is achieved when these atoms have two electrons.

octet rule ■ The tendency for certain nonmetallic atoms to achieve electronic stability by gaining or sharing a total of eight electrons

■ Atoms of elements other than hydrogen, helium, lithium, and beryllium attain the electronic structure of the noble gas nearest in atomic number. For the atoms of these elements, electronic stability is achieved by attaining a total of eight electrons in their outermost atomic orbitals. The observation that atoms other than hydrogen, helium, lithium, and beryllium achieve electronic stability by attaining a total of eight electrons is called the **octet rule**.

There are two mechanisms, called *ionic* and *covalent* bonding, by which atoms are capable of achieving these electronic structures. Each represents an extreme situation; in actuality, some degree of each description exists in all substances. The mechanisms are discussed independently in Sections 4.9 and 4.10.

valence electrons ■ An atom's outermost electrons that participate in chemical bonding to other atoms

4.8 LEWIS SYMBOLS

Although an atom of any element possesses a definite number of electrons, only some of them are involved in chemical bonding. These electrons are called the atom's **valence electrons**. The valence electrons are those electrons in an atom's outermost atomic orbital. Electrons that do not participate in bonding are called **nonbonding electrons**.

nonbonding electrons ■ An atom's electrons that do not participate in bonding with other atoms

Lewis symbol ■ The symbol of an element together with a number of dots that represent its bonding electrons

We display valence electrons by means of a **Lewis symbol**, named after an American chemist, Gilbert N. Lewis. A Lewis symbol consists of the symbol of an element together with a certain number of dots that represent the atom's valence electrons. For example, Na· is the Lewis symbol for the sodium atom. The sodium atom has only one electron that participates in chemical bonding.

FAMILY	1A	2A	3A	4A	5A	6A	7A	
	Li·	·Be·	·B·	·C·	·N·	·O·	:F·	H·
	Na·	·Mg·	·Al·	·Si·	·P·	·S·	:Cl·	
	K·	·Ca·	·Ga·	·Ge·	·As·	·Se·	:Br·	
	Rb·	·Sr·				·Te·	:I:	

TABLE 4.5 — Lewis Symbols of Some Representative Elements

Table 4.5 lists the Lewis symbols of some representative elements important to the study of hazardous materials. Note that a simple way exists for determining the number of valence electrons for any element of the A family: The number that identifies the A family on the periodic table is also the number of valence electrons possessed by elements in that family. For instance, the halogens are located in the family identified as Group 7A, and we note from Table 4.5 that each halogen has *seven* valence electrons.

4.9 IONIC BONDING

Electrons can be *transferred* from an atom of one element to an atom of another element, resulting in the formation of positive and negative ions. This phenomenon generally occurs between the atoms of metals and nonmetals. The ions that form are attracted to each other by virtue of their opposite charges. Chemists call this electrostatic force of attraction an **ionic bond**. The formation of the ionic bond is based on a fundamental law of nature by which forms of matter with like charges (+/+ or −/−) repel each other, whereas forms of matter with unlike charges (+/−) attract.

ionic bond ■ The electrostatic force of attraction between oppositely charged ions

Many atoms of the elements transfer or accept just the number of electrons that gives them eight electrons in their outermost atomic orbitals. By transferring only this number of electrons, these atoms attain the electronic stability of the noble gas nearest to them in atomic number.

For illustration, consider the ionic bonding in sodium fluoride. From Table 4.5 we learn that the Lewis symbols of sodium and fluorine are Na· and :F·, respectively. When these two elements combine at the atomic level, an atom of sodium transfers its single electron to a fluorine atom. By transferring one electron, the sodium atom electronically resembles neon. By accepting it, the fluorine atom electronically resembles neon. The atoms become charged; that is, they become ions. The process can be written schematically as follows:

$$Na· + :F· \longrightarrow Na^+ \; :F:$$

The attraction between these oppositely charged ions constitutes the ionic bond that binds the two ions together.

4.10 COVALENT BONDING

Electrons can also be *shared* by atoms of identical or different elements to form molecules of elements or compounds. This sharing of electrons is usually between nonmetal atoms. Atoms of the same nonmetal bond to one another and form molecules of the element; atoms of different nonmetals bond to one another and form molecules of a compound.

The atoms of nonmetals acquire their electronic stability by sharing electrons in a manner that permits them to resemble atoms of the noble gases nearest to them in atomic number. For the atoms in the first and second periods other than hydrogen, helium, lithium, and beryllium, this means sharing a total of eight electrons in their outermost atomic orbital. This includes the atom's valence electrons plus those it shares with another atom. Hydrogen atoms share only two electrons. One shared pair of electrons between any two atoms is called a **covalent bond**.

covalent bond ■ A shared pair of electrons between two atoms

Let's observe how two atoms of hydrogen combine to form a hydrogen molecule. H· is the Lewis symbol for hydrogen. To achieve the electronic structure of helium, the nearest noble gas, one hydrogen atom shares its only electron with the single electron from a second hydrogen atom. We represent this process as follows:

$$\text{H} \cdot + \text{H} \cdot \longrightarrow \text{H}:\text{H}$$

The pair of electrons shared between the two hydrogen atoms is a covalent bond. This manner of representing the hydrogen molecule (H:H) is called its **Lewis structure**.

Lewis structure ■ A means of displaying the bonding between the atoms of a molecule by the use of dots or dashes for shared pairs of electrons

Two chlorine atoms combine to form a chlorine molecule. :C̈l· is the Lewis symbol for chlorine. To achieve the electronic structure of argon, the nearest noble gas, each chlorine atom shares its unpaired electron. This formation of the chlorine molecule from two chlorine atoms can be represented as follows:

$$:\ddot{\text{C}}\text{l} \cdot + :\ddot{\text{C}}\text{l} \cdot \longrightarrow :\ddot{\text{C}}\text{l}:\ddot{\text{C}}\text{l}:$$

Let's consider next the combination of hydrogen and chlorine atoms. A hydrogen atom and a chlorine atom can share an electron pair to form a molecule of the substance called *hydrogen chloride*. The formation of a hydrogen chloride molecule from a hydrogen atom and a chlorine atom can be represented as follows:

$$\text{H} \cdot + :\ddot{\text{C}}\text{l} \cdot \longrightarrow \text{H}:\ddot{\text{C}}\text{l}:$$

The hydrogen atom shares an electron pair, so its electronic structure resembles that of helium, whereas the chlorine atom shares an electron pair, so that its electronic structure resembles that of argon.

An atom can also form several covalent bonds with other atoms by simply sharing more than one pair of electrons. For instance, consider the formation of the methane molecule. This molecule consists of one carbon atom and four hydrogen atoms. The Lewis symbols for carbon and hydrogen atoms are ·C̈· and H·, respectively. In the methane molecule, the carbon atom shares each of its four electrons with the electrons from four hydrogen atoms, as represented here:

$$\cdot \dot{\text{C}} \cdot + 4\text{H} \cdot \longrightarrow \begin{matrix} & \text{H} & \\ \text{H} & : \ddot{\text{C}} : & \text{H} \\ & \text{H} & \end{matrix}$$

The sharing of electrons in the methane molecule results in an electronic arrangement like that of the neon atom for the carbon atom and an electronic arrangement like that of the helium atom for each of the four hydrogen atoms.

double bond ■ A covalent bond composed of four electrons shared between two atoms

triple bond ■ A covalent bond composed of six shared electrons between two atoms

Sometimes, two nonmetallic atoms share more than one pair of electrons. This behavior is particularly characteristic of carbon atoms and results in the formation of multiple bonds. Two types of multiple bonds exist: double and triple bonds. A **double bond** consists of the sharing of two pairs of electrons (::), whereas a **triple bond** consists of the sharing of three pairs of electrons (:::). The name of the element is designated when the pairs of electrons are associated with atoms of the same element, as for example, carbon-carbon double bond and carbon-carbon triple bond.

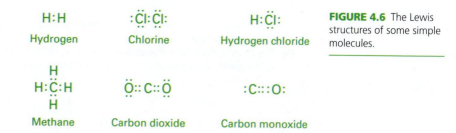

H:H   :C̈l:C̈l: H:C̈l:

Hydrogen   Chlorine   Hydrogen chloride

FIGURE 4.6 The Lewis structures of some simple molecules.

H
H:C̈:H
H

Methane   Ö::C::Ö   :C:::O:

Carbon dioxide   Carbon monoxide

The carbon dioxide molecule consists of one atom of carbon and two atoms of oxygen. The Lewis symbols for carbon and oxygen atoms are $\cdot\dot{C}\cdot$ and $\cdot\ddot{O}\cdot$, respectively. The formation of a carbon dioxide molecule from one carbon atom and two oxygen atoms can be represented as follows:

$$\cdot\ddot{O}\cdot + \cdot\dot{C}\cdot + \cdot\ddot{O}\cdot \longrightarrow O::C::O$$

Notice the existence of two pairs of electrons on each side of the carbon atom in the carbon dioxide molecule. Each of these shared pairs of electrons is a double bond. By sharing electrons in this fashion, the carbon atom and the two oxygen atoms all achieve the electronic arrangement of neon.

The formation of the carbon monoxide molecule from a carbon atom and an oxygen atom can be represented as follows:

$$\cdot\dot{C}\cdot + \cdot\ddot{O}\cdot \longrightarrow :C:::O:$$

The three shared pairs of electrons between the carbon and oxygen atoms constitute a triple bond. Once again, the carbon and oxygen atoms achieve the electronic stability of the neon atom by sharing eight electrons between them. The Lewis structures of several other molecules are illustrated in Figure 4.6.

For the sake of simplicity, Lewis structures are usually written by drawing a long dash to represent the shared pair of electrons. The dots representing all other electrons are omitted. In this notation, the compounds previously noted can be represented as follows:

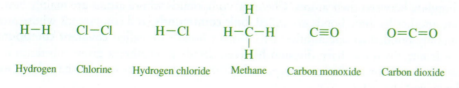

H
|
H—H   Cl—Cl   H—Cl   H—C—H   C≡O   O=C=O
|
H

Hydrogen   Chlorine   Hydrogen chloride   Methane   Carbon monoxide   Carbon dioxide

SOLVED EXERCISE 4.3

At ambient conditions, nitrogen trifluoride is a gas. It is used to manufacture items like microchips and flatscreen televisions.

(a) What is the chemical formula of nitrogen trifluoride?
(b) What is its Lewis structure?

Solution:

(a) From Table 4.9, we see that the Greek prefix *tri-* refers to three. The absence of a prefix for "nitrogen" implies that the prefix *mono-* was dropped for simplicity. This means that the nitrogen trifluoride molecule contains one nitrogen atom and three fluorine atoms. Accordingly, the chemical formula of nitrogen trifluoride is NF_3.

(b) The Lewis symbols for nitrogen and fluorine are obtained from Table 4.5 as $\cdot \ddot{N} \cdot$ and $:\ddot{F}\cdot$, respectively. Consequently, the Lewis structure of nitrogen trifluoride is represented as follows:

$$:\ddot{F}: \qquad :\ddot{F}:\ddot{N}:\ddot{F}: \quad \text{or} \quad :\ddot{F}:\ddot{N}:\ddot{F}: \qquad :\ddot{F}:$$

SOLVED EXERCISE 4.4

The chemical warfare agent called *phosgene* was responsible for approximately 80% of the casualties associated with exposures to chemical substances during World War I. It was called a "choking agent," because the troops who inhaled phosgene subsequently experienced bronchial constriction and were unable to breathe properly. The chemical formula of phosgene is $COCl_2$. What is its Lewis structure?

Solution: The Lewis symbols for the carbon, oxygen, and chlorine atoms are provided in Table 4.5. The Lewis structure of a phosgene molecule can be represented by examining its formation from one carbon atom, one oxygen atom, and two chlorine atoms as follows:

4.11 IONIC AND COVALENT COMPOUNDS AND THEIR SOLUTIONS

ionic compound ■ Any substance whose atoms are bonded together mainly by ionic bonds

covalent compound ■ Any substance whose constituent atoms are bonded together mainly by covalent bonds

solution ■ Any homogeneous mixture of a substance dissolved into another substance

solvent ■ Any substance that dissolves another substance forming a solution

solute ■ Any substance that dissolves in another substance forming a solution

Chemical compounds are classified into two groups based on the predominant nature of the bonding between their atoms. Chemical compounds whose atoms are mainly bonded to one another by ionic bonds are called **ionic compounds**, and compounds whose atoms are mainly bonded to one another by covalent bonds are called **covalent compounds**. There is not always a sharp division between deciding whether a given substance is an ionic compound or a covalent compound. Compounds are known that have both ionic and covalent character.

Ionic and covalent compounds are associated with a number of general properties that are summarized in Table 4.6. Note from this table that ionic compounds usually boil and melt at much higher temperatures than do covalent compounds. Many ionic compounds dissolve in water and conduct electricity when melted or dissolved in solution, whereas the opposite is often true of covalent compounds. This statement is meant to be a generalization that reflects the characteristics of only the majority of ionic and covalent compounds.

Many ionic and covalent compounds are used commercially in the form of homogeneous mixtures of two or more substances. They are called **solutions.** Most commonly encountered are binary solutions, that is, homogeneous mixtures of two substances only. The particular substance that is present in a larger amount in these solutions is called the **solvent** and the other substance is called the **solute.**

Hundreds of industrial solvents are encountered, having widespread application. They are used to manufacture paints, varnishes, lacquers, adhesives, synthetic textiles,

TABLE 4.6	Contrasting Properties of Ionic and Covalent Compounds
IONIC COMPOUNDS	**COVALENT COMPOUNDS**
High melting points	Low melting points
High boiling points	Low boiling points
High solubility in water	Low solubility in water
Nonflammable	Flammable
Molten substances and their water solutions conduct an electric current	Molten substances do not conduct an electric current
Exist predominantly as solids at room temperature	Exist as gases, liquids, and solids at room temperature

plastics, rubber goods, and other products. Some are used directly for such purposes as paint removers and dry-cleaning agents. Emergency responders should be wary of their presence because many solvents are flammable liquids that pose the risk of fire and explosion.

Solvents can be divided into the two groups, polar solvents and nonpolar solvents. In polar substances, the positive and negative charges in the molecules are situated such that they are asymmetrically situated about their centers. Polar substances usually dissolve readily in other polar substances, but are insoluble in nonpolar substances. The polar substance present in the larger amount is called the **polar solvent**. The most common polar solvent is water, which dissolves so many substances it is called the "universal solvent."

Similarly, in a nonpolar substance, the positive and negative charges in its molecules are situated such that symmetry exists about their centers. Nonpolar substances often are soluble in other nonpolar substances, but are insoluble in polar substances. The nonpolar substance present in the larger amount is called the **nonpolar solvent**. Benzene and *n*-hexane are common nonpolar solvents.

4.12 THE CHEMICAL FORMULA

Chemists condense information regarding the chemical composition of substances by writing a **chemical formula** for each of them. For substances consisting of molecules, the chemical formula indicates the kinds and numbers of atoms present in each molecule. For instance, we observed earlier in this chapter that hydrogen and oxygen are symbolized as H_2 and O_2, respectively; these expressions are the chemical formulas of elemental hydrogen and oxygen. The subscript 2 means that each molecule contains two atoms. These chemical formulas are read verbally as "H-two" and "O-two," respectively.

We also observed that the chemical formulas for hydrogen chloride and methane are HCl and CH_4, respectively. These formulas represent the composition of the individual molecules. One molecule of hydrogen chloride consists of one hydrogen atom and one chlorine atom, whereas a molecule of methane consists of one carbon atom and four hydrogen atoms. We verbally read these chemical formulas as "HCl" and "CH-four," respectively. Figure 4.7 further illustrates the chemical formulas of two substances based on their composition.

However, substances are not always composed of molecules. For instance, sodium fluoride, an ionic compound, is composed of ions held together by an ionic bond. Experimental evidence indicates that the components of ionic compounds are not molecules but,

polar solvent ■ Any substance consisting of molecules in which their electrical charges are separated about their centers

nonpolar solvent ■ Any substance consisting of molecules in which their electrical charges are not separated about their centers

chemical formula ■ A method of expressing the number of atoms or ions of specific types in a molecule or unit of matter through the use of chemical symbols for the elements and numbers or proportions of each kind of atom

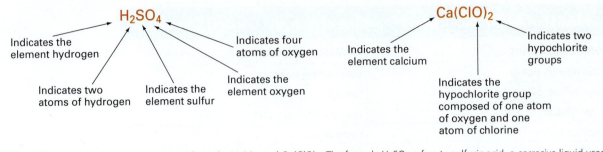

FIGURE 4.7 An explanation of the chemical formulas H_2SO_4 and $Ca(ClO)_2$. The formula H_2SO_4 refers to sulfuric acid, a corrosive liquid used to produce rayon, certain explosives, dyes, other acids, and detergents. The formula $Ca(ClO)_2$ refers to calcium hypochlorite, a constituent of some solid bleaching and disinfecting agents.

rather, aggregates of positive and negative ions. When we analyze a sample of sodium fluoride, we learn that it is composed of an equal number of sodium ions (Na^+) and fluoride ions (F^-). Chemists write its chemical formula as $Na^+\,F^-$, representing the lowest number of formula units. In common practice, the ionic charges are omitted and we simply write NaF.

In an emergency involving a hazardous material, knowing the names or chemical formulas of its constituents can be vital to the responders. When the constituents have been identified, a prompt and appropriate response can be directed toward handling the hazardous material while minimizing the risk to personnel. During emergency-response actions, the names and formulas of commercial chemical commodities can be determined in several ways. For instance, when the emergency occurs within a manufacturing plant or other workplace, the formulas are provided on MSDSs (Section 1.5-B). During a transportation mishap, the specific name of the hazardous material being transported can readily be obtained from a review of the information provided on the accompanying shipping paper. Sometimes, when a commodity is transported in bulk, its name is stenciled on the side of the transport vehicle. In Chapter 6, we review other ways to rapidly determine the nature of the hazardous materials involved in transportation mishaps.

4.13 WRITING CHEMICAL FORMULAS AND NAMING IONIC COMPOUNDS

The chemical formulas of ionic compounds are obtained from the symbols of the ions that make up a given substance. The most common ions are listed in Table 4.7. The names and symbols of these ions should be memorized. There are several simple rules you can follow.

4.13-A POSITIVE IONS

The following rules are useful for naming positive ions:

■ Alkali metals, alkaline earth metals, aluminum, zinc, and hydrogen form monatomic positive ions, each taking the name of the element from which it was derived. For the hydrogen ion and each alkali metal ion, the net charge is +1, and for each alkaline earth metal ion, the net charge is +2.

Na^+	sodium ion	Ba^{2+}	barium ion
Al^{3+}	aluminum ion	H^+	hydrogen ion

TABLE 4.7	Some Common Ions

POSITIVE IONS	NEGATIVE IONS
Ammonium, NH_4^+	Acetate, $C_2H_3O_2^-$
Copper(I) or cuprous, Cu^+	Bromide, Br^-
Hydrogen, H^+	Chloride, Cl^-
Silver, Ag^+	Chlorate, ClO_3^-
Sodium, Na^+	Chlorite, ClO_2^-
Potassium, K^+	Cyanide, CN^-
	Fluoride, F^-
Barium, Ba^{2+}	Hydrogen carbonate, or bicarbonate, HCO_3^-
Cadmium, Cd^{2+}	Hydrogen sulfate, or bisulfate, HSO_4^-
Calcium, Ca^{2+}	Hydrogen sulfite, or bisulfite, HSO_3^-
Cobalt(II) or cobaltous, Co^{2+}	Hydroxide, OH^-
Copper(II) or cupric, Cu^{2+}	Hypochlorite, ClO^-
Iron(II) or ferrous, Fe^{2+}	Iodide, I^-
Lead(II) or plumbous, Pb^{2+}	Nitrate, NO_3^-
Magnesium, Mg^{2+}	Nitrite, NO_2^-
Manganese(II) or manganous, Mn^{2+}	Perchlorate, ClO_4^-
Mercury(I) or mercurous, Hg_2^{2+}	Permanganate, MnO_4^-
Nickel, Ni^{2+}	
Strontium, Sr^{2+}	
Tin(II) or stannous, Sn^{2+}	Carbonate, CO_3^{2-}
Zinc, Zn^{2+}	Oxide, O^{2-}
	Peroxide, O_2^{2-}
Chromium(III) or chromic, Cr^{3+}	Sulfate, SO_4^{2-}
Iron(III) or ferric, Fe^{3+}	Sulfide, S^{2-}
	Sulfite, SO_3^{2-}
Tin(IV) or stannic, Sn^{4+}	
	Phosphate, PO_4^{3-}

■ If a metal forms more than one positive ion, the ion is named in either of two ways:

(a) The ion takes the English name of the metal from which it is derived, immediately followed by a Roman numeral written in parentheses that indicates the net charge on the ion. This system of naming ionic compounds is called the *IUPAC system*.

Cu^+	copper(I) ion	Sn^{2+}	tin(II) ion
Cu^{2+}	copper(II) ion	Sn^{4+}	tin(IV) ion

(b) The ion takes the Latin name of the metal from which it was derived, together with one of the suffixes *-ous* or *-ic* representing the lower and higher net ionic charge, respectively. We call this system of naming ionic compounds the *older system*.

Cu^+	cuprous ion	Sn^{2+}	stannous ion
Cu^{2+}	cupric ion	Sn^{4+}	stannic ion

- There are only two common polyatomic positive ions:

NH_4^+	ammonium ion	Hg_2^{2+}	mercury(I) or mercurous ion

4.13-B NEGATIVE IONS

There are also rules that are useful for naming negative ions:

- Monatomic negative ions are named by adding the suffix *-ide* to the stem of their names.

F^-	fluoride ion	Cl^-	chloride ion
O^{2-}	oxide ion	S^{2-}	sulfide ion

- The diatomic negative ions of oxygen are named by adding the prefixes *per-* and *super-* to the name oxide.

O_2^{2-}	peroxide ion	O_2^-	superoxide ion

- Two polyatomic negative ions are named with the suffix *-ide*.

CN^-	cyanide ion	OH^-	hydroxide ion

- The polyatomic negative ions containing carbon are named uniquely. Two polyatomic negative ions, the carbonate and acetate ions, are commonly encountered.

CO_3^{2-}	carbonate ion	$C_2H_3O_2^-$	acetate ion

- When a nonmetal forms two different negative ions containing oxygen, the suffixes *-ite* and *-ate* are used to name the ion with the lesser and greater number of oxygen atoms, respectively.

NO_2^-	nitrite ion	SO_3^{2-}	sulfite ion
NO_3^-	nitrate ion	SO_4^{2-}	sulfate ion

- When a nonmetal forms more than two different negative ions containing oxygen, the prefixes *hypo-* (meaning lower than usual) and *per-* (meaning higher than usual) are used together with the suffixes *-ite* and *-ate* to name the ions having the lowest and highest number of oxygen atoms, respectively, as follows: *hypo-_____ -ite, _____ -ite, _____ -ate,* and *per _____ -ate,* where the dashed line represents the stem of the identifying nonmetal. For example the negative ions containing one chlorine atom and one to four oxygen atoms are named as follows:

ClO^-	ClO_2^-	ClO_3^-	ClO_4^-
Hypochlorite ion	Chlorite ion	Chlorate ion	Perchlorate ion

- Negative ions containing hydrogen and oxygen atoms (other than the hydroxide and acetate ions) are named *hydrogen _____,* or alternatively, *bi _____,* where the line represents the stem of the identifying nonmetal.

HCO_3^-	hydrogen carbonate, or bicarbonate	HSO_4^-	hydrogen sulfate, or bisulfate

When we know the chemical formulas of substances, we name the associated compounds by simply listing the names of the positive and negative ions in that order. For instance, NaCl is sodium chloride, MgS is magnesium sulfide, and $BaCO_3$ is barium carbonate.

Frequently, we encounter a chemical formula containing parentheses, such as $Fe(ClO_4)_3$. To name such a formula, it is sometimes best to rewrite it and indicate the net charges on the constituent ions. In this case, we have $Fe^{3+}(ClO_4^-)_3$. This shows us that the +3 charge on the iron(III) ion balances the −3 charge on three perchlorate ions. The total positive and negative charges of the ions are equal (+3 and −3), because the compound itself is uncharged, or neutral. The compound is named iron(III) perchlorate by the IUPAC system and ferric perchlorate by the older system.

Before proceeding further, note the names of the following chemical formulas:

$Al_2(SO_3)_3$	aluminum sulfite	$(NH_4)_2CO_3$	ammonium carbonate
$Ba(CN)_2$	barium cyanide	$Cd_3(PO_4)_2$	cadmium phosphate
$Ca(MnO_4)_2$	calcium permanganate	$Ni(NO_3)_2$	nickel nitrate

Suppose next that we are interested in writing a chemical formula when we know the name of a substance. It is again best to follow two simple rules:

- If the net charge of the positive and negative ions is equal (but opposite in sign), the formula of a substance is obtained by simply listing the symbol of the positive ion first, then the symbol of the negative ion. For instance, NaF, CaS, ZnO, KOH, $NaClO_3$, $BaSO_4$, and $AlPO_4$ all have component ions with numerically equal charges. Their names are sodium fluoride, calcium sulfide, zinc oxide, potassium hydroxide, sodium chlorate, barium sulfate, and aluminum phosphate, respectively.
- If the net charges of the positive and negative ions are not numerically equal, first write the symbols of the positive and negative ions together with their respective charges. Then, simply cross the numbers representing the charges (but not the signs) to the opposite ion, as shown in the following examples:

Potassium sulfate	$K^+_2 \, SO_4^{2-},$	or	K_2SO_4	(1 is not written)
Aluminum oxide	$Al^{3+}_2 \, O^{2-}_3,$	or	Al_2O_3	
Barium iodide	$Ba^{2+} \, I^-_2,$	or	BaI_2	(1 is not written)
Ammonium sulfate	$NH_4^+{}_2 \, SO_4^{2-},$	or	$(NH_4)_2SO_4$	(1 is not written)

One special case should be noted for using this rule. When the charges are multiples of one another, such as +4 and −2, the lowest multiple is used. For instance, the chemical formula of tin(IV) oxide is SnO_2, not Sn_2O_4. In the study of hazardous materials, this exception arises only for compounds containing the tin(IV) ion.

SOLVED EXERCISE 4.5

The use of a portable chemical extinguisher containing a water solution of potassium acetate is recommended for extinguishing class K fires (Section 5.17). What is the chemical formula of potassium acetate?

Solution: By consulting Table 4.7, we determine that the potassium and acetate ions are K^+ and $C_2H_3O_2^-$, respectively. Because their charges are equal in magnitude, the chemical formula of potassium acetate is written as $KC_2H_3O_2$.

Provide acceptable names for the compounds whose chemical formulas are (**a**) $CaCO_3$, (**b**) $FeSO_4$, and (**c**) $Cd_3(PO_4)_2$.

Solution: To answer this question, we use the information in Table 4.7.

(**a**) Ca^{2+} and CO_3^{2-} are the symbols for the calcium and carbonate ions, respectively. Because their charges are equal in magnitude, $CaCO_3$ is the formula for the compound whose name is calcium carbonate.

(**b**) Iron has two ions: ferrous, or iron(II) (Fe^{2+}), and ferric, or iron(II) (Fe^{3+}); SO_4^{2-} is the symbol for the sulfate ion. Because the subscript "3" is written adjacent to the Fe in $Fe_2(SO_4)_3$, the symbol represents the ferric ion. Acceptable names for the compound whose formula is $Fe_2(SO_4)_3$ are ferric sulfate and iron(III) sulfate.

(**c**) Cd^{2+} and PO_4^{3-} are the symbols for the cadmium and phosphate ions, respectively. $Cd_3(PO_4)_2$ is the formula for the compound whose name is cadmium phosphate.

4.14 SOME CHEMICAL FORMULAS AND NAMES OF COVALENT COMPOUNDS

It is generally difficult to write the chemical formula of a covalent compound. In this text, we shall learn many formulas of covalent compounds as the occasions present themselves. Nevertheless, a great many covalent names have acquired common names that are still used as part of the chemical language. Some examples are listed in Table 4.8. These names and formulas should be memorized.

A number of covalent compounds contain only two elements, and these substances can be named in a simple fashion. We name the element that is written first in a chemical formula, preceded by a Greek prefix indicating the number of atoms of the element in the compound; then, we name the other element, also preceded by its relevant prefix to indicate the number of its atoms and modifying its ending to *-ide*. Table 4.9 lists the Greek prefixes corresponding to the numbers used to name these simple covalent compounds.

TABLE 4.8	**Common Names for Some Simple Covalent Compounds**
FORMULA OF COMPOUND	**COMMON NAME**
C_2H_2	Acetylene
NH_3	Ammonia
C_6H_6	Benzene
C_2H_6	Ethane
N_2H_4	Hydrazine
H_2S	Hydrogen sulfide
CH_4	Methane
PH_3	Phosphine
C_3H_8	Propane
H_2O	Water

TABLE 4.9		**Greek Prefixes Used in Naming Simple Covalent Compounds**
PREFIX		**COMPOUND**
Mono-	1	Carbon monoxide, CO[a,b]
Di-	2	Carbon dioxide, CO_2
Tri-	3	Phosphorus trichloride, PCl_3
Tetra-	4	Carbon tetrachloride, CCl_4
Penta-	5	Phosphorus pentachloride, PCl_5
Hexa-	6	Sulfur hexafluoride, SF_6
Hepta-	7	Dichlorine heptoxide, Cl_2O_7[a]
Octa-	8	Dichlorine octoxide, Cl_2O_8
Nona-	9	Tetraiodine nonoxide, I_4O_9
Deca-	10	Tetraphosphorus decoxide, P_4O_{10}[a]

[a]When two vowels appear next to one another, as oo in monooxide and ao in heptaoxide, the vowel from the Greek prefix is dropped for the sake of euphony.
[b]The prefix mono- is generally used only to avoid ambiguity.

TABLE 4.10	Names and Chemical Formulas of Some Oxyacids

NAME	CHEMICAL FORMULA
Sulfurous acid	H_2SO_3
Sulfuric acid	H_2SO_4
Nitrous acid	HNO_2
Nitric acid	HNO_3
Phosphorous acid	H_3PO_3
Phosphoric acid	H_3PO_4
Hypochlorous acid	$HClO$
Chlorous acid	$HClO_2$
Chloric acid	$HClO_3$
Perchloric acid	$HClO_4$

We have encountered the chemical formulas of several simple covalent compounds containing only two elements: HCl, CO, and CO_2. The compounds represented by these formulas are named hydrogen chloride, carbon *mono*xide, and carbon *di*oxide, respectively. Other examples of naming covalent compounds are provided in Table 4.10.

SOLVED EXERCISE 4.7

The fire-extinguishing agent, *Intergen*, is a mixture of three gases: carbon dioxide (8%), argon (40%), and nitrogen (52%). What are the chemical formulas of these gases?

Solution: Table 4.9 indicates that the Greek prefix *di-* refers to two. Because no prefix is provided with "carbon" in carbon dioxide, we assume the presence of a "1" despite its absence. This means that the carbon dioxide molecule contains one carbon atom and two oxygen atoms. Accordingly, the chemical formula of carbon dioxide is CO_2.

The noble gases are monatomic; hence, the chemical formula of argon is Ar.

At room conditions, the gases other than the noble gases are diatomic; hence, the chemical formula of nitrogen is N_2.

SOLVED EXERCISE 4.8

Dinitrogen tetroxide has been used to oxidize rocket fuels. What is its chemical formula?

Solution: From Table 4.9, we see that the Greek prefixes *di-* and *tetra-* refer to two and four, respectively. This means that the dinitrogen tetroxide molecule contains two nitrogen atoms and four oxygen atoms. Accordingly, the chemical formula of dinitrogen tetroxide is N_2O_4.

SOLVED EXERCISE 4.9

The gaseous substance having the chemical formula SF_4 is used to produce water and oil repellants. What is the name of this substance?

Solution: Using Table 4.9, we determine that each molecule of SF_4 is composed of one atom of sulfur and four atoms of fluorine. Consequently, the name of the substance is sulfur tetrafluoride.

4.15 NAMING ACIDS

We learn in Chapter 8 that acids are compounds that produce hydrogen ions when dissolved in water. Many commonly encountered acids contain hydrogen in their chemical structures.

4.15-A BINARY ACIDS

binary acid ■ Any acid composed of hydrogen and a nonmetal

Binary acids contain hydrogen and one other nonmetal. The chemical formula of a binary acid is HX or H_2X, where X is the symbol of the nonmetal other than hydrogen. These acids are named by placing the prefix *hydro-* before the stem of the name of an identifying nonmetal, attaching the suffix *-ic* to this stem, and adding the word *acid* to the name. There are two important binary acids whose chemical formulas should be noted:

HF   hydrofluoric acid   HCl   hydrochloric acid

The names of these binary acids are valid only when the acids are dissolved in water. As single substances, they are named as compounds of hydrogen: hydrogen fluoride and hydrogen chloride.

4.15-B OXYACIDS

oxyacid ■ Any acid composed of hydrogen, oxygen, and a nonmetal

Another important group of acids is the **oxyacids**. In addition to hydrogen and the identifying nonmetal, these acids contain one or more oxygen atoms. The chemical formula of the oxyacids is H_mXO_n, where X is the symbol of the identifying nonmetal, and m and n are numbers. There are only 10 oxyacids important to the study of hazardous materials. Their names and chemical formulas are listed in Table 4.10. They should be memorized.

The names of oxyacids correlate directly with the names of their associated polyatomic ions:

■ Polyatomic ions whose names end in *-ate* are derived from acids whose names end in *-ic*; for example, the nit*rate* ion is derived from nit*ric* acid, and the perchlor*ate* ion is derived from perchlor*ic* acid.

■ Polyatomic ions whose names end in *-ite* are derived from acids whose names end in *-ous*; for example, the nit*rite* ion is derived from nit*rous* acid, and the hypochlor*ite* ion is derived from hypochlor*ous* acid.

4.16 MOLECULAR WEIGHTS, FORMULA WEIGHTS, AND THE MOLE

molecular weight ■ The sum of the atomic weights of all the atoms in a molecular formula

The relative weight of a compound that occurs as molecules is called the **molecular weight** and is the sum of the atomic weights of all the atoms that are part of the molecule. Consider the water molecule. Using three digits after the decimal point from the compilation on the inside back cover of this text, its molecular weight is determined as follows:

$$2 \text{ hydrogen atoms} = 2 \times 1.008 = 2.016$$
$$1 \text{ oxygen atom} = 1 \times 15.999 = \underline{15.999}$$
$$\text{Molecular weight of } H_2O = 18.015$$

formula weight ■ The sum of the atomic weights of all the atoms in a chemical formula

The relative weight of a compound that occurs as formula units is called the **formula weight**. It is the sum of the atomic weights of all atoms that make up one formula unit. Consider sodium fluoride. Its formula weight is determined as follows (again selecting only the first three digits after the decimal point):

$$1 \text{ sodium ion} = 22.990$$
$$1 \text{ fluoride ion} = \underline{18.998}$$
$$\text{Formula weight of NaF} = 41.988$$

Chemists also frequently make use of a unit quantity called the **mole**. The mole is the approved SI unit for the amount of any substance. As a unit, it is abbreviated as *mol* (without the *e*). One mole of atoms, molecules, or formula units represents the amount of substance that has a mass in grams equal to its atomic weight, molecular weight, or formula weight, respectively. Thus, 1 mol of carbon is an amount of carbon that has a mass of 12.011 grams. Also, 1 mol of water is an amount of water that has a mass of 18.105 grams.

Because the concept of the mole applies to any type of particle, it is important to identify just what the particle is. For instance, 1 mol of hydrogen *atoms* has a mass of 1.0079 grams, but 1 mol of hydrogen *molecules* has a mass of 2.0158 grams.

Finally, 1 mol of atoms, molecules, or formula units contains a definite number (6.022×10^{23}) of these units. This number is called the **Avogadro number**, in honor of the scientist who first suggested its existence. The exponential notation refers to a number in which the decimal point has been moved 23 places to the right. We can also write this number without using an exponent as 602,200,000,000,000,000,000,000. The mole is analogous to units like the dozen or ream, which mean 12 and 500 items, respectively. One mol of hydrogen atoms is 6.022×10^{23} H atoms, 1 mol of hydrogen molecules is 6.022×10^{23} H_2 molecules, and 1 mol of sodium fluoride units is 6.022×10^{23} NaF units.

The Avogadro number is not just huge—it is so enormous that it is difficult to appreciate how large it is. Imagine that you are counting jelly beans at the rate of 3 beans per second. At this rate, it would take 6×10^{15} years to count 1 mol of jelly beans! This is a million times older than the age of Earth. The enormous magnitude of this number reflects the minute dimensions of atoms and molecules on the scale of "ordinary" measurements.

On October 23 each year, chemists across the world honor Avogadro's number from 6:02 a.m. to 6:02 p.m.

mole ▪ An Avogadro number of particles; the amount of a substance whose mass equals its molecular weight or formula weight, as relevant

Avogadro number ▪ The number of atoms, molecules, or other units contained in a mole of a specified substance, 6.022×10^{23}

SOLVED EXERCISE 4.10

When hydrazine burns, 148.6 kcal/mol of heat evolves to the environment. How many kcal/g of heat evolves?

Solution: From Table 4.8, we see that the chemical formula of hydrazine is N_2H_4. Using this formula, we calculate the molecular weight of hydrazine to be 32.045:

$$2 \text{ nitrogen atoms, } 2 \times 14.006 = 28.012$$
$$4 \text{ hydrogen atoms, } 4 \times 1.008 = \underline{4.032}$$
$$\text{Molecular weight} = 32.044$$

One mole of hydrazine is an amount equal to 32.044 g. When hydrazine burns, 4.5 kcal/g of heat evolves:

$$\frac{148.6 \text{ kcal/mol}}{32.044 \text{ g/mol}} = 4.6 \text{ kcal/g}$$

REVIEW EXERCISES

Elements and Compounds

1. There are 16 elements known to be essential for healthy plant growth: carbon, hydrogen, oxygen, nitrogen, phosphorus, potassium, calcium, magnesium, sulfur, boron, copper, iron, manganese, zinc, molybdenum, and chlorine. What are the symbols of these elements?

2. For the proper functioning and survival of the human organism, trace amounts of 14 metals and metalloids are essential in the diet. Their symbols are Ca, P, K, S, Na, Cl, Mg, Fe, Zn, Cu, Mo, Co, I, and Se. What are the names of these elements?

Physical and Chemical Changes

3. Classify each of the following phenomena as a physical or chemical change:
   (a) The detonation of trinitrotoluene (TNT) produces carbon monoxide, water, nitrogen, and carbon.
   (b) When exposed to an ignition source, hexane vapor ignites at −7°F (−22°C).
   (c) Battery acid is neutralized when mixed with sodium bicarbonate.

Atomic Weights of the Elements

4. Gold has only one stable isotope, gold-197. Its atomic mass is 196.96655. What is the atomic weight of gold?

5. Multiple samples of chlorine collected from sources around the world are found to consist of two stable isotopes: chlorine-35 and chlorine-37. The isotopic composition of the chlorine in one of these samples is shown below:

Isotope	Atomic mass	Natural abundance (%)
Chlorine-35	34.969	75.78
Chlorine-37	36.966	24.22

Show that the atomic weight of chlorine in this sample is 35.45.

The Periodic Table

6. Consider the element that has an atomic number of 114.
   (a) Identify the family of which this element is a member.
   (b) Identify the period of which this element is a member.
   (c) Identify the element that occupies the position immediately above it on the periodic table.

Atoms, Moles, and Ions

7. How many valence electrons do atoms of the following elements possess:
   (a) sodium   (b) carbon   (c) silicon
   (d) aluminum   (e) bromine   (f) oxygen

8. Write the Lewis symbol for the following atoms:
   (a) nitrogen   (b) calcium   (c) boron
   (d) phosphorus   (e) potassium   (f) sulfur
9. Determine the net charge of each of the following ions:
   (a) sodium   (b) oxide   (c) hydrogen
   (d) chloride   (e) magnesium   (f) sulfide

Chemical Bonding

10. Anaerobic bacteria initiate the decomposition of some forms of matter in the absence of oxygen. The phenomenon is often associated with the generation of hydrogen sulfide, a gas possessing the offensive odor of rotten eggs. What is the Lewis structure for the hydrogen sulfide molecule?
11. Dentists frequently use a 2% sodium fluoride solution in water to prevent tooth decay. Illustrate the manner by which a unit of sodium fluoride forms from atoms of the elements sodium and fluorine.
12. Fluorite is a mineral commonly used in making opalescent glass, as well as enameling cooking utensils. The principal chemical constituent of fluorite is calcium fluoride. Describe the nature of the chemical bonds in a unit of calcium fluoride and show that the chemical formula of calcium fluoride is CaF_2.

Chemical Formulas

13. Write the chemical formula of the ionic compound containing each of the following pairs of ions:
   (a) K^+ and SO_4^{2-} (b) Ca^{2+} and OH^- (c) NH_4^+ and Cl^-
   (d) Fe^{2+} and S^{2-} (e) Ag^+ and NO_3^- (f) Ni^{2+} and O^{2-}

14. Name the substance having each of the following chemical formulas:
   (a) HNO_3 (b) O_2 (c) K_2SO_4
   (d) Fe_2O_3 (e) $HgCl_2$ (f) $Al(OH)_3$
   (g) $HClO_4$ (h) Na_2SO_3 (i) $Mg(HCO_3)_2$
   (j) NH_4MnO_4 (k) $BaCO_3$ (l) C_2H_2
   (m) H_2SO_4 (n) NH_3 (o) $NiBr_2$
   (p) $Ca(NO_2)_2$ (q) H_2S (r) $Sn(C_2H_3O_2)_2$
   (s) $SrSO_3$ (t) $Zn(ClO_3)_2$ (u) CO
   (v) $Cd(ClO)_2$ (w) MnO (x) HCl (aqueous solution)

15. Give the chemical formula for each of the following substances:
   (a) nickel sulfite   (b) chlorine   (c) sulfuric acid
   (d) ammonium bromide   (e) hydrogen chloride   (f) silver bromide
   (g) barium peroxide   (h) cadmium nitrite   (i) nickel sulfate
   (j) magnesium peroxide   (k) hydrogen cyanide   (l) zinc chloride
   (m) mercuric nitrate   (n) carbon dioxide   (o) chromium(II) nitrate
   (p) magnesium bisulfite   (q) hydrogen bromide   (r) lead perchlorate
   (s) methane   (t) zinc phosphate   (u) magnesium carbonate
   (v) phosphoric acid   (w) cobalt(II) bicarbonate (x) aluminum hydroxide
   (y) strontium oxide   (z) magnesium phosphate

Molecular Weights, Formula Weights, and Moles

16. The chemical formula of α-chloroacetophenone is C_8H_7OCl. It is a riot-control agent (Section 13.12) that law-enforcement officers sometimes discharge during riots and other forms of civil unrest when attempting to control unruly mobs. What is the molecular weight of α-chloroacetophenone?

17. Argon is a monatomic noble gas that can be isolated by liquefying the components of air.
    (a) How many moles of argon are contained in 119.98 grams of argon?
    (b) How many argon molecules are contained in 119.98 grams of argon?

18. An average U.S. penny weighs 3.0 grams. Assuming that the penny is composed of 100% copper by mass, determine the number of copper atoms in an average penny.

19. When one pound of hydrogen burns in air, 61,600 Btu of heat is released to the environment. What amount of heat expressed in kilocalories per kilogram (kcal/kg) is released?

5
Principles of Chemical Reactions

Courtesy of Drägerwerk AG & Company KGaA, Lubeck.

■ Describe the scope of a balanced chemical c
Identify the types of reaction in a given example

to-energy balance in a given

■ Determine the Gauss of a fire by using information
into burning substances and required reactants and products
able to react with a chemical fuel: that are provided

■ For each of the five common types of chemical combus
fire scenario they must reactivity with the temperature
such fuel present at reduced level all fire activity
the charity some reactivity that are compatible fire limited
and having my chemical reaction and familiar fire extinguish

KEY TERMS

ABC fire extinguisher, *p. 167*

activation energy, *p. 145*

alcohol-resistant aqueous-film-forming foam, *p. 159*

aqueous-film-forming foam, *p. 158*

balanced equation, *p. 137*

catalyst, *p. 145*

chain reaction, *p. 154*

chemical equation, *p. 136*

chemical foam, *p. 158*

classes of fire, *p. 152*

combination reaction (synthesis reaction), *p. 138*

combustion, *p. 136*

complete combustion, *p. 147*

decomposition reaction, *p. 138*

double replacement reaction (double displacement reaction), *p. 139*

dry ice, *p. 161*

dry-chemical fire extinguishing agent, *p. 166*

dry-powder fire extinguishing agent, *p. 169*

dust explosion, *p. 143*

electrolysis, *p. 145*

enzyme, *p. 145*

fire, *p. 146*

fire extinguishing agent, *p. 156*

fire tetrahedron, *p. 153*

fire triangle, *p. 153*

fixed extinguishing system, *p. 157*

fossil fuel, *p. 146*

free radical, *p. 153*

fuel, *p. 146*

global warming, *p. 150*

global warming potential, *p. 151*

greenhouse gas, *p. 150*

halon, *p. 164*

heat of combustion (heat value, heat content), *p. 148*

ignition source, *p. 142*

incipient-stage fire, *p. 161*

incomplete (partial) combustion, *p. 147*

initiation, *p. 154*

kindling point, *p. 145*

law of conservation of mass and energy, *p. 137*

mechanism, *p. 154*

oxidation, *p. 140*

oxidation–reduction reaction (redox reaction), *p. 140*

oxidizer, *p. 142*

oxidizing agent, *p. 142*

phase diagram, *p. 160*

product, *p. 136*

propagation, *p. 154*

protein foam, *p. 159*

rate of reaction, *p. 142*

reactant, *p. 136*

RCRA reactivity characteristic, *p. 152*

reducing agent, *p. 142*

reduction, *p. 141*

relative humidity, *p. 156*

saponification, *p. 170*

single replacement reaction (single displacement reaction), *p. 139*

smolder, *p. 147*

soda–acid fire extinguisher, *p. 157*

spontaneous combustion, *p. 149*

streaming agent, *p. 157*

sublimation, *p. 161*

surfactant, *p. 158*

termination, *p. 155*

thermogenesis, *p. 150*

total flooding, *p. 162*

triple point, *p. 161*

wet-chemical fire extinguishing agent, *p. 169*

water mist, *p. 157*

- Describe the nature of a balanced chemical equation.
- Identify the types of simple chemical reactions.
- Identify the factors that influence the rate at which a chemical reaction occurs.
- Describe the ordinary combustion of a substance in air.
- Identify the manner in which global warming influences the workload of emergency responders.
- Identify the classes of fire that are effectively extinguished by the use of water, aqueous-film-forming foam, alcohol-resistant aqueous-film-forming foam, carbon dioxide, alkali metal bicarbonates, and dry-powder.

Once substances are properly named and their formulas written, we can begin to examine how they interact with one another. Although the chemical reactions of individual hazardous materials will be examined throughout this text, we note in this chapter some features that are common to all chemical reactions. This includes writing and balancing chemical equations and learning how certain factors affect reaction rates.

5.1 THE CHEMICAL REACTION

A substance that has undergone a chemical change is no longer the original substance. In other words, it has become one or more new substances. To describe this chemical change is to indicate that a substance "reacted" in a particular manner. For instance, we say that "dynamite detonated," "hydrogen burned," or "acid corroded." Each statement relates to a chemical change that is generally called a chemical reaction.

In chemistry, it is commonplace to summarize the result of any given reaction in the form of a **chemical equation**. An equation is a shorthand method for expressing a reaction in terms of written chemical formulas. For instance, if we wish to describe the union of elemental carbon with atmospheric oxygen, we may write the equation as follows:

$$C + O_2 \longrightarrow CO_2$$

The equation describes the **combustion** of carbon. We shall study the nature of this reaction in more detail in Section 5.6.

The chemical formulas of the substances that enter the chemical reaction are written on the left of the arrow; they are called its **reactants**. The formulas of the substances formed as a consequence of the reaction are written on the right of the arrow; they are called its **products**. The arrow itself is read as "yields," "produces," "forms," or "gives." Consequently, one way to read this equation is "carbon plus oxygen yields carbon dioxide."

Chemical equations constitute the basic language of chemistry. They should contain as much information as possible concerning the specific chemical change under consideration. We list not only the formulas of the substances reacting and forming but sometimes also their physical states under the temperature and pressure conditions of the reaction. Physical states are indicated in parentheses next to the chemical formulas of the reactants and products with the italicized letters s, l, and g, which symbolize solid, liquid, and gas, respectively. Examples of the symbols used in equations are provided in Table 5.1. Using the relevant symbols, we denote the combustion of elemental carbon by the following equation:

$$C(s) + O_2(g) \longrightarrow CO_2(g)$$

chemical equation
■ The symbolic representation of a chemical process that respectively lists the chemical formulas of its reactants and products on each side of an arrow

combustion ■ The rapid oxidation of a material in the presence of oxygen or air; the chemical process of burning

reactants ■ The substances that enter into a chemical reaction

products ■ The substances produced as a result of a chemical change

TABLE 5.1	Symbols Used in Chemical Equations	
SYMBOL	**MEANING**	**EXAMPLES**
$(g)^a$	Gaseous reactant or product	$H_2(g)$, $CO_2(g)$
(l)	Liquid reactant or product	$H_2O(l)$, $Br_2(l)$
$(s)^b$	Solid reactant or product	$Fe(s)$, $S_8(s)$
$(aq)^c$	Reactant or product dissolved in water	$NaCl(aq)$, $KNO_3(aq)$
$(conc)^d$	Reactant undiluted with water	$HCl(conc)^e$

[a] An arrow pointing upward (↑) is also used when the gas is a product of a reaction.
[b] An arrow pointing downward (↓) is also used when a solid precipitates from solution.
[c] Meaning *aqueous*.
[d] Meaning *concentrated*.
[e] Some substances that are labeled "concentrated" actually contain water. An example is concentrated hydrochloric acid (Section 8.9).

Before we can write a chemical equation, we need to know first the chemical formulas of the reactants and products. To write the more complete form of an equation, we must also know the physical state of the reactants and products. The physical states of reactants and products are determined from chemical reference books or through laboratory experimentation.

5.2 BALANCING SIMPLE EQUATIONS

Not only must an equation summarize what occurs qualitatively during a chemical change, but it must also account for other more fundamental observations. In particular, each equation must be written so that the chemical change at issue adheres to the **law of conservation of mass and energy** (Section 2.6). This law states that there is no apparent change in mass during ordinary chemical reactions. As used here, the term *ordinary* means chemical reactions occurring at the molecular or ionic level, as opposed to the reactions of atomic nuclei.

law of conservation of mass and energy ▪ The observation that the total amount of mass and energy in the universe is constant

The conservation of mass and energy requires that during a given chemical change the atoms of any element are neither created nor destroyed. This means that the number of atoms before and after a reaction remains the same. When chemical equations are written, there must be an equal number of atoms for each element on both sides of the arrow. They are said to be **balanced equations**. The simple equation written in Section 5.1 is balanced, because it has one carbon atom and two oxygen atoms on each side of the arrow.

balanced equation ▪ A chemical equation that has an equal number of like atoms on each side of the arrow

Not all equations are directly balanced after we write the chemical formulas of the reactants and products; in fact, they often are unbalanced at this point. We must select a proper coefficient to place *in front of* the appropriate formula so that the equation then becomes balanced. The correct formula of a substance must never be changed when balancing an equation. Furthermore, coefficients are never written in the middle of a formula, such as "H_23O."

Most simple equations can be balanced by inspection. Although there are no absolute rules for balancing equations by inspection, the following points are useful when first learning this process:

▪ Write the correct formula for each reactant and product and separate the reactants from the products with an arrow.

▪ If known, write the physical state of the reactants and products in parentheses after each formula.

- Choose the formula of the substance containing the greatest number of atoms of an arbitrary element. Insert a number in front of one or both formulas so that the number of atoms for this particular element is balanced.
- If polyatomic ions appear in an equation, balance them as single units only when they retain their identity on both sides of the arrow.
- Balance any remaining atoms or ions.

Let's consider an example. Suppose we wish to write the balanced chemical equation for the reaction that occurs when methane burns in air to form carbon dioxide and water vapor. First, we write the chemical formulas of the reactants and products. The formula of methane is CH_4 (from Table 4.8), oxygen is O_2, carbon dioxide is CO_2, and water is H_2O. Under the reaction conditions, each is a gas or vapor. Hence, we initially write the following:

$$CH_4(g) + O_2(g) \longrightarrow CO_2(g) + H_2O(g)$$

Next, we note that this is an unbalanced equation because there are more hydrogen atoms in CH_4 than in H_2O. We balance the hydrogen atoms by inserting a 2 in front of the formula for water, as follows:

$$CH_4(g) + O_2(g) \longrightarrow CO_2(g) + 2H_2O(g)$$

Because there are no ions in this equation, we next balance the number of the oxygen atoms by inserting a 2 in front of O_2, as follows:

$$CH_4(g) + 2O_2(g) \longrightarrow CO_2(g) + 2H_2O(g)$$

When performing such exercises it is usually best to make one final check: there are one carbon atom, four hydrogen atoms, and four oxygen atoms on each side of the arrow.

5.3 TYPES OF CHEMICAL REACTIONS

By now, one point should be apparent: an equation denoting the chemical reaction of a hazardous material cannot be written when the products of the reaction are unknown. Although identifying the reaction products is not always a simple feat, we can frequently determine them by knowing the reaction type. The reactions in which we are interested can be classified as one of four types reviewed in the next section with illustrative examples.

5.3-A COMBINATION (OR SYNTHESIS) REACTIONS

combination reaction (synthesis reaction)
- A chemical reaction involving two substances that react to form a single product

In a **combination reaction**, or **synthesis reaction**, two or more simpler substances combine to form a more complex substance. Some examples of such reactions are illustrated by the following equations:

$$2H_2(g) + O_2(g) \longrightarrow 2H_2O(l)$$
Hydrogen   Oxygen   Water

$$2Na(s) + Cl_2(g) \longrightarrow 2NaCl(s)$$
Sodium   Chlorine   Sodium chloride

$$C(s) + O_2(g) \longrightarrow CO_2(g)$$
Carbon   Oxygen   Carbon dioxide

decomposition reaction
- A chemical reaction involving the breakup of a compound into two or more simpler substances

5.3-B DECOMPOSITION REACTIONS

In a **decomposition reaction**, a relatively complex substance is broken down into several simpler substances. Some examples of decomposition reactions are illustrated by the following equations:

$$2H_2O(l) \longrightarrow 2H_2(g) + O_2(g)$$

Water   Hydrogen   Oxygen

$$Na_2CO_3(s) \longrightarrow Na_2O(s) + CO_2(g)$$

Sodium carbonate   Sodium oxide   Carbon dioxide

$$(NH_4)_2CO_3(s) \longrightarrow 2NH_3(g) + CO_2(g) + H_2O(g)$$

Ammonium carbonate   Ammonia   Carbon dioxide   Water

5.3-C SINGLE REPLACEMENT (OR SINGLE DISPLACEMENT) REACTIONS

In a **single replacement reaction**, or **single displacement reaction**, an element and a compound react so that the free element replaces an element in the compound. Some examples of replacement reactions are illustrated by the following equations:

single replacement reaction (single displacement reaction) ■ A chemical reaction in which one element replaces another within a compound

$$Mg(s) + 2HCl(aq) \longrightarrow MgCl_2(aq) + H_2(g)$$

Magnesium   Hydrochloric acid   Magnesium chloride   Hydrogen

$$Cu(s) + 2AgNO_3(aq) \longrightarrow 2Ag(s) + Cu(NO_3)_2(aq)$$

Copper   Silver nitrate   Silver   Copper(II) nitrate

$$2KI(aq) + Cl_2(g) \longrightarrow 2KCl(aq) + I_2(aq)$$

Potassium iodide   Chlorine   Potassium chloride   Iodine

5.3-D DOUBLE REPLACEMENT (OR DOUBLE DISPLACEMENT) REACTIONS

In a **double replacement reaction** or **double displacement reaction**, there is an exchange of the positively charged ions in two compounds. Some examples of this type of chemical reaction are illustrated by the following equations:

double replacement reaction (double displacement reaction) ■ A chemical reaction in which two different compounds exchange their ions to form two new compounds

$$2NaCN(s) + H_2SO_4(aq) \longrightarrow Na_2SO_4(aq) + 2HCN(g)$$

Sodium cyanide   Sulfuric acid   Sodium sulfate   Hydrogen cyanide

$$PbS(s) + 2HCl(aq) \longrightarrow PbCl_2(s) + H_2S(g)$$

Lead(II) sulfide   Hydrochloric acid   Lead(II) chloride   Hydrogen sulfide

$$Na_2SO_4(aq) + BaCl_2(aq) \longrightarrow BaSO_4(s) + 2NaCl(aq)$$

Sodium sulfate   Barium chloride   Barium sulfate   Sodium chloride

SOLVED EXERCISE 5.1

Potassium bicarbonate is a useful dry-chemical fire extinguishing agent. When heated, it produces potassium carbonate, water, and carbon dioxide. Write the balanced chemical equation for this reaction.

Solution: First, it is necessary to write the chemical formulas for the reactant and products associated with this chemical reaction. Following the methods learned in Chapter 3, the chemical formulas of the four relevant compounds are written as follows:

potassium bicarbonate, $KHCO_3(s)$ potassium carbonate, $K_2CO_3(s)$

water, $H_2O(g)$ carbon dioxide, $CO_2(g)$

The physical states of these substances are denoted parenthetically. They are known from experience.

These formulas can now be used to write an unbalanced equation representing the chemical reaction.

$$KHCO_3(s) \longrightarrow K_2CO_3(s) + H_2O(g) + CO_2(g)$$

Finally, we must balance the equation. Because oxygen atoms are more abundant in this equation than any other type of atom, we begin by balancing the number of oxygen atoms. Initially, there are three atoms of oxygen on the left side of the arrow but six on the right side (3 + 1 + 2). Balance oxygen by inserting a 2 in front of the formula for potassium bicarbonate. The 2 also balances the number of potassium atoms on each side of the arrow. The equation now looks as follows:

$$2KHCO_3(s) \longrightarrow K_2CO_3(s) + H_2O(g) + CO_2(g)$$

Performing a final check on the number of atoms, we see that there are two atoms of potassium, two atoms of hydrogen, two atoms of carbon, and six atoms of oxygen on each side of the arrow. Hence, this equation is now balanced.

5.4 OXIDATION–REDUCTION REACTIONS

oxidation–reduction reaction (redox reaction)
■ A chemical reaction between one or more oxidizing and reducing agents

Chemists also classify a chemical process in terms of whether it represents an **oxidation–reduction reaction**, frequently called a **redox reaction**. Combination, decomposition, and simple replacement reactions involve oxidation–reduction processes, whereas double replacement reactions do not. Although we study redox reactions in more depth in Chapter 11, a basic understanding is now required, because we will encounter them frequently.

5.4-A OXIDATION

oxidation ■ A chemical process during which a substance reacts as an oxidizing agent

Oxidation is any of the following processes:

■ Elements and compounds *oxidize* when they gain oxygen atoms. When a compound is oxidized, each type of atom within the compound combines with oxygen. For example, carbon, hydrogen, and methane oxidize by combining with oxygen.

$$C(s) \quad + \quad O_2(g) \quad \longrightarrow \quad CO_2(g)$$
Carbon   Oxygen   Carbon dioxide

$$2H_2(g) \quad + \quad O_2(g) \quad \longrightarrow \quad 2H_2O(l)$$
Hydrogen   Oxygen   Water

$$CH_4(g) \quad + \quad 2O_2(g) \quad \longrightarrow \quad CO_2(g) \quad + \quad 2H_2O(g)$$
Methane   Oxygen   Carbon dioxide   Water

■ Compounds also oxidize when they lose hydrogen atoms. When methanol decomposes, for instance, formaldehyde and hydrogen form.

$$CH_3OH(g) \quad \longrightarrow \quad HCHO(g) \quad + \quad H_2(g)$$
Methanol   Formaldehyde   Hydrogen

Because methanol loses hydrogen atoms, it is said to be *oxidized*.

■ An element or ion oxidizes when it becomes less affiliated with its electrons. For ionic substances, this is accomplished by the *loss* of one or more electrons.

$$Na(s) \longrightarrow Na^+(aq) + e^-$$
$$Mg(s) \longrightarrow Mg^{2+}(aq) + 2e^-$$
$$Cu(s) \longrightarrow Cu^{2+}(aq) + 2e^-$$
$$Fe^{2+}(aq) \longrightarrow Fe^{3+}(aq) + e^-$$
$$2Cl^-(aq) \longrightarrow Cl_2(g) + 2e^-$$

In the first three examples, neutral atoms of sodium, magnesium, and copper, respectively, lose either one or two electrons as indicated and become positively charged ions; in the fourth example, the iron(II) ion loses an electron and becomes the iron(III) ion; and in the fifth example, each of two chloride ions loses an electron to form a neutral molecule of chlorine.

5.4-B REDUCTION

Oxidation is always associated with the accompanying process called **reduction**. Any one of the following processes constitutes reduction:

reduction ■ A chemical process during which a substance reacts as a reducing agent

- Compounds *reduce* when they lose oxygen atoms. For example, when sodium perchlorate is heated, it loses oxygen atoms.

$$NaClO_4(s) \longrightarrow NaCl(s) + 2O_2(g)$$
Sodium perchlorate   Sodium chloride   Oxygen

Therefore, sodium perchlorate is said to be *reduced*.

- Compounds also reduce when they gain hydrogen atoms. For example, the organic compound ethene combines with hydrogen to become ethane.

$$C_2H_4(g) + H_2(g) \longrightarrow C_2H_6(g)$$
Ethene   Hydrogen   Ethane

Because it gains hydrogen atoms, ethene is *reduced*.

- Substances reduce when they become more affiliated with electrons. For ionic systems, reduction is associated with the *gain* of electrons.

$$Cl_2(g) + 2e^- \longrightarrow 2Cl^-(aq)$$
$$S_8(s) + 16e^- \longrightarrow 8S^{2-}(aq)$$
$$Fe^{3+}(aq) + e^- \longrightarrow Fe^{2+}(aq)$$
$$Fe^{2+}(aq) + 2e^- \longrightarrow Fe(s)$$

In the first two examples, neutral elements gain electrons and form negative ions; in the third example, the iron(III) ion gains an electron and becomes the iron(II) ion; and in the final example, the iron(II) ion gains two electrons and becomes an atom of elemental iron. The molecules and ions on the left of these arrows are said to be reduced.

Oxidation and reduction also occur in covalent systems, but here, an actual transference of electrons does not occur. For instance, consider the chemical reaction represented by the combination of hydrogen and chlorine.

$$H_2(g) + Cl_2(g) \longrightarrow 2HCl(g)$$
Hydrogen   Chlorine   Hydrogen chloride

In the hydrogen and chlorine molecules, the electron pairs in the covalent bonds are shared equally by their respective atoms. In the hydrogen chloride molecule, however, the chlorine atom shares the pair of bonding electrons to a greater degree than does the hydrogen atom. This unequal sharing of the electron pair is illustrated in Figure 5.1. It causes an unsymmetrical electron distribution in the molecule of hydrogen chloride. This unsymmetrical distribution of electrons is typical of oxidation in covalent systems. Hydrogen has been oxidized and chlorine has been reduced.

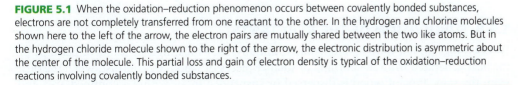

Hydrogen   Chlorine   Hydrogen chloride

FIGURE 5.1 When the oxidation–reduction phenomenon occurs between covalently bonded substances, electrons are not completely transferred from one reactant to the other. In the hydrogen and chlorine molecules shown here to the left of the arrow, the electron pairs are mutually shared between the two like atoms. But in the hydrogen chloride molecule shown to the right of the arrow, the electronic distribution is asymmetric about the center of the molecule. This partial loss and gain of electron density is typical of the oxidation–reduction reactions involving covalently bonded substances.

reducing agent ■ The substance oxidized during an oxidation–reduction reaction

oxidizing agent (oxidizer) ■ The substance reduced during an oxidation–reduction reaction

Any oxidized substance is called a **reducing agent**, and any reduced substance is called an **oxidizing agent**, or **oxidizer**. These names result from the effect that the agent has on other substances. In the combination of hydrogen and chlorine, chlorine is the oxidizing agent and hydrogen is the reducing agent.

Consider another example. Decades ago, cameras used flashbulbs to generate a brilliant blaze to lighten a darkened scene. The brilliance was associated with a chemical reaction in which metallic magnesium burned to form magnesium oxide.

$$2Mg(s) \ + \ O_2(g) \ \longrightarrow \ 2MgO(s)$$

Magnesium   Oxygen   Magnesium oxide

During this reaction, a magnesium atom loses two electrons to become a magnesium ion. It also combines with oxygen. For both reasons, magnesium is *oxidized*. Each atom of an oxygen molecule gains two electrons and becomes an oxide ion. The oxygen is *reduced*. Magnesium is the *reducing agent*, and oxygen is the *oxidizing agent*, or *oxidizer*.

5.5 FACTORS AFFECTING THE RATE OF REACTION

rate of reaction ■ The speed at which a chemical transformation occurs; the amount of a product formed, or reactant consumed, per unit of time

Each chemical reaction occurs at a definite speed called its **rate of reaction**. Sometimes the rate of reaction is referenced by correlating it to a chemical phenomenon, as in the use of terms such as the *rate of combustion*, *rate of corrosion*, or *rate of explosion*. Chemists establish these rates of reaction by experimentally noting the change in concentration of a reactant or product over time.

The speed at which a given substance undergoes a chemical change is often associated with its hazardous nature. This is clearly illustrated by the detonation of nitroglycerin. Several grams can completely decompose within a millionth of a second. Fortunately, not all chemical reactions occur as rapidly, or we would have even greater problems when responding to emergencies involving hazardous materials.

The rate of reaction depends on at least seven factors, each of which will be discussed independently in the sections that follow. When appropriate, the influence of each factor is noted as it bears on the rate of combustion.

5.5-A NATURE OF THE MATERIAL

ignition source ■ Any purposeful or incidental means by which self-sustained combustion is initiated

When exposed to air, some substances do not burn at all. Examples of such substances are water, carbon dioxide, nitrogen, and the noble gases. Other substances, like hydrogen, magnesium, and sulfur, do not begin burning in air until they are first exposed to a source of ignition. Common **ignition sources** include open flames, sparks (static, electrical, and mechanical), lightning, smoking, cutting and welding, hot surfaces, physical and chemical reactions, electrical arcs, radiant heat, and the accumulation of electrical charges (friction) generated by the movement of materials (e.g., liquids through a pipe or hose or powders through chutes or conveyors).

Still other substances burn spontaneously in air, even without exposure to an ignition source. An example is elemental white phosphorus, which bursts into flame on exposure to the air. These rates of combustion vary from zero to some finite value. It is their individual chemical nature that causes some substances not to burn at all, others to burn only when kindled, and others to burn spontaneously.

5.5-B SUBDIVISION OF THE REACTANTS

Wooden logs do not burn spontaneously. Initially, they must first be kindled, perhaps by the heat generated from the burning of smaller pieces of wood. By contrast, when the dust from the same type of wood is dispersed or suspended in air within a confined area and exposed to an ignition source, it may ignite and burn with explosive violence throughout its entire mass. The phenomenon is referred to as a **dust explosion**.[1]

Sawdust, coal dust, grain dust, flour dust, and cotton lint are examples of combustible materials that disperse in the air and undergo a dust explosion. Their rapid combustion reactions are well-acknowledged phenomena when these materials are confined within a space or enclosure. Even the static electricity generated as one particle circulates among the others can serve as an ignition source.

Why does the dust of a combustible material burn explosively, whereas bulk quantities of the same materials must be kindled before they burn? The answer to this question is associated with particle size. As the particle size of these combustible materials is reduced, the greater is the total exposed surface area of a given mass. Because more molecules become available at the particle surface to react, the likelihood of a dust explosion increases.

The factors associated with the onset of a dust explosion are displayed in the dust-explosion pentagon shown in Figure 5.2. Five factors must be present simultaneously for a dust explosion to occur:

- Combustible dust of adequate small-particle size
- Air
- Dispersion of the dust in the air
- A confined environment (such as a silo)
- An ignition source

More molecules are also available to react in configurations that provide an increased surface area. Figure 5.3 shows that a flammable liquid burns fastest in the vessel that allows it to assume the greatest surface area. In general, whenever the surface area of a given substance is increased, the substance reacts at an increased rate, because its molecules are not internally bound to one another. This lack of restriction permits them to react with the molecules of neighboring substances.

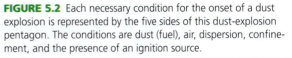

FIGURE 5.2 Each necessary condition for the onset of a dust explosion is represented by the five sides of this dust-explosion pentagon. The conditions are dust (fuel), air, dispersion, confinement, and the presence of an ignition source.

dust explosion ▪ The combustion of dust or particles of dust suspended within a confined space

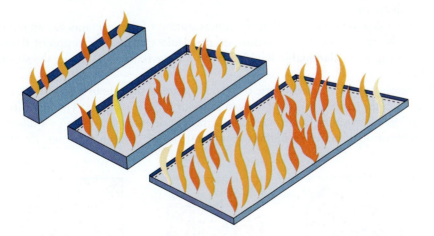

FIGURE 5.3 An equivalent amount of a flammable liquid has been added to three containers of progressively increasing size. When an ignition source is passed along their surfaces at the same instant, the liquid burns at the fastest rate within the container that provides the largest surface area.

---

[1]OSHA is in the process of propagating regulations that address the potential for generating dust explosions in the workplace.

5.5-C STATE OF AGGREGATION

The rate at which a substance reacts is also affected by its physical state of matter. This is particularly true for the rate of combustion. Generally, reactions involving gases occur much faster than reactions involving liquids or solids. These differences in reaction rates are affected by the nature of the gaseous, liquid, and solid states of matter. Molecules of gaseous substances are relatively far apart and exert small attractive forces on one another. This allows diffusion to occur very rapidly. But in the liquid and solid states of matter, the particles are in contact and held tightly together. This hinders the likelihood that they will encounter other particles. Accordingly, their reaction rates are reduced.

5.5-D CONCENTRATION OF REACTANTS

Before a chemical reaction occurs, the particles that make up the structure of the reactants must contact each other. However, particle contact does not signify that a reaction will necessarily occur. The probability that particles will contact one another increases as the number of particles in a given volume increases. In other words, if different concentrations of the same reactants are put into two vessels, the rate of the reaction is generally faster in the vessel containing the greater concentration of reactants.

Imagine that we have four containers of equal volume, such as those shown in Figure 5.4, holding different numbers of two hypothetical molecules, A and B. What is the relative number of collisions between unlike molecules? (We ignore the collisions between like molecules, such as A contacting A, or B contacting B, because these collisions do not cause a chemical reaction.)

Let's consider each container separately. The first container holds one molecule of A and one molecule of B, whereas the second container holds two molecules of A and one molecule of B. It follows that the likelihood that an A molecule will collide with a B molecule is twice as great in the second container as compared with the first one. In the third container, which holds two molecules of A and two molecules of B, the probability of collision between unlike molecules is increased to four times the probability of collision in the first container and two times that of the second one. Finally, in the fourth container, which holds two molecules of A and four molecules of B, the probability of collision between unlike molecules is increased to eight times that in the first container. This shows that an increase in the concentration of the reactants causes the rate of a given reaction to increase.

We also noted in Section 3.1 that the concentration of the reactants has an important bearing on the rate at which combustion occurs. Unless the concentration of flammable gases or the vapors of flammable liquids is within their flammable range, the substances do not burn.

The concentration of atmospheric oxygen also affects the rate of combustion. In clean air at sea level, the concentration of oxygen is about 21% by volume. The majority of the remaining 79% by volume is nitrogen gas. Atmospheric nitrogen serves as a diluent during the combustion of materials in the air and retards their combustion rates. Combustion

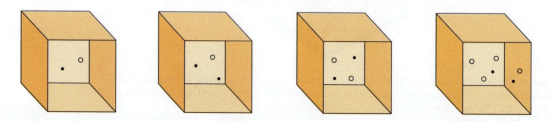

FIGURE 5.4 The probability of a chemical reaction increases as the number of reactant molecules confined within their container increases. The molecules of reactants A and B are designated here as dots and open circles, respectively.

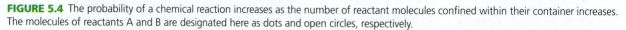

always occurs with an increased intensity within an atmosphere of pure oxygen. Some substances that are ordinarily stable in the air will burn spontaneously in an atmosphere of pure oxygen.

5.5-E ACTIVATION ENERGY

Although some substances react spontaneously on contact, most must be supplied with a minimum amount of energy before they chemically react. Consider pieces of combustible solids lying exposed to atmospheric oxygen. They do not normally burst into flame, whereas they readily ignite when exposed to an energy source. This minimum amount of energy that must be supplied to initiate the chemical reaction is called the **activation energy**.

Heat is the form of energy that is ordinarily supplied to initiate the combustion of a solid substance. The heat raises the temperature of the substance to its **kindling point**, the temperature at which the burning occurs at a sufficiently rapid rate so it may continue without the need for additional heat from an external source. In exothermic reactions, the activation energy does not need to be continuously supplied as the reaction proceeds, because it is replaced by the energy released in the process. Once combustion has been initiated, it continues until either the material or the oxygen is exhausted.

An amount of activation energy must also be supplied to initiate endothermic reactions. In this case, the chemical phenomenon ceases if the source of energy is removed. We must supply not only the activation energy but enough additional energy to replace the energy absorbed by the reactants. Consider **electrolysis**, the process of decomposing a substance through use of an electric current. Electrolysis is an endothermic phenomenon. When we disconnect the current, the electrolysis stops.

5.5-F TEMPERATURE

Heating a mixture of reactants causes its particles to move more rapidly, and this motion increases the probability that the particles will collide. As the speed of the particles increases, the temperature rises accordingly. Thus, a reaction rate increases with a rise in temperature, because more molecules become sufficiently activated at higher temperatures.

According to a classic rule in chemistry, the rate of a chemical reaction doubles for each rise of 18°F (10°C). Thus, twice as many molecules are activated when the temperature of the system is increased by 18°F (10°C); four times as many are activated when the temperature is increased 36°F (20°C); and eight times as many for a 54°F (30°C) increase in temperature.

5.5-G CATALYSIS

A **catalyst** is a substance that affects the rate of a reaction and appears to remain unchanged. A familiar example of a catalyzed reaction is the rusting of iron in the presence of atmospheric water vapor. Iron exposed to moist air corrodes much faster than iron stored in a dry atmosphere. Another example of a catalyzed reaction is the combustion of hydrogen in the presence of platinum. A mixture of hydrogen and oxygen can be kept in a vessel for years without reacting to any noticeable degree; however, if a small amount of platinum is introduced into the vessel, the mixture reacts explosively. At the end of the reaction, hydrogen and oxygen form water but the platinum remains unchanged.

During a given reaction, the catalyst itself can undergo a chemical change but then react again in a manner that causes it to return to its original condition. In the end, its overall chemical identity is unchanged. A catalyst may also be altered physically by the absorption of one or more reactants on its surface.

Catalysts are involved in controlling the rates of the reactions used to produce many of the commercial products noted in this text, such as gasoline, fertilizer, and plastics. They are also involved in controlling the rates of biochemical reactions that produce the energy and molecules needed for sustaining life. In this latter instance, the catalysts are called **enzymes**.

activation energy
■ The minimum energy that must be supplied to a group of reactants before a chemical reaction is initiated

kindling point ■ The temperature at which the combustion of a heated solid proceeds without further addition of heat from an external source

electrolysis ■ The passage of an electric current through a material, resulting in the flow of electrons and energy from an outside source

catalyst ■ A substance that increases or decreases the rate of a chemical reaction without itself being permanently altered

enzyme ■ Any substance that controls the rate of a biochemical reaction

Propane is frequently used as a fuel for heating and cooking food in areas of the country where natural gas is not delivered by pipeline to residences. When propane burns completely, carbon dioxide and water vapor are produced as combustion products. Write a balanced equation representing the complete combustion of propane.

Solution: First, we require the chemical formulas of the reactants and products. The chemical formula of propane was provided in Table 4.8 as C_3H_8. The chemical formulas of elemental oxygen, carbon dioxide, and water vapor are O_2, CO_2, and H_2O, respectively. The unbalanced equation representing the combustion of propane is then written as follows:

$$C_3H_8(g) + O_2(g) \longrightarrow CO_2(g) + H_2O(g)$$

As written, there are three atoms of carbon, eight atoms of hydrogen, and two atoms of oxygen on the left side of the arrow but only one atom of carbon, two atoms of hydrogen, and three atoms of oxygen on the right side of the arrow. To balance this equation, we write a 3 in front of the formula for carbon dioxide. This provides three atoms of carbon on each side of the arrow, but the number of hydrogen and oxygen atoms still remains unbalanced. Hence, we insert a 4 in front of the formula for water. Now, there are eight hydrogen atoms on each side of the arrow, but the number of oxygen atoms remains unbalanced: 2 on the left side and 10 on the right side of the arrow. To overcome this problem, we insert a 5 in front of the formula for oxygen, and the equation is balanced in the following final form:

$$C_3H_8(g) + 5O_2(g) \longrightarrow 3CO_2(g) + 4H_2O(g)$$

5.6 THE COMBUSTION REACTION

Combustion is a chemical reaction that releases energy to the surroundings as heat and light. The light is ordinarily visible to the naked eye and is observable as flames. During any combustion reaction, flames appear because the minute particles of the fuel or combustion products are heated to incandescence.

The flames that accompany a combustion reaction are sometimes nearly imperceptible. For example, the combustion of hydrogen or methanol is nearly impossible to detect with the naked eye. The energy released to the surroundings is still heat and light, but the light is not within the visible range. To locate these substances as they leak from storage and transport vessels during execution of an emergency response action, experts recommend the use of a thermal-imaging camera like the model shown in Figure 5.5.

Two or more substances chemically unite during a combustion reaction. Because one of them is typically oxygen, we often hear that a substance *oxidizes* when it burns. There are also combustion reactions that involve the union of one substance with another substance that is not oxygen. Elemental phosphorus, for example, burns in an oxygen environment, but it also burns in a chlorine environment. Although oxygen is not involved in the latter process, we still indicate that the phosphorus oxidizes when it combines with chlorine.

When the oxidation of a substance occurs relatively fast, the combustion process manifests itself as **fire**. For this reason, it is often described as *rapid* oxidation. The combustion process is a self-sustaining reaction; that is, unless the process is intentionally extinguished, combustion continues until the concentration of the substance falls below a minimum value.

5.6-A RELATIONSHIP BETWEEN COMBUSTION AND FIRE

The substance that burns during a fire is called the **fuel**. The fuels that originally formed in nature by the decomposition of the remnants of prehistoric organisms in oxygen-free environments are called **fossil fuels**. They consist primarily of coal, natural gas, and crude oil. The most abundant fossil fuel worldwide is coal (Section 7.6-C).

fire ▪ The physical manifestation of combustion as flames, light, and heat

fuel ▪ A material that burns; any commercial product whose combustion serves as a source of energy

fossil fuel ▪ Any energy source like coal, crude oil, and natural gas, each of which formed naturally by the decomposition of prehistoric organisms

FIGURE 5.5 These emergency responders are using a thermal-imaging camera to locate objects without hindrances from external influences such as thick smoke or darkness. The camera is also used to locate points on storage and transport vessels from which escaping vapors are burning with nearly imperceptible flames. (*Courtesy of Dräger-werk AG & Company KGaA, Lubeck.*)

When a fuel burns, its atoms are never destroyed; instead, they unite with the atoms of other substances to form one or more new substances. Because a substance typically unites with atmospheric oxygen when it burns, the products of the combustion are often metallic or nonmetallic oxides. *Oxygen itself does not burn*; it is said to *support* combustion.

Many flammable and combustible substances contain carbon in their molecular framework. When they burn in air, the constituent carbon atoms become carbon mon*oxide* or carbon di*oxide*. When carbon dioxide is produced, the burning phenomenon is said to constitute **complete combustion**; but when carbon monoxide forms, we regard the process as **incomplete** or **partial combustion**. We represent the incomplete and complete combustion of carbon as follows:

$$2C(s) + O_2(g) \longrightarrow 2CO(g)$$
Carbon Oxygen Carbon monoxide

$$C(s) + O_2(g) \longrightarrow CO_2(g)$$
Carbon Oxygen Carbon dioxide

Incomplete combustion occurs when items **smolder** or burn slowly without a visible flame. During the incomplete combustion of carbon-containing fuels, some carbon atoms do not even unite with oxygen. When this occurs, elemental carbon is produced, which is observed as the tiny particulates that constitute smoke (Section 10.9).

The prevailing oxygen concentration influences the color of the flames generated during the combustion of ordinary fuels. When adequate oxygen is available for their complete combustion, the flames are blue; but, when adequate oxygen is unavailable, they are yellow.

complete combustion ■ Combustion process involving complete oxidation

incomplete (partial) combustion ■ Combustion process involving incomplete oxidation

smolder ■ The act of burning without visible flames but usually with moderate smoke production

5.6-B EARLY ATTEMPTS TO CONTROL FIRE

Although firefighters have been trying to control fire since at least the onset of the colonial period, attempts were first achieved by others much, much longer ago. Insofar as is presently known, the use of fire in a controlled fashion first occurred at least one million years ago. This fact was revealed by the archeological discovery of burned bones and the ashes of plant material in South Africa's Wonderwerk Cave. These telltale signs of fire now serve as the oldest known evidence of an ancient hominid's attempt to control fire.[2] The progenitor was probably *Homo erectus*. Because modern man (*Homo sapiens*) evolved approximately 40,000 years ago, our ancient ancestors used and controlled fire for their benefit even before our own species had appeared on the planet.

Harnessing the power of fire brought many benefits to our ancestors, including the ability to live in colder climates, cook food over flames or in embers, and provide protection against predators. Without the ability to control fire, our ancestors would have been unable to transform the world.

Until the Atomic Age, fossil fuels were the sole source of energy. Even today, the combustion of fossil fuels generates over 80% of the energy used in North America.

5.6-C TYPES OF CHEMICAL ENERGY

As we first noted in Section 2.6, chemical energy is available in every substance. It is composed of two types:

■ Some energy is present within a substance by virtue of the motion of its atoms and molecules.
■ Energy is also stored in a substance as the result of its unique structure. This energy is associated with the strength of its chemical bonds. During any chemical reaction, some bonds break, new ones form, and any excess energy is either absorbed or released into the environment. This is the source of the energy released during the combustion process.

heat of combustion (heat value, heat content) ■ The amount of heat emitted to the surroundings when unique materials burn

As materials burn, specific amounts of energy are evolved to the surroundings as heat. The actual amount varies with their chemical nature and is called the **heat of combustion, heat value,** or **heat content.** This property is generally expressed in an energy unit per gram, pound, or mole of substance burned, for example, Btu/lb, kcal/g, kJ/kg, or kJ/mol; for flammable gases, the heat of combustion is also expressed in units like Btu/ft^3 and kJ/m^3.

Materials that burn and evolve an amount of heat greater than approximately 5000 Btu/lb (~11,600 kJ/kg) are useful commercial and industrial fuels. Some are noted in Table 5.2. When methane burns, for example, its chemical bonds are broken and new ones form as carbon dioxide and water vapor are produced. The amount of heat released during its combustion is noted in Table 5.2 as 24,100 Btu/lb, or 56,000 kJ/kg.

As an example, consider the combustion of methane. Chemists sometimes denote the complete combustion of methane with an equation like the following:

$$CH_4(g) \ + \ 2O_2(g) \ \longrightarrow \ CO_2(g) \ + \ 2H_2O(g) \ + \ 24{,}100 \text{ Btu/lb (56,000 kJ/kg)}$$

Methane   Oxygen   Carbon dioxide   Water

Methane is the fuel and oxygen is the oxidizing agent.

---

[2]R. Ferring et al., "Earliest human occupations at Dmanisi (Georgian Caucasus) dated to 1.85–1.78 Ma," *Proc. Nat. Acad. Sci. USA,* Volume 108 (2011), pp. 10432–10436.

TABLE 5.2	Heat of Combustion of Some Common Fuels[a]	
FUEL	**Btu/lb**	**kJ/kg**
Acetylene	21,600	50,200
Aviation fuel	20,420	47,600
Butane	20,900	49,510
Coal (bituminous)[b]	7300–10,000	17,000–23,250
Diesel oil	19,300	44,800
Ethanol	12,800	29,700
Hydrogen	61,000	141,790
Methane (natural gas)	24,100	56,000
Methanol	9740	22,700
Motor gasoline	20,400	47,300
Propane	21,500	50,000
Wood (dry)	6200–7500	14,400–17,400

[a]Determined at 32°F (0°C) and 1 atm (101.3 kPa).
[b]The heat of combustion for coal varies from 5500 to 14,500 Btu/lb (13,800–34,000 kJ/kg). Bituminous coal is selected here for comparison to other fuels because it is the most common rank.

5.7 SPONTANEOUS COMBUSTION

spontaneous combustion ■ Rapid oxidation initiated by the accumulation of heat in a material undergoing slow oxidation

Some substances undergo oxidation so slowly that their fires are initially imperceptible. When the oxidation occurs within a confined space where the air circulates poorly, they can absorb their own heat of reaction. The cumulative heat ultimately raises their temperatures and they self-ignite. Because a source of ignition other than the heat of reaction is absent, this process is called **spontaneous combustion**.

When exposed to the air, several animal and vegetable oils oxidize slowly. As demonstrated in Figure 5.6, they should be regarded as potential sources of spontaneous combustion. A well-known example is linseed oil (Section 13.7-C), which was formerly a component of commercial paints and other surface coatings. The improper storage or disposal of rags dampened with linseed oil is a well-acknowledged situation in which spontaneous combustion is probable. When such rags are negligently tossed into a pile in a broom closet, for instance, the heat cannot readily dissipate to the surroundings. Instead, the rags retain the heat; before long, the kindling point is reached and the oil bursts into flame.

FIGURE 5.6 Oily rags and improperly stored linseed oil, varnishes, lacquers, and other oil-based paint products are among the materials likely to undergo spontaneous combustion. Considerable loss of property is associated annually with major fires caused by spontaneous combustion.

Because the availability of oxygen is limited, a thick, black smoke usually accompanies these slow-burning fires.

Spontaneous combustion also occurs within the bulk of undried agricultural products, such as nuts, damp hay, and grass. These moist materials contain microorganisms that multiply rapidly. Their combined biological activity slowly evolves heat, a process known as **thermogenesis**. It is supplemented by chemical oxidation until the temperature rises to approximately 160°F (71°C), at which point the microorganisms are unable to survive. Although their biological activity ceases, chemical oxidation continues. When the heat evolves faster than it dissipates, the internal temperature increases until the kindling point is attained. Then, the products self-ignite.

Coal is another material that poses the risk of spontaneous combustion, especially when large amounts are stored in bins, bunkers, or piles for lengthy periods. Fires generally begin as "hot spots" deep within a reserve of coal, causing the temperature of the coal pile to elevate above that of the surroundings. Once it reaches approximately 600°F to 700°F (~310°C to 370°C), the coal self-ignites.

Spontaneous combustion is also linked with several ongoing fires in abandoned coal mines. We shall visit this subject in Section 7.6-J.

thermogenesis ■ The physiological activity of microorganisms that results in the slow liberation of heat within a material that, when absorbed by the material, can cause its spontaneous combustion

5.8 GLOBAL WARMING

Our planet is continuously bombarded by radiant energy that enters the atmosphere from the sun. Although some is emitted back into outer space, some energy is also absorbed by Earth's surface. Over the past 500,000 years, the rates at which the radiation was absorbed and emitted have been steady on average, thereby producing a surface temperature that has fluctuated between 66°F (19°C) and 80°F (27°C). Today, however, the average temperature of Earth's surface exceeds the high end of this range, because the absorbed energy has been hindered from radiating effectively back into outer space.

This climatic phenomenon is known as **global warming**. Scientists now believe that we are experiencing the effects of global warming in the form of a substantial increase in the number of extreme weather events like forest and grassland fires, heavy rainstorms, floods, and heat waves.[3] In this manner, global warming impacts the workload of emergency responders, because they must mitigate the impact of these natural disasters when they occur. Scientists also believe that the impact may become far worse than what we are experiencing today. The World Bank has shown that unless serious international attention is directed at the problem, we are on track to experience a rise in temperature of up to 7.2°F (4°C) by 2100.[4]

The enhancement in global warming in the present and recent past is believed to be caused at least in part to the presence of an elevated concentration of carbon dioxide in the atmosphere compared with its concentration in the distant past. Carbon dioxide is not the sole gas that warms the lower atmosphere. Other gases also can trap heat in Earth's atmosphere. Each is called a **greenhouse gas**.

global warming ■ The increase in the average temperature of Earth's atmosphere partly due to the greenhouse effect during the present and recent past

greenhouse gas ■ Any component of Earth's atmosphere that increases its capability to prevent the escape of heat into outer space

---

[3]Remarks released by the World Meteorological Organization at the 18th Session of the Conference of the Parties to the United Nations Framework Convention on Climate Change (Doha, Qatar: November 26–December 7, 2012).

[4]"Turn Down the Heat," A Report for the World Bank by the Potsdam Institute for Climate Impact Research and Climate Analytics (Washington, DC: World Bank, November 2012), p. 37.

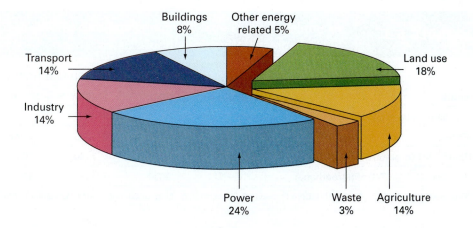

FIGURE 5.7 Global sources of carbon dioxide emissions into the atmosphere: Approximately 65% and 35% originate from energy-related and non-energy-related emissions, respectively. [*Courtesy of World Resources Institute, Climate Analysis Indicators Tool On-line Database Version 3.0 (2006), Washington, D.C. cait.wri.org*]

Carbon dioxide is the main culprit associated with global warming because prodigious amounts of it are produced when fossil fuels burn. The primary global sources that emit carbon dioxide into the atmosphere are shown in Figure 5.7. The largest amount—24%—is associated with operation of the world's fossil-fuel-fired power plants.

Although the evidence for global warming is overwhelming, not all scientists agree about its cause. Some attribute it to natural forces such as cyclic variations in solar activities. The contribution of such events to global warming is well documented: Earth has warmed and cooled every 1500 years. Regardless of the conflicting scientific positions, the vast majority of climate specialists support the prevailing theory that global warming is occurring today largely because human activities have caused the levels of carbon dioxide and other greenhouse gases in Earth's atmosphere to exceed the levels noted in the past.

As a nation, the United States has taken limited steps to curb airborne emissions of carbon dioxide. One attempt is exhibited by the requirement that American vehicle manufacturers produce cars and light-duty trucks that average 35.5 miles per gallon (15 km/L) and 54.5 miles per gallon (23 km/L) by 2016 and 2025, respectively. This requirement focuses on reducing the amount of carbon dioxide emitted during vehicular operations. It remains to be seen whether it will significantly reduce future carbon dioxide emissions.

Climatologists compare the potential ability of a substance to act as a greenhouse gas by measuring its **global warming potential,** or **GWP.** This is the amount of heat trapped in the atmosphere by one unit mass of a substance relative to that trapped by one unit mass of carbon dioxide over a specified time period. GWPs are one of the physical properties of substances. The GWPs of the greenhouse gases are published in the scientific literature and periodically reevaluated to assess the underpinnings for understanding global climate change. Some examples are provided in Table 5.3. For instance, consider methane that accumulates in the atmosphere over a time horizon of 20 years. Its GWP is designated by climatologists as 72, which means that over a 20-year period, one kilogram of methane traps 72 times more heat in the atmosphere than one kilogram of carbon dioxide.

global warming potential ■ The estimated amount of a substance that contributes to global warming compared to carbon dioxide over a specified time period

TABLE 5.3	Examples of the Global Warming Potentials (GWPs) of Some Representative Greenhouse Gases[a]			
NAME	**CHEMICAL FORMULA**	**GWP FOR INDICATED TIME HORIZON**		
		20-Y	**100-Y**	**500-Y**
Carbon dioxide	CO_2	1	1	1
Methane	CH_4	72	25	7.6
Halon 1301 (bromotrifluoromethane)	$CBrF_3$	8480	7140	2760
HFC-134a (1,1,1,2-Tetrafluoroethane)	CH_2F—CF_3	3830	1430	435

[a]Excerpted from Table 2.14, United Nations Framework Convention on Climate Change (Fourth Assessment, 2007).

5.9 THE RCRA REACTIVITY CHARACTERISTIC

The RCRA regulations (Section 1.3-C) denote a material as a hazardous waste when it exhibits one or more characteristics, one of which is reactivity. A material exhibits the **RCRA reactivity characteristic** when it possesses one or more of the following properties:

RCRA reactivity characteristic ■ For purposes of RCRA regulations, the characteristic of a waste that exhibits any property set forth at 40 C.F.R. §261.23

■ It readily undergoes violent change without detonating.
■ It reacts violently with water.
■ It forms potentially explosive mixtures with water.
■ It is a cyanide- or sulfide-bearing substance that when exposed to acid or alkaline solutions generates toxic gases, vapors, or fumes in a quantity sufficient to present a danger to human health or the environment.
■ It is capable of detonation or an explosive reaction if it is subjected to a strong initiating source or if it is heated under confinement.
■ It is readily capable of detonation, explosive decomposition, or reaction at standard temperature and pressure.
■ It is an explosive material in hazard class 1.1, 1.2, or 1.3 (Section 6.1-B).
■ It is a forbidden explosive (Section 15.4).

A waste that exhibits the RCRA reactivity characteristic is subject to EPA's treatment, storage, and disposal regulations. Its generators assign D003 as the hazardous waste number.

5.10 CLASSIFICATION OF FIRES

classes of fire ■ The five types of fire designated by NFPA, each characterized by the nature of its fuel and extinguishing agent

In its standards and codes, NFPA identifies five general **classes of fire**: class A, class B, class C, class D, and class K. Each is characterized by a designation of the nature of its fuel and the most appropriate extinguishing agent. Their individual features are as follows:

■ A class A fire results from the burning of ordinary cellulosic materials such as wood and paper as well as similar natural and synthetic materials like rubber and plastics. It is typically extinguished using water. A class A fire produces a solid residue consisting of ashes or embers.

■ A class B fire results when flammable gases and liquids burn. Firefighters generally use carbon dioxide, dry-chemical extinguishers, or foam to extinguish a class B fire fueled by flammable liquids. The use of water is typically ineffective at extinguishing a class B fire, especially when it involves the burning of bulk amounts of a combustible material.

■ A class C fire is associated with the combustion of materials occurring in or originating from energized electrical circuits, wiring, motors, and similar items. The use of carbon dioxide or dry-chemical extinguishers is recommended for extinguishing class C fires. Water should be used as a class C fire extinguishing agent only when the electrically active unit can first be deenergized.

■ A class D fire is associated with the combustion of certain metals including titanium, magnesium, aluminum, and sodium. Class D fires are extinguished with special types of fire extinguishers, such as graphite-based dry powder or sodium chloride. Water should be used as a class D fire extinguishing agent only when it can be applied in a deluging volume.

■ A class K fire consists of burning cooking materials such as vegetable and animal fats and oils. The K in class K refers to "kitchen." A class K fire may be considered a subclass of a class B fire. However, a class K fire is extinguished with special fire extinguishers, such as handheld portable wet-chemical extinguishers consisting predominantly of water solutions of either potassium acetate or potassium carbonate. Professionals do not recommend the sole use of water for extinguishing class K fires.

5.11 THE FIRE TRIANGLE AND FIRE TETRAHEDRON

The combustion process was visually represented decades ago by using the **fire triangle** shown in Figure 5.8. It served as a model for displaying the interplay between three necessary components required for a fire: fuel, oxygen, and heat. Although scientists initially believed that combustion could be described using only these three factors, it soon became apparent that the model could not adequately account for everything scientists knew about the combustion process.

Fire scientists had discovered that a fourth component—free radicals—also is important during combustion. To visually display the four components, we use the faces of the three-dimensional four-sided figure called the **fire tetrahedron**. It is shown in Figure 5.9. The message conveyed by the fire tetrahedron is that combustion occurs when adequate fuel, oxygen, heat, and free radicals are simultaneously present.

A **free radical** is a molecule that has an unpaired electron. It forms when a covalent bond is broken. Consider the methane molecule. When one of its four chemical bonds is broken or split, we represent the resulting structure as follows:

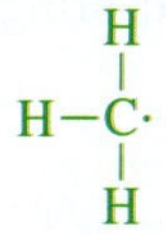

This species is called the methyl free radical. The dot represents the unpaired electron.

Once produced, a free radical rapidly seeks out an atom, a molecule, or another free radical. Because it possesses an unpaired electron, a free radical is an extraordinarily reactive, short-lived species. Nonetheless, during its transient existence, a free radical undergoes many kinds of chemical reactions, which occur in the following four ways:

■ An electron is donated to a stable molecule or atom.
■ An electron is removed from a stable molecule or atom.
■ A group of atoms is removed from a molecule.
■ A free radical adds to a molecule.

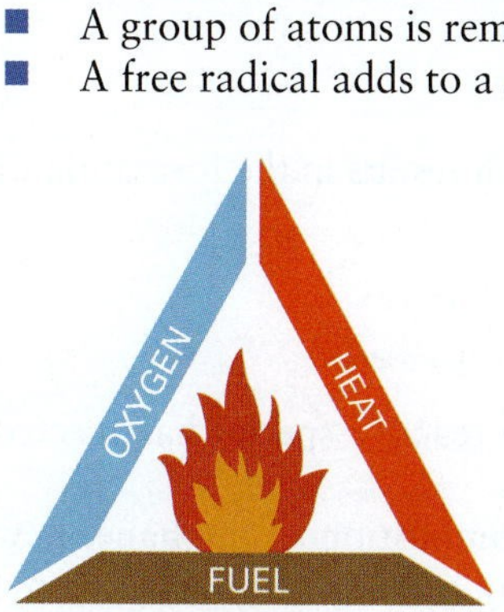

fire triangle ■ A triangle whose legs are used to designate three of the four components of a fire (fuel, oxygen, and heat)

fire tetrahedron ■ A four-sided pyramid, whose faces are used to designate the four components of a fire (fuel, oxygen, heat, and free radicals)

free radical ■ A reactive species resembling a molecule but having an unpaired electron

FIGURE 5.8 This fire triangle once was used to illustrate that the self-sustenance of an ordinary fire required the simultaneous presence of fuel, oxygen, and heat. Today, however, this concept has been broadened into the fire tetrahedron shown in Figure 5.9, which includes free radicals as a fourth component.

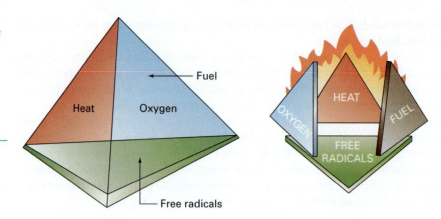

FIGURE 5.9 The fire tetrahedron, the faces of which designate fuel, oxygen, heat, and free radicals. The presence of all four components is essential for the self-sustenance of an ordinary fire.

Suppose we examine the burning of methane to illustrate the importance of free radicals during the combustion process. Chemists now view the combustion of methane as a series of individual reactions, each of which involves free radicals. As combustion proceeds, the products of one reaction activate additional molecules, which then initiate other reactions. This stepwise description of the manner by which reactants are converted into products is called the **mechanism** of the reaction.

In all, the combustion of methane can be summarized by means of a mechanism consisting of the following steps:

mechanism ■ The step-by-step depiction (initiation, propagation, and termination) of the chemical process by which reactants are converted into products

$$CH_4(g) \longrightarrow CH_3 \cdot (g) \longrightarrow HCHO(g) \longrightarrow$$
Methane   Methyl radical   Formaldehyde

$$HCO \cdot (g) \longrightarrow CO(g) \longrightarrow CO_2(g)$$
Formyl radical   Carbon monoxide   Carbon dioxide

chain reaction ■ A multistep reaction in which a reactive intermediate formed in one step reacts during a subsequent step to generate the species needed in the first step

This mechanism is an example of a **chain reaction** that is a series of steps, each of which generates a reactive substance that brings about another step. Let's examine each step more closely.

5.11-A INITIATION

The first mechanistic step involves the production of a methyl free radical and a hydrogen atom, which occurs when a methane molecule first absorbs activation energy from an ignition source.

$$CH_4(g) \longrightarrow CH_3 \cdot (g) \longrightarrow H \cdot (g)$$
Methane   Methyl radical   Hydrogen atom

initiation ■ The first mechanistic step of a chemical reaction during which reactive chemical species like free radicals are produced when heat is applied to a reactant

It is called the **initiation** step of the mechanism.

5.11-B PROPAGATION

The hydrogen atom then reacts with molecular oxygen, which results in the formation of a hydroxyl radical ($\cdot OH$) and an oxygen atom ($\cdot \ddot{O} \cdot$).

Hydrogen atom   Oxygen   Hydroxyl radical   Oxygen atom

propagation ■ The second mechanistic step of a chemical reaction during which reactant species repetitively form new species by means of multiple reactions

This reaction is an example of a **propagation**, because one reactive species has reacted and generated new ones.

During the propagation steps that occur during the combustion of methane, new reactive intermediates, atoms, and molecules are produced by reactions involving methyl radicals, hydroxyl radicals, hydrogen atoms, and oxygen atoms. Examples of these propagation steps are represented by a group of equations like the following:

$$CH_4(g) \ + \ \cdot OH(g) \ \longrightarrow \ H_2O(g) \ + \ CH_3\cdot(g)$$
Methane   Hydroxyl radical   Water   Methyl radical

$$CH_4(g) \ + \ H\cdot(g) \ \longrightarrow \ CH_3\cdot(g) \ + \ H_2(g)$$
Methane   Hydrogen atom   Methyl radical   Hydrogen

$$CH_3\cdot(g) \ + \ \cdot\ddot{O}\cdot(g) \ \longrightarrow \ H-\overset{\displaystyle O}{\overset{\|}{\underset{\underset{\displaystyle H}{|}}{C}}}(g) \ + \ H\cdot(g)$$
Methyl radical   Oxygen atom   Formaldehyde   Hydrogen atom

$$H-\overset{\displaystyle O}{\overset{\|}{\underset{\underset{\displaystyle H}{|}}{C}}}(g) \ + \ CH_3\cdot(g) \ \longrightarrow \ \cdot\overset{\displaystyle O}{\overset{\|}{\underset{\underset{\displaystyle H}{|}}{C}}}(g) \ + \ CH_4(g)$$
Formaldehyde   Methyl radical   Formyl radical   Methane

$$H-\overset{\displaystyle O}{\overset{\|}{\underset{\underset{\displaystyle H}{|}}{C}}}(g) \ + \ \cdot OH(g) \ \longrightarrow \ \cdot\overset{\displaystyle O}{\overset{\|}{\underset{\underset{\displaystyle H}{|}}{C}}}(g) \ + \ H_2O(g)$$
Formaldehyde   Hydroxyl radical   Formyl radical   Water

$$H-\overset{\displaystyle O}{\overset{\|}{\underset{\underset{\displaystyle H}{|}}{C}}}(g) \ + \ H\cdot(g) \ \longrightarrow \ \cdot\overset{\displaystyle O}{\overset{\|}{\underset{\underset{\displaystyle H}{|}}{C}}}(g) \ + \ H_2(g)$$
Formaldehyde   Hydrogen atom   Formyl radical   Hydrogen

$$H-\overset{\displaystyle O}{\overset{\|}{\underset{\underset{\displaystyle H}{|}}{C}}}(g) \ + \ \cdot\ddot{O}\cdot(g) \ \longrightarrow \ \cdot\overset{\displaystyle O}{\overset{\|}{\underset{\underset{\displaystyle H}{|}}{C}}}(g) \ + \ \cdot OH(g)$$
Formaldehyde   Oxygen atom   Formyl radical   Hydroxyl radical

$$\cdot\overset{\displaystyle O}{\overset{\|}{\underset{\underset{\displaystyle H}{|}}{C}}}(g) \ \longrightarrow \ CO(g) \ + \ H\cdot(g)$$
Formyl radical   Carbon monoxide   Hydrogen atom

$$CO(g) \ + \ \cdot OH(g) \ \longrightarrow \ CO_2(g) \ + \ H\cdot(g)$$
Carbon monoxide   Hydroxyl radical   Carbon dioxide   Hydrogen atom

The combination of the reactions illustrated by these equations allows combustion to continue until the supply of the reactants or the reactive intermediates is exhausted.

5.11-C TERMINATION

Finally, the free radicals and atoms combine in **termination** steps similar to the reactions denoted by the following equations:

$$H\cdot(g) \ + \ \cdot OH(g) \ \longrightarrow \ H_2O(g)$$
Hydrogen atom   Hydroxyl radical   Water

$$CO(g) \ + \ \cdot\ddot{O}\cdot(g) \ \longrightarrow \ CO_2(g)$$
Carbon monoxide   Oxygen atom   Carbon dioxide

termination ■ The third mechanistic step of a chemical reaction during which certain reactive species formed during the propagation step unite to produce the product

During the termination steps of the mechanism, two reactive intermediates combine. Because new reactive species are not generated, the combustion process slows or ceases.

5.12 WATER AS A FIRE EXTINGUISHING AGENT

fire extinguishing agent ■ A substance that suppresses or extinguishes fire or the propagation of flame

Water is the most common **fire extinguishing agent**, but its effectiveness generally is limited to class A fires. Nonetheless, it can also be used for extinguishing class B fires in circumstances like the following:

■ Water is effective as a fire extinguishing agent when it is immiscible with a burning liquid and floats on its surface. The water extinguishes the fire by absorbing heat and preventing the escape of vapor from the liquid into the atmosphere.

■ Water is effective as a fire extinguishing agent when it is soluble in the burning fuel. If a flammable liquid is soluble in water, it becomes nonflammable when a sufficient amount of water has been mixed with it. Water extinguishes these fires primarily by reducing the rate at which the fuel vaporizes.

Water is often selected as a fire extinguishing agent because it is usually available in relatively large quantities at most locations. Yet, water is not an all-purpose fire extinguishing agent for the following reasons:

■ The application of water to fires often causes considerable damage; in fact, the damage resulting from the use of water frequently surpasses that caused directly by the fire. Water is not a "clean" fire extinguishing agent, because it leaves a messy residue following its application.

■ The fluidity of water causes it to be highly inefficient as a fire extinguishing agent. Most water that is applied to a fire drains from the scene into adjacent areas, where its exposure to the heat generated by the fire may be so limited that it does not rapidly vaporize.

■ Many parts of the world are cold enough for water to freeze when it is discharged. Although antifreeze agents can be added to keep the water liquid at temperatures as low as –60°F (–51°C), these water solutions usually are corrosive to metals and require the use of special equipment for mixing and discharging them. Antifreeze agents burn; hence, when an insufficient volume of the antifreeze/water mixture is applied to a fire, the water evaporates and the antifreeze agent bursts into flame.

■ Water often is denser than the burning liquids that constitute the fuels of class B fires. When water is applied to burning liquids that are confined within containers or tanks, it sinks below their surface, where it is incapable of absorbing the evolved heat.

■ Water ruins delicate electronic circuitry; thus, it is not recommended as an extinguisher of class C fires. Water containing dissolved mineral salts conducts electricity, which can put firefighters at an unreasonable risk of being electrocuted.

■ Water often reacts with the burning metals that constitute the fuels of class D fires. The products of these reactions may ignite and aid in sustaining the combustion of the metals rather than extinguishing their fires.

relative humidity ■ A measurement in percent by volume at a given temperature of the amount of water vapor present in the air compared with saturation at 100%

Airborne water also acts as a fire extinguishing agent when the moisture content of the air is relatively high. Fires burn less vigorously in this climate compared to when the water content of the air is low. Although the air always contains some water vapor, there is a limit to the amount of water that it contains at a given temperature. When the air attains this limit, it is said to be *saturated*. A measurement of the **relative humidity** indicates the percentage of water vapor present in the air compared with the amount the air would hold if it were 100% saturated at a given temperature. Weather forecasters warn that grassland fires are most likely to occur when the grass is dry, the wind velocity is high, and the relative humidity is low.

5.12-A DISCHARGING WATER AS A FIRE EXTINGUISHING AGENT

Water may be discharged in two ways to extinguish or control fires. When used on small fires, the water may be discharged as either a **streaming agent** or **water mist** from a portable handheld fire extinguisher. For large fires, the water is similarly discharged, but when it is discharged as a mist, the water is supplied through a hose fitted with a misting nozzle or from a **fixed extinguishing system**. The misting nozzle can be selected to provide a water-spray pattern throughout a range of angles from 10 to 90 degrees. Fixed extinguishing systems typically are located in factories and other workplaces to provide exposure protection for personnel, equipment, buildings, storage containers, tanks, and other articles.

The efficiency of water as a fire extinguishing agent is significantly improved when the water is discharged as a mist. As illustrated in Figure 5.10, the increased surface area of the particles causes the water to vaporize more rapidly. Heat is subsequently removed faster from the fire scene because the rate at which the water may vaporize has been increased.

Deionized water is often used as the fire extinguishing agent in portable, handheld fire extinguishers. The use of deionized water provides a degree of safety against experiencing an electrical shock, because deionized water does not conduct an electric current.

In some fire extinguishers, a chemical reaction produces carbon dioxide, which generates an internal pressure sufficient to discharge water from the extinguishers. For example, carbon dioxide was produced by chemical action in the once-popular **soda–acid fire extinguisher**. (*Baking soda* is the common name for sodium bicarbonate.) Separate solutions of baking soda and sulfuric acid were arranged so that when the assembly was inverted or tilted, the two solutions mixed together. The resulting chemical reaction produced carbon dioxide and generated a pressure that forced water through the nozzle.

$$2NaHCO_3(aq) \ + \ H_2SO_4(aq) \longrightarrow Na_2SO_4(aq) \ + \ 2H_2O(l) \ + \ 2CO_2(g)$$

Sodium bicarbonate   Sulfuric acid   Sodium sulfate   Water   Carbon dioxide

It was the water that extinguished the fire, not the carbon dioxide. The standard 2½-gallon soda–acid fire extinguisher shown in Figure 5.11 provided 2½ gallons of water. Because sodium bicarbonate solutions slowly deteriorate over time, soda-acid extinguishers required periodic recharging with fresh solutions.

streaming agent ■ Any fire suppressant, typically liquid, that is applied directly on an unobstructed fire in a stream

water mist ■ Water used as a fire extinguishing agent in the physical form of a mist, fine spray, or fog

fixed extinguishing system ■ A permanently installed system from which a fire extinguishing agent (usually water or carbon dioxide) is discharged upon demand to extinguish or suppress the intensity of a fire

soda–acid fire extinguisher ■ A portable handheld fire extinguisher from which water is expelled at a high rate due to the generation of carbon dioxide by chemical action

Stream   Fog

FIGURE 5.10 A comparison between two ways by which firefighters may apply water when extinguishing a fire. On the left, the water is discharged from a hose as a solid stream, and on the right, it is discharged from a hose as a fog. In the latter case, a larger surface area is provided by the droplets of water, which are able to absorb heat from the fire more effectively compared to the use of the solid stream of water.

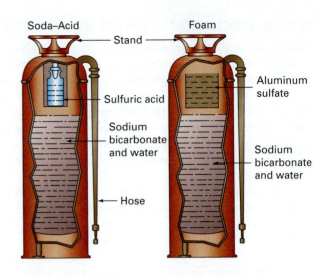

FIGURE 5.11 On the left is the soda–acid fire extinguisher, now relatively obsolete. The extinguisher held sodium bicarbonate and sulfuric acid solutions. When it was inverted, the two solutions mixed and chemically reacted. The carbon dioxide that was produced forced the solution mixture from the nozzle, which the user directed at a fire. On the right is a modification of the soda–acid fire extinguisher, which uses solutions of sodium bicarbonate and aluminum sulfate to produce the carbon dioxide.

SOLVED EXERCISE 5.3

For quenching a fire, discharging a fine mist of water is more effective than discharging water streams on burning items. Why is this so?

Solution: Due to their smaller sizes, the droplets in a water mist evaporate more readily than the water discharged as a stream. These droplets are converted faster into steam, which absorbs a large amount of heat (970 Btu/lb) and cools the burning items below the temperatures at which they ignite.

5.12-B AQUEOUS-FILM-FORMING FOAM

chemical foam ■ Any chemical mixture that produces a gel or foam in which a fire extinguishing agent is physically sealed

aqueous-film-forming foam ■ An aqueous foam prepared by mixing a commercially available concentrate with water, intended for use as a fire extinguishing agent on fires that involve burning flammable, water-insoluble liquids

surfactant ■ A substance that reduces the interfacial tension between two immiscible liquids

To improve the efficiency of a fire extinguishing agent, the agent may be sealed into a **chemical foam**, such as an **aqueous-film-forming foam** (AFFF). Several AFFF concentrates are commercially available for producing the chemical foam on demand. One component of the concentrate is a **surfactant**, a compound capable of increasing the wetting and penetrating ability of water. One AFFF surfactant is perfluorooctanoic acid, whose properties are noted in Section 13.6-B.

An AFFF concentrate can be mixed with fresh, sea, or brackish water in specialized equipment. Firefighters typically use an AFFF to extinguish class B fires involving a bulk volume of a flammable liquid that is water-insoluble, like gasoline and diesel oil. The use of AFFF to fight flight-deck fires aboard ships that carry aircraft is well known. The use of aqueous-film-forming foam is also recommended for extinguishing some stubborn class A fires in upholstery, bedding, paper, hay, and brush.

AFFF is an effective fire extinguishing agent mainly because it functions as a temporary blanketing agent over the burning fuel. Because it produces a barrier between the fuel and the atmosphere, AFFF prevents the fuel's vapors from contacting the atmospheric oxygen needed for combustion; that is, it serves as a vapor-sealing film over the surface of the fuel.

Foam concentrate often is stored in a bladder-tank system. Water pressure squeezes the bladder, which then provides the concentrate at the same pressure. Although normally little maintenance of the system is needed, firefighters soon learn that considerable skill must be developed to effectively use AFFF for extinguishing a major fire. They must apply the foam across the surface of a burning fuel rather than plunging it into the fuel. When it is applied improperly, the foam cannot isolate the burning fuel from atmospheric oxygen, and the fire continues to burn.

There has been some apprehension about using certain aqueous-film-forming foams, because their use can release hydrofluoric acid (Section 8.11) and other fluorides. Their use can also lead to groundwater contamination and failure of the wastewater treatment systems to which they are discharged.

5.12-C ALCOHOL-RESISTANT AQUEOUS-FILM-FORMING FOAM

A versatile fire extinguishing agent is **alcohol-resistant aqueous-film-forming foam** (AR-AFFF). It is widely used for extinguishing fires fueled by polar solvents. Several AR-AFFF concentrates are available commercially. When they are mixed with water and properly dispensed, the resulting foams are used to extinguish class B fires involving either a water-soluble or a water-insoluble flammable liquid. The AR-AFFF fire extinguishing agents are also available in fire extinguishers, in which the AR-AFFF concentrate is already mixed with water for immediate discharge. In this instance, the fire extinguishers are primarily useful for extinguishing relatively small fires. Two examples are illustrated in Figure 5.12.

alcohol-resistant aqueous film-forming foam ■ An aqueous foam prepared by mixing a commercially available concentrate with water, intended for use as a fire extinguishing agent on flammable liquid fires

5.12-D PROTEIN FOAM

Another example of a fire extinguishing foam is **protein foam**. It is prepared from a natural protein material such as soybeans, fish meal, horn-and-hoof meal, or feather meal. It typically contains from 3% to 6% by weight of a protein concentrate as well as a stabilizing agent to provide permanence. Protein foam has a lubricating nature; thus, it is used on airport runways to assist disabled aircraft during landing.

Several types of protein foam are commercially available: regular protein foam, or P; fluoroprotein foam, or FP; alcohol-resistant fluoroprotein foam, or AR-FP; film-forming fluoroprotein foam, FFFP; and alcohol-resistant film fluoroprotein foam, AR-FFFP. Each is used to bring fires of various flammable materials under control.

protein foam ■ A film-forming foam prepared by mixing natural protein materials with water and used to extinguish fires

5.12-E PYROCOOL FIRE EXTINGUISHING FOAM

It is sometimes possible to circumvent the environmental problems associated with the use of ordinary firefighting foams by using products such as Pyrocool FEF. This is a special

FIGURE 5.12 These portable AR-AFFF fire extinguishers are useful when combating relatively small fires within laboratory, research, or industrial settings. Their operation is simple: remove the locking pin, aim the horn at the base of the fire, squeeze the lever to discharge the foam, and release the lever to stop the discharge. (*Courtesy of Badger Fire Protection, Charlottesville, Virginia.*)

foaming agent consisting of a blend of surface-active agents that reduce the interfacial tensions between two liquids or between a liquid and a solid, thus permitting the foam to flow freely and rapidly over a liquid or solid compared with the rate at which water flows.

There are two significant features of Pyrocool FEF that distinguish it from other fire extinguishing foams:

- On application to a burning surface, a slippery foam is produced that spreads quickly and coats the surface. This allows the heat of the fire to transfer quickly to the water so the fire is extinguished rapidly.
- Pyrocool FEF contains additives that absorb the ultraviolet waves emitted from a fire and then reemits them at a lower energy. Thus, application of the foam effectively serves to lower the available heat to sustain the fire.

The use of Pyrocool FEF is purported to be environmentally benign. Following its application to a fire, no hazardous substances are produced that could pollute the nearby environment. In this sense, its application can differ significantly from the use of aqueous-film-forming foam.

5.13 CARBON DIOXIDE AS A FIRE EXTINGUISHING AGENT

At ordinary temperatures and pressures, carbon dioxide is encountered as a colorless, nonflammable gas. Its important physical properties are cited in Table 5.4. Carbon dioxide is often used as a fire extinguishing agent on class B, class C, and even some class A fires, but it is rarely recommended on class D or class K fires.

The effectiveness of carbon dioxide as a fire extinguishing agent is linked with the magnitude of its vapor density. Because carbon dioxide has a vapor density of 1.52, it is about one and a half times denser than air. Thus, when it is applied to a fire, carbon dioxide prevents contact between the burning material and atmospheric oxygen. It also causes a reduction in the concentration of oxygen and/or the gaseous phase of the fuel. Their displacement causes the fire to be smothered.

5.13-A THE PRESSURE–TEMPERATURE RELATIONSHIP FOR CARBON DIOXIDE

phase diagram ■ A plot of the temperature versus pressure of a substance indicating the conditions at which it exists as a gas, liquid, and solid

Figure 5.13 illustrates the pressure–temperature relationship for carbon dioxide in a graphical format called a **phase diagram.** A phase diagram provides the temperature and pressure conditions at which a substance exists as a gas, liquid, and solid. This phase diagram illustrates that at normal room conditions, carbon dioxide exists either as a solid or vapor but not as a liquid. This is a relatively rare property for a substance.

Carbon dioxide may exist as a liquid but not at atmospheric pressure. The lowest pressure and temperature at which carbon dioxide can exist as a liquid within a container

TABLE 5.4	Physical Properties of Carbon Dioxide
Melting point	−70°F (−57°C) (sublimates)
Boiling point at 5.12 atm	−70°F (−57°C)
Specific gravity (gas) at 32°F (0°C) and 1 atm	1.97
Specific gravity (liquid) at 68°F (20°C) and 56 atm	0.770
Specific gravity (dry ice) at −109.3°F(−78.5°C) and 1 atm	1.562
Vapor density (air = 1)	1.52

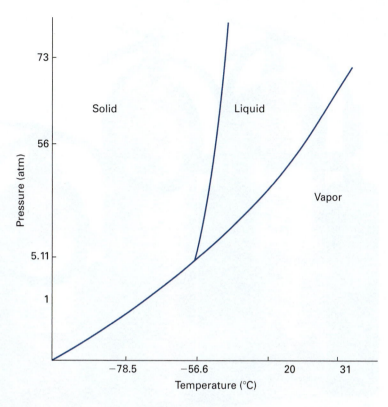

FIGURE 5.13 This phase diagram of carbon dioxide (not to scale) shows that at 1 atmosphere (101.3 kPa), carbon dioxide may exist only as either its vapor or solid called dry ice. Liquid carbon dioxide exists only at pressures equal to or greater than 5.11 atmospheres (517 kPa) and temperatures between −69.9°F (−56.6°C) and approximately 88°F (31°C).

are 3884 psia (518 kPa) and −69.9°F (−56.6°C), respectively. These temperature and pressure conditions are referred to as the **triple point**. At the triple point, the solid, liquid, and gaseous phases of a confined substance coexist in equilibrium. Liquid carbon dioxide can be used not only as a fire extinguishing agent, but also as a refrigerant and dry-cleaning agent.

Whereas most solids liquefy at atmospheric pressure before they vaporize, solid carbon dioxide, called **dry ice**, changes directly from the solid state into its vapor at normal atmospheric conditions. The physical transformation of a substance directly from the solid state to the gaseous state without becoming liquid is called **sublimation**. The substance is said to *sublimate*, or *sublime*. Carbon dioxide sublimates at all pressures less than 14.7 psi (101.3 kPa), regardless of the prevailing temperature.

5.13-B GENERAL FEATURES OF CARBON DIOXIDE AS A FIRE EXTINGUISHING AGENT

When intended for use as a fire extinguishing agent, carbon dioxide is often packaged as a compressed gas within portable extinguishers like those shown in Figure 5.14. Portable handheld extinguishers containing compressed carbon dioxide in capacities from 2 to 25 pounds (0.9 to 10 kg) are available commercially. They are used primarily to suppress or extinguish **incipient-stage fires**.

For fire extinguishing purposes, carbon dioxide may also be discharged from a fixed extinguishing system as a gas from storage tanks in which it is confined as a liquid under pressure. When discharged as the gas, the carbon dioxide is much colder than the surrounding environment. Thus, it functions as a fire extinguishing agent not only by smothering the burning material but also by cooling it.

Unlike water, carbon dioxide is considered a clean fire extinguishing agent. Because it is a colorless gas at normal temperature and pressure conditions, carbon dioxide dissipates into the surrounding atmosphere after use without leaving a visible residue. Nonetheless, when carbon dioxide is discharged from a high-pressure system for total flooding

triple point ■ The temperature and pressure at which the solid, liquid, and gaseous phases of matter coexist in equilibrium for a given confined substance

dry ice ■ Carbon dioxide in the solid phase of matter

sublimation ■ A physical change during which a substance passes from the solid directly to the gaseous state of matter, without first liquefying

incipient-stage fire ■ A fire occurring in its initial or beginning stage

FIGURE 5.14 The use of these portable carbon dioxide fire extinguishers quickly delivers smothering action to flames and suffocates fires supported by atmospheric oxygen. Their operation is simple: remove the locking pin, aim the horn at the base of the fire, squeeze the lever to discharge carbon dioxide, and release the lever to stop the discharge. (*Courtesy of Walter Kidde, Mebane, North Carolina.*)

(Section 5.13-B), the force of the discharge may scatter burning materials to the adjacent environment, where they may kindle secondary fires.

Carbon dioxide generally is applied to fires that occur inside buildings where the atmosphere is calm. Even under this circumstance, however, carbon dioxide may need to be applied repeatedly to fires that have been initially extinguished. The heat generated by the hot material causes the gas to dissipate into the atmosphere. Unless it has cooled sufficiently, the material may reignite upon contacting atmospheric oxygen.

5.13-C CARBON DIOXIDE IN TOTAL-FLOODING SYSTEMS

total flooding ■ The automatic discharge of a fire extinguishing agent from a fixed extinguishing system into an enclosed structure in order to achieve a predetermined concentration for the purpose of fire extinguishment or control

Carbon dioxide often is stored at industrial plants in a fixed extinguishing system for potential use as a fire extinguishing agent by total flooding. The term **total flooding** means that the affected area is literally flooded on demand with carbon dioxide. The following two systems involving the use of carbon dioxide for total flooding are commercially available:

■ In the *high-pressure system*, liquid carbon dioxide is stored under a pressure of approximately 850 psia (5900 kPa) within steel cylinders.
■ In the *low-pressure system*, liquid carbon dioxide is stored in tanks maintained at 0°F (−18°C) by means of refrigeration at a pressure of approximately 300 psia (2100 kPa).

Figure 5.15 illustrates that the carbon dioxide is discharged as the gas even though the substance is stored as the liquid.

The principal danger posed by the use of carbon dioxide during total flooding is the threat that individuals are likely to experience the adverse health effects noted in Table 5.5. Individuals experience these effects at the lower concentrations (1–5% by volume) not because they inhale carbon dioxide *per se*, but because insufficient oxygen is available in the

FIGURE 5.15 Low-pressure carbon dioxide is discharged into an evacuated newspaper pressroom to serve as a method of fire protection. (*Courtesy of Chemetron Fire Systems, Matteson, Illinois.*)

air they breathe for the body to function properly. By contrast, the ill effects resulting from breathing air with carbon dioxide concentrations exceeding 5% are caused directly by the exposure to carbon dioxide.

When discharged into a confined area, a carbon dioxide plume replaces the ground-level air. Because individuals who inhale air containing an elevated concentration of carbon dioxide within the area could lose consciousness, it is critical to first evacuate everyone from the area *before* the carbon dioxide is discharged.

TABLE 5.5	Ill Effects Caused by Inhaling Air Containing Elevated Levels of Carbon Dioxide[a]
PERCENT CARBON DIOXIDE BY VOLUME	**SIGNS AND SYMPTOMS**
1	Breathing rate increases slightly; drowsiness upon prolonged exposure
2	Increased blood pressure and pulse rate; prolonged exposure may cause headache and tiredness; reduced hearing
3	Breathing rate increases to two times the normal rate and becomes labored; impaired hearing; mild headache; sweating; increased blood pressure and pulse rate
4–5	Breathing rate increases to four times the normal rate; stimulation of respiration; dizziness; confusion; sweating; nausea
5–10	Very labored breathing; headache; visual impairment; ringing in the ears; restlessness; disorientation; impaired judgment, followed by loss of consciousness
10–100	Tremors; convulsions; loss of consciousness after 15-min exposure, resulting in death

[a]Adapted in part from "Health Effects of Carbon Dioxide Gas," Canadian Centre for Occupational Health and Safety (2012).

Chapter 5 Principles of Chemical Reactions   **163**

In the workplace, OSHA requires employers to adhere to the following five egress and fire-protection standards when carbon dioxide is used as a fire extinguishing agent in a total-flooding system:

- 29 C.F.R. §1910.36(b)(1) – The extinguishing components and agents in a fixed extinguishing system shall be designed and approved for use on the specific fire hazards they are expected to suppress or extinguish.

- 29 C.F.R. §1910.160(b)(5) – The employer shall post hazard warning or caution signs at the entrance to, and inside of, areas protected by a fixed extinguishing system that use agents in concentrations known to be hazardous to employee safety and health.

- 29 C.F.R. §1910.160(c)(1) – The employer shall assure that inspection and maintenance dates are recorded on the container, or a tag attached to the container, or in a central location. A record of the last semiannual check shall be maintained until the container is checked again or for the life of the container, whichever is less.

- 29 C.F.R. §1910.160(c)(3) – On all total-flooding systems, the employer shall provide an OSHA-approved pre-discharge employee alarm that is capable of being perceived above ambient light or noise levels before the system discharges, to give employees time to safely exit from the discharge area prior to system discharge.

- 29 C.F.R. §1910.160(c)(4) – The employer shall provide automatic actuation of total-flooding systems by means of an approved fire-detection device installed and interconnected with a predischarge employee alarm system to give employees time to safely exit from the discharge area prior to system discharge.

5.13-D CARBON DIOXIDE PRODUCTION BY CHEMICAL ACTION

Carbon dioxide can also be contained as bubbles within a chemical foam. This is another example of a film-forming foam. Here, sodium bicarbonate and aluminum sulfate are stored as separate water solutions or powders. When they are mixed, these substances react to produce carbon dioxide.

$$6NaHCO_3(aq) \; + \; Al_2(SO_4)_3(aq) \; \longrightarrow \; 3Na_2SO_4(aq) \; + \; 2Al(OH)_3(s) \; + \; 6CO_2(g)$$

Sodium bicarbonate · · · · · Aluminum sulfate · · · · · · · · · · · Sodium sulfate · · · · · Aluminum hydroxide · · · Carbon dioxide

Sometimes an extract of licorice root is mixed with these solutions to help keep the bubbles from collapsing. In combination with aluminum hydroxide, a tough coating is formed around each bubble of carbon dioxide. The entire mass emerges as a foam and acts as a wet blanket. When the foam is applied to a fire, it prevents the burning material from contacting atmospheric oxygen.

Some large petroleum refineries and/or their outlying tank farms employ a two-solution wet-foam system for smothering fires that occur in or around their tanks. Solutions of aluminum sulfate and sodium bicarbonate are stored in separate tanks equipped with foam-mixing chambers, usually located at the top ring or roof of these tanks, where they may be combined to produce the chemical foam. When spread on the surface of a burning petroleum product, the foam starves the fire by preventing its contact with atmospheric oxygen.

5.14 HALONS AS FIRE EXTINGUISHING AGENTS

halon ■ A fire suppressant primarily composed of a substance whose molecules have one or two carbon atoms and four to six fluorine, chlorine, or bromine atoms

The **halons** are nonflammable organic compounds whose molecules possess one or two carbon atoms and from four to six halogen atoms. They are halogenated derivatives of methane and ethane, that is, compounds in which one or more halogen atoms have been substituted for the hydrogen atoms in the molecules of methane (CH_4) and ethane (C_2H_6).

Halon 1301 and Halon 1211 are two representative halons. They are halogenated derivatives of methane having the chemical formulas CF_3Br and CF_2ClBr, respectively. In the first formula, three hydrogen atoms in the methane molecule have been substituted with three fluorine atoms and one has been substituted with a bromine atom. In the second, the four hydrogen atoms have been substituted with one chlorine atom, one bromine atom, and two fluorine atoms. Both compounds were once very popular fire extinguishing agents.

The word *halon* was first devised by the U.S. Army Corps of Engineers as an abbreviation for <u>hal</u>ogenated hydrocarb<u>on</u>, a class of organic compounds we visit again in Chapter 12. The commercial halons are named by using the word *Halon* and a three- or four-digit number. A halon whose formula is $C_aF_bCl_cBr_d$ is named commercially *Halon abcd*, where *a*, *b*, *c*, and *d* are integers. When the numerical value of *d* is 0, it is dropped.

The most broadly used halons were Halon 1301 and Halon 1211. Both were more effective at extinguishing class A, class B, and class C fires than either water or carbon dioxide. Halon 1301 was primarily discharged for total flooding from a fixed extinguishing system, whereas Halon 1211 was mainly discharged from portable, handheld fire extinguishers as a streaming agent.

Chemists name the halons as halogenated derivatives of methane or ethane. The halogen atoms are named by replacing the *-ine* suffix on the name of the halogen with *-o*. The number of halogen atoms is indicated by the use of *mono-*, *di-*, *tri-*, and *tetra-*, as relevant, for 1, 2, 3, and 4, respectively, and when multiple halogens are part of the same substance, they are named alphabetically. The prefix *mono* is often avoided when the number of halogen atoms is clearly one. Thus, the chemical names of the halons having the chemical formulas CF_3Br and CF_2ClBr are bromotrifluoromethane and bromochlorodifluoromethane, respectively.

Bromotrifluoromethane
(Halon 1301)

Bromochlorodifluoromethane
(Halon 1211)

Collectively, the halons possess two properties that caused them to be effective fire extinguishing agents:

- Their vapors are much denser than air. This means that when a halon is discharged on a fire, its vapor acts as a fire extinguishing agent by smothering it.
- The halons produce free radicals and atoms at the elevated temperatures experienced during fires. These reactive species scavenge the combustion intermediates produced during class B fires. For instance, when Halon 1301 is applied to a fire, trifluoromethyl radicals and bromine atoms are produced.

$$CF_3Br(g) \longrightarrow F-\underset{F}{\overset{F}{\vert}}C\cdot(g) \ + \ Br\cdot(g)$$

These bromine atoms have a far greater affinity for reacting with combustion intermediates than the combustion intermediates have for reacting with one another.

The halons were once popular fire extinguishing agents for the following reasons:

- They effectively extinguished class A, class B, and class C fires.
- They performed as clean fire extinguishing agents by dissipating into the atmosphere following their use without leaving a residue, thereby avoiding secondary property damage. This was a highly desirable feature when assessing the protection of delicate and expensive electronic equipment, computer circuitry, and aircraft interiors.
- They could not conduct electricity.
- Human exposure to elevated concentrations of the halons may cause dizziness, impaired coordination, and a reduction in mental sharpness. When discharged with proper exposure controls, however, the halons posed only a minimum health risk to firefighters and other personnel. This was a highly desirable feature when considering their use within enclosed structures like aircraft interiors.

The halons were especially suitable for extinguishing fires in vaults, museums, libraries, hospitals, large computer and data storage facilities, and the engine rooms aboard ships. They were also used in military and civilian planes to extinguish fires in engines, cargo bays, cockpits, passenger compartments, and washrooms. Onboard aircraft, Halon 1301 and Halon 1211 were primarily discharged from portable, handheld fire extinguishers to suppress fires in cargo compartments, engine nacelles, and washrooms.

Using the authority of the Clean Air Act, EPA banned the manufacture and import of virgin halons in January 1994, but does not prohibit the import of *reclaimed* halons into the United States. The use of reclaimed halon fire extinguishing agents is now severely limited for all but the most critical needs although their use is still permitted in any fire-protection system as long as there is no safe and effective substitute for their application and the recharged system is properly maintained.

Recycled halons and inventories manufactured before January 1994 are now the sole sources for supplying and recharging fixed extinguishing systems. The most common contemporary users of recycled halon fire extinguishing agents are the aviation, military, and aerospace industries. When the halon fire extinguishing agents are used in the workplace in fixed extinguishing systems, OSHA requires employers to adhere to the regulations published at 29 C.F.R. §1910.162.

Because the halons were such extraordinarily effective fire extinguishing agents, why has their use been severely curtailed under federal law? The answer to this question is linked with their ability to deplete the ozone that occupies a stratospheric layer lying at the edge of Earth's atmosphere. Ozone is a form of elemental oxygen having three atoms of oxygen per molecule instead of the usual two. As we examine in more detail in Section 7.1-M, the ozone that exists as a blanket around Earth protects life from overexposure to harmful ultraviolet radiation.

When most substances are released into the atmosphere, they ultimately undergo chemical reactions that convert them into relatively innocuous substances. The halons, however, are very stable compounds that do not readily undergo conversion reactions. When released into the environment, their vapors migrate upward and, ultimately, react with the stratospheric ozone. The combination of countless reactions between ozone and compounds like the halons has resulted in depletion of the stratospheric ozone. Scientists fear that this depletion has allowed excessive ultraviolet radiation to penetrate to the surface of our planet, where it contributes to the incidence of skin cancer and cataracts and causes untold harm in other ways.

As noted earlier, EPA banned the production and manufacture of halons as fire extinguishing agents since January 1994. Although EPA permits the supplies of halons that existed before the ban to be recycled and reused, the ban has substantially reduced the worldwide use of halons as fire extinguishing agents.

Although EPA has also approved the use of certain substances and blends as halon alternatives (Section 12.15-C), international efforts continue to curb the use of the halons and other substances that contribute to depletion of stratospheric ozone. Most likely, generations will pass before this threat to the ozone layer terminates.

5.15 DRY-CHEMICAL FIRE EXTINGUISHING AGENTS

dry-chemical fire extinguishing agent
■ Any solid substance that functions by removing one or more fire components from the fire scene

The following chemical substances generally are used in the solid state as **dry-chemical fire extinguishing agents**:

- ■ Sodium bicarbonate ($NaHCO_3$)
- ■ Potassium bicarbonate ($KHCO_3$)
- ■ Sodium chloride ($NaCl$)
- ■ Potassium chloride (KCl)
- ■ Monoammonium phosphate ($NH_4H_2PO_4$)

Some dry-chemical fire extinguishing agents effectively extinguish or suppress the intensity of class B and class C fires; others are useful on class D fires; and some extinguish class A, class B, and class C fires. They are not clean fire extinguishing agents, because messy residues remain following their application.

Dry-chemical fire extinguishing agents are corrosive to aluminum and other metals; furthermore, once discharged, they impair visibility. Consequently, their use during aircraft fires is not warranted unless no other fire extinguishing agent is immediately available. The use of dry-chemical fire extinguishing agents during aircraft fires also damages the airplane's structure and sensitive electrical and electronic systems.

The dry-chemical fire extinguishing agents are commercially available for use in portable cylinders and fixed extinguishing systems. When they are intended for use in the workplace in fixed extinguishing systems, OSHA requires employers to adhere to the regulations published at 29 C.F.R. §1910.161.

5.15-A ALKALI METAL BICARBONATES

Fire extinguishers containing sodium bicarbonate and potassium bicarbonate are sometimes referred to as *regular* dry-chemical fire extinguishers. Potassium bicarbonate is the fire extinguishing agent in the commercial product *Purple K*. Although potassium bicarbonate is a white compound, Purple K assumes its color from a violet-colored pigment additive.

Although ineffective on class A fires, both sodium bicarbonate and potassium bicarbonate effectively suppress and extinguish class B flammable liquid fires. They are also used during initial tactics when fighting a fire involving a flammable gas.

The effectiveness of the alkali metal bicarbonates as fire extinguishing agents is related in part to their ability to produce carbon dioxide when heated.

$$2NaHCO_3(s) \longrightarrow Na_2CO_3(s) + CO_2(g) + H_2O(g)$$

Sodium bicarbonate   Sodium carbonate   Carbon dioxide   Water

They use the heat that evolves from the fire to decompose. The carbon dioxide that is produced smothers the fire.

In addition, sodium atoms are produced when these compounds are exposed to high temperatures. They react with other reactive chemical species to produce water. For example, during fires involving petrochemical fuels, among the reactive species produced are hydroxyl radicals ($\cdot$OH). When alkali metal bicarbonates are used to extinguish the fires, sodium atoms react with these hydroxyl radicals by the following stepwise process:

$$Na\cdot(g) + \cdot OH(g) \longrightarrow NaOH(g)$$

Sodium atoms   Hydroxyl radicals   Sodium hydroxide

$$NaOH(g) + \cdot H(g) \longrightarrow H_2O(l) + Na\cdot(g)$$

Sodium hydroxide   Hydrogen atoms   Water molecules   Sodium atoms

The overall reaction produces water.

$$H\cdot(g) + \cdot OH(g) \longrightarrow H_2O(l)$$

Hydrogen atoms   Hydroxyl radicals   Water molecules

The sodium atoms serve to catalyze the production of water and remove the reactive free radicals.

5.15-B ABC FIRE EXTINGUISHING AGENT

Another dry-chemical fire extinguishing agent is monoammonium phosphate, also known as ammonium dihydrogen phosphate. This substance is employed in the multipurpose **ABC fire extinguisher** shown in Figure 5.16. It has also been used since the 1950s for

ABC fire extinguisher ▪ A multipurpose extinguisher containing monoammonium phosphate, which functions primarily by removing heat and free radicals from the scene of an class A, class B, or class C fire

FIGURE 5.16 Easily portable, this ABC dry-chemical fire extinguisher containing monoammonium phosphate is recommended for use on fires within the home and small manufacturing and process facilities. The letters *ABC* imply that its use effectively suppresses most class A, class B, and class C fires. The user holds the unit upright, pulls the ring pin, and aims the hose at the base of the fire while squeezing the lever and sweeping the spray from side to side. (*Courtesy of Walter Kidde, Mebane, North Carolina.*)

fighting major forest fires. In the latter case, a solution of monoammonium phosphate and other additives is applied aerially either ahead of the fire line or directly upon the fire. Figure 5.17 shows that aircraft are used to apply this fire extinguisher, often in areas otherwise inaccessible to ground-based firefighters. The solution is dyed red to aid pilots in selecting the spot where it can be dropped for most efficient use.

The reason for the effectiveness of monoammonium phosphate as a fire extinguishing agent is twofold:

■ When applied to a burning material, monoammonium phosphate decomposes endothermically as follows:

$$4NH_4H_2PO_4(s) \longrightarrow P_4O_{10}(s) + 4NH_3(g) + 6H_2O(g)$$

Monoammonium phosphate   Tetraphosphorus decoxide   Ammonia   Water

Because it absorbs heat as it decomposes, monoammonium phosphate cools the combustible material below the minimum temperature at which it can burn.

■ The ammonia generated by the thermal decomposition of monoammonium phosphate reacts with certain atoms and free radicals. For example, ammonia reacts with hydroxyl radicals produced when petroleum fuels burn. The reaction effectively removes the free radicals.

Despite its general effectiveness, monoammonium phosphate is not a clean fire extinguishing agent because a messy residue remains following its use. Furthermore, if the agent contacts water, the subsequent reaction forms phosphoric acid (Section 8.12), a corrosive material that damages aluminum and other metals.

$$NH_4H_2PO_4(s) + H_2O(l) \longrightarrow H_3PO_4(aq) + NH_4OH(aq)$$

Monoammonium phosphate   Water   Phosphoric acid   Ammonium hydroxide

As noted earlier, it is the corrosive nature of the ABC fire extinguisher and other dry-chemical fire extinguishing agents that warrants its nonuse on aircraft fires.

FIGURE 5.17 Helicopters and air tankers are used to aerially apply fire extinguisher on unburned vegetation during range and forest fires. Because the retardant is dyed red, air crews are able to apply it as directed. The retardant decreases the rate of combustion of trees, brush, and grass. (*Courtesy of Andrea Booher/Federal Emergency Management Agency, Washington, DC.*)

5.16 FIRE EXTINGUISHING AGENTS FOR CLASS D FIRES

Commercial chemical products known as **dry-powder fire extinguishing agents** have been manufactured specifically to effectively suppress class D fires. Most are composed primarily of graphite. The graphite-based dry powders extinguish the fires of combustible metals such as magnesium and aluminum. They are applied on the burning metal by using a scoop or shovel.

Although graphite-based fire extinguishing agents are successful for suppressing class D fires, caution must be exercised when they are used. When applied to a combustible metal fire, graphite produces water-reactive metallic carbides like magnesium carbide and aluminum carbide. On exposure to water, the metallic carbides generate flammable gases including methane and acetylene. For this reason the water reactivity of metallic carbides must be considered when removing the residues following the use of graphite-based dry powders on combustible metal fires. We revisit this issue in Chapter 9.

dry-powder fire extinguishing agent ■ Any extinguisher usually containing graphite powder that functions by removing heat from the scene of a class D fire

5.17 FIRE EXTINGUISHING AGENTS FOR CLASS K FIRES

As we first noted in Section 5.10, a mixture of either potassium acetate or potassium carbonate with water is recommended for use as a fire extinguishing agent on kitchen fires. It is an example of a **wet-chemical fire extinguishing agent**. A commercially available dispenser of this mixture is shown in Figure 5.18.

Kitchen fires are class K fires when they involve the burning of fats, oils, and grease within cooking appliances. The addition of potassium acetate or potassium carbonate converts these fats, oils, and greases into noncombustible potassium soaps and glycerol.

wet-chemical fire extinguishing agent ■ An extinguisher consisting of a substance dissolved in water that is capable of forming a soapy foam blanket when applied to burning oil

FIGURE 5.18 The use of this wet-chemical fire extinguisher is recommended on class K fires. The user holds the unit upright, pulls the ring pin, and aims the hose at the base of the fire while squeezing the lever and discharging the contents on the burning surface. (*Courtesy of Walter Kidde, Mebane, North Carolina.*)

saponification ■ The conversion of triglycerides into soaps

This type of chemical reaction is called **saponification**. Fats and oils contain complex organic compounds called *triglycerides*. When they saponify, glycerol and soap are produced by means of two chemical reactions that occur simultaneously. Using potassium carbonate as an example, potassium hydroxide is first produced.

$$K_2CO_3(aq) \ + \ 2H_2O(l) \ \longrightarrow \ 2KOH(aq) \ + \ CO_2(g) \ + \ H_2O(g)$$

Potassium carbonate · · · Water · · · Potassium hydroxide · Carbon dioxide · · · Water

Then, the triglycerides react with the potassium hydroxide to form soaps. Although the latter chemical reactions are relatively complex, we can simplify their nature as follows:

$$\text{triglycerides} \ + \ KOH(aq) \ \longrightarrow \ \begin{array}{c} CH_2{-}OH \\ | \\ CH{-}OH(l) \\ | \\ CH_2{-}OH \end{array} \ + \ \text{soaps}$$

Potassium hydroxide · · · · Glycerol

The potassium soaps extinguish class K fires because they do not burn. Regardless of their success, however, wet-chemical fire extinguishing agents leave messy residues following their application and are not considered clean fire extinguishing agents.

Writing and Balancing Chemical Equations

1. Carbon dioxide is removed from the air onboard the International Space Station by using lithium hydroxide filters. Write the equation that describes this reaction.
2. Convert the information described below into balanced equations:
   (a) When a mixture of copper(II) oxide and carbon is heated, elemental copper forms and carbon monoxide evolves into the atmosphere.
   (b) When a concentrated solution of sodium hydroxide is added to an aqueous solution of lead acetate, lead hydroxide forms as a solid dispersed in a solution of sodium acetate.
   (c) When aqueous solutions of cupric chloride and sodium phosphate are mixed, solid cupric phosphate separates from a solution of sodium chloride.
   (d) When heated in an aqueous solution of ammonium acetate, calcium oxide reacts to form calcium acetate, ammonia, and water.
3. Balance each of the following equations:
   (a) $CaCl_2(aq) + Na_2CO_3(aq) \longrightarrow CaCO_3(s) + NaCl(aq)$
   (b) $Br_2(l) + H_2SO_3(aq) + H_2O(l) \longrightarrow H_2SO_4(aq) + HBr(aq)$
   (c) $C_2H_2(g) + O_2(g) \longrightarrow CO_2(g) + H_2O(g)$
   (d) $NH_3(g) + CuO(s) \longrightarrow N_2(g) + Cu(s) + H_2O(g)$
   (e) $HClO_4(conc) \longrightarrow Cl_2(g) + O_2(g) + H_2O(g)$
   (f) $CrCl_3(aq) + KCN(aq) \longrightarrow Cr(CN)_3(s) + KCl(aq)$
   (g) $HgSO_4(s) + Hg(l) + NaCl(s) \longrightarrow Hg_2Cl_2(s) + Na_2SO_4(s)$
   (h) $Fe_2O_3(aq) + HNO_3(aq) \longrightarrow Fe(NO_3)_3(aq) + H_2O(l)$
   (i) $HNO_3(conc) \longrightarrow NO_2(g) + O_2(g) + H_2O(l)$
   (j) $NH_3(g) + O_2(g) \longrightarrow NH_4NO_2(s) + H_2O(g)$
   (k) $P_4O_{10}(s) + H_2O(l) \longrightarrow H_3PO_4(aq)$
4. Supply the name of the reactant or product in the blanks that appear in each of the following word equations, and convert them into balanced equations:
   (a) Sulfur + _____ $\longrightarrow$ disulfur dichloride
   (b) Calcium hypochlorite + _____ $\longrightarrow$ calcium carbonate + sodium hypochlorite
   (c) Potassium oxide + _____ $\longrightarrow$ potassium hydroxide
   (d) _____ + hydrochloric acid $\longrightarrow$ iron(III) chloride + hydrogen
   (e) _____ + strontium nitrate $\longrightarrow$ strontium hydroxide + sodium nitrate
   (f) Magnesium + sulfuric acid $\longrightarrow$ magnesium sulfate + _____
   (g) Silver nitrate + _____ $\longrightarrow$ silver sulfide + nitric acid
   (h) Copper(II) sulfite $\longrightarrow$ _____ + sulfur dioxide
5. Provide the formula for each substance noted below and balance the equations:
   (a) Phosphine + oxygen $\longrightarrow$ phosphoric acid
   (b) Lead acetate + barium chloride $\longrightarrow$ lead chloride + barium acetate
   (c) Ammonia + hydrogen sulfide $\longrightarrow$ ammonium sulfide
   (d) Sulfur tetrachloride + water $\longrightarrow$ sulfurous acid + hydrogen chloride
   (e) Bismuth + sulfuric acid $\longrightarrow$ bismuth sulfate + sulfur dioxide + water
   (f) Zinc + nitric acid $\longrightarrow$ zinc nitrate + nitrogen monoxide + water

Factors Affecting Reaction Rates

6. Among the various factors that influence reaction rates, identify which is involved in each of the following observations:
   (a) When heated in a flame and then placed in oxygen, iron wire ignites, whereas when an iron rod having a diameter of 0.4 inch (1 cm) is also heated to the same temperature and placed in oxygen, it does not ignite.
   (b) When heated in a flame and then placed in oxygen, iron wire ignites, whereas when heated in a flame and then exposed to the air, it does not ignite.
   (c) Gasoline ignites more easily in the combustion chambers of an automobile when the ambient surroundings are warm than when they are cold.
   (d) A stack of wooden logs lying exposed to the air does not burn until a single log in the stack first has been kindled.

Combustion Phenomena

7. Charcoal briquettes consist almost entirely of elemental carbon. When they burn in the open air, carbon monoxide and carbon dioxide are produced. Write balanced chemical equations illustrating the production of each combustion product.
8. Sooty plumes generally billow from petroleum tank fires. What does the presence of these sooty plumes generally indicate about the nature of the combustion phenomena?
9. Why should vendors exercise special caution when storing bulk quantities of nuts, especially pistachios, in confined containers?

Global Warming

10. In what major way has global warming affected firefighting in the western United States?

Classification of Fires

11. Identify the class of fire associated with the ignition and burning of each of the following:
    (a) Vegetable oil
    (b) Automobile tires
    (c) Aluminum powder
    (d) Energized electrical circuit

Chemistry of Fire Extinguishing Agents

12. The OSHA regulation at 29 C.F.R. §1910.162(b)(5) stipulates that employers must provide a predischarge-employee alarm system within the workplace to alert employees if a concentration of 4% or greater of carbon dioxide is released from a fixed extinguishing system. What is the most likely reason OSHA enacted this regulation?
13. At 14 C.F.R. §121.265, FAA permits the use of carbon dioxide as a fire extinguishing agent onboard aircraft, provided that provisions are made to prevent harmful concentrations from entering any personnel compartment. Use Table 5.5 to identify the adverse health effects likely to be experienced by a crew member who inhales air containing 8% carbon dioxide for a 5-minute period?
14. The chemical formula of methyl bromide is CH_3Br. Why does methyl bromide perform effectively as a fire extinguishing agent?

CHAPTER

6

Use of the DOT Hazardous Materials Regulations by Emergency Responders

Courtesy of the U.S. Department of Transportation, Washington, DC.

KEY TERMS

OBJECTIVES

- Illustrate how shippers of hazardous materials use the Hazardous Materials Table at 49 C.F.R. §172.101 to prepare a shipping description.
- Identify the specific information emergency responders obtain from hazardous material shipping descriptions.
- Memorize the colors and inscriptions on the DOT labels and placards corresponding to all hazard classes and divisions.

- Identify the information that DOT requires shippers to mark on the packaging used to transport hazardous materials.
- Memorize the hazard classes and divisions of the hazardous materials whose transport vehicles always require placarding; and those that require placarding only when the amount of the materials equals or exceeds 1001 pounds (454 kg).
- Demonstrate how first-on-the-scene responders use UN/NA identification numbers and DOT's *Emergency Response Guidebook* to mitigate the impact of a transport mishap involving hazardous materials.

The Pipeline and Hazardous Materials Safety Administration, a division of the U.S. Department of Transportation (DOT), regulates hazardous material shipments for the following major reasons:

- To protect transportation personnel and equipment from exposure to hazardous materials, and
- To identify the nature of the shipment to emergency-response personnel for their use during a transportation mishap.

The regulations promulgated by DOT now conform to the generally accepted international standards represented by the United Nation's requirements of the Globally Harmonized System of Classification and Labeling of Chemicals, or *GHS* (Section 1.9). Consequently, their proper implementation potentially aids emergency-response personnel, the public, and transportation workers worldwide.

The material presented in this chapter aims to demonstrate how emergency responders may use the hazardous materials regulations when they encounter a spill or leak of a hazardous material during a transportation mishap. Because the regulations are subject to change from time to time, the parties affected by their use should always consult 49 C.F.R. Parts 171 through 199 for the latest revisions and amendments.

6.1 DOT HAZARDOUS MATERIALS AND THEIR PROPER SHIPPING NAMES

A **hazardous material** is uniquely defined by DOT as follows:

> a designated substance or material that has been determined by the Secretary of Transportation to be capable of posing an unreasonable risk to health, safety, and property when transported in **commerce**.

The hazardous materials have designated names that are listed in the **Hazardous Materials Table** at 49 C.F.R. §172.101, available via the Internet at www.gpoaccess.gov/cfr/index.html. The table has more than 16,000 entries. In addition, the term *hazardous material* includes hazardous wastes (Section 1.3-C), hazardous substances (Section 1.3-D), marine pollutants and severe marine pollutants (Section 6.2-C), and elevated-temperature materials (Section 6.5-B). The majority of the regulated hazardous materials are identified in the table by their chemical names, but some are identified generically with expressions like "flammable liquids, toxic, n.o.s." (not otherwise specified) and "elevated-temperature liquid, flammable, n.o.s." or with **technical names** like "tear gas substances, liquid, n.o.s." and "organophosphorus pesticides, solid, toxic." DOT may regulate their transportation in certain instances only when they are shipped either domestically or internationally (but not both), or by means of aircraft or watercraft.

Although there are 10 columnar entries in the Hazardous Materials Table, only the entries in columns 1 through 7 are important to emergency responders. An excerpt of the Hazardous Materials Table listing the information in the first seven columns only is provided in Appendix C.

hazardous material ■ A substance or material in any form that constitutes a present or potential threat to human health, safety, welfare, or the environment when improperly stored, treated, transported, disposed or used, or otherwise managed

commerce ■ Trade, traffic, transportation, importation, exportation, and similar activities engaged in by shippers and carriers

Hazardous Materials Table ■ The compilation of hazardous materials whose transportation is regulated by DOT at 49 C.F.R. §172.101

technical name ■ Any recognized name in current use in scientific and technical handbooks and textbooks

6.1-A PROPER SHIPPING NAMES

The **proper shipping name** of a hazardous material appears in column 2 of the Hazardous Materials Table. The proper shipping name is the name that appears in plain, nonitalic type, *not* the name sometimes listed in italics or any other name used to describe the material. DOT allows shippers to select a proper shipping name from several general descriptions in the event that the correct technical name of a material is neither listed in the table nor entirely accurate. The general descriptions often use the letters *n.o.s., n.o.i.* (not otherwise indexed), or *n.o.i.b.n.* (not otherwise indexed by name).

Shippers describe a particular hazardous material by selecting the name from the Hazardous Materials Table that most accurately describes the material to be transported. Suppose a shipper desires to transport ten 55-gallon drums containing industrial heating oil. DOT requires the shipper to identify this commodity as "Fuel Oil" on a shipping paper accompanying the shipment, because the Hazardous Materials Table provides only this name as the closest description of the commodity. An arbitrary selection of "Oil," "Heating Oil," or any other similar term does not comply with the intent of these regulations. DOT does allow the shipper to include additional information parenthetically, but only *in addition to* the proper shipping name. In this example, the commodity may be correctly noted as "Fuel Oil (Fuel Oil No. 2)" when this parenthetical entry constitutes the fuel's closest commercial description. On occasion, the word "forbidden" appears in column 3 of the Hazardous Materials Table. The entry indicates that the hazardous material is a **forbidden material**, that is, a substance whose transportation is prohibited by DOT by any mode. Most forbidden materials are explosive materials, readily combustible solids, self-heating materials, or self-reactive materials.

6.1-B HAZARD CLASSES AND DIVISIONS

Each hazardous material is associated with a specific hazard class that is characterized by a unique defining criterion. There are nine hazard classes, each symbolized by a number 1 through 9.

Several hazard classes are further divided into divisions. Each division is symbolized by two numbers: the hazard class number and a digit separated by a period, like 1.5. Table 6.1 lists the DOT hazard class numbers and their divisions, if any, that DOT assigns to hazardous materials as well as the names of the classes and divisions associated with them.

There is also an additional unnumbered hazard class that is represented as **other regulated materials**, or **ORM**. When shippers intend to transport them domestically, they use the designation **ORM-D**. These materials pose a limited hazard during transportation due to their forms, quantities, and packaging. They are generally **consumer commodities** like drugs, medicines, hair spray, and other items used for personal care in the home.

Although many hazardous materials are characterized by a single hazard or division class, some hazardous materials comply with the defining criteria of more than one hazard or division class. In such instances, DOT refers to their major hazard class as the **primary hazard class** or **division**. All other hazards are called their **subsidiary hazards**.

The primary and subsidiary hazard classes or divisions of a hazardous material are noted in columns 3 and 6 of the Hazardous Materials Table. Column 6 is headed "Label Codes." We note in Section 6.4 that these codes also identify the specific warning labels that shippers affix to hazardous material packaging.

Let's consider an example. As noted in column 3 of Appendix C, compressed oxygen—a nonflammable gas—has a primary division number of 2.2. However, oxygen also supports combustion, which is the defining criterion of an oxidizer. Hence, compressed oxygen has a subsidiary division number of 5.1. Both the primary and subsidiary division numbers are listed in column 6 of Appendix C.

proper shipping name
■ For purposes of DOT regulations, the name in Roman print assigned to a hazardous material listed in the Hazardous Materials Table published at 49 C.F.R. §172.101

forbidden material
■ For purposes of DOT regulations, a hazardous material listed in the Hazardous Materials Table at 49 C.F.R. §172.101 with the word "forbidden" in column 3, signifying that the material may not be transported by any mode

other regulated material (ORM) ■ For purposes of DOT regulations, a hazardous material that does not conform to the definition of any of the nine hazard classes but is nonetheless considered to possess a dangerous property that could pose a risk to transportation personnel and equipment when it is transported

ORM-D ■ For purposes of DOT regulations, any "other regulated material" whose transportation is subject to certain regulations when transported domestically

consumer commodity ■ Any material that is packaged and distributed in a form intended or suitable for sale through retail sales agencies for utilization by individuals for the purpose of personal care or household use

primary hazard class (division) ■ For purposes of DOT regulations, any of the nine categories of hazard assigned to a hazardous material

TABLE 6.1 Hazard Classes and Divisions of Hazardous Materials[a]

HAZARD CLASS NUMBER	DIVISION NUMBER (IF ANY)	NAME OF CLASS OR DIVISION
1	1.1	Explosives (with a mass explosion hazard)
1	1.2	Explosives (with a projection hazard)
1	1.3	Explosives (with predominantly a fire hazard)
1	1.4	Explosives (with no significant blast hazard)
1	1.5	Very insensitive explosives; blasting agents
1	1.6	Extremely insensitive detonating substances
2	2.1	Flammable gas
2	2.2	Nonflammable compressed gas
2	2.3	Poison gas
3		Flammable and combustible liquid
4	4.1	Flammable solid
4	4.2	Spontaneously combustible material
4	4.3	Dangerous-when-wet material
5	5.1	Oxidizer
5	5.2	Organic peroxide
6	6.1	Poisonous materials
6	6.2	Infectious substance
7		Radioactive material
8		Corrosive material
9		Miscellaneous hazardous material
		Other regulated material (ORM)

[a]40 C.F.R. §173.2.

subsidiary hazard ■ For purposes of DOT regulations, any hazard of a material (other than its primary hazard) whose hazard class numbers are listed in column 6 of the Hazardous Materials Table published at 49 C.F.R. §172.101

UN/NA identification number ■ For purposes of DOT regulations, the number assigned to a hazardous material and listed in column 4 of the Hazardous Materials Table published at 49 C.F.R. §172.101

Emergency Response Guidebook ■ A DOT publication that provides emergency actions for implementation at hazardous materials transportation mishaps

6.1-C UN/NA IDENTIFICATION NUMBERS

DOT uses a **UN/NA identification number** to categorize each hazardous material. It is a four-digit number preceded by either the prefix "UN" (United Nations) or "NA" (North America). The identification numbers preceded by "UN" are used when transporting hazardous materials domestically *and* internationally, whereas identification numbers preceded by "NA" are used when transporting certain hazardous materials domestically and within Canada *only*. The "UN/NA" identification number is listed in column 4 for each hazardous material denoted in the Hazardous Materials Table.

The "UN/NA" identification number is assigned to a hazardous material to rapidly convey important information concerning the consignment to emergency-response teams. When the team members know the "UN/NA" identification number, they may quickly identify the physical and chemical nature of a hazardous material and respond effectively to the incident in which the material is involved by referring to a guidebook published by DOT, the ***Emergency Response Guidebook*** (ERG). In this guidebook, each "UN/NA" identification number is referenced to a guide number that provides specific instructions relating to appropriate response-actions for a given listing. We examine the use of this guidebook in Section 6.7.

6.1-D PACKING GROUPS

In most instances, DOT assigns a **packing group** to each hazardous material. This term represents a grouping of hazardous materials according to the collective degree of hazard that is posed by exposure to them. There are only three packing groups. Packing Group I is assigned to those hazardous materials associated with a great degree of danger; Packing Group II, a medium degree of danger; and Packing Group III, a minor degree of hazard.

The packing group of a hazardous material appears as the Roman numeral I, II, or III in column 5 of the Hazardous Materials Table. The packing group affects the nature of the packaging that shippers select when they transport a hazardous material. Packing groups for hazard classes 2 and 7, Division 6.2 (other than regulated medical wastes), and ORM-D do not have packing groups, because hazardous materials in these hazard classes require their own specially approved packaging.

DOT allows certain hazardous materials within a single hazard class or division number to be shipped in packaging associated with each of the three packing groups. For example, shippers or carriers may transport flammable liquids (hazard class 3) in Packing Group I, II, or III in accordance with the following flashpoint and boiling point data:

Packing Group I	Initial boiling point equal to or less than 95°F (≤35°C)
Packing Group II	Flashpoint less than 73°F (≤23°C); initial boiling point greater than 95°F (>35°C)
Packing Group III	Flashpoint equal to or greater than 73°F (>23°C) but equal to or less than 140°F (≤60°C); initial boiling point greater than 95°F (>35°C)

6.1-E PACKAGING

Packaging is defined by DOT as the receptacle and any other components or materials necessary for the receptacle to perform its containment function and to ensure compliance with packing requirements. It represents the **container** used to handle a hazardous material for a designated purpose, like movement from one location to another. A container is the bag, barrel, bottle, box, can, cylinder, drum, reaction vessel, tank (other than a fixed storage tank), or other device into which a hazardous material is confined. The packaging together with its contents is called the **package**.

There are three types of performance-oriented containers for transporting hazardous materials:

- **Nonbulk containers,** such as bags, bottles, boxes, cans, drums, barrels, carboys, cylinders, and jerricans, sometimes are encountered as inner packages that have been overpacked in crates to provide an additional measure of safety during transit. A nonbulk container has a maximum capacity of 119 gallons (450 L) as a receptacle for a liquid; a maximum net mass less than 882 pounds (<400 kg) and a maximum capacity of 119 gallons (450 L) as a receptacle for a solid; or a water capacity less than 1000 pounds (<454 kg) as a receptacle for a gas. Some examples of nonbulk containers commonly used to transport liquid hazardous materials are illustrated in Figure 6.1.
- **Intermediate bulk containers,** or **IBCs,** such as the forms of rigid or portable packaging other than a cylinder or portable tank that has been designed for mechanical handling and has a capacity equal to or less than 0.08 cubic feet (≤3 m³). An example is illustrated in Figure 6.2.

packing group ■ For purposes of DOT regulations, a classification based on the degree of danger (great, medium, and minor) presented by hazardous materials containers

For purposes of DOT regulations, a portable device in which a material is stored, transported, treated, disposed of, or otherwise handled

packaging ■ For purposes of DOT regulations, the receptacle and any other components or materials necessary for the receptacle to perform its containment function and to ensure conformance with prescribed minimum requirements

package ■ For purposes of DOT regulations, the means of containment used to transport a hazardous material from one location to another

nonbulk container ■ For purposes of DOT regulations, packaging that has a (a) maximum capacity of 119 gallons (450 L) or less as a liquid receptacle; (b) maximum net mass of 882 pounds (400 kg) or less and a maximum capacity of 119 gallons (450 L) or less as a solid receptacle; or (c) water

intermediate bulk container (IBC) ■ For purposes of DOT regulations, a rigid or flexible portable packaging other than a cylinder or portable tank designed for mechanical handling and having an intermediate volume between nonbulk and bulk packaging

FIGURE 6.1 Examples of nonbulk containers used for the transportation of liquid and solid hazardous materials. (*Adapted with permission of Avantor Performance Materials, Inc., Center Valley, Pennsylvania, from "Packaging Systems: Innovative Design Drives Productivity".*)

Four-oz (114-mL) amber glass bottle	0.6-gal (2.5-L) glass bottle	Lined steel drum
Polyethylene bottle	5-gal (19-L) polyethylene hedpak	Polyethylene pail
Polyethylene drum	Polypropylene bottle	211-gal (200-L) polyethylene drum
55-gal (208-L) steel drum	5-gal (19-L) steel pail	5.3-gal (20-L) lined polyethylene drum

bulk container ■ For purposes of DOT regulations, packaging that has a (a) maximum capacity of greater than 119 gallons (450 L) for a liquid receptacle; (b) maximum net mass greater than 882 pounds (400 kg) for a solid receptacle; *or* (c) water capacity greater than 1000 pounds (454 kg) as a gas receptacle

transport vehicle ■ A cargo-carrying vehicle used for the transportation of cargo by any mode

freight container ■ A reusable container having a volume equal to or greater than 64 cubic feet (1.8 m³), designed and constructed so it may be lifted with its contents and intended primarily for the containment of packages (in unit form) during transportation

■ **Bulk containers,** such as any transport vehicle, ton cylinder, or freight container. A **transport vehicle** is any vehicle that is approved for carrying cargo such as an automobile, van, tractor, truck, semi-trailer, tankcar, or railcar used for the transportation of cargo by any mode. A **freight container** is any type of bulk packaging because it is used primarily for containment of packages in unit form during transportation. Bulk packaging has a maximum capacity greater than 119 gallons (>450 L) as a receptacle for a liquid; a maximum net mass greater than 882 pounds (>400 kg) and a maximum capacity greater than 119 gallons (>450 L) as a receptacle for a solid; or a water capacity greater than 1000 pounds (>454 kg) or less as a receptacle for a gas.

All forms of packaging intended for containerization of a hazardous material is subjected to rigorous testing procedures prescribed by DOT to ensure that they will retain their integrity when subjected to normal conditions of transportation. There are five principal tests for each type of packaging: drop test; leakproofness test; hydrostatic test; stacking test; and vibration test. DOT publishes the minimum requirements for compliance with these tests at 49 C.F.R. Subpart M.

A shipper may petition DOT to containerize a hazardous material for transportation in an unconventional fashion by submitting relevant information to demonstrate that the packaging retains its integrity when subjected to normal conditions of transportation. When DOT agrees, it assigns the shipper an approved special permit as "DOT-SP-" followed by the assigned special permit number. The assignment of an approved special permit allows carriers to legally transport a hazardous material in unauthorized packaging for a specified period, after which the shipper must reapply for a new special permit.

FIGURE 6.2 This intermediate bulk container may be used to transport sulfuric acid. *(Courtesy of Avantor Performance Materials, Inc., Center Valley, Pennsylvania, "Packaging Systems: Innovative Design Drives Productivity".)*

6.1-F SPECIAL PROVISIONS

DOT identifies certain special provisions that apply to the transportation of unique hazardous materials. Each provision is identified by a code. We are primarily concerned here only with the provisions that apply directly to the needs of emergency-response personnel.

When shippers and carriers transport a gaseous or liquid hazardous material that poses a **health hazard by inhalation**, DOT requires them to establish its **hazard zone**. For gases, the hazard zone is any of the designations Zone A, Zone B, Zone C, or Zone D; and for liquids, it is either Zone A or Zone B. The greatest degree of health hazard is posed by exposure to hazardous materials that have been assigned to Zone A.

The code for the hazard zone of a hazardous material that poses health hazard by inhalation appears in column 7 of the Hazardous Materials Table. When DOT lists any of the codes 1, 2, 3, or 4 in column 7, shippers and carriers associate them with Zone A, Zone B, Zone C, and Zone D, respectively.

6.2 THE SHIPPING PAPER

With few exceptions, DOT requires a hazardous material to be properly described for transportation on a **shipping paper**. There are several forms of a shipping paper, such as a shipping order, bill of lading, manifest, railroad waybill, or similar document. Most resemble the example shown in Figure 6.3, a portion of which we shall use as a template throughout the remainder of this text. For emergency responders, the information entered in the section noted as "Shipping Description" is especially important.

6.2-A SHIPPING DESCRIPTIONS OF HAZARDOUS MATERIALS

DOT requires shippers to provide the **shipping description** of each hazardous material in a given consignment on the accompanying shipping paper. The shipping description of a hazardous material is composed in part of a composite listing of the following components in the sequence noted.

- "UN/NA" Identification number
- Proper shipping name (including the technical name in parentheses, when applicable)
- Primary hazard class or division

health hazard by inhalation ▪ The expression used to describe the property of certain gases and liquid vapors that could cause health problems when inhaled

hazard zone ▪ For purposes of DOT regulations, any of the four hazard levels assigned at 49 C.F.R. §§173.116(a) and 173.133(a), respectively, to poison gases and either of the two levels (Zones A and B) assigned to liquids whose vapors pose a health hazard by inhalation

shipping paper ▪ The shipping order, bill of lading, air bill, dangerous-cargo manifest, or similar shipping document issued by a shipper or carrier as required by DOT at 49 C.F.R. §§172.202, 172.203, and 172.204

Non-Negotiable
BILL OF LADING

No.	
	_____OF_____

Date Received:	Dispatch/Pro No:	Driver:	Truck No:	Trailer No:

SHIPPER	CONSIGNEE	CHARGES

SHIPPER	CONSIGNEE	
Name	Name	
Street Address	Street Address	
City, State	City, State	
Notify/Contact Phone	Notify/Contact Phone	

PARTICULARS FURNISHED BY SHIPPER

Units	HM	Shipping Description (Identification Number, Proper Shipping Name, Primary Hazard Class or Division, Subsidiary Hazard Class or Division, and Packing Group)	Weight (lb)

	Placards Required:	Emergency Telephone:	ERG No.

For hazardous materials transported by vessel: Shipper declares that the packing/loading of freight containers and/or transport vehicles containing hazardous materials has been carried out in accordance with the provisions of 49 C.F.R. §176.27(c). Signature: _____	This is to certify that the above-named materials are properly classified, described, packaged, marked and labeled, and are in proper condition for transportation according to the applicable regulations of the Department of Transportation (49 C.F.R. §172.204). Signature:_____

LIABILITY FOR LOSS, DAMAGE, ETC. TO GOODS

Shipper's attention is directed to Section 11 on the reverse side of this Bill of Lading. All goods shall have an agreed release valuation of $0.10 per pound unless Shipper declares a higher value and Carrier accepts that valuation in the space below. For water carriage, see Section 5 on the reverse side of this Bill of Lading.
Shipper's initials:_____ Higher value: $_____ per pound Carrier's acceptance:_____

ORIGIN   **DESTINATION**

time in :	time in :	
time out:	time out:	Received in good order, count and condition unless otherwise noted above.
date :	date :	

SHIPPER   **CONSIGNEE**

Authorized Signature Date	Authorized Signature Date	Authorized Signature Date

FIGURE 6.3 One form of a shipping paper used in commerce.

■ Subsidiary hazard class or division in parentheses, if any
■ Packing group (when applicable)

DOT requires shippers to provide this information in English on a shipping paper.

DOT also requires the total number of packaging units and the total amount of each hazardous material in a consignment to be communicated in the shipping description. The total number of units is indicated by such expressions as 4 boxes, 17 cylinders, 5 steel drums, 3 glass bottles, 1 tank truck, and so forth. The total amount is typically provided as a gross aggregate mass or volume. Abbreviations are used to express the common units of measurement.

DOT lists six symbols in column 1 of the Hazardous Materials Table. When a G is listed, DOT requires shippers to provide the technical name parenthetically with the hazardous material's shipping description. When a + symbol is listed, the proper shipping name, hazard class, and packing group may not be altered for any reason. When a D or I is listed, the proper shipping name is appropriately described for domestic and international transportation, respectively. Finally, when an A or W is listed, DOT regulates the transportation of the given hazardous material only by aircraft and watercraft, respectively.

When a consignment comprises hazardous materials and other materials whose transportation is not regulated by DOT, the proper shipping descriptions of the hazardous materials must be entered *first* on the relevant shipping paper, or a contrasting color or an X in the "HM" column can be entered to identify the entries that are hazardous materials. For example, gasoline, nitrogen, and a flammable solid are listed on the following portion of a shipping paper *before* the materials that are not subject to the DOT hazardous materials regulations. The hazardous materials are specifically identified by an X in the column headed "HM."

shipping description
■ For purposes of DOT regulations, the following minimum information for a hazardous material: proper shipping name (including the technical name, when required); hazard class or division; identification number; packing group; total quantity by mass, volume, or as otherwise appropriate; and the number and types of packages

UNITS	HM	SHIPPING DESCRIPTION (IDENTIFICATION NUMBER, PROPER SHIPPING NAME, PRIMARY HAZARD CLASS OR DIVISION, SUBSIDIARY HAZARD CLASS OR DIVISION, AND PACKING GROUP)	WEIGHT (lb)
10 steel drums (UN1A1)	X	UN1203, Gasoline, 3, PG II	4500
40 cylinders (DOT-3AA1800)	X	UN1066, Nitrogen, compressed, 2.2	800
1 steel drum (UN1A1)	X	UN2926, Flammable solids, toxic, organic, n.o.s. (Cloth/Paper containing 2,4-dinitrophenol), 4.1, (6.1), PG II	452
4 boxes (UN4D)		Advertising materials, paper, n.o.i.	60
1 roll		Paper, printing, newsprint	690
12 sets		Carbon paper	22
		EMERGENCY CONTACT: (000) 000-0000	

This example also illustrates that the shipper must provide an emergency-contact telephone number on a shipping paper for each hazardous material within the consignment. In this instance, there are multiple hazardous materials listed on the same paper, and the single telephone number applies to all of them. During a transportation mishap, emergency responders secure the shipping paper and provide this number to CHEMTREC.

When providing the shipping description of a hazardous material, shippers must be aware of other DOT regulatory requirements that are relevant to identifying the given consignment by a selected mode of transportation. Examples of these regulations are provided in Table 6.2. They illustrate that emergency-response personnel obtain valuable information about the nature of a unique hazardous materials consignment from the information that has been entered on shipping papers.

TABLE 6.2	Information Required in Shipping Descriptions[a,b]
EXAMPLE (SELECT PHRASES)[c]	**RELEVANT REGULATION**
Transportation of a Hazardous Material by All Modes	
RQ, UN1052, Hydrogen fluoride, anhydrous, 8, (6.1), PG I (Poison - Inhalation Hazard, Zone C) *or* UN1052, Hydrogen fluoride, anhydrous, 8, (6.1), PG I, RQ (Poison - Inhalation Hazard, Zone C)	When a shipper or carrier transports a hazardous substance in an amount that exceeds its reportable quantity in a single package, the letters RQ are included before or after the shipping description.
RQ, UN1428, Sodium, 4.3, PG I (Dangerous When Wet)	When a hazardous material, by chemical interaction with water, is liable to become spontaneously flammable or give off flammable gases in dangerous quantities, the words "Dangerous When Wet" are included with the shipping description.
UN1831, Sulfuric acid, fuming, 8, (6.1), PG I (Poison - Inhalation Hazard, Zone B)	When a hazardous material possesses multiple hazards, its subsidiary hazard class or division numbers are entered parenthetically immediately following the primary hazard class.
UN1760, Corrosive liquids, n.o.s. (Valeric acid), 8, PG I	When a hazardous material is described with an n.o.s. entry or other generic description in the Hazardous Materials Table, the shipping description includes the name of the substance in parentheses.
UN1992, Flammable liquids, toxic, n.o.s. (contains xylene and methanol), 3, (6.1), PG I	When a mixture or solution of two or more hazardous materials described by an n.o.s. entry in the Hazardous Materials Table is to be transported, the technical names of at least two components most predominantly contributing to the hazards of the mixture or solution are parenthetically entered in the shipping description.
UN1202, Diesel fuel solution, 3, PG III *or* UN1202, Diesel fuel mixture, 3, PG III	When a mixture or solution of a hazardous material not listed by name in the Hazardous Materials Table and a non-hazardous material is to be transported, the proper name of the hazardous material is entered in the shipping description along with the word "solution" or "mixture."
UN1987, Alcohol, n.o.s. (contains 2-octanol), 3, PG III	When the name of a hazardous material is not specifically listed in the Hazardous Materials Table, the entry that most appropriately describes the chemical class of the material is used, not its hazard class.
UN2672, Ammonia solutions (with 12.8% ammonia in water), 8, PG III	When a concentration range is a component of a proper shipping name, the actual concentration of the hazardous material, if it is within the indicated range, is used in lieu of the range.
UN2789, Acetic acid glacial, 8, (3), PG II (with >80% acid by mass) *or* UN2789, Acetic acid, glacial (containing more than 80% acid by mass), 8, (3), PG II	Technical and chemical group names may include an appropriate modifier like "contains" or "containing" which is entered parenthetically between the shipping name and the hazard class *or* following the shipping description.

[a]49 C.F.R. §§172.202 and 172.203.
[b]Not provided in this table is the unique information that DOT requires in the shipping descriptions of organic peroxides, explosives, and radioactive materials. This information is discussed independently in Sections 13.19-A, 15.4, and 16.10, respectively.
[c]The component of the shipping description that is applicable to the regulation is provided in blue print.

TABLE 6.2 Information Required in Shipping Descriptions (*continued*)

EXAMPLE (SELECT PHRASES)	RELEVANT REGULATION
RQ, UN2029, Hydrazine anhydrous, 8, (3, 6.1), UN2029, PG I (Poison) *or* RQ, UN2029, Hydrazine anhydrous, 8, (3, 6.1), PG I (Toxic)	Regardless of the hazard class to which a hazardous material is assigned, if a liquid or solid material in a package meets the definition of a Division 6.1, PG I or II, and the fact that it is a poison is not disclosed in its shipping name or class entry, either of the words "Poison" or "Toxic" is entered in the shipping description.
UN2783, Organophosphorus pesticides, solid, toxic (Ciodrin), 6.1, PG II *or* UN2783, Organophosphorus pesticides, solid, toxic, 6.1, PG I (Ciodrin)	Notwithstanding the hazard class to which a hazardous material is assigned, if the technical name of the compound or principal constituent that causes a material to meet the definition of a Division 6.1, PG I or II, is not included in the proper shipping name for the material, the technical name is entered in parentheses in the shipping description.
RQ, UN2199, Phosphine, 2.3, (2.1) (Poison - Inhalation Hazard, Zone A) *or* RQ, UN2199, Phosphine, 2.3 (2.1) (Toxic - Inhalation Hazard, Zone A)	For materials that pose an inhalation hazard, either of the expressions "Poison - Inhalation Hazard" or "Toxic - Inhalation Hazard" and the applicable words "Zone A," "Zone B," "Zone C," or "Zone D" for gases or "Zone A" or "Zone B" for liquids, are entered on the shipping paper immediately following the shipping description.
Flammable solids, n.o.s. (sodium), 4.3, UN1325, PG II UN3082, Hazardous waste, liquid, n.o.s., 9, (contains trichloroethylene), PG III (F001)[d] NA3082, Hazardous waste, liquid, n.o.s., (contains toluene and methanol), 9, PG III (EPA ignitability) *or* NA3082, Hazardous waste, liquid, n.o.s., (contains toluene and methanol), 9, PG III (D001)	If a hazardous material is also a hazardous substance other than a radioactive material, and the name of the hazardous material does not identify the hazardous substance by name, one of the following descriptions is entered in parentheses in the shipping description of the hazardous material: **(a)** the name of the hazardous substance; *or* **(b)** for waste streams, the hazardous waste number; *or* **(c)** for waste streams that exhibit an EPA characteristic, the letters "EPA" followed by the characteristic or the EPA hazardous waste number.[d,e]
RQ, UN1621, London purple (contains diarsenic trioxide and aniline), 6.1, PG II (Marine Pollutant) *or* UN1621, London purple (contains diarsenic trioxide and aniline), 6.1, PG II, (Marine Pollutant) RQ	If a hazardous material is also a hazardous substance other than a radioactive material, and the name of the hazardous material does not identify the hazardous substance by name, the letters "RQ" are entered either before or after the shipping description.
UN1193, Residue: Last contained ethyl methyl ketone, 3, PG II	The shipping description for the residue of a hazardous material contained in packaging other than a rail tankcar *may include* the words "Residue: Last contained _____ (name of hazardous material in packaging before it was emptied)."

[d]The EPA hazardous waste numbers for listed hazardous wastes are prefixed with one of the capital letters F, K, P, or U. The nature of the listed hazardous wastes is published at 40 C.F.R. §§261.31, 261.32, 261.33(e) and 261.33(f).

[e]Four EPA characteristics of a hazardous waste are identified at 40 C.F.R. §§261.21–261.24: ignitability (Section 3.2); corrosivity (Section 8.15); reactivity (Section 5.9); and toxicity (Section 10.1-B). Hazardous wastes exhibiting these characteristics have been assigned the EPA hazardous waste numbers D001, D002, D003, and D004 through D043, respectively.

(continued)

TABLE 6.2 | Information Required in Shipping Descriptions (*continued*)

EXAMPLE (SELECT PHRASES)	RELEVANT REGULATION
RQ, UN1052, Hydrogen fluoride, anhydrous, 8, (6.1), PG I (Poison - Inhalation Hazard, Zone C) (DOT-SP- ***) (*** is replaced with the appropriate special permit number)	When a shipper has been assigned an approved special permit from DOT to containerize a hazardous material in an unconventional fashion, the shipping description includes the notation "DOT-SP-" followed by the assigned special permit number and so located that the notation is clearly associated with the description to which the exemption applies.
UN2214, RQ, HOT Phthalic anhydride, 8, PG III	If a liquid hazardous material is to be transported at an "elevated temperature," and the fact that it is an elevated-temperature material is not disclosed in the shipping name, the word "HOT" immediately precedes the proper shipping name of the hazardous material.
6 cyl (DOT-3AA1800), RQ, UN1076, Phosgene, 2.3, (8), (Poison - Inhalation Hazard, Zone A) *or* 5 steel drums (UN1A1), UN1193, Ethyl methyl ketone, 3, PG II	The number and type of each nonbulk package with its DOT specification number are included with the shipping description.
Transportation of a Hazardous Material by Passenger-Carrying Railroad or Aircraft UN1827, Stannic chloride, anhydrous, 8, PG II (Limited Quantity) *or* UN1827, Stannic chloride, anhydrous, 8, PG II (Ltd Qty)	When DOT authorizes the transportation of a hazardous material in a limited quantity by passenger-carrying railroad or aircraft, the shipping description includes the words "Limited Quantity" or "Ltd Qty."
Transportation of a Hazardous Material by Cargo Aircraft UN2363, Ethyl mercaptan, 3, PG I (Cargo Aircraft Only) (Limited Quantity)	When DOT authorizes the transportation of a hazardous material by cargo aircraft, the words "Cargo Aircraft Only" are entered in the shipping description.
UN2363, Ethyl mercaptan, 3, PG I (Cargo Aircraft Only) (Limited Quantity) *or* UN2363, Ethyl mercaptan, 3, PG I (Cargo Aircraft Only) (Ltd Qty)	When DOT authorizes the transportation of a hazardous material in a limited quantity by cargo aircraft, the shipping description includes the words "Limited Quantity" or "Ltd Qty" in addition to "Cargo Aircraft Only."
Nonbulk Transportation of a Marine Pollutant on Water and Bulk Transportation of a Marine Pollutant in All Modes UN1381, Phosphorus, white, under water, 4.2, (6.1), PG I (Marine Pollutant) (Poison)	When a shipper or carrier transports a nonbulk quantity of a marine pollutant onboard watercraft or a bulk quantity of a marine pollutant by any mode, the words "Marine Pollutant" are included with the shipping description.
UN1621, London Purple (contains diarsenic trioxide and aniline), 6.1, PG II (Marine Pollutant) (Poison)	When the name of the component causing a material to be a marine pollutant is not evident, the name is identified parenthetically in the shipping description.
Environmentally hazardous substance, solid, n.o.s. (Zinc bromide) (Marine Pollutant)	If a marine pollutant is not listed by name in the Hazardous Materials Table and it does not meet the defining criteria of classes 1 through 8, its shipping description is either "Environmentally hazardous substance, liquid, n.o.s." or "Environmentally hazardous substance, solid, n.o.s.," as relevant

TABLE 6.2	Information Required in Shipping Descriptions (*continued*)

EXAMPLE (SELECT PHRASES)	RELEVANT REGULATION
RQ, UN3082, Environmentally hazardous substances, liquid, n.o.s. (sodium cyanide and cupric cyanide), 9, PG III (Marine Pollutant) (Poison)	When a material designated with an n.o.s. entry in the Hazardous Materials Table is a marine pollutant, the names of at least two of the components most predominantly contributing to the marine pollutant designation are entered in parentheses in the shipping description.
Bulk Transportation of a Hazardous Material on Water	
UN1175, Ethylbenzene, 3, PG II (flashpoint = 15°C)	When a hazardous material possesses a flashpoint equal to or less than 142°F (61°C) and is transported on water, the flashpoint in degrees Celsius (closed cup) is entered with the shipping description.
UN2796, Battery fluid, acid, 8, PG II (IMDG Code 1 – Acids)	When a hazardous material is designated with an n.o.s. entry in the Hazardous Materials Table and is not listed in section 3.1.4 of the IMDG Code,[f] the applicable IMDG Code segregation group is entered with the shipping description.
Bulk Transportation of a Hazardous Material by Railway	
UN1962, Ethylene, compressed, 2.1, (DOT-113) (Do Not Hump or Cut Off Car While in Motion) Placarded: FLAMMABLE GAS (HYDX11111)	The shipping description of a flammable gas or its residue contained within a DOT-113 rail tankcar contains an appropriate notation such as "DOT-113" and the statement "Do Not Hump or Cut Off Car While in Motion."
UN2214, HOT Phthalic anhydride, 8, UN2214, III, RQ (Maximum Operating Speed 15 mph) (LCTX111111)	When DOT allows the transportation by railroad of an elevated-temperature material pursuant to certain exceptions, the shipping description contains an appropriate notation such as "Maximum Operating Speed 15 mph."
UN1193, RESIDUE: Last contained ethyl methyl ketone, 3, PG II Placarded: FLAMMABLE (ACTX11111)	The shipping description for the residue of a hazardous material contained in a rail tankcar *must include* the words "Residue: Last contained _____ (name of hazardous material in a rail tankcar before it was emptied)."
RQ, UN1076, Phosgene, 2.3, (8), (Poison - Inhalation Hazard, Zone A) Placarded: POISON GAS (DUPX11111)[g]	When hazardous materials are transported by rail, the accompanying shipping paper bears the notation "Placarded:_____," in which the name of the placard displayed on the railcar is inserted.
RQ, UN1075, Liquefied petroleum gas, 2.1, (Noncorrosive) (CSXT111111)[g]	When hazardous materials are transported by railway, the accompanying shipping paper lists the reporting mark and number[g] displayed on the bulk packaging used for shipment.
Bulk Transportation of Anhydrous Ammonia by Highway	
RQ, UN1005, Ammonia, anhydrous, 2.2, (0.2 percent water) (Inhalation Hazard)	When a carrier transports anhydrous ammonia having a water content equal to or greater than 0.2% by mass in a DOT Specification MC330 or MC331 tank truck, the shipping description includes "0.2 percent water" to indicate the suitability for shipping anhydrous ammonia in a cargo tank constructed from quenched and tempered steel.
RQ, UN1005 Ammonia, anhydrous, 2.2, (Not for Q and T tanks) (Inhalation Hazard)	When a carrier transports anhydrous ammonia having a water content less than 0.2% by mass in a DOT Specification MC330 or MC331 tank truck, the shipping description includes the phrase "Not for Q and T tanks."

[f]The International Maritime Dangerous Goods (IMDG) Code governs the vast majority of the shipments of hazardous materials by water.
[g]See Section 3.7-A. DUPX and CSXT are the reporting marks of E. I du Pont de Nemours & Company and CSX Transportation Company, respectively.

(continued)

| TABLE 6.2 | Information Required in Shipping Descriptions (*continued*) |

EXAMPLE (SELECT PHRASES)	RELEVANT REGULATION
RQ, UN1005, Ammonia, anhydrous, 2.2, (0.2 percent water) (Inhalation Hazard)	When a carrier transports anhydrous ammonia or ammonia solutions, *relative density less than 0.880 at 15 degrees C in water*, the shipping description includes the words "Inhalation Hazard" without either of the words "Poison" or "Toxic" or the designation of a hazard zone.
Bulk Transportation of Liquefied Petroleum Gas by Highway UN1075, Liquefied petroleum gas, 2.1 (Noncorrosive) *or* UN1075, Liquefied petroleum gas, 2.1 (Noncor)	When a carrier transports liquefied petroleum gas in a DOT Specification MC330 or MC331 tank truck, the shipping description includes "Noncorrosive" (or "Noncor") to indicate the suitability of transporting noncorrosive liquefied petroleum gas in a cargo tank made of quenched-and-tempered steel.
Bulk Transportation of Oil by Highway or Railroad UN1263, Paint-related material, 3, PG III (oil)	When petroleum oil is transported in packaging having a capacity of 3500 gal (13,250 L) or more, *or* when any oil is transported in packaging in a quantity of 42,000 gal (159,000 L) or more, the word "oil" is included on the shipping paper.

Let's consider another example. Suppose a fuel company desires to use a tank truck to move aviation gasoline from a petroleum storage facility to an airport terminal. DOT obligates the company to select the particular entry in the Hazardous Materials Table that most accurately describes the material to be transported. In this instance, the correct selection is "Fuel, aviation, turbine engine." This combination of information is properly entered on a shipping paper as follows:

hazardous substance
■ For purposes of DOT regulations, a substance listed in Appendix A at 40 C.F.R. §172.101 in an amount that exceeds its reportable quantity and is transported in a single package

UNITS	HM	SHIPPING DESCRIPTION (IDENTIFICATION NUMBER, PROPER SHIPPING NAME, PRIMARY HAZARD CLASS OR DIVISION, SUBSIDIARY HAZARD CLASS OR DIVISION, AND PACKING GROUP)	VOLUME (gal)
1 tank truck	X	UN1863, Fuel, aviation, turbine engine, 3, PG I	1500

When shippers offer aviation fuel for transportation, DOT requires them to enter only this description of the commodity on a shipping paper. The selection of any other name constitutes an illegal entry.

6.2-B REPORTABLE QUANTITIES

reportable quantity (RQ) ■ For purposes of DOT regulations, the amount listed at 40 C.F.R. §302.4 and 49 C.F.R. §172.101 of a "hazardous substance" within the meaning of DOT and CERCLA, the release of which triggers mandatory notification to the National Response Center

On occasion, the letters *RQ* are entered with the shipping description of a hazardous material. "RQ" refers to the amount of a **hazardous substance** that constitutes a **reportable quantity**. For the purposes of the Superfund law(Section 1.3-D), any substance listed at 40 C.F.R. §302.4 is a hazardous substance, regardless of the amount at issue; but for the purposes of the DOT regulations, only a substance listed at 49 C.F.R. §172.101, Appendix A, in an amount equal to or exceeding the reportable quantity is a hazardous substance. There are only five reportable quantities: 5000 lb (2270 kg); 1000 lb (454 kg); 100 lb (45.4 kg); 10 lb (4.54 kg); and 1 lb (0.454 kg). Examples of several hazardous substances and their DOT reportable quantities are provided in Table 6.3.

	REPORTABLE QUANTITY	
HAZARDOUS SUBSTANCE	**POUNDS**	**KILOGRAMS**
Acetone	5000	2270
Acetonitrile	5000	2270
Ammonia solutions	1000	454
Asbestos, white[b]	1	0.454
Barium cyanide	10	4.54
Cupric cyanide	10	4.54
Chlorine	10	4.54
Chlorobenzene	100	45.4
Dichloromethane	1000	454
Ethyl methyl ketone	5000	2270
Hydrazine	1	0.454
Hydrogen fluoride, anhydrous	100	45.4
Hydrogen sulfide	100	45.4
Methyl isobutyl ketone	5000	2270
Methyl parathion	100	45.4
Nickel carbonyl	10	4.54
Nitric acid	1000	454
Phosgene	10	4.54
Phosphine	100	45.4
Phosphorus	1	0.454
Phthalic anhydride	5000	2270
Sodium	10	4.54
Selenium disulfide	10	4.54
Sodium hydroxide	1000	454
Sodium hypochlorite	100	45.4
Toluene	1000	454
1,1,1-Trichloroethane	1000	454
Xylene(s)	100	45.4

TABLE 6.3 Some Hazardous Substances and their Reportable Quantities[a]

[a]40 C.F.R. §302.4 and 49 C.F.R. §172.101, Appendix A, Table 1.
[b]Section 10.19-A.

When a shipper intends to transport a hazardous substance in an amount equal to or greater than its reportable quantity, DOT requires "RQ" to be entered on the shipping paper either before or after its shipping description. Suppose, for instance, that a shipper intends to transport domestically by tank truck 4500 pounds of an aqueous ammonia solution containing 12% ammonia having a gross weight of 4500 pounds. Although four shipping descriptions for ammonia solutions are listed in Appendix C, the description having the identification number UN2672 most adequately describes the commodity.

Because the shipper intends to transport an amount of an ammonia solution exceeding the reportable quantity listed in Table 6.3 as 1000 pounds (454 kg), "RQ" must be included either before or after its proper shipping description on the accompanying waybill. If the shipper elects to enter "RQ" before the shipping description, the complete shipping description is noted as follows:

UNITS	HM	SHIPPING DESCRIPTION (IDENTIFICATION NUMBER, PROPER SHIPPING NAME, PRIMARY HAZARD CLASS OR DIVISION, SUBSIDIARY HAZARD CLASS OR DIVISION, AND PACKING GROUP)	WEIGHT (lb)
1 tank truck	X	RQ, UN2672, Ammonia solution (contains 12% ammonia in water), 8, PG III	4500

SOLVED EXERCISE 6.1

A shipper offers 2500 pounds of phosgene to E. I. du Pont de Nemours & Company, Inc. for transportation in a rail tankcar marked with the number 11112. What shipping description does DOT require on the shipping paper? (DUCX is du Pont's reporting mark.)

Solution: Table 6.3 indicates that phosgene is a hazardous substance. From this knowledge and the information listed in Appendix C, the shipping description is provided as follows:

UNITS	HM	SHIPPING DESCRIPTION (IDENTIFICATION NUMBER, PROPER SHIPPING NAME, PRIMARY HAZARD CLASS OR DIVISION, SUBSIDIARY HAZARD CLASS OR DIVISION, AND PACKING GROUP)	WEIGHT (lb)
1 c/t	X	RQ, UN1076, Phosgene, 2.3, (8), (Poison - Inhalation Hazard, Zone A) (Placarded: POISON GAS) (DUCX11112)	2500

SOLVED EXERCISE 6.2

The following shipping description of a hazardous material is provided on a transportation manifest:

UNITS	HM	SHIPPING DESCRIPTION (IDENTIFICATION NUMBER, PROPER SHIPPING NAME, PRIMARY HAZARD CLASS OR DIVISION, SUBSIDIARY HAZARD CLASS OR DIVISION, AND PACKING GROUP)	WEIGHT (gal)
1 tankcar	X	RQ, NA3082, Hazardous waste liquid, n.o.s. (contains toluene and xylene), 9, PG III (EPA ignitability) (UTLX42888)	10,000

How is this shipping description useful to an emergency-response team?

Solution: There are at least four pieces of information in this shipping description that are useful to emergency-response personnel:

- The notation "RQ" informs first-on-the-scene responders that the commodity is a hazardous substance, the release of which could adversely affect public health and the environment. The emergency-response crew should proceed to dam or dike the area to confine the leak and prevent its widespread release into the environment.
- The reference to "EPA ignitability" indicates that the hazardous waste liquid exhibits the RCRA characteristic of ignitability (Section 3.2). This designation informs emergency-response personnel that the hazardous waste liquid possesses a flashpoint equal to or less than 140°F (60°C). Thus, it poses a risk of fire and explosion.
- Packing Group III indicates that the relative degree of hazard possessed by the commodity is minor. A minor degree of hazard means that although the material ignites, other groups of hazardous materials are far more hazardous.
- In Section 6.7, we shall note that identification numbers direct emergency-response personnel to certain recommended procedures for properly responding to an incident involving the release of this hazardous waste liquid.

6.2-C MARINE POLLUTANTS AND SEVERE MARINE POLLUTANTS

DOT regulates the transportation of a variety of environmentally hazardous substances including **marine pollutants** and **severe marine pollutants,** or **SMP**. These are substances or mixtures that negatively impact aquatic life or tend to bioaccumulate in seafood when released into a waterway. Protection of the marine environment is not an immediate mandate of DOT; nonetheless, DOT provides a listing of marine pollutants at 49 C.F.R. §172.101, Appendix B, an excerpt of which is reproduced in Table 6.4.

Marine pollutants that potentially damage the marine environment more severely than others are called severe marine pollutants. In Table 6.4, severe marine pollutants are identified with the capital letters *PP*, meaning "pollution potential," in the column before their listing. The listed substances that are not identified as severe marine pollutants are referred to as marine pollutants.

When a substance or mixture is designated as a marine pollutant or severe marine pollutant, the words "Marine Pollutant" are included parenthetically with its shipping description. For example, when a glass bottle containing 10 pounds (4.54 kg) of

marine pollutant ■ For purposes of DOT regulations, any substance denoted in Appendix B, 49 C.F.R. §172.101, and, when in a solution or mixture of one or more marine pollutants, is packaged in a concentration that equals or exceeds: (1) 10% by weight of the solution or mixture for materials identified in the appendix as marine pollutants; or (2) 1% by weight of the solution or mixture for materials identified in the appendix as severe marine pollutants

severe marine pollutants (SMP) ■ For purposes of DOT regulations, any substance denoted in Appendix B, 49 C.F.R. §172.101, and indexed with the capital letters PP

TABLE 6.4		Examples of Some Marine Pollutants and Severe Marine Pollutants[a]		
	MARINE POLLUTANT		**MARINE POLLUTANT**	
	Anisole		Isobutyl isobutyrate	
	Barium cyanide		Isobutyl propionate	
	Butanedione		Lead acetate	
	Butylbenzenes		London purple	
	Butyraldehyde	PP	Mercuric oxide	
PP	Cadmium compounds	PP	Nickel carbonyl	
	Carbon tetrachloride		1-Octanol	
	Chlorine	PP	Phosphorus, white or yellow, dry, under water, or in solution	
	Cresols (*o-, m-, p-*)		α-Pinene	
	Cumene	PP	Polychlorinated biphenyls (PCBs)	
PP	Cupric sulfate		Propionaldehyde	
PP	1,3,5-Cyclododecatriene		*n*-Propylbenzene	
PP	Cymenes (*o-, m-, p-*)		Sodium cyanide, solid or solution	
PP	Dichlorobenzene (*o-, m-, p-*)		Styrene monomer, inhibited	
	Diethylbenzenes		1,1,2,2-Tetrachloroethane	
	Epichlorohydrin		Tetrachloroethylene	
	Ethyl acrylate, inhibited		1,3,5-Trimethylbenzene	
	Ferric arsenate		Triethylbenzene	
PP	*n*-Heptaldehyde		Turpentine	
PP	Hexachlorobutadiene		*n*-Valeraldehyde	
	Isobutyraldehyde		Xylenols	
	Isobutyl butyrate		Zinc bromide	

[a]49 C.F.R. §172.101, Appendix B.

mercury(II) oxide is overpacked in a fiberboard box and transported by watercraft, the shipper conveys that the substance is a marine pollutant on the accompanying **dangerous-cargo manifest** as follows:

UNITS	HM	SHIPPING DESCRIPTION (IDENTIFICATION NUMBER, PROPER SHIPPING NAME, PRIMARY HAZARD CLASS OR DIVISION, SUBSIDIARY HAZARD CLASS OR DIVISION, AND PACKING GROUP)	WEIGHT (lb)
Fiberboard box (UN6G)	X	UN1641, Mercury oxide, 6.1, PG II (Marine Pollutant) (Toxic)	10

The dangerous-cargo manifest also lists the specific location at which the hazardous material is stowed aboard watercraft. When the name of the commodity does not include the name of the component that causes it to be a marine pollutant, DOT requires the shipper to parenthetically identify the name of the component. An example is provided in Table 6.2.

6.3 LOCATION OF THE SHIPPING PAPER DURING TRANSIT

During a transportation mishap, where may emergency-response personnel expect to locate the shipping paper? To avoid any confusion over its location, DOT requires the shipping paper to be carried on transport vehicles in a specific location. Because it is important to quickly locate this document when responding to incidents involving the release of a hazardous material, these requirements are briefly summarized next.

DOT requires the drivers of motor vehicles transporting hazardous materials and each carrier using the vehicles to clearly distinguish the appropriate shipping paper from other papers. When the driver is at the vehicle's controls, DOT requires the shipping paper to be within immediate reach and readily visible to any person entering the driver's compartment. For example, as shown in Figure 6.4, it may be stored in a holder mounted to the inside of the door on the driver's side of the motor vehicle. When the driver is not at the vehicle's controls, DOT requires the shipping paper to be located either in the holder or on the driver's seat inside the vehicle.

When hazardous materials are transported by railroad, DOT requires a member of the train crew, usually the conductor, to be in charge of the freight waybill that describes

FIGURE 6.4 When the driver of a motor vehicle is at the vehicle's controls, DOT requires the shipping paper to be kept within immediate reach inside a holder mounted to the inside of the door on the driver's side of the vehicle.

Shipping papers

the consignments that are transported. This person also retains the **consist**, a document indicating the position in the train of each loaded and placarded car containing a hazardous material. The consist usually is a computer printout that lists each cargo tank, boxcar, and other transport vehicle as a separate entry in ascending order from the front of the train, or alternatively, in descending order from the rear of the train, by its position, reporting mark (Section 3.7-A), and placard.

When hazardous materials are transported by aircraft, DOT requires the aircraft carrier to provide a shipping paper describing the consignment to the pilot in command. The pilot usually retains the shipping paper in the cockpit until the plane reaches its destination.

When hazardous materials are transported by watercraft, DOT requires the carrier to prepare a dangerous-cargo manifest, list, or stowage plan and submit it to the master of the vessel. This document must be contained within a designated holder on or near the vessel's bridge. On barges, it is kept in the pilot house of the tugboat.

6.4 DOT LABELING REQUIREMENTS

In most instances involving the transportation of a hazardous material, DOT requires shippers to affix prescribed warning **labels** corresponding to the primary and subsidiary hazard classes of the material directly to the outside surfaces of the packaging. The required labels are identified by the code entries in column 6 of the Hazardous Materials Table. Additional labeling requirements for hazard classes 1 and 7 are provided in Sections 15.4-D and 16.10-B, respectively.

The DOT labels that shippers affix to packaging containing a hazardous material are displayed in Figure 6.5 and their names are listed in Table 6.5. Each label is diamond-shaped and color-coded to signify a specific hazard class or division and may also include a pictograph to rapidly signify the potential hazards of the contents. Each label also bears the hazard class or division number in its lower corner.

Two points regarding the nature of the EXPLOSIVE labels should be noted:

- The uppercase letters that follow the division numbers for the EXPLOSIVE labels designate the compatibility groups of explosives, whose characteristics are discussed in Section 15.4-B.
- The label at the far right of Hazard Class 1 is the EXPLOSIVE subsidiary label. It is affixed to packaging when the code listed in column 6 of the Hazardous Materials Table indicates that the hazardous material has an explosive subsidiary hazard.

DOT requires shippers and carriers to affix the specified labels on the surface of the package near the spot where they indicate the proper shipping name. For instance, when shippers and carriers transport gasoline in drums, DOT requires them to affix the FLAMMABLE LIQUID label to the outside surface of each drum adjacent to the spot where it is marked "Gasoline."

When cylinders or other packages have irregular surfaces on which labels cannot be satisfactorily affixed, DOT allows the labels to be secured on a tag, unless the packages contain radioactive materials. Because of the irregularity of their surfaces, gas cylinders are labeled on shoulder tags like that shown in Figure 6.6. When the tags are embossed on the cylinders, OSHA/GHS hazard and precautionary statements may be cited along with the posting of the relevant transportation pictograms, or DOT labels.

DOT also requires shippers to affix *duplicate* labels corresponding to their applicable codes to certain packaging types. Duplicate labels are displayed on at least two sides or two ends other than the bottom. For multiunit car tanks, they are displayed on each

consist ▪ A shipping document that lists the location of one or more locomotives coupled to one or more railcars loaded with hazardous materials as they are placed, one by one, in a train, sometimes from the rear forward

label ▪ For purposes of DOT regulations, any of several diamond-shaped and color-coded warning signs, with or without a pictogram, affixed on the packages of hazardous materials to identify the specific hazard class or division of their contents

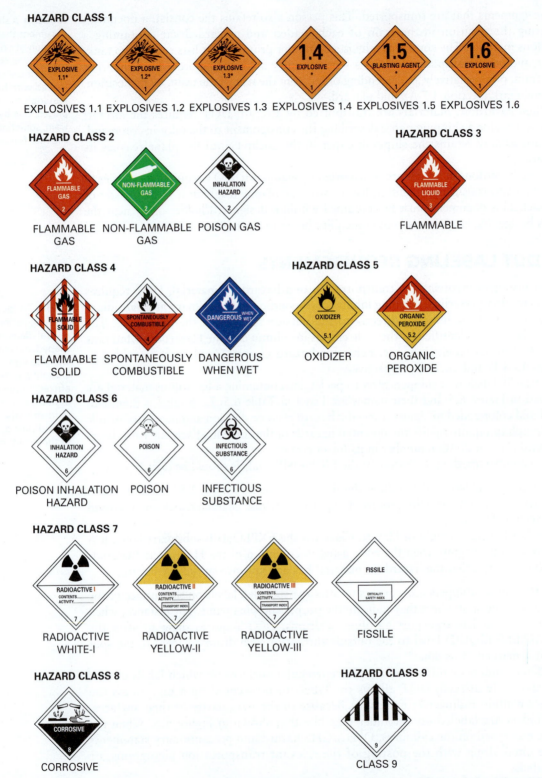

FIGURE 6.5 DOT requires shippers to affix warning labels that identify the primary and subsidiary hazard classes of a hazardous material on the exterior surface of its packaging. The asterisks in the orange EXPLOSIVE labels are replaced by the compatibility group (Section 15.4-B), or left blank if the explosive hazard is the subsidiary risk associated with the product. The INFECTIOUS SUBSTANCE label bears the following words (not shown here on label due to space constraints): "In case of Damage or Leakage, Immediately Notify Public Health Authority. In USA, Notify Director CDC, Atlanta, GA (800) 232-0124."

TABLE 6.5	DOT Warning Labels[a]		
HAZARD CLASS OR DIVISION	**LABEL NAME**	**HAZARD CLASS OR DIVISION**	**LABEL NAME**
1.1	EXPLOSIVE 1.1	5.1	OXIDIZER
1.2	EXPLOSIVE 1.2	5.2	ORGANIC PEROXIDE
1.3	EXPLOSIVE 1.3	6.1 (Inhalation hazard, Zone A or B)	POISON INHALATION HAZARD
1.4	EXPLOSIVE 1.4	6.1 (other than inhalation hazard, Zone A or B)	POISON
1.5	EXPLOSIVE 1.5	6.2	INFECTIOUS SUBSTANCE
1.6	EXPLOSIVE 1.6	7	RADIOACTIVE WHITE-I
2.1	FLAMMABLE GAS	7	RADIOACTIVE YELLOW-II
2.2	NON-FLAMMABLE GAS	7	RADIOACTIVE YELLOW-III
2.3	POISON GAS	7 (fissile material)	FISSILE
3	FLAMMABLE LIQUID	7 (empty packages)[b]	EMPTY
4.1	FLAMMABLE SOLID	8	CORROSIVE
4.2	SPONTANEOUSLY COMBUSTIBLE	9	CLASS 9
4.3	DANGEROUS WHEN WET		

[a]49 C.F.R. §172.400.
[b]Section 16.10-B.

end. For a freight container or aircraft unit load device, they are displayed on or near the closure. The regulations pertaining to use of duplicate labels are published at 49 C.F.R. §172.406.

When shippers offer for transportation packages containing hazardous materials authorized solely for transport on **cargo aircraft**, or when the packages contain net quantities that exceed those authorized on passenger-carrying aircraft, DOT requires them at 49 C.F.R. §172.448 to display the CARGO AIRCRAFT ONLY label in Figure 6.7 in addition to any other required labels.

cargo aircraft ■ An air-transport vehicle designed solely to transport cargo but not paying passengers

FIGURE 6.6 At 49 C.F.R. §172.406(b)(2), DOT allows the use of tags to provide both marking and labeling information when they are affixed to the surface of gas cylinders. As required by OSHA/GHS, hazard and precautionary statements are provided on the left of this shoulder tag and the transportation pictogram (or DOT label) is provided on the right. This particular tag is permanently affixed to the outside surface of a propane cylinder. This shoulder tag provides the required marking information on the left and the warning label on the right.

FIGURE 6.7 At 49 C.F.R. §172.448, DOT requires shippers to affix the CARGO AIRCRAFT ONLY label to the exterior surface of the hazardous material packaging when its transportation is solely approved on board cargo aircraft or the net quantity of the hazardous material approved for transportation on passenger-carrying aircraft is exceeded. The printing on the CARGO AIRCRAFT ONLY label is black on an orange background.

6.5 DOT MARKING REQUIREMENTS

marking (mark) ■ For purposes of DOT regulations, a descriptive name, identification number, instructions, caution, weight, specification, "UN" marks, or any combination thereof required on outer packaging of hazardous materials

DOT requires several **markings**, or **marks**, to be written or affixed on packaging containing a hazardous material when it is offered for transportation. Some markings are written precautionary statements like the expressions "INHALATION HAZARD" or "MOLTEN ALUMINUM"; other markings resemble labels with pictograms with or without written precautionary statements.

There are two general types of these markings. The first type, called the *specification marking*, is provided by the manufacturer of the packaging. In compliance with DOT regulations, the manufacturer marks certain information on packaging before it is used by shippers or carriers. It usually consists of professionally printed information on an outer exterior surface of the packaging indicating that the packaging was constructed and tested in compliance with applicable DOT specifications and standards. The specification marking also includes the name and address or symbol of the packaging manufacturer or approval agency and the letters and numerals that identify the standard or specification. DOT's requirements relating to specification marking for gases transported in cylinders and rail tankcars were previously reviewed in Sections 3.3-B and 3.7-A, respectively.

The second type of marking requires shippers and carriers of hazardous materials to legibly mark certain other information on an outer exterior surface of the packaging used for shipment. These markings are written in English and are displayed on the packaging, unobstructed by the presence of labels, advertising, and other information. In particular, DOT requires shippers to mark on an outer exterior surface of nonbulk packaging certain information published at 49 C.F.R. §172.301. This information consists of the name and address of the shipper and receiver and the proper shipping name and UN/NA identification number of the hazardous material. When the proper shipping name in the hazardous materials table is designated by an n.o.s. entry, DOT requires shippers to mark the outer exterior surface of the packaging with the technical name of the material in parentheses immediately following or below the proper shipping name.

Some commonly encountered DOT marking requirements for shippers and carriers of hazardous materials are provided in the sections that immediately follow. Sections 6.5-A through 6.5-M, and Sections 6.5-O through 6.5-W address the requirements for nonbulk and bulk packaging, respectively. Special marking requirements applicable to hazard classes and divisions 1, 6.2, and 7 are noted later, in Sections 15.4-D, 10.21-D, and 16.10-C, respectively.

6.5-A MARKING NONBULK PACKAGING TO DENOTE PERFORMANCE TEST FEATURES

In compliance with global harmonization (Section 1.9), DOT requires manufacturers at 49 C.F.R. §178.503 to provide the following performance-oriented specification markings in the presented sequence on nonbulk outer packaging used to transport liquid and solid hazardous materials:

- The "un" mark shown in the left margin (the lowercase letters *u* and *n* inscribed within a circle) or the capital letters *UN*.
- A packaging identification code designating the type of package and the material from which it was constructed. This code is a combination of a first number, capital letter, and second number in this sequence. Their significance is provided in Table 6.6.
- One of the capital letters *X*, *Y*, or *Z*, each of which designates the performance standard for which the packaging was successfully tested, as follows:

  X   For packages meeting PG I, PG II, and PG III tests

  Y   For packages meeting PG II and PG III tests

  Z   For packages meeting PG III tests only

- A designation of the specific gravity of liquids, or the mass in kilograms of solids, for which the packaging design has been successfully tested without an inner lining
- The letter *S* for single and composite packaging intended to contain liquids, the test pressure in kilopascals rounded to the nearest 10 kilopascals of the hydrostatic pressure test that the packaging design type has successfully passed, and for packages intended to contain solids or inner packagings
- The last two digits of the year during which the package was manufactured
- The letters *USA* indicating that the package was marked pursuant to DOT's standards
- The symbol and identification number of the manufacturer

When the packaging and its contents weigh more than 66 pounds (30 kg), the markings are applied on an exterior surface of the top and a side of the packaging; otherwise they appear on the bottom of the packaging. They may be applied on a single line or multiple lines, and slash marks are generally used to separate the information into sections. Examples of these markings are provided in Figure 6.8.

Suppose 110 pounds (50 kg) of cupric cyanide is transported within several glass bottles that are individually cushioned in a fiberboard box. DOT requires the specification markings shown in Figure 6.9 to be embossed on an exterior surface of the box, the majority of which are provided by its manufacturer.

6.5-B MARKING CYLINDERS CONTAINING COMPRESSED GASES AND LIQUEFIED COMPRESSED GASES

We noted in Section 3.3-B that DOT requires gas cylinders to be marked so as to identify their service pressures. In addition, DOT requires gas cylinders to be periodically tested and retested to determine their mechanical integrity by authorized persons. As with the service pressures, the results of these testing activities and the identification number of the tester and retester must be marked on the outside surfaces of the cylinders.

CODE[a,b]	MEANING	CODE	MEANING
1A1	Steel drum, nonremovable head	5H3	Woven plastic bag, water-resistant
1A2	Steel drum, removable head	5H4	Plastic film bag
1B1	Aluminum drum, nonremovable head	5L1	Textile bag, unlined or noncoated
1B2	Aluminum drum, removable head	5L2	Textile bag, silt-proof
1D	Plywood drum	5L3	Paper bag, water-resistant
1G	Fiber drum	5M1	Paper bag, multiwall
1H1	Plastic drum, nonremovable head	5M2	Paper bag, multiwall water-resistant
1H2	Plastic drum, removable head	6HA1	Plastic receptacle within a protective steel drum
1N1	Metal drum, nonremovable head	6HA2	Plastic receptacle within a protective steel crate or box
1N2	Metal drum, removabale head	6HB1	Plastic receptacle within a protective aluminum drum
2C1	Wooden barrel, bung type	6HB2	Plastic receptacle within a protective aluminum crate or box
2C2	Wooden barrel, slack type, removable head	6HC	Plastic receptacle within a protective wooden box
3A1	Steel jerrican, nonremovable head	6HD1	Plastic receptacle within a protective plywood drum
3A2	Steel jerrican, removable head	6HD2	Plastic receptacle within a protective plywood box
3B1	Aluminum jerrican, nonremovable head	6HG1	Plastic receptacle within a protective fiber drum
3B2	Aluminum jerrican, removable head	6HG2	Plastic receptacle within a protective fiberboard box
3H1	Plastic jerrican, nonremovable head	6HH1	Plastic receptacle within a protective plastic drum
3H2	Plastic jerrican, removable head	6HH2	Plastic receptacle within a protective plastic box
4A	Steel box	6PA1	Glass, porcelain, or stoneware receptacles within a protective steel drum
4B	Aluminum box	6PA2	Glass, porcelain, or stoneware receptacles within a protective steel crate or box
4C1	Wooden box, ordinary	6PB1	Glass, porcelain, or stoneware receptacles within a protective aluminum drum
4C2	Wooden box, silt-proof walls	6PB2	Glass, porcelain, or stoneware receptacles within a protective aluminum crate or box
4D	Plywood box	6PC	Glass, porcelain, or stoneware receptacles within a protective wooden box
4F	Reconstituted wooden box	6PD1	Glass, porcelain, or stoneware receptacles within a protective plywood drum
4G	Fiberboard box	6PD2	Glass, porcelain, or stoneware receptacles within a protective wickerwork hamper
4H1	Plastic box, expanded	6PG1	Glass, porcelain, or stoneware receptacles within a protective fiber drum
4H2	Plastic box, solid	6PG2	Glass, porcelain, or stoneware receptacles within a protective fiberboard box
5H1	Woven plastic bag, unlined or uncoated	6PH1	Glass, porcelain, or stoneware receptacles within a protective expanded plastic packaging
5H2	Woven plastic bag, silt-proof	6PH2	Glass, porcelain, or stoneware receptacles within a protective solid plastic packaging

[a]49 C.F.R. §§178.504–178.521.
[b]49 C.F.R. §§178.522 and 178.523.

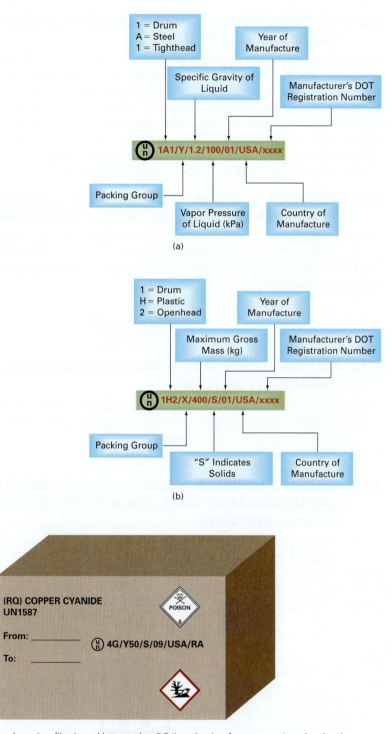

FIGURE 6.8 Examples of package identification markings embossed on (a) a steel tighthead drum containing a liquid chemical product and (b) an openhead plastic drum containing a solid chemical product.

(a) 1A1/Y/1.2/100/01/USA/xxxx

1 = Drum
A = Steel
1 = Tighthead

Specific Gravity of Liquid

Year of Manufacture

Manufacturer's DOT Registration Number

Packing Group

Vapor Pressure of Liquid (kPa)

Country of Manufacture

(b) 1H2/X/400/S/01/USA/xxxx

1 = Drum
H = Plastic
2 = Openhead

Maximum Gross Mass (kg)

Year of Manufacture

Manufacturer's DOT Registration Number

Packing Group

"S" Indicates Solids

Country of Manufacture

(RQ) COPPER CYANIDE
UN1587

From: _____

To: _____

4G/Y50/S/09/USA/RA

FIGURE 6.9 This outer package is a fiberboard box used as PG II packaging for transporting glass bottles containing copper cyanide, the proper shipping name for cupric cyanide. "RQ" must be included with the proper shipping name, because the reportable quantity for this commodity is 10 pounds. "4G" is provided in Table 6.6 as the UN identification code for a fiberboard box; "Y" designates the performance standard to determine whether the box retained its integrity when 50 kilograms of material was contained therein and subjected to packaging performance tests for Packing Group II; "S" designates that cupric cyanide is a solid; "09" means the box was manufactured in 2009; and "USA/RA" designates that the box was manufactured in the United States and marked pursuant to DOT standards by "RA," the symbol of the box's manufacturer. DOT also requires the shipper to affix the MARINE POLLUTANT mark and POISON labels to the box.

6.5-C MARKING NONBULK PACKAGING CONTAINING HAZARDOUS SUBSTANCES

When shippers transport a hazardous substance in nonbulk packaging in an amount that exceeds its reportable quantity, DOT requires them to mark the outer exterior surface of the packaging with the letters *RQ* in association with the proper shipping name. If the proper shipping name does not specifically identify the hazardous substance, DOT also requires them to indicate the name of the hazardous substance on the packaging in parentheses in association with the proper shipping name. If more than one hazardous substance is contained within the packaging, DOT requires them to indicate at least the two hazardous substances with the lowest reportable quantities.

6.5-D MARKING NONBULK PACKAGING CONTAINING HAZARDOUS MATERIALS THAT POSE AN INHALATION HAZARD

When shippers offer for transportation in nonbulk packaging a hazardous material that poses a health hazard by inhalation, DOT requires them at 49 C.F.R. §172.102.13 to mark "INHALATION HAZARD" in association with the proper shipping name and identification number.

6.5-E MARKING NONBULK PACKAGING CONTAINING CLASS 6 HAZARDOUS MATERIALS

When a poisonous material is transported within nonbulk plastic packaging, DOT requires shippers to *permanently* mark the packaging with the word "POISON."

POISON

6.5-F MARKING NONBULK PACKAGING CONTAINING LIQUID HAZARDOUS MATERIALS OTHER THAN LIQUEFIED COMPRESSED GASES

When shippers transport liquid hazardous materials other than liquefied compressed gases in nonbulk packaging, DOT requires them at 49 C.F.R. §172.312 to assure that the closure is upward, and that the packages are marked with package orientation markings on the outer exterior surface of the packaging on two opposing sides with arrows pointing in the correct upright direction, as shown in the following illustration:

limited quantity (Ltd Qty) ■ For purposes of DOT regulations, the maximum amount of a specific hazardous material for which there may be labeling or packaging exceptions

6.5-G MARKING NONBULK PACKAGING CONTAINING A LIMITED QUANTITY OF A HAZARDOUS MATERIAL FOR SHIPMENT BY RAIL OR AIR

DOT authorizes the transportation of certain hazardous materials in a **limited quantity (Ltd Qty)** by passenger-carrying aircraft, rail, or cargo aircraft. DOT provides this authorization in column (9) of the Hazardous Materials Table published at 49 C.F.R. §172.101.

When the transportation of a hazardous material is authorized in a limited quantity by rail, DOT requires the shipper at 49 C.F.R. §172.315 to mark its package on at least one side and one end as shown on the left below. When the transportation of a hazardous material is authorized in a limited quantity by passenger-carrying or cargo aircraft, DOT requires them to mark the package on at least one side and one end as shown on the right.

The top and bottom triangular portions of both markings, their borders, and the letter *Y* are black and their centers are white.

6.5-H MARKING NONBULK PACKAGING REQUIRING A SPECIAL PERMIT

When DOT assigns a special permit to a shipper to containerize a hazardous material in an unconventional fashion, it requires the shipper to mark the packaging with the notation "DOT-SP-" followed by the assigned special permit number.

6.5-I MARKING NONBULK PACKAGING CONTAINING HAZARDOUS WASTES

When an RCRA-regulated hazardous waste (Section 1.3-C) is transported in containers having a capacity of 119 gallons (450 L) or less, EPA requires the hazardous waste generator to affix the HAZARDOUS WASTE marking in Figure 6.10 on an outer surface of the containers.

6.5-J MARKING NONBULK PACKAGING OF MARINE POLLUTANTS OR SEVERE MARINE POLLUTANTS

When shippers intend to transport a marine pollutant or severe marine pollutant by watercraft in nonbulk packaging, DOT requires them to affix the marine pollutant mark shown in Figure 6.11 on an outer exterior surface of the packaging along with the following information:

- When the proper shipping name does not specifically identify the name of the marine pollutant, DOT requires them to indicate the name on the packaging in parentheses in association with the proper shipping name.

HAZARDOUS WASTE

STATE & FEDERAL LAW PROHIBITS IMPROPER DISPOSAL.

IF FOUND, CONTACT THE NEAREST POLICE OR PUBLIC SAFETY
AUTHORITY OR THE U.S. ENVIRONMENTAL PROTECTION AGENCY
OR THE CALIFORNIA DEPARTMENT OF TOXIC SUBSTANCES CONTROL.

GENERATOR INFORMATION:
NAME _____
ADDRESS _____ PHONE _____
CITY _____ STATE _____ ZIP _____
EPA /MANIFEST
ID NO. / DOCUMENT NO. _____ / _____
EPA
WASTE NO. _____ CA. WASTE NO. _____ ACCUMULATION START DATE _____
CONTENTS COMPOSITION _____

PHYSICAL STATE: HAZARDOUS PROPERTIES ☐ FLAMMABLE ☐ TOXIC
☐ SOLID ☐ LIQUID   ☐ CORROSIVE ☐ REACTIVITY ☐ OTHER

D.O.T. PROPER SHIPPING NAME AND UN OR NA NO. WITH PREFIX

HANDLE WITH CARE!
CONTAINS HAZARDOUS OR TOXIC WASTES

FIGURE 6.10 When a container having a capacity of 119 gallons (450 L) or less is used to transport hazardous wastes (Section 1.3-C), DOT and EPA require their generators at 40 C.F.R. §262.32 to affix this HAZARDOUS WASTE marking on the containers together with the generator's name and address and the relevant manifest number.

FIGURE 6.11 When a marine pollutant is offered for transportation in nonbulk packaging by watercraft or bulk packaging by any mode, DOT requires shippers and carriers at 49 C.F.R. §172.322 to post this MARINE POLLUTANT marking on its packaging. The pictogram on this DOT marking is identical to the GHS Environment (Aquatic Toxicity) pictogram shown in Table 1.2. It displays a dead tree and dead fish, and is printed in black on a white background or other contrasting color.

- If two or more marine pollutants are contained within packaging, DOT requires shippers to indicate parenthetically at least the two components most predominantly contributing to the marine pollutant designation in association with the proper shipping name.
- When a nonbulk container is inserted into a transport vehicle or freight container for shipment by watercraft, shippers must affix the MARINE POLLUTANT marking on each side and end of the transport vehicle or freight container.
- When DOT has not listed a marine pollutant in the Hazardous Materials Table at 49 C.F.R. §172.101 and when the marine pollutant does not meet the definition of any class 1 through 8 material, DOT requires shippers to designate by either of the following class 9 entries, as relevant: "UN3082, Environmentally hazardous substance, liquid, n.o.s., 9, PG II" or "UN3077, Environmentally hazardous substance, solid, n.o.s., 9, PG III."

6.5-K MARKING OVERPACKED NONBULK PACKAGING

To provide protection or convenience in handling hazardous materials, DOT allows a single consignor to **overpack** two or more nonbulk packages on a pellet and secure them by strapping, shrink-wrapping, or placement in a box or crate. When the overpacked consignment is offered for transportation, DOT requires the consignor at 49 C.F.R. §173.25(a)(4) to mark the packaging with the word *OVERPACK* on an outer exterior surface.

overpack ▪ For purposes of DOT regulations, an enclosure other than a transport vehicle, freight container, or aircraft unit-loading device that is used by a single consignor of a hazardous material to provide protection or convenience in handling or to consolidate two or more packages of hazardous materials

OVERPACK

The packaging must also be marked with the proper shipping name, identification number, and orientation arrows, when applicable, and be labeled as required for each hazardous material contained therein.

6.5-L MARKING NONBULK PACKAGING CONTAINING AN EXCEPTED QUANTITY

excepted quantity ▪ For purposes of DOT regulations, an authorized amount of a hazardous material in certain hazard classes that may be offered for transportation in accordance with regulations published at 49 C.F.R. §173.4

DOT allows hazardous materials in certain hazard classes to be shipped as **excepted quantities.** These are very small amounts that a shipper may offer for transportation by a given mode without being subjected to DOT's shipping paper, label, and placard requirements. When an excepted quantity of a hazardous material is offered for transportation, DOT requires its shipper at 49 C.F.R. §173.4(g) to affix the EXCEPTED QUANTITY marking shown at the top of the following page or on an outer exterior surface of the packaging. The * and ** are used to identify the relevant hazard class/division number and the name of the shipper or consignee, respectively. The crosshatched border and pictogram are red on a white background. The lettering for the specific hazard class/division number and the name of the shipper or consignee are printed in black.

6.5-M MARKING NONBULK PACKAGING CONTAINING ORM-DS

When a material classified as ORM-D is offered for transportation, DOT requires shippers to mark its nonbulk packaging with the "ORM-D" designation placed within a rectangle. When the package is intended for shipment by air, the "ORM-D" designation is changed to "ORM-D-AIR."

DOT requires shippers to mark "ORM" or "ORM-D-AIR," as relevant, on at least one side and one end on the surface of their packaging or on an attached tag.

6.5-N MARKING ALL FORMS OF BULK PACKAGING CONTAINING HAZARDOUS MATERIALS

At 49 C.F.R. §172.302(a), DOT requires carriers to mark bulk packaging containing a hazardous material with the relevant identification number on each side and each end if the packaging has a capacity of 1000 gallons (3785 L) or more, *or* on two opposing sides if the packaging has a capacity less than 1000 gallons (3785 L).

6.5-O MARKING TRANSPORT VEHICLES AND FREIGHT CONTAINERS LOADED WITH HAZARDOUS MATERIALS IN NONBULK PACKAGING

When an aggregate gross weight of more than 8820 pounds (4000 kg) of a *single* hazardous material in nonbulk packaging is loaded at one facility and is to be transported in a transport vehicle or freight container, DOT requires the carrier at 49 C.F.R. §172.301(a)(3) to mark the transport vehicle or freight container on each side and each end with the proper shipping name and identification number of the hazardous material.

When an aggregate gross weight of more than 2205 pounds (1000 kg) of a single hazardous material in hazard class 6.1, Zones A or B, is loaded at one facility and is to be transported in a transport vehicle or freight container, DOT requires the carrier at 49 C.F.R. §172.313(c) to mark the packaging with the identification number of the material on each side and each end if the packaging has a capacity equal to or greater than 1000 gallons (3785 L); *or* on two opposing sides if the packaging has a capacity of less than 1000 gallons (3785 L).

6.5-P MARKING FREIGHT CONTAINERS CONTAINING FUELED ITEMS

To draw attention to the potential for explosive ignition, DOT requires shippers and carriers at 49 C.F.R. §176.905 to mark freight containers holding a motor vehicle or mechanical equipment with fuel in its tanks. The EXPLOSIVE-MIXTURE marking resembles the following:

6.5-Q MARKING BULK PACKAGING CONTAINING HAZARDOUS MATERIALS THAT POSE AN INHALATION HAZARD

When shippers transport in bulk packaging a hazardous material that poses a health hazard by inhalation, DOT requires them at 49 C.F.R. §172.102.13 to mark the words *INHALATION HAZARD* on two opposing sides of the bulk packaging.

> **INHALATION HAZARD**

6.5-R MARKING BULK PACKAGING CONTAINING ELEVATED-TEMPERATURE MATERIALS

elevated-temperature material ■ For purposes of the DOT regulations, a hazardous material that when offered for transportation or transported in a bulk packaging exists as a liquid at a temperature equal to or greater than 212°F (100°C); a liquid with a flashpoint equal to or greater than 100°F (37.8°C) that is intentionally heated and offered for transportation or transported at or above its flashpoint; *or* a solid at a temperature equal to or greater than 464°F (240°C)

The DOT regulations refer to an **elevated-temperature material** as any of the following:

- A liquid at a temperature equal to or greater than 212°F (100°C);
- A liquid with a flashpoint equal to or greater than 100°F (37.8°C); *or*
- A solid at a temperature equal to or greater than 464°F (240°C)

When carriers transport bulk packaging containing an elevated-temperature material by highway or rail, DOT requires them at 49 C.F.R. §172.325 to display the HOT marking shown below on two opposing sides of the packaging. For example, when they transport a liquid whose identification number is 3256, they post the following HOT marking together with an orange panel displaying the liquid's identification number and a FLAMMABLE placard:

DOT designed the HOT marking as a white square-on-point diamond with black lettering and border.

DOT also allows carriers to post the identification number of the elevated-temperature material on a HOT marking within a hatched rectangle below the word *HOT* in a white square-on-point diamond, or at the upper corner of the same square-on-point diamond. For example, a liquid whose identification number is 3256 may be transported with the following HOT marking that displays the liquid's identification number and a FLAMMABLE placard:

6.5-S MARKING BULK PACKAGING CONTAINING MARINE POLLUTANTS OR SEVERE MARINE POLLUTANTS

When carriers transport a marine pollutant or severe marine pollutant in bulk packaging by any mode, DOT requires them to affix the MARINE POLLUTANT marking on at least two opposing sides *or* two ends other than the bottom if the packaging has a capacity of less than 1000 gallons (3785 L), *or* on each side *and* each end if the packaging has a capacity equal to or exceeding 1000 gallons (3785 L). When carriers transport a bulk quantity of a marine pollutant or severe marine pollutant in a transport vehicle or freight container, DOT requires them to mark "MARINE POLLUTANT" on each of its sides and ends.

6.5-T MARKING EMPTIED BULK PACKAGING THAT CONTAINED HAZARDOUS MATERIALS

DOT requires at 49 C.F.R. §172.302(d) bulk packaging that has been marked with a proper shipping name, common name, or identification number to remain marked when it is emptied unless it is sufficiently cleaned of residue and purged of vapors to remove any potential hazard, *or* refilled with a material requiring different markings or no markings, such that any residue remaining in the packaging is no longer hazardous.

6.5-U MARKING PORTABLE TANKS CONTAINING HAZARDOUS MATERIALS

When a hazardous material is transported in a portable tank, DOT requires shippers at 49 C.F.R. §172.326 to mark the portable tank on two opposing sides with the proper shipping name of its contents. For example, when a solution of a metallic hypochlorite (Section 11.6) is transported in a portable tank, DOT requires the tank to be marked in the following manner:

> **HYPOCHLORITE SOLUTION**

6.5-V MARKING BULK PACKAGING CONTAINING CERTAIN HAZARDOUS MATERIALS FOR SHIPMENT BY HIGHWAY OR RAIL

When transporting a hazardous material in bulk packaging by highway or rail, DOT requires carriers at 49 C.F.R. §§172.332, 172.334, and 172.336 to identify the identification number of the hazardous material by displaying it on each side and each end of the

TABLE 6.7	Key Words of Proper Shipping Names That Are Marked on Each Side of a Tankcar for Shipment by Highway or Rail[a]	
Acrolein, stabilized	Hydrogen cyanide, stabilized (less than 3% water)
Ammonia, anhydrous, liquefied	Hydrogen fluoride
Ammonia solutions (more than 50% ammonia)	Hydrogen peroxide, aqueous solutions (greater than 10% hydrogen peroxide)
Bromine *or* Bromine solutions	Hydrogen peroxide, stabilized
Bromine chloride	Hydrogen peroxide and peroxyacetic acid mixtures
Chloroprene, stabilized	Nitric acid (other than red fuming)
Dispersant gas *or* Refrigerant gas	Phosphorus (amorphous)
Division 2.1 materials	Phosphorus, white dry *or* Phosphorus, white, under water *or* Phosphorus white, in solution, *or* Phosphorus, yellow dry *or* Phosphorus, yellow under water *or* Phosphorus, yellow in solution
Division 2.2 materials (DOT-107 tank-cars only)	Phosphorus, white, molten
Division 2.3 materials	Potassium nitrate and sodium nitrate mixtures
Formic acid	Potassium permanganate
Hydrocyanic acid, aqueous solutions	Sulfur trioxide, stabilized
	Sulfur trioxide, uninhibited

[a]49 C.F.R. §172.330(a)(ii).

bulk packaging across the centers of placards, white square-on-point diamonds, or orange panels as shown in (a), (b), and (c), respectively, of the following figure:

(a)   (b)   (c)

When transporting gasoline, the FLAMMABLE placard is placed adjacent to both the orange panel and the white square-on-point diamond.

When carriers transport more than one hazardous material separately in compartmented portable tanks, cargo tanks, or tankcars, DOT requires them to display identification numbers of the hazardous materials on the ends of the tank *and* on the sides in the same sequence as the compartments containing the materials they identify.

When shippers transport a bulk quantity of the hazardous materials listed in 7 in a tankcar by highway or railroad, DOT requires them to mark the key word from the applicable proper shipping name on each side of the tankcar.

6.5-W MARKING BULK PACKAGING DURING OR FOLLOWING THEIR FUMIGATION

When the lading in a railcar, freight container, truck body, or trailer has been fumigated or is undergoing fumigation, DOT requires the carrier at 49 C.F.R. §173.9 to display the FUMIGANT marking shown in Figure 6.12 so it is observable when a person attempts to

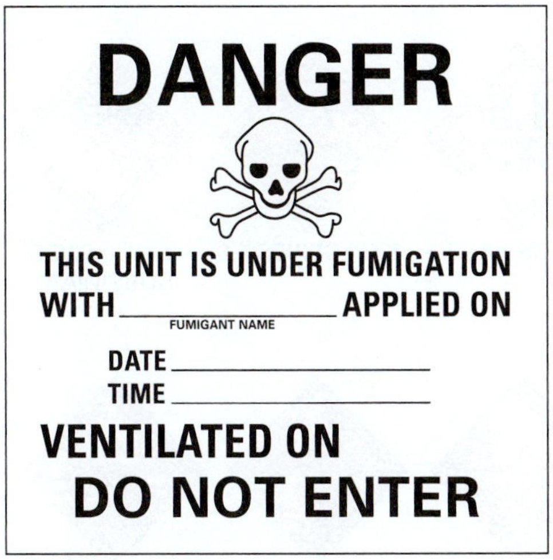

DANGER

☠

THIS UNIT IS UNDER FUMIGATION WITH_____APPLIED ON
FUMIGANT NAME

DATE_____
TIME_____

VENTILATED ON
DO NOT ENTER

FIGURE 6.12 A fumigant is a pesticide that is a vapor or gas, or forms a vapor or gas on application, and whose method of pesticidal action is through inhalation of the gaseous state. DOT regards a rail tankcar, freight container, truck body, or trailer in which lading has been fumigated, or is undergoing fumigation, as a package containing a hazardous material. DOT requires carriers at 49 C.F.R. §173.9(c) to display this FUMIGANT marking in such a manner that it may be readily observed by a person attempting to inspect the lading. When a FUMIGANT marking is displayed, DOT requires carriers to include on the marking the name of the fumigant, the date and time of fumigation, and the date on which ventilation occurred. The printing may be either black or red on a white background.

enter it or otherwise comes in contact with the **fumigated lading**. DOT also requires the carrier to inscribe the technical name of the fumigant and the date and time of its application on the face of the marking.

6.6 DOT PLACARDING REQUIREMENTS

DOT requires carriers at 49 C.F.R. §172.504 to display in plain view one or more warning **placards** on each side and each end of the bulk packaging, freight container, unit load device, transport vehicle, or rail tankcar used to transport hazardous materials. These DOT placards are illustrated in Figure 6.13. The features of the individual placards are similar to those of the corresponding labels. Both are diamond-shaped and color-coded to signify a specific hazard class or division and may also include a pictograph.

The following points regarding the nature of DOT labels and placards are relevant:

- Although there is a COMBUSTIBLE placard, there is not an analogous label. Furthermore, although there is an INFECTIOUS SUBSTANCE label, there is not an analogous placard.
- The word *GASOLINE* may be used in lieu of the word *FLAMMABLE* on a placard that is displayed on a cargo tank or a portable tank used to transport gasoline by highway.
- The words *FUEL OIL* may be used in lieu of the word *COMBUSTIBLE* on a placard that is displayed on a cargo tank used to transport fuel oil that is not classed as a flammable liquid by highway.
- When solely transporting oxygen, carriers may display OXYGEN placards (Section 7.1-F) in lieu of NON-FLAMMABLE GAS placards. There is not an analogous OXYGEN label.
- As with the EXPLOSIVE labels, the uppercase letters that follow the division designations for the EXPLOSIVES placards are examples of the designations for the compatibility groups of explosives, whose nature is discussed in Section 15.4-B.

Although shippers and carriers are mutually responsible for the choice of the placards that are displayed on transport vehicles, the carriers typically bear the responsibility for posting them; i.e., shippers and carriers select the applicable placards by referring to the codes listed in column 3 of the Hazardous Materials Table, and the carriers display them as follows:

- Carriers *always* display the applicable placards when transporting materials whose hazard classes are listed in Table 6.8, regardless of the amounts transported. For example, carriers display RADIOACTIVE placards on each side and each end of a motor van used to transport wooden boxes on which RADIOACTIVE YELLOW-III labels have been affixed, regardless of the amount of hazard-class-7 material in the boxes.

fumigated lading ■ A material in transit or intended for transportation, such as grain, that has been treated with a fumigant to destroy rodents, insects, and germs

placard ■ For purposes of DOT regulations, a sign displayed by the carrier on bulk packaging, freight containers, transport vehicles, unit containment devices, or railcars to rapidly communicate hazard information relating to the hazardous material being transported

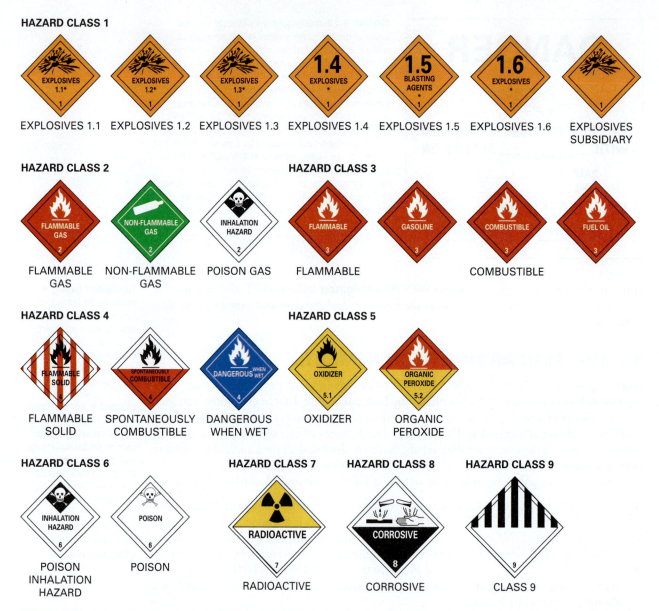

FIGURE 6.13 When carriers transport hazardous materials, DOT requires them to post warning placards on their bulk packaging. The placard at the far right in Hazard Class 1 is the EXPLOSIVES subsidiary placard. It is posted on bulk packaging when the number *1* is listed in column 6 of the Hazardous Materials Table as a subsidiary hazard. The asterisks in the orange EXPLOSIVES placards are replaced by the compatibility group (Section 15.4-B), or left blank if the explosive hazard is the subsidiary risk associated with the product.

■ Carriers display the applicable placards on the same transport vehicle when they transport a material whose hazard class is listed in Table 6.9 and when the aggregate gross mass of the hazardous materials in any one hazard class equals or exceeds 1001 pounds (454 kg). For example, carriers display FLAMMABLE placards on each side and each end of a rail boxcar used to transport steel drums containing acetonitrile, a flammable liquid, only when their aggregate gross mass equals or exceeds 1001 pounds (454 kg).

■ When carriers transport multiple hazardous materials including at least one of whose mass exceeds 1001 pounds (454 kg), they display the applicable placards for *each* hazardous material on the transport vehicle. For example, when carriers transport more than 1001 pounds (454 kg) of acetonitrile with 500 pounds (227 kg) of sulfuric acid, a

TABLE 6.8	DOT Hazard Classes That Always Require Placarding[a]
HAZARD CLASS NUMBER, DIVISION NUMBER, AND ADDITIONAL INFORMATION	**PLACARD NAME**
1.1	EXPLOSIVES 1.1
1.2	EXPLOSIVES 1.2
1.3	EXPLOSIVES 1.3
2.3	POISON GAS
4.3	DANGEROUS WHEN WET
5.2 (Organic peroxide, Type B, liquid or solid, temperature-controlled)	ORGANIC PEROXIDE
6.1 (Inhalation hazard, Zone A or B)	POISON INHALATION HAZARD
7 (RADIOACTIVE YELLOW-III label only)[b]	RADIOACTIVE

[a]49 C.F.R. §172.504, Table 1.
[b]Section 16.10-B.

corrosive material, DOT requires them to display *both* FLAMMABLE and CORROSIVE placards on each side and each end of the transport vehicle.

■ When carriers transport one or more hazardous materials in a placarded motor vehicle, DOT requires the operator of the vehicle to have been issued a commercial driver's license with a **hazardous materials endorsement,** the standards of which are provided at 49 C.F.R. §383.121. These standards largely require that the driver know the hazardous materials regulations.

hazardous materials endorsement ■ For purposes of DOT regulations, a recognition by the U.S. Transportation Security Administration that an applicant has knowledge of the standards published at 49 C.F.R. §383.121 and is unlikely to pose a security risk when transporting hazardous materials in a placarded motor vehicle

TABLE 6.9	DOT Hazard Classes That Require Placarding Only Under Certain Conditions[a]
HAZARD CLASS NUMBER, DIVISION NUMBER, AND ADDITIONAL INFORMATION	**PLACARD NAME**
1.4	EXPLOSIVES 1.4
1.5	EXPLOSIVES 1.5
1.6	EXPLOSIVES 1.6
2.1	FLAMMABLE GAS
2.2	NON-FLAMMABLE GAS
3	FLAMMABLE
Combustible liquid	COMBUSTIBLE
4.1	FLAMMABLE SOLID
4.2	SPONTANEOUSLY COMBUSTIBLE
5.1	OXIDIZER
5.2 (other than organic peroxide, type B, liquid or solid, temperature-controlled)	ORGANIC PEROXIDE
6.1 (other than inhalation hazard, Zone A or Zone B)	POISON
6.2	(None)
8	CORROSIVE
9	CLASS 9
ORM-D	(None)

[a]49 C.F.R. §172.504, Table 2.

Identify the placard(s), if any, that DOT requires a carrier to display on a motor van used to transport six cylinders of compressed fluorine gas, each of which weighs 2.5 pounds.

Solution: By referring to the entry in column 3 of Appendix C, we determine that the hazard class for compressed fluorine is 2.3. By referring to Table 6.8, we determine that when a carrier transports a hazardous material whose hazard class is 2.3, placarding of the transport vehicle is *always* required, irrespective of the amount transported. Consequently, DOT requires the carrier to display POISON GAS placards on each side and each end of the transport vehicle.

6.6-A PLACARDING REQUIREMENTS WHEN SHIPPING MULTIPLE PACKAGES OF MATERIALS WHOSE HAZARD CLASSES ARE LISTED SOLELY IN TABLE 6.9

When carriers transport multiple nonbulk packages containing materials requiring the posting of several placards listed in Table 6.9, DOT permits at 49 C.F.R. §172.504 the display of the DANGEROUS placard shown in the left margin in lieu of the separate hazard class placards.

However, DOT prohibits the use of DANGEROUS placards when an aggregate gross mass equal to or exceeding 2205 pounds (1000 kg) of materials having a single hazard class listed in Table 6.9 is loaded at a single location. In this instance, carriers are obligated to display the applicable placards specified in Table 6.10 subject to the following conditions:

■ When the hazardous materials in Table 6.10 are transported by highway or rail, placards are not required when the transport vehicle or freight container contains less than 1001 pounds (454 kg) of a hazardous material listed in Table 6.9.

■ When the hazardous materials are transported in a railcar loaded with transport vehicles or freight containers, none of which is required to be placarded, placards are not required.

DOT never allows carriers to display DANGEROUS placards in lieu of the required placards when they transport hazardous materials whose hazardous class or division is listed in Table 6.8; nor does DOT permit carriers to display UN/NA identification numbers across the center face of any posted DANGEROUS placard.

A carrier loads nonbulk containers of the following hazardous materials into a rail boxcar for delivery from a chemical warehouse to a customer's plant site: 1500 pounds (682 kg) of acetone, 500 pounds (227 kg) of ethyl mercaptan, 500 pounds (227 kg) of barium cyanide, and 1000 pounds (454 kg) of methyl isobutyl ketone. Which placards does DOT require the carrier to display on the boxcar?

Solution: We note from Appendix C that the hazard class codes of acetone, ethyl mercaptan, barium cyanide, and methyl isobutyl ketone are 3, 3, 6.1, and 3, respectively. The gross aggregate mass of the hazardous materials in hazard class 3 is 1500 pounds + 500 pounds + 1000 pounds, or 3000 pounds. Because DOT authorizes carriers to display DANGEROUS placards on the boxcar in lieu of the separate hazard class placards *only* when the aggregate gross mass of the materials in any one hazard class does not exceed 2205 pounds (1000 kg), DOT prohibits the carrier from displaying DANGEROUS placards and requires the posting of FLAMMABLE placards on each side and each end of the boxcar. The posting of POISON placards is not required since the boxcar is loaded with less than 1001 pounds (454 kg) of barium cyanide.

6.6-B PLACARDING FOR SUBSIDIARY HAZARDS

In the following three circumstances, DOT requires carriers to display warning placards to account for a subsidiary hazard *in addition to* displaying other required placards:

- When carriers transport a hazardous material whose subsidiary hazard is 2.3 or 6.1, DOT requires them at 49 C.F.R. §172.505 to display POISON GAS or POISON INHALATION HAZARD placards, as relevant, on each side and each end of the transport vehicle, freight container, portable tank, unit load device, or railcar used for transportation. This regulatory requirement applies to all hazardous materials that pose an inhalation hazard, regardless of their unique hazard zones.
- When carriers transport a hazardous material whose subsidiary hazard is dangerous-when-wet, DOT requires them to display DANGEROUS WHEN WET placards on each side and each end of the transport vehicle, freight container, portable tank, unit load device, or railcar used for transportation.
- When carriers transport 1001 lb (454 kg) or more of uranium hexafluoride (UF_6), DOT requires them to display CORROSIVE and RADIOACTIVE placards on each side and each end of the transport vehicle, portable tank, or freight container used for transportation. This subject is noted again in Section 16.9-D.

When a hazardous material possesses subsidiary hazards, DOT requires carriers to display multiple placards to account for the subsidiary hazards. However, DOT prohibits them from displaying the UN/NA identification number across the center face of a subsidiary hazard placard.

6.6-C SPECIAL PLACARDING REQUIREMENTS WHEN HAZARDOUS MATERIALS ARE TRANSPORTED IN SEPARATE COMPARTMENTED PORTABLE TANKS, CARGO TANKS, OR TANKCARS

When carriers transport more than one hazardous material in separate compartmented portable tanks, cargo tanks, or tankcars, DOT requires the carrier to display the applicable placard on the ends of the tank *and* on the sides in the same sequence as the compartments containing the materials they identify. On the sides, the placards are positioned near the location of each compartment as shown in Figure 6.14.

FIGURE 6.14 When a multicompartment cargo tank is used to transport more than one hazardous material by highway, DOT requires its carrier to position the appropriate placard corresponding to each hazardous material on the two sides near the location of each separate compartment. On the two ends, DOT requires the carrier to position them in the same sequence as the compartments containing the materials they identify. This three-compartment cargo tank is placarded to simultaneously transport gasoline, an ethanol and gasoline mixture, and denatured alcohol, each of which DOT designates in hazard class 3. Because these hazardous materials have only a single hazard class, DOT requires the carrier to post only a single FLAMMABLE placard on each side and each end of the vehicle. *(Courtesy of the U.S. Department of Transportation, Washington, DC.)*

6.6-D SPECIAL PLACARDING REQUIREMENTS WHEN HAZARDOUS MATERIALS IN CERTAIN DIVISIONS ARE TRANSPORTED BY RAIL

When carriers transport hazardous materials in the following hazard classes by rail, DOT requires them at 49 C.F.R. §172.510 to post the relevant placard on a square having a white background and black border:

- Division 1.1 explosive materials
- Division 1.2 explosive materials
- Division 2.3 poisonous gases, Zone A
- Division 6.1 poisonous materials, Zone A, to which Packing Group I is assigned
- Division 2.1 flammable gases when transported in a DOT-113 rail tankcar, including tankcars containing only the residues of Division 2.1 flammable gases.

By requiring the posting of these placards on a white square with a black border, DOT directs a railcar-switching crew to exercise extraordinary precaution to prevent the tankcar from being cut off while in motion. These five placards resemble the following:

6.7 RESPONDING TO INCIDENTS INVOLVING THE RELEASE OF HAZARDOUS MATERIALS

When shippers and carriers have correctly complied with the DOT regulations, the UN/NA identification number of a hazardous material can be readily located during an emergency-response action by any of the following means:

- Listed on a shipping paper as a component of the shipping description of the hazardous material
- Marked on the packaging containing the hazardous material
- Displayed on bulk packaging within an orange panel or across the center area of a placard or a white square-on-point diamond

Having located the UN/NA identification number, how may first-on-the-scene responders use it during an incident involving the release of a hazardous material? The answer to this question involves consulting the *Emergency Response Guidebook* (*ERG*).[1] This is a DOT manual that primarily provides direction to emergency-response crews during the initial phases of a response action. OSHA and EPA require training for emergency responders regarding the use of the ERG at 29 C.F.R. §1910.120 and 40 C.F.R. Part 311, respectively.

The guidebook serves as the primary reference book for first-on-the-scene personnel. It directs them to a guide, which provides suggested actions that should be considered during the initial response phase of a transportation mishap. When properly implemented, each guide listed within the guidebook provides emergency-response personnel with vital information on how to initially deal with an incident involving the release of a unique hazardous material until more specific information is obtained from the shipper, carrier, manufacturer, or CHEM-

---

[1]In the United States, the *Emergency Response Guidebook* is provided free of charge to state and local safety authorities. It is also available from private commercial companies; the Research and Special Programs Administration, Hazardous Materials Transportation Bureau, U.S. Department of Transportation, Washington, DC 20590; the U.S. Government Printing Office, Superintendent of Documents, Mail Stop SSOP, Washington, DC 20402-9328; and the American Chemistry Council's Bookstore at (301) 617-7842. From government agencies, request ISBN 0-16-042938-2.

TREC. Each guide provides information on the potential hazards of the situation (fire and health), public safety comments for first-responders (initial instructions, protective clothing, and evacuation), and emergency response directions (fire, spill or leak, and first aid).

The *Emergency Response Guidebook* is organized into the following sections:

■ **YELLOW SECTION**, is so called because it has yellow-bordered pages. It provides a listing of the individual UN/NA identification numbers and the hazardous materials having these numbers. A guide number that refers the user to the Orange Section is also provided.

■ **BLUE SECTION** This section has blue-bordered pages. It lists alphabetically the names of the hazardous materials with their corresponding UN/NA identification numbers. Like the Yellow Section, it provides a guide number that refers the user to the Orange Section.

■ **ORANGE SECTION** This section has orange-bordered pages. It contains 172 individual guide numbers, each of which constitutes a two-page summary of information concerning the potential hazards of the referenced hazardous material, recommended actions that relate to public safety, and recommended emergency-response actions to be implemented during a transportation mishap involving the material in question.

■ **GREEN SECTION** This section has green-bordered pages. It provides initial isolation and protective-action distances for small and large spills of hazardous materials that are highlighted in the yellow and blue sections. "Small spills" refer to the release of the contents from a single, small package (e.g., a small gas cylinder or a drum whose liquid capacity is 55 gallons [208 L] or less) or a small leak from a large package. "Large spills" consist of a spill from a large package (e.g., a rail tankcar, a highway tankcar or trailer, or a ton-container) or multiple spills from many small packages. An exception applies to the release of chemical warfare agents (Section 13.11), for which "small spills" include releases up to 4.4 pounds (2 kg) and "large spills" include releases up to 55 pounds (25 kg).

The Green Section also provides container-specific guidance for first-on-the-scene responders relating to the release of the six most common toxic substances transported by rail or highway. The containers are rail tankcars; highway tank trucks or trailers; multiple ton cylinders; and multiple small cylinders or single ton cylinders. The toxic substances are chlorine, hydrogen chloride, ammonia, hydrogen fluoride, sulfur dioxide, and ethylene oxide.

The guidebook should always be accessible for immediate use within emergency-response vehicles, so that responding personnel may locate the appropriate guide number and implement the recommended actions. DOT also requires shippers and carriers to have emergency-response information available to workers during all phases of the transportation of a hazardous material. This requirement is often met by having available a copy of the *Emergency Response Guidebook* during hazardous material loading, unloading, and transfer operations.

Suppose that information is needed concerning the appropriate response action to be taken at a transportation incident involving an initially unidentified hazardous material being transported by motor carrier on a crowded highway. On arriving at the scene, personnel note that CORROSIVE placards are displayed on the exterior surface of the vehicle, adjacent to which are orange panels in which the number *1805* is inscribed. On securing the shipping paper from the driver, personnel also note that a number of items not regulated by DOT are components of the consignment as well as a hazardous material having the following shipping description:

UNITS	HM	SHIPPING DESCRIPTION (IDENTIFICATION NUMBER, PROPER SHIPPING NAME, PRIMARY HAZARD CLASS OR DIVISION, SUBSIDIARY HAZARD CLASS OR DIVISION, AND PACKING GROUP)	WEIGHT (lb)
10 drums (UN1A1)	X	UN1805, Phosphoric acid solution, 8, PG III	585

FIGURE 6.15 *Guide 154 from* Emergency Response Guidebook (*Washington, DC: U.S. Department of Transportation, 2012*), pp. 246–247. (*Courtesy of the U.S. Department of Transportation, Washington, DC.*)

GUIDE 154	SUBSTANCES - TOXIC AND/OR CORROSIVE (NON-COMBUSTIBLE)	ERG2012

POTENTIAL HAZARDS

HEALTH

- TOXIC; inhalation, ingestion or skin contact with material may cause severe injury or death.
- Contact with molten substance may cause severe burns to skin and eyes.
- Avoid any skin contact.
- Effects of contact or inhalation may be delayed.
- Fire may produce irritating, corrosive and/or toxic gases.
- Runoff from fire control or dilution water may be corrosive and/or toxic and cause pollution.

FIRE OR EXPLOSION

- Non-combustible, substance itself does not burn but may decompose upon heating to produce corrosive and/or toxic fumes.
- Some are oxidizers and may ignite combustibles (wood, paper, oil, clothing, etc.).
- Contact with metals may evolve flammable hydrogen gas.
- Containers may explode when heated.
- For UN3171, if Lithium ion batteries are involved, also consult GUIDE 147.

PUBLIC SAFETY

- **CALL EMERGENCY RESPONSE Telephone Number on Shipping Paper first. If Shipping Paper not available or no answer, refer to appropriate telephone number listed on the inside back cover.**
- As an immediate precautionary measure, isolate spill or leak area in all directions for at least 50 meters (150 feet) for liquids and at least 25 meters (75 feet) for solids.
- Keep unauthorized personnel away.
- Stay upwind.
- Keep out of low areas.
- Ventilate enclosed areas.

PROTECTIVE CLOTHING

- Wear positive pressure self-contained breathing apparatus (SCBA).
- Wear chemical protective clothing that is specifically recommended by the manufacturer. It may provide little or no thermal protection.
- Structural firefighters' protective clothing provides limited protection in fire situations ONLY; it is not effective is spill situations where direct contact with the substance is possible.

EVACUATION

Spill
- See Table 1 - Initial Isolation and Protective Action Distances for highlighted materials. For non-highlighted materials, increase, in the downwind direction, as necessary, the isolation distance shown under "PUBLIC SAFETY".

Fire
- If tank, rail car or tank truck is involved in a fire, ISOLATE for 800 meters (1/2 mile) in all directions; also, consider initial evacuation for 800 meters (1/2 mile) in all directions.

The orange panel and shipping description provide the UN/NA identification number of the hazardous material as UN1805. The shipping description also provides the precise name of the hazardous material contained within the drums: phosphoric acid solution.

In the *Emergency Response Guidebook*, the number *1805* directs the reader to Guide 154, which is reproduced in Figure 6.15. This guide number provides general information concerning this substance. Most important, it notes that the hazardous material is toxic and/or corrosive, but noncombustible. Responders may then dismiss any concern as to whether the substance may ignite.

EMERGENCY RESPONSE

FIRE

Small Fire
- Dry chemical, CO_2 or water spray.

Large Fire
- Dry chemical, CO_2, alcohol-resistant foam or water spray.
- Move containers from fire area if you can do it without risk.
- Dike fire-control water for later disposal; do not scatter the material.

Fire involving Tanks or Car/Trailer Loads
- Fight fire from maximum distance or use unmanned hose holders or monitor nozzles.
- Do not get water inside containers.
- Cool containers with flooding quantities of water until well after fire is out.
- Withdraw immediately in case of rising sound from venting safety devices or discoloration of tank.
- ALWAYS stay away from tanks engulfed in fire.

SPILL OR LEAK

- ELIMINATE all ignition sources (no smoking, flares, sparks or flames in immediate area).
- Do not touch damaged containers or spilled material unless wearing appropriate protective clothing.
- Stop leak if you can do it without risk.
- Prevent entry into waterways, sewers, basements or confined areas.
- Absorb or cover with dry earth, sand or other non-combustible material and transfer to containers.
- DO NOT GET WATER INSIDE CONTAINERS.

FIRST AID

- Move victim to fresh air.
- Call 911 or emergency medical service.
- Give artificial respiration if victim is not breathing.
- **Do not use mouth-to-mouth method if victim ingested or inhaled the substance; give artificial respiration with the aid of a pocket mask equipped with a one-way valve or other proper respiratory medical device.**
- Administer oxygen if breathing is difficult.
- Remove and isolate contaminated clothing and shoes.
- In case of contact with substance, immediately flush skin or eyes with running water for at least 20 minutes.
- For minor skin contact, avoid spreading material on unaffected skin.
- Keep victim warm and quiet.
- Effects of exposure (inhalation, ingestion or skin contact) to substance may be delayed.
- Ensure that medical personnel are aware of the material(s) involved and take precautions to protect themselves.

SOLVED EXERCISE 6.6

When a firefighting team arrives at the scene of a highway transportation mishap, they encounter an overturned motor van bearing POISON INHALATION HAZARD placards on its visible sides and ends. On cautiously examining the van's interior from a distance, they also observe broken boxes to which POISON INHALATION HAZARD labels are affixed. The boxes are marked "Nickel carbonyl, UN1259," "INHALATION HAZARD, ZONE A," and "MARINE POLLUTANT." Punctured cylinders of the hazardous material are scattered outside their protective boxes, and their liquid contents have spilled on the underlying floor of the van. Securing the shipping paper from the driver, the team captain notes the relevant portion of the shipping description of this consignment as follows:

UNITS	HM	SHIPPING DESCRIPTION (IDENTIFICATION NUMBER, PROPER SHIPPING NAME, PRIMARY HAZARD CLASS OR DIVISION, SUBSIDIARY HAZARD CLASS OR DIVISION, AND PACKING GROUP)	WEIGHT (lb)
Eight steel cylinders over-packed within wooden boxes (UN4C)	X	UN1259, Nickel carbonyl, 6.1, (3), PG I (Poison - Inhalation Hazard, Zone A) (Marine Pollutant)	8

What precautionary actions should be implemented by these first-on-the-scene responders?

Solution: From the shipping description, labeling, marking, and placarding information, the team members learn that nickel carbonyl is a toxic and flammable liquid. Because the liquid is not confined within its containers, its vapor poses a pronounced risk of inhalation toxicity *and* fire. To protect public health and the environment, they should immediately cordon off the area, direct the public from the scene, and prevent entrance of the liquid into a waterway or sewer. Even though the total quantity of this nonbulk shipment is only 8 pounds, emergency-response personnel should don total-encapsulating suits with self-contained breathing apparatus. Only then should they remove the intact boxes from the van and set them aside.

For more specific guidance in responding to this incident, CHEMTREC should be contacted. The team members should also consult the *Emergency Response Guidebook* for immediate assistance. The identification number 1259 refers the responders to Guide 131. Here, personnel are provided with directions such as absorbing the spilled liquid and eliminating all potential sources of ignition. Guide 131 also provides first-aid information to help the responders who experience ill effects from inadvertent exposure to nickel carbonyl vapor.

6.8 REPORTING THE RELEASE OF A HAZARDOUS SUBSTANCE

At 40 C.F.R. §302.6, EPA uses the legal authority of CERCLA to require persons in charge of facilities (including transport vehicles) to notify the National Response Center (NRC) (Section 1.13) when a hazardous substance has been released into the environment in an amount equal to or greater than its reportable quantity within a 24-hour period. At 40 C.F.R. §110.3, EPA uses the legal authority of FWPPCA to require persons in charge of facilities to notify the NRC when oil is released into a waterway or when the release of oil may affect the quality of a waterway based on use of the following criteria:

- The discharge causes a sheen or discoloration on the surface of a body of water.
- The discharge violates applicable water-quality standards, and
- The discharge causes a sludge or emulsion to be deposited beneath the surface of the water or on adjoining shorelines

Notice to the NRC is also required when oil is released into a waterway or when the release of oil may affect the quality of a waterway.

In addition, DOT requires at 49 C.F.R. §171.15 that notice be given to the NRC whenever any of the following circumstances result:

- In connection with the release of one or more hazardous materials, at least one of the following is true:
  - A person is killed.
  - A person receives injuries requiring admittance to a hospital.

- The general public is evacuated for 1 hour or more.
- A major transportation artery or facility is closed or shut down for 1 hour or more.
- The operational flight pattern or routine of aircraft is altered.
- Fire, breakage, spillage, or suspected radioactive contamination occurs.
- Fire, breakage, spillage, or suspected contamination involving an infectious substance (Section 10.21-A) occurs. (Notice of the release of an infectious substance may be reported to the Centers for Disease Control and Prevention, U.S. Public Health Service, Atlanta, Georgia, at (800) 232-0124 in lieu of notice to the NRC.)
- A marine pollutant is released in a quantity exceeding 119 gallons (450 L) for a liquid or 882 pounds (400 kg) for a solid.
- Even when these criteria are not met, the transportation mishap is so severe that, in the judgment of the carrier, it should be reported nonetheless.

The notice provided to the NRC includes the following:

- The name of the reporter
- The name and address of the person represented by the reporter
- The telephone number where the reporter can be contacted
- The date, time, and location of the release incident
- The class or division, proper shipping name, and quantity of the hazardous material involved in the release, if such information is available
- The type of incident and nature of the hazardous material involvement and whether a continuing danger to life exists at the scene

As previously noted in Section 1.13, the NRC can initially be contacted by telephone at (800) 424-8802 or, in the District of Columbia, at (202) 426-2675. EPA and DOT require the submission of a follow-up written report to the following person:

Director, Office of Hazardous Materials Regulations
Materials Transportation Bureau
Department of Transportation
Washington, DC 20590

The informational requirements of the follow-up report are published at at 49 C.F.R. §171.16.

6.9 TRANSPORTATION SECURITY PLANS

At 49 C.F.R. Part 172, Subpart I, DOT requires carriers to develop and implement a **transportation security plan** when they intend to transport in commerce the specific hazardous materials listed in Table 6.10 in quantities that exceed the published values. This requirement is applicable to carriers who transport the hazardous materials in a motor vehicle, freight container, or railcar. Its purpose is to assess security risks during the transit of hazardous materials to deter terrorists and other illegal acts involving them. A copy of the plan must be made available to the DOT or the Department of Homeland Security upon their request.

At 40 C.F.R. §172.802, DOT requires carriers to address the following matters in their transportation security plans:

- Personnel Security. Measures confirming information provided by each job applicant hired for a position that involves access to and handling of the hazardous materials covered by the security plan. This information includes the applicant's references, employment history, and immigration status.
- Unauthorized Access. Measures addressing the assessed risk that unauthorized persons may gain from access to the hazardous materials covered by the security plan or transport conveyances being prepared to transport them. Such measures include identification

transportation security plan ■ For purposes of DOT regulations, a document that addresses personnel security, unauthorized access, and enroute security of the hazardous materials identified at 49 C.F.R. §172.800 in amounts equal to or exceeding threshold values

TABLE 6.10	Hazardous Materials and Their Quantities Whose Transportation in Commerce Requires DOT's Acceptance of a Transportation Security Plan[a]

HAZARDOUS MATERIAL	QUANTITY
Division 1.1, 1.2, or 1.3 materials	Any quantity
Division 1.4, 1.5, or 1.6 materials	A quantity requiring placarding
Division 2.1 materials	792 gal (3000 L) in a bulk container
Division 2.2 materials with a subsidiary hazard of 5.1	792 gal (3000 L) in a bulk container
Materials that pose an inhalation health hazard	Any quantity
Class 3 materials in a packaging meeting the criteria for Packing Group I or II	792 gal (3000 L)
Class 3 and Division 4.1 materials containing desensitized explosives	A quantity requiring placarding
Division 4.2 materials meeting the criteria of Packing Groups I or II	6614 lb (3000 kg) for solids or 792 gal (3000 L) for liquids in a bulk container
Division 4.3 materials meeting the criteria of Packing Groups I or II	A quantity requiring placarding
Division 5.1 materials in packaging meeting the criteria of Packing Groups I or II; metallic perchlorates; ammonium nitrate, ammonium nitrate fertilizers, or ammonium nitrate emulsions, suspensions, or gels	6614 lb (3000 kg) for solids or 792 gal (3000 L) for liquids
Temperature-controlled organic peroxides, type B, liquid or solid (Section 13.9-A)	Any quantity
Division 6.1 materials other than those that pose an inhalation health hazard	6614 lb (3000 kg) for solids or 792 gal (3000 L) for liquids in a bulk container
A select agent or toxin regulated by the Centers for Disease Control and Prevention (Section 10.20-A)	
Uranium hexafluoride	A quantity requiring placarding
"Highway route–controlled quantity" (Section 16.10-G) of a class 7 hazardous material	
Class 8 materials meeting the criteria for Packing Group I	792 gal (3000 L) in a bulk container

[a]49 C.F.R. §172.800.

of the measures to be taken that prevent unauthorized persons from gaining access to a shipment when it is stopped. These measures aim to prevent tampering or other illegal activity by securing closures of containers and locking them inside their transport vehicles.

■ Enroute Security. Measures addressing the assessed security risk of shipping the hazardous materials covered by the security plan from origin to destination, including shipments stored incidental to their movement. Such measures include identification of the preferred and alternate route plans that the carrier anticipates using during transit and an assessment of how the selected route favors other routes. They also include, when warranted, the reporting of suspicious incidents or events to local law enforcement officials and the nearest FBI field office and the contacting of local fire department and emergency-rescue personnel.

hazardous materials safety permit ■ For purposes of DOT regulations, a document issued by the Federal Motor Carrier Safety Administration that confers authority upon motor carriers to transport in commerce the hazardous materials identified at 49 C.F.R. §385.403 in amounts equal to or exceeding threshold values

6.10 HAZARDOUS MATERIALS SAFETY PERMITS

At 49 C.F.R. §385.400, DOT requires motor carriers that transport certain hazardous materials within interstate or intrastate commerce to obtain and maintain a **hazardous materials safety permit**. To obtain this permit, DOT requires the motor carrier to have achieved a "satisfactory" safety rating issued either by the Federal Motor Carrier Safety Administration (FMCSA) or the state in which the carrier has its principal place of

TABLE 6.11	Hazardous Materials and Their Quantities Whose Transportation in Commerce by Motor Carrier Requires DOT's Issuance of a Hazardous Materials Safety Permit[a]
HAZARDOUS MATERIAL	**QUANTITY**
Division 1.1, 1.2, or 1.3 materials	More than 55 lb (25 kg)
Division 1.5 materials	A quantity requiring placarding
Compressed or liquefied natural gas, or other liquefied gas with a methane concentration of at least 85% in bulk packaging (Section 12.5)	3500 gal (13,248 L) in bulk packaging
Materials that pose an inhalation health hazard and that meet the criteria for Zone A (Section 10.7)	1.08 qt (1 L) per package
Materials that pose an inhalation health hazard and that meet the criteria for Zone B (Section 10.7)	119 gal (450 L) in bulk packaging
Materials that pose an inhalation health hazard and that meet the criteria for Zones C or D (Section 10.7)	3500 gal (13,248 L)
Class 7 hazardous materials	Highway route-controlled quantity (Section 16.10-G)

[a]49 C.F.R. §385.403.

business. DOT further requires the motor carrier to certify that its security program complies with requirements published at 49 C.F.R. §385.407. In addition, DOT requires motor carriers to keep a copy of the permit or other document showing the permit number within the vehicle and provide it to federal, state, or local authorities upon their request.

The issuance of a hazardous materials safety permit confers authority on the motor carrier to transport in commerce the hazardous materials denoted in Table 6.11 in amounts exceeding the listed threshold values. It also requires the carrier to prepare and implement a transportation security plan that complies with the requirements noted in Section 6.9 and ensures the security of the selected route plan.

Shipping Papers

1. A liquor company intends to transport twenty 25-gallon wooden barrels of Scotch whiskey by motor van from its distillery to cellars where the barrels will be stored during aging. Use Appendix C to identify the shipping description that DOT requires on the accompanying shipping paper.

2. Potentially dangerous quantities of flammable hydrogen may be generated when lithium hydride reacts with water. Use Appendix C and Table 6.2 to identify the shipping description that DOT requires a shipper to enter on the accompanying shipping paper when transporting by motor van 25 pounds (11 kg) of lithium hydride in a metal can cushioned in a fiberboard box.

3. Phillips Petroleum Company (PPC) intends to transport 4500 cubic meters of liquefied petroleum gas, equally distributed in five DOT-113A rail tankcars bearing the numbers PPRX111111, PPRX111112, PPRX111113, PPRX111114, and PPRX111115, respectively. Using Appendix C and Table 6.2, identify the shipping description that DOT requires the shipper to provide on the accompanying waybill. (PPRX is PPC's reporting mark.)

DOT Labeling, Marking, and Placarding Requirements

4. A chemical manufacturer offers a single 25-liter glass bottle of ethyl mercaptan cushioned with bubble wrap in a wooden box to an airline carrier for domestic shipment by cargo aircraft.
   (a) Use Appendix C and Table 6.2 to identify the shipping description that DOT requires the shipper to provide on the accompanying shipping paper. (DOT authorizes the transportation of ethyl mercaptan in the limited quantity of 30 L by cargo aircraft.)
   (b) Use Appendix C to identify the warning labels, if any, that DOT requires the carrier to affix to the packaging.

5. What markings does DOT require a box manufacturer to provide on the exterior surface of a plastic box containing 5 pounds (2.3 kg) of the poisonous solid barium cyanide, when it is offered to a carrier as freight by motor van? (The plastic box was successfully tested to meet the performance standards for PG I packaging, and its manufacturer complied with all other DOT standards.)

6. What information is available to emergency responders when they observe each of the following markings on nonbulk packaging used for the transportation of a hazardous material?
   (a) UN 1H1/Y1.8/100/98/USA/JG
   (b) UN 1B1/Y1.4/150/83/USA/VL

7. A shipper offers a single cylinder of compressed fluorine gas for transportation by private motor carrier, and enters the following information on the accompanying shipping paper:

UNITS	HM	SHIPPING DESCRIPTION (IDENTIFICATION NUMBER, PROPER SHIPPING NAME, PRIMARY HAZARD CLASS OR DIVISION, SUBSIDIARY HAZARD CLASS OR DIVISION, AND PACKING GROUP)	WEIGHT (lb)
1 cylinder (DOT-3A1000/UN7A)	X	UN1045, Fluorine, compressed, 2.3, (5.1, 8), (Poison - Inhalation Hazard, Zone A)	2.5

   (a) What warning labels does DOT require the shipper to affix on the cylinder?

   (b) What placards does DOT require the carrier to post on the motor vehicle?

   (c) What licensing requirement is the driver of the motor vehicle obligated to fulfill?

8. Use Table 6.7 to identify the unique marking that DOT requires a rail carrier to display on the exterior surface of a DOT-105A rail tankcar used for the transportation of anhydrous ammonia.

9. A paint manufacturer retains CSX Transportation Company to transport 5000 gallons of a solvent mixture consisting of 15% toluene, 30% methyl ethyl ketone, 25% ethylbenzene, 15% ethyl acetate and 15% xylene in a rail tankcar.

   (a) Using Appendix C and Table 6.2, determine the proper shipping description that DOT requires the shipper to enter on the accompanying waybill. (CSXT is the reporting mark of CSX Transportation Company.)

   (b) What information concerning this shipment does the train crew document?

   (c) What information does DOT require to be marked on the exterior surface of the tankcar?

10. Show that the DOT regulations do not allow carriers to display DANGEROUS placards on a motor van used to transport the mixed load of hazardous materials obtained from a single facility and noted on the following bill of lading:

UNITS	HM	SHIPPING DESCRIPTION (IDENTIFICATION NUMBER, PROPER SHIPPING NAME, PRIMARY HAZARD CLASS OR DIVISION, SUBSIDIARY HAZARD CLASS OR DIVISION, AND PACKING GROUP)	WEIGHT (lb)
4 glass bottles (UN6P)	X	RQ, UN1565, Barium cyanide, 6.1, PG I (toxic)	100
3 drums (UN1A1)	X	UN1245, Methyl isobutyl ketone, 3, PG II	1100
2 boxes (UN4G)	X	UN1504, Sodium peroxide, 5.1, PG I	50
42 drums (UN1A1)	X	RQ, UN1294, Toluene, 3, PG II	16,800

Emergency-Response Actions at Transportation Mishaps

11. A train crew notes that a liquid is leaking from the piping connected to a rail tankcar containing a liquid hazardous material. This causes the engineer to stop the train and contact the local firefighting department. Upon arrival of the firefighters at the scene, a member of the train crew provides a waybill to the team captain. The relevant portion of the waybill gives the next shipping description of the leaking liquid:

UNITS	HM	SHIPPING DESCRIPTION (IDENTIFICATION NUMBER, PROPER SHIPPING NAME, PRIMARY HAZARD CLASS OR DIVISION, SUBSIDIARY HAZARD CLASS OR DIVISION, AND PACKING GROUP)	WEIGHT (lb)
1 tankcar	X	RQ, UN1134, Chlorobenzene, 3, PG III (Placarded FLAMMABLE) (DUPX1111)	5,000

The captain also notes that the presence of the following orange panel on each side and end of the tankcar adjacent to a FLAMMABLE placard:

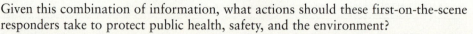

Given this combination of information, what actions should these first-on-the-scene responders take to protect public health, safety, and the environment?

12. A firefighting team responds to a fire involving an overturned motor van on which DANGEROUS placards have been displayed. The intensity of the fire prevents the team from approaching the vehicle and retrieving the shipping paper. Despite the fact that the DANGEROUS placards are visible, why is it difficult for the team to determine the best action to take at this scene to protect public health, safety, and the environment?

13. Arriving at the scene of a train wreck, the members of a firefighting team note the presence of the following orange panel on at least three sides of an overturned rail tankcar that is now leaking its contents.

The placards and markings on the tankcar are otherwise obscured by dense smoke. Using the *Emergency Response Guidebook*, determine each of the following:

(a) The name of the hazardous material having the identification number 2795
(b) The guide number to which the identification number 2795 is referenced
(c) Whether the relevant guide number indicates that the hazardous material is flammable or poisonous

14. On entering a burning garage owned by a landscaping company, the captain of a firefighting team observes the presence of five portable tanks against an inner wall. Although the captain is unable to determine information relating to their contents through the dense smoke, each tank appears to be intact and placarded as shown here:

To protect public health, safety, and the environment, what information about the contents should the captain provide to the team members?

7

Chemistry of Some Common Elements

Courtesy of the Chlorine Institute, Arlington, Virginia and Andrew Johnson, iMed Design, Inc., Reno, Nevada.

KEY TERMS

- Associate the physical and health hazards of the elements noted in this chapter with the information provided by their hazard diamonds and GHS pictograms.
- Identify the primary industries that use the elements noted in this chapter.
- Describe how exposure to ground-level ozone may diminish the quality of life.
- Describe how the depletion of ozone in the stratosphere negatively affects life on Earth.
- Identify the common metals that react with acids and/or water to produce hydrogen.
- Identify the principal health hazards associated with exposure to chlorine and phosphorus vapor.
- Illustrate that sulfur and carbon act as reducing agents in chemical reactions.
- Identify the labels, markings, and placards that DOT requires on packaging of the elements noted in this chapter and the transport vehicles used for their shipment.
- Identify the response actions to be executed when the elements noted in this chapter are released from their packaging into the environment.

In this chapter, we examine the characteristic properties of six elements: oxygen, hydrogen, chlorine, phosphorus, sulfur, and carbon. These particular elements have been selected for two primary reasons: Not only do they possess hazardous features of which emergency responders should be aware, but they are also used more widely than other elements. This latter distinction increases the likelihood that these elements will be encountered during emergency response efforts involving hazardous materials.

7.1 OXYGEN

What image is conjured up in your mind when someone mentions oxygen? Most individuals first think of the air. It is true that oxygen is a constituent of the air, yet oxygen makes up only about 21% of its volume. The major component—78% by volume—is nitrogen. Although oxygen is the most abundant element on Earth, most of it is not found in the free state. Instead, it is combined with other elements in chemical compounds such as water and carbon dioxide. Many oxygen compounds also exist in the chemical structures of the rocks and minerals that make up Earth's crust. A review of Table 4.1 shows that oxygen accounts for approximately 50% by mass of all substances in Earth's crust, the atmosphere, oceans, and other bodies of water.

As is well known, oxygen is needed for our survival. Each time we take a breath of air, we use the oxygen and exhale carbon dioxide. Although atmospheric oxygen is constantly consumed by living organisms, it is also replenished during **photosynthesis**, a chemical process in which plants convert carbon dioxide and water, in the presence of chlorophyll and sunlight, into oxygen and carbohydrates like cellulose (Section 14.5-A) and glucose.

photosynthesis ■ The chemical process by which plants use carbon dioxide and water to produce cellulosic substances with the aid of sunlight and chlorophyll

$$6CO_2(g) \ + \ 6H_2O(l) \ \longrightarrow \ C_2H_{12}O_6(aq) \ + \ 6H_2O(l) \ + \ 6O_2(g)$$

Carbon dioxide   Water   Glucose   Water   Oxygen

During the carboniferous age, 300 million years ago, the life processes of plants increased the atmospheric oxygen concentration to an amazing 35%, but this concentration later was reduced by an evolving ecosystem to today's 21%.

Scientists estimate that approximately 70% of the world's supply of atmospheric oxygen is now generated by trees within the huge span of the Amazonian rainforest in South America. Removal of the rainforest could place all of civilization in peril. Much of the other 30% originates during the life processes of **phytoplankton**, one-celled plants that thrive in

phytoplankton ■ One-celled organisms, mainly algae, whose biological processes produce oxygen

TABLE 7.1	Physical Properties of Elemental Oxygen
Melting point	−360°F (−218°C)
Boiling point	−297°F (−183°C)
Specific gravity at 68°F (20°C)	1.43
Specific gravity (liquid) at boiling point	1.14
Vapor density (air = 1)	1.11
Liquid-to-gas expansion ratio	875

the oceans. These land and ocean plant masses coexist and maintain the concentration of atmospheric oxygen at 21% by volume. Without them, we would be unable to survive.

At the ambient temperatures encountered on Earth, oxygen is a colorless, odorless, and tasteless gas that is only slightly denser than air. At very low temperatures and under pressure, oxygen exists as a pale blue liquid or solid and has the curious feature of being attracted to a magnet. Other physical properties of oxygen are listed in Table 7.1. Chemists represent the chemical formula of elemental oxygen as O_2.

7.1-A LIQUID OXYGEN

Oxygen is usually obtained for commercial use by liquefying air. Confined air liquefies when it is compressed at a pressure exceeding 545 psia (3760 kPa) while being simultaneously cooled below −318°F (−194°C). To obtain these temperature and pressure conditions, air is passed through a series of compression, expansion, and cooling operations.

By volume, liquid air is essentially a mixture of approximately 43% nitrogen, 54% oxygen, and less than 3% by mass of argon and other gases. When this mixture is allowed to boil at atmospheric pressure, the vapor above the liquid becomes enriched in nitrogen while, simultaneously, the remaining liquid slowly becomes enriched in oxygen. The oxygen-enriched mixture is called *liquid oxygen, cryogenic oxygen,* or *LOX*. When it vaporizes, LOX becomes gaseous oxygen, sometimes denoted as *GOX*. The vaporization of 1 liter of liquid oxygen at room temperature produces approximately 875 liters of gaseous oxygen.

Facilities using large quantities of oxygen often store it as the cryogenic liquid within a specialized tank like that shown in Figure 7.1. Although oxygen may be discharged from the tank as either the liquid or gas, the release of gaseous oxygen generally is more desirable from the standpoint of regulating its temperature and pressure.

7.1-B COMMERCIAL USES OF OXYGEN

Oxygen generally is used in connection with combustion reactions. The combustion process is *supported* by oxygen; that is, when a substance burns, it unites with oxygen. This means that oxygen is the oxidizing agent in ordinary combustion reactions. Earth is the only known planet where a fire can burn, because the atmospheres of other planets do not contain ample oxygen to support combustion.

Huge volumes of liquid oxygen are used by the aerospace industry to oxidize rocket fuels and propellants. In today's rockets, liquid oxygen and fuel are pumped into a combustion chamber, in which the mixture is ignited. The combustion reaction produces a high-pressure, high-velocity stream of hot gases that flows through the nozzle and accelerates the rocket to the tremendous velocities needed to reach outer space. Although other oxidizers could be used to oxidize rocket fuels, none has been shown to perform as effectively and economically as liquid oxygen.

LOX

GOX

FIGURE 7.1 Large volumes of medical oxygen often are stored at hospitals and clinics as its cryogenic liquid (LOX) in tanks. Although it is stored as the liquid, the oxygen is generally first vaporized before it is delivered into hospital lines for subsequent use. (*Courtesy of Eugene Meyer.*)

Oxygen is also used in a number of less esoteric ways. For example, it is used by cold storage and food processing plants, hospitals, metal fabrication plants, electrical power plants, and the steel industry. Within hospitals, oxygen-enriched air is delivered into oxygen tents and hoods for therapeutic uses. Gaseous oxygen under increased pressure, called *hyperbaric oxygen*, is also discharged within an enclosed chamber to treat a patient's circulatory and respiratory problems, thermal and radiation burns, cerebral palsy, brain injuries, and carbon monoxide poisoning.

Compressed oxygen may be encountered in steel cylinders of various sizes whose volumes range from 40 cubic feet (1.13 m^3) to 400 cubic feet (11.33 m^3). It is also confined in portable aluminum cylinders for use by individuals who require a supplemental supply of oxygen. Within a hospital or clinic, emergency responders may encounter these cylinders within a specialized cabinet like the one shown in Figure 7.2, in which the cylinders have been nested for storage to reduce or eliminate the likelihood of their movement.

FIGURE 7.2 This cabinet is specifically marked to identify it as an area to be utilized solely for the storage of aluminum cylinders containing compressed oxygen. (*Courtesy of A & A Sheet Metal Products, Inc., SE-Cur-All® Products, LaPorte, Indiana.*)

7.1-C RESPIRATION

In human and animal organisms, oxygen sustains the life process. The oxygen obtained from breathing air is assimilated through the lungs, from which it is carried by the bloodstream to the body's cells where it is used to oxidize food nutrients. The product of this oxidation is carbon dioxide, which is carried back to the lungs and exhaled into the atmosphere. Carbon dioxide is produced at approximately the same rate as oxygen is consumed. The overall phenomenon, called **respiration**, releases energy that humans and animals use to maintain life. For example, during respiration, the nutrient glucose is converted into carbon dioxide and water.

$$C_6H_{12}O_6(aq) \ + \ 6H_2O(l) \ + \ 6O_2(g) \ \longrightarrow \ 6CO_2(g) \ + \ 12H_2O(l) \ + \ \text{energy}$$

Glucose　　　　　Water　　　　Oxygen　　　　Carbon dioxide　　　Water

respiration ■ The chemical process whereby living organisms inhale oxygen and use it to convert organic substances into carbon dioxide, water, and energy

Without sufficient oxygen, our survival is impossible.

Despite the unconscious act of breathing, an individual may breathe air improperly. This usually occurs during situations requiring physical or mental stress because too much carbon dioxide has been lost from the bloodstream. This imbalance causes the condition known as **hyperventilation**, during which the individual breathes at a depth and rate greater than the body requires. The symptoms of hyperventilation are dizziness, chest pain, lightheadedness, or pins-and-needles sensations in the lips or mouth, fingers, or toes. Breathing into a paper bag can eliminate the imbalance.

hyperventilation ■ The act of improper breathing, causing an increase in the rate and depth of respiration and a deficiency in the carbon dioxide concentration in the blood

Earth's atmosphere has not always been oxygenated sufficiently to support life. Although the age of our planet is 4.5 billion years, oxygen did not begin to significantly enter Earth's atmosphere until roughly 2.2 billion years ago. Only after the atmosphere was oxygenated during the primeval past could animals and humans irrevocably establish themselves.

Anaerobic microorganisms like cyanobacteria (blue-green algae) may have been initially responsible for changing the chemical nature of the atmosphere. They accomplished this feat by using energy from the sun to produce oxygen as a byproduct. As oxygen accumulated in the atmosphere, the evolving zooplankton used it for survival. Over the subsequent millennia, more complex animals also used atmospheric oxygen as they evolved. When hominids finally arrived on Earth, perhaps 4 to 8 million years ago, plants had been producing oxygen for some time, ultimately providing an atmosphere that contained oxygen at a concentration of 21% by volume—the concentration in today's atmosphere.

When obliged to breathe concentrations lower or higher than 21%, the human organism responds to the situation. Consider, for instance, breathing an atmosphere containing oxygen at a concentration more than 23.5%. At 29 C.F.R. §1910.146, OSHA defines this situation as an **oxygen-enriched atmosphere**. Elevated oxygen levels are commonly administered to patients with respiratory problems for therapeutic purposes. For example, an oxygen concentration between 25% and 50% by volume often is delivered to patients inside oxygen tents, hoods, and chambers. Breathing this elevated concentration typically helps them return to good health more rapidly.

oxygen-enriched atmosphere ■ An environment containing more than 23.5% oxygen by volume

Yet, long-term breathing of an elevated concentration of oxygen does not always provide a positive outcome. Clinical observations demonstrate that persons may experience adverse health effects by inhaling elevated oxygen concentrations for extended periods. For instance, individuals who breathe an atmosphere containing 80% oxygen for more than 12 hours often suffer coughing, nasal stuffiness, sore throat, chest pain, and other respiratory problems.

It is from the mass of the overlying air that humans extract oxygen for survival. As first noted in Section 2.8, the mass of air exerts an average atmospheric pressure on planet Earth of 14.7 psi, or 101.3 kPa. At elevated heights from Earth's surface, the air still contains 21% oxygen, but the atmospheric pressure is reduced. This means that there is less pressure to push oxygen into the lungs. Thus, when breathing the air at lower-than-normal pressure, it takes more effort to inhale enough oxygen for survival. This occurs when mountaineers climb to summits and experience **altitude sickness**. They experience heavy breathing, difficulty in

altitude sickness ■ The phenomenon associated with the inability of the human organism to extract sufficient oxygen for survival from air at elevated altitudes

TABLE 7.2	Adverse Health Effects Associated with Breathing Reduced Levels of Oxygen

PERCENT OXYGEN BY VOLUME	SIGNS AND SYMPTOMS
21 (normal atmospheric concentration)	No signs or symptoms
15–19	Decreased ability to perform tasks; may induce early symptoms in persons with heart, lung, or circulatory problems
12–15	Deeper respiration; faster pulse; muscular coordination for skilled movements is lost
10–12	Giddiness; faulty judgment; muscular effort leads to rapid fatigue; slightly blue lips
8–10	Nausea; vomiting; unconsciousness; ashen face; fainting; mental failure
6–8	Death in 8 min; 50% death and 50% recovery with treatment in 6 min; 100% recovery with treatment in 4 to 5 min
<4	Coma in 40 s; convulsions; respiration ceases; death in less than 1 min

concentrating, headaches, and visual problems. Because they are unable to obtain sufficient oxygen from the air, the use of supplemental oxygen may be necessary.

Aviators constitute another group of individuals who may experience altitude sickness. Pilots who fly in unpressurized airplanes at altitudes between 8000 feet (2438 m) and 12,000 feet (3658 m) without supplemental oxygen make more procedural errors than do pilots who use supplemental oxygen. Pilots also experience a residual effect from flying without adequate oxygen by continuing to make errors at lower altitudes on descent and approach. To reduce or eliminate the likelihood of experiencing altitude sickness onboard operating aircraft, flight crews and aircraft occupants must abide by Federal Aviation Administration regulations published at 14 C.F.R. §91.211.

Conversely, adverse health effects may also be experienced from breathing oxygen at an *increased* pressure, even for short periods of time. This is the reason clinicians routinely administer hyperbaric oxygen therapy to patients for no more than ½ to 1 hour.

Firefighters are more likely to encounter incidents in which the oxygen concentration has been *reduced* below 21% by volume. At 29 C.F.R. §1910.146, OSHA defines an environment containing less than 19.5% oxygen as an **oxygen-deficient atmosphere**. As Table 7.2 illustrates, individuals gasp for breath when the oxygen concentration in the air falls below 17%. This occurs when materials burn within an enclosed area. Without an alternative air supply, emergency responders could collapse and die if compelled to remain within such an environment for even a few minutes.

Compressed oxygen and LOX are concentrated forms of the element. When they are released from their containers at fire scenes, the oxygen enhances the rate at which nearby combustible materials burn. Put simply, matter burns more rapidly in an atmosphere of pure oxygen than it does in the air. This observation is consistent with our general understanding from Section 5.5-D, in which we noted that the rate of a reaction increases when the concentration of a reactant is increased.

Liquid oxygen increases the rate of combustion in the following circumstances:

■ *The combustion of ordinary matter.* Some substances burn slowly in air after exposure to an ignition source; but, they may burn spontaneously when they contact liquid oxygen. Fuels, oils, greases, tar, asphalt, paper, and textiles are examples of matter that

oxygen-deficient atmosphere ■ An environment containing less than 19.5% oxygen by volume

do not easily ignite without first being kindled. When they have been exposed to liquid oxygen, however, these materials are likely to ignite spontaneously.

■ *The combustion of metals.* Certain metals are likely to burn on contact with liquid oxygen. The combination of magnesium shavings and liquid oxygen, for example, is so chemically reactive that these elements explode on contact. The components of aluminum pump parts also react with liquid oxygen, especially when the reaction is initiated by friction. This phenomenon is the reason for avoiding the use of aluminum in equipment intended for the storage and handling of cryogenic oxygen.

■ *The combustion of porous materials.* Wood, concrete, and asphalt are examples of porous materials. When liquid oxygen contacts them, it is readily absorbed into their pores and produces media that can be shock-sensitive. An example is an asphalt surface on which liquid oxygen has spilled. This oxygen-enriched asphalt may ignite explosively when subjected to a mechanical impact.

7.1-D CHEMICAL OXYGEN GENERATORS

Chemical oxygen generators are portable devices that can be chemically actuated to produce oxygen on demand. They are used in mines and other places with limited accessibility in which the available oxygen supply may be limited. They were once used to escape from noxious or poisonous atmospheres, but this practice is now obsolete.

chemical oxygen generator ■ Any portable device in which oxygen is produced upon demand by chemical reaction

One type of chemical oxygen generator consists of an apparatus with two separate compartments containing powdered iron and sodium chlorate, respectively. Oxygen is generated when these substances are mixed and heated.

$$Fe(s) \ + \ NaClO_3(s) \ \longrightarrow \ FeO(s) \ + \ NaCl(s) \ + \ O_2(g)$$
Iron　　Sodium chlorate　　Iron(II) oxide　Sodium chloride　Oxygen

A percussion cap activated by a hammer provides the heat needed to initiate the reaction. The heat that is evolved by this exothermic reaction permits it to be self-sustaining until either of the reactants is depleted.

Another type of oxygen generator utilizes the thermal decomposition of an alkali metal perchlorate. Lithium perchlorate, for example, decomposes when heated as follows:

$$LiClO_4(s) \ \longrightarrow \ LiCl(s) \ + \ 2O_2(g)$$
Lithium perchlorate　　Lithium chloride　　Oxygen

The reaction is initiated through use of an igniter that is struck by a firing pin.

7.1-E WORKPLACE REGULATIONS INVOLVING BULK OXYGEN SYSTEMS

OSHA defines a **bulk oxygen system** as the assembly of equipment (storage tank, pressure regulators, safety devices, vaporizers, manifolds, and interconnecting piping) that possesses an oxygen storage capacity at normal temperature and pressure of more than 13,000 cubic feet (368 m^3) when connected for service, or more than 25,000 cubic feet (708 m^3) when available as an unconnected reserve.

bulk oxygen system ■ For purposes of OSHA regulations, the assembly of equipment that has an oxygen storage capacity at normal temperature and pressure of more than 13,000 cubic feet (368 m^3) when connected for service, or more than 25,000 cubic feet (708 m^3) when available as an unconnected reserve

When bulk quantities of oxygen are stored at the premises of industrial or institutional consumers, OSHA requires their owners to comply with certain regulations to prevent or minimize the risk of fire and explosion. These regulations pertain to the location of the system, its elevation, its accessibility to authorized personnel, measures associated with leakage from the system, diking, required distances between the system and nearby exposures, and other safety issues. For example, OSHA requires the location of bulk oxygen storage systems to be either outdoors aboveground *or* within a building of noncombustible construction that is adequately vented and used for that purpose exclusively. The selected location must be such that containers and associated equipment are not exposed to electric power lines, flammable gas lines, or flammable liquid lines.

OSHA requires the location of bulk oxygen storage systems to be either outdoors aboveground *or* within a building of noncombustible construction that is adequately vented and used for that purpose exclusively. The selected location must be such that containers and associated equipment are not exposed to electric power lines, flammable or combustible liquid lines, or flammable gas lines. Despite the fact that oxygen is a nonflammable gas, what is the most likely reason OSHA requires bulk oxygen storage systems to be installed in this fashion?

Solution: The most likely reason OSHA enacted this regulation is to minimize a pronounced risk of fire and explosion. Although oxygen is a nonflammable gas, it *supports* the combustion of many common materials. Because these materials burn at increased rates in an oxygen-enriched environment, it is prudent to prevent the accumulation of oxygen near flammable or combustible materials. This is accomplished by installing bulk oxygen storage systems in the indicated fashion.

OSHA advises the owners of bulk oxygen systems to locate them on ground higher than flammable liquid storage tanks in nearby storage. When it is necessary to locate a bulk oxygen system on ground lower than adjacent flammable liquid storage tanks, OSHA requires the owner to provide a suitable means of diking, diversion curbing, or grading such that in the event of a release from these tanks, the liquids do not accumulate under the bulk oxygen system. Because flammable liquids burn at increased rates in an oxygen-enriched environment, it is prudent to prevent their accumulation near the components of a bulk oxygen system.

To warn workers of the presence of a bulk oxygen system within a workplace, OSHA also requires its owner at 29 C.F.R. §1910.104(b)(8)(viii) to permanently post a placard that reads as follows:

7.1-F TRANSPORTING OXYGEN

When shippers offer GOX or LOX for transportation, DOT requires them to identify it as shown in Table 7.3 on the accompanying shipping paper. DOT also requires them to affix NON-FLAMMABLE GAS and OXIDIZER labels to the cylinders or other packaging.

When carriers transport 1001 pounds (454 kg) or more of GOX or LOX, DOT requires them to display NON-FLAMMABLE GAS placards on the bulk packaging used

TABLE 7.3	Shipping Descriptions of Oxygen and Oxygen Generators
OXYGEN	**SHIPPING DESCRIPTION**
Compressed oxygen	UN1072, Oxygen, compressed, 2.2, (5.1)
Cryogenic oxygen	UN1073, Oxygen, refrigerated liquid, 2.2, (5.1)
Oxygen generator, chemical	UN3356, Oxygen generator, chemical, 5.1, PG II
Spent oxygen generator, chemical	NA3356, Oxygen generator, chemical spent, 9, PG III

for the shipment; or, in lieu of posting NON-FLAMMABLE GAS placards, DOT allows carriers to display OXYGEN placards that look like the following:

The motor van shown in Figure 7.3 has been placarded for the transportation of cylinders and tanks containing both oxygen and nitrogen.

When carriers transport oxygen within bulk packaging, DOT also requires them to display the relevant identification number—1072 or 1073—on orange panels, across the center area of the NON-FLAMMABLE GAS or OXYGEN placards, or on white square-on-point diamonds affixed on each side and each end of the packaging. For example, any of the following means may be used to display 1072, the identification number of oxygen:

When oxygen is transported in bulk by highway or rail, DOT requires carriers to display the name *OXYGEN* on two opposing sides of the tankcar used for shipment.

OXYGEN

DOT requires shippers who transport chemical oxygen generators or spent chemical oxygen generators to identify them as shown in Table 7.3 on the accompanying shipping

FIGURE 7.3 DOT authorizes carriers at 49 C.F.R. §172.504(f)(7) to display OXYGEN placards in lieu of NON-FLAMMABLE GAS placards on vehicles used for the sole shipment of oxygen in nonbulk cylinders or other containers. To emphasize that oxygen is an oxidizer, the background color of the OXYGEN placard is yellow, and the color of the symbol, text, hazard class number, and border is black. When carriers transport oxygen and other nonflammable gases in nonbulk cylinders or other containers within the same vehicle, DOT permits them to display NON-FLAMMABLE GAS placards *or* both OXYGEN and NON-FLAMMABLE GAS placards on the vehicle. (*Courtesy of Airgas Inc., Radnor, Pennsylvania.*)

paper. When the generators are equipped with an attached means of initiation, DOT requires their carriers to obtain approval for shipment by demonstrating that they have been outfitted with two positive means of preventing their unintentional actuation.

7.1-G RESPONDING TO INCIDENTS INVOLVING A RELEASE OF OXYGEN

When oxygen is first released from a storage or transportation device, its concentration becomes elevated within the immediate area compared to its level in the air. Then, secondary fires are likely to occur, especially when the oxygen contacts fuels. An attempt should be made to combat secondary fires only when they are located far from the area where the oxygen is released or when the flow of oxygen from its containment vessel can be stopped without exposing personnel to unnecessary risk.

In the event of a transportation mishap involving the release of oxygen to the environment, DOT recommends an initial downwind evacuation of unauthorized persons to a distance of at least 0.33 miles (500 m) from the scene of the mishap. If a rail tankcar or tank truck is involved in a fire, DOT advises evacuation to a distance of 0.5 miles (800 m).

7.1-H OZONE, THE ALLOTROPE OF OXYGEN

allotrope ■ A form of the same element having its own unique set of physical and chemical properties

Chemists refer to ozone as an allotrope of oxygen. An **allotrope** is a unique variation of an element possessing physical and chemical properties that substantially differ from those of any other form of the element. Allotropic forms are exhibited by certain elements that occupy Groups 4A through 6A on the periodic table. The allotropes of four elements are noted in this chapter: oxygen, phosphorus, sulfur, and carbon.

Ozone is a form of elemental oxygen having three atoms of oxygen per molecule instead of the usual two; thus, its chemical formula is represented as O_3 and its Lewis structure is represented as follows:

$$\ddot{\text{:}}\ddot{O}\text{:}\ddot{O}\text{::}\ddot{O}\text{:} \text{ or } O-O=O$$

Ozone is produced when either an electrical discharge or ultraviolet radiation is passed through oxygen. Although it is reasonably stable at relatively low temperatures, ozone decomposes rapidly at room temperature; that is, it spontaneously reverts into "ordinary" oxygen. It is this relative instability that accounts for the fact that ozone is generally encountered at very low concentrations.

At ordinarily encountered temperatures and pressures, ozone exists as a pale blue gas. With a vapor density of 1.7 relative to air, ozone is denser than normal oxygen. It possesses a pleasing smell in low concentrations, but has an irritating, pungent, "metallic" odor at moderate to high concentrations. The fresh, clean, spring-rain smell sometimes detected around operating electric motors or following a lightning storm usually can be attributed to the presence of ozone.

Two characteristics are responsible for ozone's reputation as one of the most hazardous materials known: a pronounced chemical reactivity and a distinct toxicity. Ozone is a powerful oxidizing agent—considerably more reactive than oxygen itself. For example, ozone converts lead sulfide rapidly into lead sulfate, whereas oxygen reacts so slowly with lead sulfide that the reaction is virtually imperceptible.

$$3PbS(s) \quad + \quad 4O_3(g) \quad \longrightarrow \quad 3PbSO_4(s)$$
Lead sulfide   Ozone   Lead sulfate

Because ozone is a toxic substance, precaution should be undertaken to avoid inhalation exposure. When it is inhaled, ozone damages the scavenger cells of the immune system. These cells, called *macrophages*, customarily destroy foreign bacteria, but when they are damaged, they are unable to effectively accomplish this task. Individuals whose macrophages have been damaged are more susceptible to contracting diseases.

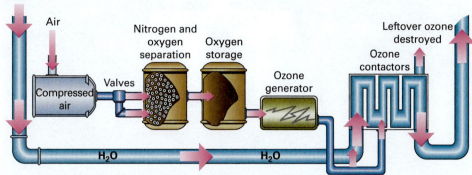

FIGURE 7.4 A simplified schematic route that illustrates the production of ozone from air and its subsequent use for producing pure water.

7.1-I COMMERCIAL USES OF OZONE

Ozone is used commercially for several purposes. It is used to bleach undesirable colors from oils, fats, textiles, and sugar solutions. It is also used as a **microbicide** at drinking water and wastewater treatment plants. A microbicide is a substance that kills disease-causing microorganisms. Ozone is an especially effective microbicide. Most importantly, it is capable of destroying the parasite *Cryptosporidium parvum* that is sometimes found in chlorinated drinking water. Chlorine alone does not destroy this parasite (Section 7.3-B).

When water contaminated with *Cryptosporidium parvum* is consumed, the parasite causes gastrointestinal diseases that can be fatal. In 1993, for example, consumption of water in the greater Milwaukee area caused 403,000 people to experience stomach cramps, fever, diarrhea, and dehydration.[1] Moreover, the deaths of 104 people were attributed to the outbreak of diseases caused by the presence of *Cryptosporidium parvum* in their drinking water. Fortunately, such incidents are now rare in the United States and other developed countries that choose to ozonize their drinking water supplies.

When ozone is used for water treatment, the processes illustrated in Figure 7.4 are implemented. Air is first compressed to separate oxygen and nitrogen. Then, the oxygen is zapped with electricity to produce ozone, which is passed into less-than-pristine water under pressure. The ozone reacts with the impurities in the water, thereby killing its constituent microorganisms. The water is then filtered and pumped into the municipal drinking water supply.

The use of ozone is also advantageous for treating wastewater. Ozone converts the constituent hydrogen sulfide (sewer gas, Section 10.13) into sulfuric acid. At the concentration produced, the sulfuric acid is benign.

$$3H_2S(s) \ + \ 4O_3(g) \ \longrightarrow \ 3H_2SO_4(aq)$$
Hydrogen sulfide   Ozone   Sulfuric acid

When ozone is used for treating drinking water or wastewater, the process is called **ozonation**.

Because ozone is unstable and extremely poisonous, it is always synthesized at its intended point of use in minutely low concentrations. This synthesis is accomplished within an ozone generator, an apparatus that supplies an electrical current to oxygen or air. The ozone generator shown in Figure 7.5 may be familiar to firefighters, because its use often is encouraged within buildings damaged by fire. While the buildings are enclosed and vacated, ozone produced by the generator oxidizes the components of smoke. This removes the offensive odors from furniture, clothing, and other items damaged by fire. Ozone generators are also used to produce ozone for removal of musty odors that persist within walls and carpeting damaged by mold or mildew.

microbicide ■ A chemical substance capable of destroying microorganisms

ozonation ■ The chemical reaction involving the addition of ozone to a substance; the treatment of contaminated drinking water to kill the microorganisms that cause waterborne diseases

---

[1] W. R. Mac Kenzie et al., "A massive outbreak in Milwaukee of *Cryptosporidium* infection transmitted through the public water supply," *N. Engl. J. Med.*, Vol. 331 (1994) pp. 161–167.

7.1-J GROUND-LEVEL OZONE

ground-level ozone (tropospheric ozone) ■ The ozone that forms in the troposphere (the lower atmosphere) by photochemically catalyzed reactions between VOCs and the nitrogen oxides

volatile organic compounds (VOCs) ■ Certain vapors that undergo photochemical reactions in the atmosphere to form ozone and other air pollutants

In certain metropolitan regions of the United States, **ground-level ozone**, or **tropospheric ozone**, often is simultaneously present in the air with nitric oxide, nitrogen dioxide (Section 10.14) and certain substances called **volatile organic compounds (VOCs)**. The VOCs are components of the air emissions from petroleum refineries, fuel dispensers, chemical plants, and other industrial facilities. They typically are regarded as organic compounds that boil at temperatures less than approximately 392°F (200°C). Hence, the substances in gasoline vapor are examples of VOCs. Figure 7.6 shows that the majority of the atmospheric VOCs originate when nonchemical industrial processes are conducted.

The oxides of nitrogen are generated during the operation of automobiles. They are components of vehicular exhaust and the plumes emanating from the smokestacks of fossil-fuel-fired power plants. Within the troposphere—or lower atmosphere immediately above the ground—these oxides react photochemically with the VOCs to produce ozone. At ground level where we live and breathe, ozone and the nitrogen oxides often are the *primary* components of polluted air, especially during the daylight hours of summertime, when ample sunlight catalyzes ozone production.

When the ground-level ozone concentration exceeds approximately 100 ppb, weather forecasters declare an *ozone alert*, which signifies that the ozone concentration in the air is approaching an unhealthful condition. At this concentration, affected individuals—especially the elderly—struggle for breath and feel dizzy. Their eyes, noses, throats, and lungs become irritated, and they experience fatigue, a lethargic feeling, coughing, wheezing, and hoarseness.

The inhalation of low concentrations of ozone is also likely to exacerbate the illnesses of individuals suffering from heart ailments, emphysema, or chronic bronchitis. These people are likely to cough more frequently and with greater intensity, experience chest pain and sinus congestion, and suffer severe headaches. Short-term exposure to ozone has been linked with premature death, and long-term cumulative exposure has been linked with an increased

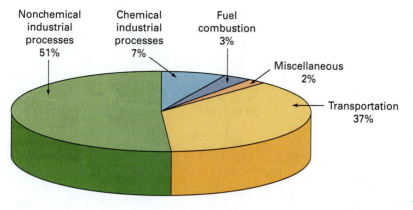

Nonchemical industrial processes 51%

Chemical industrial processes 7%

Fuel combustion 3%

Miscellaneous 2%

Transportation 37%

FIGURE 7.6 Ground-level ozone is produced when the volatile organic compounds (VOCs) in the air react with the nitrogen oxides discharged in vehicular exhaust. VOCs are contained primarily within the gaseous discharges of power plants, motor vehicles, and petroleum refineries. (*Courtesy of United States Environmental Protection Agency.*)

risk of fatalities from respiratory causes. This risk rises up to 4% for every 10 ppb increase in exposure to ozone.[2]

7.1-K WORKPLACE REGULATIONS INVOLVING OZONE

When the use of ozone is needed in the workplace, OSHA requires employers to limit employee exposure to an ozone concentration in air of 0.1 ppm, averaged over an 8-hour workday.

7.1-L ENVIRONMENTAL REGULATIONS INVOLVING GROUND-LEVEL OZONE

EPA has evaluated the adverse health effects that result from exposure to ground-level ozone. Using the legal authority of the Clean Air Act, EPA regulates ground-level ozone as a criteria air pollutant (Section 1.3-A) by setting its primary and secondary national ambient air quality standards at 0.12 parts per million (235 $\mu g/m^3$) as a 1-hour average and 0.079 parts per million (157 $\mu g/m^3$) as an 8-hour average.

SOLVED EXERCISE 7.2

On a hot, humid day, the ozone concentration in urban air can exceed 100 parts per billion. What actions to limit the adverse health effects associated with breathing ozone should be taken by persons who suffer from emphysema?

Solution: Emphysema is a lung ailment associated with the swelling of the alveoli and connecting tissues in the lungs. Emphysema sufferers cough frequently, experience headaches, and have trouble breathing. These ill effects are exacerbated when they inhale air contaminated with ozone. To limit undue distress when the ozone concentration exceeds 100 parts per billion, emphysema sufferers are advised to remain in an air-conditioned environment, avoid heavy work and exercise, and breathe oxygen from a portable source.

7.1-M STRATOSPHERIC OZONE

Although the presence of ozone in the troposphere can be a troublesome factor for maintaining good health, the presence of ozone in the stratosphere provides an exceptional benefit to the inhabitants of Earth. Approximately 10 to 19 miles (16 to 30 km) above Earth's surface, ozone occurs naturally in the region of the stratosphere called the **ozone layer**. As shown in Figure 7.7, oxygen is bombarded within the stratosphere by high-energy particles that originate in the sun. The bombardment causes some oxygen molecules to dissociate into their individual atoms ($\cdot \overset{..}{\underset{..}{O}} \cdot$). The oxygen within the ozone

ozone layer ■ The region of the upper atmosphere that is especially plentiful in ozone

---

[2]Michael Jerrett et al., "Long-term ozone exposure and mortality," *N. Engl. J. Med., Vol.* 360 (2009), pp. 1085–95.

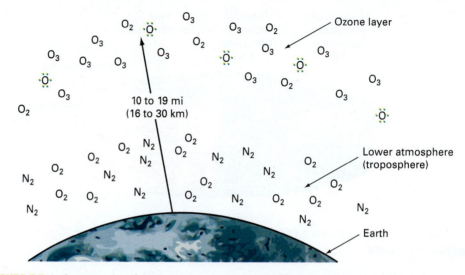

FIGURE 7.7 At the outer edge of the stratosphere is a region called the ozone layer, where oxygen is constantly converted into ozone. The ozone layer prevents much of the sun's ultraviolet radiation from reaching Earth. Atmospheric scientists have discovered that in modern times the amount of stratospheric ozone has been depleted compared with the amount in past times. This depletion allows more harmful ultraviolet radiation to penetrate to Earth's surface, increasing the instances of skin cancer and cataracts and weakening immune systems.

layer exists as a mixture of oxygen atoms and oxygen molecules. These oxygen species collide to initiate a chemical reaction producing ozone.

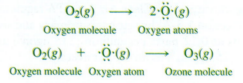

When molecules of stratospheric ozone absorb low-energy ultraviolet radiation, some revert to the ordinary form of oxygen. Under normal conditions, the continuous formation and photodecomposition of ozone compete favorably with one another. This chemical activity produces a steady-state condition in which the rate at which the stratospheric ozone forms from oxygen equals the rate at which it undergoes photodecomposition into oxygen.

In the past, the presence of the stratospheric ozone layer has protected Earth, its occupants, and plant life from overexposure to the harshness of ultraviolet radiation. Today, however, the former use of the halon fire extinguishing agents (Section 5.14), Freon refrigerants, foam-blowing agents and coolants (Section 12.15), and other halogen-containing compounds continues to adversely affect the environmental quality of our planet. Released into the environment, these compounds diffuse upward to the ozone layer, where their exposure to ultraviolet radiation produces halogen atoms. We note more fully in Section 12.15 that these atoms catalyze the decomposition of ozone and contribute to its overall stratospheric depletion.

In 1985, atmospheric scientists began to note a precipitous drop in stratospheric ozone, first over Antarctica and then over the Arctic. The areas of the ozone layer in which this thinning of the ozone layer has been observed are referred to as **ozone holes**. Due to their existence, more ultraviolet radiation from the sun now penetrates Earth's atmosphere compared with the amount that reached Earth before 1985. The concern among scientists is that this increased radiation exposure is the cause of certain health and environmental problems such as the following:

ozone hole ■ A thin spot in the ozone layer resulting from destruction of stratospheric ozone by its chemical reaction with chlorofluorocarbons and other substances

■ Radiation weakens plant life and contributes to lower agricultural outputs.
■ Radiation may cause climatic changes that could ultimately disturb the balance of aquatic and land ecosystems.

- The overexposure to ultraviolet radiation in humans has dramatically increased the incidence of skin cancer and cataracts and weakened immune systems.
- In aquatic systems, radiation harms marine life.

7.1-N ENVIRONMENTAL REGULATIONS INVOLVING STRATOSPHERIC OZONE

To ensure the continued survival of life on planet Earth, 43 representatives of the world's industrialized nations agreed in 1987 to coordinate the control of ozone-depleting substances by phasing out their manufacture and use. This agreement, sponsored by the United Nations, is known as the Montréal Protocol on Substances That Deplete the Ozone Layer.

The original **Montréal Protocol** has now been ratified by 196 nations including the United States. It banned the parties from producing and consuming most ozone-depleting substances by January 1, 2010. Even with adherence to the Montréal Protocol, however, scientists project that it will take until 2050 for the ozone layer to completely recover from the world's misuse of these hazardous substances.

Although the United States is a party to the Montréal Protocol, EPA has also used the legal authority of the Clean Air Act to regulate the manufacture, use, and importation of substances that deplete stratospheric ozone. Among these substances are methyl bromide, carbon tetrachloride, 1,1,1-trichloroethane, the halon fire extinguishing agents (Section 5.14), and certain chlorofluorocarbons (Section 12.15).

Montréal Protocol ■
The international agreement that provides for phasing out the worldwide production, manufacture, and use of chlorofluorocarbons and other ozone-depleting substances

7.2 HYDROGEN

Although Table 4.1 shows that hydrogen ranks ninth in natural abundance by mass on Earth, only traces of this element exist naturally in the free state. Hydrogen is such a reactive element that it is found on Earth only in compounds like water, acids, and hydrocarbons. Throughout the entire universe, however, hydrogen is the most abundant element, 72% by mass.

Elemental hydrogen is an odorless, colorless, tasteless, and nontoxic substance. Its chemical formula is H_2. Although hydrogen exists as a gas at ordinary temperatures and pressures, the gas liquefies when compressed under a pressure greater than 294 psi (2030 kPa) and at a temperature less than $-390°F$ ($-234.5°C$). Liquid hydrogen is sometimes denoted as *LH2*.

7.2-A COMMERCIAL USES OF HYDROGEN

Elemental hydrogen is commercially available as both the compressed gas and the cryogenic liquid. The chemical industry uses it as a raw material for the production and manufacture of other chemical substances. The petroleum industry uses it to process crude petroleum fractions (Section 12.13-C) during the production of petroleum fuels, and the aerospace industry uses hydrogen as a fuel for propelling rockets into space.

Hydrogen is chosen as a rocket fuel by aerospace engineers, because the combustion of a very small mass produces a prodigious amount of energy (see below). It is within the aerospace industry that massive volumes of hydrogen are likely to be encountered in storage. For example, at the John F. Kennedy Space Center, Cape Canaveral, Florida, a staggering 800,000 gallons (3000 m^3) of cryogenic hydrogen is held in a single storage tank.

For more than half a century, physicists and engineers have also been investigating ways to successfully develop a thermonuclear reactor, in which the nuclei of hydrogen atoms fuse and produce heavier nuclei by converting mass into energy. Nuclear fusion is the phenomenon that powers the sun and other stars; it was also the means used on Earth to detonate the hydrogen bomb pictured in Figure 7.8. However, the energy released during the detonation

Hydrogen gas

LH2

of a hydrogen bomb is essentially uncontrolled. To serve as a commercially viable energy source, self-sustaining fusion reactions must be controlled on a large scale. If a thermonuclear reactor could be successfully developed, hydrogen would most likely become *the* energy source of the future.[3] However, serious technical obstacles in achieving a sustainable, controlled nuclear power source now exist. Commercialized nuclear fusion is not anticipated to begin until late in this century, if at all.

When hydrogen burns in an atmosphere of pure oxygen, the accompanying heat of combustion is 61,000 Btu/lb (141,790 kJ/kg). This is the most energy for the least mass that evolves when any substance burns. This energy is used for launching spacecraft and cutting and welding metals and glass. Jewelry manufacturers, for example, use the oxyhydrogen torch illustrated in Figure 7.9 to craft rings and bracelets made of platinum, a metal that melts at 3191°F (1755°C). Compressed hydrogen and oxygen are separately stored within steel cylinders and then pressure-fed through tubing into a mixing chamber within the torch in a 2:1 ratio. This mixture is then discharged from the nozzle, where combustion of these gases occurs.

---

[3]Research relating to the development of a fusion-demonstration reactor is now being conducted in a multinational effort called the International Thermonuclear Experimental Reactor Consortium. The United States, European Union, Russia, India, Japan, South Korea, and China are members of the consortium that funds the project. In 2006, representatives of these seven partner nations agreed to build the plant at Cadarache in southern France, where it is now under construction and slated to be finished in 2018 and begin operation by 2020.

The underlying technology in demonstrating fusion involves the use of a doughnut-shaped magnetic bottle called a *tokamak*, which confines the hydrogen as it is heated to 180 million degrees. At this superheated temperature, the hydrogen exists as plasma (see footnote in Section 2.1) and its nuclei fuse to form helium, neutrons, and sufficient energy that can be used to generate electricity. If the initial tests are successful, the consortium estimates that the construction of a commercial-scale fusion power plant could begin by midcentury.

In the United States, fusion research is also underway at a number of government facilities, most notably the National Ignition Facility, Livermore, California, and Los Alamos National Laboratory, Los Alamos, New Mexico.

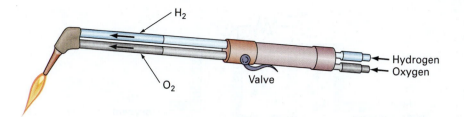

FIGURE 7.9 Hydrogen and oxygen separately enter the mixing chamber of this oxyhydrogen torch from the right. A 2:1 mixture discharges from the nozzle at the left at a rate that causes the flame to burn at the tip of the torch. When this mixture is ignited, temperatures ranging between 3300 and 4400°F (1800 and 2400°C) are achieved.

Hydrogen is also used commercially in connection with the chemical process called **hydrogenation**. At elevated temperatures and pressures, and generally in the presence of a catalyst, hydrogen combines with certain organic compounds to form commercially useful products. Vegetable oils are hydrogenated, for example, for use as shortening and raw materials for the manufacture of soaps and lubricants.

hydrogenation ■ The addition of hydrogen to a substance

Within the chemical industry, hydrogen is used to produce metallic and nonmetallic hydrides. Hydrogen combines with elemental sodium, for example, to produce sodium hydride.

$$2Na(s) \quad + \quad H_2(g) \quad \longrightarrow \quad 2NaH(s)$$
Sodium   Hydrogen   Sodium hydride

It is also used to produce compounds such as hydrogen chloride and ammonia.

$$H_2(g) \quad + \quad Cl_2(g) \quad \longrightarrow \quad 2HCl(g)$$
Hydrogen   Chlorine   Hydrogen chloride

$$3H_2(g) \quad + \quad N_2(g) \quad \longrightarrow \quad 2NH_3(g)$$
Hydrogen   Nitrogen   Ammonia

7.2-B HYDROGEN AS AN ALTERNATIVE MOTOR FUEL

In the mid-2000s, there were persuasive reasons for believing that compressed hydrogen would become a popular fuel for powering automobiles and other surface vehicles. In particular, the prospective use of hydrogen as a vehicular fuel would reduce air pollution and our reliance on fossil fuel resources from foreign countries. The use of hydrogen as a vehicular fuel would truly revolutionize the automotive industry and theoretically produce a cleaner environment.

In hydrogen-powered automobiles, compressed hydrogen directly replaces the gasoline used in gasoline-powered vehicles. It is an example of an **alternative motor fuel**. In 1984, Daimler Benz first demonstrated the use of hydrogen to power the Mercedes Benz 280 Te automobile.

alternative motor fuel ■ Any of the substances or mixtures that can power vehicles as an alternative to the use of motor gasoline or diesel oil

Hydrogen may also be encountered as an alternative motor fuel in the form of a metallic hydride, or metal hydride. A metal powder is generally contained in an aluminum cylinder with a pressure-relief valve and a coupling for connecting to a system for potential use. When hydrogen is charged into the cylinder, it is absorbed into the metal and produces the corresponding metal hydride. Because the hydrogen is only loosely bonded to the metal, it is desorbed readily for use upon demand. In this sense, it performs similarly to the storage and withdrawal of electricity from a battery.

Although the energy from burning hydrogen may be used directly to power motor vehicles, hydrogen may also be used indirectly in the form of **fuel cells**. These are devices that produce electricity through a chemical reaction between a source fuel and

fuel cell ■ Any device that produces electricity through a chemical reaction between a source fuel and an oxidizer

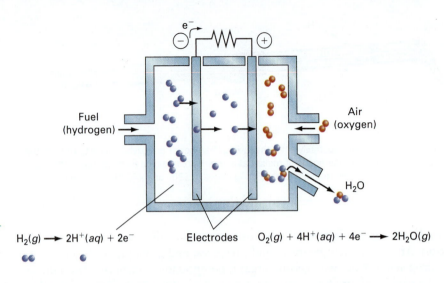

FIGURE 7.10 Hydrogen and oxygen react to produce energy in a hydrogen fuel cell. These reactants are provided to the cell from external sources. The flow of electrons in the cell produces electricity, which is the source of power. When the electricity is used to power a motor vehicle, the hydrogen is regarded as an alternative motor fuel.

Fuel (hydrogen)

Air (oxygen)

H_2O

$H_2(g) \longrightarrow 2H^+(aq) + 2e^-$ Electrodes $O_2(g) + 4H^+(aq) + 4e^- \longrightarrow 2H_2O(g)$

hydrogen fuel cell ■ A device that produces energy by the reaction of hydrogen and oxygen

an oxidizer. The cross section of a **hydrogen fuel cell** is shown in Figure 7.10. Hydrogen fuel cells are manufactured tiny enough to power a cell phone and large enough to power a car or truck. The reactants in a hydrogen fuel cell are hydrogen and atmospheric oxygen, which react to produce water and energy. The water is discharged to the atmosphere and the energy is converted into electricity, which powers the vehicle. Although the sales of vehicles powered by hydrogen fuel cells has annually increased in the United States, gasoline-powered vehicles are still the most widespread form of transportation.

The direct use of hydrogen gas for powering motor vehicles is now limited to areas equipped with hydrogen-refueling stations like the one shown in Figure 7.11. These stations are primarily located in California, where the current use of hydrogen as an automotive motor fuel is greatest. Each refueling station has equipment from which the hydrogen

FIGURE 7.11 At a hydrogen refueling station, hydrogen fuel is stored either as a liquid or a gas under pressure. As hydrogen fuel is dispensed from the pump, its pressure typically ranges from 5000 psig (34,456 kPa) to 10,000 psig (68,911 kPa). (*Courtesy of Air Products, Allentown, Pennsylvania.*)

fuel is dispensed to customers. At 16 C.F.R. §306.12, the U.S. Federal Trade Commission requires retail hydrogen distributors to affix an orange-and-black label on the hydrogen dispenser that resembles the following:

This label identifies hydrogen as the alternative motor fuel and provides its minimum fuel rating as 98% by volume. The U.S. Federal Trade Commission also requires new-vehicle manufacturers and used-vehicle dealers to affix the label on a visible surface of each vehicle that is powered by hydrogen.

7.2-C PRODUCTION OF HYDROGEN

Hydrogen is produced for commercial use by the following methods:

■ In the first method, steam is first passed over red-hot coke or coal under high temperature and pressure. This process, called **coal gasification**, produces a mixture of gases called **water gas**, **synthesis gas**, or **syngas**. This mixture consists mainly of carbon monoxide and hydrogen.

$$C(s) \quad + \quad H_2O(g) \quad \longrightarrow \quad CO(g) \quad + \quad H_2(g)$$

Carbon   Water   Carbon monoxide   Hydrogen

coal gasification ■ The chemical process that produces a mixture of carbon monoxide and hydrogen when steam is exposed to coal at high temperature and pressure

The hydrogen is isolated from the carbon monoxide, or the carbon monoxide is converted to carbon dioxide, by passing the water gas with additional steam over a catalyst such as iron(III) oxide.

$$CO(g) \quad + \quad H_2O(l) \quad \longrightarrow \quad CO_2(g) \quad + \quad H_2(g)$$

Carbon monoxide   Water   Carbon dioxide   Hydrogen

water gas (synthesis gas; syngas) ■ The mixture of carbon monoxide and hydrogen produced by blowing steam through a bed of red-hot coke

The carbon dioxide in the resulting mixture is passed through an alkaline solution.

■ The second industrial method of producing hydrogen involves the chemical action of steam on methane at high temperatures.

$$CH_4(g) \quad + \quad H_2O(g) \quad \longrightarrow \quad 3H_2(g) \quad + \quad CO(g)$$

Methane   Water   Hydrogen   Carbon monoxide

The water gas produced by the chemical reaction is treated by either of the methods previously noted for isolating carbon monoxide–free hydrogen.

■ Hydrogen is also produced by a multistep, complex reaction between propane and steam that is denoted by the following overall equation:

$$C_3H_8(g) \quad + \quad 6H_2O(g) \quad \longrightarrow \quad 3CO_2(g) \quad + \quad 10H_2(g)$$

Propane   Water   Carbon dioxide   Hydrogen

The carbon dioxide in the resulting mixture is passed through an alkaline solution.

■ Hydrogen is also produced by the reaction of steam on methanol vapor using a copper oxide/zinc oxide catalyst as follows:

$$CH_3OH(g) \quad + \quad H_2O(g) \quad \longrightarrow \quad CO_2(g) \quad + \quad 3H_2(g)$$

Methanol   Water   Carbon dioxide   Hydrogen

Once again, the carbon dioxide is passed through an alkaline solution.

TABLE 7.4	Physical Properties of Elemental Hydrogen
Melting point	−434.5°F (−259.2°C)
Boiling point	−423.0°F (−252.8°C)
Specific gravity (gas) at 68°F (20°C)	0.071
Vapor density (air = 1)	0.069
Autoignition point	1058°F (570°C)
Lower flammable limit	4% by volume
Upper flammable limit	75% by volume
Liquid-to-gas expansion ratio	848

7.2-D PROPERTIES OF HYDROGEN

Elemental hydrogen possesses several interesting physical properties, some of which are listed in Table 7.4. A notable property is its vapor density—only 0.07 compared with air. Because this value is so low, hydrogen possesses a natural **lifting power** or buoyancy potential. When expressed in metric units, the lifting power of hydrogen is equal to the difference in mass between 1 liter of air and 1 liter of hydrogen at a given temperature and pressure. At 32°F (0°C) and 14.7 psi (101.3 kPa), the mass of 1 liter of air is 1.2930 grams, whereas the mass of 1 liter of hydrogen is 0.0899 grams; hence, the lifting power of hydrogen in air at this temperature and pressure is 1.2930 g/L − 0.0899 g/L, or 1.2031 g/L. As we note in Section 7.2-H, this lifting power once was used to keep dirigibles aloft.

lifting power ■ The natural ability of a gas that is less dense than air to move upwards

Another important property of hydrogen is associated with the tiny size of its molecules. Hydrogen molecules are extremely small when compared with other molecules. They are so tiny that they easily leak through openings such as pipe joints and valve connections. To prevent loss during storage and transport, the steel cylinders, tubes, and tanks used to hold the compressed gas must be uniquely designed.

Another important property of hydrogen is the relatively high rate at which it diffuses into the air. When released from a tank or container, hydrogen diffuses into the air more rapidly than any other substance; that is, at the same temperature, hydrogen moves faster than other gases or vapors.

Hydrogen is a flammable gas that burns when its concentration in air is between 4% and 75% by volume. A concentration within this wide range can readily be achieved when hydrogen is released from its container or tank. Nonetheless, because it dissipates so rapidly, unconfined hydrogen does not remain in any single area for long.

Hydrogen burns in air to form water vapor.

$$2H_2(g) \ + \ O_2(g) \ \longrightarrow \ 2H_2O(g)$$
$$\text{Hydrogen} \qquad \text{Oxygen} \qquad\qquad \text{Water}$$

Given its low vapor density and high diffusion rate, the combustion reaction occurs as the gas moves upward. Hydrogen burns with an almost nonluminous flame that is especially difficult to observe during daylight hours.

7.2-E HYDROGEN AND THE RISK OF FIRE AND EXPLOSION

The physical properties of hydrogen give rise to two potential scenarios relating to its risk of fire and explosion:

■ When released indoors or into an enclosure where the gas can accumulate, the presence of the hydrogen poses a pronounced risk of fire and explosion.

■ When released outdoors or in a manner that enables the hydrogen to readily dissipate, the likelihood that it will form a flammable mixture is comparatively small.

Hydrogen diffuses into the air at the rate of 0.098 in.²/s (0.634 cm²/s) at 32°F (0°C). What does the magnitude of this rate reveal about the likelihood that a flammable mixture of hydrogen and air will be produced under most accident scenarios that occur outside buildings?

Solution: This information indicates that hydrogen diffuses very rapidly into the surrounding air. In fact, unconfined hydrogen moves more rapidly into air than any other gas. Hydrogen is also flammable over a wide concentration range (4% to 75% by volume). Because hydrogen does not remain localized for long, a flammable mixture of hydrogen and air is not ordinarily produced under most accident scenarios that occur outside buildings.

Because hydrogen rises in the air, hydrogen-sensing devices often are installed near the ceilings of enclosures in which hydrogen could be inadvertently released, as from a leaking storage tank into a room. These devices activate exhaust fans capable of evacuating the hydrogen into the outside air at very rapid speeds. When emergency response crews are called to such incidents, they should enter the room only after the hydrogen has been dispelled and its concentration measures less than 4% by volume.

7.2-F CHEMICAL REACTIONS THAT GENERATE HYDROGEN

Although hydrogen may be encountered in cylinders and storage tanks, it also may be inadvertently generated as the product of chemical reactions. Certain metals generate hydrogen by displacing it from water and acid solutions. These displacement reactions occur at rates that depend on the nature of the metal.

Most alkali metals and alkaline earth metals react with water and acids to produce hydrogen. Certain other metals react much more slowly with water and acids. Nonetheless, these reactions still can pose the risk of fire and explosion, because the hydrogen produced may absorb the heat of reaction and burst spontaneously into flame.

Many metals possess the capability of releasing hydrogen from acids. For instance, metallic tin and aluminum displace hydrogen from aqueous solutions of hydrochloric acid and sulfuric acid, respectively, as follows:

$$\underset{\text{Tin}}{Sn(s)} \; + \; \underset{\text{Hydrochloric acid}}{2HCl(aq)} \; \longrightarrow \; \underset{\text{Tin(II) chloride}}{SnCl_2(aq)} \; + \; \underset{\text{Hydrogen}}{H_2(g)}$$

$$\underset{\text{Aluminum}}{2Al(s)} \; + \; \underset{\text{Sulfuric acid}}{3H_2SO_4(aq)} \; \longrightarrow \; \underset{\text{Aluminum sulfate}}{Al_2(SO_4)_3(aq)} \; + \; \underset{\text{Hydrogen}}{3H_2(g)}$$

The rates at which metals displace hydrogen from water and acidic solutions can be determined experimentally. When the metals are then arranged by decreasing reaction rates, the compilation in Table 7.5 is obtained. This arrangement of the metals is called an **activity series**. Its significance is summarized as follows:

activity series ■ An arrangement of the metals in order of their decreasing ability to generate hydrogen by chemical reaction with water and acids

- The metals listed at the top of the table are so chemically reactive that they react with *both* water and acids.
- The metals below magnesium release hydrogen from steam and acid solutions, but not from liquid water.
- The metals below iron in the series do not displace hydrogen from steam, even when the temperature is elevated.
- The metals below lead possess insufficient chemical reactivity to release hydrogen from either water or acids.

Hydrogen is also displaced by certain metals from solutions of sodium hydroxide. The common metals exhibiting this chemistry are aluminum, zinc, and lead, but these reactions occur slowly.

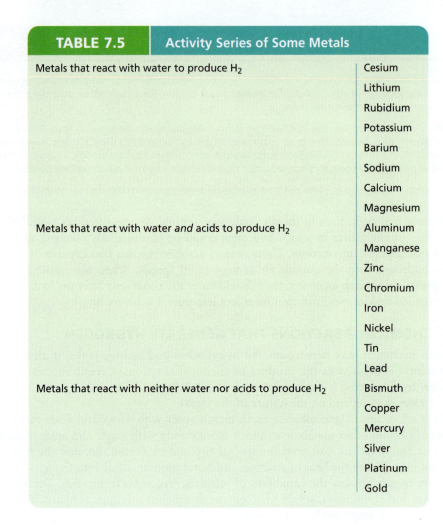

TABLE 7.5	Activity Series of Some Metals	
Metals that react with water to produce H_2		Cesium
		Lithium
		Rubidium
		Potassium
		Barium
		Sodium
		Calcium
		Magnesium
Metals that react with water *and* acids to produce H_2		Aluminum
		Manganese
		Zinc
		Chromium
		Iron
		Nickel
		Tin
		Lead
Metals that react with neither water nor acids to produce H_2		Bismuth
		Copper
		Mercury
		Silver
		Platinum
		Gold

7.2-G HYDROGEN GENERATION WHEN CHARGING LEAD–ACID STORAGE BATTERIES

lead–acid storage battery ■ The collective group of individual cells composed of lead and lead oxide plates and immersed in a solution of sulfuric acid having a specific gravity ranging from 1.25 to 1.30

Most of us are familiar with the **lead–acid storage battery** as the power source in virtually all motor vehicles, including electric hybrid vehicles. A battery differs from a fuel cell in that its reactants are generated by chemical action. On the other hand, each fuel-cell reactant must be continuously provided to a fuel cell from an external source.

A lead-acid storage battery produces electric current by means of two reversible chemical reactions commonly referred to as charging and discharging. A battery is said to be charged when an outside current has been delivered through it; then, as the battery discharges, it produces electricity by means of chemical reactions involving the substances that were formed during the charging process. Hydrogen is one of them. When the battery is being charged, bubbles of hydrogen accumulate near the negative plate; simultaneously, bubbles of oxygen accumulate near its positive plate. As they are produced, both gases dissipate into the atmosphere.

When banks of lead–acid storage batteries like those shown in Figure 7.12 are simultaneously charged within an enclosure, however, the concentration of hydrogen is likely to exceed its lower flammable limit. Within an enclosed room, for instance, hydrogen rises and concentrates along the length of the ceiling. Because a concentration of only 4% by volume of hydrogen is sufficient to produce a flammable mixture in air, adequate ventilation should always be provided within rooms containing banks of storage batteries to

FIGURE 7.12 This battery-charging station is a location at which banks of individual batteries are simultaneously charged. Hydrogen gas is produced during the battery-charging process. When the battery-charging station is located inside a large ventilated room, the production of hydrogen is a relatively small problem, but when it is located inside an enclosure, the hydrogen may reach a concentration within its flammable range. Then, the accumulation of hydrogen poses the risk of fire and explosion. *(Photo by Eugene Meyer and courtesy of Interstate Batteries of Las Vegas, Las Vegas, Nevada.)*

prevent the accumulation of hydrogen and minimize an unreasonable risk of fire and explosion. OSHA requires the owners and operators of enclosures in which lead-acid storage batteries are charged to post warning signs like the following:

7.2-H THE *HINDENBURG*

Perhaps the best-known incident involving the burning of hydrogen was the destruction of the German dirigible named the *Hindenburg*. Although the combustion of diesel fuel powered its engines, hydrogen stored in large gas bags held this airship aloft. Germany had intended to use the *Hindenburg* to inaugurate a new era in fast transatlantic travel, but the airship mysteriously caught fire and exploded while attempting its landing approach at Lakehurst, New Jersey. Given the presence of hydrogen onboard, the immediate sentiment linked a hydrogen leak with the burning of the airship. In Section 9.3-D, we shall examine another likely cause.

The tragedy of the *Hindenburg* caused Germany and other nations to discontinue the use of hydrogen as a buoyant gas in airships. Today, helium is used to provide lifting power in airships, and it is a safer choice because it is a nonflammable gas. It possesses about 93% of the lifting power of hydrogen.

7.2-I WORKPLACE REGULATIONS INVOLVING HYDROGEN

When it is intended for use at manufacturing and processing plants, hydrogen is generally encountered as a confined gas within cylinders or storage tanks. The capacities of the tanks used to store hydrogen range from less than 3000 cubic feet (85 m^3) to over 15,000 cubic feet (425 m^3). At 29 C.F.R. §§1910.103(b)(2)(1)(a)–(d), OSHA regulates their locations in relation to the position of buildings and other storage tanks so they are readily accessible to delivery equipment and authorized personnel. Hydrogen storage systems must be located aboveground, but not beneath electric power lines, and away from piping for other flammable gases and flammable liquids. To alert individuals to the presence of this flammable gas, OSHA also requires the tanks to be permanently placarded with a sign that reads as follows:

7.2-J TRANSPORTING HYDROGEN

When shippers intend to transport compressed or cryogenic hydrogen, DOT requires them to identify it as shown in Table 7.6 on the accompanying shipping paper. All labeling, marking, and placarding requirements apply.

When hydrogen is transported in bulk by highway or rail, its name must be displayed on two opposing sides of the tankcar used for shipment. When it is transported by rail in a DOT-113 tankcar, DOT requires the carrier to display FLAMMABLE GAS placards on squares having a white background and black border.

To meet demand in high-use areas of the United States, hydrogen is also transferred by pipeline from production plants to user facilities. For example, hydrogen is supplied to the petroleum refineries and petrochemical plants located along the coast of the Gulf of

TABLE 7.6	Shipping Descriptions of Hydrogen
HYDROGEN	**SHIPPING DESCRIPTION**
Compressed hydrogen	UN1049, Hydrogen, compressed, 2.1
Cryogenic hydrogen	UN1966, Hydrogen, refrigerated liquid, 2.1
Hydrogen in a metal hydride storage system	UN3468, Hydrogen in a metal hydride storage system, 2.1 *or* UN3468, Hydrogen in a metal hydride storage system contained in equipment, 2.1 *or* UN3468, Hydrogen in a metal hydride storage system packed with equipment, 2.1

Mexico from a supply pipeline that stretches from the Houston Ship Channel to New Orleans and connects to a 600-mile (965-km) pipeline network.

7.2-K RESPONDING TO INCIDENTS INVOLVING A RELEASE OF HYDROGEN

As a practical matter, combating an ongoing fire involving hydrogen is rarely successful unless the flow of hydrogen from its containment vessel can be stopped without exposing personnel to undue risk. For this reason, it is often best to permit a hydrogen fire to burn without interference until the fuel is entirely exhausted. Because a hydrogen-fueled fire is likely to cause secondary fires, appropriate emergency response actions should include prevention of its spread.

When LH2 has been released, the use of a water fog is generally warranted. It prevents or reduces the possibility that hydrogen will concentrate in the nearby atmosphere and ignite.

7.3 CHLORINE

Elemental chlorine is not found naturally, but chlorine is found on Earth to the extent of 0.19% by mass in a variety of compounds including sodium chloride, potassium chloride, calcium chloride, and magnesium chloride.

At room conditions, chlorine exists as a yellow-green gas with a characteristic penetrating and irritating odor. It is encountered as a gas and a liquefied compressed gas. The element is about 2½ times heavier than air and is highly poisonous when inhaled. Several other physical properties of chlorine are noted in Table 7.7.

Elemental chlorine should not be confused with solid "chlorine" products used to treat the water in residential and municipal swimming pools. Although the latter products frequently are called chlorine, they actually are oxidizing agents that generate chlorine by chemical action within the pools. Their properties are more appropriately discussed in Chapter 11.

7.3-A PRODUCTION AND COMMERCIAL USES OF ELEMENTAL CHLORINE

For commercial use, elemental chlorine is prepared by passing an electric current through either molten sodium chloride or an aqueous solution of sodium chloride or magnesium chloride. When aqueous sodium chloride is used, sodium hydroxide and hydrogen are simultaneously produced.

$$2NaCl(aq) \ + \ 2H_2O(l) \ \longrightarrow \ 2NaOH(aq) \ + \ H_2(g) \ + \ Cl_2(g)$$

Sodium chloride   Water   Sodium hydroxide   Hydrogen   Chlorine

Chlorine

TABLE 7.7	Physical Properties of Elemental Chlorine
Melting point	−150°F (−101°C)
Boiling point	−30°F (−35°C)
Specific gravity (gas) at 68°F (20°C)	1.56
Specific gravity (liquid) at 6.86 atm	1.41
Vapor density (air = 1)	2.49
Liquid-to-gas expansion ratio	457.6

TABLE 7.8	Adverse Health Effects Associated with Breathing Chlorine[a]

TIME EXPOSURE FOR CONCENTRATION	ADVERSE HEALTH EFFECT
1–8 hr exposure at < 0.5 ppm	No signs or symptoms of adverse effects
1-hr exposure at 0.5–2 ppm	Strong odor; slight irritation of nose, throat, and eyes
1-hr exposure at 2–20 ppm	Burning of eyes or throat; coughing and choking sensations
1-hr exposure at > 20 ppm	Sense of suffocation; chest pain; shortness of breath; nausea; vomiting; and hoarseness
1-hr exposure at ≥ 34 ppm	Pulmonary edema; sudden death; bronchospasm (closure of the larynx)

[a]U.S. Army Center for Health Promotion and Preventative Medicine (USACHPPM) Technical Guide 230, "Environmental Health Risk Assessment and Chemical Exposure Guidelines for Deployed Military Personnel" (U.S. Army Public Health Command, June 2010).

Throughout the civilized world, large volumes of elemental chlorine are required annually for the following commercial applications:

- A raw material for the production and manufacture of a wide range of chlorine-containing compounds used as solvents, pesticides, dyes, bleaching agents, plastics, refrigerants, and other commercial products
- A microbicide for treatment of drinking water and wastewater to prevent waterborne infectious diseases
- A bleaching agent of paper pulp and certain textiles

7.3-B ILL EFFECTS CAUSED BY INHALING CHLORINE

Exposure to elemental chlorine poses the threat of inhalation toxicity, its principal risk. The exposure initially causes coughing, dizziness, nausea, headache, and severe inflammation of the eyes, nose, and throat. Prolonged (>1 hr) exposure to moderate concentrations potentially causes congestion of the lungs, which can give rise to the onset of **pulmonary edema**, obstruction of the airways, and painful and difficult breathing. Other ill effects are listed in Table 7.8.

pulmonary edema ■ The excessive accumulation of fluid within the lungs

7.3-C CHLORINE AS A CHEMICAL WARFARE AGENT

On a somber note, Germany used chlorine during World War I as a chemical warfare agent (Section 13.11). The worst incident occurred at Ypres, Belgium in 1915. Taking advantage of its highly poisonous nature, the Germans unleashed the gas as a weapon of mass destruction. The chlorine was discharged from 5730 pressurized cylinders into the wind, which carried the gas into the trenches. As a poisonous gas with a vapor density of 2.49 (air = 1), the gas caused mass casualties by maintaining ground-level concentrations capable of causing death when inhaled.

The reported loss of life at Ypres ranged from 7000 to 15,000 people. The British retaliated by discharging chlorine from 5100 cylinders at German troops during the Battle of Loos. However, meteorological conditions caused the chlorine gas to move back toward the British troops, resulting in 2632 British casualties, including 7 deaths.

In 1925, the League of Nations took the first step toward eliminating the use of poisonous gases during warfare between civilized nations. As a consequence of its efforts, over 130 nations ratified the **Geneva Protocol**, which prohibits the use of chlorine and other poisonous gases as chemical warfare agents. Known formally as the *Protocol for the Prohibition of the Use in War of Asphyxiating, Poisonous, or Other Gases, and of Bacteriological Methods of Warfare*, it has successfully prevented civilized countries from using poisonous gases against their wartime enemies.

Geneva Protocol ■
The international agreement that bans the use of chemical weapons against wartime enemies

A rail tankcar contains 1 ton (0.9 t) of chlorine as a liquefied compressed gas. The following information is imprinted on its exterior surface:

PROX28829		
DOT-105A500W		
LD LMT	5000 LB	2273 KG
LT WT	2000 LB	909 KG
NEW	08 97	

Use the data in Table 7.7 to calculate the approximate volume in cubic feet of gaseous chlorine that is generated when the contents of this tank are suddenly released into the atmosphere during a transportation mishap, and ascertain how this amount impacts emergency responders when the wind speed is <6 mi/hr (<10 km/hr).

Solution: In Section 3.7-A, we learned that "LD LMT" and "LT WT" are the abbreviations for the load limit and light weight of a rail tankcar, respectively. Hence, the imprinted information indicates that the tankcar weighs 2000 pounds (909 kg) when empty and may be used to safely transport up to 5000 pounds (2273 kg) of liquid chlorine.

Using the specific gravity of liquid chlorine in Table 7.7, we calculate the density of the liquid as follows:

$$1.41 \times 62.4 \frac{lb}{ft^3} = 88.0 \ lb/ft^3$$

The volume occupied by 1 ton of liquid chlorine is then calculated to be 22.7 cubic feet:

$$1 \ \overline{tn} \times 2000 \frac{\overline{lb}}{\overline{tn}} \times \frac{1 \ ft^3}{88.0 \ \overline{lb}} = 22.7 \ ft^3$$

Using its liquid-to-gas expansion ratio, we determine that 22.7 cubic feet of liquid chlorine expands to a gaseous volume of 10,388 cubic feet.

$$22.7 \ ft^3_{(liquid)} \times 457.6 = 10,388 ft^3_{(gas)}$$

One ton of chlorine per 10,388 cubic feet is equivalent to an undiluted concentration in air of 0.19 lb/ft^3, or 3000 parts per million.[4] Because the wind speed is relatively low, this concentration remains virtually unchanged during the response action. Table 7.8 indicates that the inhalation of 34 parts per million is likely to cause sudden death. The inhalation of 3000 parts per million is unquestionably fatal.

7.3-D CHEMICAL REACTIVITY OF CHLORINE

Like oxygen, elemental chlorine is a nonflammable gas capable of supporting combustion. The oxidizing ability of chlorine is apparent from its reaction with hydrogen, which burns in a chlorine atmosphere to form hydrogen chloride.

$$H_2(g) \quad + \quad Cl_2(g) \quad \longrightarrow \quad 2HCl(g)$$

Hydrogen   Chlorine   Hydrogen chloride

---

[4]A concentration of 0.19 lb/ft^3 is converted to its equivalent in part per million through use of the program at http://www.unitconversion.org.

Elements other than hydrogen also burn when exposed to an atmosphere of chlorine. For example, exposure of finely divided copper, arsenic, antimony, phosphorus, and sulfur to chlorine causes them to burn with incandescence.

$$Cu(s) \quad + \quad Cl_2(g) \quad \longrightarrow \quad CuCl_2(s)$$
Copper   Chlorine   Copper(II) chloride

$$2As(s) \quad + \quad 3Cl_2(g) \quad \longrightarrow \quad 2AsCl_3(s)$$
Arsenic   Chlorine   Arsenic trichloride

$$2Sb(s) \quad + \quad 3Cl_2(g) \quad \longrightarrow \quad 2SbCl_3(s)$$
Antimony   Chlorine   Antimony trichloride

$$P_4(s) \quad + \quad 6Cl_2(g) \quad \longrightarrow \quad 4PCl_3(l)$$
Phosphorus   Chlorine   Phosphorus trichloride

$$S_8(l) \quad + \quad 10Cl_2(g) \quad \longrightarrow \quad 2S_2Cl_2(l) \quad + \quad 4SCl_4(l)$$
Sulfur   Chlorine   Disulfur dichloride   Sulfur tetrachloride

In these instances, the chlorine acts as an oxidizing agent.

Chlorine also supports the combustion of certain organic compounds. These reactions occur slowly unless the mixture of the compound and chlorine is exposed to light. For instance, a mixture of elemental chlorine and gasoline vapor is essentially unreactive in the dark, but when the mixture is exposed to light, the reaction occurs instantaneously. In such reactions, the light acts catalytically.

chlorination ■ A chemical reaction that involves the addition of chlorine to a substance; the treatment of contaminated drinking water to kill the microorganisms that cause dysentery, cholera, typhoid, hepatitis, and other diseases

When chlorine atoms are incorporated into a compound, the associated chemical phenomenon is referred to as the **chlorination** of the compound. This term is also used to describe the treatment of drinking water and wastewater with elemental chlorine and chlorine-containing oxidizing agents. When used in this fashion, chlorine acts as a microbicide by killing undesirable microorganisms.

In the United States, there are nearly 60,000 municipal water treatment facilities that provide drinking water to over 260 million people. The use of chlorine for treating drinking water saves millions of lives annually from waterborne diseases like cholera and dysentery. Notwithstanding this fact, we noted in Section 7.1-I that chlorine is not a totally effective microbicide, because it does not effectively destroy the *Cryptosporidium parvum* bacterium.

When it is dissolved in water, chlorine reacts to form hypochlorous acid, which is even more powerful than chlorine as an oxidizing agent.

$$Cl_2(g) \quad + \quad H_2O(l) \quad \longrightarrow \quad HClO(aq) \quad + \quad HCl(aq)$$
Chlorine   Water   Hypochlorous acid   Hydrochloric acid

During the chlorination of drinking water and wastewater, it is the presence of hypochlorous acid that aids in making chlorine an effective microbicide.

7.3-E WORKPLACE REGULATIONS INVOLVING CHLORINE

When the use of elemental chlorine is necessary in the workplace, OSHA requires employers to provide their workers with respiratory protective gear. When employees are exposed to a relatively small amount of chlorine, they should wear goggles and impermeable gloves, and whenever possible, their activities should be conducted within a fume hood. When working on bulk chlorine storage systems, however, employees should wear fully encapsulating suits and use self-contained breathing apparatus.

To avoid or minimize the ill effects associated with exposure to chlorine, OSHA requires employers to limit employee exposure in the workplace to a maximum chlorine concentration in air no greater than 1 part per million, averaged over an 8-hour workday.

FIGURE 7.13 When carriers transport any amount of chlorine in a rail tankcar, DOT requires them to post POISON GAS placards on each of its sides and to mark *CHLORINE* and *INHALATION HAZARD* on two opposing sides. DOT also requires them to display the identification number 1017 on each side. In this instance, the carrier chose to display 1017 across the center area of the POISON GAS placards. HOKX 7718 is the reporting mark and number of Occidental Chemical Corporation, Dallas, Texas. At the time this tank was photographed, DOT did not require the MARINE POLLUTANT marking to appear on two opposing sides. (*Courtesy of the Chlorine Institute, Arlington, Virginia and Andrew Johnson, iMed Design, Inc., Reno, Nevada.*)

7.3-F TRANSPORTING CHLORINE

Chlorine is transported as a liquefied compressed gas in steel cylinders, ton-containers, cargo tanks, and rail tankcars at 84 psi (580 kPa) at 70°F (21°C). The ton-container is a welded tank having a maximum loaded mass of 3700 pounds (1680 kg).

When shippers intend to transport chlorine, DOT requires them to identify it on the accompanying shipping paper as follows:

UN1017, Chlorine, 2.3, (5.1), (8) (Marine Pollutant) (Poison - Inhalation Hazard, Zone B)

The rail tankcar in Figure 7.13 is an example of a type of bulk packaging used to transport chlorine. DOT requires carriers to display POISON GAS placards and MARINE POLLUTANT on the tankcar. DOT also requires them to display the identification number 1017 on orange panels, across the center area of the POISON GAS placards, or on white square-on-point diamonds, and to mark the expression INHALATION HAZARD on the bulk packaging. One means of complying with these DOT requirements is noted below:

When chlorine is transported in bulk by highway or rail, DOT requires carriers to display its name on two opposing sides of the tankcar or cargo tank used for shipment. When it is transported by rail, DOT requires carriers to display POISON GAS placards on squares having a white background and black border (Section 6.6-D).

The DOT-approved tanks and cylinders used for transporting chlorine are equipped with fusible plugs designed to melt in the temperature range of 158 to 165°F (70 to 74°C). The plugs are located on the valve just below the valve seat. When exposed to the temperatures routinely associated with fire conditions, the melting of these plugs permits chlorine to be slowly released to the environment and prevents the vessels from rupturing.

DOT also regulates the construction and fabrication of rail tankcars and tank trucks used to transport chlorine. To maintain the confined chlorine primarily in the liquid state, DOT requires a minimum of 4 inches (10 cm) of insulation about them. As noted in Section 3.7-A, thermally insulated tankcars constructed to DOT specifications have been designed to allow external heat to be slowly transferred to the contents under the normal conditions of transport. When exposed to heat, however, these tanks lose their integrity and rupture.

7.3-G RESPONDING TO INCIDENTS INVOLVING A RELEASE OF CHLORINE

Because elemental chlorine poses an inhalation health hazard, emergency responders cannot be too wary when they encounter chlorine during transportation incidents. In 2005, after two freight trains collided in Graniteville, South Carolina, a pressurized tankcar ruptured and released more than 9200 gallons (35 m^3) of liquid chlorine into the environment. This sole incident caused the deaths of nine people, the hospitalization of more than 250 individuals, and the evacuation of more than 5400 people.[5]

When emergency responders are called to a scene at which chlorine tanks or containers have ruptured or could potentially rupture, it is vital to acknowledge the toxicity hazard associated with elemental chlorine. Because inhaling chlorine may be fatal, the use of fully encapsulated suits and self-contained breathing apparatus is absolutely essential.

The responsibilities of a first-on-the-scene team responding to a large spill of chlorine begin with the immediate isolation and evacuation of all unauthorized persons to a distance that relates to the amount of chlorine that has been or could potentially be released into the environment. As initial guidance,[6] DOT recommends the *isolation* of unauthorized persons 3000 feet (1000 m) in all directions when a release of chlorine from a rail tankcar, highway tank truck, or trailer is involved; 1250 feet (400 m) from multiple ton cylinders; and 800 feet (250 m) from multiple small cylinders or a single ton cylinder. As additional protection during a large spill of chlorine, DOT also recommends the *evacuation* of unauthorized persons to a distance ranging from 0.5 mile (0.8 km) to 7+ miles (11+ km) during day and nighttime hours, respectively, from the same transport vessels. The latter distances are selected by consideration of the prevailing wind speed [low (<6 mph, or <10 km/hr), moderate (6–12 mph, or 10–20 km/hr), or high (>12 mph, or 20 km/hr)] for the relevant type of transport vessel.

Emergency responders must also identify those who have been exposed to chlorine from leaks or ruptured containers. These individuals should be moved downwind to fresh air where, if necessary, artificial respiration can be supplied. They should also be kept warm with blankets and provided with immediate first-aid attention. To minimize skin

---

[5]David van Sickle, "Acute health effects after exposure to chlorine gas released after a train derailment," *Amer. J. Emerg. Med.,* Vol. 27 (2009), pp. 1–7.

[6]Table 3, *Emergency Response Guidebook* (Washington, DC: U.S. Department of Transportation, 2012), p. 353.

burns, persons exposed to chlorine should remove their clothing and shower thoroughly. To prevent the impairment of vision, they should irrigate their eyes with flowing water for approximately 30 minutes. Finally, they should be transported to an emergency medical facility for follow-up examinations by physicians.

An emergency response team may be required to examine the physical condition of a chlorine container or rail tankcar. Unruptured vessels should be cooled with water as the team members pinpoint the specific spots from which chlorine is leaking or could potentially leak. Because containers and transport vessels contain *liquid* chlorine, it is often the liquid that drips from valves, fittings, or openings. Liquid chlorine is substantially more concentrated than its gas, and when unconfined, it readily evaporates. One volume of liquid chlorine evaporates into approximately 460 volumes of gas. Consequently, the immediate area surrounding a chlorine leak becomes a highly toxic environment within seconds.

Locating a chlorine leak is not always a simple matter, especially when the gas has been escaping from its container or transport vessel for some time and the atmosphere is heavily laden with chlorine. One method of detecting a chlorine leak is based on the chemical reaction between chlorine and ammonia. These two substances react to form ammonium chloride and ammonium hypochlorite, each of which is a white solid.

$$2NH_3(g) \;+\; Cl_2(g) \;+\; H_2O(l) \;\longrightarrow\; NH_4Cl(s) \;+\; NH_4ClO(s)$$

Ammonia   Chlorine   Water   Ammonium chloride Ammonium hypochlorite

The method consists of tying a rag soaked with household ammonia to a broomstick and then passing the stick along the surface of the chlorine container or tank. Ammonia vaporizes from the liquid and reacts with the chlorine at the point from which it is escaping. At this location, a white cloud drifts into the air. By observing the cloud's formation, the source of the leak is easily identified. This simple procedure is ineffective when the surrounding atmosphere is heavily laden with chlorine, because under these conditions, the ammonium compounds are produced virtually throughout the area.

When chlorine is leaking from a tank or container, the exit points must be closed or sealed. When sealing an opening is impossible or impractical, an attempt should be made to prevent the further escape of liquid chlorine into the environment. This can sometimes be accomplished by rolling the vessel so that the opening points upward. Although chlorine gas continues to escape from the opening, the more concentrated liquid remains confined within the vessel.

When emergency responders arrive at the scene of an incident involving the release of chlorine, specialized equipment should be available for their immediate use. An essential item is a kit that contains a clamping device to seal a leak at the fusible plug and a patching device to seal a leak in the cylinder sidewall. They are components of the so-called *Chlorine Institute Emergency Kit "A,"* which is available from several commercial outlets.[7] There are also "B" and "C" kits for use on ton cylinders and tanks, respectively.

To improve the speed and effectiveness of a response action at an emergency involving the release of chlorine, the Chlorine Institute formalized an action known as the **Chlorine Emergency Plan**, or **CHLOREP**. Under this plan, the United States and Canada are divided into regional sectors in which specially trained teams are located. In the event of an emergency involving the release of chlorine, or when CHEMTREC (Section 1.12) is contacted, the caller is put into immediate contact with the closest CHLOREP team, which then oversees the handling of the incident.

Chlorine Emergency Plan (CHLOREP) ■ An action plan for aiding emergency responders during incidents involving the release of chlorine during transportation mishaps or at user locations

---

[7]The Chlorine Institute is a trade association consisting primarily of company representatives interested in the safe production, distribution, and use of chlorine and other substances associated with the chlor-alkali industry. Their offices are located at 1300 Wilson Boulevard, Arlington, Virginia 22209.

TABLE 7.9	Physical Properties of Elemental Phosphorus	
	WHITE PHOSPHORUS	**RED PHOSPHORUS (AMORPHOUS)**
Melting point	111°F (44°C)	1094°F (590°C) at 43 atm (4400 kPa)
Boiling point	535°F (280°C)	781°F (416°C) (sublimates)
Specific gravity at 68°F (20°C)	1.82	2.34
Vapor density (air = 1)	4.42	4.77
Autoignition point	86°F (30°C) (spontaneous ignition in dry air)	500°F (260°C)

7.4 PHOSPHORUS

Elemental phosphorus has several allotropes, two of which are especially important: *white phosphorus* and *red phosphorus*. White phosphorus is the unstable form of the element at room conditions. On standing, it acquires a yellow coloration due to the partial conversion of the white allotrope to the more stable red allotrope. For this reason, white phosphorus is also called *yellow phosphorus*.

The physical properties of white and red phosphorus are provided in Table 7.9. Their properties are so strikingly different that we discuss them here separately. Both are used to manufacture important chemical products such as special alloys (e.g., phosphor bronze), rodenticides, fireworks, matches, phosphoric acid, and metallic phosphides. In the past, both were also used as the active agent in incendiary bombs. When dropped from military aircraft, these bombs set fires to objects or caused burn injuries to persons through the action of flames and heat. The composition of some incendiary bombs containing red phosphorus was 50% of the incendiary mixture.

White phosphorus has also been used in incendiary bombs to illuminate the battlefield during nighttime. However, the addition of white phosphorus in munitions is now regarded as an unwarranted wartime practice, because flaming droplets from exploded ordnance can become embedded beneath skin and seriously burn civilians.

White phosphorus has also been used in tracer, smoke, and signaling systems. The mixture of tetraphosphorus hexoxide and tetraphosphorus decoxide produced by the combustion of white phosphorus possesses the best obscuring power of any known smoke-producing material.

In 1983, many of the world's civilized countries agreed to ban the use of incendiary weapons during certain warfare operations by acceptance of the **Convention on Certain Conventional Weapons**. Known formally as the *Convention on Prohibitions or Restrictions of the Use of Certain Conventional Weapons Which May Be Deemed to Be Excessively Injurious or to Have Indiscriminate Effects*, it prohibits the use of incendiary weapons against civilian populations and restricts their use against military targets located within a concentration of civilians, but it does not restrict the use of incendiary bombs in illumination, tracer, smoke, or signaling systems.

Although the United States is a signatory to the Convention, it has not ratified the nonuse of incendiary weapons on military targets located in civilian-populated areas. In 2005, during the U.S.-led assault on Fallujah, Iraq, incendiary munitions containing white phosphorus were used to flush enemy troops from covered positions.

Convention on Certain Conventional Weapons ■ An international treaty that aims to restrict or prohibit the use of certain conventional weapons during warfare, including those containing incendiary agents

White phosphorus

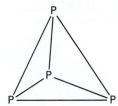

7.4-A PRODUCTION AND PROPERTIES OF WHITE PHOSPHORUS

White phosphorus is a waxy, translucent solid at ambient conditions. Each of its molecules is tetratomic and has the tetrahedral shape like that shown in the left margin. For these reasons, the chemical formula of white phosphorus is P_4.

White phosphorus is industrially prepared by heating calcium phosphate rock with sand and coke in an electric furnace. The principal components of sand and coke are silicon dioxide and carbon, respectively. The production of white phosphorus is denoted as follows:

$$2Ca_3(PO_4)_2(s) \quad + \quad 6SiO_2(s) \quad + \quad 10C(s) \quad \longrightarrow \quad 6CaSiO_3(s) \quad + \quad 10CO(g) \quad + \quad P_4(g)$$

Calcium phosphate   Silicon dioxide   Carbon   Calcium silicate   Carbon monoxide   Phosphorus

The phosphorus vapors are vented from the furnace and condensed under water to produce a white solid.

Because the autoignition temperature of white phosphorus is only 86°F (30°C), white phosphorus spontaneously ignites when exposed to air. The autoignition temperature is so low that body heat can serve as an ignition source. To reduce or eliminate the risk of its spontaneous combustion, white phosphorus is stored under water or a blanket of nitrogen.

White phosphorus is often said to possess the offensive odor of a mixture of garlic and rotten fish, but this comparison is misleading. This specific odor is more likely associated with the presence of phosphine (Section 9.6-B), a toxic gas slowly produced by the reaction between phosphorus and cold water.

$$P_4(s) \quad + \quad 6H_2O(l) \quad \longrightarrow \quad 3H_3PO_2(aq) \quad + \quad PH_3(g)$$

Phosphorus   Water   Hypophosphorous acid   Phosphine

The combustion of white phosphorus produces two oxides, tetraphosphorus hexoxide and tetraphosphorus decoxide, as follows:

$$P_4(s) \quad + \quad 3O_2(g) \quad \longrightarrow \quad P_4O_6(s)$$

Phosphorus   Oxygen   Tetraphosphorus hexoxide

$$P_4(s) \quad + \quad 5O_2(g) \quad \longrightarrow \quad P_4O_{10}(s)$$

Phosphorus   Oxygen   Tetraphosphorus decoxide

As implied by these equations, tetraphosphorus hexoxide and tetraphosphorus decoxide are the products of incomplete and complete combustion, respectively. Both are white compounds. Consequently, when white phosphorus burns, billows of dense white, choking smoke are produced. This luminous dense smoke accounts for the use of white phosphorus as a component of military smoke and signaling systems.

Although spontaneous combustion is the principal hazard associated with white phosphorus, its vapor also is highly poisonous. Prolonged exposure to phosphorus vapor causes **phossy jaw**, or **phosphorus necrosis**, which in the worst instances causes disintegration of the jawbone. In addition, white phosphorus burns the skin and produces wounds that are extremely painful and slow to heal. For this reason, users must always handle white phosphorus with gloves.

phossy jaw (phosphorus necrosis) ■ The disfiguring affliction caused by overexposure to phosphorus vapor

7.4-B TRANSPORTING WHITE PHOSPHORUS

When shippers offer solid or molten white phosphorus for transportation in bulk, DOT requires them to enter the relevant shipping description shown in Table 7.10 on the accompanying shipping paper. When the element is molten, HOT markings must be displayed.

DOT requires carriers to display the relevant identification number—1381 or 2447—on orange panels or across the center area of SPONTANEOUSLY COMBUSTIBLE placards or white square-on-point diamonds. For example, any of the following may be used to display the identification number 1381 on bulk packaging:

TABLE 7.10	Shipping Descriptions of White Phosphorus
WHITE PHOSPHORUS	**SHIPPING DESCRIPTION**
White phosphorus, solid	UN1381, Phosphorus, white, dry, 4.2, (6.1), PG I (Marine Pollutant) (Poison)
White phosphorus, molten	HOT, UN2447, Phosphorus, white, molten, 4.2, (6.1), PG I (Marine Pollutant) (Poison)

When white phosphorus is transported in bulk by highway or rail, DOT requires carriers to display a name such as *PHOSPHORUS, WHITE, DRY* or *PHOSPHORUS, WHITE, MOLTEN* on two opposing sides of the transport vehicle used for shipment. All other labeling, marking, and placarding requirements apply.

7.4-C RESPONDING TO INCIDENTS INVOLVING A RELEASE OF WHITE PHOSPHORUS

White phosphorus generally is burning when first-on-the-scene responders arrive at a scene involving its release. Although the fire can be effectively extinguished with water, experts advise firefighters to avoid its use due to the formation of toxic phosphine. Instead, they recommend the use of dry sand, because it can blanket the element, prevent reignition, and minimize the potential exposure to phosphine.

When small quantities of white phosphorus are burning, it is best to segregate them from nearby combustible materials. This may not be a simple matter, because phosphorus melts at a relatively low temperature and flows as it burns into nearby low areas. For this reason, appropriate action should be taken by constructing dams or dikes.

Firefighters responding to incidents involving elemental phosphorus should wear protective gear and use self-contained breathing apparatus. They should be particularly cautious to avoid inhaling the fumes from a phosphorus fire. Fumes contain particulates of the phosphorus oxides, which when inhaled, can seriously irritate the nose, mouth, throat, and lungs.

7.4-D RED PHOSPHORUS

Red phosphorus

The red allotrope of phosphorus is a dark red solid whose molecules consist of long chains of P_4 tetrahedra, each having an undefined length. The following example shows only three interlinked tetrahedra, but this number is generally much larger.

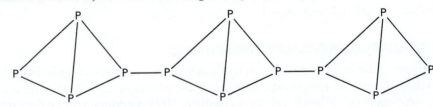

The chemical formula of red phosphorus is usually denoted as either P or P_∞. Due to its undefined structure, red phosphorus is also known as amorphous (unstructured) phosphorus. It is produced industrially by heating white phosphorus at 482°F (250°C) in an iron container from which air has been excluded.

When compared to white phosphorus, the red allotrope is sluggish in chemical reactivity. Small quantities are not spontaneously combustible. They generally are stored in closed containers without an overlying layer of water or nitrogen.

Bulk quantities of red phosphorus, however, are spontaneously combustible, forming a mixture of tetraphosphorus hexoxide and tetraphosphorus decoxide. The red allotrope combines spontaneously with atmospheric oxygen, but the combustion reaction occurs very slowly. When it is stored in bulk, the accumulated heat of combustion triggers the burning of the entire mass. For this reason, bulk quantities of red phosphorus are rarely stored.

Red phosphorus is nonpoisonous, but like the white allotrope, it reacts with water to form phosphine.

7.4-E TRANSPORTING RED PHOSPHORUS

When shippers intend to transport red phosphorus, DOT requires them to identify it on the accompanying shipping paper as follows:

UN1338, Phosphorus, amorphous, 4.1, PG III

When red phosphorus is transported in bulk by highway or rail, DOT requires carriers to display the name *PHOSPHORUS (AMORPHOUS)* on two opposing sides of the transport vehicle used for shipment. All other labeling, marking, and placarding requirements apply.

7.4-F RESPONDING TO INCIDENTS INVOLVING A RELEASE OF RED PHOSPHORUS

Bulk quantities of red phosphorus are unlikely to be encountered, but containers holding small quantities have been involved in fires. These fires can be extinguished by the application of dry sand, foam, or dry chemicals, but not water. Because phosphine is produced when water is applied to burning red phosphorus, the use of dry sand is advised for fire extinguishment.

7.5 SULFUR

Elemental sulfur occurs naturally, particularly in countries bordering the Gulf of Mexico and in Japan, Mexico, and Italy. Sulfur also occurs in minerals and ores too numerous to mention, in which it is combined with metals and other nonmetals. Sulfur accounts for 0.06% by mass of all the elements found on Earth.

7.5-A PRODUCTION AND PROPERTIES OF SULFUR

Most of the world's supply of elemental sulfur comes from natural deposits of the element frequently called *brimstone*. These deposits are often located near hot springs and volcanoes. To isolate the sulfur from them, hot water under pressure is pumped into the subterranean brimstone-bearing deposits, whereupon the sulfur melts and is brought to the surface by an airlift. Brimstone often has an offensive odor because it contains hydrogen sulfide, a toxic gas having the odor of rotten eggs (Section 10.13-A). Their containers and tanks must be handled with special care, as the gas can evolve and accumulate in the vent spaces above the liquid.

By contrast, pure sulfur is an odorless solid. When it is heated at atmospheric pressure, it vaporizes, but the vapor can be condensed on a cold surface, producing a fine dust called **flowers of sulfur**. It is used as a commercial fungicide and acaricide (Section 7.5-B).

The solid state of sulfur occurs in numerous allotropic forms, the two most common of which are *orthorhombic sulfur* and *monoclinic sulfur*. Orthorhombic sulfur is the stable allotrope of solid sulfur at ambient conditions. It is a yellow, crystalline solid whose physical properties are noted in Table 7.11. When orthorhombic sulfur is maintained at a temperature between 205 and 235°F (96 and 113°C), it changes into a mass of long transparent needles composed of monoclinic sulfur. Because monoclinic sulfur is not the stable allotrope, it slowly changes back into the orthorhombic form.

Sulfur, solid

Sulfur, liquid

flowers of sulfur ■ The finely divided powder of elemental sulfur

TABLE 7.11	Physical Properties of Elemental Sulfur (Orthorhombic)
Melting point	248°F (120°C)
Boiling point	832°F (445°C)
Specific gravity at 68°F (20°C)	2.07
Vapor density (air = 1)	8.9
Vapor pressure at 475°F (246°C)	10 mmHg
Flashpoint	405°F (207°C)
Autoignition point	450°F (232°C)
Lower flammable limit (as dust)[a]	0.002 lb/ft^3 (35 g/m^3)
Upper flammable limit (as dust)[a]	0.09 lb/ft^3 (1400 g/m^3)

[a]The lower and upper flammable limits of sulfur dust vary with its particle size and depth of dispersion.

Both solid allotropes exist as molecules having eight sulfur atoms bonded together in the puckered-ring arrangement shown below:

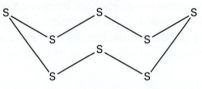

Molten sulfur has a highly complex molecular arrangement. Chemists denote it as S_x, where x is a relatively small but undefined number. The chemical formula of sulfur vapor at its boiling point is also represented as S_8, but when the vapor is further heated, the cyclic arrangement breaks down, and the sulfur molecules assume the formula S_2. To avoid ambiguity, S_8 is denoted in this text as the formula of liquid and solid sulfur, and S_2 is represented as the formula of gaseous sulfur.

When elemental sulfur is exposed to an ignition source, it first melts and the liquid sulfur burns with a blue flame before it vaporizes. The combustion produces sulfur dioxide, a poisonous gas having a suffocating, choking odor (Section 10.12).

$$S_8(l) \ + \ 8O_2(g) \longrightarrow 8SO_2(g)$$

Sulfur   Oxygen   Sulfur dioxide

Elemental sulfur also combines with most metals. For example, when a mixture of mercury and iron is heated, the elements unite to form mercury(II) sulfide and iron(II) sulfide, respectively.

$$8Hg(l) \ + \ S_8(l) \longrightarrow 8HgS(s)$$

Mercury   Sulfur   Mercury(II) sulfide

$$8Fe(s) \ + \ S_8(l) \longrightarrow 8FeS(s)$$

Iron   Sulfur   Iron(II) sulfide

Elemental sulfur is likely to spontaneously ignite under the following conditions:

■ When flowers of sulfur are dispersed into air, a potentially explosive mixture is produced. The spontaneous ignition of this mixture is triggered by the static electricity generated by the movement of the sulfur particles within the air. The potential for a dust explosion may be markedly reduced by electrically grounding the vessel in which the sulfur is confined.

FIGURE 7.14 Elemental sulfur is a constituent of many industrial and domestic products including gunpowder, matches, insecticides, fertilizers, and vulcanized rubber. When the sulfur burns, it is converted into the toxic gas sulfur dioxide.

■ Elemental sulfur reacts with many oxidizing agents. When the mixture is activated, the resulting heat of reaction is likely to cause the ignition of the residual sulfur. Consequently, all mixtures of elemental sulfur and oxidizing agents pose the risk of fire and explosion. Every effort should be employed to keep sulfur and oxidizing agents segregated.

7.5-B COMMERCIAL USES OF SULFUR

Sulfur is one of the world's most important raw materials. There is hardly a segment of the chemical industry that does not use elemental sulfur or one of its compounds in manufacturing or production processes. Approximately 80% of the elemental sulfur produced in the United States is used as a raw material for the manufacture of sulfuric acid (Section 8.7). As Figure 7.14 illustrates, sulfur is also used to produce vulcanized rubber products (Section 14.11-A), fertilizers, dyes and other chemical substances, drugs and other pharmaceuticals, black gunpowder, fireworks, pesticides, and matches.

Elemental sulfur is used as a fungicide (for killing fungi) and an acaricide (for killing mites and ticks). For agricultural use, it is commercially available as a dust, paste, and wettable powder. The dust is applied undiluted, but the paste and wettable powder formulations may contain pulverized clay, which deposit and stick to the surfaces of leaves. They kill by direct contact.

Sulfur is also used as a raw material by the chemical industry to produce other chemical products. For example, elemental sulfur is united with carbon and fluorine to produce carbon disulfide (Section 13.10) and sulfur hexafluoride, respectively:

■ Carbon disulfide is produced when sulfur vapor is passed over very hot carbon in the absence of air. The presence of air is avoided to reduce or prevent the likelihood that carbon disulfide will ignite.

$$C(s) \ + \ S_2(g) \ \longrightarrow \ CS_2(g)$$

Carbon   Sulfur   Carbon disulfide

TABLE 7.12	Shipping Descriptions of Sulfur
SULFUR	**SHIPPING DESCRIPTION**
Sulfur, elemental, solid	NA1350, Sulfur, 9, PG III
Sulfur, elemental, molten	NA2448, Sulfur, molten, 9, PG III
	or
	UN2448, Sulfur, molten, 4.1, PG III

- Sulfur hexafluoride is the predominant product formed when sulfur combines with fluorine.

$$S_8(s) \quad + \quad 24F_2(g) \quad \longrightarrow \quad 8SF_6(g)$$

Sulfur   Fluorine   Sulfur hexafluoride

It is used as an insulator in high-voltage electrical equipment.

7.5-C TRANSPORTING SULFUR

When shippers offer sulfur for transportation, DOT requires them to identify the relevant entry in Table 7.12 on the accompanying shipping paper. DOT regulates the transportation of sulfur for domestic and international transportation as a class 9 and a flammable solid, respectively. For domestic transportation, DOT notes at 49 C.F.R. §172.102.30 that shippers are not subject to its labeling requirements if the sulfur is transported in nonbulk packaging, or if it is formed in a specific shape like prills, granules, pellets, pastilles, or flakes.

At 49 C.F.R. §172.102.30, DOT also notes that the CLASS 9 placard is not required on bulk packaging when molten sulfur is transported domestically, as long as the packaging is marked with the identification number 2448 on orange panels, white square-on-point diamonds, or HOT markings. At 49 C.F.R. §172.325, DOT also requires the packaging to be marked with the expression *MOLTEN SULFUR*.

When molten sulfur is transported internationally in bulk packaging, DOT requires its carriers to display FLAMMABLE SOLID placards on the packaging. DOT also requires the HOT marking and the expression *MOLTEN SULFUR* to be displayed on the packaging. Alternatively, the identification number may be displayed across the center of the FLAMMABLE SOLID placard, but the HOT marking and the expression *MOLTEN SULFUR* are still displayed on the packaging.

7.5-D RESPONDING TO INCIDENTS INVOLVING A RELEASE OF SULFUR

The following practices should be implemented at fire scenes involving burning sulfur:

- Firefighters may effectively extinguish sulfur fires by applying water to them as a fog. The fog not only removes heat from the fire scene, but it also avoids the potential buildup of steam that could cause the hot material to splatter.

TABLE 7.13	Physical Properties of Graphite and Diamond	
	GRAPHITE	**DIAMOND**
Melting point	6606–6687°F (3652–3697°C)	6422°F (3550°C)
Boiling point	7592°F (4200°C)	8726°F (4830°C)
Specific gravity at 68°F (20°C)	2.20–2.35	3.52
Vapor density (air = 1)	0.4	
Flashpoint	>200°F (>93°C)	
Autoignition point	1346°F (730°C)	

■ To avoid inhaling toxic levels of sulfur dioxide, firefighters responding to incidents involving burning sulfur must wear fully encapsulating suits with self-contained breathing apparatus.

■ Because sulfur readily melts under fire conditions and flows into lower adjacent areas where it can ignite secondary fires, firefighters must take appropriate action to segregate the molten material from combustible materials by constructing dams or dikes.

■ When firefighters respond to transport incidents involving molten brimstone, they must understand that hydrogen sulfide may evolve and accumulate in the vent spaces of the transport vessels. Inhalation of the gas may be deadly.

7.6 CARBON

Elemental carbon occurs naturally as its two primary allotropes, graphite and diamond.[8] It also occurs in soot, coal, coke, charcoal, and carbon black. In these two primary allotropes, the elemental carbon is represented by its chemical symbol, C.

The natural abundance of carbon is only 0.08% by mass on Earth, yet carbon ranks much higher in terms of importance, because it is a constituent of the compounds needed by all living organisms for survival.

7.6-A COMMON ALLOTROPES OF CARBON

Neither diamond nor graphite is generally considered hazardous. Each possesses substantially different physical properties, some of which are compared in Table 7.13. Graphite (the "lead" of pencils) is a black material that feels slippery, but diamond can be cut and polished to a crystalline, transparent luster. Although graphite is one of the softest known substances, diamond is the hardest substance found in nature.

Graphite

---

[8]There are several other carbon allotopes whose properties are not discussed in this text. In 1962, a one-atom-thick sheet of carbon called *graphene* was first isolated from graphite. Its structure resembles a densely packed hexagonal honeycomb of carbon atoms. It is regarded as a distinct allotrope of carbon because it possesses a set of unique properties, even different from those of graphite.

In 1985, the carbon allotrope having 60 carbon atoms per molecule (C_{60}) was discovered by vaporizing graphite with a laser. It is called *buckminsterfullerene*. The name is derived from the observation that its hollow, soccer-ball-shaped molecules resemble the celebrated geodesic domes designed by American architect and engineer, R. Buckminster Fuller. Each C_{60} molecule is colloquially called a "buckeyball." Its symmetry consists of 32 interlocking rings (20 hexagons and 12 pentagons). This geometrical shape is called a truncated icosahedron.

In 1991, the carbon allotrope called a carbon *nanotube* was discovered. It takes the physical form of cylindrical carbon molecules with at least one end typically capped with a hemisphere of the buckeyball structure. The name is derived from its size, because the diameter of a nanotube is of the order of a few nanometers but may be several millimeters in length. A carbon nanotube exhibits extraordinary strength and is currently finding usefulness in diverse areas including food, cosmetics, electronics, and medicine.

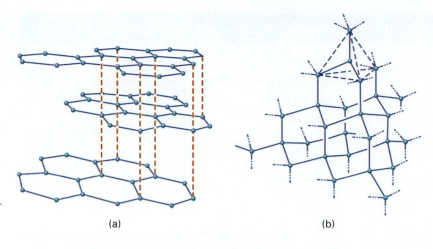

FIGURE 7.15 The structures of two carbon allotropes, diamond and graphite. In the graphite structure shown in (a), planar hexagonal rings covalently bond to one another in successive sheets. In the diamond structure shown in (b), each carbon atom is covalently bonded to four other carbon atoms in a tetrahedral arrangement.

(a)   (b)

The physical properties of graphite and diamond are linked with the structures of the substances. In the graphite structure shown in Figure 7.15(a), each carbon atom is bonded to other carbon atoms in planar hexagonal rings joined to one another in successive sheets. In the diamond structure shown in Figure 7.15(b), each carbon atom is symmetrically bonded to four other carbon atoms in a tetrahedral arrangement.

The diamond structure is produced in nature when carbon-containing materials are subjected to a pressure of approximately 0.8 million pounds per square inch (5.6 million kPa) within the hot mantle of Earth, over 75 miles (120 km) below its surface. The highest gem-quality diamonds are discovered in ancient volcanic pipes in a greenish rock, kimberlite, in South Africa, Canada, Arkansas, Siberia, and elsewhere.

For decades, scientists and engineers have attempted to produce diamonds in laboratories by replicating Nature's process. Limited success was initially experienced. The diamonds produced were small and useful only as low-grade industrial diamonds. They were too dark in color to be suitable for gemstones. Today, however, significant advances in producing gemstone-quality diamonds have been made by passing carbon vapor over diamond seeds inside a vacuum chamber at an approximate temperature of 2000°F (1093°C). Using this technique, the ability to produce 10-carat diamonds has been mastered.[9] By comparison, the largest polished (faceted) natural diamond is the *Golden Jubilee Diamond*, which weighs 545.67 carats (109.13 g). It is among the crown jewels of the royal family of Thailand.

At moderate temperatures and pressures, graphite is the stable allotrope of carbon. On Earth's surface, diamond converts into graphite at an imperceptibly slow rate. However, when diamond is heated to approximately 3583°F (1700°C) in the absence of air, it rapidly converts into graphite.

$$C_{(diamond)} \longrightarrow C_{(graphite)}$$

Although graphite and diamond are stable substances at the conditions experienced on Earth's surface, both burn in air. The products of their incomplete and complete combustion are carbon monoxide and carbon dioxide, respectively.

7.6-B COMMERCIAL USES OF DIAMOND AND GRAPHITE

Diamond companies advertise that a diamond is "a girl's best friend," the supreme token of love and affection. In many parts of the world, a woman's acceptance of a

---

[9]Lauren K. Wolf and Carmen Drahl, "Carbon Goes Deep," *Chem. Eng. News*, Vol. 90 (March 12, 2012), pp. 54–58.

gemstone-quality diamond from her fiancé serves as a formal sign of their betrothal. Gemstone-quality diamonds primarily serve in this exalted role, because their dazzling sparkle is so appealing. Industrial diamonds have more mundane uses in saw blades used to cut marble, in wire-drawing dies, and in drill bits, grinding wheels, and hacksaw blades.

Graphite and diamond undergo phase changes from solid to liquid carbon at exceptionally high temperatures. For example, Table 7.13 shows that diamond melts at 6422°F (3550°C). This is the highest melting point of any element that occurs naturally.

Graphite and diamond also possess an important anomalous property when compared with other nonmetals: They are extraordinarily good conductors of heat. If you are fortunate enough to own a 5-carat diamond, hold it to the tip of your tongue. The diamond feels cold, because it conducts the heat from your tongue. Diamond possesses the highest thermal conductivity of all substances.

The properties of graphite give rise to many industrial applications, of which the following are representative:

- Graphite is molded into crucibles that are used to hold molten steel and other high-melting metals.
- Graphite is used to line the walls of furnaces and other vessels where high-temperature operations are conducted. The graphite protects the underlying metal from melting or softening.
- The nose and leading edges of aircraft wings generally are coated with graphite to protect the underlying metal from the heat of friction generated when the aircraft travels at high speeds through the air.
- Graphite-based dry powder is an effective fire extinguishing agent on class D fires, because the carbon conducts heat away from the burning metal.

Carbon in the form of coke (Section 7.6-E) is also used in the chemical and metallurgical industries as a reducing agent. This chemical property is put to use at foundries during the production of iron and other metals from their naturally occurring ores.

7.6-C COAL

Almost without exception, all fossil fuels found in nature can be traced to the giant plants that grew during the carboniferous age. Three hundred million years ago, plants grew much more luxuriantly than they do today. During this period, the dominant plants were tree ferns, which grew 30 feet (9.1 m) in height with crowns of large, feathery fronds. As Earth evolved over the subsequent millennia, the remains of the ferns and other plants ultimately were buried at great depths below the planet's surface, where intense temperature and pressure compacted, hardened, and chemically altered them into fossil fuels: coal, natural gas (Section 12.5), and crude oil (Section 12.13-A). This conversion process is called the **carbonization** of vegetable matter.

Coal is a readily combustible black rock whose composition consists of a mixture of carbonaceous material and organic and inorganic compounds. It is extracted from the earth by surface mining as well as from seams that exist in deep, underground natural deposits. In the United States, coal is abundant far more than either natural gas or crude petroleum. Natural sources of coal occur primarily in West Virginia, Pennsylvania, Kentucky, and Wyoming. They are also plentiful in areas outside the United States, especially Australia, India, and China. On the international scale, China is coal's major consumer.

Coal is largely mined to operate factories and power plants that generate electricity. In the United States, coal is used as the fuel at 460 power plants where electricity is generated. The worldwide burning of coal at coal-fired power plants accounts for approximately 24% of the carbon dioxide found in the atmosphere (Figure 5.6). Coal and other

carbonization ■ The process of converting an organic compound into carbon or a carbon-containing residue, typically conducted under intense temperature and pressure

coal ■ Any black or brownish combustible rock that formed naturally by the partial decomposition of plant life at increased pressure and temperature

fossil fuels are nonrenewable natural resources; once used, only their ash and other by-products remain. As we continue to advance into the twenty-first century, the impact of EPA regulations could trigger a decline in the use of coal for energy production; yet, it is more likely that coal will remain the United States' primary fuel source for electricity generation until 2035.[10]

An underground coal mine itself poses a potentially hazardous environment. Flammable methane seeps from the coal seams into the surrounding environment where it poses the risk of fire and explosion (Section 12.5-A). In addition, the atmosphere of a coal mine often is laden with coal dust. When the dust is inhaled over long periods, it causes respiratory diseases including **black lung disease**, or **pneumoconiosis**. These miners have lungs coated with coal dust, the presence of which severely restricts the exchange of oxygen between the lungs and the blood. Their lungs ultimately become scarred, and they experience emphysema, chronic bronchitis, shortness of breath, disability, and premature death. There is no effective treatment or cure for those who have contracted black lung disease.

To limit the amount of coal dust that workers could potentially inhale within a mine, regulations promulgated pursuant to the **Federal Coal Mine Health and Safety Act of 1969**, or **Coal Act**, require mine operators to reduce coal-mine dust within active work areas to less than 2 mg/m^3 of air. Although this regulation produced a decline in the number of workers who now contract black lung disease, approximately 1000 American workers still contract the disease annually.

The hazards of coal dust are not limited to its potential health hazards. It is also a readily combustible material. Coal dust ignites readily when exposed to heat, sparks, or other ignition sources. When dispersed in air, its lower flammable limit is greater than 0.05 oz/ft^3 (>50 g/m^3). Precaution always needs to be exercised by the generators and users of coal dust to prevent its ignition.

7.6-D RANKS OF COAL

Coal occurs naturally in several forms, each differentiated from the others by its **rank**, or **grade**. The ranks of coal are determined by the extent to which carbonization has occurred. Six major ranks of coal are recognized: peat, lignite, subbituminous, bituminous, semianthracite, and anthracite. As we read from left to right in this list, each rank is progressively older and denser than those before it. Lignite and anthracite, for example, have specific gravities of 1.29 and 1.47, respectively. Given this wide range, these are sometimes referred to as *soft coal* and *hard coal*, respectively.

Some representative information about the individual ranks of coal is provided in Table 7.14. Each rank is characterized by its heat content and carbon content. The carbon occurs both as the element and in the form of numerous compounds that become locked within the complex structure of coal. The compounds that are volatile evolve from coal as it is crushed and pulverized; they burn when their vapors are exposed to an ignition source. Before they ignite, however, a sufficient energy of activation generally must be provided to first vaporize and release them from the inner structure of coal. The evolved heat serves to produce more flammable vapor, which subsequently ignites and evolves heat. In this fashion, coal fires are self-sustaining until the flammable compounds in the coal have been entirely exhausted.

Table 7.14 shows that sulfurous compounds also are components of coal. When coal burns, sulfur dioxide is produced.

After the flammable components of coal have burned, a solid residue or ash remains. This ash consists of a mixture of the oxides of arsenic, barium, beryllium, boron, cadmium,

black lung disease (pneumoconiosis) ■ The lung disease caused by long-term inhalation of coal-mine dust

Federal Coal Mine Health and Safety Act of 1969 ■ The federal statute that empowers the Mine Safety and Health Administration within the U.S. Department of Interior to regulate health and safety conditions within coal mines

rank (grade) ■ Any class or type of coal distinguished by its carbon content and density

---

[10]U.S. Energy Information Administration, *Today in Energy* (March 9, 2012).

TABLE 7.14	Some Properties of the Different Ranks of Coal			
RANK OF COAL	**HEAT CONTENT**		**CARBON CONTENT**	**SULFUR CONTENT**
	Btu/lb	**kJ/kg**	**(%)**	**(%)**
Peat	5500–8800	13,800–20,500	25–35	0.5–3
Lignite	7000	16,300	25–35	0.5–3
Bituminous	7300–10,000	17,000–23,250	45–86	0.8–5.0
Subbituminous	9000	20,960	35–45	0.6–1.8
Semianthracite	11,500–14,000	26,700–32,500		
Anthracite	14,000–14,500	32,500–34,000	40	0.4–1.9

flyash ■ The mixture of ultralight, fine particles generated during the combustion of coal

coal gas ■ The flammable and toxic mixture of gases and vapors produced when coal is strongly heated in the absence of air

coal tar ■ The condensed black, viscous liquid produced by heating coal in the absence of air

distillation ■ A physical process in which a substance or group of substances is separated from a mixture by heating the mixture to a specified temperature or range of temperatures, thereby converting one or more components into a vapor that is subsequently condensed and collected

chromium, iron, mercury, thorium, uranium, zinc, and other elements. Although components of this mixture were formerly ejected as **flyash** into the atmosphere from the smokestacks of coal-fired plants, today's environmental regulations require its capture using air-pollution-control equipment. However, there are no federal regulations that address the ultimate fate of the flyash generated during the burning of coal. Most of it ends up in storage near the plants in huge piles, reservoirs, or impoundments.

7.6-E CHEMICAL PRODUCTS OBTAINED FROM COAL

A mixture of volatile gases evolves when coal is heated in the absence of air within a simple closed assembly like that shown in Figure 7.16. The mixture is called **coal gas**. It consists of ammonia, carbon monoxide, hydrogen sulfide, hydrogen cyanide, and methane, none of which condenses when exposed to the temperature of a cold water bath. Although coal gas once was used to heat and illuminate homes and other buildings, its use is now obsolete.

Coal gas

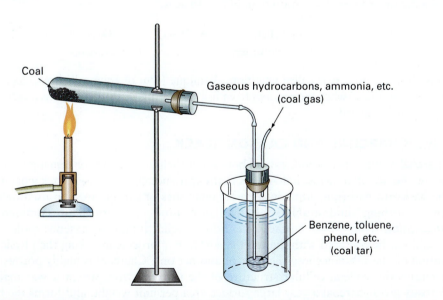

FIGURE 7.16 In this laboratory demonstration, coal is strongly heated in the absence of air. In this manner, two flammable or combustible mixtures are isolated: coal gas and coal tar. When coal is exposed to an ignition source, the chemical components of these mixtures ignite and burn.

Labels within Figure 7.16: Coal; Gaseous hydrocarbons, ammonia, etc. (coal gas); Benzene, toluene, phenol, etc. (coal tar)

Coal tar

The heating of coal also produces a viscous liquid called **coal tar**, a commercial product used for waterproofing when sealing roofing or pavements and coating underground pipelines. It is also used as a binder during the construction of carbon electrodes for the production of elemental aluminum.

Coal tar may be subjected to a separation process called **distillation**, during which it is heated within specified temperature ranges to vaporize its components. When these components are subsequently condensed, the resulting materials are called **coal tar distillates**. Although the latter term is loosely defined, three distillates are recognized commercially:

- The "light" oil is so named because it is a mixture of compounds like benzene (Section 12.11-A) and toluene (Section 12.11-B), whose densities are less than the density of water; hence, it floats on water. Light oil boils near 392°F (200°C).
- The "middle" oil boils between 392 and approximately 518°F (200 and 270°C). It is a raw material from which the chemical industry obtains naphthalene (Section 12.12-A), phenol (Section 13.2-I), and cresols (Section 13.2-J).
- The "heavy" oil boils between 518 and 662°F (270 and 350°C). It contains anthracene and other polynuclear aromatic hydrocarbons (Section 12.12-E). A commercially important product produced from heavy coal tar distillate is **creosote oil**, a viscous liquid widely used to preserve railroad cross ties and utility poles against decay and to cut asphalt so it is suitable for application as a road and roofing tar.

Creosote oil

The viscous residue remaining after coal tar is heated to 662°F (350°C) is called **coal tar pitch**. Products made from this residue are used chiefly as sealants and roofing- and road-paving compounds.

The solid residue that remains when coal and coal tar pitch are heated in the absence of air is called **coke**. The heating process is conducted in industrial ovens known as beehives. Coke is the form of carbon used by metallurgists to reduce the ores of arsenic, tin, copper, iron, zinc, phosphorus, and other elements. It is also used in the steel industry for reducing the iron oxide in iron ore in blast furnaces.

$$\text{C}(s) \quad + \quad \text{FeO}(s) \quad \longrightarrow \quad \text{Fe}(s) \quad + \quad \text{CO}(g)$$
Carbon   Iron(II) oxide   Iron   Carbon monoxide

This chemical process is called **smelting**. A unique form of coke called petroleum coke is similarly produced when the residues from heating certain crude oil fractions (Section 12.14) are thermally treated.

7.6-F CHARCOAL AND CARBON BLACK

Coal tar pitch

Charcoal results when wood, animal bones, nut shells, corn cobs, or peach pits are heated in the absence of air. Most individuals first experience charcoal briquettes when they fuel the backyard barbeque, but considerable quantities of carbon are also used industrially for "adsorbing" undesirable substances from products destined for commercial use. **Adsorption** refers to the surface retention of solid, liquid, or gaseous molecules and should not be confused with the term *absorption*, a process involving the physical penetration of one substance into the bulk of another one. Charcoal is highly porous because it retains the skeletal cellular structure of the material from which it was made. This porosity gives charcoal a very large surface area per unit weight, and forms the basis for its effectiveness as an adsorbing agent.

Charcoal may also be heated in the absence of air at 1472 to 1652°F (800 to 900°C) to produce **activated charcoal**, or **activated carbon**. This material often is chosen as an adsorbing medium, because its average internal surface area is 284,000 ft^2/oz (929 m^2/g). Medical personnel use activated charcoal as an antidote for the quick treatment of potential poisoning caused by the consumption of certain drugs and pesticides. It is also used industrially as an adsorbing agent to purify atmospheric emissions from exhaust stacks by reducing or eliminating the concentration of undesirable toxic gases. On a smaller scale, it is used in gas mask canisters and cigarette filter tips for the same purpose.

When coal tar is burned in a furnace with a limited amount of air, a finely divided form of carbon called **carbon black** is produced. The soot that forms during petroleum fires is composed primarily of carbon black. Different grades are available commercially that are distinguished primarily by their particle size. They are used as components of a number of consumer products including inks, paints, plastics, and tires, belts, and other abrasion-resistant rubber products. Carbon black is an ideal component in these products due to its high surface-area-to-volume ratio.

7.6-G CONSUMER PRODUCT REGULATIONS INVOLVING CHARCOAL

To inform the public that carbon monoxide is produced when charcoal burns, CPSC requires charcoal manufacturers to affix the label shown in Figure 7.17 to charcoal packaging. In addition to the written warning, the label contains a pictograph of a grill situated inside a tent, home, and vehicle. These drawings are enclosed in a circle with an X through it. They serve to convey the message that burning charcoal in enclosed areas should be avoided, because deadly concentrations of carbon monoxide could accumulate and kill the occupants.

7.6-H TRANSPORTING CARBON-BASED PRODUCTS

When shippers intend to transport coal tar distillates, DOT requires them to identify the distillates on the accompanying shipping paper as shown in Table 7.15. All other labeling, marking, and placarding requirements apply.

When shippers offer hot coal tar pitch for transportation in a kettle or other bulk container as an elevated-temperature material, DOT requires them to display the HOT marking on both sides and ends of the container.

DOT also regulates the transportation of coal dust, charcoal, activated carbon, coke, and carbon black. DOT requires shippers who offer these products for transportation to identify it as shown in Table 7.15 on the accompanying shipping paper. For domestic transportation, DOT allows the use of other appropriate shipping names. All other labeling, marking, and placarding requirements apply.

coal tar distillate ■ Any fraction obtained by distilling coal tar

creosote oil ■ An oily liquid obtained from the distillation of coal tar

coal tar pitch ■ The residue remaining after coal is heated to approximately 662°F (350°C) in the absence of air

coke ■ The final residue remaining after coal or coal tar pitch is heated in the absence of air

smelting ■ A chemical process used to isolate an element from its naturally occurring ore by reacting the ore with coke

charcoal ■ The residue remaining after wood, animal bones, nut shells, corn cobs, or peach pits are heated in the absence of air

adsorption ■ A physical phenomenon characterized by the adherence or occlusion of atoms, ions, or molecules of a gas or liquid to the surface of another substance

activated charcoal (activated carbon) ■ The amorphous form of carbon characterized by a high absorptivity for certain gases and vapors

carbon black ■ The finely divided form of carbon produced when coal tar burns in a furnace with limited air

⚠WARNING **CARBON MONOXIDE HAZARD**

Burning charcoal inside can kill you. It gives off carbon monoxide, which has no odor.

NEVER burn charcoal inside homes, vehicles or tents.

FIGURE 7.17 At 16 C.F.R. §1500.14(b)(6), CPSC requires charcoal manufacturers to affix this label to the front and back panels of bags holding charcoal briquettes and other forms of charcoal that are intended for retail sale and use in cooking or heating.

TABLE 7.15	Shipping Descriptions of Carbon-Based Chemical Products
CARBON-BASED PRODUCT	**SHIPPING DESCRIPTION**
Activated carbon	UN1362, Carbon, activated, 4.2, PG III
Carbon black	UN1361, Carbon (carbon black), 4.2, PG II
	or
	UN1361, Carbon (carbon black), 4.2, PG III
Charcoal	UN1361, Charcoal, 4.2, PG III
Coal dust	UN1361, Carbon (coal dust), 4.2, PG II
Coal gas	UN1023, Coal gas, compressed, 2.3 (2.1)
Coal tar distillates	UN1136, Coal tar distillates, flammable, 3, PG II
	or
	UN1136, Coal tar distillates, flammable, 3, PG III
Coal tar pitch	UN3257, Elevated-temperature liquid, n.o.s., 9, PG III
	or
	UN3258, Elevated-temperature solid, n.o.s., 9, PG III
	or
	UN1999, Tars, liquid, 3, PG II
	or
	UN1999, Tars, liquid, 3, PG III
Coke	UN1361, Carbon (coke), 4.2, PG II
	or
	UN1361, Carbon (coke), 4.2, PG III

7.6-I RESPONDING TO INCIDENTS INVOLVING A RELEASE OF COAL

Most fires involving coal may be effectively extinguished with water. When bulk quantities of coal are burning, it is essential to use a deluging volume of water for the following reasons:

- When fires are extinguished only on the surfaces of coal, combustion continues within the interior of the reserve. Because the liberated heat cannot easily dissipate, the temperature of the entire bulk increases, and the coal erupts into flame once again.
- When water contacts coal, coke, or charcoal, the mixture may chemically react to form carbon monoxide and hydrogen. This is the "water gas" previously noted in Section 7.2-C. When water gas ignites with in the confinement of a coal mine, it serves to rekindle coal fires.

7.6-J UNCONTROLLED COAL FIRES

Uncontrolled coal fires typically are confined to natural coal beds and abandoned coal mines. They represent potentially destructive phenomena, not only because they consume a valuable natural resource, but because they transform landscapes and generate dozens of toxic air pollutants—the constituents of coal gas and coal tar. In essence, coal fires provide the conditions that can cause environmental catastrophes. The residents in surrounding communities can be swallowed into sinkholes or valleys, or be sickened by inhaling pollutants in the dominant atmosphere, perhaps fatally.

When coal fires exist in coal beds, the coal was probably ignited by means of lightning strikes or spontaneous combustion. A notable example of an uncontrolled fire in a coal bed is Burning Mountain in New South Wales, Australia. This fire has been burning for 6000 years by some estimates and is credited as the oldest known ongoing coal fire.

In the United States, there are at least 112 documented out-of-control underground fires in abandoned coal mines,[11] but most have been burning "only" for decades. Worldwide, however, there are thousands of documented fires in abandoned coal mines.[12] These fires were most likely initiated when methane (natural gas) seeped from the coal, accumulated in the surrounding confined environment, and ignited; then, the burning methane kindled the burning of the coal.

Fires burn downward in coal mines, acquiring oxygen from the air that passes through the fissures in surrounding rock. Although attempts are sometimes made to extinguish them, most abandoned coal mines are nearly impossible to access. Because entering them is a highly dangerous undertaking, firefighters cannot deluge the fires with water. Furthermore, it is also impossible to starve the fires of oxygen, because the burning coal is exposed almost continuously to new sources of air from the countless boreholes driven into the working mines to provide ventilation for miners. Consequently, the fires in abandoned coal mines are only rarely extinguished. Many quietly smolder from one generation into the next.

---

[11]Kristin Ohlson, "Earth on Fire," *Discover* (July/August 2011), pp. 60–65.

[12]G. B. Stracher, editor, "Coal Fires Burning around the World: A Global Catastrophe," *Int. J. Coal Geol.*, Vol. 59 (2004), pp. 7–17.

REVIEW **EXERCISES**

Oxygen

1. Complete and balance each of the following equations and provide an acceptable name for each product:
   (a) $Mg(s) + O_2(g) \longrightarrow$
   (b) $H_2(g) + O_2(g) \longrightarrow$
   (c) $C(s) + O_2(g) \longrightarrow$
2. The OSHA regulation at 29 C.F.R. §1910.104(b)(2)(iii) requires owners of bulk oxygen systems to provide noncombustible surfacing in areas where liquid oxygen might leak during operation of the system or during the filling of a storage container. Are owners of bulk oxygen systems in compliance with this regulation if they provide an asphalt surface in areas on which oxygen could potentially leak?

Ozone

3. When ozone generators are used to treat smoke-damaged items following a residential fire, why is it essential to conduct the treatment while the home is unoccupied?

Hydrogen

4. Complete and balance each of the following equations, and provide an acceptable name for each product:
   (a) $Li(l) + H_2(g) \longrightarrow$
   (b) $C(s) + H_2(g) \longrightarrow$
   (c) $S_8(s) + H_2(g) \longrightarrow$
   (d) $FeO(s) + H_2(g) \longrightarrow$
5. The OSHA regulation at 29 C.F.R. §1910.103(b) requires employers to arrange safety-relief devices on all containers of hydrogen in the workplace, other than containers having a capacity of 2 cubic feet (0.06 m^3) or less, so that they discharge upward into the open air in an unobstructed fashion and in such a manner as to prevent the impingement of escaping gas on its container, adjacent structures, or personnel.
   (a) What is the most likely reason OSHA requires employers to arrange the safety-relief devices so that they discharge *upward* into the open air?
   (b) What is the most likely reason that OSHA requires employers to arrange the safety-relief devices so that the escaping hydrogen does not impinge on its container, adjacent structures, or personnel?

Chlorine

6. Complete and balance each of the following equations, and provide an acceptable name for each product:
   (a) $Fe(s) + Cl_2(g) \longrightarrow$
   (b) $H_2(g) + Cl_2(g) \longrightarrow$
   (c) $Al(s) + Cl_2(g) \longrightarrow$

7. Identify the GHS pictograms that chlorine manufacturers, distributors, and importers display on the labels attached to steel cylinders containing chlorine.

8. When chlorine is transported in a DOT-105 rail tankcar, does DOT require the carrier to post POISON GAS placards on squares having a white background and a black border on each side and each end of the railcar?

Phosphorus

9. Complete and balance each of the following equations, and provide an acceptable name for each product:
   (a) $P_4(s) + O_2(g) \longrightarrow$
   (b) $P_4(s) + Ca(s) \longrightarrow$
   (c) $P_4(s) + Cl_2(g) \longrightarrow$

10. Why should firefighters avoid the use of water to extinguish phosphorus fires?

11. Identify the GHS pictograms that manufacturers, distributors, and importers of white and red phosphorus display on the labels affixed to their nonbulk containers.

Sulfur

12. Complete and balance each of the following equations, and provide an acceptable name for each product:
   (a) $C(s) + S_8(s) \longrightarrow$
   (b) $Cl_2(g) + S_8(s) \longrightarrow$
   (c) $Cu(s) + S_8(s) \longrightarrow$

13. Why are secondary fires likely to result near burning bulk quantities of sulfur?

Carbon

14. Complete and balance each of the following equations, and provide an acceptable name for each product:
   (a) $C(s) + F_2(g) \longrightarrow$
   (b) $C(s) + CuO(s) \longrightarrow$
   (c) $C(s) + H_2O(g) \longrightarrow$
   (d) $C(s) + SnO_2(s) \longrightarrow$

15. The DOT regulation at 49 C.F.R. §174.450 stipulates that when fire occurs in a domestic rail shipment of charcoal in transit, water should not be used if it is practicable to locate and remove the burning charcoal. Any charcoal that becomes wet while extinguishing a fire must be removed from the car and not reshipped, and the remainder of the charcoal must be held under observation in a dry place for at least 5 days before forwarding. What is the most likely reason that DOT regulates the domestic transportation of charcoal in this fashion?

16. The flashpoint of creosote oil ranges from 165 to 185°F (74 to 85°C). When 5000 pounds (2270 kg) of creosote oil is transported at ambient temperature by means of a motorized cargo tank, which placards does DOT require the carrier to post on the vehicle?

8

Chemistry of Some Corrosive Materials

2012

Courtesy of Bulk Transportation, Inc., Walnut, California.

KEY TERMS

acid, *p. 272*

acidic, *p. 274*

acidic anhydride (nonmetallic oxide), *p. 277*

alkaline, *p. 274*

anhydride, *p. 277*

aqua regia, *p. 285*

base, *p. 272*

basic (caustic), *p. 274*

basic anhydride (metallic oxide), *p. 277*

concentrated acid, *p. 274*

corrosive material, *p. 271*

corrosive substance, *p. 271*

diluted acid, *p. 274*

hydronium ion, *p. 272*

hypocalcemia, *p. 292*

mineral acid, *p. 273*

nonoxidizing acid, *p. 274*

oleum (fuming sulfuric acid), *p. 282*

organic acid (carboxylic acid), *p. 274*

oxidizing acid, *p. 274*

pH, *p. 274*

pickling, *p. 278*

RCRA corrosivity characteristic, *p. 300*

salt, *p. 276*

slaking, *p. 299*

strong acid, *p. 273*

strong base, *p. 273*

weak acid, *p. 273*

weak base, *p. 273*

OBJECTIVES

- Associate the physical and health hazards of the corrosive materials noted in this chapter with the information provided by their hazard diamonds and GHS pictograms.
- Identify the primary industries that use the corrosive materials noted in this chapter.
- Identify the concentrated acids that vaporize at room temperature and cause ill effects when inhaled.
- Identify the labels, markings, and placards that DOT requires on packaging of corrosive materials and the transport vehicles used for their shipment.
- Identify the response actions to be executed when corrosive materials are released from their packaging into the environment.

The most common examples of corrosive materials are substances known chemically as acids and bases. They are compounds capable of causing irritation, burns, or more serious damage to tissue, depending on their strength. Their unique properties and the manner by which acids and bases corrode matter are the principal subjects of this chapter.

8.1 THE NATURE OF CORROSIVITY

Corrosivity is a chemical property exhibited by certain substances. In everyday practice, we say that these substances "eat into" and destroy the quality of metals, minerals, and body tissues. Government regulations, however, provide more specificity, as the following definitions demonstrate:

- CPSC defines a **corrosive substance** as any consumer product that destroys living tissue such as skin or eyes by chemical action. At 16 C.F.R. §1500.41, a product is considered corrosive to the skin if, when tested on the intact skin of the albino rabbit, the structure of the tissue at the site of contact is destroyed or changed irreversibly in 24 hours or less.

- DOT defines a **corrosive material** as a liquid or solid that causes full-thickness destruction of human skin at the site of contact within a specified period, or a liquid that chemically reacts with steel or aluminum surfaces at a rate exceeding 0.25 inches/year (6.25 mm/y) at a test temperature of 130°F (54°C) when measured in accordance with prescribed testing procedures.

During emergency response actions, corrosive materials often are encountered in non-bulk containers like lined steel drums (1A1 or 1A2), plastic drums (1H1 or 1H2), and plastic carboys (3H1 or 3H2). As a minimum, OSHA requires manufacturers, distributors, and importers of chemical products to post the corrosion-symbol pictogram on the labels of corrosive materials, along with an appropriate hazard signal word and hazard and precautionary statements.

Corrosive materials are transported in bulk in rail tankcars or cargo tanks. Both the DOT-111 rail tankcar and MC-312 cargo tank are low-pressure [<75 psi (517 kPa)] transport vessels manufactured from steel and generally lined with rubber. Their design and construction features are published at 49 C.F.R. §§173.242 and 178.343, respectively. To provide an element of safety, the MC-312 cargo tank is equipped with stiffening rings and rollover protection.

When carriers transport 1001 pounds (454 kg) or more of a corrosive material, DOT requires them to display CORROSIVE placards on the bulk packaging used for shipment. When more than one corrosive material is transported in a compartmented tankcar or cargo tank, DOT requires carriers to placard each compartment separately as shown in Figure 8.1.

When carriers transport corrosive materials by highway or rail, DOT requires them to display the relevant identification number on orange panels or across the center area of the CORROSIVE placards or white square-on-point diamonds. For example, when sodium hydroxide (Section 8.14), a corrosive material, is transported by highway or rail, carriers may use any of the following means to display its identification number 1824:

corrosive substance ■ For purposes of CPSC regulations, any substance that causes the destruction of living tissue by chemical action

corrosive material ■ For purposes of DOT regulations, a liquid or solid that causes full-thickness destruction of human skin at the site of contact within a specified period; or a liquid that chemically reacts with steel or aluminum surfaces at a rate exceeding 0.25 inches/year (6.25 mm/y) at a test temperature of 130°F (54°C) when measured in accordance with prescribed testing procedures

FIGURE 8.1 DOT requires the front and rear ends of a compartmented cargo tank to be placarded in same sequence as the compartments are being transported. This three-compartment cargo tank is placarded to transport caustic soda solution, formic acid, and an "other regulated liquid substance." The latter is a class 9 hazardous material, whereas caustic soda solution and formic acid are class 8 corrosive materials. To comply with DOT marking requirements, the carrier has posted identification numbers across the center of each placard. (*Courtesy of Bulk Transportation, Inc., Walnut, California.*)

8.2 THE NATURE OF ACIDS AND BASES

acid ■ A compound that forms hydrated hydrogen ions, $H^+(aq)$, when dissolved in water

Several theories have been proposed to account for the properties of acids and bases, but for simplicity's sake, we use only one of them in this text. In 1887, Swedish chemist Svante Arrhenius first advocated a theory that has been modernized here to reflect current scientific knowledge. Arrhenius proposed that an **acid** is any substance that generates hydrogen ions (H^+) when dissolved in water. Hydrogen ions are hydrogen atoms that have been stripped of their electrons. A single hydrogen ion is the same as the nucleus of a hydrogen atom, or a single proton.

hydronium ion ■ The simplest hydrated hydrogen ion, denoted as $H(H_2O)^+$, or H_3O^+

Today, chemists know that free hydrogen ions cannot exist alone in aqueous solution due to their high charge density. Instead, they rapidly become "solvated"; that is, they bond loosely to water molecules. These solvated hydrogen ions are very complex in the manner in which they interact with water. The simplest ion is $H(H_2O)^+$, or H_3O^+; it is called the **hydronium ion**. We collectively represent here all solvated hydrogen ions by the notation $H^+(aq)$.

For instance, when hydrogen chloride dissolves in water, solvated hydrogen ions and chloride ions are generated.

$$HCl(aq) \longrightarrow H^+(aq) + Cl^-(aq)$$
Hydrogen chloride　　　Hydrogen ion　　　Chloride ion

This solution of hydrogen chloride in water is called hydrochloric acid. Its chemical formula is $HCl(aq)$.

base ■ A hydroxide of the alkali and alkaline earth metals and certain other substances whose aqueous solutions have a pH greater than 7

Arrhenius further proposed that a **base** is a substance that produces hydroxide ions (OH^-) when it is dissolved in water. For example, when sodium hydroxide dissolves in water, solvated sodium and hydroxide ions are generated. We represent them as $Na^+(aq)$ and $OH^-(aq)$, respectively.

$$NaOH(aq) \longrightarrow Na^+(aq) + OH^-(aq)$$
Sodium hydroxide　　　Sodium ion　　　Hydroxide ion

The solution of sodium hydroxide in water is represented as $NaOH(aq)$.

8.2-A STRONG AND WEAK ACIDS AND BASES

It is the ability of acids and bases to form ions in water that gives rise to their corrosive nature. The relative strength of an acid or base refers to the tendency of an individual substance to form hydrated hydrogen ions and hydroxide ions, respectively, when dissolved in water.

Acids and bases that yield a relatively high concentration in water of hydrated hydrogen and hydroxide ions are called **strong acids** and **strong bases**, respectively. For instance, hydrochloric acid is an example of a strong acid, because the hydrogen chloride almost completely ionizes in water. Likewise, sodium hydroxide is an example of a strong base, because it almost completely ionizes in water.

In contrast, some substances essentially retain their unit formulas when they are dissolved in water. Such substances yield relatively low concentrations of hydrogen or hydroxide ions and are called **weak acids** and **weak bases**, respectively. Acetic acid is an example of a weak acid, because it primarily exists as molecules of acetic acid when it is dissolved in water, although some hydrogen and acetate ions are also produced. Ammonium hydroxide is an example of a weak base. When ammonia is dissolved in water, it continues to exist primarily as molecular ammonia and does not appreciably from ammonium and hydroxide ions.

Each acid in the group listed in Table 8.1 is ranked as a strong or a weak acid. Phosphoric acid is regarded as a moderately strong acid, whereas the acids above and below phosphoric acid in this listing are considered strong acids and weak acids, respectively.

strong acid ■ Any acid that predominantly forms hydrated hydrogen ions when dissolved in water

strong base ■ Any base that predominantly forms hydrated hydroxide ions when dissolved in water

weak acid ■ Any acid that predominantly retains its molecular identity when dissolved in water

weak base ■ Any base that predominantly retains its unit identity when dissolved in water

8.2-B MINERAL ACIDS AND ORGANIC ACIDS

Acids may be classified as either mineral acids or organic acids. **Mineral acids** consist of molecules having atoms of hydrogen; an identifying nonmetal like chlorine, sulfur, or phosphorus; and sometimes oxygen. Mineral acids most likely are so named because they were initially produced from minerals existing in naturally occurring ores.

mineral acid ■ Any inorganic acid, chiefly hydrochloric acid, sulfuric acid, nitric acid, perchloric acid, phosphoric acid, and hydrofluoric acid

TABLE 8.1	Relative Strengths of Some Common Acids in Water
NAME OF ACID	**CHEMICAL FORMULA**
Perchloric acid	$HClO_4(aq)$
Sulfuric acid	$H_2SO_4(aq)$
Hydrochloric acid	$HCl(aq)$
Nitric acid	$HNO_3(aq)$
Phosphoric acid	$H_3PO_4(aq)$
Nitrous acid	$HNO_2(aq)$
Hydrofluoric acid	$HF(aq)$
Acetic acid	$CH_3COOH(aq)$
Carbonic acid[a]	$CO_2(aq)$
Hydrocyanic acid	$HCN(aq)$
Boric acid	$H_3BO_3(aq)$

[a]The chemical formula of carbonic acid sometimes appears in the literature as $H_2CO_3(aq)$. An acid having this chemical composition, however, has never been isolated or identified. The chemical formula of carbonic acid is correctly denoted as $CO_2(aq)$. Notwithstanding the fact that carbonic acid has never been discovered, solutions of carbon dioxide in water are acidic in nature.

organic acid (carboxylic acid) ▪ An organic compound containing one or more of the group of atoms

$$-\overset{\displaystyle O}{\overset{\|}{C}}\diagdown_{OH}$$

The mineral acids are distinguished from **organic acids**, or **carboxylic acids**, substances whose molecules usually possess carbon, hydrogen, and oxygen atoms only. All organic acids have molecular structures that contain the following group of atoms:

$$-\overset{\displaystyle O}{\overset{\|}{C}}\diagdown_{OH}, \text{ or } -COOH.$$

Acetic acid is the only organic acid noted in this chapter, but in Section 13.6, we will visit other organic acids. Collectively, they are weak acids.

8.2-C OXIDIZING AND NONOXIDIZING ACIDS

oxidizing acid ▪ Any acid capable of reacting as an oxidizing agent

nonoxidizing acid ▪ Any acid incapable of reacting as an oxidizing agent

An acid chemically reacts as either an oxidizing acid or a nonoxidizing acid. An **oxidizing acid** participates in chemical reactions as an oxidizing agent. A **nonoxidizing acid** participates in a chemical reaction by some means other than oxidation.

Hot, concentrated sulfuric acid, nitric acid, and perchloric acid are oxidizing acids, but hydrochloric acid, hydrofluoric acid, phosphoric acid, and acetic acid are nonoxidizing acids. The degree to which the oxidizing acids corrode depends on how powerfully they participate as oxidizing agents. We note this characteristic when we examine their individual properties.

8.2-D CONCENTRATED AND DILUTED ACIDS

concentrated acid ▪ The term routinely applied to the commercially available form of an acid containing the greatest concentration of the substance

diluted acid ▪ Any form of an acid that has been produced by mixing a concentrated acid with water

Chemists often refer to an acid as either diluted or concentrated. When they refer to a **concentrated acid**, chemists generally mean the commercially available acid that has the greatest concentration. When referring to a **diluted acid**, they mean a solution produced by adding water to the concentrated acid.

This is not meant to imply that water is absent from a concentrated acid. A concentrated acid may actually contain some amount of water. For example, concentrated hydrochloric acid consists of approximately 36% to 38% hydrogen chloride by mass in water; 62% to 64% of the solution is water. Diluted hydrochloric acid is any solution that results when additional water is added to this concentrated acid. Although concentrated hydrochloric acid is a corrosive material, diluted hydrochloric acid may or may not exhibit corrosiveness depending on its strength. When highly diluted with water, all acids lose their corrosive character.

8.3 THE pH SCALE

pH ▪ A numerical scale from 0 to 14 used to quantify the acidity of alkalinity of a solution with neutrality indicated as 7

acidic ▪ The property of aqueous solutions that have a pH ranging from 0 to 7; the property of any substance that corrodes steel or destroys tissue at the site of contact

The **pH** is a number ranging from 0 to 14 that denotes the acidity or alkalinity of an aqueous solution. Aqueous solutions of substances that have a pH less than 7 are said to be **acidic**. Let's consider a simple example. In pure water, there is a very small hydrogen ion concentration derived by the dissociation of the water molecules. This is represented as follows:

$$H_2O(l) \longrightarrow H^+(aq) + OH^-(aq)$$
Water   Hydrogen ion   Hydroxide ion

When the hydrogen ion concentration in pure water is determined experimentally, we find that it is only 0.0000001 (or 10^{-7}) moles/liter. Instead of writing all these zeros, we express this concentration by indicating that the pH equals 7; that is, the pH is the degree of the negative exponent. When the hydrogen ion concentration of an aqueous solution is 0.01 (or 10^{-2}) moles/liter, the pH of the solution equals 2.

basic (caustic) ▪ The property of aqueous solutions that have a pH ranging from 8 to 14

alkaline ▪ Basic, as opposed to acidic; an aqueous solution or other liquid whose pH is greater than 7

Aqueous solutions having a pH greater than 7 are referred to as **basic**, **caustic**, or **alkaline**. Because water is neither acidic nor basic, the pH of pure water is 7. Table 8.2 lists the pH values of some common solutions and mixtures.

A unit change in a pH value represents a 10-fold difference in the hydrogen ion concentration of a solution, and a difference of two pH units represents a 100-fold difference. This

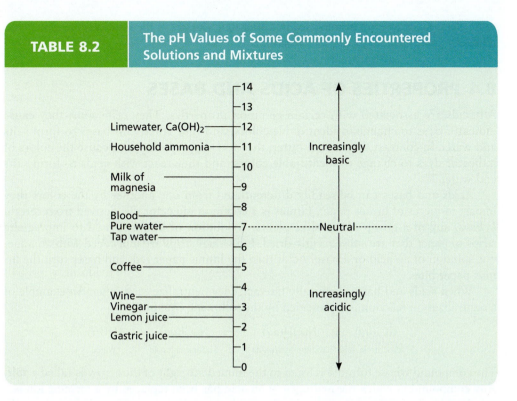

means that the hydrogen ion concentration of an aqueous solution having a pH of 4 is 100 times greater than the hydrogen ion concentration of a solution having a pH of 6; and it is 1000 times greater than one having a pH equal of 7. In everyday practice, we say that a solution having a pH of 4 is 100 times more acidic than a solution having a pH of 6 and 1000 times more acidic than a solution having a pH of 7. Also, a solution with a pH of 8 is 10 times as alkaline as one with a pH of 7; a solution having a pH of 9 is 10 times as alkaline as one having a pH of 8 and 100 times as alkaline as one having a pH of 7; and so on.

The pH is frequently determined by means of an electrometric apparatus called a pH meter. The pH meter illustrated in Figure 8.2 is a voltage-measuring device that is connected

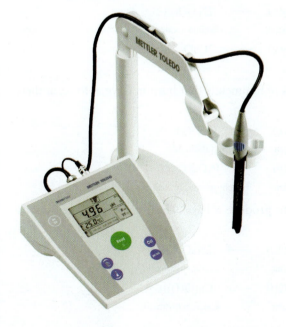

FIGURE 8.2 The pH of an aqueous solution is determined by using a pH meter. The user immerses the electrodes in a sample of the solution, whose pH is read on the display monitor. (*Courtesy of Mettler-Toledo International, Inc., Columbus, Ohio.*)

to an electrode whose tip is immersed in a solution. The pH of the solution is readily determined by simply reading a display monitor.

8.4 PROPERTIES OF ACIDS AND BASES

All acids are associated with certain common properties. They taste sour; they cause indicator dyes to change to identifiable colors; and they react with bases to form salts and water. In contrast, bases taste bitter; they feel slippery; they also cause the colors of indicator dyes to change to identifiable colors; and they react with acids to form salts and water.

Acids and bases can be readily differentiated from one another by the colors they impart to pieces of litmus paper. Litmus is a common indicator dye derived from certain *lichens*, any of a group of mosslike plants. A solution of litmus is used to impregnate strips of paper that are subsequently dried. Individual strips are moistened with an aqueous solution of an acid or a base. Acids turn the litmus paper red, and bases turn the litmus paper blue.

When acids and bases chemically interact, they neutralize each other. An example of a neutralization reaction is represented by the following equation:

$$\text{HCl}(aq) \ + \ \text{NaOH}(aq) \ \longrightarrow \ \text{NaCl}(aq) \ + \ \text{H}_2\text{O}(l)$$

Hydrochloric acid   Sodium hydroxide   Sodium chloride   Water

salt ■ A compound in which the hydrogen ion from an acid has been substituted with a metallic ion

The compound whose formula is listed to the immediate right of the arrow is called a **salt**. Any compound in which the hydrogen in an acid has been replaced by a metallic ion is a salt. NaCl is the chemical formula of sodium chloride, which we commonly know as ordinary table salt. It has properties that are dissimilar from those of both hydrochloric acid and sodium hydroxide.

8.5 THE ANHYDRIDES OF ACIDS AND BASES

When a metallic or nonmetallic element combines with oxygen, the compounds produced are metallic oxides and nonmetallic oxides, respectively. Sodium and calcium are two examples of metals that burn in oxygen to form their corresponding metallic oxides.

$$4\text{Na}(s) \ + \ \text{O}_2(g) \ \longrightarrow \ 2\text{Na}_2\text{O}(s)$$

Sodium   Oxygen   Sodium oxide

$$2\text{Ca}(s) \ + \ \text{O}_2(g) \ \longrightarrow \ 2\text{CaO}(s)$$

Calcium   Oxygen   Calcium oxide

Sulfur and phosphorus are two examples of nonmetals that burn in oxygen to form their corresponding nonmetallic oxides.

$$\text{S}_8(l) \ + \ 8\text{O}_2(g) \ \longrightarrow \ 8\text{SO}_2(g)$$

Sulfur   Oxygen   Sulfur dioxide

$$\text{P}_4(s) \ + \ 3\text{O}_2(g) \ \longrightarrow \ \text{P}_4\text{O}_6(s)$$

Phosphorus   Oxygen   Tetraphosphorus hexoxide

A metallic oxide reacts with water to produce a base. These combination reactions are denoted as follows:

$$\text{Na}_2\text{O}(s) \ + \ \text{H}_2\text{O}(l) \ \longrightarrow \ 2\text{NaOH}(aq)$$

Sodium oxide   Water   Sodium hydroxide

$$\text{CaO}(s) \ + \ \text{H}_2\text{O}(l) \ \longrightarrow \ \text{Ca(OH)}_2(aq)$$

Calcium oxide   Water   Calcium hydroxide

TABLE 8.3		Acidic and Basic Anhydrides	
ANHYDRIDE	CHEMICAL FORMULA	ACID/BASE	CHEMICAL FORMULA
Dinitrogen trioxide	$N_2O_3(g)$	Nitrous acid	$HNO_2(aq)$
Dinitrogen pentoxide	$N_2O_5(g)$	Nitric acid	$HNO_3(aq)$
Sulfur dioxide	$SO_2(g)$	Sulfurous acid	$H_2SO_3(aq)$
Sulfur trioxide	$SO_3(g)$	Sulfuric acid	$H_2SO_4(aq)$
Tetraphosphorus hexoxide	$P_4O_6(s)$	Phosphorous acid	$H_3PO_3(aq)$
Tetraphosphorus decoxide	$P_4O_{10}(s)$	Phosphoric acid	$H_3PO_4(aq)$
Calcium oxide	$CaO(s)$	Calcium hydroxide	$Ca(OH)_2(aq)$
Magnesium oxide	$MgO(s)$	Magnesium hydroxide	$Mg(OH)_2(aq)$
Potassium oxide	$K_2O(s)$	Potassium hydroxide	$KOH(aq)$
Sodium oxide	$Na_2O(s)$	Sodium hydroxide	$NaOH(aq)$

A nonmetallic oxide reacts with water to produce an acid. These combination reactions are represented by the following equations:

$$SO_2(g) \ + \ H_2O(l) \ \longrightarrow \ H_2SO_3(aq)$$
Sulfur dioxide   Water   Sulfurous acid

$$P_4O_6(s) \ + \ 6H_2O(l) \ \longrightarrow \ 4H_3PO_3(aq)$$
Tetraphosphorus hexoxide   Water   Phosphorous acid

The word **anhydride** refers to a substance from which the elements of water have been extracted. It is either a metallic or a nonmetallic oxide. A metallic oxide that combines with water to produce a base is called a **basic anhydride**, and a nonmetallic oxide that combines with water to produce an acid is called an **acidic anhydride**. Sodium oxide and calcium oxide are examples of basic anhydrides, and sulfur dioxide and tetraphosphorus hexoxide are examples of acidic anhydrides. Some common acids and bases and their respective acidic and basic anhydrides are listed in Table 8.3.

anhydride ▪ Any metallic or nonmetallic oxide

basic anhydride ▪ Any metallic oxide that chemically combines with water to produce a base

acidic anhydride ▪ Any nonmetallic oxide that chemically combines with water to produce an acid

SOLVED EXERCISE 8.1

Write equations for the following chemical phenomena:

(a) Bubbles of hydrogen are generated when small chunks of metallic calcium are dropped into an aqueous solution of hydrochloric acid.

(b) A surface coating of zinc oxide is removed by an aqueous solution of sulfuric acid.

(c) Bubbles of carbon dioxide are generated when potassium carbonate is mixed into an aqueous solution of phosphoric acid.

Solution: The phenomena in (a), (b), and (c) are representative of the chemical reactions of acids with metals, metallic oxides, and metallic carbonates, respectively.

(a) Acids react with common metals other than copper, silver, gold, and mercury to produce hydrogen and a salt of the metal. Consequently, metallic calcium reacts with hydrochloric acid to produce hydrogen and calcium chloride according to the following equation:

$$Ca(s) \ + \ 2HCl(aq) \ \longrightarrow \ CaCl_2(aq) \ + \ H_2(g)$$
Calcium   Hydrochloric acid   Calcium chloride   Hydrogen

(b) Acids react with metallic oxides to produce water and a salt of the metal. Consequently, zinc oxide reacts with sulfuric acid to produce zinc sulfate and water.

$$ZnO(s) \;+\; H_2SO_4(aq) \longrightarrow ZnSO_4(aq) \;+\; H_2O(l)$$

Zinc oxide   Sulfuric acid   Zinc sulfate   Water

(c) Acids react with metallic carbonates to produce carbon dioxide, water, and a salt of the metal. Consequently, the reaction between potassium carbonate and phosphoric acid produces carbon dioxide, water, and potassium phosphate.

$$3K_2CO_3(s) \;+\; 2H_3PO_4(aq) \longrightarrow 3CO_2(g) \;+\; 3H_2O(l) \;+\; 2K_3PO_4(aq)$$

Potassium carbonate   Phosphoric acid   Carbon dioxide   Water   Potassium phosphate

8.6 ACIDS AND BASES AS CORROSIVE MATERIALS

Acids act as corrosive materials by reacting with metals, metallic oxides, metallic carbonates, and skin tissue. Bases act as corrosive materials by reacting with metals and skin tissue. We consider these phenomena separately.

8.6-A REACTIONS OF ACIDS AND METALS

Diluted acids react with all the commonly encountered metals other than copper, silver, gold, and mercury. These are simple displacement reactions that produce hydrogen and a salt of the metal. Two examples of such reactions are illustrated by the following equations:

$$Mg(s) \;+\; 2HCl(aq) \longrightarrow MgCl_2(aq) \;+\; H_2(g)$$

Magnesium   Hydrochloric acid   Magnesium chloride   Hydrogen

$$2Al(s) \;+\; 3H_2SO_4(aq) \longrightarrow Al_2(SO_4)_3(aq) \;+\; 3H_2(g)$$

Aluminum   Sulfuric acid   Aluminum sulfate   Hydrogen

Because acids and metals are chemically incompatible, it is unsafe to store acids in metal containers. This is why nonbulk quantities of acids are usually stored in glass or plastic containers. When they are stored in metal drums or pails, the containers must be rubber-lined.

8.6-B REACTIONS OF ACIDS AND METALLIC OXIDES

Acids react with metallic oxides to form a salt of the metal and water. Examples of this type of double-displacement reaction are illustrated by the following equations:

$$FeO(s) \;+\; 2HCl(aq) \longrightarrow FeCl_2(aq) \;+\; H_2O(l)$$

Iron(II) oxide   Hydrochloric acid   Iron(II) chloride   Water

$$Al_2O_3(s) \;+\; 6HNO_3(aq) \longrightarrow 2Al(NO_3)_3(aq) \;+\; 3H_2O(l)$$

Aluminum oxide   Nitric acid   Aluminum nitrate   Water

pickling ■ The combination of chemical reactions associated with removing impurities from the surfaces of metals by immersing them in an acid bath

An acid is sometimes beneficially used to remove metallic oxides and other impurities from the surface of metals. When used in this fashion, the acid is commonly called *pickle liquor*, and the associated phenomenon is called **pickling**. The steel industry uses large volumes of pickle liquor during the manufacture of such products as wire, rod, nuts, and bolts. The common acids used in pickle liquors are sulfuric acid, hydrochloric acid, and phosphoric acid.

8.6-C REACTIONS OF ACIDS AND METALLIC CARBONATES

Acids react with metallic carbonates to produce carbon dioxide, water, and a salt of the metal. Examples of these reactions are illustrated by the following equations:

$$CaCO_3(s) \; + \; 2HCl(aq) \; \longrightarrow \; CaCl_2(aq) \; + \; CO_2(g) \; + \; H_2O(l)$$

Calcium carbonate   Hydrochloric acid   Calcium chloride   Carbon dioxide   Water

$$ZnCO_3(s) \; + \; H_2SO_4(aq) \; \longrightarrow \; ZnSO_4(aq) \; + \; CO_2(g) \; + \; H_2O(l)$$

Zinc carbonate   Sulfuric acid   Zinc sulfate   Carbon dioxide   Water

8.6-D REACTIONS OF ACIDS WITH SKIN TISSUE

The nature of the corrosive effect caused by prolonged exposure of skin tissue to an acid depends on the concentration of the acid. When the skin contacts a diluted acid, the site of contact may appear only reddened, whereas exposure to a concentrated acid for the same duration could cause the skin to blister. In the worst incidents, prolonged exposure of the skin to a concentrated acid causes irreversible tissue damage and permanent disfigurement at the site of contact. In either situation, the skin tissue is said to be "burned," because its appearance visually resembles a thermal burn.

8.6-E REACTIONS OF BASES AND METALS

Three common metals react with the concentrated solutions of strong bases, namely, aluminum, zinc, and lead. During the chemical reactions, hydrogen and a complex compound of the metal are produced. The following equations illustrate the chemical behavior of these three metals with a concentrated solution of sodium hydroxide:

$$2Al(s) \; + \; 6NaOH(aq) \; \longrightarrow \; 2Na_3AlO_3(aq) \; + \; 3H_2(g)$$

Aluminum   Sodium hydroxide   Sodium aluminate   Hydrogen

$$Zn(s) \; + \; 2NaOH(aq) \; \longrightarrow \; Na_2ZnO_2(aq) \; + \; H_2(g)$$

Zinc   Sodium hydroxide   Sodium zincate   Hydrogen

$$Pb(s) \; + \; 2NaOH(aq) \; \longrightarrow \; Na_2PbO_2(aq) \; + \; H_2(g)$$

Lead   Sodium hydroxide   Sodium plumbite   Hydrogen

The reaction between sodium hydroxide and aluminum is especially important when examining the corrosive nature that strong bases exhibit with metals.

8.6-F REACTION OF BASES WITH SKIN TISSUE

Aqueous solutions of bases corrode skin tissue in a fashion that is associated with the concentration of the base. When skin is exposed to a diluted solution, it appears reddened at the site of contact. Although wounds of this type heal rapidly, when the skin has been exposed to a more concentrated solution of the same base, it becomes sticky, soapy, and soft in texture. Prolonged exposure causes the development of deep wounds that turn black and leathery in texture and are very slow to heal. The tissue damage may be irreversible and cause permanent disfigurement at the site of contact.

Exposure of the eyes to solutions of caustic substances is particularly worrisome. It causes injurious changes in the structure of the cornea, ultimately leading to complete opacification (clouding).

Sulfuric acid

8.7 SULFURIC ACID

Sulfuric acid is a mineral acid whose chemical formula is H_2SO_4. It is an odorless, colorless, oily liquid having a density approximately twice that of water. Impure or spent sulfuric acid is brown to black in color. An industrial grade of sulfuric acid is sometimes

TABLE 8.4	Physical Properties of Concentrated Sulfuric Acid
Melting point	50°F (10°C)
Boiling point	640°F (338°C)
Specific gravity at 68°F (20°C)	1.84
Vapor density (air = 1)	2.8
Vapor pressure at 68°F (20°C)	<0.001 mmHg
Solubility in water	Infinitely soluble

encountered as a clear-to-brownish-colored liquid. The brown color reflects the presence of dissolved salts. Some of the important physical properties of sulfuric acid are noted in Table 8.4.

If we were to list chemical substances by the amounts produced and consumed annually within the United States, we would find sulfuric acid near the top of these lists for the past five decades. In the United States alone, well over 40 million tons (36 million t) of sulfuric acid are produced annually. Sulfuric acid is so important commercially that its production and consumption rates have been used by economists to estimate the extent to which a country has industrialized.

The layperson is generally aware of sulfuric acid through its use as the electrolyte in the lead–acid storage battery shown in Figures 1.6 and 8.3. As previously noted in Section 7.2-G, this battery is the electrical source in virtually all motor vehicles. However, sulfuric acid has countless other uses. In the chemical industry alone, sulfuric acid is used to manufacture explosives, fertilizers, drugs, and dozens of other compounds—including other acids. Given its widespread usage, sulfuric acid has been called the workhorse of the industrial world. The popularity of its use implies that sulfuric acid is likely to be encountered more frequently than other corrosive materials when responding to emergencies involving hazardous materials.

As first noted in Section 1.5-A, OSHA requires every manufacturer, distributor, and importer of a chemical substance to ensure that appropriate hazard warnings are provided on their containers. For example, OSHA requires them to provide the GHS hazard information on the label shown in Figure 8.4 when the label is affixed to containers of concentrated sulfuric acid. The information written under the heading DANGER consists of GHS hazard and precautionary statements. The symbols required by WHMIS also are shown.

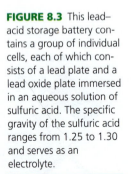

FIGURE 8.3 This lead–acid storage battery contains a group of individual cells, each of which consists of a lead plate and a lead oxide plate immersed in an aqueous solution of sulfuric acid. The specific gravity of the sulfuric acid ranges from 1.25 to 1.30 and serves as an electrolyte.

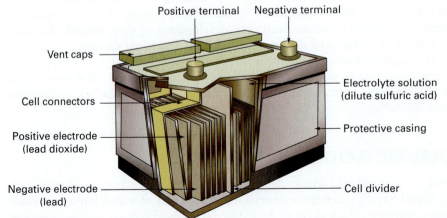

Positive terminal   Negative terminal

Vent caps

Cell connectors

Positive electrode (lead dioxide)

Negative electrode (lead)

Electrolyte solution (dilute sulfuric acid)

Protective casing

Cell divider

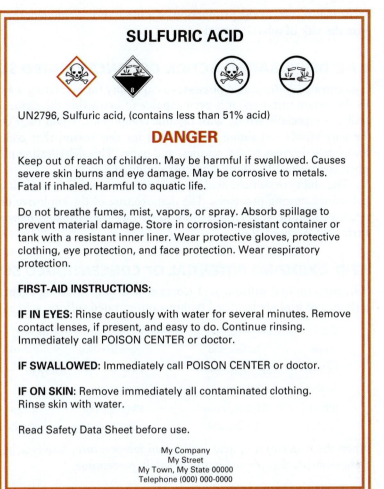

FIGURE 8.4 This label affixed to nonbulk containers of sulfuric acid complies with OSHA regulations published at 29 C.F.R. §1910.1200(f).

The label reads:

SULFURIC ACID

UN2796, Sulfuric acid, (contains less than 51% acid)

DANGER

Keep out of reach of children. May be harmful if swallowed. Causes severe skin burns and eye damage. May be corrosive to metals. Fatal if inhaled. Harmful to aquatic life.

Do not breathe fumes, mist, vapors, or spray. Absorb spillage to prevent material damage. Store in corrosion-resistant container or tank with a resistant inner liner. Wear protective gloves, protective clothing, eye protection, and face protection. Wear respiratory protection.

FIRST-AID INSTRUCTIONS:

IF IN EYES: Rinse cautiously with water for several minutes. Remove contact lenses, if present, and easy to do. Continue rinsing. Immediately call POISON CENTER or doctor.

IF SWALLOWED: Immediately call POISON CENTER or doctor.

IF ON SKIN: Remove immediately all contaminated clothing. Rinse skin with water.

Read Safety Data Sheet before use.

My Company
My Street
My Town, My State 00000
Telephone (000) 000-0000

8.7-A PRODUCTION OF SULFURIC ACID

Sulfuric acid is prepared at industrial plants by first burning sulfur to produce sulfur dioxide, and then oxidizing the sulfur dioxide further in the presence of a catalyst to produce sulfur trioxide. This is the acidic anhydride of sulfuric acid. It reacts with water to form sulfuric acid.

However, sulfur trioxide does not unite with pure water readily. When sulfuric acid is manufactured industrially, the sulfur trioxide is absorbed into a solution of sulfuric acid containing 97% sulfuric acid by mass instead of pure water. Sulfur trioxide readily dissolves in this sulfuric acid solution. The final solution boils at 640°F (338°C) at 14.7 psia (101.3 kPa) and contains 98.3% sulfuric acid by mass. In commerce, this solution is identified as concentrated sulfuric acid. Emergency responders may encounter it in a variety of containers such as those illustrated in Figures 6.1 and 6.2.

8.7-B THE HAZARD ASSOCIATED WITH DILUTING SULFURIC ACID WITH WATER

When concentrated sulfuric acid is diluted with water, considerable heat is released to the environment. Because this reaction is highly exothermic, extreme caution must always be exercised when diluting concentrated sulfuric acid with water. The recommended practice is to slowly pour the acid into water while stirring. When concentrated sulfuric acid is diluted in the reverse manner, localized boiling and violent spattering occur that results in the production of a fume. When inhaled, the sulfuric acid fume can pose the risk of inhalation toxicity.

8.7-C DEHYDRATING ACTION OF CONCENTRATED SULFURIC ACID

Concentrated sulfuric acid possesses a capability for extracting water from certain materials. In certain instances, it is even capable of extracting the elements of water. This latter ability is especially strong with organic compounds. Upon their contact, carbon is often the only visibly remaining residue. It is for this reason that concentrated sulfuric acid completely destroys wood, textiles, and paper. This dehydrating phenomenon also occurs when concentrated sulfuric acid burns body tissues.

The ability of sulfuric acid to extract water is put to beneficial use during many chemical manufacturing processes. The manufacture of the explosive nitroglycerin, for example, uses sulfuric acid catalytically to extract the elements of water from glycerol and nitric acid (Section 15.6).

8.7-D OXIDIZING POTENTIAL OF CONCENTRATED SULFURIC ACID

Hot, concentrated sulfuric acid reacts as a strong oxidizing agent. For example, copper, carbon, and lead are oxidized by hot, concentrated sulfuric acid as follows:

$$\text{Cu}(s) + 2\text{H}_2\text{SO}_4(conc) \longrightarrow \text{CuSO}_4(aq) + \text{SO}_2(g) + 2\text{H}_2\text{O}(g)$$

Copper   Sulfuric acid   Copper(II) sulfate   Sulfur dioxide   Water

$$\text{C}(s) + 2\text{H}_2\text{SO}_4(conc) \longrightarrow \text{CO}_2(g) + 2\text{SO}_2(g) + 2\text{H}_2\text{O}(g)$$

Carbon   Sulfuric acid   Carbon dioxide   Sulfur dioxide   Water

$$\text{Pb}(s) + 3\text{H}_2\text{SO}_4(conc) \longrightarrow \text{Pb}(\text{HSO}_4)_2(s) + \text{SO}_2(g) + 2\text{H}_2\text{O}(g)$$

Lead   Sulfuric acid   Lead(II) bisulfate   Sulfur dioxide   Water

When the concentrated acid is at room temperature, however, it reacts so slowly with these elements that the oxidation is barely perceptible.

8.7-E ILL EFFECTS CAUSED BY INHALING THE VAPORS, MISTS, AND FUMES OF SULFURIC ACID

The repeated inhalation of sulfuric acid vapors, mists, and fumes by workers in occupational settings has been linked with the onset of larynx, paranasal sinus, and lung cancer. These airborne forms of sulfuric acid are generated during the manufacture and use of sulfuric acid, especially the use of the acid for pickling. Given its potential to cause cancer, all work with sulfuric acid should be conducted only within a workplace that provides adequate ventilation.

8.7-F WORKPLACE REGULATIONS INVOLVING SULFURIC ACID

In the workplace, OSHA requires employees to limit employee exposure to a maximum sulfuric acid vapor concentration of 1 milligram/cubic meter, averaged over an 8-hour workday.

8.7-G OLEUM (FUMING SULFURIC ACID)

Sulfur trioxide dissolves in concentrated sulfuric acid, producing a thick, fuming yellow liquid. When this liquid contains a higher proportion of sulfur trioxide than is found in ordinary sulfuric acid, the resulting material is called **oleum**, or **fuming sulfuric acid**. This acid is available commercially with varying concentrations of sulfur trioxide up to 99.9%. Although oleum was formerly used in the petroleum refining industry, hydrofluoric acid (Section 8.11) now serves the same purpose. Today, oleum is used primarily within the chemical industry.

Oleum

oleum (fuming sulfuric acid) ■ Concentrated sulfuric acid containing additional dissolved sulfur trioxide

TABLE 8.5	Shipping Descriptions of Sulfuric Acid
FORM OF SULFURIC ACID	SHIPPING DESCRIPTION[a]
Sulfuric acid (contains less than 51% sulfuric acid)	UN2796, Sulfuric acid, 8, PG II
Sulfuric acid (contains more than 51% sulfuric acid)	UN1830, Sulfuric acid, 8, PG II
Sulfuric acid, spent	UN1832, Sulfuric acid, spent, 8, PG II
Sulfuric acid, fuming (contains less than 30% sulfur trioxide)	UN1831, Sulfuric acid, fuming, 8, PG I
Sulfuric acid, fuming (contains 30% or more sulfur trioxide)	UN1831, Sulfuric acid, fuming, 8, (6.1), PG I (Poison - Inhalation Hazard, Zone B)

[a]When shippers know the actual concentration of sulfuric acid or sulfur trioxide in the acid that is transported, DOT requires them to identify the concentration in lieu of a concentration range.

The chemical formula of oleum is often denoted as $xH_2SO_4 \cdot ySO_3$, where x and y are the number of moles of sulfuric acid and sulfur trioxide, respectively. For example, oleum that contains 1 mole of sulfur trioxide (80.1 g) for each mole of sulfuric acid (98.1 g) has the formula $H_2S_2O_7$, because in this instance, $x = y = 1$. Oleum containing 65% sulfur trioxide by mass is expressed by the formula $4H_2SO_4 \cdot 9SO_3$.

Because oleum contains dissolved sulfur trioxide, it has an elevated vapor pressure compared to the value for concentrated sulfuric acid (<0.001 mmHg). The vapor pressure for oleum varies according to the concentration of dissolved sulfur trioxide but ranges from 342 mmHg to 433 mmHg at 68°F (20°C). This means that fuming sulfuric acid poses the risk of inhalation toxicity.

Like sulfuric acid, oleum is a corrosive material. It severely burns the skin, which generally heals to produce ugly scars. In addition, oleum spontaneously releases toxic sulfur trioxide vapor, which poses the potential risk of inhalation toxicity. Consequently, oleum should be stored and used only in a well-ventilated location.

8.7-H TRANSPORTING SULFURIC ACID

DOT regulates the transportation of four forms of sulfuric acid: a liquid having 51% or more acid; a liquid with less than 51% acid; a spent sulfuric acid solution; and fuming sulfuric acid. When these acids are offered for transportation, DOT requires their shippers to provide the relevant shipping description shown Table 8.5 on the accompanying shipping paper. The Hazardous Materials Table at 49 C.F.R. §172.101 lists several concentrations of sulfuric acid. DOT requires shippers to parenthetically include an appropriate modifier like "contains" or "containing" between the shipping name and the hazard class *or* following the shipping description. All other labeling, marking, and placarding requirements apply.

8.8 NITRIC ACID

Nitric acid is second among the acids most commonly used throughout the United States. It is the raw material used for the manufacture of ammonium nitrate fertilizers, explosives, dyes, perfumes, drugs, and nitrated organic compounds. Its use is also required for the production of patent leather and related fabrics. The chemical formula of nitric acid is HNO_3. Some of its important physical properties are provided in Table 8.6.

Nitric acid

TABLE 8.6	Physical Properties of Concentrated Nitric Acid	
Melting point	–44°F (–42°C)	
Boiling point	187°F (86°C)	
Specific gravity at 68°F (20°C)	1.50	
Vapor density (air = 1)	3.2	
Vapor pressure at 68°F (20°C)	47.9 mmHg	
Solubility in water	Infinitely soluble	

Pure nitric acid is a colorless liquid. It is encountered commercially in concentrated and diluted forms. Concentrated nitric acid is an aqueous solution consisting of 68.2% nitric acid by mass. All concentrations containing less than 68.2% acid are forms of diluted nitric acid.

When encountered, nitric acid is often yellow to red-brown in color, which indicates that nitrogen dioxide is present. Nitrogen dioxide is a red-brown gas produced by the slow decomposition of nitric acid, a reaction catalyzed by sunlight.

$$4HNO_3(l) \longrightarrow 4NO_2(g) + 2H_2O(l) + O_2(g)$$

Nitric acid   Nitrogen dioxide   Water   Oxygen

8.8-A PRODUCTION OF NITRIC ACID

Almost all nitric acid is industrially manufactured from ammonia by means of a series of reactions. Gaseous ammonia is first mixed with about 10 times its volume of air, and then exposed to platinum gauze. The platinum increases the rate of the reaction that converts ammonia into nitric monoxide (NO). Then, additional air is permitted to enter the reaction system so the nitrogen monoxide can be further oxidized to nitrogen dioxide.

$$4NH_3(g) + 5O_2(g) \longrightarrow 4NO(g) + 6H_2O(g)$$

Ammonia   Oxygen   Nitrogen monoxide   Water

$$2NO(g) + O_2(g) \longrightarrow 2NO_2(g)$$

Nitrogen monoxide   Oxygen   Nitrogen dioxide

The nitrogen dioxide is then reacted with water to produce nitric acid.

$$3NO_2(g) + H_2O(l) \longrightarrow 2HNO_3(l) + NO(g)$$

Nitrogen dioxide   Water   Nitric acid   Nitrogen monoxide

The excess nitrogen monoxide produced in this reaction is recycled through the system.

8.8-B OXIDATION OF METALS BY NITRIC ACID

Like hot, concentrated sulfuric acid, nitric acid is an oxidizing acid, even when it is diluted with water at room temperature. When metals are chemically attacked by nitric acid, they are oxidized to their corresponding positive ions and the nitric acid is reduced to one or more of the following: nitrogen, nitrogen monoxide, nitrogen dioxide, dinitrogen monoxide, or the ammonium ion. Nitric acid reacts with some metals to form each of these products under specific conditions. For example, it reacts with zinc to form nitrogen, nitrogen monoxide, nitrogen dioxide, dinitrogen monoxide, or ammonium nitrate in separate reactions.

$$5Zn(s) \;+\; 12HNO_3(aq) \longrightarrow 5Zn(NO_3)_2(aq) \;+\; 6H_2O(l) \;+\; N_2(g)$$

Zinc   Nitric acid   Zinc nitrate   Water   Nitrogen

$$3Zn(s) \;+\; 8HNO_3(aq) \longrightarrow 3Zn(NO_3)_2(aq) \;+\; 4H_2O(l) \;+\; 2NO(g)$$

Zinc   Nitric acid   Zinc nitrate   Water   Nitrogen monoxide

$$Zn(s) \;+\; 4HNO_3(conc) \longrightarrow Zn(NO_3)_2(aq) \;+\; 2H_2O(l) \;+\; 2NO_2(g)$$

Zinc   Nitric acid   Zinc nitrate   Water   Nitrogen dioxide

$$4Zn(s) \;+\; 10HNO_3(aq) \longrightarrow 4Zn(NO_3)_2(aq) \;+\; 5H_2O(l) \;+\; N_2O(g)$$

Zinc   Nitric acid   Zinc nitrate   Water   Dinitrogen monoxide

$$4Zn(s) \;+\; 10HNO_3(aq) \longrightarrow 4Zn(NO_3)_2(aq) \;+\; 3H_2O(l) \;+\; NH_4NO_3(aq)$$

Zinc   Nitric acid   Zinc nitrate   Water   Ammonium nitrate

It is the combination of these reactions that defines the corrosivity of nitric acid.

When nitric acid oxidizes a metal, the most common products formed are nitrogen monoxide and nitrogen dioxide. In general, *diluted* nitric acid oxidizes metals to produce nitrogen monoxide, and *concentrated* nitric acid oxidizes metals to produce nitrogen dioxide.

On occasion, chemists use a nitrating solution to oxidize metals that are ordinarily considered unreactive. The most common solution used for this purpose is a 1:3 mixture of concentrated nitric and hydrochloric acids called **aqua regia**. It is one of a limited number of mixtures known to react with metallic gold. The nitric acid oxidizes the hydrochloric acid, as follows:

aqua regia ■ A 1:3 mixture of concentrated nitric acid and hydrochloric acid

$$HNO_3(aq) \;+\; 3HCl(aq) \longrightarrow NOCl(g) \;+\; Cl_2(g) \;+\; 2H_2O(l)$$

Nitric acid   Hydrochloric acid   Nitrosyl chloride   Chlorine   Water

Nitrosyl chloride has a reddish-yellow color; hence, aqua regia is also reddish-yellow.

8.8-C OXIDATION OF NONMETALS BY NITRIC ACID

Hot nitric acid corrodes nonmetals. For example, carbon and sulfur are oxidized by hot, concentrated nitric acid as depicted by the following equations:

$$C(s) \;+\; 4HNO_3(conc) \longrightarrow CO_2(g) \;+\; 4NO_2(g) \;+\; 2H_2O(l)$$

Carbon   Nitric acid   Carbon dioxide   Nitrogen dioxide   Water

$$S_8(s) \;+\; 48HNO_3(conc) \longrightarrow 8H_2SO_4(l) \;+\; 48NO_2(g) \;+\; 16H_2O(l)$$

Sulfur   Nitric acid   Sulfuric acid   Nitrogen dioxide   Water

8.8-D OXIDATION OF ORGANIC COMPOUNDS BY NITRIC ACID

Nitric acid oxidizes many flammable organic compounds, sometimes at explosive rates. This results in their subsequent ignition. For example, the organic compounds turpentine, acetic acid, acetone, ethanol, nitrobenzene, and aniline react so vigorously when mixed with hot, concentrated nitric acid that they burst into flame.

8.8-E REACTIONS OF NITRIC ACID WITH CELLULOSIC MATERIALS

Nitric acid is capable of initiating the spontaneous ignition of wood, excelsior, and other cellulosic materials, especially when these materials have been finely divided. It is for this reason that bottles of nitric acid are not cushioned with a cellulosic material when they are transported.

8.8-F ILL EFFECTS CAUSED BY EXPOSURE TO NITRIC ACID

Nitric acid corrodes body tissues by reacting with the complex proteins that make up their structures. The exposure of nitric acid to the skin results in ugly, yellow burns that heal very slowly. The chemistry associated with this phenomenon involves the production of a yellow-brown substance called *xanthoproteic acid*. The discoloration of the skin typically wears away in two to three weeks.

Fuming nitric acid

8.8-G FUMING NITRIC ACID

The oxides of nitrogen are readily soluble in concentrated nitric acid. When the acid contains a higher proportion of nitrogen oxides than is contained in ordinary nitric acid, the resulting material is called fuming nitric acid, of which there are two varieties:

■ *Red fuming nitric acid* contains more than 85% nitric acid, less than 5% water, and from 6% to 15% nitrogen oxides. Red fuming nitric acid containing 70% acid and 6% to 15% nitrogen oxides has a vapor pressure of 49 mmHg at 68°F (20°C).

■ *White fuming nitric acid* contains more than 97.5% nitric acid, less than 2% water, and less than 0.5% nitrogen oxides. It has a vapor pressure of 57 mmHg at 77°F (25°C). This second form is not widely encountered because it slowly reverts to the red form.

Contact of the skin with fuming nitric acid is highly irritating. In addition, the red form spontaneously releases toxic nitrogen oxide vapor, which poses the risk of inhalation toxicity. For this reason, fuming nitric acid should always be segregated from other chemical substances and stored and used in well-ventilated locations.

8.8-H WORKPLACE REGULATIONS INVOLVING NITRIC ACID

In the workplace, OSHA requires employers to limit employee exposure to a maximum nitric acid vapor concentration of 2 ppm, averaged over an 8-hour workday, and 4 ppm for a short-term exposure.

8.8-I TRANSPORTING NITRIC ACID

When shippers offer nitric acid or fuming nitric acid for transportation, DOT requires them to identify the appropriate substance with the relevant entry shown in Table 8.7 on the accompanying shipping paper. When carriers transport nitric acid by highway or rail, DOT requires them to display the name *NITRIC ACID* on two opposing sides of the transport vehicle. All other labeling, marking, and placarding requirements apply.

TABLE 8.7	Representative Shipping Descriptions of Nitric Acid
FORM OF NITRIC ACID	**SHIPPING DESCRIPTION**
Nitric acid, other than red fuming, with not more than 70% nitric acid	UN2031, Nitric acid, 8, PG I
Nitric acid, other than red fuming, with more than 70% nitric acid	UN2031, Nitric acid, 8, (5.1), PG II
Nitric acid mixtures, spent, with more than 50% nitric acid	UN1796, Nitrating acid mixtures, spent, 8, (5.1), PG I
Nitric acid mixtures, spent, with not more than 50% nitric acid	UN1826, Nitrating acid mixtures, spent, 8, PG II
Nitric acid mixtures with more than 50% nitric acid	UN1796, Nitric acid mixtures, 8, (5.1), PG I
Nitric acid mixtures with not more than 50% nitric acid	UN1796, Nitric acid mixtures, 8, PG II
Nitric acid, other than red fuming, with not more than 20% nitric acid	UN2031, Nitric acid, 8, PG II
Red fuming nitric acid with not more than 70% nitric acid	UN2032, Nitric acid, red fuming, 8, (5.1, 6.1), PG I (Poison - Inhalation Hazard, Zone B)
Nitric acid, other than red fuming, with more than 70% nitric acid	UN2031, Nitric acid, 8, (5.1), PG I
Nitric acid, other than red fuming, with not more than 20% nitric acid	UN2031, Nitric acid, 8, PG II

8.9 HYDROCHLORIC ACID

Hydrochloric acid is another commercially important acid. It is most familiar to the general public as a constituent of certain household cleaning products like Lysol Toilet Bowl Cleaner. It also is the liquid used to maintain the proper acidity of the water in residential and public swimming pools.

Outside the home, hydrochloric acid has many other uses. As a component of pickle liquor, it is used for galvanizing, tinning, and enameling. In the food industry, hydrochloric acid is used as a processing agent during the production of certain food products such as corn syrup. It is used in the petroleum industry to activate petroleum wells, and it is used in the chemical industry to manufacture and produce dozens of important compounds. Hydrochloric acid is frequently encountered in educational and research facilities, where it usually stored in nonbulk plastic containers like those shown in Figure 6.1.

Pure hydrochloric acid is a colorless, fuming, and pungent-smelling liquid composed of hydrogen chloride dissolved in water. The chemical formulas of hydrochloric acid and hydrogen chloride are $HCl(aq)$ and $HCl(g)$, respectively. Some physical properties of the concentrated acid and its vapor are noted in Table 8.8.

8.9-A PRODUCTION OF HYDROCHLORIC ACID

Hydrochloric acid is prepared by dissolving anhydrous hydrogen chloride in water. The concentrated acid contains from 36% to 38% hydrogen chloride by mass and is colorless. When purity is not a factor, the industrial grade of hydrochloric acid called *muriatic acid* often is used. It is a dilute solution of undefined concentration that is slightly yellow due to the presence of metal impurities like iron salts.

Hydrochloric acid

TABLE 8.8	Physical Properties of Hydrochloric Acid and Its Vapor	
	HCl (*conc*)	HCl (*g*)
Melting point	−101°F (−74°C)	−174°F (114°C)
Boiling point	127°F (53°C)	−121°F (−85°C)
Specific gravity at 68°F (20°C)	1.18	1.27
Vapor density (air = 1)	1.3	1.3
Vapor pressure at 68°F (20°C)	150 mmHg	30,780 mmHg
Solubility in water	85 g/100 g of H_2O at 68°F (20°C)	67% at 68°F (20°C)

8.9-B ILL EFFECTS CAUSED BY INHALING HYDROGEN CHLORIDE

The principal hazardous feature of hydrochloric acid is associated with inhaling the hydrogen chloride vapor released spontaneously from the concentrated acid. Hydrogen chloride is a poisonous gas. When concentrated hydrochloric acid spills or leaks from its container, the pungent odor of its toxic vapor is immediately detectable. Because the vapor is approximately one-fifth heavier than air, it lingers in low areas, where it can pose a risk to health and safety.

When hydrogen chloride vapor is encountered, individuals experience the symptoms noted in Table 8.9. In the worst situations, prolonged exposure to massive amounts of hydrogen chloride causes pulmonary edema (Section 7.3-B), which can completely deteriorate the tissue cells within the respiratory tract and destroy its lining.

TABLE 8.9	Ill Effects Caused by Inhaling Hydrogen Chloride
HYDROGEN CHLORIDE (ppm)	SYMPTOMS
1–5	Threshold limit for detection of smell
5–10	Mild irritation of mucous membranes, eyes, nose, and throat
35	Distinct irritation of mucous membranes, eyes, nose, and throat
5–100	Barely tolerable effects including severe coughing and panic; possible injury to the bronchial region
1000	Danger of pulmonary edema after 24-hr exposure; potentially fatal

SOLVED EXERCISE 8.3

Use the data in Tables 8.4 and 8.8 to determine whether each of the following acids poses the greater hazard by inhalation toxicity at 68°F (20°C) when all other factors are considered equally: **(a)** concentrated sulfuric acid or concentrated hydrochloric acid; **(b)** oleum or concentrated hydrochloric acid.

Solution:

(a) To pose an inhalation toxicity hazard, the liquid acid must produce sufficient vapor at 68°F (20°C) to cause illness or death when inhaled. In Table 8.4, the vapor pressure of concentrated sulfuric acid is listed as <0.001 mmHg. This low value signifies that virtually no vapor is produced at this temperature. Accordingly, exposure to concentrated sulfuric acid poses a very low risk of inhalation toxicity. However, the vapor pressure of hydrochloric acid is listed in Table 8.8 as 150 mmHg. This elevated value indicates that when inhaled, hydrochloric acid produces sufficient vapor at 68°F (20°C) to cause the ill effects noted in Table 8.9.

(b) Oleum, or fuming sulfuric acid, has a vapor pressure ranging from 342 mmHg to 433 mmHg at 68°F (20°C) since it contains free sulfur trioxide. Hence, it poses a greater risk of inhalation toxicity than does concentrated hydrochloric acid. Nonetheless, some caution must be exercised with the use of the phrase "all other factors considered equally." In particular, sulfur trioxide is a cancer-causing substance whereas hydrogen chloride is not. On this basis alone, health officials are likely to consider that oleum poses a greater risk of inhalation toxicity than does concentrated hydrochloric acid.

8.9-C REACTIONS OF HYDROCHLORIC ACID WITH OXIDIZING AGENTS

Hydrochloric acid is chemically incompatible with oxidizing agents such as metallic chlorates, metallic dichromates, and metallic permanganates. These reactions result in the production of toxic chlorine.

$$KClO_3(s) \ + \ 6HCl(conc) \ \longrightarrow \ KCl(aq) \ + \ 3H_2O(l) \ + \ 3Cl_2(g)$$

Potassium chlorate Hydrochloric acid Potassium chloride Water Chlorine

$$K_2Cr_2O_7(s) \ + \ 14HCl(conc) \ \longrightarrow \ 2KCl(aq) \ + \ 2CrCl_3(aq) + 7H_2O(l) + 3Cl_2(g)$$

Potassium dichromate Hydrochloric acid Potassium chloride Chromium(III) chloride Water Chlorine

$$2KMnO_4(s) \ + \ 16HCl(conc) \ \longrightarrow \ 2MnCl_2(aq) \ + \ 2KCl(aq) + 8H_2O(l) + 5Cl_2(g)$$

Potassium permanganate Hydrochloric acid Manganese(II) chloride Potassium chloride Water Chlorine

To reduce or eliminate the likelihood of an unwanted reaction between hydrochloric acid and chlorine-producing pool chemicals, acid manufacturers often label their containers with a message like the following:

DO NOT STORE NEAR CHLORINE-PRODUCING POOL CHEMICALS

8.9-D ANHYDROUS HYDROGEN CHLORIDE

Anhydrous hydrogen chloride itself is a commercial chemical product that is prepared by the direct combination of elemental hydrogen and chlorine.

$$H_2(g) \ + \ Cl_2(g) \ \longrightarrow \ 2HCl(g)$$

Hydrogen Chlorine Hydrogen chloride

Hydrogen chloride is also produced as a by-product during the chlorination of organic compounds.

8.9-E WORKPLACE REGULATIONS INVOLVING HYDROGEN CHLORIDE

In the workplace, OSHA requires employers to limit employee exposure to a maximum hydrogen chloride vapor concentration of 5 ppm, which should not be exceeded during any part of the working exposure.

8.9-F TRANSPORTING HYDROCHLORIC ACID AND ANHYDROUS HYDROGEN CHLORIDE

When shippers offer hydrochloric acid or anhydrous hydrogen chloride for transportation, DOT requires them to provide the relevant shipping description shown in Table 8.10 on the accompanying shipping paper. When hydrogen chloride is transported in bulk by highway or rail, DOT also requires carriers to mark the bulk packaging on two opposing sides with the words *INHALATION HAZARD*. All labeling, marking, and placarding requirements apply.

Anhydrous hydrogen chloride

TABLE 8.10	Shipping Descriptions of Hydrochloric Acid and Hydrogen Chloride

HYDOCHLORIC ACID/HYDROGEN CHLORIDE	SHIPPING DESCRIPTION
Hydrochloric acid	UN1789, Hydrochloric acid, 8, PG II
	or
	UN1789, Hydrochloric acid, 8, PG III
Hydrogen chloride, anhydrous	UN1050, Hydrogen chloride, anhydrous, 2.3, (8), (Poison - Inhalation Hazard, Zone C)
Hydrogen chloride, cryogenic liquid	UN2186, Hydrogen chloride, refrigerated liquid, 2.3, (8), (Poison - Inhalation Hazard, Zone C)

Perchloric acid

8.10 PERCHLORIC ACID

Perchloric acid is an acid used primarily by the chemical, electroplating, and incendiary (fireworks) industries. Its chemical formula is $HClO_4$.

8.10-A PRODUCTION OF PERCHLORIC ACID

Concentrated perchloric acid is prepared by distilling a mixture of potassium perchlorate and sulfuric acid at reduced pressure (i.e., lower than atmospheric pressure). The chemical reaction associated with its production follows:

$$2KClO_4(s) \ + \ H_2SO_4(aq) \ \longrightarrow \ 2HClO_4(conc) \ + \ K_2SO_4(aq)$$

Potassium perchlorate Sulfuric acid Perchloric acid Potassium sulfate

The acid is a colorless, aqueous liquid having an approximate composition of 72% by mass. This is the composition of the concentrated perchloric acid available commercially. Although solutions of perchloric acid having an acid concentration greater than 72% are known, they are not routinely encountered, because they are explosively unstable. Some physical properties of concentrated perchloric acid are noted in Table 8.11.

8.10-B THERMAL DECOMPOSITION OF PERCHLORIC ACID

Concentrated perchloric acid may be safely heated to 194°F (90°C), but above this temperature, it is likely to explosively decompose as follows:

$$4HClO_4(conc) \ \longrightarrow \ 2Cl_2(g) \ + \ 7O_2(g) \ + \ 2H_2O(g)$$

Perchloric acid Chlorine Oxygen Water

TABLE 8.11	Physical Properties of Concentrated Perchloric Acid
Melting point	0°F (−18°C)
Boiling point	397°F (203°C)
Specific gravity at 68°F (20°C)	1.70
Vapor density (air = 1)	3.5
Vapor pressure at 68°F (20°C)	6.75 mmHg
Solubility in water	Very soluble

TABLE 8.12	Shipping Descriptions of Perchloric Acid
FORM OF PERCHLORIC ACID	SHIPPING DESCRIPTION
Perchloric acid (contains not more than 50% acid by mass)	UN1802, Perchloric acid, 8, (5.1), PG II
Perchloric acid (contains more than 50% acid but not more than 72% acid by mass)	UN1873, Perchloric acid, 5.1, (8), PG I

8.10-C OXIDIZING POTENTIAL OF PERCHLORIC ACID

Hot, concentrated perchloric acid is a powerful oxidizing acid, but when it is diluted with water, perchloric acid reacts as a weak oxidizing agent even when hot. The hot, concentrated perchloric acid reacts violently with organic compounds including cellulosic materials such as sawdust. In fact, a mixture of perchloric acid and sawdust ignites spontaneously. Every measure should be taken to segregate perchloric acid and organic compounds, because their combination is regarded as a fire and explosion hazard.

8.10-D TRANSPORTING PERCHLORIC ACID

When shippers offer perchloric acid for transportation, DOT requires them to identify it on the accompanying shipping paper in one of the ways shown in Table 8.12, as relevant. All labeling, marking, and placarding requirements apply.

8.11 HYDROFLUORIC ACID

Hydrogen fluoride is a colorless, fuming liquid or vapor having the chemical formula $HF(l)$ or $HF(g)$, respectively. Its anhydrous form is typically represented by the acronym AHF. Solutions of hydrogen fluoride in water are denoted by the chemical formula $HF(aq)$.

Hydrofluoric acid solutions are used for a variety of purposes. In the household, they are found as components of rust removers and aluminum-cleaning products. In the glass industry, they are used to polish, etch, and frost glass. In the metallurgical and steel industries, they are used to pickle brass, copper, and certain steel alloys. In the computer industry, they are used to etch silicon wafers during the manufacture of computer chips. In the chemical industry, they are used as catalysts and fluorinating agents.

Hydrofluoric acid

8.11-A PRODUCTION OF HYDROFLUORIC ACID

Hydrofluoric acid is prepared by reacting sulfuric acid with calcium fluoride, a constituent of the naturally occurring ores fluorspar and fluorite. This production reaction is represented as follows:

$$CaF_2(s) \ + \ H_2SO_4(conc) \ \longrightarrow \ CaSO_4(s) \ + \ 2HF(aq)$$

Calcium fluoride   Sulfuric acid   Calcium sulfate   Hydrofluoric acid

The concentrated hydrofluoric acid of commerce is a liquid solution containing 70% hydrogen fluoride by mass. A solution containing 49% hydrofluoric acid is also available commercially. The physical properties of these three forms of hydrofluoric acid are noted in Table 8.13.

The calcium sulfate is filtered from the resulting mixture, and the solution of hydrofluoric acid is boiled until the desired concentration is achieved.

8.11-B ILL EFFECTS CAUSED BY EXPOSURE TO HYDROFLUORIC ACID

Although the data in Table 8.1 show that hydrofluoric acid is a weak acid, both the liquid and its vapor permeate the skin and produce severe burns that heal very slowly. Although the skin damage is generally noticeable as external destruction to the outermost layer, the underlying tissues may also be severely damaged.

TABLE 8.13	Physical Properties of Hydrofluoric Acid		
	ANHYDROUS HF(l) (AHF)	**HF(aq) (49%)**	**HF(aq) (70%)**
Melting point	−118°F (−84°C)	−34°F (−37°C)	−96°F (−71°C)
Boiling point	67°F (20°C)	224°F (106°C)	151°F (66°C)
Specific gravity at 70°F (21°C)	0.97	1.16	1.225
Vapor density (air = 1) 70°F (21°C)	2.21	1.175	1.76
Vapor pressure at 70°F (21°C)	776 mmHg	27 mmHg	110 mmHg
Solubility in water	Infinitely soluble	Infinitely soluble	Infinitely soluble

hypocalcemia ■ The adverse health condition associated with low concentrations of calcium in the bloodstream

Exposure of the skin to hydrofluoric acid may be fatal because it can cause **hypocalcemia**, during which calcium fluoride precipitates in the body's tissues and removes calcium from the bloodstream. Calcium is essential for blood clotting and normal muscle and nerve functions. Its absence in the bloodstream may cause cardiac arrest. To avoid or reduce the impact from the ill effects caused by exposure to hydrofluoric acid, it is essential to follow the first-aid instructions noted in Figure 8.5.

The pain associated with exposure to hydrofluoric acid may not be immediately experienced, and the visible signs of corrosion may be unapparent until hours after the initial exposure. At this point, the area of contact may appear blanched and bloodless. Gangrene may develop if the wound is left unattended. The acid may penetrate into the bones, where decalcification can also occur. Specialized medical treatment following exposure to the acid or its fumes is often necessary. This specialized medical treatment includes injection of calcium-containing substances into burned areas or into the bloodstream.

8.11-C REACTIONS OF HYDROFLUORIC ACID WITH SILICON COMPOUNDS

The most distinguishing chemical property of concentrated hydrofluoric acid is its ability to slowly react with silicon compounds to produce gaseous silicon tetrafluoride. Silicon compounds are components of ordinary glass. Hydrofluoric acid reacts with them to produce silicon tetrafluoride. For example, hydrofluoric acid reacts with the sodium silicate and calcium silicate in glass as follows:

Anhydrous hydrogen fluoride

$$Na_2SiO_3(s) \ + \ 6HF(conc) \ \longrightarrow \ 2NaF(aq) \ + \ SiF_4(g) \ + \ 3H_2O(l)$$
Sodium silicate   Hydrofluoric acid   Sodium fluoride   Silicon tetrachloride   Water

$$CaSiO_3(s) \ + \ 6HF(conc) \ \longrightarrow \ CaF_2(s) \ + \ SiF_4(g) \ + \ 3H_2O(l)$$
Calcium silicate   Hydrofluoric acid   Calcium fluoride   Silicon tetrachloride   Water

Because hydrofluoric acid reacts with the components of glass, the acid is routinely stored and transported in polyethylene or other hydrofluoric acid–resistant plastic bottles and drums.

8.11-D ANHYDROUS HYDROGEN FLUORIDE

Anhydrous hydrogen fluoride (AHF), sometimes called anhydrous hydrofluoric acid, is itself a commercial chemical product. Like its solutions, it is prepared from calcium fluoride and sulfuric acid. The physical properties listed in Table 8.13 illustrate that anhydrous hydrogen fluoride readily vaporizes near room temperature.

Anhydrous hydrogen fluoride is used mainly in the chemical industry to produce chlorofluorocarbons, hydrofluorocarbons, and hydrochlorofluorocarbons (Section 12.15). In the petroleum refining industry, it is used as an alkylate catalyst for the production of high-octane fuels (Section 12.13-E). It is also used during uranium-enrichment processes

FIRST AID
FOR HYDROFLUORIC ACID
EXPOSURE

SEEK IMMEDIATE MEDICAL ATTENTION
CALL 911

SERIOUS TISSUE DAMAGE
WITH DELAYED ONSET

SKIN CONTACT

- **IMMEDIATELY** (within seconds) proceed to the **NEAREST SAFETY SHOWER** and wash affected area **FOR 5 MINUTES**.
- **REMOVE** all contaminated **CLOTHING** while in the shower.
- **WITH NITRILE DOUBLE-GLOVED HANDS MASSAGE CALCIUM GLUCONATE GEL** into the affected area. If calcium gluconate gel is **not available, wash area for at least 15 minutes** or until emergency medical assistance arrives.
- **REAPPLY CALCIUM GLUCONATE GEL** and massage it into affected area **EVERY 15 MINUTES** until medical assistance arrives or pain disappears.

EYE CONTACT

- **IMMEDIATELY** (within seconds) proceed **TO THE NEAREST EYEWASH STATION**.
- Thoroughly **WASH EYES WITH WATER FOR AT LEAST 15 MINUTES** while holding eyelids open.
- **DO NOT APPLY CALCIUM GLUCONATE GEL TO EYES**.

INHALATION

- **GET MEDICAL ASSISTANCE** by calling 911.

FIGURE 8.5 This poster provides first-aid instructions for individuals who have been exposed to hydrofluoric acid. (*Courtesy of the Department of Environmental Health & Safety, University of Washington, Seattle, Washington.*)

(Section 16.9-D) to prepare uranium hexafluoride, a gas essential for the production of uranium fuel for the nuclear power industry.

Exposure to even low concentrations (<15 ppm) of anhydrous hydrogen fluoride is irritating to the eyes, skin, and the bronchial passageways and lung surfaces. Prolonged exposure to the vapor may cause the development of pulmonary edema.

As with its solutions, contact with anhydrous hydrogen fluoride severely corrodes skin tissue and produces severe burns deep beneath the skin. The tissue corrosion may become

TABLE 8.14	Shipping Descriptions of Hydrofluoric Acid and Hydrogen Fluoride

HYDROFLUORIC ACID/HYDROGEN FLUORIDE	SHIPPING DESCRIPTION
Hydrofluoric acid (contains not more than 60% strength)	UN1790, Hydrofluoric acid solution, 8 (6.1), PG II
Hydrofluoric acid (contains more than 60% strength)	UN1790, Hydrofluoric acid solution, 8, (6.1), PG I
Hydrogen fluoride, anhydrous	UN1052, Hydrogen fluoride, anhydrous, 8, (6.1), PG I (Poison - Inhalation Hazard, Zone C)

evident only hours following the initial exposure with no immediate experience of pain. Vapor burns to the eyes may result in the formation of lesions or may cause blindness.

8.11-E WORKPLACE REGULATIONS INVOLVING HYDROGEN FLUORIDE

In the workplace, OSHA requires employers to limit employee exposure to a maximum hydrogen fluoride vapor concentration of 3 parts per million, averaged over an 8-hour workday.

8.11-F TRANSPORTING HYDROFLUORIC ACID AND ANHYDROUS HYDROGEN FLUORIDE

Hydrofluoric acid and anhydrous hydrogen fluoride are transported as liquids. When shippers offer these compounds for transportation, DOT requires them to provide the relevant shipping description shown in Table 8.14 on the accompanying shipping paper. All labeling, marking, and placarding requirements apply.

Phosphoric acid

8.12 PHOSPHORIC ACID

Although there are at least eight mineral acids containing phosphorus, phosphoric acid is the only one that is commonly encountered. Its chemical formula is H_3PO_4.

Phosphoric acid is most familiar to the general public as a constituent of certain household cleaning products like Lime-Away. In the chemical industry, it is used as a raw material for manufacturing organophosphate pesticides (Section 10.20) and a number of commercially important metallic phosphates. For instance, phosphoric acid is used to produce superphosphate fertilizer, a synthetic fertilizer consisting of a mixture of calcium dihydrogen phosphate and calcium sulfate. Phosphoric acid is also used to produce monoammonium phosphate, the dry-chemical fire extinguisher (Section 5.15-B). It is also used to prepare the surface of steel sheets before painting.

8.12-A PRODUCTION OF PHOSPHORIC ACID

Phosphoric acid is manufactured from phosphate rock, an ore containing calcium phosphate, by either of the following methods:

■ In the first method, phosphate rock is reacted with sulfuric acid.

$$Ca_3(PO_4)_2(s) \ + \ 3H_2SO_4(aq) \ \longrightarrow \ 3CaSO_4(s) \ + \ 2H_3PO_4(aq)$$
Calcium phosphate   Sulfuric acid   Calcium sulfate   Phosphoric acid

The calcium sulfate is filtered from the acid.

■ In the second method, elemental phosphorus is produced from phosphate rock; then, the phosphorus is burned to form tetraphosphorus decoxide, which is reacted with water.

TABLE 8.15	Physical Properties of Concentrated Phosphoric Acid
Melting point	108°F (42°C)
Boiling point	500°F (260°C)
Specific gravity at 68°F (20°C)	1.69
Vapor density (air = 1)	3.4
Vapor pressure at 68°F (20°C)	0.0285 mmHg
Solubility in water	Very soluble

$$P_4(s) \ + \ 5O_2(g) \ \longrightarrow \ P_4O_{10}(s)$$

Phosphorus   Oxygen   Tetraphosphorus decoxide

$$P_4O_{10}(s) \ + \ 6H_2O(l) \ \longrightarrow \ 4H_3PO_4(aq)$$

Tetraphosphorus decoxide   Water   Phosphoric acid

The reaction between tetraphosphorus decoxide and water is especially violent and releases considerable heat, 41 kilojoules/mole.

Various grades of phosphoric acid are commercially available. When an aqueous solution of phosphoric acid is allowed to boil at atmospheric pressure, a syrupy solution containing about 85% phosphoric acid by mass is produced. This is the concentrated phosphoric acid of commerce, some physical properties of which are noted in Table 8.15. In addition, a commercially available food-grade phosphoric acid is used as an ingredient of certain soft drinks and other food products.

Phosphoric acid exhibits the hazardous features of all corrosive materials: It reacts with metals, metallic oxides, and metallic carbonates, and it damages skin tissue.

8.12-B PHOSPHORIC ANHYDRIDE

Tetraphosphorus decoxide, or phosphoric anhydride, is the acidic anhydride of phosphoric acid. Its chemical formula is P_4O_{10}. This substance has such a strong affinity for water that it is used industrially as a drying agent.

Because its reaction with water releases 41 kilojoules/mole into the surroundings, phosphoric anhydride should be segregated from combustible matter to prevent its inadvertent ignition. This amount of heat also causes thermal burns when the oxide comes in contact with the skin.

8.12-C TRANSPORTING PHOSPHORIC ACID AND PHOSPHORIC ANHYDRIDE

When shippers offer phosphoric acid or phosphoric anhydride for transportation, DOT requires them to provide the relevant shipping description shown in Table 8.16 on the accompanying shipping paper. All labeling, marking, and placarding requirements apply.

Phosphoric anhydride

TABLE 8.16	Shipping Descriptions of Phosphoric Acid and Phosphoric Anhydride
PHOSPHORIC ACID/PHOSPHORUS ANHYDRIDE	**SHIPPING DESCRIPTION**
Phosphoric acid, solid	UN1805, Phosphoric acid, solid, 8, PG III
Phosphoric acid solutions	UN1805, Phosphoric acid solution, 8, PG III
Phosphoric anhydride	UN1807, Phosphorus pentoxide, 8, PG II

Glacial acetic acid

8.13 ACETIC ACID

Acetic acid is the most commonly encountered organic acid. Its chemical formula is CH_3COOH, or $HC_2H_3O_2$. Each molecule of acetic acid is represented by the following Lewis structure:

Acetic acid is the substance responsible for the sour taste and sharp odor of vinegar, the common food product containing from 3% to 6% acetic acid by volume. It is used primarily by the chemical industry as a raw material for the synthesis of ethyl acetate, vinyl acetate, cellulose acetate, and other chemical and pharmaceutical products.

8.13-A PRODUCTION OF ACETIC ACID

Chemical manufacturers produce acetic acid by a number of methods, the most popular of which involves the gas-phase combination of methanol and carbon monoxide.

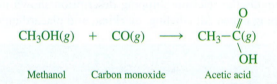

When aqueous solutions of acetic acid are cooled to temperatures near 61°F (16°C), a mixture of liquid and solid phases is produced. The liquid phase contains impurities and is recycled or discarded, but the solid phase typically contains more than 99% acetic acid by mass. This component is the concentrated acetic acid of commerce, called glacial acetic acid. When encountered, it is generally stored in glass and plastic containers.

TABLE 8.17	Physical Properties of Concentrated Acetic Acid
Melting point	61°F (17°C)
Boiling point	244°F (118°C)
Specific gravity at 68°F (20°C)	1.05
Vapor density (air = 1)	2.1
Vapor pressure at 68°F (20°C)	11 mmHg
Flashpoint	109°F (43°C)
Autoignition point	800°F (426°C)
Lower flammable limit	4% by volume
Upper flammable limit	19.9% by volume
Solubility in water	Infinitely soluble

At room temperature, glacial acetic acid is a colorless, pungent liquid. Some of its important physical properties are noted in Table 8.17. Other varieties are also available commercially containing from 80% to 99% acid by mass.

8.13-B VAPORIZATION OF ACETIC ACID

Concentrated acetic acid releases a vapor, which, when inhaled, is choking and suffocating and can readily damage the bronchial tract. Exposure to other body tissues, particularly the eyes, results in severe burns.

8.13-C COMBUSTIBLE NATURE OF ACETIC ACID

Concentrated acetic acid is an OSHA category 3 flammable liquid (NFPA class II combustible liquid), but aqueous solutions of acetic acid containing less than 80% acid are nonflammable. The complete combustion of acetic acid vapor is represented as follows:

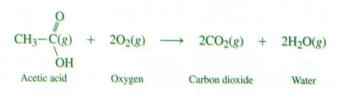

$$CH_3-\overset{\overset{O}{\|}}{\underset{OH}{C}}(g) \ + \ 2O_2(g) \ \longrightarrow \ 2CO_2(g) \ + \ 2H_2O(g)$$

Acetic acid   Oxygen   Carbon dioxide   Water

Fires involving concentrated acetic acid may be extinguished by diluting the acid with water.

8.13-D WORKPLACE REGULATIONS INVOLVING ACETIC ACID

In the workplace, OSHA requires employers to limit employee exposure to a maximum acetic acid vapor concentration of 10 parts per million, averaged over an 8-hour workday.

8.13-E TRANSPORTING ACETIC ACID

When shippers offer acetic acid or its solutions for transportation, DOT requires them to provide the relevant shipping description shown in Table 8.18 on the accompanying shipping paper. All labeling, marking, and placarding requirements apply.

TABLE 8.18 | **Shipping Descriptions of Acetic Acid**

FORM OF ACETIC ACID	SHIPPING DESCRIPTION
Acetic acid (contains more than 10% but not more than 50% acid by mass)	UN2790, Acetic acid solution, 8, PG III
Acetic acid solution (contains not less than 50% but not more than 80% acid by mass)	UN2790, Acetic acid solution, 8, PG II
Acetic acid, glacial (contains more than 80% acid by mass)	UN2789, Acetic acid, glacial, 8, (3), PG II
Acetic acid solution (contains more than 80% acid by mass)	UN2789, Acetic acid solution, 8, (3), PG II

Sodium hydroxide

Potassium hydroxide

8.14 SODIUM HYDROXIDE, POTASSIUM HYDROXIDE, AND CALCIUM HYDROXIDE

There are three commercially important corrosive materials that are alkaline: sodium hydroxide, potassium hydroxide, and calcium hydroxide. Their chemical formulas are NaOH, KOH, and $Ca(OH)_2$, respectively. Their physical properties are noted collectively in Table 8.19.

To the layperson, sodium hydroxide is best known as a constituent of consumer products such as Drano and Liquid-Plumr, which are used to unclog drainage pipes, and oven cleaners like Easy-Off. Sodium hydroxide solutions are also used as industrial cleaners at car- and truck-washing facilities and garages, and they are used at wastewater facilities to isolate metallic compounds as water-insoluble metallic hydroxides.

Sodium hydroxide is industrially used in countless applications including the purification of petroleum products, the reclaiming of rubber, and the processing of textiles and paper. The chemical industry uses large amounts of sodium hydroxide as a raw material during the manufacturing of soap, rayon, cellophane, and numerous commercial chemical substances.

Potassium hydroxide is used mainly by the chemical industry for the production of compounds like fertilizers, soft soaps, and pharmaceutical products. It is also the electrolyte in alkaline storage batteries. Potassium hydroxide was formerly used in liquid drain cleaners, but at 16 C.F.R. §1500.17, CPSC now bans the manufacture and sale of cleaners containing potassium hydroxide at a concentration of 10% or more by weight.

Calcium hydroxide is primarily used as a raw material for the production of mortar, plaster, and cement, but it is also widely used commercially for other purposes. Emergency responders use it to neutralize acids when they have been unintentionally released from their containers. A food grade of calcium hydroxide is a component of some antacids.

TABLE 8.19	**Physical Properties of Several Metallic Hydroxides**		
	SODIUM HYDROXIDE	POTASSIUM HYDROXIDE	CALCIUM HYDROXIDE
Melting point	599°F (315°C)	680°F (360°C)	1076°F (580°C)
Boiling point	2534°F (1390°C)	2408°F (1320°C)	Decomposes
Specific gravity at 68°F (20°C)	2.13	2.04	2.50
Solubility in 100 g of water	42 g	107 g	0.18 g

8.14-A PRODUCTION OF SODIUM HYDROXIDE AND POTASSIUM HYDROXIDE

Sodium hydroxide and potassium hydroxide are generally manufactured by passing an electric current through solutions of sodium chloride (brine) and potassium chloride, respectively.

$$2NaCl(aq) \ + \ 2H_2O(l) \ \longrightarrow \ 2NaOH(aq) \ + \ H_2(g) \ + \ Cl_2(g)$$

Sodium chloride   Water   Sodium hydroxide   Hydrogen   Chlorine

$$2KCl(aq) \ + \ 2H_2O(l) \ \longrightarrow \ 2KOH(aq) \ + \ H_2(g) \ + \ Cl_2(g)$$

Potassium chloride   Water   Potassium hydroxide   Hydrogen   Chlorine

At room temperature, they are commercially available as flakes, pellets, sticks, granules, and concentrated aqueous solutions. The industrial forms of sodium hydroxide are also known as *lye* and *caustic soda*, whereas the forms of potassium hydroxide are called *caustic potash* and *potash lye*.

8.14-B PRODUCTION OF CALCIUM HYDROXIDE

Calcium hydroxide is industrially prepared by reacting calcium oxide and water as follows:

$$CaO(s) \ + \ H_2O(l) \ \longrightarrow \ Ca(OH)_2(s)$$

Calcium oxide   Water   Calcium hydroxide

This reaction is called **slaking**, which gives rise to *slaked lime*, the industrial name of calcium hydroxide. Calcium oxide is called *unslaked lime* and *quicklime*.

slaking ■ The chemical process associated with reacting calcium oxide (unslaked lime) with water to form calcium hydroxide

8.14-C TRANSPORTING SODIUM HYDROXIDE, POTASSIUM HYDROXIDE, AND CALCIUM HYDROXIDE

When shippers intend to transport sodium hydroxide, potassium hydroxide, and calcium hydroxide, DOT requires them to provide the relevant shipping description in Table 8.20 on the accompanying shipping paper. All labeling, marking, and placarding requirements apply.

DOT regulates the transportation of calcium oxide by air only. Hence, shippers are obligated to affix a CARGO AIRCRAFT ONLY label in addition to the CORROSIVE label to its packaging.

When carriers transport 1001 pounds (454 kg) or more of these substances, DOT requires them to display CORROSIVE placards on the bulk packaging used for shipment as shown in Figure 8.6.

DOT does not regulate the transportation of calcium hydroxide.

TABLE 8.20	Shipping Descriptions of Some Alkaline Substances
ALKALINE SUBSTANCE	**SHIPPING DESCRIPTION**
Calcium oxide (unslaked lime)	UN1910, Calcium oxide, 8, PG III
Potassium hydroxide, solid	UN1813, Potassium hydroxide, solid, 8, PG II
Potassium hydroxide solution	UN1814, Potassium hydroxide, solution, 8, PG II
Sodium hydroxide, solid	UN1823, Sodium hydroxide solid, 8, PG II
Sodium hydroxide, solution	UN1824, Sodium hydroxide, solution, 8, PG II

FIGURE 8.6 When carriers transport a sodium hydroxide solution, DOT requires them to display the identification number 1824 across the center area of the CORROSIVE placards posted on the transport vehicle. DOT also permits the identification number to be displayed on orange panels or white square-on-point diamonds. *(Courtesy of Bulk Transportation, Inc., Walnut, California.)*

8.15 RCRA CORROSIVITY CHARACTERISTIC

As first noted in Section 1.3-C, EPA uses the legal authority of RCRA to regulate the treatment, storage, and disposal of hazardous wastes that exhibit certain characteristics, one of which is corrosivity.

At 40 C.F.R. §261.22, a waste exhibits the **RCRA corrosivity characteristic** when it is either of the following:

- An aqueous liquid that has a pH less than or equal to 2 or greater than or equal to 12.5.
- A liquid that corrodes steel at a rate greater than 0.25 inches/year (6.25 mm/y) at a test temperature of 130°F (55°C) using a specified test method.

A waste that exhibits the RCRA corrosivity characteristic is assigned the hazardous waste number D002.

8.16 WORKPLACE REGULATIONS INVOLVING CORROSIVE MATERIALS

Among its other responsibilities, OSHA is charged with protecting workers from the ill effects caused by exposure to corrosive materials. To accomplish this aim, it requires employers to display accident-prevention tags, warning labels, and worded signs that signal the presence of corrosive materials in the workplace when exposure could damage property or cause accidental injury to workers. Some examples that relate to identifying corrosive acids and bases are shown in Figure 8.7.

RCRA corrosivity characteristic ■ For purposes of RCRA regulations, the characteristic of either liquid waste noted at 40 C.F.R. §261.22

FIGURE 8.7 OSHA requires employers to post worded signs that warn employees of the presence of corrosive materials in the workplace. OSHA also requires employers at 29 C.F.R. §1200(h)(3)(iv) to affix accident-prevention tags or warning labels to their in-plant containers of corrosive materials.

Corrosive materials are frequently used in science laboratories and in certain work environments. In all areas where individuals may be exposed to an injurious corrosive material, OSHA requires the availability of suitable facilities for quick drenching and flushing of the eyes and body. An example of a suitable eyewash station and shower is shown in Figure 8.8.

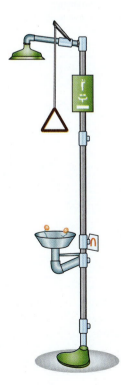

FIGURE 8.8 OSHA requires employers at 29 C.F.R. §1910.51(c) to provide emergency-use eyewash units and drench showers in areas where employees may be exposed to corrosive materials. Each eyewash unit and shower should deliver 4 gallons/minute and 20 gallons/minute of water, respectively, during a 15-minute period.

8.17 RESPONDING TO INCIDENTS INVOLVING A RELEASE OF A CORROSIVE MATERIAL

First-on-the-scene responders may generally identify the presence of a corrosive material at the emergency scene by reference to the NFPA hazard diamond affixed or imprinted on its storage vessel. Any of the expressions *ACID*, *ALK*, or *CORR* may appear in the bottom quadrant of the hazard diagram. They rapidly convey the message that a corrosive material is contained within the vessel.

At a transportation mishap, emergency responders identify the presence of a corrosive material by observing the following:

- The number *8* as a component of a shipping description of a hazardous material listed on a shipping paper
- The word *CORROSIVE* and the number *8* printed on black-and-white labels affixed to containment devices
- The word *CORROSIVE* and the number *8* printed on black-and-white placards displayed on each side and each end of a transport vehicle containing 1001 pounds (454 kg) or more of a corrosive material

At an emergency scene involving a corrosive material, the members of the response crew should conduct their work only while wearing fully encapsulated entry suits with clear face shields. Because these suits are fabricated from a material through which the corrosive material cannot penetrate, they protect their wearers from bodily contact with it. The firefighter shown in Figure 8.9 is wearing an "Entry 1-suit," which provides sufficient protection to work without the fear of exposure to a corrosive material.

FIGURE 8.9 This firefighter is wearing a fully encapsulating body suit for responding to an emergency incident involving the release of a corrosive material into the environment. The chemical nature of the fabric prevents bodily contact with the corrosive material. Self-contained breathing apparatus is also essential when responding to emergencies in which the corrosive material possesses a significant vapor pressure. (*Courtesy of Lakeland Industries, Inc., Ronkonkoma, New York; Image © 2012, All Rights Reserved.*)

Although fully encapsulated entry suits are not themselves designed to protect their users from inhaling toxic vapors or fumes, apparel equipped with self-contained breathing apparatus (SCBA) is marketed for use by emergency responders. When a bulk quantity of a corrosive material that has a significant vapor pressure is encountered, the responders must use SCBA in addition to wearing these suits. Examples of acids that spontaneously emit harmful vapors are oleum, fuming nitric acid, concentrated hydrochloric acid, and concentrated acetic acid.

When they are called to a scene involving the release of a corrosive material, emergency responders should consider the following actions:

■ Dilute the corrosive material with an approximate volume of water equal to at least 10 times the volume of the material that has been released into the environment. This is usually an adequate response when relatively small quantities of an acid or base have spilled or leaked from a container or storage tank. For instance, suppose that a 1-gallon container of liquid sulfuric acid has inadvertently been spilled on a laboratory floor and has flowed toward a drain leading to an off-site wastewater treatment plant. In this situation, dilution of the spilled acid with a copious volume of water lessens the acid's corrosive nature.

Special care should be exercised to avoid inhaling the fumes that arise when diluting fuming sulfuric acid, fuming nitric acid, concentrated hydrochloric acid, and concentrated acetic acid. The inhalation of these fumes poses the risk of inhalation toxicity because the fumes can seriously damage the respiratory system.

■ Neutralize the corrosive material. This action is recommended when a crew responds to the environmental release of a relatively large volume of a corrosive material such as a leak of 10,000 gallons (38 m^3) of an acid from a storage tank or during a transportation mishap. It generally is impractical to dilute such a large volume of a corrosive material with water, because an even larger volume of water is needed to effectively reduce its corrosiveness.

A large volume of an acid may be effectively neutralized with solid substances such as either slaked lime or soda ash, the common names for calcium hydroxide and anhydrous sodium carbonate, respectively. As noted in Section 8.3, an acid reacts with a base to produce a salt of the acid and water; and an acid reacts with a metallic carbonate to produce a salt of the acid, water, and carbon dioxide.

When slaked lime or soda ash is used to neutralize a spill of hydrochloric acid, the resulting chemical reactions may be represented by the following equations:

$$\underset{\text{Calcium hydroxide}}{Ca(OH)_2(s)} + \underset{\text{Hydrochloric acid}}{2HCl(aq)} \longrightarrow \underset{\text{Calcium chloride}}{CaCl_2(aq)} + \underset{\text{Water}}{2H_2O(l)}$$

$$\underset{\text{Sodium carbonate}}{Na_2CO_3(s)} + \underset{\text{Hydrochloric acid}}{2HCl(aq)} \longrightarrow \underset{\text{Sodium chloride}}{2NaCl(aq)} + \underset{\text{Carbon dioxide}}{CO_2(g)} + \underset{\text{Water}}{H_2O(l)}$$

The use of slaked lime or soda ash results in reducing or eliminating the corrosive nature of hydrochloric acid by chemically converting it into a group of relatively benign substances.

When first-on-the-scene emergency responders are called to a transportation mishap involving the release a bulk shipment of hydrogen chloride or hydrogen fluoride, they must acknowledge that the vapor could cause serious respiratory damage to the team, transportation personnel, and the general public. The *Emergency Response Guidebook* recommends isolation and evacuation distances for large spills of these substances from multiple small cylinders or single ton cylinders, multiple ton cylinders, highway tank trucks or trailers, and rail tankcars (for hydrogen chloride) and multiple small cylinders or single ton cylinders, highway tank trucks or trailers, and rail tankcars (for hydrogen fluoride).[1]

---

[1]Table 3, *Emergency Response Guidebook* (Washington, DC: U.S. Department of Transportation, 2012), pp. 354–355.

First-on-the-scene responders arriving at a domestic transportation mishap observe that a 5000-gallon (19-m^3) overturned tank truck is leaking its liquid contents. The transportation manifest indicates that the consignment consists solely of concentrated hydrochloric acid. The highway on which the truck was traveling is located approximately 250 feet (76.2 m) from the edge of a lake. What procedures should these first responders implement to save lives, property, and the environment?

Solution: Hydrochloric acid is a corrosive material. Consequently, first responders must don an acid-impervious suit that has been designed to prevent bodily contact with a corrosive material. Another hazard associated with concentrated hydrochloric acid is exposure to the vapor that spontaneously evolves from the liquid. To avoid inhaling this vapor, emergency responders must use self-contained breathing apparatus.

To save lives, property, and the environment, first-on-the-scene responders should consider implementation of the following procedures:

- Use self-contained breathing apparatus.
- Wear special protective clothing.
- Use water fog or foam to reduce toxic gas fumes in the air.
- Dike or dam the spilled material to prevent its spread.
- Use slaked lime to neutralize the acid within the diked area.

8.18 RESPONDING TO INCIDENTS INVOLVING ACID AND ALKALI POISONING

Paramedic teams are often called to assist in situations in which individuals have inadvertently been exposed to corrosive materials. In such incidents, a member of the paramedic team should immediately contact the American Association of Poison Control Centers.[2] In addition, the following actions are appropriate:

■ Because corrosive materials can irreversibly alter skin tissue at the site of contact, the affected area should be thoroughly flushed with water.

■ When a corrosive material has been inadvertently splashed into an individual's eyes, pain, swelling, corneal erosion, and blindness can rapidly ensue. Consequently, it is vital to immediately flush the eyes with a gentle stream of running water for at least 30 minutes (lifting the upper and lower lids occasionally). If the individual is wearing contact lenses, the eyes should first be irrigated for several minutes, and then, the lenses should be removed and the eyes again irrigated. The individual should be advised to promptly contact an ophthalmologist for professional eye treatment.

■ When a corrosive material has been ingested, the individual may experience difficulty in swallowing, nausea, intense thirst, shock, difficulty in breathing, and death. Vomiting should *not* be induced unless advised by a physician. The stomach wall is relatively tough and normally capable of withstanding the presence of gastric juices having a pH of less than 2. Vomiting should not be induced, because the individual's stomach contents could inadvertently be channeled into the bronchial tract, where they could cause serious damage. However, paramedics may attempt to neutralize the individual's stomach contents. *Milk of magnesia*, a white suspension of magnesium hydroxide in water, may be used to neutralize acids. A popular brand is Phillips' Milk of Magnesia, which is often found in home medicine cabinets. Acidic foods such as vinegar and citrus fruit juices may be used to neutralize bases. To ensure that the corrosive material has been completely neutralized, the individual should consume a volume of the neutralizing agent at least equal to the amount ingested.

---

[2]The American Association of Poison Control Centers can be accessed by telephone at (800) 222–1222. It is essentially a library staffed by nonmedical personnel. Be prepared to give the name of the poisonous product involved in the emergency and any information provided on the label. The poison hotline may also be contacted when an individual overdoses on medication.

Chemical Nature of Acids and Bases

1. Complete and balance each of the following equations, and provide an acceptable name for each reactant and product:
   (a) $Al_2O_3(s) + H_2SO_4(aq) \longrightarrow$
   (b) $PbCO_3(s) + HCl(aq) \longrightarrow$
   (c) $Fe(s) + HClO_4(aq) \longrightarrow$
   (d) $NaOH(aq) + H_3PO_4(aq) \longrightarrow$

pH Scale

2. A solution having a pH of 4 is how many times more acidic than a solution having a pH of 5?

Corroding Nature of Substances

3. Write an equation that illustrates the corrosive effect, if any, that occurs when each substance listed under the A heading reacts with the substance listed under the B heading:

A	B
(a) sulfuric acid	iron
(b) hydrochloric acid	cadmium oxide
(c) sulfuric acid	chromium(III) carbonate

4. How can the storage of a diluted acid within a steel tank connected to metal fittings and pipes constitute a fire and explosion hazard?

Sulfuric Acid

5. Complete and balance each of the following equations, and provide an acceptable name for each reactant and product:
   (a) $Ba(CN)_2(s) + H_2SO_4(aq) \longrightarrow$
   (b) $K(s) + H_2SO_4(aq) \longrightarrow$
   (c) $Cu(s) + H_2SO_4(conc, hot) \longrightarrow$
   (d) $NaCl(s) + H_2SO_4(conc) \longrightarrow$

6. Why does an aqueous solution of sulfuric acid conduct an electric current, but concentrated sulfuric acid does not?

Nitric Acid

7. Complete and balance each of the following equations, and provide an acceptable name for each reactant and product:
   (a) $Li_2CO_3(s) + HNO_3(aq) \longrightarrow$
   (b) $Fe_2O_3(s) + HNO_3(aq) \longrightarrow$
   (c) $CuS(s) + HNO_3(aq) \longrightarrow$
   (d) $Ag(s) + HNO_3(conc) \longrightarrow$

8. Following work at a transportation scene involving the release of nitric acid, emergency responders note a yellowing of the skin on their feet. What is the most likely cause of the discoloration?

Hydrochloric Acid

9. Complete and balance each of the following equations, and provide an acceptable name for each reactant and product:
   (a) $BaO(s) + HCl(aq) \longrightarrow$
   (b) $K_2SO_3(s) + HCl(aq) \longrightarrow$
   (c) $SnO_2(s) + HCl(aq) \longrightarrow$

Perchloric Acid

10. Complete and balance each of the following equations, and provide an acceptable name for each reactant and product:
    (a) $CaO(s) + HClO_4(aq) \longrightarrow$
    (b) $Al_2O_3(s) + HClO_4(aq) \longrightarrow$
    (c) $CdCO_3(s) + HClO_4(aq) \longrightarrow$
    (d) $Zn(s) + HClO_4(aq) \longrightarrow$

Hydrofluoric Acid

11. Complete and balance each of the following equations, and provide an acceptable name for each reactant and product:
    (a) $SnO(s) + HF(aq) \longrightarrow$
    (b) $Mg(s) + HF(aq) \longrightarrow$
    (c) $Al_2(CO_3)_3(s) + HF(aq) \longrightarrow$

12. Why do physicians often prescribe calcium injections for individuals who have been exposed to hydrofluoric acid?

Phosphoric Acid

13. Complete and balance each of the following equations, and provide an acceptable name for each reactant and product:
    (a) $SrO(s) + H_3PO_4(aq) \longrightarrow$
    (b) $Sn(s) + H_3PO_4(aq) \longrightarrow$
    (c) $CuCO_3(s) + H_3PO_4(aq) \longrightarrow$

Acetic Acid

14. Complete and balance each of the following equations, and provide an acceptable name for each reactant and product:
    (a) $CdO(s) + CH_3COOH(aq) \longrightarrow$
    (b) $Na(s) + CH_3COOH(aq) \longrightarrow$
    (c) $MgCO_3(s) + CH_3COOH(aq) \longrightarrow$

Alkaline Metal Hydroxides

15. When responding to a domestic transportation mishap involving a spill of caustic soda near a lake, why should emergency responders take precaution to prevent the spilled material from entering the lake?

Transporting Corrosive Materials

16. Nalco Chemical Company (NCC) agrees to transport 5000 pounds of caustic soda solution in a rail tankcar from Chicago, Illinois, to a customer in Baltimore, Maryland. What shipping description does DOT require NCC to enter on the accompanying waybill? (NALX is NCC's reporting mark.)

Responding to Incidents Involving a Release of Corrosive Materials

17. When first-on-the-scene responders arrive at a transportation mishap, they discover that a motor van has overturned, and liquids are leaking from a ruptured stainless-steel drum and a broken plastic jerrican. The accompanying bill of lading reads as follows:

UNITS	HM	SHIPPING DESCRIPTION (IDENTIFICATION NUMBER, PROPER SHIPPING NAME, PRIMARY HAZARD CLASS OR DIVISION, SUBSIDIARY HAZARD CLASS OR DIVISION, AND PACKING GROUP)	VOLUME (gal)
1 drum (UN1A1)	X	UN2031, Nitric acid, other than red fuming, with more than 70% nitric acid, 8, (5.1), PG I	55
1 plastic jerrican (UN3H)	X	UN1805, Phosphoric acid solution, 8, PG III	5

What procedures should be implemented by these responders to save lives, property, and the environment?

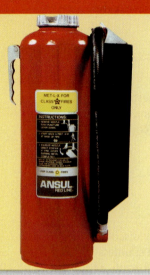

CHAPTER 9

Chemistry of Some Water- and Air-Reactive Substances

KEY TERMS

air-reactive substance
(pyrophoric substance), *p. 309*

alloy, *p. 313*

aluminum alkyl, *p. 324*

amalgam, *p. 313*

chlorosilane, *p. 334*

combustible metal, *p. 316*

dangerous-when-wet
material, *p. 310*

ductile, *p. 321*

flammable solid, *p. 310*

galvanize, *p. 323*

hydrolysis, *p. 309*

ionic hydride, *p. 328*

malleable, *p. 320*

metal fume fever, *p. 318*

metallic carbide, *p. 331*

metallic phosphide, *p. 330*

metallic superoxide (metallic
hyperoxide), *p. 314*

silane, *p. 334*

spontaneously combustible
material, *p. 310*

thermite, *p. 321*

thermite reaction, *p. 321*

water-reactive substance, *p. 309*

Ziegler–Natta catalyst, *p. 325*

OBJECTIVES

- Associate the physical and health hazards of the water- and air-reactive materials noted in this chapter with the information provided by their hazard diamonds and GHS pictograms.
- Identify the industries that use the water- and air-reactive materials noted in this chapter.
- Identify the labels, markings, and placards that DOT requires on packaging of water- and air-reactive materials and the transport vehicles used for their shipment.
- Identify the response actions to be executed when water- and air-reactive materials are released from their packaging into the environment.

The members of several classes of substances are likely to pose special problems when they are encountered by emergency responders. Two such classes are water-reactive and air-reactive (pyrophoric) substances.

- A **water-reactive substance** is an element or compound that reacts with water to produce either flammable gases that ignite spontaneously, or toxic or corrosive compounds that may endanger one's health upon exposure.
- An **air-reactive substance (pyrophoric substance)** is an element or compound that ignites spontaneously upon exposure to the oxygen or moisture in the ambient air, typically posing the risk of fire and explosion.

When a substance is water- or air-reactive, the use of water as a fire extinguisher is not only inappropriate, but could be dangerous. Fires involving these substances usually must be fought using special fire extinguishers, not water. As we progress through this chapter, these special extinguishers will be noted.

water-reactive substance ■ A substance that, by its chemical reaction with water, is likely to become spontaneously flammable, emit flammable or toxic gases, or generate sufficient heat to self-ignite or cause the ignition of nearby combustible materials

air-reactive substance (pyrophoric substance) ■ A substance that ignites spontaneously upon exposure to the air

9.1 WATER- AND AIR-REACTIVE SUBSTANCES

When water reacts with another substance, the chemical phenomenon is called **hydrolysis**. In chemistry, this process is represented by the following general equation:

$$A + H_2O(l) \longrightarrow C + D$$

Here, A represents a water-reactive substance, and C and D are the substances produced when A reacts with water. Exposure to the hydrolysis products can be harmful, because C and D may be flammable, corrosive, or toxic. The application of water should always be avoided during emergency response actions at which A is present, especially when either C or D is a flammable or toxic substance.

Consider the metals displayed in Figure 9.1. They react with water, including atmospheric moisture, to produce flammable hydrogen; others are so chemically reactive that they spontaneously ignite in air without exposure to an ignition source. As the hydrogen forms, it absorbs the heat of reaction, self-ignites, and triggers the combustion of the metals. These metals constitute the fuels of class D fires.

As noted, the hydrolysis of a substance may result in the formation of a solution that is corrosive. For example, ferric chloride reacts with water to produce ferric hydroxide and hydrochloric acid.

hydrolysis ■ The chemical reaction between a substance and water

Ferric chloride

$$\underset{\text{Ferric chloride}}{\text{FeCl}_3(aq)} + \underset{\text{Water}}{3\text{H}_2\text{O}(l)} \longrightarrow \underset{\text{Ferric hydroxide}}{\text{Fe(OH)}_3(s)} + \underset{\text{Hydrochloric acid}}{3\text{HCl}(aq)}$$

1A	2A											3A	4A	5A	6A	7A	8A
3 Li	4 Be											5 B	6 C	7 N	8 O	9 F	10 Ne
11 Na	12 Mg	3B	4B	5B	6B	7B		8B		1B	2B	13 Al	14 Si	15 P	16 S	17 Cl	18 Ar
19 K	20 Ca	21 Sc	22 Ti	23 V	24 Cr	25 Mn	26 Fe	27 Co	28 Ni	29 Cu	30 Zn	31 Ga	32 Ge	33 As	34 Se	35 Br	36 Kr
37 Rb	38 Sr	39 Y	40 Zr	41 Nb	42 Mo	43 Tc	44 Ru	45 Rh	46 Pd	47 Ag	48 Cd	49 In	50 Sn	51 Sb	52 Te	53 I	54 Xe

FIGURE 9.1 The symbols of the metals whose background shading are blue and yellow, respectively, represent the alkali metals and combustible metals noted in this chapter. Their finely divided physical forms may self-ignite upon exposure to air.

Ferric hydroxide is insoluble; hence, the character of the solution is provided by the acid. For this reason, aqueous solutions of ferric chloride are corrosive. Relatively large volumes of them are transported in tank trucks to facilities that treat water and wastewater. DOT requires carriers to display the UN identification number 2582 on orange panels, across the center of CORROSIVE placards, or on white square-on-point diamonds posed on each side and each end of the trucks.

When the hydrolysis of a substance produces a toxic vapor, emergency responders need to be especially cautious to avoid inhaling it. Some dangerous-when-wet substances produce sufficient toxic vapor when they undergo hydrolysis that they pose an inhalation health risk to individuals who are located 0.3 to 6.0 miles (0.5–10 km) downwind. A list of these substances is provided in DOT's *Emergency Response Guidebook*, and several representative dangerous-when-wet materials are reproduced in Table 9.1. When these substances are involved in emergencies, first-on-the scene responders should give special attention to them and select appropriate actions to reduce or eliminate the potential for their inhalation.

9.1-A IDENTIFYING AIR-REACTIVE (PYROPHORIC) SUBSTANCES

Pyrophoric substances pose the risk of fire and explosion because they ignite rapidly when exposed to atmospheric oxygen. This inherent hazard may be initiated when they react with the atmospheric moisture encountered as they are released from their containers. These substances not only burn spontaneously, but their fires are so exothermic that they pose a unique challenge to firefighting efforts. To prevent their premature ignition, pyrophoric substances sometimes are stored and processed under oil or other nonaqueous liquids or within enclosed, oxygen-free, dry atmospheres. To avoid their premature ignition during storage and transportation, manufacturers seal them hermetically in airtight containers.

Fortunately, few pyrophoric substances are used as commercial chemical products, and they are almost always stored in nonbulk amounts. OSHA requires their manufacturers, distributors, and importers to post the GHS flame pictogram on the labels of pyrophoric liquids or solids, and the flame *and* explosive pictograms on the labels of substances or mixtures that form flammable gases upon contact with water.

9.1-B TRANSPORTING WATER-REACTIVE SUBSTANCES

When water-reactive substances are transported, DOT regulates their transportation as **dangerous-when-wet materials, spontaneously combustible materials, flammable solids,** or corrosive materials. Individual water-reactive substances may be members of one or more of these hazard classes. DOT requires their shippers and carriers to comply with relevant labeling, marking, and placarding requirements.

dangerous-when-wet material ■ For purposes of DOT regulations, a material that by interaction with water is likely to become spontaneously flammable or to release a flammable or toxic gas or vapor at a rate greater than 28 in.^{3}/lb (1 L/kg) per hour when subjected to prescribed test procedures

spontaneously combustible material ■ For purposes of DOT regulations, either a pyrophoric material or a self-heating material

flammable solid ■ For purposes of DOT regulations, any of the following types of materials: wetted explosives; thermally unstable compounds that can undergo a strongly exothermic decomposition even without the participation of atmospheric oxygen; and readily combustible solids

SOLVED EXERCISE 9.1

What information regarding water reactivity is immediately conveyed to responding firefighters by a hazard diamond displayed on the exterior wall of a burning shed?

Solution: Information regarding the water reactivity of a substance is conveyed on a hazard diamond in two ways. As first noted in Section 1.11, a substance's relative degree of the health, fire, and instability hazards is conveyed through numbers in the three topmost quadrants of a diamond. The relative degree of water reactivity is conveyed by the number that appears in the rightmost yellow quadrant, as follows:

- "3" means that the substance reacts explosively with water.
- "2" means that the substance may react violently with water or may form potentially explosive mixtures with water.
- "1" means that the substance may react with water with some release of energy, but not violently.
- "0" means that the substance does not react with water.

In addition, the presence of a capital letter W with a line through its center (W̶) in the bottommost quadrant of the diamond serves to signal firefighters that they should avoid applying water when fighting a fire in the shed. Responding firefighters use this combination of information to select an action appropriate to the incident at hand.

TABLE 9.1	Some Classes of Water-Reactive Substances[a]		
CLASS OF SUBSTANCE	**EXAMPLE**	**CHEMICAL FORMULA**	**HAZARDOUS HYDROLYSIS PRODUCT**
Acetyl halides	Acetyl bromide	$CH_3-C\!\!\underset{Br}{\overset{O}{\parallel}}$	Hydrogen bromide
	Acetyl chloride	$CH_3-C\!\!\underset{Cl}{\overset{O}{\parallel}}$	Hydrogen chloride
Acids	Fluorosulfonic acid	$F-SO_2-OH$	Hydrogen fluoride
	Nitrosylsulfuric acid	$O{=}N-O-SO_2-O-H$	Nitrogen dioxide
Chlorosilanes	Methyldichlorosilane	$CH_3-Si-Cl_2H$	Hydrogen chloride
	Methyltrichlorosilane	$CH_3-Si-Cl_3$	Hydrogen chloride
	Trichlorosilane	Cl_3SiH	Hydrogen chloride
Metallic amides	Lithium amide	$LiNH_2$	Ammonia
	Magnesium diamide	$Mg(NH_2)_2$	Ammonia
Metallic halides	Aluminum bromide, anhydrous	$AlBr_3$	Hydrogen bromide
	Aluminum chloride, anhydrous	$AlCl_3$	Hydrogen chloride
	Antimony pentafluoride, anhydrous	SbF_5	Hydrogen fluoride
Metallic hypochlorites	Calcium hypochlorite	$Ca(ClO)_2$	Chlorine, hydrogen chloride
	Lithium hypochlorite	$LiClO$	Chlorine, hydrogen chloride
Metallic nitrides	Lithium nitride	Li_3N	Ammonia
Metallic oxychlorides	Chromium oxychloride	$Cr(OCl)_3$	Hydrogen chloride
Metallic phosphides	Aluminum phosphide	AlP	Phosphine
	Calcium phosphide	Ca_3P_2	Phosphine
	Magnesium aluminum phosphide	$Mg_3P_2 \cdot AlP$	Phosphine
	Magnesium phosphide	Mg_3P_2	Phosphine
	Potassium phosphide	K_3P	Phosphine
	Sodium phosphide	Na_3P	Phosphine
	Zinc phosphide	Zn_3P_2	Phosphine
Nonmetallic halides	Iodine pentafluoride	IF_5	Hydrogen fluoride
	Phosphorus pentachloride	PCl_5	Hydrogen chloride
	Silicon tetrachloride	$SiCl_4$	Hydrogen chloride
	Thionyl chloride	$SOCl_2$	Hydrogen chloride, sulfur dioxide
Sulfides	Ammonium hydrosulfide	NH_4HS	Hydrogen sulfide, ammonia
	Ammonium sulfide	$(NH_4)_2S$	Hydrogen sulfide, ammonia
Others	Chlorine dioxide (hydrate)[b]	ClO_2	Chlorine
	Uranium hexafluoride	UF_6	Hydrogen fluoride

[a]Adapted in part from Table 2, *Emergency Response Guidebook* (Washington, DC: U.S. Department of Transportation, 2012).
[b]Section 11.8.

TABLE 9.2	Physical Properties of the Alkali Metals		
	LITHIUM	SODIUM	POTASSIUM
Melting point	354°F (179°C)	208°F (98°C)	147°F (64°C)
Boiling point	2437°F (1337°C)	1618°F (881°C)	1425°F (774°C)
Specific gravity at 68°F (20°C)	0.53	0.97	0.86
Autoignition point	352°F (178°C)	250°F (121°C)	

9.2 ALKALI METALS

We consider the properties of three alkali metals in this section: lithium, sodium, and potassium. The properties of these three metals illustrate the uniqueness of their reactions. Some of their physical properties are noted in Table 9.2.

The alkali metals spontaneously ignite. Furthermore, they displace hydrogen from water as the following equations illustrate:

$$2Li(s) \ + \ 2H_2O(l) \ \longrightarrow \ 2LiOH(aq) \ + \ H_2(g)$$
Lithium   Water   Lithium hydroxide   Hydrogen

$$2Na(s) \ + \ 2H_2O(l) \ \longrightarrow \ 2NaOH(aq) \ + \ H_2(g)$$
Sodium   Water   Sodium hydroxide   Hydrogen

$$2K(s) \ + \ 2H_2O(l) \ \longrightarrow \ 2KOH(aq) \ + \ H_2(g)$$
Potassium   Water   Potassium hydroxide   Hydrogen

When metallic lithium reacts with water, the hydrogen produced does not immediately ignite; but when metallic sodium and potassium react with water, the hydrogen bursts spontaneously into flame as it is produced.

Chemical manufacturers display the GHS flame pictogram on labels affixed to containers holding the alkali metals.

9.2-A METALLIC LITHIUM

Lithium metal

Lithium is a soft, silvery metal and is the least dense solid element at normal conditions. Metallic lithium is so light that pieces of it float even in low-density petroleum products like kerosene and gasoline.

Metallic lithium and lithium compounds are valuable raw materials used to manufacture porcelain, ceramics, castings, batteries, zero-expansion glass, fungicides, bleaching agents, pharmaceuticals, and greases. In contemporary times, they have become increasingly popular within the chemical industry as raw materials used to synthesize organic compounds. Metallic lithium itself is a component of a lightweight magnesium alloy.

There are two types of lithium batteries, called primary and secondary lithium batteries. Both types pose the risk of fire. They differ as follows:

■ *Primary lithium-metal batteries.* These are disposable, non-rechargeable batteries. Although they have variable compositions, the most common primary lithium battery uses metallic lithium and manganese dioxide as its electrodes, and lithium perchlorate dissolved in propylene carbonate and dimethoxyethane as the electrolyte. Primary lithium-metal batteries are encountered mainly in the coin or button cells used in watches and digital cameras. Upon contact with water, they produce hydrogen.

■ *Secondary lithium-ion batteries.* These are rechargeable batteries, also having variable compositions. A typical type uses a lithium alloy as the positive plate, graphite as the negative plate, and lithium hexafluorophosphate ($LiPF_6$) dissolved in an organic solvent as the electrolyte. Examples of the organic solvents are diethyl carbonate and ethylene carbonate. Secondary lithium-ion batteries are used in laptop computers, cell phones, power tools, and all-electric automobiles. When damaged, the liquid electrolyte in secondary lithium-ion batteries may ignite.

Both primary and secondary lithium batteries may short-circuit and ignite when they have been improperly packaged or damaged. Sufficient heat is generated to cause fire when they discharge suddenly during short-circuiting. They are also potentially hazardous because the electrolytes used in them may ignite when exposed to the heat generated during short-circuiting.

When lithium metal reacts with water, the hydrogen is slowly displaced from the water. The hydrogen dissipates into the surrounding environment without ever achieving a concentration equal to or greater than 4% by volume, its lower flammable limit.

During its reaction with water, metallic lithium remains in the solid state of matter. Although the heat of the reaction initially is absorbed by the metal, it is transmitted to the surrounding water, and the metal's temperature remains below the boiling point of water.

When metallic lithium is left exposed to the air at room conditions, it does not spontaneously ignite. Although the metal oxidizes in the air, it does so very slowly. Even molten lithium oxidizes so slowly that it can be poured from a container in the open air without losing its bright luster.

In an atmosphere of absolutely dry air, lithium metal does not spontaneously burn. When exposed to an ignition source, however, the metal burns in the air with a characteristic crimson color, forming a mixture of lithium oxide and lithium nitride.

$$4Li(s) \ + \ O_2(g) \longrightarrow 2Li_2O(s)$$
Lithium   Oxygen   Lithium oxide

$$6Li(s) \ + \ N_2(g) \longrightarrow 2Li_3N(s)$$
Lithium   Nitrogen   Lithium nitride

9.2-B METALLIC SODIUM

Like metallic lithium, sodium also is a soft, silvery bright metal. It is the most commonly encountered alkali metal and the only one produced in bulk. Sodium metal generally is available for commercial use in the form of solid bricks.

The majority of the metallic sodium produced in the United States formerly was used as a raw material for the manufacture of the vehicular fuel additives tetraethyllead and tetramethyllead. This use of metallic sodium was sharply curtailed in 1975, when EPA banned the use of leaded gasoline in vehicular fuels. Today, metallic sodium is used primarily as a raw material for the production of highly reactive sodium compounds such as sodium peroxide and sodium hydride. In addition, metallic sodium is used as a catalyst during the production of certain types of synthetic rubber.

Metallic sodium sometimes is encountered commercially in alloys such as sodium/potassium alloys, sodium/lead alloys, and sodium amalgams. An **alloy** is a solid mixture of two or more elements, none of which can be separated by mechanical means. An **amalgam** is a special alloy in which one of these elements is elemental mercury. When a sodium alloy or amalgam is used instead of sodium alone, sodium reacts less vigorously. Consequently, chemical manufacturers often use sodium alloys and amalgams when the rate of a reaction requires careful control; notwithstanding this fact, the use of all amalgams has lost its popularity owing to the toxicity of mercury.

Metallic sodium reacts rapidly with water. In fact, the reaction occurs so rapidly that the hydrogen produced is unable to dissipate before it ignites. Instead, it concentrates in the immediate vicinity of the metal, where, induced by the heat of reaction, it self-ignites and spontaneously burns. The metallic sodium absorbs the heat of reaction and melts, thereby exposing an underlying surface of the solid metal for further reaction.

Sodium metal

alloy ■ A solid mixture of two or more mechanically inseparable elements

amalgam ■ An alloy of mercury with one or more elements

Unlike lithium, metallic sodium does not react with atmospheric nitrogen. Metallic sodium burns in an atmosphere of oxygen, producing a mixture of sodium oxide and sodium peroxide.

$$4Na(s) \; + \; O_2(g) \; \longrightarrow \; 2Na_2O(s)$$
Sodium   Oxygen   Sodium oxide

$$2Na(s) \; + \; O_2(g) \; \longrightarrow \; Na_2O_2(s)$$
Sodium   Oxygen   Sodium peroxide

Because sodium peroxide is a powerful oxidizer (Section 11.16-A), it itself is a hazardous material.

Although bulk sodium is not pyrophoric, nonbulk pieces of metallic sodium ignite spontaneously at room temperature with a characteristic yellow flame. In absolutely dry air, however, this oxidation does not occur at an appreciable rate, which suggests that the oxidation of metallic sodium in air is triggered by the reaction of metallic sodium and atmospheric water vapor. To reduce its potential for ignition, sodium generally is stored under kerosene.

9.2-C METALLIC POTASSIUM

Potassium metal

Potassium is a soft, silvery metal. Although it formerly was used with sodium as a heat-exchanger fluid in nuclear reactors, metallic potassium now has so few commercial uses that it is rarely encountered.

The combustion of potassium metal in air is associated with the production of a characteristic purple flame. The combustion product is primarily potassium oxide.

$$4K(s) \; + \; O_2(g) \; \longrightarrow \; 2K_2O(s)$$
Potassium   Oxygen   Potassium oxide

When potassium burns in an atmosphere of pure oxygen, however, a mixture of potassium oxide, potassium peroxide, and potassium superoxide is produced.

$$2K(s) \; + \; O_2(g) \; \longrightarrow \; K_2O_2(s)$$
Potassium   Oxygen   Potassium peroxide

$$K(s) \; + \; O_2(g) \; \longrightarrow \; KO_2(s)$$
Potassium   Oxygen   Potassium superoxide

metallic superoxide (metallic hyperoxide)

■ An inorganic compound composed of metallic and superoxide ions

A superoxide, more properly called a *hyperoxide,* is a compound containing the superoxide ion, whose chemical formula is O_2^-. **Metallic superoxides (metallic hyperoxides)** are extraordinarily reactive oxidizing agents (Section 11.16-B).

Metallic potassium reacts with water even more rapidly than does sodium. The vigorous nature of this reaction most likely is due to the presence of minute amounts of potassium superoxide produced when potassium oxidizes. The hydrogen produced by the reaction of potassium and water initially concentrates around the metal, where it self-ignites; then, the heat of reaction triggers the burning of the potassium metal. The reaction between potassium superoxide and atmospheric moisture occurs with such ease that it has been employed commercially as a means of supplying oxygen in self-contained breathing apparatus.

9.2-D TRANSPORTING ALKALI METALS AND PRIMARY LITHIUM BATTERIES

When shippers offer an alkali metal or its alloys, amalgams, or dispersions for transportation, DOT requires them to identify the appropriate material on the accompanying shipping paper. The shipping descriptions of some representative examples are listed in Table 9.3. DOT also requires shippers and carriers to comply with all applicable labeling, marking, and placarding requirements.

TABLE 9.3	Shipping Descriptions of Some Representative Alkali Metals, Their Amalgams, Dispersions, Alloys, and Products
ALKALI METAL OR ITS AMALGAMS, DISPERSIONS, ALLOYS, OR PRODUCTS	**SHIPPING DESCRIPTION**
Alkali metal alloys (liquid)	UN1421, Alkali metal alloy, liquid, n.o.s., 4.3, PG I (Dangerous When Wet)
Alkali metal dispersions	UN1391, Alkali metal dispersion, flammable, 4.3, PG I (Dangerous When Wet)
Alkaline earth metal alloys	UN1393, Alkaline earth metal alloy, n.o.s., 4.3, PG II
Lithium	UN1415, Lithium, 4.3, PG I (Dangerous When Wet)
Primary lithium metal battery	UN3090, Lithium battery, 9, PG II (Dangerous When Wet)
Primary lithium metal batteries, contained in equipment	UN3091, Lithium battery, contained in equipment, 9, PG II (Dangerous When Wet)
Primary lithium metal batteries, packed with equipment	UN3091, Lithium battery, packed with equipment, 9, PG II (Dangerous When Wet)
Potassium	UN2257, Potassium, 4.3, PG I (Dangerous When Wet)
Sodium	UN1428, Sodium, 4.3, PG I (Dangerous When Wet)

The transportation of primary lithium batteries is subject to additional regulations published at 49 C.F.R. §173.185. DOT requires manufacturers, shippers, and carriers to implement certain safety precautions when offering lithium-metal batteries for transportation. For example, shippers must package individual lithium-metal batteries in an inner packaging, separated by a divider and surrounded by noncombustible, nonconductive cushioning that prevents contact of the battery terminals with other batteries, metal objects, or conductive surfaces. Strong outer packaging or containment that complies with Packing-Group-II performance standards also is required.

When transporting primary lithium batteries domestically, DOT requires shippers to include the following statement on the shipping paper:

This shipment contains primary lithium batteries. Do not damage or mishandle the packages. If the package is damaged, flammability hazard may exist; batteries must be quarantined, inspected, and repacked.

When shippers intend to transport lithium-metal batteries by aircraft, DOT also requires them at 49 C.F.R. §172.102.188 to affix the lithium-battery-handling label shown in Figure 9.2 on two opposing sides or ends (other than the bottom) of the packaging, and to provide either of the following markings on its surface:

PRIMARY LITHIUM BATTERIES - FORBIDDEN FOR TRANSPORT ABOARD PASSENGER AIRCRAFT

LITHIUM METAL BATTERIES - FORBIDDEN FOR TRANSPORT ABOARD PASSENGER AIRCRAFT

When shippers intend to transport lithium-metal batteries by cargo aircraft, DOT also requires the CLASS 9 and CARGO AIRCRAFT ONLY labels shown in Figures 6.5 and 6.6,

FIGURE 9.2 When packages of lithium-metal batteries are transported by cargo aircraft, DOT requires shippers to affix this lithium-battery handling label on opposing sides of each package adjacent to a CLASS 9 and a CARGO AIRCRAFT ONLY label. The printing on the lithium-battery-handling label is black with a red-hatching border on a contrasting background.

respectively, to be affixed adjacent to each other on two opposing sides or ends (other than the bottom) of the packaging. DOT also regulates the nature of the packaging to reduce the likelihood of short circuiting and damage to battery terminals.

The Federal Aviation Administration does not permit large palletized shipments of primary lithium batteries on cargo *or* passenger aircraft. Furthermore, although DOT permits the transportation of primary lithium batteries contained in electronic equipment, it does not permit air transport of loose lithium batteries in checked baggage.

9.3 COMBUSTIBLE METALS

Magnesium, titanium, zirconium, aluminum, and zinc possess a common hazardous feature. Although bulk pieces of these metals typically are difficult to ignite, their finely divided forms may self-ignite in air without exposure to an ignition source. They are examples of **combustible metals**, some of which are also pyrophoric at elevated temperatures. These metals represent the fuels of class D fires. Some of their relevant physical properties are provided in Table 9.4.

Chemical manufacturers display the GHS flame pictogram on labels affixed to containers holding hazardous forms of the combustible metals.

The finely divided forms of some combustible metals are regarded as water- and air-reactive substances to varying degrees. They include dusts, powders, chips, turnings, flakes, punchings, borings, ribbons, and shavings. These forms are commonly produced during metal-forging and metal-machining operations. Often, the heat retained from these processes is sufficient to cause the metals to spontaneously ignite. The finely divided forms of metals generated during machining, grinding, boring, and other fabrication processes are also likely to be coated with the cutting oils used as lubricants, which can ignite as the primary fuel.

The finely divided forms of combustible metals react with water to produce hydrogen. The spontaneous ignition of the hydrogen kindles the burning of the underlying metal. The rate of hydrogen production is affected by a number of factors including the particle size, distribution and dispersion, purity, and ignition temperature of the metal, as well as the moisture content of the surrounding atmosphere. Some relevant physical properties of these metals are provided in Table 9.4.

combustible metal ■ Any metal whose distinct particles or pieces, regardless of size or shape, can readily ignite to produce an NFPA class D fire

TABLE 9.4	Physical Properties of Several Combustible Metals		
	MAGNESIUM	**ZIRCONIUM**	**TITANIUM**
Melting point	1200°F (649°C)	3326°F (1830°C)	3034°F (1668°C)
Boiling point	2012°F (1100°C)	7911°F (4377°C)	5948°F (3260°C)
Specific gravity at 68°F (20°C)	1.74	6.49	4.51
Autoignition point	883°F (472.78°C) (powder) 950°F (510°C) (ribbons and shavings) 1202°F (650°C) (massive chunks)	662°F (350°C) (powder)	482°F (250°C) (3.175-mm-thick plate); >2192°F (>1200°C) (6.35-mm-diameter rod)
	ALUMINUM	**ZINC**	
Melting point	1220°F (660°C)	786°F (419°C)	
Boiling point	4221°F (2327°C)	1665°F (907°C)	
Specific gravity at 68°F (20°C)	2.70	7.14	
Autoignition point	1400°F (760°C) (powder)	860°F (460°C) (powder)	

9.3-A METALLIC MAGNESIUM

As previously noted in Table 4.1, magnesium occurs to the extent of 1.9% by mass on Earth's surface, where it typically is found in such ores as *magnesite*, *dolomite*, *soapstone*, and *brucite*. Magnesium is also found extensively in underground brines, mainly as magnesium chloride. It is also present in seawater as magnesium chloride and magnesium sulfate.

Magnesium metal is produced mainly for commercial use by the electrolysis of a molten mixture of anhydrous magnesium chloride and potassium chloride.

$$MgCl_2(l) \longrightarrow Mg(l) + Cl_2(g)$$
Magnesium chloride   Magnesium   Chlorine

The potassium chloride increases the conductivity of the salt mixture and reduces its melting point. At the temperature of the electrolytic cell, molten magnesium floats on the salt mixture and is periodically removed through a trough and poured into molds.

Magnesium is an exceptionally lightweight metal. It therefore is often employed in the construction of aircraft, racing cars, transportable machinery, engine parts, automobile frames and bumpers, wheel rims, and other items for which the mass of the object is pertinent. Because of its popularity, magnesium is commercially available in a variety of sizes ranging from a dust or powder to massive ingots.

As illustrated by the following examples, magnesium is a very reactive metal:

- Although it reacts slowly with cold water, magnesium reacts rapidly with warm and hot water, producing hydrogen.

$$Mg(s) + 2H_2O(l) \longrightarrow Mg(OH)_2(s) + H_2(g)$$
Magnesium   Water   Magnesium hydroxide   Hydrogen

- Metallic magnesium is also a strong reducing agent. This property is put to use in the metal manufacturing industry, where molten metallic magnesium is used to reduce the metals in certain ores such as those containing titanium and zirconium compounds. Magnesium powder is also used as a reducing agent in many fireworks, in which its reactions contribute to the production of brilliant displays of light.

Magnesium metal turnings

The most well-known chemical property of magnesium metal is its combustibility. As elemental magnesium burns, approximately 75% combines with atmospheric oxygen to form magnesium oxide.

$$2Mg(s) \ + \ O_2(g) \ \longrightarrow \ 2MgO(s)$$

Magnesium   Oxygen   Magnesium oxide

The remaining 25% combines with atmospheric nitrogen to form magnesium nitride.

$$3Mg(s) \ + \ N_2(g) \ \longrightarrow \ Mg_3N_2(s)$$

Magnesium   Nitrogen   Magnesium nitride

Magnesium ribbon once was used in one-time-use photo flashbulbs to illuminate scenes with a brilliant flash of light.

The burning of bulk pieces of magnesium is hazardous. When raised to a temperature of 1200°F (649°C), massive ingots, castings, and other bulk forms of metallic magnesium melt and burn vigorously with the production of brilliant, blinding white flames. Burning as it flows, the liquid metal drops from its initial location to lower levels, where it ignites combustible materials encountered in its pathway.

Individuals exposed to magnesium fumes are at risk of contracting **metal fume fever**. This disease is characterized by a rise in body temperature, cough, sore throat, chest tightness, headache, fever, metallic taste, nausea, vomiting, and blurred vision.

metal fume fever ■ The occupational disease associated with inhaling metal fumes and dusts, especially of magnesium and zinc

9.3-B METALLIC TITANIUM

As noted in Table 4.1, titanium occurs on Earth's surface to the extent of 0.58% by mass. The element occurs primarily as titanium(IV) oxide. One such ore, *rutile*, is abundant in beach sands in Australia, South Africa, and Sri Lanka.

The titanium manufacturing process consists of the following two steps:

■ First, rutile or a similar ore is reacted with chlorine and carbon at approximately 1112°F (600°C) to produce titanium(IV) chloride.

$$TiO_2(s) \ + \ 2C(s) \ + \ 2Cl_2(g) \ \longrightarrow \ TiCl_4(g) \ + \ 2CO(g)$$

Titanium(IV) oxide   Carbon   Chlorine   Titanium(IV) chloride   Carbon monoxide

■ Then, titanium(IV) chloride is reacted with molten magnesium within a steel vessel under an atmosphere of an inert gas like helium or argon at approximately 1472°F (800°C).

$$TiCl_4(g) \ + \ 2Mg(l) \ \longrightarrow \ Ti(s) \ + \ 2MgCl_2(s)$$

Titanium(IV) chloride   Magnesium   Titanium   Magnesium chloride

The magnesium chloride formed as a by-product is leached from the reaction mixture, leaving the basic form of the metal known as *titanium sponge*, so called because its physical appearance resembles the shape of a sponge.

These production steps are costly, which currently hinders the widespread use of titanium.

Although it is 45% lighter in mass than steel, metallic titanium is just as strong as steel. Because titanium possesses this combination of lightness and strength, it often is alloyed with aluminum and vanadium, which then is used to manufacture aircraft and automotive parts, jet engines, and missiles. In modern-day commercial and military jet aircraft, titanium is replacing aluminum in blades, discs, rings, and engine cases, as well as bulkheads, tail sections, landing gear, wing supports, and fasteners.

In the automotive industry, titanium metal now is used in several consumer car and motorcycle applications. It is used primarily for exhaust systems, suspension springs, engine valves, connecting rods, and turbocharger compressor wheels.

Titanium metal powder

Metallic titanium is also found in everyday items such as jewelry, skis, and golf equipment. Titanium prostheses often are selected by orthopedic surgeons for hip and knee replacements.

Titanium metal has a great affinity for oxygen. Once exposed to air, a thin layer of titanium(IV) oxide quickly deposits on the surface of the metal and protects the underlying metal from further chemical attack. Then, the metal is highly resistant to corrosion by most acids, chlorine, oxidizing agents, and seawater.

The resistance of titanium to chemical attack has been put to good use in various ways, including the following:

- Because it is lightweight and resistant to corrosion by seawater, metallic titanium is used in the structure of underwater machinery. Most likely, its corrosion resistance was an influencing factor that prompted Russia in 2007 to mount a flag made from titanium on the ocean floor at the North Pole from a deep-submergence vehicle.
- Because metallic titanium is resistant to corrosion by seawater, the U.S. Office of Naval Research ordered its first ship hull constructed entirely from it in 2012.
- Because it is resistant to corrosion by most acids and chlorine, metallic titanium is also used in the construction of vessels in which these raw materials will be stored, transported, or reacted. One notable exception to this general observation, however, is hydrofluoric acid, which reacts with metallic titanium.

Although the bulk forms of titanium are not considered hazardous, Table 9.4 shows that titanium has the lowest auto-ignition point of the combustible metals. Finely divided titanium poses a dangerous risk of fire and explosion. It is generated when fabrication operations are conducted on titanium and its pieces are cut, formed, and welded.

When metallic titanium burns in air, a mixture of titanium(IV) oxide and titanium(III) nitride is produced.

$$\underset{\text{Titanium}}{Ti(s)} \quad + \quad \underset{\text{Oxygen}}{O_2(g)} \quad \longrightarrow \quad \underset{\text{Titanium(IV) oxide}}{TiO_2(s)}$$

$$\underset{\text{Titanium}}{2Ti(s)} \quad + \quad \underset{\text{Nitrogen}}{N_2(g)} \quad \longrightarrow \quad \underset{\text{Titanium(III) nitride}}{2TiN(s)}$$

The reactions may be initiated by the combustion of the hydrogen produced when the finely divided metal reacts with atmospheric moisture.

$$\underset{\text{Titanium}}{Ti(s)} \quad + \quad \underset{\text{Water}}{2H_2O(g)} \quad \longrightarrow \quad \underset{\text{Titanium(IV) oxide}}{TiO_2(s)} \quad + \quad \underset{\text{Hydrogen}}{2H_2(g)}$$

9.3-C METALLIC ZIRCONIUM

Metallic zirconium is produced primarily from *zirconia*, a naturally occurring ore containing zirconium(IV) oxide (ZrO_2), by the method just described for production of titanium. Today, the metal is used almost exclusively in the nuclear and steel industries. In the nuclear industry, tubes made of a zirconium alloy are used to hold the uranium(IV) oxide pellets needed as fuel for use in reactor cores (Section 16.9-C), and in the steel industry, zirconium is used to remove oxygen from molten steel. Zirconium dust formerly was used as the active component of specialized camera photoflash bulbs, and it has also been used militarily as an incendiary agent.

Zirconium dust constitutes a risk of fire and explosion. When it is transferred from one container to another, the dust particles absorb the heat generated by friction as the particles move against one another. When these particles are hot, as when they are first produced, the temperature of the dust can exceed its autoignition point, whereupon it spontaneously ignites.

Zirconium metal powder/dust

The ease of ignition of zirconium dust is associated with its former use as an incendiary agent in a shotgun round known as *Dragon's Breath*. The round, when inserted into the magazine of a shotgun, burst into flame when the gun was fired and shot from the gun's barrel like a flamethrower. Zirconium dust, however, is extremely expensive. Despite its fearsomeness in application, the routine military use of Dragon's Breath is cost-prohibitive.

When zirconium dust burns in air, the resulting fire provides an exceedingly brilliant, white flame. The combustion results in the production of a mixture of zirconium oxide and zirconium nitride.

$$Zr(s) \ + \ O_2(g) \ \longrightarrow \ ZrO_2(s)$$
Zirconium   Oxygen   Zirconium oxide

$$2Zr(s) \ + \ N_2(g) \ \longrightarrow \ 2ZrN(s)$$
Zirconium   Nitrogen   Zirconium nitride

The reactions may be initiated by the combustion of hydrogen, produced when the dust reacts with hot atmospheric moisture.

$$Zr(s) \ + \ 2H_2O(g) \ \longrightarrow \ ZrO_2(s) \ + \ 2H_2(g)$$
Zirconium   Water   Zirconium oxide   Hydrogen

Water does not react with the zirconium alloy used in nuclear reactors; in fact, under normal operating conditions, it cools their fuel assemblies. When zirconium or its alloy is red-hot, however, the metal decomposes water to form hydrogen. When hydrogen is produced at malfunctioning nuclear reactor sites, it must be vented to the outside environment to prevent the confinement vessel from exploding.

Aluminum metal powder/dust

9.3-D METALLIC ALUMINUM

Table 4.1 lists aluminum as the most abundant metal on Earth's surface, 7.5% by mass. As the element, aluminum is too reactive to be found in an uncombined form in nature. Instead, it is found in minerals like cryolite and in such materials as clay and feldspar, in which it is combined with silicon and oxygen.

Aluminum metal is produced by the electrolysis of aluminum oxide dissolved in fused cryolite, which serves as the electrolyte.

$$2Al_2O_3(s) \ \longrightarrow \ 4Al(l) \ + \ 3O_2(g)$$
Aluminum oxide   Aluminum   Oxygen

The aluminum metal is denser than molten cryolite; therefore, it collects at the bottom of the electrolytic cell, from which it is tapped and cast into ingots.

Aluminum is an example of a **malleable** metal; that is, it can be rolled or hammered into a relatively thin sheet or foil. Aluminum foil is a popular kitchen item. Firefighters encounter aluminum foil bonded to fabric in aluminized protective suits that are worn when they must approach fires that release exceptionally high levels of radiant heat.

Because it is lightweight and durable, aluminum sheeting is used to produce a wide variety of commercial products including the common soda can. In building construction, aluminum sheeting is used as siding, eaves, screens, and window and door frames. During building fires, these aluminum components can melt and collapse, because the temperature attained often exceeds the melting point of aluminum, 1220°F (660°C).

The principal material employed as the metal skin of most standard aircraft is an aluminum or aluminum alloy sheeting. The metal cannot be used as the outer skin of supersonic aircraft, however, because it becomes too hot and softens from the friction generated by the fast movement through the air.

malleable ■ The property or capability of being rolled or hammered into shapes

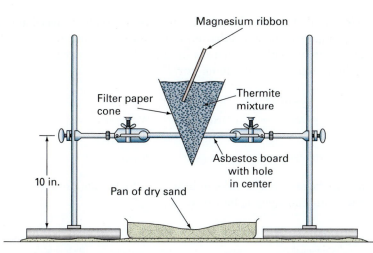

FIGURE 9.3 In this laboratory demonstration, a mixture of powdered aluminum and iron(III) oxide is inserted into a cone over a pan of dry sand (which protects the tabletop from possible damage). A magnesium ribbon is inserted into the thermite mixture and ignited. The heat of combustion initiates the reaction between aluminum and iron(III) oxide. Molten iron spits from the reaction mixture and drips into the sand.

Aluminum is also a **ductile** substance; that is, it can be drawn into wires. Although aluminum wire is twice as effective as copper wire for conducting electricity, the use of aluminum electrical wiring is undesirable. This is because the heavy deposit of aluminum oxide produced on the surface of aluminum wiring restricts the flow of electrical current and causes the metal to become overheated. This situation constitutes a fire hazard.

ductile ■ The property of metals associated with their capability of being stretched into wires

The deposition of aluminum oxide on the surface of aluminum wiring is associated with the extraordinary affinity that aluminum and oxygen have for each other. Aluminum exposed to air is covered with a thin, tenacious coating of aluminum oxide that gives the metal a dull, white luster. Although this oxide coating protects the underlying metal from further oxidation, the coating does not protect the aluminum from other forms of chemical attack. Seawater, for instance, corrodes metallic aluminum.

This chemical affinity of metallic aluminum for oxygen is evident from the chemical reaction noted in Figure 9.3 involving powdered aluminum and iron(III) oxide. The mixture of 27% powdered aluminum and 73% iron(III) oxide is commonly called **thermite**. When the mixture is activated by a magnesium fuse, a reaction producing molten iron and aluminum oxide occurs.

thermite ■ The mixture of 27% powdered aluminum and 73% iron(III) oxide

$$2Al(s) \ + \ Fe_2O_3(s) \ \longrightarrow \ 2Fe(l) \ + \ Al_2O_3(s)$$

Aluminum   Ferric oxide   Iron   Aluminum oxide

This phenomenon is called the **thermite reaction**. It releases such considerable heat that temperatures of approximately 3990°F (2199°C) result. Because this temperature is above the melting point of iron [2800°F (1538°C)], it is produced by this chemical reaction as a molten, white-hot liquid. The thermite reaction cannot be stopped with water.

thermite reaction ■ The chemical reaction used in some welding operations and incendiary weapons during which elemental iron is produced by the reduction of iron(III) oxide with elemental aluminum

In the days of the old West, the thermite reaction was used to weld rails together during the construction of railroads. During World War II, thermite was used extensively as the incendiary agent in bombs, especially against the British during the London Blitz. In contemporary warfare, however, the use of thermite in incendiary weapons is essentially banned by Protocol III of the Convention on Certain Conventional Weapons (Section 7.3). Protocol III limits the use of all incendiary weapons against civilian targets.

Although elemental aluminum is stable in the form of foil and sheets, aluminum dust and powder are pyrophoric materials that pose the risk of fire and explosion. The aluminum burns violently in air with an intensely bright, white and orange flame producing a mixture of aluminum oxide and aluminum nitride.

$$4Al(s) \ + \ 3O_2(g) \ \longrightarrow \ 2Al_2O_3(s)$$

Aluminum Oxygen Aluminum oxide

$$2Al(s) \ + \ N_2(g) \ \longrightarrow \ 2AlN(s)$$

Aluminum Nitrogen Aluminum nitride

These reactions may be initiated by the combustion of hydrogen, produced when the dust and powder react with atmospheric moisture.

$$2Al(s) \ + \ 3H_2O(l) \ \longrightarrow \ Al_2O_3(s) \ + \ 3H_2(g)$$

Aluminum Water Aluminum oxide Hydrogen

Powdered aluminum burns spontaneously on contact with liquid oxygen. Aluminum oxide is the sole product of combustion.

The reactivity of aluminum powder is put to use in the formulations of many fireworks, in which the metal, when activated, burns to produce a brilliant display of orange light. It is also incorporated into certain paints and varnishes for its decorative and heat-reflective features; but consideration must be given to their use, because these coatings may behave as flammable solids once the paint solvent has evaporated. Aluminum powder is also a component of solid rocket fuels, in which it is mixed with ammonium nitrate and ammonium perchlorate. The mixture of powdered aluminum and ammonium nitrate is an explosive called *ammonal*.

The catastrophe of the German dirigible *Hindenburg* may have been linked with the combustion of aluminum powder. The exterior surface of the dirigible consisted of a cloth cover impregnated with a doping mixture of aluminum powder and ferric oxide. The presence of aluminum powder provided a surface having high reflectivity. The cover was intended to serve an important purpose: The aluminum particles reflected heat off the vessel and prevented the hydrogen from expanding. The prevailing theory is that the aluminum powder first caught fire at an isolated location, perhaps triggered by static electricity or lightning. Once initiated, the fire then rapidly spread across the entire covering, ultimately igniting the reserves of hydrogen. The resulting inferno consumed the vessel.

In circumstances where the temperature is substantially elevated compared with the norm, even bulk aluminum acts as a fast-burning fuel. The skin of shuttle aircraft, for example, must be armored with heat shielding to protect the shuttle when it reenters Earth's atmosphere from outer space, experiencing temperatures in excess of 3000°F (1650°C). If this shielding is pierced in any way, the underlying aluminum becomes superheated. Aluminum melts at 1220°F (660°C) and vaporizes at 4221°F (2327°C). At these temperatures, aluminum fires occur when oxygen is available to support the combustion.

In 2003, the space shuttle *Columbia* disintegrated on reentry into Earth's atmosphere, killing the seven astronauts onboard. The shuttle was covered with more than 20,000 interlocking ceramic tiles designed to protect the aluminum alloy shell from the heat of reentry. Experts who examined debris from the accident wreckage observed droplets of aluminum and stainless steel. This observation suggests that the cause of the accident was linked with the loss of the thermal protective system on the left wing, especially along its leading edge. Without its protective covering, the underlying aluminum alloy most likely burned, ultimately destroying the entire shuttle.

9.3-E METALLIC ZINC

Zinc is produced primarily by means of the following two-step thermal process:

Zinc metal powder/dust

■ First, the zinc sulfide ore *sphalerite*, or *zinc blende*, is roasted in air to produce zinc oxide.

$$2ZnS(s) \;+\; 3O_2(g) \;\longrightarrow\; 2ZnO(s) \;+\; 2SO_2(g)$$
Zinc sulfide   Oxygen   Zinc oxide   Sulfur dioxide

■ Then, the oxide is reduced with carbon monoxide.

$$ZnO(s) \;+\; CO(g) \;\longrightarrow\; Zn(g) \;+\; CO_2(g)$$
Zinc oxide   Carbon monoxide   Zinc   Carbon dioxide

The zinc vapor produced by the reaction is then distilled, condensed, and cast into ingots. The zinc deposits on the walls of the distillation apparatus as a gray, finely divided powder known as zinc dust.

The zinc manufacturing process is complicated by the presence of the metal impurities silver, lead, copper, arsenic, antimony, and manganese, all of which occur naturally in sphalerite. These metals are removed by a combination of chemical processes. The manufacturing process is also complicated by the simultaneous production of the pollutant sulfur dioxide (Section 10.12), which must be scrubbed from the off-gas plume generated during the roasting process.

Metallic zinc is used for several purposes. The metal is coated on iron products to protect them from corrosion by the air. This zinc-coated iron is said to be **galvanized**. Zinc is also used as a component of several alloys; for example, zinc and copper are combined in the molten state to produce brass. Metallic zinc is also used in the manufacture of dry-cell batteries and a variety of structural materials. Zinc dust is a component of certain primers and rust-resistant paints.

galvanize ■ The process of coating a metal with a protective layer of elemental zinc

Zinc is hazardous only as its dust. Especially when it is hot, zinc dust is a pyrophoric material that poses a fire and explosion hazard. It ignites spontaneously in air with a green flame, producing zinc oxide as the sole combustion product.

$$2Zn(s) \;+\; O_2(g) \;\longrightarrow\; 2ZnO(s)$$
Zinc   Oxygen   Zinc oxide

The reaction may be initiated by the combustion of hydrogen, produced when the dust reacts with atmospheric moisture.

$$Zn(s) \;+\; H_2O(l) \;\longrightarrow\; ZnO(s) \;+\; H_2(g)$$
Zinc   Water   Zinc oxide   Hydrogen

9.3-F TRANSPORTING COMBUSTIBLE METALS

When shippers offer a combustible metal for transportation, DOT requires them to identify the appropriate material on the accompanying shipping paper. Some examples for several representative combustible metals are listed in Table 9.5. DOT also requires shippers and carriers to comply with all labeling, marking, and placarding requirements.

When molten aluminum is transported in bulk packaging by highway or rail, DOT requires carriers at 49 C.F.R. §172.325 to mark the packaging with the expression *MOLTEN ALUMINUM* and the identification number 9260 on orange panels, white square-on-point diamonds, or HOT markings. The following examples illustrate the nature of these markings:

TABLE 9.5	Shipping Descriptions of Some Representative Combustible Metals
COMBUSTIBLE METALS	**SHIPPING DESCRIPTION**
Aluminum, molten	NA9260, Aluminum, molten, 9, PG I
Aluminum powder	UN1309, Aluminum powder, coated, 4.1, PG II *or* UN1396, Aluminum powder, uncoated, 4.3, PG II (Dangerous When Wet)
Magnesium (with more than 50% magnesium in pellets, turnings, or ribbons)	UN1869, Magnesium, 4.1, PG III
Magnesium alloys (with more than 50% magnesium in pellets, turnings, or ribbons)	UN1869, Magnesium alloys, 4.1, PG III
Magnesium granules (particle size not less than 149 microns)	UN2950, Magnesium granules, coated, 4.3, PG III (Dangerous When Wet)
Magnesium powder	UN1418, Magnesium powder, 4.3, (4.2), PG I (Dangerous When Wet) *or* UN1418, Magnesium powder, 4.3, (4.2), PG II (Dangerous When Wet) *or* UN1418, Magnesium powder, 4.3, (4.2), PG III (Dangerous When Wet)
Titanium powder	UN2546, Titanium powder, dry, 4.2, PG I
Titanium (powder), wetted with not less than 25% water (a visible excess of water must be present) **(a)** mechanically produced, particle size less than 53 microns; **(b)** chemically produced, particle size less than 840 microns	UN1352, Titanium powder, wetted, 4.1, PG II
Titanium sponge granules	UN2878, Titanium sponge granules, 4.1, PG III
Titanium sponge powders	UN2878, Titanium sponge powders, 4.1, PG III
Zinc dust	UN1436, Zinc dust, 4.3, (4.2), PG I (Dangerous When Wet)
Zinc powder	UN1436, Zinc powder 4.3, (4.2), PG I (Dangerous When Wet)
Zirconium, dry (finished sheets, strip, or coil wire)	UN2008, Zirconium, dry, 4.1 PG III
Zirconium powder, wetted with not more than 25% water [(a visible excess of water must be present) **(a)** mechanically produced, particle size less than 53 microns; **(b)** chemically produced, particle size less than 840 microns]	UN1358, Zirconium powder, wetted, 4.1, PG II

9.4 ALUMINUM ALKYL COMPOUNDS AND THEIR DERIVATIVES

Organometallic substances are compounds whose molecules have one or more metal atoms covalently bonded directly to a nonmetal atom. Examples of organometallic substances include the **aluminum alkyls**, whose molecules have an aluminum atom covalently bonded to three carbon atoms. An example of an aluminum alkyl compound is triethylaluminum, whose chemical formula is $Al(CH_2CH_3)_3$, or $Al(C_2H_5)_3$.

aluminum alkyl ■ A compound whose molecules are composed of an aluminum atom covalently bonded to three carbon atoms, each of which is a component of an alkyl group

$$CH_2CH_3$$
$$|$$
$$CH_3CH_2-Al-CH_2CH_3$$

Triethylaluminum
(TEA)

In this instance, the alkyl group is named *ethyl*, which has the formula $-CH_2CH_3$. In the chemical industry, triethylaluminum is often designated as TEA. Its properties are representative of aluminum alkyl compounds.

Two special groups of aluminum alkyl compounds are the aluminum alkyl halides and aluminum alkyl hydrides. These compounds are the halide and hydride derivatives of aluminum alkyl compounds, respectively, in which one or two halide or hydrogen atoms substitute for an alkyl group. Examples of these derivatives are diethylaluminum chloride and diisobutylaluminum hydride, whose formulas are $(C_2H_5)_2AlCl$ and $[(CH_3)_2CHCH_2]_2AlH$, respectively.

$$\underset{\underset{\text{(DEAC)}}{\text{Diethylaluminum chloride}}}{CH_3CH_2 - \overset{\overset{\text{Cl}}{|}}{Al} - CH_2CH_3} \qquad \underset{\underset{\text{(DIBAH)}}{\text{Diisobutylaluminum hydride}}}{(CH_3)_2CHCH_2 - \overset{\overset{\text{H}}{|}}{Al} - CH_2CH(CH_3)_2}$$

The alkyl group having the formula $(CH_3)_2CHCH_2-$ is named *isobutyl*. In the chemical industry, these compounds are sometimes designated as *DEAC* and *DIBAH*, respectively. In this section, we consider them as representative of the halide and hydride derivatives of all aluminum alkyl compounds.

Table 9.6 provides some physical properties of triethylaluminum, diethylaluminum chloride, and diisobutylaluminum hydride. Chemical manufacturers display the flame pictogram on labels affixed to containers holding the aluminum alkyls and their halide and hydride derivatives.

9.4-A COMMERCIAL USES OF THE ALUMINUM ALKYL COMPOUNDS AND THEIR DERIVATIVES

The aluminum alkyls are used by the chemical industry primarily as polymerization catalysts, one of which is a mixture of titanium(IV) chloride and an aluminum alkyl. It is called a **Ziegler–Natta catalyst**, after Karl Ziegler and Giulio Natta, the chemists who first discovered its catalytic capability. Aluminum alkyl halides and aluminum alkyl hydrides are also primarily used as catalysts in the chemical industry.

Aluminum alkyl compounds have also been used by the military, albeit rarely, as incendiary agents. For example, triethylaluminum has been used as the active component in flamethrowers. Trimethylaluminum has also been used to produce luminous trails in the upper atmosphere for tracking the location of rockets.

Ziegler–Natta catalyst
■ Any of a group of compounds produced from titanium tetrachloride and an aluminum alkyl compound that is used mainly as a catalyst

TABLE 9.6	Physical Properties of an Aluminum Alkyl Compound and Two Metal Alkyl Derivatives		
	TRIETHYLALUMINUM	DIETHYLALUMINUM CHLORIDE	DIISOBUTYLALUMINUM HYDRIDE
Melting point	−62°F (−52°C)	−121°F (−85°C)	−112°F (−80°C)
Boiling point	367°F (186°C)	417°F (214°C)	237°F (114°C)[a]
Specific gravity	0.837[b]	0.961[c]	0.798[c]
Vapor pressure	0.0147 mmHg[b]	0.17 mmHg[c]	
Flashpoint	−63°F (−53°C)	−9.4°F (−23°C)	
Autoignition temperature	Spontaneously flammable in air	Spontaneously flammable in air	Spontaneously flammable in air

[a]At 3 mmHg (0.3 kPa).
[b]At 68°F (20°C).
[c]At 77°F (25°C).

9.4-B PROPERTIES OF THE ALUMINUM ALKYL COMPOUNDS AND THEIR DERIVATIVES

The aluminum alkyl compounds and their derivatives are spontaneously combustible, pyrophoric, violently water-reactive, and highly toxic liquids. They are commercially available as individual compounds and solutions in which they are dissolved in organic solvents. When triethylaluminum, diethylaluminum chloride, and diisobutylaluminum hydride spontaneously ignite, their combustion reactions are represented as follows:

$$2(C_2H_5)_3Al(l) \ + \ 21O_2(g) \ \longrightarrow \ Al_2O_3(s) \ + \ 12CO_2(g) \ + \ 15H_2O(g)$$

Triethylaluminum (TEA)   Oxygen   Aluminum oxide   Carbon dioxide   Water

$$2(C_2H_5)_2AlCl(l) \ + \ 14O_2(g) \ \longrightarrow \ Al_2O_3(s) \ + \ 8CO_2(g) \ + \ 9H_2O(g) \ + \ 2HCl(g)$$

Diethylaluminum chloride (DEAC)   Oxygen   Aluminum oxide   Carbon dioxide   Water   Hydrogen chloride

$$2[(CH_3)_2CHCH_2]_2AlH(s) \ + \ 27O_2(g) \ \longrightarrow \ Al_2O_3(s) \ + \ 16CO_2(g) \ + \ 19H_2O(g)$$

Diisobutylaluminum hydride (DIBAH)   Oxygen   Aluminum oxide   Carbon dioxide   Water

When triethylaluminum and diethylaluminum chloride react with water, the flammable gas ethane (Section 12.2) is produced as a hydrolysis product.

$$Al(C_2H_5)_3(l) \ + \ 3H_2O(l) \ \longrightarrow \ Al(OH)_3(s) \ + \ 3C_2H_6(g)$$

Triethylaluminum (TEA)   Water   Aluminum hydroxide   Ethane

$$(C_2H_5)_2AlCl(l) \ + \ 3H_2O(l) \ \longrightarrow \ Al(OH)_3(s) \ + \ 2C_2H_6(g) \ + \ HCl(g)$$

Diethylaluminum chloride (DEAC)   Water   Aluminum hydroxide   Ethane   Hydrogen chloride

Diisobutylaluminum hydride, however, is a reducing agent. When it reacts with water, the flammable gases isobutene and hydrogen are produced.

$$[(CH_3)_2CHCH_2]_2AlH(s) \ + \ 3H_2O(l) \ \longrightarrow \ Al(OH)_3(s) \ + \ 2(CH_3)_2CH{=}CH_2(g) \ + \ 2H_2(g)$$

Diisobutylaluminum hydride (DIBAH)   Oxygen   Aluminum hydroxide   Isobutene   Hydrogen

When water is applied to these reactive substances, the gaseous hydrolysis products immediately burst into flame as they are generated. The considerable heat evolved to the environment often triggers secondary fires. Bulk quantities of aluminum alkyl compounds burn so vigorously and persistently that they pose an especially dangerous risk of fire and explosion. The heat of combustion that evolves necessitates that firefighters wear special protective gear like the *silvers* shown in Figure 9.4 when combating these fires.

To prevent their accidental ignition, the aluminum alkyl compounds and their halide and hydride derivatives often are stored within electrically grounded containers under an atmosphere of nitrogen in a cool, well-ventilated area.

9.4-C TRANSPORTING ALUMINUM ALKYL COMPOUNDS AND THEIR DERIVATIVES

When shippers offer an aluminum alkyl compound or a halide or hydride derivative for transportation, DOT requires them to identify it with the proper shipping name, "organometallic substance," on the accompanying shipper paper. The Hazardous Materials Table at 49 C.F.R. §172.101 lists several shipping names for organometallic substances. Because triethylaluminum is both water- and air-reactive, its most appropriate shipping description is the following:

UN3394, Organometallic substance, liquid, pyrophoric, water-reactive (triethylaluminum), 4.2, (4.3), PG I (Dangerous When Wet).

FIGURE 9.5 When a carrier transports an aluminum alkyl compound or its halide or hydride derivative in an amount exceeding 1001 pounds (454 kg), DOT requires SPONTANEOUSLY COMBUSTIBLE and DANGEROUS WHEN WET placards to be displayed on the transport vehicle. AKZX is the reporting mark of AKZO Nobel Chemicals, Inc., Chicago, Illinois, a distributor of triethylaluminum and other class 4 compounds.

When transporting aluminum alkyl compounds or their halide or hydride derivatives, shippers and carriers must also comply with all applicable labeling, marking, and placarding requirements. Figure 9.5 illustrates that DOT requires carriers to display DANGEROUS WHEN WET placards on the bulk packaging used for shipment regardless of the amount transported.

SOLVED EXERCISE 9.2

When 387 gallons (829 L) of liquid diisobutylaluminum hydride is transported in a 400-gallon (857-L) portable tank by highway:

(a) What shipping description does DOT require the shipper to enter on the accompanying shipping paper?
(b) How does DOT require the carrier to placard and mark the tank?

Solution:

(a) There are two regulations in Table 6.2 that are pertinent to preparing the shipping description. First, when a hazardous material is described with a generic description in the Hazardous Materials Table, shippers must include the name of the substance in parentheses in the shipping description. Second, when a hazardous material, by chemical interaction with water, is liable to become spontaneously flammable or give off flammable gases in dangerous quantities, the words "Dangerous When Wet" must be included with the shipping description. Consequently, the shipping description of diisobutyla-luminum hydride is entered on a shipping paper as follows:

UNITS	HM	SHIPPING DESCRIPTION (IDENTIFICATION NUMBER, PROPER SHIPPING NAME, PRIMARY HAZARD CLASS OR DIVISION, SUBSIDIARY HAZARD CLASS OR DIVISION, AND PACKING GROUP)	VOLUME (gal)
1 portable tank	X	UN3394, Organometallic substance, liquid, pyrophoric, water-reactive (diisobutylaluminum hydride), 4.2, (4.3), PG I (Dangerous When Wet)	387

(b) Since the amount transported exceeds 1001 pounds, DOT requires carriers to display side by side a SPONTANEOUSLY COMBUSTIBLE *and* a DANGEROUS WHEN WET placard on each side and each end of the cargo tank. Because the tank has a capacity of less than 1000 gallons (3785 L), DOT requires them to mark the tank with the identification number 3394 on two opposing sides on orange panels, across the center area of the SPONTANEOUSLY COMBUSTIBLE placards, or on white square-on-point diamonds.

9.5 IONIC HYDRIDES

ionic hydride ■ A compound composed of a metallic ion and a simple or complex hydride ion

Approximately ten **ionic hydrides** are encountered commercially. They are compounds consisting of metallic ions bonded to simple or complex hydride ions. Some metallic hydrides are not ionic hydrides. For example, although tin(IV) hydride is a metallic hydride, it is composed of molecules. Each molecule consists of a tin atom *covalently* bonded to four hydrogen atoms. Its chemical formula is SnH_4.

Ionic hydrides are used as powerful reducing agents by the chemical industry. They can be classified according to their general chemical composition as simple ionic hydrides, ionic borohydrides, and ionic aluminum hydrides.

9.5-A SIMPLE IONIC HYDRIDES

Sodium hydride

Simple ionic hydrides are compounds consisting of metallic ions bonded to hydride ions (H^-). They are lithium hydride, sodium hydride, calcium hydride, magnesium hydride, and aluminum hydride, whose chemical formulas are LiH, NaH, CaH_2, MgH_2, and AlH_3, respectively. They are produced by reactions between the corresponding metal and hydrogen. For example, sodium hydride is a simple ionic hydride produced by the union of sodium metal and hydrogen.

$$2Na(s) \ + \ H_2(g) \ \longrightarrow \ 2NaH(s)$$
Sodium   Hydrogen   Sodium hydride

9.5-B IONIC BOROHYDRIDES

Ionic borohydrides are ionic hydrides in which metallic ions are bonded to borohydride ions (BH_4^-). The commercially important ionic borohydrides are lithium borohydride, sodium borohydride, and aluminum borohydride, whose chemical formulas are $LiBH_4$, $NaBH_4$, and $Al(BH_4)_3$, respectively. The ionic borohydrides are produced by relatively complex chemical reactions.

9.5-C IONIC ALUMINUM HYDRIDES

Ionic aluminum hydrides are ionic hydrides in which metallic ions are bonded to aluminum hydride ions (AlH_4^-). Two commercially important ionic aluminum hydrides are lithium aluminum hydride and sodium aluminum hydride, whose chemical formulas are $LiAlH_4$ and $NaAlH_4$, respectively. They are produced by reacting the relevant ionic hydride and anhydrous aluminum chloride (Section 9.8-A).

9.5-D WATER REACTIVITY OF THE IONIC HYDRIDES

Although the ionic hydrides are relatively stable compounds, they possess several common hazardous features. Of special interest here is the fact that they react with water to produce flammable hydrogen.

To prevent their contact with atmospheric moisture, all ionic hydrides are stored in tightly sealed containers. When encountered commercially, they are often covered with petroleum oil. The presence of the oil lends an element of safety when handling and storing them. However, these compounds are also encountered as ethereal solutions; that is, they are dissolved in diethyl ether, a highly flammable liquid. The combination of diethyl ether and an ionic hydride poses the risk of fire and explosion.

The following equations illustrate the water reactivity of several representative ionic hydrides:

$$\underset{\text{Lithium hydride}}{\text{LiH}(s)} + \underset{\text{Water}}{\text{H}_2\text{O}(l)} \longrightarrow \underset{\text{Lithium hydroxide}}{\text{LiOH}(aq)} + \underset{\text{Hydrogen}}{\text{H}_2(g)}$$

$$\underset{\text{Sodium borohydride}}{3\text{NaBH}_4(s)} + \underset{\text{Water}}{6\text{H}_2\text{O}(l)} \longrightarrow \underset{\text{Sodium borate}}{3\text{NaBO}_2(aq)} + \underset{\text{Hydrogen}}{12\text{H}_2(g)}$$

$$\underset{\text{Lithium aluminum hydride}}{\text{LiAlH}_4(s)} + \underset{\text{Water}}{4\text{H}_2\text{O}(l)} \longrightarrow \underset{\text{Aluminum hydroxide}}{\text{Al(OH)}_3(s)} + \underset{\text{Lithium hydroxide}}{\text{LiOH}(aq)} + \underset{\text{Hydrogen}}{4\text{H}_2(g)}$$

$$\underset{\text{Aluminum borohydride}}{\text{Al(BH}_4)_3(s)} + \underset{\text{Water}}{12\text{H}_2\text{O}(l)} \longrightarrow \underset{\text{Aluminum hydroxide}}{\text{Al(OH)}_3(s)} + \underset{\text{Boric acid}}{3\text{H}_3\text{BO}_3(aq)} + \underset{\text{Hydrogen}}{12\text{H}_2(g)}$$

As these ionic hydrides react with water, the evolved hydrogen absorbs the heat of reaction and spontaneously bursts into flame.

Because the ionic hydrides are water-reactive substances, precautions should be exercised to avoid exposing them to humid air or other potential sources of water. Experts recommend that firefighters use water as a fire extinguisher only when they encounter small spills of these substances.

9.5-E TRANSPORTING IONIC HYDRIDES

When shippers offer an ionic hydride for transportation, DOT requires them to provide the appropriate shipping description on the accompanying shipping paper. Table 9.7 provides

Sodium borohydride

Lithium aluminum hydride

TABLE 9.7	Shipping Descriptions of Some Representative Ionic Hydrides
IONIC HYDRIDE	**SHIPPING DESCRIPTION**
Aluminum borohydride	UN2870, Aluminum borohydride, 4.2, (4.3), PG I (Dangerous When Wet) *or* UN2870, Aluminum borohydride in devices, 4.2, (4.3), PG I (Dangerous When Wet)
Calcium hydride	UN1404, Calcium hydride, 4.3, PG I (Dangerous When Wet)
Lithium aluminum hydride	UN1410, Lithium aluminum hydride, 4.3, PG I (Dangerous When Wet)
Lithium aluminum hydride dissolved in ether	UN1411, Lithium aluminum hydride, ethereal, 4.3, (3), PG I (Dangerous When Wet)
Lithium borohydride	UN1413, Lithium borohydride, 4.3, PG I (Dangerous When Wet)
Lithium hydride	UN1414, Lithium hydride, 4.3, PG I (Dangerous When Wet)
Sodium aluminum hydride	UN2835, Sodium aluminum hydride, 4.3, PG II (Dangerous When Wet)
Sodium borohydride	UN1426, Sodium borohydride, 4.3, PG I (Dangerous When Wet)
Sodium hydride	UN1427, Sodium hydride, 4.3, PG I (Dangerous When Wet)

Chapter 9 Chemistry of Some Water- and Air-Reactive Substances   **329**

some representative examples. When the shipping description is not listed at 49 C.F.R. §172.101, DOT requires them to identify the commodity generically and include the name of the specific compound parenthetically. DOT also requires shippers and carriers to comply with all applicable labeling, marking, and placarding requirements.

9.6 METALLIC PHOSPHIDES

metallic phosphide
■ An inorganic compound composed of metallic and phosphide ions

Metallic phosphides are produced by combination reactions in which a given metal unites with elemental phosphorus. Calcium phosphide, for example, is formed by heating calcium and phosphorus.

$$6Ca(s) + P_4(s) \longrightarrow 2Ca_3P_2(s)$$

Calcium   Phosphorus   Calcium phosphide

These compounds once were popular fumigants used on grain and other postharvest crops, but in the United States, their use is not nearly as popular now as it was in the past.

The metallic phosphides function as fumigants by reacting with atmospheric moisture to produce the toxic gas phosphine.

$$Ca_3P_2(s) + 6H_2O(l) \longrightarrow 3Ca(OH)_2(s) + 2PH_3(g)$$

Calcium phosphide   Water   Calcium hydroxide   Phosphine

When calcium phosphide is applied within an enclosure used for the storage of crops, it is the phosphine produced by hydrolysis that actually kills mice and other unwanted pests.

Calcium phosphide

9.6-A TRANSPORTING METALLIC PHOSPHIDES

When shippers offer a metallic phosphide for transportation, DOT requires them to identify the appropriate material on the accompanying shipping paper. Some examples of the shipping descriptions for several representative metallic phosphides are listed in Table 9.8. DOT also requires shippers and carriers to comply with all applicable labeling, marking, and placarding requirements.

SOLVED EXERCISE 9.3

What is the most likely reason that DOT requires shippers to affix DANGEROUS WHEN WET and POISON labels to packages of stannic phosphide?

Solution: DOT assigns two hazard codes, 4.3 and 6.1, to stannic phosphide because the properties of this substance comply with the defining criteria for both a dangerous-when-wet substance and a poisonous material (Section 10.1-C). Stannic phosphide is a solid compound that reacts with water to produce phosphine, a flammable and toxic gas.

$$Sn_3P_4(s) + 12H_2O(l) \longrightarrow 3Sn(OH)_4(s) + 4PH_3(g)$$

Stannic phosphide   Water   Stannic hydroxide   Phosphine

When shippers affix DANGEROUS WHEN WET and POISON labels to packages of stannic phosphide, these hazard warning labels quickly inform emergency responders that the substance is simultaneously water-reactive and toxic.

9.6-B PHOSPHINE

As noted previously, phosphine is generated when metallic phosphides react with water. This substance possesses the following hazardous features:

■ Phosphine is a poisonous gas. Toxicity is its primary hazard. The gas may be detected by its exceptionally offensive odor, which has been described as a mixture of garlic and rotten fish. The odor threshold for phosphine is only 0.15 part per million.

TABLE 9.8	Shipping Descriptions of Some Representative Metallic Phosphides	
METALLIC PHOSPHIDE	**SHIPPING DESCRIPTION**	
Aluminum phosphide	UN1397, Aluminum phosphide, 4.3, (6.1), PG I (Dangerous When Wet) (Poison)	
Aluminum phosphide (pesticides)	UN3048, Aluminum phosphide pesticides, 6.1, PG I (Poison)	
Calcium phosphide	UN1360, Calcium phosphide, 4.3, (6.1), PG I (Dangerous When Wet) (Poison)	
Magnesium phosphide	UN2011, Magnesium phosphide, 4.3, (6.1), PG I (Dangerous When Wet) (Poison)	
Potassium phosphide	UN2012, Potassium phosphide, 4.3, (6.1), PG I (Dangerous When Wet) (Poison)	
Sodium phosphide	UN1432, Sodium phosphide, 4.3, (6.1), PG I (Dangerous When Wet) (Poison)	

When inhaled, it primarily attacks the cardiovascular and respiratory systems, causing pulmonary edema (Section 7.3-B) and massive destruction of the lung tissues. Long-term exposure to lesser concentrations causes the bones to soften. Exposure to a concentration of 50 parts per million is immediately dangerous to an individual's life and health.

■ Phosphine is also a spontaneously flammable gas. Flammability is considered its secondary risk. Its autoignition temperature is only 100°F (37.8°C), a value readily attained in most environments. Phosphine burns in air to produce a dense white cloud of tetraphosphorus decoxide and water vapor.

$$4PH_3(g) \ + \ 8O_2(g) \ \longrightarrow \ P_4O_{10}(s) \ + \ 6H_2O(g)$$

Phosphine   Oxygen   Tetraphosphorus decoxide   Water

We examine the methods recommended for extinguishing fires involving gases that are simultaneously flammable and poisonous in Chapter 10.

9.6-C WORKPLACE REGULATIONS INVOLVING PHOSPHINE

When phosphine is present in the workplace, OSHA requires employers to limit employee exposure to a concentration of 0.3 part per million, averaged over the 8-hour workday.

9.6-D TRANSPORTING PHOSPHINE

When shippers offer phosphine for transportation, DOT requires them to identify the gas on the accompanying shipping paper as follows:

UN2199, Phosphine, 2.3, (2.1) (Poison - Inhalation Hazard, Zone A)

DOT also requires shippers and carriers to comply with all applicable labeling, marking, and placarding requirements.

9.7 METALLIC CARBIDES

Metals bond with carbon to form compounds having either ionic or covalent units. Our concern here is solely with the compounds composed of metallic ions and carbon ions that exist as C_2^{2-} or C^{4-}. They are called **metallic carbides**. There are only two commercially important metallic carbides: aluminum carbide and calcium carbide. Both are water-reactive substances.

metallic carbide ■ An inorganic compound composed of metallic and carbide ions

Phosphine

TABLE 9.9	Shipping Descriptions of Metallic Carbides
METALLIC CARBIDE	**SHIPPING DESCRIPTION**
Aluminum carbide	UN1394, Aluminum carbide, 4.3, PG II (Dangerous When Wet)
Calcium carbide	UN1402, Calcium carbide, 4.3, PG I (Dangerous When Wet)

9.7-A ALUMINUM CARBIDE

Aluminum carbide

Aluminum carbide is prepared by heating aluminum oxide with coke in an electric furnace. It is used as a catalyst by the chemical industry. When aluminum carbide reacts with water, flammable methane is produced as a hydrolysis product.

$$Al_4C_3(s) \ + \ 12H_2O(l) \ \longrightarrow \ 4Al(OH)_3(s) \ + \ 3CH_4(g)$$

Aluminum carbide   Water   Aluminum hydroxide   Methane

Aluminum carbide that has been exposed to humid air poses a flammable and explosive hazard.

9.7-B CALCIUM CARBIDE

Calcium carbide

Calcium carbide is manufactured by heating a mixture of coke and lime at an elevated temperature in an electric furnace.

$$CaO(s) \ + \ 3C(s) \ \longrightarrow \ CaC_2(s) \ + \ CO(g) \quad [t > 3600°F \ (1982°C)]$$

Calcium oxide   Carbon   Calcium carbide   Carbon monoxide

The production process is very energy-intensive.

    Calcium carbide formerly was used as a raw material for the production of industrial-grade acetylene.

$$CaC_2(s) \ + \ 2H_2O(l) \ \longrightarrow \ Ca(OH)_2(s) \ + \ C_2H_2(g)$$

Calcium carbide   Water   Calcium hydroxide   Acetylene

Today, this method of producing and manufacturing acetylene is used to a very limited extent because it produces a huge amount of calcium hydroxide slag. Although some of this slag is incorporated into cement, acetylene manufacturers often choose to avoid dealing with the negative environmental impact altogether.

    Because of its water-reactive nature, calcium carbide that has been exposed to humid air poses a flammable and explosive hazard. For this reason, it must be stored and handled in a dry environment that is free of ignition sources.

9.7-C TRANSPORTING METALLIC CARBIDES

Anhydrous aluminum chloride

When shippers offer aluminum carbide or calcium carbide for transportation, DOT requires them to identify it as shown in Table 9.9 on the accompanying shipping paper. DOT also requires shippers and carriers to comply with all applicable labeling, marking, and placarding requirements.

9.8 WATER-REACTIVE SUBSTANCES THAT PRODUCE HYDROGEN CHLORIDE

Certain substances react with water to produce hydrogen chloride vapor or hydrochloric acid as products of their hydrolysis. When they are encountered, these substances routinely have the suffocating, pungent odor of hydrogen chloride, which fumes in air and limits visibility. As first noted in Chapter 8, hydrogen chloride is a toxic, irritating gas, and hydrochloric acid is a corrosive liquid. In this section, the hydrolysis product is denoted solely as hydrogen chloride vapor.

9.8-A ALUMINUM CHLORIDE, ANHYDROUS

Anhydrous aluminum chloride is a white-to-yellow solid whose chemical formula is $AlCl_3$. In the chemical industry substantial quantities are used as catalysts and as a raw material for the production of aluminum alkyl compounds and lithium aluminum hydride. It is also used to produce antiperspirants.

Anhydrous aluminum chloride reacts violently with water to produce hydrogen chloride.

$$2AlCl_3(s) \quad + \quad 3H_2O(l) \quad \longrightarrow \quad Al_2O_3(s) \quad + \quad 6HCl(g)$$

Aluminum chloride   Water   Aluminum oxide   Hydrogen chloride

For this reason, the manufacturers and distributors of this substance caution potential users that it irritates the skin, eyes, and respiratory tract.

9.8-B PHOSPHORUS OXYCHLORIDE

Phosphorus oxychloride, also called *phosphoryl chloride*, is a colorless, fuming liquid whose chemical formula is $POCl_3$. It is used primarily by the chemical industry as a chlorinating agent.

Phosphorus oxychloride reacts violently with water to produce hydrogen chloride.

$$POCl_3(l) \quad + \quad 3H_2O(l) \quad \longrightarrow \quad H_3PO_4(aq) \quad + \quad 3HCl(g)$$

Phosphorus oxychloride   Water   Phosphoric acid   Hydrogen chloride

9.8-C PHOSPHORUS PENTACHLORIDE

This substance is a yellow-to-green solid whose chemical formula is PCl_5. It is primarily used as a chlorinating and dehydrating agent by the chemical industry.

Phosphorus pentachloride is decomposed by water in a multistep process, summarized in the following equation:

$$PCl_5(s) \quad + \quad 4H_2O(l) \quad \longrightarrow \quad H_3PO_4(aq) \quad + \quad 5HCl(g)$$

Phosphorus pentachloride   Water   Phosphoric acid   Hydrogen chloride

9.8-D PHOSPHORUS TRICHLORIDE

This substance is a colorless, fuming liquid whose chemical formula is PCl_3. It is used in the chemical industry as a chlorinating agent and catalyst. Phosphorus trichloride is used, for example, as a raw material for producing acetyl chloride (Section 9.9-C).

Phosphorus trichloride reacts with water to form phosphorous acid and hydrogen chloride.

$$PCl_3(l) \quad + \quad 3H_2O(l) \quad \longrightarrow \quad H_3PO_3(aq) \quad + \quad HCl(g)$$

Phosphorus trichloride   Water   Phosphorous acid   Hydrogen chloride

9.8-E SILICON TETRACHLORIDE

Silicon tetrachloride is a colorless, fuming liquid whose chemical formula is $SiCl_4$. It is used as a raw material to manufacture liquid or semisolid silicon-containing polymers known as *silicones*, substances widely used in electrical insulation. Silicon tetrachloride is also used in the semiconductor manufacturing industry.

Silicon tetrachloride reacts vigorously with water to form silicic acid and hydrogen chloride.

$$SiCl_4(l) \quad + \quad 4H_2O(l) \quad \longrightarrow \quad H_4SiO_4(aq) \quad + \quad 4HCl(g)$$

Silicon tetrachloride   Water   Silicic acid   Hydrogen chloride

Phosphorus oxychloride

Phosphorus pentachloride

Phosphorus trichloride

9.8-F SULFURYL CHLORIDE

Sulfuryl chloride, also called *sulfonyl chloride*, is a colorless liquid whose chemical formula is SO_2Cl_2. Sulfuryl chloride is used mainly in the chemical industry as a chlorinating and dehydrating agent.

Sulfuryl chloride reacts slowly with water to form sulfuric acid and hydrogen chloride.

$$SO_2Cl_2(l) \ + \ 2H_2O(l) \ \longrightarrow \ H_2SO_4(aq) \ + \ 2HCl(g)$$
Sulfuryl chloride   Water   Sulfuric acid   Hydrogen chloride

9.8-G THIONYL CHLORIDE

Thionyl chloride is a red-to-yellow liquid whose chemical formula is $SOCl_2$. Thionyl chloride is used mainly within the chemical industry.

Thionyl chloride reacts vigorously with water to form sulfurous acid and hydrogen chloride.

$$SOCl_2(l) \ + \ 2H_2O(l) \ \longrightarrow \ H_2SO_3(aq) \ + \ 2HCl(g)$$
Thionyl chloride   Water   Sulfurous acid   Hydrogen chloride

9.8-H TIN(IV) CHLORIDE, ANHYDROUS

Tin(IV) chloride, also known as *stannic tetrachloride*, is a colorless, fuming liquid whose chemical formula is $SnCl_4$. It is used to manufacture blueprint and similarly sensitized types of paper.

Anhydrous tin(IV) chloride reacts slowly with water to form tin(IV) oxide and hydrogen chloride.

$$SnCl_4(l) \ + \ 2H_2O(l) \ \longrightarrow \ SnO_2(s) \ + \ 4HCl(g)$$
Tin(IV) chloride   Water   Tin(IV) oxide   Hydrogen chloride

9.8-I TITANIUM(IV) CHLORIDE, ANHYDROUS

Titanium(IV) chloride, also called *titanium tetrachloride*, is a colorless, volatile liquid whose chemical formula is $TiCl_4$. Titanium(IV) chloride is the intermediate compound produced during the production of metallic titanium and the white paint pigment titanium dioxide. The mixture of titanium(IV) chloride and an aluminum alkyl compound is an important polymerization catalyst (Section 9.4).

Titanium(IV) chloride reacts slowly with water to produce titanium(IV) dioxide and hydrogen chloride.

$$TiCl_4(l) \ + \ 2H_2O(l) \ \longrightarrow \ TiO_2(s) \ + \ 4HCl(g)$$
Titanium(IV) chloride   Water   Titanium(IV) oxide   Hydrogen chloride

9.8-J CHLOROSILANES

The hydrides of silicon are called **silanes**. They consist of a class of compounds having the chemical formula Si_nH_{2n+2}, where n is a nonzero integer. The simplest silane, itself called *silane*, is a substance having the formula SiH_4. When one or more of the hydrogen atoms in a silane molecule is replaced with a chlorine atom, the resulting substances are called **chlorosilanes**. All chlorosilanes are water-reactive substances.

The chlorosilanes are used predominantly in the polymer industry. For example, trichlorosilane is used to manufacture the polysilicon employed for the production of solar cells and solar wafers. Some of its physical properties are noted in Table 9.10. It is a water-reactive, flammable, and corrosive liquid having the chemical formula $SiHCl_3$.

silane ■ An organic compound whose molecules are composed of silicon and hydrogen atoms

chlorosilane ■ A chlorinated derivative of silane (SiH_4)

TABLE 9.10 | Physical Properties of Some Chlorosilanes

	METHYLDICHLOROSILANE	METHYLTRICHLOROSILANE	TRICHLOROSILANE
Melting point	−135°F (−93°C)	−130°F (−90°C)	−196°F (−127°C)
Boiling point	106°F (41°C)	149°F (66.4°C)	89°F (32°C)
Specific gravity at 68°F (20 °C)	1.1	1.27	1.33
Vapor density (air = 1)	3.97	5.2	4.7
Vapor pressure at 68°F (20°C)	321 mmHg	134 mmHg	500 mmHg
Flashpoint	−18°F (−28°C)	14°F (−10°C)	7°F (−14°C)
Autoignition point	471°F (244°C)	759°F (404°C)	219 °F (104°C)
Lower flammable limit	3.4% by volume	3.4% by volume	1.2% by volume
Upper flammable limit	55% by volume	>55% by volume	90.5% by volume

Trichlorosilane reacts violently with water, producing choking vapors of hydrogen chloride and trihydroxysilane.

$$H{-}SiCl_3(l) \ + \ 3H_2O(l) \ \longrightarrow \ H{-}Si(OH)_3(s) \ + \ 3HCl(g)$$

Trichlorosilane   Water   Trihydroxysilane   Hydrogen chloride

Other chlorosilanes, including methyldichlorosilane and methyltrichlorosilane, possess similar hazardous properties. Their physical properties are included in Table 9.10.

9.8-K TRANSPORTING SUBSTANCES THAT REACT WITH WATER TO PRODUCE HYDROGEN CHLORIDE VAPOR

When shippers transport any substance noted in this section, DOT requires them to provide the relevant shipping description shown in Table 9.11 on the accompanying shipping paper. DOT also requires shippers and carriers to comply with all applicable labeling, marking, and placarding requirements.

Sulfuryl chloride

TABLE 9.11	Shipping Descriptions of Substances That Produce Hydrogen Chloride Upon Reacting with Water
WATER-REACTIVE SUBSTANCE	**SHIPPING DESCRIPTION**
Aluminum chloride, anhydrous	UN1726, Aluminum chloride, anhydrous, 8, PG II
Methyldichlorosilane	UN1242, Methyldichlorosilane, 4.3, (8), (3), PG I (Dangerous When Wet)
Methyltrichlorosilane	UN1250, Methyltrichlorosilane, 3, (8), PG I
Phosphorus oxychloride	UN1810, Phosphorus oxychloride, 6.1, (8), PG II (Poison - Inhalation Hazard, Zone B)
Phosphorus pentachloride	UN1806, Phosphorus pentachloride, 8, PG II
Phosphorus trichloride	UN1809, Phosphorus trichloride, 6.1, (8), PG I (Poison - Inhalation Hazard, Zone B)
Silicon tetrachloride	UN1818, Silicon tetrachloride, 8, PG II
Stannic chloride, anhydrous	UN1827, Stannic chloride, anhydrous, 8, PG II
Sulfuryl chloride	UN1834, Sulfuryl chloride, 6.1, (8), PG I (Poison - Inhalation Hazard, Zone A)
Thionyl chloride	UN1836, Thionyl chloride, 8, PG I
Titanium tetrachloride, anhydrous	UN1838, Titanium tetrachloride, 6.1, (8), PG II (Poison - Inhalation Hazard, Zone B)
Trichlorosilane	UN1295, Trichlorosilane, 4.3, (8), (3), PG I (Dangerous When Wet)

Why should emergency responders use protective gear including self-contained breathing apparatus when investigating a transportation mishap involving a massive spill of liquid methyldichlorosilane?

Solution: As indicated by the shipping description in Table 9.11, methyldichlorosilane is a water-reactive, flammable, and corrosive liquid. When it reacts with water, methyldichlorosilane forms hydrogen chloride vapor. Table 8.9 shows that the inhalation of this vapor causes exposed individuals to experience a variety of adverse health effects. Because hydrogen chloride poses a health hazard by inhalation, the use of self-contained breathing apparatus is essential when emergency responders are investigating a transportation mishap involving methyldichlorosilane.

Thionyl chloride

Anhydrous Tin(IV) chloride

DOT requires carriers to display DANGEROUS WHEN WET placards on the bulk packaging or transport vehicle used to transport trichlorosilane, and when transporting an amount exceeding 1001 pounds (454 kg), to display FLAMMABLE and CORROSIVE placards as well.

9.9 WATER-REACTIVE COMPOUNDS THAT PRODUCE ACETIC ACID VAPOR

When acetic acid is formed as a product of a chemical reaction, its highly irritating and pungent odor is immediately evident. This can pose serious consequences, as the inhalation of acetic acid vapor is suffocating, and exposure to the eyes and nose is severely irritating.

We briefly note here two organic compounds that react with water to produce acetic acid vapor: acetic anhydride and acetyl chloride. Some physical properties of these compounds are provided in Table 9.12.

Acetic anhydride causes severe burns on contact with the skin or eyes. Inhalation of its vapor is suffocating and causes irritation of the respiratory tract. Before the advent of the GHS, chemical manufacturers often marked acetic anhydride containers as follows to warn users of these potential hazards:

> **DANGER: CORROSIVE**
>
> CAUSES BURNS TO ANY AREA OF CONTACT
> FLAMMABLE LIQUID AND VAPOR WATER REACTIVE
> HARMFUL IF SWALLOWED OR INHALED
> VAPOR CAUSES RESPIRATORY TRACT
> IRRITATION AND SEVERE EYE IRRITATION

9.9-A ACETIC ANHYDRIDE

Acetic anhydride is a colorless, fuming liquid whose condensed chemical formula is $(CH_3CO)_2O$. Acetic anhydride is used principally by the chemical, pharmaceutical, and polymer industries for the manufacture of aspirin, cellulose acetate (Section 14.5-A), and related products.

When acetic anhydride combines with water, the sole product produced is acetic acid.

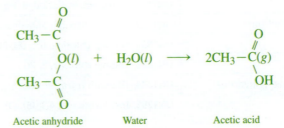

Acetic anhydride Water Acetic acid

TABLE 9.10 — Physical Properties of Some Chlorosilanes

	METHYLDICHLOROSILANE	METHYLTRICHLOROSILANE	TRICHLOROSILANE
Melting point	−135°F (−93°C)	−130°F (−90°C)	−196°F (−127°C)
Boiling point	106°F (41°C)	149°F (66.4°C)	89°F (32°C)
Specific gravity at 68°F (20 °C)	1.1	1.27	1.33
Vapor density (air = 1)	3.97	5.2	4.7
Vapor pressure at 68°F (20°C)	321 mmHg	134 mmHg	500 mmHg
Flashpoint	−18°F (−28°C)	14°F (−10°C)	7°F (−14°C)
Autoignition point	471°F (244°C)	759°F (404°C)	219 °F (104°C)
Lower flammable limit	3.4% by volume	3.4% by volume	1.2% by volume
Upper flammable limit	55% by volume	>55% by volume	90.5% by volume

Trichlorosilane reacts violently with water, producing choking vapors of hydrogen chloride and trihydroxysilane.

$$H-SiCl_3(l) \quad + \quad 3H_2O(l) \quad \longrightarrow \quad H-Si(OH)_3(s) \quad + \quad 3HCl(g)$$

Trichlorosilane　　　　　Water　　　　　Trihydroxysilane　　　Hydrogen chloride

Other chlorosilanes, including methyldichlorosilane and methyltrichlorosilane, possess similar hazardous properties. Their physical properties are included in Table 9.10.

9.8-K TRANSPORTING SUBSTANCES THAT REACT WITH WATER TO PRODUCE HYDROGEN CHLORIDE VAPOR

When shippers transport any substance noted in this section, DOT requires them to provide the relevant shipping description shown in Table 9.11 on the accompanying shipping paper. DOT also requires shippers and carriers to comply with all applicable labeling, marking, and placarding requirements.

Sulfuryl chloride

TABLE 9.11 — Shipping Descriptions of Substances That Produce Hydrogen Chloride Upon Reacting with Water

WATER-REACTIVE SUBSTANCE	SHIPPING DESCRIPTION
Aluminum chloride, anhydrous	UN1726, Aluminum chloride, anhydrous, 8, PG II
Methyldichlorosilane	UN1242, Methyldichlorosilane, 4.3, (8), (3), PG I (Dangerous When Wet)
Methyltrichlorosilane	UN1250, Methyltrichlorosilane, 3, (8), PG I
Phosphorus oxychloride	UN1810, Phosphorus oxychloride, 6.1, (8), PG II (Poison - Inhalation Hazard, Zone B)
Phosphorus pentachloride	UN1806, Phosphorus pentachloride, 8, PG II
Phosphorus trichloride	UN1809, Phosphorus trichloride, 6.1, (8), PG I (Poison - Inhalation Hazard, Zone B)
Silicon tetrachloride	UN1818, Silicon tetrachloride, 8, PG II
Stannic chloride, anhydrous	UN1827, Stannic chloride, anhydrous, 8, PG II
Sulfuryl chloride	UN1834, Sulfuryl chloride, 6.1, (8), PG I (Poison - Inhalation Hazard, Zone A)
Thionyl chloride	UN1836, Thionyl chloride, 8, PG I
Titanium tetrachloride, anhydrous	UN1838, Titanium tetrachloride, 6.1, (8), PG II (Poison - Inhalation Hazard, Zone B)
Trichlorosilane	UN1295, Trichlorosilane, 4.3, (8), (3), PG I (Dangerous When Wet)

Why should emergency responders use protective gear including self-contained breathing apparatus when investigating a transportation mishap involving a massive spill of liquid methyldichlorosilane?

Solution: As indicated by the shipping description in Table 9.11, methyldichlorosilane is a water-reactive, flammable, and corrosive liquid. When it reacts with water, methyldichlorosilane forms hydrogen chloride vapor. Table 8.9 shows that the inhalation of this vapor causes exposed individuals to experience a variety of adverse health effects. Because hydrogen chloride poses a health hazard by inhalation, the use of self-contained breathing apparatus is essential when emergency responders are investigating a transportation mishap involving methyldichlorosilane.

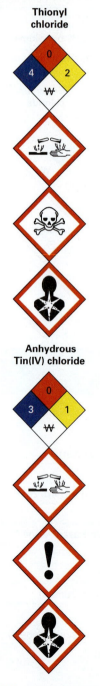

Thionyl chloride

Anhydrous Tin(IV) chloride

DOT requires carriers to display DANGEROUS WHEN WET placards on the bulk packaging or transport vehicle used to transport trichlorosilane, and when transporting an amount exceeding 1001 pounds (454 kg), to display FLAMMABLE and CORROSIVE placards as well.

9.9 WATER-REACTIVE COMPOUNDS THAT PRODUCE ACETIC ACID VAPOR

When acetic acid is formed as a product of a chemical reaction, its highly irritating and pungent odor is immediately evident. This can pose serious consequences, as the inhalation of acetic acid vapor is suffocating, and exposure to the eyes and nose is severely irritating.

We briefly note here two organic compounds that react with water to produce acetic acid vapor: acetic anhydride and acetyl chloride. Some physical properties of these compounds are provided in Table 9.12.

Acetic anhydride causes severe burns on contact with the skin or eyes. Inhalation of its vapor is suffocating and causes irritation of the respiratory tract. Before the advent of the GHS, chemical manufacturers often marked acetic anhydride containers as follows to warn users of these potential hazards:

> **DANGER: CORROSIVE**
>
> CAUSES BURNS TO ANY AREA OF CONTACT
> FLAMMABLE LIQUID AND VAPOR WATER REACTIVE
> HARMFUL IF SWALLOWED OR INHALED
> VAPOR CAUSES RESPIRATORY TRACT
> IRRITATION AND SEVERE EYE IRRITATION

9.9-A ACETIC ANHYDRIDE

Acetic anhydride is a colorless, fuming liquid whose condensed chemical formula is $(CH_3CO)_2O$. Acetic anhydride is used principally by the chemical, pharmaceutical, and polymer industries for the manufacture of aspirin, cellulose acetate (Section 14.5-A), and related products.

When acetic anhydride combines with water, the sole product produced is acetic acid.

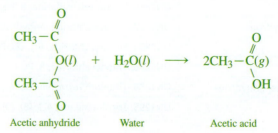

Acetic anhydride   Water   Acetic acid

TABLE 9.12	Physical Properties of Two Water-Reactive Organic Compounds	
	ACETIC ANHYDRIDE	**ACETYL CHLORIDE**
Melting point	−99°F (−73°C)	−170°F (−112°C)
Boiling point	284°F (140°C)	124°F (51°C)
Specific gravity at 68°F (20°C)	1.08	1.10
Vapor density (air = 1)	3.52	2.7
Vapor pressure at 68°F (20°C)	4 mmHg	249 mmHg
Flashpoint	130°F (54°C)	40°F (4.44°C)
Autoignition point	734°F (390°C)	734°F (390°C)
Lower flammable limit	2.7% by volume	7.3% by volume
Upper flammable limit	10.3% by volume	19% by volume

9.9-B WORKPLACE REGULATIONS INVOLVING ACETIC ANHYDRIDE

When acetic anhydride is used in the workplace, OSHA requires employers to limit employee exposure to a concentration of 5 parts per million, averaged over the 8-hour workday.

9.9-C ACETYL CHLORIDE

Acetyl chloride is a colorless, fuming liquid with the chemical formula $CH_3-C\begin{smallmatrix}O\\\\||\\\\\backslash\\Cl\end{smallmatrix}$. Acetyl chloride is used principally in the chemical industry. It is produced by various means, one of which involves the reaction between phosphorus trichloride and acetic acid.

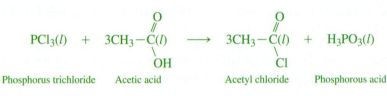

$$PCl_3(l) \; + \; 3CH_3-\overset{\displaystyle O}{\underset{\displaystyle OH}{C}}(l) \; \longrightarrow \; 3CH_3-\overset{\displaystyle O}{\underset{\displaystyle Cl}{C}}(l) \; + \; H_3PO_3(l)$$

Phosphorus trichloride   Acetic acid   Acetyl chloride   Phosphorous acid

On contact with water, acetyl chloride reacts violently, producing acetic acid and hydrochloric acid vapor. This vapor poses the risk of inhalation toxicity.

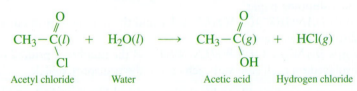

$$CH_3-\overset{\displaystyle O}{\underset{\displaystyle Cl}{C}}(l) \; + \; H_2O(l) \; \longrightarrow \; CH_3-\overset{\displaystyle O}{\underset{\displaystyle OH}{C}}(g) \; + \; HCl(g)$$

Acetyl chloride   Water   Acetic acid   Hydrogen chloride

Table 9.12 reveals that acetyl chloride has a relatively low flashpoint. Consequently, acetyl chloride also poses a dangerous risk of fire and explosion.

9.9-D TRANSPORTING ACETIC ANHYDRIDE AND ACETYL CHLORIDE

When shippers offer either acetic anhydride or acetyl chloride for transportation, DOT requires them to identify it as shown in Table 9.13 on the accompanying shipping paper. DOT also requires shippers and carriers to comply with all applicable labeling, marking, and placarding requirements.

Anhydrous titanium(IV) chloride

Methyldichloro-silane

Methyltrichloro-silane

Trichloro-silane

Acetic anhydride

TABLE 9.13	Shipping Descriptions of Organic Compounds That Produce Acetic Acid Vapor Upon Reacting With Water
WATER-REACTIVE COMPOUND	**SHIPPING DESCRIPTION**
Acetic anhydride	UN1715, Acetic anhydride, 8, (3), PG II
Acetyl chloride	UN1717, Acetyl chloride, 3, (8), PG II

9.10 RESPONDING TO INCIDENTS INVOLVING THE RELEASE OF A MATERIAL IN HAZARD CLASSES 4.1, 4.2, AND 4.3

When a flammable solid, spontaneously combustible material, or water-reactive substance is involved in a transportation mishap, first-on-the-scene responders may readily identify it by noting the following, as relevant:

(a) A flammable solid
- The number 4.1 as a component of a shipping description of a hazardous material listed on a shipping paper
- The words *FLAMMABLE SOLID* and the number 4 printed on white-and-red-striped labels affixed to packaging
- The words *FLAMMABLE SOLID* and the number 4 printed on white-and-red-striped placards displayed on each side and each end of a transport vehicle containing 1001 pounds (454 kg) or more of a flammable solid

(b) A spontaneously combustible material
- The number 4.2 as a component of a shipping description of a hazardous material listed on a shipping paper
- The words *SPONTANEOUSLY COMBUSTIBLE* and the number 4 printed on white-and-red labels affixed to packaging
- The words *SPONTANEOUSLY COMBUSTIBLE* and the number 4 printed on white-and-red placards displayed on each side and each end of a transport vehicle containing 1001 pounds (454 kg) or more of a spontaneously combustible material

(c) A water-reactive material
- The number 4.3 as a component of a shipping description of a hazardous material listed on a shipping paper
- The words *DANGEROUS WHEN WET* and the number 4 printed on blue labels affixed to their packages
- The words *DANGEROUS WHEN WET* and the number 4 printed on blue placards displayed on each side and each end of the transport vehicle

Although it is always desirable to know the chemical identity of a hazardous material involved in any transportation mishap, this statement has special meaning to the first responders arriving at a scene involving a substance that could spontaneously burst into flame or readily react with water. These responders are confronted with a unique challenge, because the use of most common extinguishers would exacerbate the ongoing emergency.

The *Emergency Response Guidebook* provides emergency responders with special information regarding the following hazardous materials: gases and volatile liquids that pose a hazard by inhalation toxicity; chemical warfare agents; and water-reactive materials that produce toxic gases upon contact with water. The *ERG* identifies these hazardous materials by their DOT identification numbers and by green highlighting in both the yellow- and blue-bordered sections. The *ERG* further directs emergency responders to implement the following actions:

■ When there is no fire, go directly to the green-bordered pages, locate the identification number and name of the hazardous material, and identify the initial isolation and protective-action distances.

■ When a fire is involved, also consult the orange guide and, if applicable, apply the evacuation information shown under PUBLIC SAFETY.

DOT provides two entries in the green-bordered pages of the *ERG*, one for land-based spills and the other for water-based spills. When a water-reactive material is not a hazardous material that poses a hazard by inhalation toxicity, and when it is not spilled in water, emergency responders should consult the safety distances in the orange-bordered pages.

Because special procedures are required when responding to fires involving the hazardous materials noted in this chapter, we briefly note the recommended practices for the following groups of substances.

Acetyl chloride

9.10-A ALKALI METALS

Fires involving the alkali metals are extremely difficult to extinguish. When first-on-the-scene responders consider a response action involving an alkali metal fire, it is appropriate to recall that these elements react with two common fire extinguishers, water and carbon dioxide. Alkali metal fires cannot be extinguished with water, because the alkali metals displace flammable hydrogen from water. Furthermore, alkali metal fires cannot be extinguished with carbon dioxide, because the alkali metal reacts with it to produce carbon particulates.

$$4Na(s) \ + \ CO_2(g) \ \longrightarrow \ 2Na_2O(s) \ + \ C(s)$$

Sodium   Carbon dioxide   Sodium oxide   Carbon

Because this reaction is exothermic, the underlying metal usually erupts into flame as the carbon dioxide dissipates.

The use of a dry-chemical or dry-powder fire extinguisher frequently is recommended for extinguishing or controlling the spread of alkali metal fires. Nonetheless, caution needs to be exercised when using graphite-based dry powder to extinguish an alkali metal fire. Graphite effectively extinguishes the fire by a smothering action, thereby limiting the amount of atmospheric oxygen and moisture available to the metal. At the high temperatures accompanying alkali metal fires, however, the graphite may react with the metal to produce metallic carbides. As previously noted in Section 9.7, these compounds are water-reactive substances. Even the moisture in the air could cause the alkali metal to reignite.

When combating fires involving nonbulk quantities of lithium, two fire extinguishing agents are uniquely recommended for use: lithium chloride, a dry-chemical fire extinguisher; and LITH-X, the graphite-based dry-powder extinguisher illustrated in Figure 9.6. Both effectively function by their smothering action. The use of LITH-X is also recommended for extinguishing magnesium, sodium, potassium, and zirconium fires.

When primary lithium batteries are involved in a fire, large volumes of water can be used to consume the lithium, but experts consider the use of LITH-X to be preferable.

9.10-B COMBUSTIBLE METALS

Firefighters frequently are warned against using water on combustible metal fires. Notwithstanding this generally sound advice, water effectively extinguishes combustible metal fires when the following two conditions are met:

■ The water is discharged in a volume that totally deluges the fire scene and cools the metal.
■ The water is discharged rapidly soon after the metal first ignites.

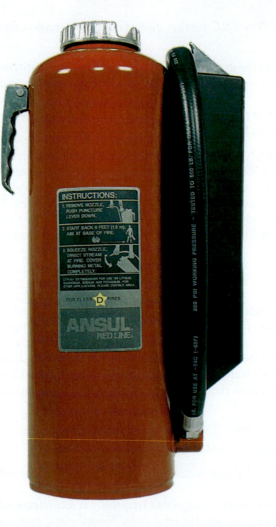

FIGURE 9.6 LITH-X dry powder, a graphite-based fire extinguisher. Although it was initially developed to extinguish lithium fires, LITH-X also extinguishes magnesium, sodium, potassium, and zirconium fires. (*Courtesy of Tyco Fire Protection Products, Lansdale, Pennsylvania.*)

Firefighters must consider not only whether enough water is available at the fire scene but also whether an appropriate means is available to rapidly apply it to the fire.

When a deluging volume of water is unavailable for extinguishing a combustible metal fire, experts recommend the use of dry sand, earth, dry-chemical extinguishers, or dry powders. The application of a special extinguishing agent is recommended for use on specific combustible metal fires. For example, the use of MET-L-X is recommended for extinguishing fires involving nonbulk quantities of metallic magnesium, titanium, zirconium, and aluminum.

The fire-extinguishing agent used in the MET-L-X extinguisher, shown in Figure 9.7, is primarily sodium chloride containing a plastic additive. The metal fire causes the plastic to melt, thereby forming a crust on the surface of the burning metal. This effectively prevents contact between the burning metal and atmospheric oxygen. It is the displacement of the oxygen that causes the fire to be smothered.

The application of carbon dioxide is not recommended for extinguishing combustible metal fires, because these hot metals react with carbon dioxide. Magnesium, for example, reacts with carbon dioxide to produce a sooty plume of carbon.

$$2Mg(s) \ + \ CO_2(g) \ \longrightarrow \ 2MgO(s) \ + \ C(s)$$
$$\text{Magnesium} \quad \text{Carbon dioxide} \quad \text{Magnesium oxide} \quad \text{Carbon}$$

Because this reaction is exothermic, the use of carbon dioxide does not cool the burning metal, and the magnesium fire is not extinguished.

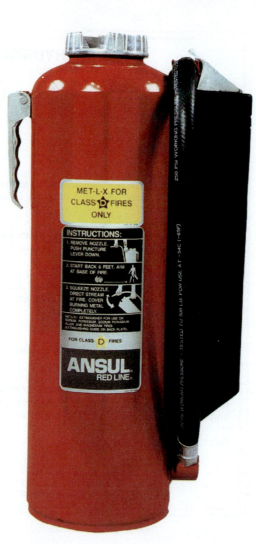

FIGURE 9.7 MET-L-X, a sodium chloride–based fire extinguisher intended for use on NFPA class D fires, especially magnesium fires. (*Courtesy of Tyco Fire Protection Products, Lansdale, Pennsylvania.*)

The environment of a class D fire can be extremely caustic because of the formation of the corresponding metallic oxides, hydroxides, and carbonates. The particulates of such compounds are constituents of the smoke accompanying alkali metal and combustible metal fires. Their inhalation can cause adverse health effects ranging from minor irritation and congestion of the nose, throat, and bronchi to severe lung injury. Because combustible metal fires frequently burn with exceptionally brilliant flames, firefighters should be aware that the evolved radiant energy could damage the retinas of their eyes. They should also avoid breathing the smoke evolved during these fires, as it contains tiny particulates of caustic metallic oxides. When inhaled, exposure to these particulates causes considerable discomfort and localized injury to the respiratory tract and inflammation of the eyes.[1]

9.10-C ALUMINUM ALKYL COMPOUNDS AND METALLIC HYDRIDES, PHOSPHIDES, AND CARBIDES

When first-on-the-scene responders encounter the release of an aluminum alkyl compound or a metallic hydride, phosphide, or carbide from its packaging, experts recommend the use of vermiculite, dry sand, or dry powder pressurized with nitrogen to extinguish fires.

---

[1]NFPA 484, Standard for Combustible Metals (Quincy, Massachusetts: National Fire Protection Association, 2012).

FIGURE 9.8 This portion of the label affixed to containers of sodium hydride communicates hazard information that includes caution against exposing the contents to water.

SODIUM HYDRIDE

UN1427, Sodium hydride

DANGER

In contact with water, releases flammable gases that may ignite spontaneously. Causes mild skin irritation. Causes serious eye irritation.

Keep away from possible contact with water because of violent reaction and possible flash fire. Handle under inert gas. Protect from moisture.

In case of fire: Use dry sand, dry chemical or alcohol-resistant foam for extinction. Store contents under inert gas.

FIRST-AID INSTRUCTIONS:

IF ON SKIN: Remove immediately all contaminated clothing. Rinse with water/shower.

IF IN EYES: Rinse cautiously with water for several minutes. Remove contact lenses, if present, and easy to do. Continue rinsing. Immediately call POISON CENTER or doctor.

Read Safety Data Sheet before use.

My Company
My Street
My Town, My State 00000
Telephone (000) 000-0000

The container label in Figure 9.8 shows that caution should be exercised to avoid exposing sodium hydride to humid air or other potential sources of water. Experts recommend the use of water as a fire extinguisher only when they encounter very small spills of these substances.

9.10-D WATER-REACTIVE SUBSTANCES THAT GENERATE HYDROGEN CHLORIDE

When responding to a transportation mishap involving the release of any substance listed in Table 9.13, firefighters should use water only sparingly and cautiously. Due to the corrosiveness of these products and their potential to form hydrogen chloride, responders must wear fully-encapsulated protective clothing and use self-contained breathing apparatus.

9.10-E ACETIC ANHYDRIDE AND ACETYL CHLORIDE

To extinguish fires involving acetic anhydride or acetyl chloride, experts recommend the use of carbon dioxide or dry chemical. When emergency responders are called to a scene involving a release of acetic anhydride or acetyl chloride, they should wear fully-encapsulated protective clothing, use self-contained breathing apparatus, and totally avoid the use of water as a fire extinguisher.

Metallic Lithium

1. Upon being called to a U.S. postal facility, emergency responders discover the release of smoke from a partially crushed fiberboard box marked on two opposing sides as follows:

> **SURFACE MAIL ONLY**
>
> **PRIMARY LITHIUM BATTERIES - FORBIDDEN FOR TRANSPORT ABOARD PASSENGER AIRCRAFT**

   What immediate action should they take to protect lives, property, and the environment?

Metallic Sodium

2. When metallic sodium is briefly exposed to moist air, three white compounds form on its surface. Identify them and write balanced equations that illustrate their formation.

Metallic Potassium

3. Why does metallic potassium react at an explosive rate with kerosene, whereas metallic lithium and sodium can be covered with kerosene during storage in bottles and cans without incident?

Metallic Magnesium

4. When firefighters inadvertently inhale the fumes released from metallic magnesium fires, they generally exhibit influenza-type symptoms. What substances most likely cause these symptoms to develop?

Metallic Titanium

5. Most combustible materials can be stored in an enclosed area containing an atmosphere of nitrogen to prevent their ignition. Nonetheless, this practice is inadvisable for storing freshly generated titanium chips, turnings, and other fines as they are generated. Why are these forms of titanium metal still likely to ignite when stored in an enclosed area containing nitrogen?

Metallic Zirconium

6. Why are zirconium strips frequently fabricated with the metal submerged under water?

Metallic Aluminum

7. The paint formerly used on some storage tanks contained aluminum powder, which reflected heat from the surface of the tanks. Why were these tanks grounded and used for the sole storage of noncombustible liquids?

Metallic Zinc

8. Which is more hazardous as a flammable solid: zinc dust at room temperature or zinc dust as it is removed after its production within a clay retort?

Aluminum Alkyl Compounds and Their Halide and Hydride Derivatives

9. The DOT regulation at 49 C.F.R. §173.181 requires shippers and carriers to transport pyrophoric liquids in specific types of packaging such as metal cans and steel drums that remain sealed by means of friction-free closures until use. What is the most likely reason that DOT requires friction-free closures on the packaging used for the domestic shipment of triethylaluminum?

Metallic Hydrides

10. What is the most likely reason that DOT limits the quantity of sodium aluminum hydride transported on domestic cargo aircraft to 25 pounds (11 kg)?

Metallic Phosphides

11. What is the most likely reason that DOT prohibits railroad carriers from accepting for domestic transport any quantity of aluminum phosphide within a passenger-carrying railcar?

Metallic Carbides

12. What is the most likely reason that DOT prohibits the domestic transportation of any quantity of calcium carbide within a passenger-carrying railcar?

Substances That React with Water to Produce Hydrogen Chloride

13. Anhydrous titanium tetrachloride is used in offensive and defensive operations as a means of concealing the movement of troops and installations in a combat zone.
    (a) Describe how the use of anhydrous titanium tetrachloride effectively deceives the enemy and prevents them from locating the military's presence. (*Hint:* Titanium dioxide is a white solid.)
    (b) When using anhydrous titanium tetrachloride, why are military personnel concerned about the humidity of the air in the combat zone?
    (c) Which compound is responsible for the acrid smell in the air when anhydrous titanium tetrachloride is used?
14. What is the most logical reason that DOT prohibits airline carriers from accepting for transport any quantity of phosphorus oxychloride by passenger-carrying aircraft?

Substances That React with Water to Produce Acetic Acid

15. Why do chemical manufacturers recommend grounding a rail tankcar containing acetyl chloride before transferring the product into a storage tank?

10
Chemistry of Some Toxic Substances

Courtesy of The Fertilizer Institute, Washington, DC.

OBJECTIVES

- Associate the physical and health hazards of the toxic substances noted in this chapter with the information provided by their hazard diamonds and GHS pictograms.
- Identify the common means by which toxic substances may enter the body and adversely impact human health.
- Describe generally the ways in which a toxic substance may adversely affect one's health.
- Identify the factors that affect the degree of toxicity resulting from chemical exposure.
- Identify the mechanisms by which carbon monoxide and hydrogen cyanide interfere with the proper transfer of oxygen to the cells of the body.
- Identify how on-duty firefighters may determine whether they have been overexposed to carbon monoxide and hydrogen cyanide.
- Identify the chemical features of commercial products that produce sulfur dioxide, hydrogen sulfide, and nitrogen oxides when the products smolder or burn.
- Identify the primary industries that are likely to use anhydrous ammonia.
- Identify the labels, markings, and placards that DOT requires on the packaging of toxic substances and the transport vehicles used for their shipment.
- Identify the response actions to be executed when toxic substances are released from their packaging into the environment.
- Describe how emergency responders use DOT's *Emergency Response Guidebook* to establish the initial isolation and protective-action zones associated with large and small spills of a toxic substance.

toxic substance (toxicant; poison) ■ Any substance that negatively impacts an individual exposed to relatively small concentrations, resulting in death, incapacitation, or the onset of disease or other forms of harm

toxicology ■ The study of toxic substances that adversely impact the health of humans and other organisms, as well as the ways the impact may be counteracted

epidemiology ■ The study of the adverse effects that toxic substances and diseases pose to human populations

Certain substances cause death, illness, injury, or incapacitation when the body is exposed to them in relatively small quantities. They are called **toxic substances**, **toxicants**, or **poisons**. Throughout the remainder of this text, these terms are used interchangeably.

The study of toxic substances is called **toxicology.** In particular, it investigates the manner in which how toxic substances adversely affect a living organism, the diseases they may cause, the concentrations at which the onset of the adverse effects is noted, and ways to prevent or minimize them. The study of the adverse effects that toxic substances and diseases have on human populations is called **epidemiology.**

Toxic substances are routinely encountered by firefighters and other emergency responders. For example, carbon monoxide gas is produced by the assorted materials that burn, and asbestos and lead are encountered in the dust that is generated as old buildings burn or are otherwise wracked by the forces associated with fires. It is for this reason that a need exists for emergency responders to study the commonly encountered toxic substances in some detail.

10.1 TOXIC SUBSTANCES AND GOVERNMENT REGULATIONS

Although the layperson generally understands the meaning of a *poison*, it is important to acknowledge the specific way the term is defined and used in government regulations. We briefly review how toxicity affects workplace, environmental, and transportation regulations.

10.1-A WORKPLACE REGULATIONS INVOLVING TOXIC SUBSTANCES

Among its responsibilities, OSHA requires employers to provide their employees with a place of employment free of conditions or activities that are recognized as hazardous and could cause their death or serious physical harm, when there is a feasible method to abate the hazards. Beyond this general duty obligation, when toxic substances are stored or used in the workplace, OSHA also requires employers to ensure that employee exposure to toxic substances is limited to no more than certain maximum allowable concentrations averaged over the 8-hour workday.

OSHA also requires employers to tag, label, or otherwise identify the presence of toxic substances in the workplace by the use of accident-prevention tags and worded signs. These tags and signs warn employees that they should exercise adequate precaution to reduce or eliminate their exposure to toxic substances. Most bear the word *POISON* in addition to the imprint of a skull and crossbones, which has long served as the internationally acknowledged symbol for poison.

As we first noted in Section 1.9, OSHA also requires chemical manufacturers, distributors, and importers to mark container labels with signal words, GHS pictograms, and appropriate hazard and precautionary statements. The health hazards are identified by marking the container labels with the skull-and-crossbones, health hazard, and exclamation point pictograms.

10.1-B RCRA TOXICITY CHARACTERISTIC

EPA uses RCRA to regulate the treatment, storage, and disposal of hazardous waste that exhibits certain characteristics, one of which is toxicity. As noted at 40 C.F.R. §261.24, a chemical analyst determines whether a representative sample of a waste exhibits the **RCRA toxicity characteristic** by performing a laboratory test designed to simulate the environment within a landfill. In practice, the analyst conducts certain specified test procedures to determine whether one or more of the waste constituents listed at 40 C.F.R. §261.24 leaches from the sample at concentrations equal to or greater than their relevant threshold levels. Some representative waste constituents and their threshold levels are listed in Table 10.1. Each listed substance adversely affects public health and the environment. When the analyst determines that at least one constituent leaches from the sample at a concentration equal to or exceeding the threshold level, the waste is said to exhibit the RCRA toxicity characteristic.

10.1-C TRANSPORTING TOXIC SUBSTANCES

In Chapter 6, we first noted that DOT regulates the transportation of several types of toxic substances: **gases poisonous-by-inhalation (poison gases), poisonous materials**, and infectious substances. These toxic substances are designated as hazardous materials in hazard classes 2.3, 6.1, and 6.2, respectively. All hazardous materials in hazard classes 2.3 and 6.1 that are designated by a 1, 2, 3, or 4 in column 7 of the Hazardous Materials Table pose a health threat by inhalation.

When toxic substances are transported, DOT requires shippers and carriers to comply with applicable labeling, marking, and placarding requirements. When preparing the

RCRA toxicity characteristic ■ For purposes of RCRA regulations, the characteristic of a waste that contains one or more of listed constituents at concentrations equal to or greater than the threshold levels designated at 40 C.F.R. §261.24

gas poisonous-by-inhalation (poison gas) ■ For purposes of DOT regulations, a gas at 68°F (20°C) or less and a pressure of 14.7 psi$_a$ (101.3 kPa) that is known to be so toxic to humans as to pose a health hazard during transportation; *or* in the absence of adequate data on human toxicity, is presumed to be toxic to humans because when tested on laboratory animals, it has an LC$_{50}$ of less than 5000 mL/m^3

poisonous material ■ For purposes of DOT regulations, a material other than a gas that is known to be so toxic to humans as to afford a hazard to health during transportation, *or that* in the absence of adequate data on human toxicity, is presumed to be toxic to humans because of data obtained from prescribed oral, dermal, and inhalation toxicity tests performed on animals; *or* is an irritating material with properties similar to those of tear gas that causes extreme irritation, especially in confined spaces

TABLE 10.1	Threshold Concentrations of Some Representative Waste Contaminants Subject to the RCRA Toxicity Characteristic[a]	
EPA HAZARDOUS WASTE NUMBER	**HAZARDOUS WASTE CONTAMINANT**	**THRESHOLD LEVEL (mg/L)**
D004	Arsenic	5.0
D005	Barium	100.0
D018	Benzene	0.5
D006	Cadmium	1.0
D019	Carbon tetrachloride	0.5
D007	Chromium	5.0
D025	*p*-Cresol	200.0
D028	1,2-Dichloroethane	0.5
D029	1,1-Dichloroethylene	0.7
D008	Lead	5.0
D009	Mercury	0.2
D010	Selenium	1.0
D011	Silver	5.0
D039	Tetrachloroethylene	0.7
D040	Trichloroethylene	0.5

[a]40 C.F.R. §261.24.

shipping description of a toxic substance, DOT requires shippers to include the expression "Poison - Inhalation Hazard" and the relevant hazard zone in the description. DOT also requires shippers to affix the applicable POISON GAS or POISON INHALATION HAZARD labels to the relevant packaging.

DOT requires carriers to display the POISON GAS or POISON INHALATION HAZARD placards shown in Figure 6.13 as follows:

■ DOT requires carriers to display the applicable POISON GAS or POISON IN-HALATION HAZARD placards on both sides and both ends of the transport vehicle, freight container, portable tank, unit load device, or rail tankcar used for shipment of a hazardous material having the hazard code 2.3 or 6.1, respectively.

■ DOT requires carriers to display POISON GAS placards when transporting any quantity of a substance that has been assigned a hazard class number 2.3.

■ DOT requires carriers to display POISON INHALATION HAZARD placards when transporting any quantity of a substance that has been assigned a hazard class number 6.1 and a hazard zone classification as Zone A or Zone B.

■ As noted in Section 6.6-D, when carriers transport a hazardous material whose hazard class is 2.3 or 6.1 by rail, DOT requires them to post POISON GAS or POISON INHALATION HAZARD placards on white squares with black borders when the hazard zone for the hazardous material is Zone A.

■ When carriers transport multiple toxic substances whose hazard classes are designated 2.3 *and* 6.1 in nonbulk packages in the same transport vehicle, DOT allows them to post POISON GAS placards only on the vehicle.

■ When the subsidiary hazard class of a hazardous material is 6.1, DOT requires carriers to display POISON INHALATION HAZARD placards on its packaging.

When a material that poses a health hazard by inhalation is shipped in bulk packaging, DOT also requires the shipper to mark the packaging on two opposing sides with

TABLE 10.2	Shipping Descriptions of Some Representative Hazardous Wastes Exhibiting the EPA Toxicity Characteristic
HAZARDOUS WASTE	**SHIPPING DESCRIPTION**
Liquid hazardous waste that exhibits the toxicity characteristic for arsenic	NA3082, Hazardous waste, liquid, n.o.s. (arsenic), 9, PG III (EPA toxicity) *or* NA3082, Hazardous waste, liquid, n.o.s. (arsenic), 9, PG III (D004)
Solid hazardous waste that exhibits the toxicity characteristic for mercury	NA3077, Hazardous waste, solid, n.o.s. (mercury), 9, PG III (EPA toxicity) *or* NA3077, Hazardous waste, solid, n.o.s. (mercury), 9, PG III (D009)

the words INHALATION HAZARD. Even when a hazardous material in division 6.1 does not pose a health hazard by inhalation, DOT may obligate shippers and carriers to disclose that it is poisonous. They do so by entering "Poison" or "Toxic" in the shipping description, affixing POISON labels to the packaging, marking POISON or TOXIC on the packaging, and displaying POISON placards on the transport vehicle when the aggregate gross mass equals or exceeds 1001 pounds (454 kg).

When carriers transport more than 1.06 quart (1 L) per package of a material that poses a health hazard by inhalation and meets the criteria for Zone A, DOT also requires them to prepare and implement a *security plan* whose components comply with the requirements of 49 C.F.R. §172.802. Motor carriers are also required to obtain a hazardous materials safety permit (Section 6.10) before transporting any of the following:

- A material that poses an inhalation health hazard and meets the criteria for Zone A in an amount more than 1.08 pounds (1 L) per package
- A material that poses an inhalation health hazard and meets the criteria for Zone B in bulk packaging [capacity greater than 119 gallons (460 L)]
- A material that poses an inhalation health hazard and meets the criteria for Zone C or Zone D in packaging having a capacity equal to or greater than 3500 gallons (13,248 L)

Table 6.2 notes that DOT requires shippers to identify any hazardous waste that exhibits the toxicity characteristic in its shipping description by using the term "EPA toxicity" or the relevant EPA hazardous waste number listed in Table 10.1. For example, when shippers offer for domestic transportation a hazardous waste that exhibits the toxicity characteristic owing to its arsenic or mercury concentration, DOT requires the waste characteristic to be parenthetically noted on the accompanying waste manifest by use of either "EPA toxicity" or the EPA hazardous waste number of the hazardous constituent. Some representative examples are cited in Table 10.2.

10.2 HOW TOXIC SUBSTANCES ENTER THE BODY

A toxic substance may enter the body by various routes including direct injection into the bloodstream, but we are concerned here with the three means of most concern to emergency responders: oral **ingestion**, **skin absorption**, and **inhalation**.

ingestion ■ The receipt of a substance through the mouth and into the digestive system

skin absorption ■ The passage through the epidermis into the dermis or subcutaneous tissue

inhalation ■ The breathing of a substance in the form of a gas, vapor, fume, mist, or dust into the respiratory system

FIGURE 10.1 The major components of the human digestive system. A substance taken orally passes through the mouth, into the esophagus, and then into the stomach. Partial degradation or complete chemical alteration of the substance occurs in the mouth and stomach. Molecules of the substance then may pass through the stomach wall directly into the bloodstream. More generally, however, absorption into the bloodstream occurs after the molecules of the substance or its degradation products pass into the small intestine.

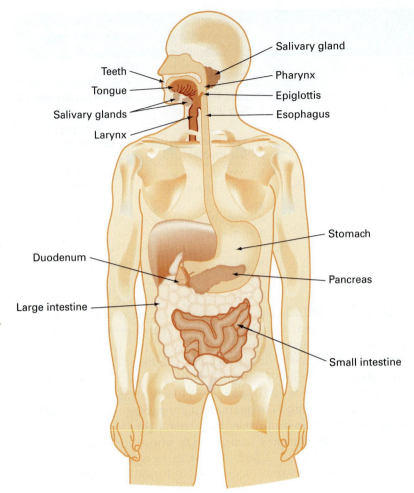

Salivary gland

Teeth

Tongue

Salivary glands

Larynx

Pharynx

Epiglottis

Esophagus

Duodenum

Large intestine

Stomach

Pancreas

Small intestine

10.2-A ORAL INGESTION

As shown in Figure 10.1, oral ingestion refers to the swallowing of a substance through the mouth into the stomach and its subsequent movement through the gastrointestinal tract. Once it is ingested, components of the substance may pass through the intestinal walls and into the circulatory system, where their molecules are further disseminated to the organs and tissues of the body. The substance may undergo chemical changes in the cells of these organs and tissues. The combination of these chemical reactions is called **metabolism**. Although the individual organs of the digestive system metabolize many substances, metabolic processes also occur within the liver.

metabolism ■ The sum total of the chemical processes that occur in the body's cells to break down absorbed foods and ingested substances

10.2-B SKIN ABSORPTION

The skin, a cross section of which is illustrated in Figure 10.2, constitutes the largest single organ of the human body. The skin of an average adult has an area of 21 square feet (2 m^2), weighs 9 pounds (4 kg), and contains more than 11 miles (18 km) of blood vessels. The skin helps the body maintain a normal temperature. It also protects the internal organs and prevents direct contact between them and foreign substances.

Although the skin acts as an organ of defense, it also can act as a permeable membrane. This is especially true when toxic substances directly contact the mucous membranes and the eyes. Some foreign substances also penetrate the epidermis, the outermost skin layer,

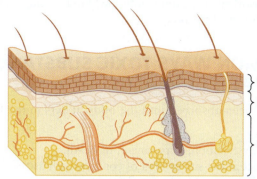

Epidermis

Dermis

Subcutaneous

FIGURE 10.2 The cross section of human skin showing some of its principal components. The outer layer of the skin is the epidermis, a thin surface membrane of dead cells. Before a substance may be absorbed into the bloodstream, it must first penetrate past the epidermis and enter the dermis, a collection of cells that collectively act as a porous medium. From the dermis, the substance is absorbed into the bloodstream.

and enter the underlying dermis or subcutaneous tissue, from which they may be further absorbed into the circulatory system and spread throughout the body.

10.2-C INHALATION

Inhalation is the route responsible for the movement of gases, vapors, and fumes through the components of the respiratory system. Emergency responders are most commonly injured by this route of exposure. As illustrated in Figure 10.3, the respiratory system consists of the nasal passageways, pharynx, larynx, trachea, bronchi, and lungs. When an inhaled substance enters the lungs, it is exposed to blood vessels that cover an average surface area of approximately 90 square yards (75 m²). Given this massive area of

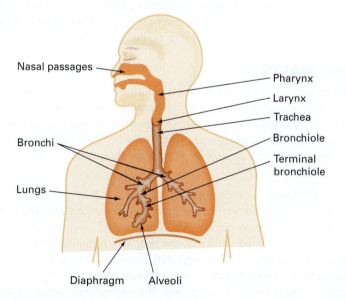

Nasal passages

Pharynx

Larynx

Trachea

Bronchiole

Bronchi

Terminal bronchiole

Lungs

Diaphragm   Alveoli

FIGURE 10.3 The major components of the human respiratory system. Inhaled gases, fumes, and vapors—including air—first pass through either the mouth or nose, through the pharynx, and into the trachea (commonly called the windpipe) at the larynx. These inhaled substances then enter either of two bronchi, each of which leads to a lung. The individual divisions of each bronchus are called *bronchioles*.

exposure, the substance is absorbed very rapidly into the bloodstream and then circulated to all the other organs and tissues of the body.

10.3 SOME COMMON WAYS TOXIC SUBSTANCES ADVERSELY AFFECT HEALTH

Once a toxic substance has been absorbed into the body, it acts in the following general ways:

- ■ It may cause immediate dysfunction, impairment, or death.
- ■ It may cause a **local effect** at the site or area of contact, such as the onset of pulmonary edema following inhalation of hydrogen fluoride.
- ■ It may affect the body as a whole, and thereby, cause an ailment that is not localized to one spot or area. This is a **systemic effect**. Human exposure to certain poisons, for example, impacts the entire nervous system and causes survivors to experience unrestricted neurobehavioral changes.
- ■ It may target a specific organ that undergoes localized dysfunction or impairment. When a toxic substance targets a specific organ, the substance is denoted accordingly as one or more of the following:
  - ■ **Hemotoxicant,** a substance that decreases the function of the blood's hemoglobin and deprives the tissues of oxygen
  - ■ **Hepatotoxicant,** a substance that causes liver damage
  - ■ **Nephrotoxicant,** a substance that causes kidney damage
  - ■ **Neurotoxicant,** a substance that adversely affects the central nervous system (the brain and spinal cord)
  - ■ **Respiratory toxicant,** a substance that adversely affects the nasal passages, pharynx, trachea, bronchi, and lungs
  - ■ **Reproductive toxicant,** a substance that adversely affects sexual function and fertility in adults, as well as development in their offspring
- ■ Its metabolic by-products may cause the host victim to experience ill effects. For instance, when an individual ingests methanol and ethylene glycol, neither substance causes ill effects directly; nonetheless, their metabolites are toxic substances and may cause the individual to experience ill effects, including death.

10.3-A ASPHYXIANTS

One way that a gas or vapor adversely affects health is by inducing unconsciousness when inhaled. The gas or vapor that affects the human body in this manner is called an **asphyxiant**. The most common asphyxiants are carbon monoxide, carbon dioxide, nitrogen, methane, and the noble gases. Asphyxiants may be toxic or nontoxic.

There are three common means by which an asphyxiant causes unconsciousness in humans:

- ■ *Simple asphyxiants.* As first noted in Section 2.5-B, a gas having a vapor density greater than 1 displaces the air within the lower regions of an enclosure. An example of such a gas is carbon dioxide. A person who inhales carbon dioxide can lose consciousness, because the body is denied sufficient oxygen. When provided fresh air or oxygen in a timely fashion, consciousness can be regained, but if unattended, the individual usually dies from asphyxiation or experiences long-term neurological damage.

The reduced oxygen concentration, not the asphyxiant concentration, is usually used to establish the negative impact that is likely to result from inhaling an asphyxiant. The signs and symptoms associated with breathing a reduced-oxygen atmosphere were previously noted in Section 7.1.

local effect ■ Any ailment occurring at the site or area that a toxic substance initially contacted

systemic effect ■ Any ailment occurring anywhere in the body, including throughout the entire body.

hemotoxicant ■ Any substance that decreases the function of the blood's hemoglobin and deprives the tissues of oxygen

hepatotoxicant ■ Any substance that causes liver damage

nephrotoxicant ■ Any substance that causes kidney damage

neurotoxicant ■ Any substance that adversely affects the central nervous system

respiratory toxicant ■ Any substance capable of inflaming the airways or otherwise decreasing lung function

reproductive toxicant ■ Any substance that adversely affects sexual function and fertility in adults, as well as normal development in their offspring

asphyxiant ■ Any gas or vapor that dilutes or displaces air and, when inhaled, causes unconsciousness or death

Use Table 7.2 to identify the adverse health effects most likely to be experienced by an emergency responder who breathes an atmosphere containing 42% methane and 9% oxygen by volume.

Solution: When it is inhaled, methane acts as an asphyxiant. Based on the information in Table 7.2, any individual who inhales air containing only 9% oxygen by volume is likely to experience nausea, vomiting, unconsciousness, an ashen face, fainting, and mental failure.

■ *Action as a hemotoxicant.* Asphyxiation can also be caused by chemical action. Substances that chemically impair the body's use of oxygen are chemical asphyxiants. For example, carbon monoxide causes unconsciousness as a chemical asphyxiant by reacting with the blood's hemoglobin. This chemical action prevents the normal transmission of oxygen in the bloodstream. Although consciousness can be regained when fresh air or oxygen is provided, we shall learn in Section 10.10 that a short-term exposure to certain concentrations of carbon monoxide is fatal.

■ *Inability to utilize cellular oxygen.* Another way that an asphyxiant causes unconsciousness is by hindering the cells' ability to utilize oxygen. This occurs when an individual inhales hydrogen cyanide. In Section 10.11-A, we shall learn that the inhalation exposure renders an important enzyme inactive. Individuals who have inhaled hydrogen cyanide may have an adequate oxygen level in their blood, but the inactivity of the enzyme prevents them from effectively utilizing it. Subsequent exposure to fresh air or oxygen may not revive them, and survival may be possible only when they are promptly treated with an appropriate **antidote**.

antidote ■ An agent that counteracts the ability of a substance to act as a poison

10.3-B IRRITANTS

The inhalation of a gas or vapor may also affect the body as an **irritant** by injuring the tissues that it contacts. When a low concentration of an irritant is inhaled, the result is often minor inflammation of the tissues that line the respiratory passageways. This happens when a relatively small amount of ammonia vapor is inhaled. However, inhalation of an elevated concentration of the same substance may cause respiratory failure. Whether a substance causes respiratory irritation or failure is often associated with the degree to which it corrodes the nose, pharynx, larynx, trachea, bronchi, and lungs.

irritant ■ Any substance capable of causing a reversible inflammatory effect at a site upon immediate, prolonged, or repeated contact with living tissue

Not all irritants are gases or vapors. Skin exposure to a solution of a weak acid, for example, may cause the development of a rash or other skin disorder at the site of contact in susceptible individuals, especially after prolonged or repeated contact.

CPSC regulates exposure to irritants in consumer products. For its purposes, an irritant is defined as any noncorrosive substance, which, on immediate, prolonged, or repeated contact with living tissue, induces a local inflammatory reaction.

The GHS exclamation-point pictogram is affixed on the container labels of chemical products holding skin and eye irritants unless the corrosion or health hazard pictogram also is affixed.

10.4 TYPES OF TOXICOLOGICAL EFFECTS

The harmful health effects associated with exposure to substances are often classified as acute, chronic, short-term, or latent. When these substances are constituents of commercial chemical products, these effects are noted by the manufacturer on labels and SDSs.

10.4-A ACUTE HEALTH EFFECT

An **acute health effect** is an injury, disease, or death that develops rapidly (in minutes, hours, or days) and quickly comes to a crisis following exposure to a substance for less than 24 hours. An example of an acute health effect is tearing of the eyes upon exposure to ammonia vapor, because the injury occurs virtually immediately from a short-term exposure.

10.4-B CHRONIC HEALTH EFFECT

A **chronic health effect** is an injury, disease, or death that develops slowly over a few days, weeks, years, or decades following an exposure or repeated exposure to a substance. Most types of **coronary heart disease** are examples of chronic health effects.

10.4-C SHORT-TERM HEALTH EFFECT

A **short-term health effect** is manifested when an injury or disease of relatively short duration *and* from which recovery occurs rapidly is caused by exposure to a substance. For example, when an individual is temporarily asphyxiated from carbon dioxide exposure but recovers when provided with ample oxygen or air, the asphyxiation is called a short-term health effect.

10.4-D LATENT HEALTH EFFECT

A **latent health effect** is manifested when an injury or disease caused by exposure to a substance occurs *only* after an extensive time period has passed. For example, the onset of certain cancerous and noncancerous diseases following exposure to asbestos fibers is regarded as a latent health effect, because these diseases usually emerge 20 to 40 years after an individual's initial exposure to the fibers.

10.5 FACTORS AFFECTING THE DEGREE OF TOXICITY

The human body is a very complex and delicately balanced system. Its cells assimilate nutrients from foods, resist biological attack, reproduce, and generate the substances that the body needs for its survival. Cellular reactions are responsible for life itself; but when toxic substances are absorbed across cellular membranes, they can upset these delicately controlled biochemical processes. When the cells malfunction, the host victim experiences impaired health, disease, or death.

Fortunately, the body has natural mechanisms for protecting itself against the impact of foreign substances. One such mechanism involves the specific actions of its organs. For instance, the liver is particularly effective at converting many harmful substances into harmless ones or into substances that can be rapidly excreted. As a result of these metabolic changes, toxic substances may be modified in chemical structure, temporarily stored in specific organs, and/or directly eliminated from the body.

Aside from these natural protective mechanisms, the body is sometimes incapable of protecting itself from invasion by toxic substances. Whether exposure to a toxic substance actually causes death, disease, or injury depends on several factors, the most important of which are the following: the quantity of substance, the duration of exposure; the rate at which a substance is absorbed into the bloodstream; the age, sex, ethnicity, and health of the afflicted person; and individual sensitivities. A parent's exposure to toxic substances may also affect the health of his or her children, including unborn children. We examine each factor independently.

10.5-A QUANTITY OF SUBSTANCE

Although all substances have the potential to be toxic, we are concerned with those substances that cause an adverse impact on an organism in relatively small amounts. The amount of the substance administered or absorbed per body weight is called the **dose**. Some substances are so toxic that exposure to the tiniest dose is lethal. For instance, the toxin *Clostridium botulinum*, the causative agent of botulism, is a single cell that releases a bacterium so potent that just 1 gram can cause the death of more than 100,000 individuals. Fortunately, there are relatively few substances that exhibit this pronounced degree of toxicity.

A specific substance impacts the body as a function of its dose. For example, consider how the consumption of alcoholic beverages affects the body. When an individual drinks enough alcohol to achieve a blood alcohol concentration of 50 mg/dL, he or she may feel slightly subdued, relaxed, perhaps even elevated in spirit; but when sufficient alcohol is consumed to achieve a blood alcohol concentration of 150 mg/dL, the person is patently intoxicated. [One deciliter (dL) is 100 milliliters, or one-tenth of a liter.]

10.5-B DURATION OF EXPOSURE

As a general policy, frequent exposures to a toxic substance cause individuals to experience ill effects. Consider the approximately 4000 substances that have been identified as constituents of tobacco smoke, about 200 of which are toxic. Exposure to them causes many tobacco smokers to develop **emphysema, chronic bronchitis**, and/or cancers of the lungs, mouth, larynx, kidney, or bladder; yet, the exposure appears to impact other smokers as if they were nonsmokers. If 200 toxic substances in tobacco smoke cause health problems, why do some smokers appear to be shielded from contracting these diseases?

The answer to this question is connected to the fact that the initiation, promotion, and manifestation of a specific disease occur stepwise. Although many random factors can accelerate or delay the progression of these steps, the possibility that ill effects will result from exposure to toxic substances increases when individuals are exposed more and more frequently to them.

Also, individuals do not respond in an identical fashion when exposed to a given concentration of a toxic substance over the same length of time. Some individuals are more capable than others of tolerating toxic substances in their environment. Some even appear to adapt to their presence. Individuals respond to toxic substances over a given length of time in the following ways:

- Although some individuals experience ill effects from exposure to a toxic substance, others experience the same effects only after they have been repeatedly exposed to the substance in the aggregate for a longer period of time.
- Although some individuals experience adverse health effects from exposure to a given concentration of a substance, they may not experience the same effects on subsequent exposure to the same concentration of the same substance.
- Some individuals who do not experience ill effects from a short-term exposure to one concentration of a substance may be unable to tolerate a long-term exposure to a smaller concentration of the same substance. These individuals require this longer period of exposure to reach their *body burden*, that is, the concentration that causes them to experience ill effects.

10.5-C RATE AT WHICH A SUBSTANCE ABSORBS INTO THE BLOODSTREAM

When a toxic substance is inhaled, the adverse effect caused by its presence may not be evident immediately. Inhaling a lethal concentration of chlorine or carbon monoxide, for example, causes death within seconds or minutes, whereas inhaling a lethal concentration of hydrogen chloride or nitrogen dioxide may result in death hours or days later. These observations demonstrate that the manifestation of an effect caused by exposure to a toxic substance depends not only on the concentration to which an individual is exposed but on the rate of the mechanism that causes the impairment or death.

dose ▪ The quantity of a substance per body weight administered directly to and absorbed by an organism

emphysema ▪ A disease in which the alveoli in the lungs become irreversibly distended, decreasing the efficiency of respiration and causing breathlessness and wheezy coughing

chronic bronchitis ▪ A disease in which the bronchi of the lungs become permanently inflamed, causing breathlessness and chronic coughing

When a toxic substance is ingested or absorbed through the skin, sufficient time must then elapse for the substance to be absorbed into the bloodstream. When ingested, the substance must usually pass from the stomach into the small intestine, where it is then absorbed into the bloodstream. When exposed to skin, the substance must first pass into the dermis, where it is then absorbed into the bloodstream.

It is fortuitous that time is required for the absorption of a substance into the bloodstream. The time needed for this process to occur provides a window of opportunity during which emergency-medical personnel can administer an antidote, induce vomiting, or implement other appropriate first-aid measures.

10.5-D AGE, SEX, ETHNICITY, AND THE GENERAL HEALTH OF INDIVIDUALS

The degree to which toxic substances affect individuals often depends on their age, sex, ethnicity, or general state of health. Young children and the elderly typically are more susceptible than middle-aged individuals to the effects of exposure to toxic substances. Children are particularly vulnerable, because their average body weight is relatively low and their immune systems are still developing. A weak immune system leaves everyone more vulnerable to contracting diseases and experiencing other health-related complications. Newborn infants are at the highest risk, because they must rely heavily on the immune factors acquired from their mothers before their birth.

The sex and ethnicity of a person also affect the susceptibility to contracting certain diseases. For example, even though the physiology of the circulatory system is the same in both men and women, women are more prone to contract certain types of heart disease that are fundamentally different from the types contracted by men. Another example is evident when examining the diseases contracted by white and black people. Even though the composition of the blood is the same in all races, black people of African descent are more prone to developing sickle-cell anemia than are white people. Why is one sex or one ethnic group more prone to be stricken with certain diseases? Although the answer to this question has not been identified, genetics most likely plays a significant role.

Likewise, those persons already weakened from disease are more likely than healthy people to be susceptible to the effects resulting from exposure to toxic substances. This sensitivity is particularly evident in individuals suffering from heart and respiratory ailments. Individuals suffering from heart ailments, for example, may be incapable of tolerating exposure to low concentrations of ground-level ozone, but healthy people may experience no symptoms at all when exposed to the same concentration.

A well-acknowledged group of unhealthy individuals are tobacco smokers. As a general rule, nonsmokers can tolerate an exposure to toxic substances better than smokers because the immune systems of smokers have been compromised. Smoking increases the potential that particulate matter will adhere to lung tissue and delay the rate of removal of hazardous substances. This may explain why smokers are more likely to suffer from asbestos- or radon-induced cancers than nonsmokers for the same exposure.

10.5-E INDIVIDUAL SENSITIVITIES

Sensitizers are a unique group of substances that cause only certain exposed individuals to experience an allergic reaction. An exposure to a sensitizer can also influence the development of a more serious disease. For instance, exposure to the dusts and metal fumes of beryllium can promote the allergic reaction called **beryllium sensitization**, whose symptoms include coughing, shortness of breath, fatigue, fever, and night sweats. Contracting beryllium sensitization is the first step towards developing **beryllosis**, or **chronic beryllium disease**, a painful scarring of lung tissue for which there is no cure.

To explain why only certain people adversely respond to sensitizers is not straightforward. The explanation may have something to do with one's inherited traits; that is, our

sensitizer ■ Any substance that causes a substantial proportion of exposed people or animals to develop an allergic reaction after repeated exposures

beryllium sensitization ■ The allergic response caused by inhalation of beryllium dusts and fumes

beryllosis (chronic beryllium disease) ■ The debilitating disease characterized by scarring of the lung tissue following the inhalation of beryllium dusts and fumes

genes may play a role in explaining why some individuals tolerate exposure to a given concentration of a sensitizer whereas other individuals experience ill effects from exposure to the same or a lesser concentration.

Three federal agencies regulate exposure to sensitizers:

■ CPSC regulates exposure to the sensitizers listed at 16 C.F.R. §1500.13 when they are components of consumer products.

■ FDA regulates exposure to foods containing peanuts, tree nuts, eggs, milk, shellfish, fish, wheat, and any food whose treatment produces a sulfite (SO_3^{2-}) concentration equal to or greater than 10 parts per million.

■ OSHA regulates exposure to sensitizers that are stored and used in the workplace. The manufacturers, distributors, and importers of respiratory and skin sensitizers are required by OSHA to affix health hazard and exclamation point GHS pictograms on their respective container labels.

10.5-F ADVERSE HEALTH IMPACTS ON A DEVELOPING FETUS

When a pregnant woman has been exposed to a toxic substance, her placenta may shield the developing fetus from exposure to a toxic substance. The placenta's primary function is to provide the fetus with oxygen and nutrients, but it can also function as a barrier by preventing the toxic substance from passing into the mother's womb.

Notwithstanding this statement, the placenta does not always block the passage of toxic substances into the womb. Some substances appear to move unhindered through the placenta and enter the fetus's blood, after which they may cause harm by negatively influencing growth and development or contributing to an illness or disability.

The adverse health effects faced by a developing fetus are vividly evident when a pregnant woman is exposed to illegal drugs. A pregnant woman who uses cocaine or heroin, for example, puts her unborn child at risk because the mother is likely to experience premature labor, miscarriage, and stillbirth. Babies who survive their mother's exposure are not only born undeveloped, but they may be addicted and require treatment for withdrawal from the drug. A woman's use of illegal drugs during her pregnancy may also weaken the placenta and allow other toxic substances to cross more readily into her womb.

A developing fetus is also affected when a woman smokes tobacco products during her pregnancy. Inhaling the toxic substances in tobacco smoke nearly doubles her risk of giving birth to an undeveloped baby and increases her risk of a preterm delivery. The risk to a baby's health is extended beyond the nine months during which the fetus is carried in its mother's womb. When a woman smokes during her pregnancy, her newborn child is likely to experience more illnesses and disabilities compared to a child that is born to a nonsmoking mother. Because its lungs are often undeveloped, the newborn child may experience difficulty in breathing. Later in life, the child is likely to experience disabilities such as mental retardation and learning problems.

We shall note later that a developing fetus may also be negatively impacted when its mother is exposed during her pregnancy to ethanol (Section 13.2-E), Bisphenol-A (Section 13.2-O), ethylene glycol alkyl ethers (Section 13.3-E), and diethylhexyl phthalate (Section 13.7-B).

10.6 MEASURING TOXICITY

Toxicologists have devised procedures for measuring the concentration of a substance that causes an organism to experience injury, disease, or death. These procedures involve exposing groups of laboratory animals to concentrations of a substance, observing the effects by the exposure, and extrapolating the results to humans. When exposed to a unique concentration of a substance, animals and humans do not always respond in the same way. Even different animals respond differently to an exposure. Consequently, these parameters may have only limited relevance.

In the workplace, OSHA and NIOSH require employers to reduce or eliminate the likelihood that employees will experience adverse health effects from exposure to toxic substances. To do so, employers consider three relevant exposure limits: the short-term exposure limit, the permissible exposure limit, and the immediately-dangerous-to-life-and-health level. Firefighters may also use these limits to evaluate the impact posed by inhaling a known concentration of an airborne toxic substance.

The relative differences among several of the measurements noted in this section are listed in Table 10.3 for some common gases and vapors.

TABLE 10.3		Toxicity Measurements of Some Common Gases and Vapors		
TOXIC SUBSTANCE	LC$_{50}$ (PPM)	EXPOSURE LIMIT	SHORT-TERM EXPOSURE LIMIT	IMMEDIATELY-DANGEROUS-TO-LIFE-AND-HEALTH LIMIT (PPM)
Ammonia	4000	25 ppm (18 mg/m^3) (NIOSH)	50 ppm (35 mg/m^3) (OSHA) 35 ppm (27 mg/m^3) (NIOSH)	300
Carbon monoxide	3760	50 ppm (55 mg/m^3) (OSHA) 35 ppm (40 mg/m^3) Ceiling 200 ppm (229 mg/m^3) (NIOSH)		1200
Chlorine	293	Ceiling 1 ppm (3 mg/m^3) (OSHA) Ceiling 0.5 ppm (1.45 mg/m^3), 15 min (NIOSH)		10
Fluorine	185	0.1 ppm (0.2 mg/m^3) (OSHA/NIOSH)		25
Hydrogen chloride	2810	Ceiling 5 ppm (7 mg/m^3) (OSHA/NIOSH)		50
Hydrogen cyanide		10 ppm (11 mg/m^3), skin absorption (OSHA)	4.7 ppm (5 mg/m^3), skin absorption (NIOSH)	50
Hydrogen sulfide	712	Ceiling 20 ppm (28.4 mg/m^3) 50 ppm/10-min maximum peak (OSHA) 10 ppm (15 mg/m^3), 10-min maximum peak (NIOSH)		100
Nitric oxide	115	25 ppm (30 mg/m^3) (OSHA/NIOSH)		100
Nitrogen dioxide	115	Ceiling 5 ppm (9 mg/m^3) (OSHA)	1 ppm (1.8 mg/m^3) (NIOSH)	20
Phosgene	5	0.1 ppm (0.2 mg/m^3) (OSHA) 0.1 ppm (0.4 mg/m^3) Ceiling 0.2 ppm (0.8 mg/m^3), 15-min exposure (NIOSH)		2
Phosphine	20	0.3 ppm (0.4 mg/m^3) (OSHA/NIOSH)	1 ppm (1 mg/m^3) (NIOSH)	50
Sulfur dioxide	2520	5 ppm (13 mg/m^3) (OSHA) 2 ppm (5 mg/m^3) (NIOSH)	5 ppm (13 mg/m^3) (NIOSH)	100

10.6-A LETHAL DOSE, 50% KILL

The **lethal dose, 50% kill**, or **LD$_{50}$**, is the amount of a substance that kills half of a group of laboratory animals to which the substance is administered during a preestablished time. The LD$_{50}$ is expressed in milligrams of administered substance per body mass of the animal in kilograms (mg/kg). When a substance affects a specific organ, the LD$_{50}$ may be measured by considering the mass of that organ.

The lethal dose of a substance to humans is calculated using the LD$_{50}$ measurement obtained from animal studies. For an average person having a mass of w kilograms, the lethal dose is the product of the LD$_{50}$ and the mass:

$$\text{Lethal dose} = \text{LD}_{50} \times w$$

lethal dose, 50% kill (LD$_{50}$) ▪ The dose of a substance that is lethal to 50% of a group of laboratory animals tested during a specified time

---

SOLVED EXERCISE 10.2

The LD$_{50}$ (rabbit) of phenol is 630 mg/kg by skin contact. Use this measurement as a technical basis to estimate the lethal dose of phenol for a 200-pound emergency responder by skin contact.

Solution: The lethal dose of phenol for a 200-pound person by skin contact is calculated as follows:

$$200 \, \text{lb} \times \frac{1 \, \text{kg}}{2.2 \, \text{lb}} \times 630 \times \frac{\text{mg}}{\text{kg}} \times \frac{1 \, \text{g}}{1000 \, \text{mg}} = 57 \, \text{g}$$

When a 200-pound emergency responder absorbs 57 g or more of phenol through the skin, the exposure is likely to be lethal.

10.6-B LETHAL CONCENTRATION, 50% KILL

The **lethal concentration, 50% kill**, or **LC$_{50}$**, is the concentration of a substance that kills half of a group of laboratory animals to which the substance is administered during a preestablished time. A LC$_{50}$ is typically expressed in parts per million (ppm) by volume. These measurements are used in different disciplines for a variety of purposes. DOT uses LC$_{50}$ values to characterize materials in hazard classes 2.3 and 6.1 to establish the hazard zone for a substance that poses a health hazard by inhalation. We review this process in Section 10.8.

The lethal concentration of a substance to humans is calculated using the LC$_{50}$ measurement obtained from animal studies. For an average person having a mass of w kilograms, the lethal concentration is the product of the LC$_{50}$ and the mass:

$$\text{Lethal dose} = \text{LD}_{50} \times w$$

lethal concentration, 50% kill (LC$_{50}$) ▪ The concentration of a substance that is lethal to 50% of a group of laboratory animals tested during a specified time

---

10.6-C THRESHOLD LIMIT VALUE

The **threshold limit value**, or **TLV**, is the upper limit of a concentration to which an average healthy person can be repeatedly exposed on an all-day, everyday basis without suffering adverse health effects. The TLV for airborne gaseous substances is usually expressed in parts per million. The TLV for airborne fumes, mists, and particulates is expressed in milligrams per cubic meter (mg/m^3). The TLV is the level of exposure at which the probability of the occurrence of adverse health effects is deemed negligible. These values are established by the American Conference of Governmental Industrial Hygienists (ACGIH), which recommends them to OSHA for its consideration of the adverse impact on a worker's long-term health.[1]

threshold limit value (TLV) ▪ A guideline standard established by ACGIH for exposure to an airborne concentration of a substance to which an average worker may be repeatedly exposed, day after day, without experiencing adverse health effects

---

[1]TLVs are published by the American Conference of Governmental Industrial Hygienists in *Documentation of the Threshold Limit Values and Biological Exposure Indices*, Cincinnati, Ohio (2012).

10.6-D PERMISSIBLE AND CEILING EXPOSURE LIMITS

The **permissible exposure limit**, or **PEL**, is the time-weighted average level of exposure to a toxic substance to which workers can be exposed continuously for either an 8-hour workday or a 40-hour workweek in conformance with relevant OSHA regulations.

The **ceiling limit**, or **C**, for some substances is also recommended by NIOSH to OSHA. This value represents the maximum concentration to which workers may safely be exposed without experiencing ill effects. When workers are exposed to a concentration of a substance that exceeds its ceiling limit, their safety is considered to be in imminent danger.

10.6-E SHORT-TERM EXPOSURE LIMIT

The **short-term exposure limit**, or **STEL**, is the maximum concentration of a substance to which workers may be exposed continuously for a time period of 15 minutes without suffering irritation, chronic or irreversible tissue damage, or narcosis sufficient to increase the risk of accidental injury, impair self-rescue, or materially reduce work-related efficiency. The STEL should not be exceeded more than four times throughout a working day with at least 60 minutes between exposures. STELs for specific substances are published when toxicological effects from relatively elevated short-term exposures to either humans or animals have been reported. They are established by NIOSH.[2]

10.6-F IMMEDIATELY-DANGEROUS-TO-LIFE-AND-HEALTH LIMIT

The **immediately-dangerous-to-life-and-health limit**, or **IDLH**, is the airborne concentration of any substance that poses an immediate threat to life, causes irreversible or delayed adverse health effects, or interferes with an individual's ability to escape during a 15-minute period from a dangerous atmosphere. These values are established by NIOSH.

10.6-G RECOMMENDED EXPOSURE LIMIT

NIOSH also recommends to OSHA the concentrations of certain substances that a facility may use as guidelines for protecting its workers against an injurious outcome. Each concentration is called a **recommended exposure limit**, or **REL**. They account for the safety of an employee during his or her tenure at a facility when they are used in combination with recommended protective equipment such as goggles and masks, posted signs that warn of potential dangers from exposure, monitoring of exposure levels, and other prudent safety measures.

NIOSH develops and publishes RELs that are based on a time-weighted average concentration for a 10-hour workday.

10.7 CPSC CRITERIA OF A TOXIC SUBSTANCE

At 16 C.F.R. §1500.3(c)(1), CPSC defines a consumer product as toxic if it can produce personal injury or illness to humans when it is inhaled, swallowed, or absorbed through the skin. Certain tests are conducted on animals to determine whether a product can cause immediate injury. CPSC also regards a product as toxic if it can cause long-term chronic effects like cancer, birth defects, or neurotoxicity.

---

[2]STELs, PELs, and IDLHs are published by the National Institute of Occupational Safety and Health in *NIOSH Pocket Guide to Chemical Hazards*, U.S. Department of Health and Human Services, Public Health Service, Centers for Disease Control and Prevention, Washington, DC (2010). PELs are also components of the standards for general industry workers published at 29 C.F.R. §1910.1000, Table Z-1; for construction industry workers at 29 C.F.R. §1926.55, Appendix A; and for maritime workers at 29 C.F.R. §1915.1000, Table Z.

At 16 C.F.R. §1500.3(c)(2)(i), CPSC defines a consumer product as highly toxic if it complies with any of the following conditions:

- It produces death within 14 days in half or more than half of a group of 10 or more laboratory white rats, each weighing between 200 and 300 grams, at a single dose of 50 mg/kg or less when orally administered; or
- It produces death within 14 days in half or more than half of a group of 10 or more laboratory white rats, each weighing between 200 and 300 grams, when inhaled continuously for a period of 1 hour or less at an atmospheric concentration of 200 parts per million by volume or less of gas or vapor or 2 mg/L by volume or less of mist or dust, provided such concentration is likely to be encountered by humans when the substance is used in any reasonably foreseeable manner; or
- It produces death within 14 days in half or more than half of a group of 10 or more rabbits tested in a dosage of 200 mg/kg of body weight, when administered by continuous contact with the bare skin for 24 hours or less

10.8 THE HAZARD ZONE

In Chapter 6, it was noted that DOT assigns one of four hazard zones to gases and one of two hazard zones to liquids that pose a health hazard by inhalation of their vapors. DOT determines the hazard zones for all toxic gases and liquids by reference to their LC_{50}s as follows:

- When the LC_{50} is equal to or less than 200 parts per million, DOT assigns Zone A to the gas or liquid.
- When the LC_{50} is greater than 200 parts per million but equal to or less than 1000 parts per million, DOT assigns Zone B to the gas or liquid.
- When the LC_{50} is greater than 1000 parts per million but equal to or less than 3000 parts per million, DOT assigns Zone C to the gas.
- When the LC_{50} is greater than 3000 parts per million but equal to or less than 5000 parts per million, DOT assigns Zone D to the gas.

Emergency responders use the hazard zone designation to determine the degree of toxicity of a given gas or liquid. Although exposure to a gas with a Zone A designation is likely to be immediately dangerous to life, exposure to a gas with a Zone D designation could be momentarily tolerable for some individuals. This distinction notwithstanding, the warning to emergency responders at transportation mishaps involving the spill or leak of a gas or liquid to which any of the four hazard zones has been assigned is to don fully-encapsulating protective clothing and use self-contained breathing apparatus before executing a response action.

10.9 TOXICITY OF THE FIRE SCENE

The fire scene is usually an extremely dangerous environment. Even if all its physical hazards could be eliminated, the atmosphere would still be filled with smoke, dust, toxic gases, and other hazardous substances. The potential for illness and death at a fire scene most often results when individuals are exposed to the smoke and toxic gases produced during the fire.

Put simply, **smoke** is the gray-to-black plume of matter consisting of an airborne dispersion of finely divided **particulate matter** composed of carbon particles whose diameters range from 0.01 to 10 micrometers. During most fires, carbon is initially produced as a product of incomplete combustion as microscopic particles, which rapidly agglomerate into black particulates collectively referred to as **soot**. When they are inhaled, the larger particulates in smoke are usually filtered in the nasal passageways, but the smaller ones can be drawn into the bronchi and lungs. When soot is produced during the combustion of petroleum products, it is primarily carbon black (Section 7.6-F).

smoke ■ The air dispersion of particles of carbon and other solids and liquids of incomplete combustion

particulate matter ■ Solid and liquid particles suspended in the atmosphere

soot ■ The agglomeration of particulate matter generated during the incomplete combustion of carbonaceous materials

EPA divides particulate matter into two classes, PM–2.5 and PM–10. PM–2.5 refers to ultrafine particulate matter having particle diameters equal to or less than 2.5 micrometers. PM–10 refers to particulate matter having particle diameters ranging from 2.5 to 10 micrometers. The PM–2.5 particles pose the greater risk to one's health, because they lodge deeply in the lungs and are not cleared by coughing. Trapped by the surrounding tissue of the lungs, they obstruct their proper functioning and contribute to the onset of pulmonary illnesses and premature death.

The long-term inhalation of particulate matter has been linked with an increase in cardiovascular and pulmonary diseases in susceptible individuals. In particular, the inhalation of particulate matter hastens the deaths of the sick and elderly by contributing to the premature onset of heart attacks, strokes, and emphysema. It is logical to infer that soot particulates can also contribute to the inception of adverse health effects in those firefighters who inhale soot regularly while combating fires.

Figure 10.4 demonstrates that the production of smoke is directly linked with the incomplete combustion of matter. When produced at a fire scene, the smoke and combustion products like carbon monoxide and hydrogen cyanide often represent a greater hazard to life and a more serious hindrance to firefighting efforts than the fire itself. Individuals who are unable to escape from a fire scene often die from smoke inhalation, even before the fire reaches them.

Fires are not the sole source of soot particles, although they generally are the indirect source. Other sources include the flyash in stack emissions from fossil-fuel-fired power plants and industrial manufacturing plants and the tailpipe emissions from heavy-duty vehicles and machinery.

10.9-A THE IMPACT OF SMOKE ON VISION

lacrimator ■ Any substance that causes the eyes to involuntarily tear and close

When the eyes are exposed to the irritant components of smoke, they sting and involuntarily tear and close due to the presence of certain irritants in smoke called **lacrimators** (Section 13.12). Their vapors act on the sensitive nerve endings of the mucous membrane of the eyes and cause an excessive, involuntary unleashing of a flood of tears. Two examples of lacrimators found in wood smoke are formaldehyde and acrolein. In the most severe instances, exposure to them causes zero visibility, especially when smoke cannot readily escape from a burning structure.

10.9-B ILL EFFECTS CAUSED BY INHALING SMOKE

The most immediate ill effect caused by inhaling smoke is heat damage to the tissues of the respiratory tract. This damage is usually limited to the tissues of the mouth and upper throat, but in the most egregious instances, it can cause pulmonary edema of the lungs (Section 7.3-B).

cilia ■ Fine hairlike projections such as those along the exterior of the respiratory tract that move in unison to help prevent the passage of fluids and fine particulate matter into the lungs

Another serious problem caused by smoke inhalation is deposition of carbon particulates on the surfaces of the respiratory passages. The trachea and bronchi are lined with tiny hairlike projections called **cilia** that act in a wavelike manner to force deposited particles upward towards the esophagus, where they generally are swallowed or deposited in phlegm. Inhaled smoke, however, greatly impairs the ability of the cilia to effectively remove carbon particulates from the respiratory passages. Cilia that have been damaged by heat cause an individual to choke, gag, and experience labored breathing. Then, oxygen levels in the blood drop to seriously low levels, and fatalities are more likely.

Breathing particulate matter also increases the likelihood of contracting *lung cancer*, the form of cancer characterized by the emergence of malignant tumors on the lungs. Individuals who regularly inhale particulate matter are 8% more likely to develop lung cancer than nonexposed individuals. The onset of lung cancer is typically exacerbated by the simultaneous exposure to certain compounds that are adsorbed on smoke particulates. These compounds include the cancer-causing polynuclear aromatic hydrocarbons, which we visit in Section 12.12.

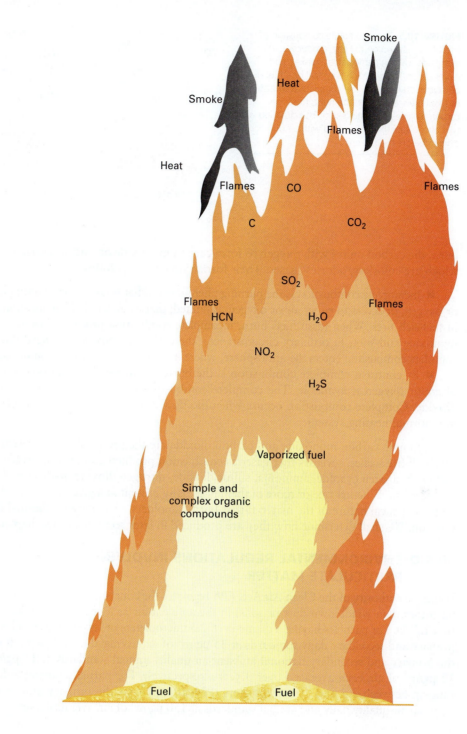

Smoke

Heat

Smoke

Flames

Heat

Flames

CO

C

CO₂

Flames

Flames

SO₂

Flames

Flames

HCN

H₂O

NO₂

H₂S

Vaporized fuel

Simple and
complex organic
compounds

Fuel   Fuel

FIGURE 10.4 Smoke, carbon monoxide, and water vapor are typically produced during the incomplete combustion of fossil fuels. Depending on the nature of the fuel, the atmosphere surrounding a fire may also contain vapors of the unburned fuel, its decomposition products, and other oxidation products. The inhalation of this combination of substances generally is more injurious to the health of firefighters and fire victims than are the accompanying flames or heat.

10.9-C THE ADSORPTION OF GASES ON THE SURFACES OF CARBON PARTICULATES

The carbon particulates in smoke act as adsorbents for gases. This implies that when we inhale smoke, it isn't just the carbon particulates that are drawn into our bronchi and lungs.

Which gases adsorb to the particulates? To answer this question, it is necessary to revisit two processes introduced earlier: incomplete and complete combustion. Their nature is schematically shown in Figure 10.5. When carbon-rich materials ignite and

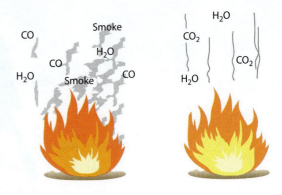

FIGURE 10.5 The incomplete combustion of fossil fuels, shown on the left, produces carbon monoxide, carbon particulates (smoke), and water vapor. This situation is typical of a fuel-rich or oxygen-starved fire. The complete combustion of fossil fuels, shown on the right, produces carbon dioxide and water vapor. This situation is typical of a fuel that was mixed with air or oxygen before its ignition.

burn, the carbon unites with oxygen to form carbon monoxide or carbon dioxide, both of which are colorless, odorless gases. Their formation occurs as follows:

■ Incomplete combustion is the burning phenomenon that occurs when the supply of air or oxygen, or access to either air or oxygen, is limited; that is, incomplete combustion occurs in fuel-rich fires. When fossil fuels burn, incomplete combustion processes produce carbon monoxide and soot. In ordinary circumstances, the carbon monoxide produced during incomplete combustion enters the atmosphere, where it slowly oxidizes to carbon dioxide.

■ By contrast, complete combustion is the burning process that occurs when plenty of air or oxygen is available. The complete combustion of fossil fuels produces carbon dioxide. Complete combustion occurs when the flow of fuel and air can be regulated, as in a normal heating system.

Although carbon monoxide and carbon dioxide are the principal gases present at virtually all fire scenes, several other gaseous combustion products can also be present. They include hydrogen cyanide, ammonia, sulfur dioxide, nitrogen dioxide, hydrogen chloride, and acrolein. Whether one or more of them is actually present at a given fire scene depends on the chemical nature of the material that burns, smolders, or undergoes thermal decomposition. We shall determine how they are produced in later sections of this chapter.

Carbon monoxide

10.9-D ENVIRONMENTAL REGULATIONS INVOLVING PARTICULATE MATTER

Using the authority of the Clean Air Act, EPA regulates the concentration of particulate matter (other than windblown dust or soils) in the ambient air as a criteria air pollutant (Section 1.3-A). For particles having a diameter of 2.5 micrometers or less (PM–2.5), EPA set the primary and secondary daily standards at 35 $\mu g/m^3$ of air, averaged over 3 years. It also set the primary and secondary national ambient air-quality annual standards at 15 $\mu g/m^3$ and 12 $\mu g/m^3$ of air, each averaged over 3 years, respectively. For particles having a diameter ranging between 2.5 and 10 micrometers (PM–10), EPA set the primary and secondary national air-quality daily and annual standards at 150 $\mu g/m^3$ of air, averaged over 3 years.

10.10 CARBON MONOXIDE

Carbon monoxide is an odorless, colorless, tasteless, and nonirritating gas at ordinary room conditions. It is also the major poison most likely to be encountered by emergency responders. Some other important physical properties are provided in Table 10.4.

As a commercial chemical product, carbon monoxide is largely used by the chemical industry for the production and manufacture of other substances like methanol (Section 13.2-A). Because it is a component of water gas (Section 7.2-C), carbon monoxide is produced in large volumes by the chemical industry during the manufacture of hydrogen.

TABLE 10.4	Physical Properties of Carbon Monoxide
Melting point	$-341°F$ $(-207°C)$
Boiling point	$-314°F$ $(-192°C)$
Specific gravity at 68°F (20°C)	0.81
Vapor density (air = 1)	0.969
Autoignition point	1128°F (609°C)
Lower flammable limit	12.5% by volume
Upper flammable limit	74.2% by volume

In the metallurgical industry, carbon monoxide is used to reduce metallic oxides to their corresponding metals. For example, the following equation shows that metallic copper is produced by the reduction of copper(II) oxide:

$$CuO(s) \ + \ CO(g) \ \longrightarrow \ Cu(s) \ + \ CO_2(g)$$

Copper(II) oxide   Carbon monoxide   Copper   Carbon dioxide

10.10-A PRODUCTION OF CARBON MONOXIDE

The isolation of carbon monoxide from water gas serves as the primary means by which carbon monoxide is produced for commercial use.

10.10-B ILL EFFECTS CAUSED BY INHALING CARBON MONOXIDE

Inhalation toxicity is the primary risk associated with exposure to carbon monoxide. To understand why this gas is poisonous, we must first examine the chemistry that occurs during respiration.

When air is inhaled into the lungs, a supply of atmospheric oxygen is assimilated into the bloodstream. The oxygen is carried throughout the body by a complex component of the blood called **hemoglobin**. Each red blood cell contains about 300 million hemoglobin molecules, each of which contains iron in the form of the ferrous ion (Fe^{2+}). The molecular structure of each hemoglobin molecule is very complex; hence, we represent it here by the symbol Hb.

When hemoglobin unites with oxygen, the compound called **oxyhemoglobin** is produced. Because its molecules are also complex, we represent them as O_2Hb. The importance of its formation during respiration is represented as follows:

$$Hb(aq) \ + \ O_2(aq) \ \longrightarrow \ O_2Hb(aq)$$

Hemoglobin   Oxygen   Oxyhemoglobin

hemoglobin ■ The component of red blood cells that transports oxygen to the tissues of the body

oxyhemoglobin ■ The compound that forms when oxygen combines with the blood's hemoglobin

As oxyhemoglobin molecules move about the circulatory system, they arrive at the various tissues and organs of the body, where the oxygen is released at the cellular level and they again become hemoglobin molecules. The hemoglobin then returns through the circulatory system to the lungs, where they secure a new supply of oxygen. This process occurs over and over with every breath we take.

When carbon monoxide is inhaled into the lungs, it interrupts normal respiration by bonding to the blood's hemoglobin molecules. This union produces the substance called **carboxyhemoglobin**, represented as COHb.

$$Hb(aq) \ + \ CO(g) \ \longrightarrow \ COHb(aq)$$

Hemoglobin   Carbon monoxide   Carboxyhemoglobin

carboxyhemoglobin ■ The compound that forms when carbon monoxide reacts with hemoglobin

TABLE 10.5	Ill Effects Associated with Various Carboxyhemoglobin Concentrations in Human Blood[a]

PERCENT CARBOXYHEMOGLOBIN	INHALATION AND TOXIC SYMPTOMS
0–10	No symptoms
10–20	Tightness across forehead; possible headache
20–30	Headache; throbbing in the temples; slowing of reflexes; drowsiness
30–40	Severe headache; weakness; dizziness; dim vision; nausea and vomiting; collapse
40–50	Same as immediately above with greater possibility of collapse or fainting; increased respiration and pulse
50–60	Fainting, increased respiration and pulse; coma with intermittent convulsions
60–70	Coma with intermittent convulsions; depressed respiration and cardiac function; possible death
70–80	Weak pulse and slowed respiration; respiratory failure and death
80–90	Death
90–100	Death

[a]Adapted in part from Laurence Brunton, Bruce Chabner, and Bjorn Knollman, *Goodman and Gilman's The Pharmacological Basis of Therapeutics*, 12th edition (New York, NY: McGraw-Hill Book Company, 2002), p. 1881.

When hemoglobin is bound to carbon monoxide as COHb, it is unable to perform its normal bodily function. This is particularly grave because hemoglobin's chemical affinity for carbon monoxide is roughly 210 times greater than its affinity for oxygen. Because COHb forms so readily when elevated concentrations of carbon monoxide are inhaled, the situation may be immediately deadly.

The ill effects caused by inhaling carbon monoxide are not solely associated with the production of COHb. Carbon monoxide inhaled into the lungs does not always bond to the blood's hemoglobin; instead, it assimilates into the cells of the body's tissues, where it may interfere with enzymatic processes. For this reason, long-term exposure to a low concentration of carbon monoxide may cause ill effects that are not otherwise experienced by a short-term exposure to the same concentration.

Individuals who inhale carbon monoxide experience the symptoms listed in Table 10.5, but there are multiple ways to acquire lower-than-normal blood oxygen levels. Firefighters, for example, are rarely exposed at fire scenes to carbon monoxide alone. It usually is produced with other combustion products like hydrogen cyanide and nitrogen oxides. Consequently, when firefighters inhale these mixtures of noxious gases, they may experience ill effects that are dissimilar from those listed in Table 10.5 for carbon monoxide alone.

When the symptoms are experienced by individuals who have inhaled carbon monoxide, they are linked with the conversion of hemoglobin into the biologically useless carboxyhemoglobin. It is the presence of COHb in the bloodstream that hinders the transport of oxygen and causes the headaches, fainting, throbbing in the temples, and other symptoms listed in Table 10.5.

pulse CO-oximeter ■ A device that may be used to measure the oxyhemoglobin, carboxyhemoglobin, and methemoglobin concentrations in a person's blood

10.10-C CARBON MONOXIDE AT FIRE SCENES AND OTHER LOCATIONS

Carbon monoxide becomes a component of the atmosphere at virtually every fire scene. To assure that firefighters do not inhale a lethal concentration of carbon monoxide, experts recommend their use of a **pulse CO-oximeter.** This device has been constructed to estimate the COHb concentration in the blood of the user.[3] Current models measure not only the concentration of COHb, but also the concentrations of oxyhemoglobin and methemoglobin (Section 10.14-B). These measurements are essential for facilitating the on-scene recovery of firefighters. Monitoring for poisoning by pulse CO-oximetry or other available methods is mandated by NFPA[4] for "any firefighter exposed to carbon monoxide or presenting with headache, nausea, shortness of breath, or gastrointestinal symptoms."

When a person's COHb concentration exceeds 50%, physicians recommend immediate medical attention for the individual. Even when the subsequent treatment is successful, however, survivors may still experience long-term cardiac, neurocognitive, and neuropsychiatric damage (such as the onset of dementia).

Exposure to carbon monoxide is linked with the majority of the illnesses and deaths that arise in connection with fires. However, carbon monoxide is not associated solely with the burning of buildings and other structures. As illustrated in Figure 10.6, the

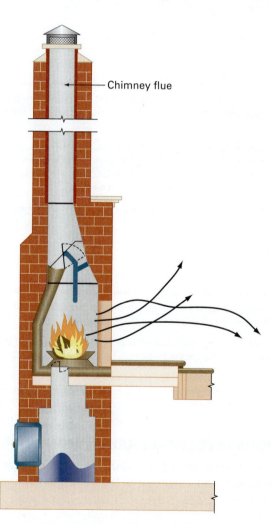

Chimney flue

FIGURE 10.6 The combustion products produced within an operating fireplace may enter the home when a chimney flue becomes blocked. Under this circumstance, the fireplace serves as a potential source of carbon monoxide poisoning. To prevent the development of a poisonous atmosphere in the home, occupants should always confirm that chimney flues are unblocked, so the noxious combustion products properly vent up the flue and into the outside atmosphere.

---

[3]An analogous device is the blood-oxygen-saturation meter called an *oxygen pulse oximeter*. It measures blood-oxygen saturation, i.e., the ratio of the total oxyhemoglobin concentration to the total hemoglobin concentration in the blood. Normal readings range from 95% to 100% at sea level. Values below 90% are considered low. Experts recommend that pilots use an oxygen pulse oximeter when flying unpressurized planes at altitudes between 8000 feet (2438 m) and 12,000 feet (3658 m) to determine whether they require supplemental oxygen to improve work-related tasks.

[4]NFPA 1584, *Standard on the Rehabilitation Process for Members During Emergency Operations and Training Exercises* (Quincy, Massachusetts: National Fire Protection Agency, 2008).

inhalation of carbon monoxide generated during the use of faulty furnaces, stoves, space heaters, blocked chimneys, and flues is considered the leading cause of accidental poisoning in the American home.

Women who are pregnant must be especially wary of exposure to carbon monoxide, because the gas is capable of crossing the placental barrier and dissolving in the blood of the unborn child. Because the fetus receives less oxygen, it is subjected to an increased risk of acquiring birth defects, particularly when the mothers have inhaled sufficient carbon monoxide to cause them to lose consciousness.

Although the primary risk associated with exposure to carbon monoxide is inhalation toxicity, bulk quantities of the gas also pose the risk of fire and explosion. When ignited, carbon monoxide oxidizes to carbon dioxide.

$$2CO(g) \quad + \quad O_2(g) \quad \longrightarrow \quad 2CO_2(g)$$

Carbon monoxide   Oxygen   Carbon dioxide

SOLVED EXERCISE 10.3

The carbon monoxide concentration in the exhaust from a car without a catalytic converter is approximately 8% by volume when the car is idling, but is reduced to approximately 4% when the car is moving. Why are these concentrations so dissimilar?

Solution: When the supply of air is restricted, as inside a car's combustion chambers, incomplete combustion occurs and carbon monoxide is produced when its fuel burns. A comparatively smaller volume of air is drawn into the car's combustion chambers during idling than when the car is moving. This causes a correspondingly larger amount of carbon monoxide to be produced when the car is idling compared to when it is moving.

SOLVED EXERCISE 10.4

Using a pulse CO-oximeter, a paramedic establishes that the blood of an on-duty firefighter contains carboxyhemoglobin at a concentration of 62% by volume. Is this concentration considered life-threatening?

Solution: Table 10.5 indicates that a carboxyhemoglobin concentration in the range of 60% to 70% could be fatal. Based on this information, it is prudent to conclude that a carboxyhemoglobin concentration of 62% is life-threatening.

10.10-D WORKPLACE REGULATIONS INVOLVING CARBON MONOXIDE

When the use of carbon monoxide is necessary in the workplace, OSHA requires employers to limit employee exposure to a maximum concentration of 50 parts per million (55 mg/m^3), averaged over an 8-hour workday.

10.10-E ENVIRONMENTAL REGULATIONS INVOLVING CARBON MONOXIDE

Relatively low concentrations of carbon monoxide are natural components of polluted air. Using the authority of the Clean Air Act, EPA regulates the carbon monoxide concentration in the ambient air as a criteria air pollutant by setting the primary national ambient air-quality standard for carbon monoxide at 9 parts per million (10 mg/m^3) as an 8-hour average and 35 parts per million (40 mg/m^3) as a 1-hour average.

FIGURE 10.7 To minimize or eliminate exposure to carbon monoxide from the use of portable generators, CPSC requires at 16 C.F.R. 1407.3 to affix this label to portable generators and their packaging.

10.10-F CONSUMER PRODUCT REGULATIONS INVOLVING CARBON MONOXIDE

The death toll from inhaling carbon monoxide in the home is considerable. This fact has caused the CPSC to inform the public that exposure to the gas is potentially lethal. The CPSC uses labeling to warn people about two relatively common ways of generating carbon monoxide: the burning of charcoal indoors in grills, hibachis, and similar items; and the operation indoors of portable generators. The production of carbon monoxide in both instances may kill a home's occupants in minutes.

As noted earlier in Section 7.6-G, CPSC requires charcoal manufacturers to affix the label shown in Figure 7.17 on charcoal packaging. CPSC and FEMA also require portable generator manufacturers to affix the label shown in Figure 10.7 on portable generators to warn the unsuspecting public that using these devices indoors could lead to fatalities.

10.10-G TRANSPORTING CARBON MONOXIDE

Carbon monoxide is commercially available as a nonliquefied compressed gas and cryogenic liquid. When shippers offer either commodity for transportation, DOT requires them to provide the relevant shipping description shown in Table 10.6 on the accompanying shipping paper. DOT also requires shippers and carriers to comply with all applicable labeling, marking, and placarding requirements.

TABLE 10.6	Shipping Descriptions of Carbon Monoxide
CARBON MONOXIDE	**SHIPPING DESCRIPTION**
Carbon monoxide, compressed gas	UN1016, Carbon monoxide, compressed, 2.3, (2.1), (Poison - Inhalation Hazard, Zone D)
Carbon monoxide, cryogenic liquid	NA9202, Carbon monoxide, refrigerated liquid, 2.3, (2.1), (Poison - Inhalation Hazard, Zone D)

10.10-H RESPONDING TO INCIDENTS ASSOCIATED WITH EXPOSURE TO CARBON MONOXIDE

When individuals lose consciousness from inhaling carbon monoxide, they should be moved swiftly to an open space where a means of artificial respiration and fresh oxygen can be administered. Breathing 100% oxygen for one hour reduces the concentration of carbon monoxide in the blood to approximately half its initial concentration.

When emergency responders have been excessively exposed to carbon monoxide, they are often transported to a facility where arrangements have been made to administer **hyperbaric oxygen therapy** under a physician's care. This treatment process involves subjecting them to a higher-than-normal-oxygen atmosphere inside a pressurized chamber of the type shown in Figure 10.8. The intermittent inhalation of 100% oxygen at a pressure of 2 atmosphere (202.6 kPa) for ½ to 1 hour can be life-saving because it results in the faster conversion of carboxyhemoglobin to oxyhemoglobin compared with the use of 100% oxygen at 1 atmosphere (101.3 kPa) for the same period. When hyperbaric oxygen therapy is used to treat individuals exposed to carbon monoxide, the oxygen acts as an antidote.

The use of a hyperbaric chamber was first introduced during the nineteenth century to aid divers suffering from work-related disorders. When divers move deeper and deeper

hyperbaric oxygen therapy ■ The inhalation of oxygen under increased pressure within a sealed steel chamber

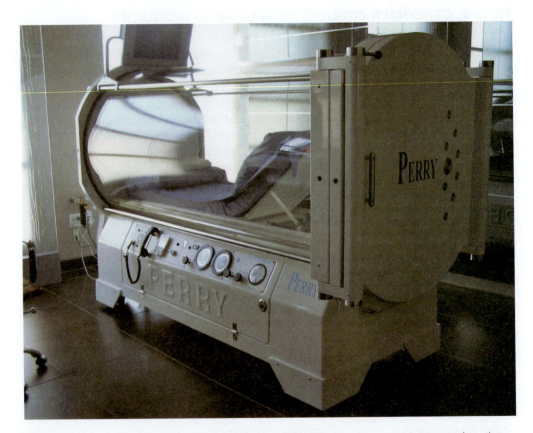

FIGURE 10.8 Hyperbaric oxygen therapy involves administering 100% oxygen at a pressure up to three times the normal atmospheric pressure for 30 to 60 minutes to patients while they lie inside enclosed chambers. The use of hyperbaric oxygen therapy at 2 atmospheres (202.6 kPa) accelerates the removal of carbon monoxide from the bloodstream. For this reason, it can potentially save the lives of emergency responders who have been overexposed to carbon monoxide. (*Courtesy of Perry Baromedical Corporation, Riviera Beach, Florida.*)

underwater, the increased pressure of the surroundings causes more nitrogen and oxygen to dissolve in their blood compared to the amount dissolved at the water's surface. When the divers return to the surface too rapidly, these gases form bubbles that release rapidly into their bloodstream and tissues. This compresses nerves and obstructs arteries, veins, and lymphatic vessels, causing excruciating pain, especially in the joints. This painful condition is commonly called "the bends." To avoid it, divers must return to the surface very slowly or take advantage of the use of a hyperbaric chamber. The hyperbaric chamber allows the divers to experience the same pressure that they experienced at underwater depths. Under this condition, the gases escape naturally.

Hyperbaric chambers have been used not only to treat divers and individuals exposed to carbon monoxide, but also individuals subjected with certain other ailments. Physicians have successfully treated diabetic foot ulcers, bone infections, thermal burns, and radiation burns to bone and soft tissue. The success of hyperbaric oxygen therapy for treating these injuries occurs for two major reasons:

■ The use of a hyperbaric chamber increases the amount of oxygen that the blood delivers to a patient's organs and tissues.

■ Breathing oxygen under increased pressure allows the blood to carry oxygen deeper into the body's organs and tissues. Under this increased pressure, the capillaries diffuse the blood farther, allowing oxygen-rich blood to be carried to areas that may otherwise be inaccessible.

Although the use of hyperbaric oxygen therapy has been successful for treating individuals exposed to carbon monoxide, the same degree of success has not been achieved when it is used for the treatment of individuals exposed to other poisonous gases like hydrogen cyanide (Section 10.11) and hydrogen sulfide (Section 10.13).

10.11 HYDROGEN CYANIDE

At temperatures above 79°F (26°C), hydrogen cyanide is a colorless, toxic gas. Some of its physical properties are provided in Table 10.7. In the chemical industry, it is used primarily to produce methacrylates (Section 14.6-E). In the past, hydrogen cyanide was used as a fumigant to kill rodents and insects, especially onboard ships, but given the accompanying human danger, this practice is no longer popular.

Hydrogen cyanide gas possesses the distinctive odor of bitter almonds, which normally is perceived at a concentration of 0.2–5.0 parts per million. Due to a unique genetic predisposition, however, some people cannot perceive the odor of bitter almonds.

Hydrogen cyanide

TABLE 10.7	Physical Properties of Hydrogen Cyanide
Melting point	7°F (−14°C)
Boiling point	79°F (26°C)
Specific gravity at 68°F (20°C)	0.69
Vapor density (air = 1)	0.938
Flashpoint	0°F (−18°C)
Autoignition point	~1000°F (~538°C)
Lower flammable limit	6% by volume
Upper flammable limit	41% by volume

As shown in the following illustration, there are two theoretical Lewis structures for a substance whose molecules contain only one carbon, hydrogen, and nitrogen atom:

$$H-C\equiv N: \qquad H-N\equiv C:$$

The compounds represented by these distinctly different structures are inseparable. For historic reasons, the substance called hydrogen cyanide is represented by the chemical formula HCN.

10.11-A PRODUCTION OF HYDROGEN CYANIDE

Hydrogen cyanide is produced as a commercial chemical product by the catalytic reaction between ammonia and air with natural gas.

$$\underset{\text{Ammonia}}{2NH_3(g)} + \underset{\text{Oxygen}}{3O_2(g)} + \underset{\text{Methane}}{2CH_4(g)} \longrightarrow \underset{\text{Hydrogen cyanide}}{2HCN(g)} + \underset{\text{Water}}{6H_2O(g)}$$

The hydrogen cyanide molecules produced by the reaction tend to react at an explosive rate with other hydrogen cyanide molecules. To prevent this reaction, the commercial grades of hydrogen cyanide are stabilized with water and 0.05% phosphoric acid.

10.11-B ILL EFFECTS CAUSED BY INHALING HYDROGEN CYANIDE

Exposure to hydrogen cyanide occurs primarily when the gas is inhaled, as a result of which the body experiences the ill effects listed in Table 10.8. The initial symptoms include dizziness, headache, diarrhea, and anemia, which are also characteristic of exposure to most poisons. Exposure to hydrogen cyanide can also occur by absorption through the skin.

Exposure to hydrogen cyanide generally is associated with the onset of **cyanosis**, a bluish coloration in the fingernail beds, lips, ear lobes, conjunctive, mucous membranes, and tongue. Cyanosis is caused when the lack of oxygen in the blood (*hypoxia*) is severe. The general public is most familiar with cyanosis as the cause of the *blue-baby syndrome*. Cyanosis is not solely symptomatic of hydrogen cyanide exposure. It also results when certain other harmful gases are inhaled.

cyanosis ■ The combination of physical symptoms associated with a reduced oxygen concentration in the blood

TABLE 10.8	Ill Effects Caused by Inhaling Hydrogen Cyanide[a]
HYDROGEN CYANIDE (PPM)	**SIGNS AND SYMPTOMS**
0.2–5.0	Odor threshold
10	Threshold limit value
18–36	Dizziness, headache, diarrhea, and anemia after several hours
50–60	Tolerated for 20 min to 1 hour without difficulty
120–150	May be fatal after 30-min exposure
181	Likely to be fatal after 10-min exposure
300	Immediately fatal

[a]Adapted in part from Fina Petrova Simeonova and Lawrence Fishbein, "Hydrogen cyanide and cyanides: Human health aspects," Concise International Chemical Assessment Document 61 (Geneva, Switzerland, World Health Organization, 2004), p. 30.

The mechanism by which hydrogen cyanide kills individuals is linked with its ability to inhibit the normal biological activity of cytochrome c oxidase, an enzyme essential for cellular respiration and energy production. Cyanide exposure renders the enzyme inactive. Although the supply of oxygen in the cells of the body's tissues may be plentiful, the hydrogen cyanide inhibits the ability of the cells to use the oxygen effectively.

Anaerobic cellular metabolism occurs when the body is incapable of using oxygen. This process forms cytotoxic byproducts, whose accumulation in the bloodstream contributes to cyanide toxicity.

10.11-C USE OF HYDROGEN CYANIDE FOR LEGAL EXECUTIONS

Hydrogen cyanide formerly was used as a means for carrying out the death penalties declared by certain judicial courts. Today, only six states—Arizona, California, Maryland, Mississippi, Missouri, and Wyoming—still authorize the use of a lethal gas under certain conditions for conducting judicially mandated executions. When hydrogen cyanide is the lethal gas, it is generated by dropping metallic cyanide pellets into a container of acid within an enclosed, airtight gas chamber.

$$NaCN(s) \quad + \quad HCl(aq) \quad \longrightarrow \quad NaCl(aq) \quad + \quad HCN(g)$$
Sodium cyanide   Hydrochloric acid   Sodium chloride   Hydrogen cyanide

In contemporary times, most states have condemned the use of a lethal gas for conducting death sentences. In fact, there exists today a nationwide inclination towards abolishing the death sentence altogether—by any means. In states where capital punishment is still authorized, the use of a lethal gas has been replaced by lethal injection as the preferred means of execution.

10.11-D HYDROGEN CYANIDE AS A CHEMICAL WARFARE AGENT

During the Holocaust in World War II, Nazi Germany used hydrogen cyanide as a chemical warfare agent for the taking of human life. The Germans had developed a method for adsorbing hydrogen cyanide on calcium sulfate pellets, and sealing them in steel cans for potential use as a pesticide called *Zyklon B*, where the "B" refers to the German word *Blausäure*, meaning "hydrocyanic." Hydrogen cyanide was generated when the pellets were exposed to the atmosphere.

Zyklon B was used by the Nazis as a weapon of mass destruction to systematically exterminate European Jews, non-Jewish Poles, political opponents, gypsies, homosexuals, disabled, and other people deemed to be "undesirable." At the Nazi death camps, pellets of the poison were dropped into the vents of locked rooms that served as gas chambers. The hydrogen cyanide that evolved killed the occupants within 20 minutes.

The most infamous Nazi death camp was located in Auschwitz, Poland. Historians estimate that exposure to hydrogen cyanide at this single camp alone caused the deaths of 1 to 2 million people. Their murders serve as one of the worst instances of genocide in the twentieth century. In all, it has been estimated that 6 million people died in the death camps.

In modern times, there have been allegations that domestic terrorists attempted to use hydrogen cyanide as a weapon of mass destruction. In 2003, the U.S. Department of Homeland Security issued a warning to law enforcement personnel that al-Qaida operatives planned to use hydrogen cyanide within the confines of the New York City subway system. Although this incident was never confirmed, it highlights the fact that cyanide could be used as a chemical weapon against Americans.[5]

Given the unorthodox ways by which terrorists may potentially use hazardous materials to kill massive numbers of people, law enforcement agencies should be provided with

[5]Ron Suskind, *The One Percent Doctrine* (New York, NY: Simon & Schuster, 2007), pp. 194–196.

a jurisdictional accounting of the cyanide products that terrorists may obtain for clandestine purposes.

10.11-E HYDROGEN CYANIDE AT FIRE SCENES

Hydrogen cyanide is generated at fire scenes by the thermal decomposition of consumer products manufactured from organic compounds whose molecules have one or more cyanide groups ($-C\equiv N$). These products include fabrics, carpeting, and mattresses manufactured from polyacrylonitrile and polyurethane. Hydrogen cyanide is released into the surrounding environment as a product of their decomposition.

Although hydrogen cyanide may form during a fire, scientists generally believe that it does not ultimately survive because its lower flammable limit is only 6% by volume. Hydrogen cyanide readily ignites by incomplete and complete combustion, as demonstrated by the following equations:

$$4HCN(g) \quad + \quad 5O_2(g) \quad \longrightarrow \quad 4CO(g) \quad + \quad 4NO(g) \quad + \quad 2H_2O(g)$$

Hydrogen cyanide   Oxygen   Carbon monoxide   Nitric oxide   Water

$$4HCN(g) \quad + \quad 9O_2(g) \quad \longrightarrow \quad 4CO_2(g) \quad + \quad 4NO_2(g) \quad + \quad 2H_2O(g)$$

Hydrogen cyanide   Oxygen   Carbon dioxide   Nitrogen dioxide   Water

The ease of combustion implies that hydrogen cyanide is likely to be present at its highest concentration during the early stages of a fire, when sufficient heat is present to initiate thermal decomposition but before the gas oxidizes.

Notwithstanding its ease of combustion, hydrogen cyanide may survive in the atmosphere of a fire scene when textiles smolder in confined-space environments having low levels of oxygen. On-scene firefighters must be wary because they may inhale the gas before it oxidizes. Because of this possibility, they are often treated as a precautionary measure even when it is unclear whether they actually inhaled the gas.

Firefighters who have survived acute exposure to hydrogen cyanide are treated by administration of an antidote. Today, many physicians choose to administer hydroxocobalamine.[6] When injected intravenously, preferably by paramedics at the fire scene, this substance binds the cyanide to form cyanocobalamine, which subsequently is excreted from the body in urine.

10.11-F WORKPLACE REGULATIONS INVOLVING HYDROGEN CYANIDE

When the use of hydrogen cyanide is needed in the workplace, OSHA requires employers to limit employee exposure to a concentration of 10 parts per million (11 mg/m^3) averaged over an 8-hour workday.

10.11-G TRANSPORTING HYDROGEN CYANIDE

When shippers offer hydrogen cyanide for transportation, DOT requires them to provide the relevant shipping description shown in Table 10.9 on the accompanying shipping paper. DOT also requires shippers and carriers to comply with all applicable labeling, marking, and placarding requirements. As noted in Section 6.6-D, when hydrogen cyanide is transported by rail in bulk packaging, DOT requires the carriers to display the POISON INHALATION HAZARD placards on white squares with black borders.

---

[6]Stephen W. Borron, Frédéric J. Baud, Bruno Mégarbane, and Chantal Bismuth, "Hydroxocobalamin for severe acute cyanide poisoning by ingestion or inhalation," *Amer. J. Emerg. Med.*, Vol. 25 (2007), pp. 551–558.

TABLE 10.9	Shipping Descriptions of Hydrogen Cyanide
HYDROGEN CYANIDE	**SHIPPING DESCRIPTION**
Hydrogen cyanide, stabilized (contains less than 3% water)[a]	UN1051, Hydrogen cyanide, stabilized, 6.1, (3), PG I (Marine Pollutant) (Poison Inhalation Hazard, Zone A)
Hydrogen cyanide, stabilized (contains less than 3% water; absorbed into a porous inert material)[a]	UN1614, Hydrogen cyanide, stabilized, 6.1, PG I (Marine Pollutant)

[a]Prior to its shipment, DOT requires the addition of a substance to hydrogen cyanide to inhibit its autopolymerization and thereby promote stabilization (Section 14.3).

10.11-H HYDROCYANIC ACID

Hydrogen cyanide dissolves in water to form a colorless solution known as *hydrocyanic acid*, or *prussic acid*. The acid solution is so weak it is incapable of turning litmus paper red. Solutions containing more than 20% hydrogen cyanide are so volatile that they have been used as fumigants and rodenticides in ships, warehouses, and greenhouses.

10.11-I TRANSPORTING HYDROCYANIC ACID

Three hydrocyanic acid solutions are available in commerce. When shippers intend to transport a hydrocyanic acid solution in bulk, DOT requires them to identify the appropriate solution as shown in Table 10.10 on an accompanying shipping paper. DOT also requires shippers and carriers to comply with all applicable labeling, marking, and placarding requirements.

10.11-J METALLIC CYANIDES

Metallic cyanides are ionic compounds composed of metallic and cyanide ions. Sodium cyanide and potassium cyanide are available commercially as solids and aqueous solutions, whereas copper(II) cyanide and zinc cyanide are available solely as solids. The aqueous solutions of these latter two substances are used to electroplate copper and zinc, respectively.

One of the major commercial uses of sodium cyanide is connected with the leaching of precious metals from their ores. As the price of gold and silver began to increase throughout the 2000s, mining for precious metals became profitable. Accordingly, the commercial demand for sodium cyanide has grown.

Sodium cyanide

TABLE 10.10	Shipping Descriptions of Hydrocyanic Acid
FORM OF HYDROCYANIC ACID	**SHIPPING DESCRIPTION**
Hydrocyanic acid, aqueous solutions containing not more than 20% hydrogen cyanide	UN1613, Hydrocyanic acid aqueous solutions, 6.1, PG I (Marine Pollutant) (Poison - Inhalation Hazard, Zone B)
Hydrocyanic acid, aqueous solutions containing less than 5% hydrogen cyanide	NA1613, Hydrocyanic acid aqueous solutions, 6.1, PG II
Hydrogen cyanide, solution in alcohol containing not more than 45% hydrogen cyanide	UN3294, Hydrogen cyanide solution in alcohol, 6.1, (3), PG I (Marine Pollutant) (Poison Inhalation Hazard, Zone B)

TABLE 10.11	Shipping Descriptions of Some Representative Metallic Cyanides
METALLIC CYANIDE	**SHIPPING DESCRIPTION**
Potassium cyanide, solid	UN1680, Potassium cyanide, solid, 6.1, PG I (Marine Pollutant) (Poison)
Sodium cyanide, solid	UN1689, Sodium cyanide, solid, 6.1, PG I (Marine Pollutant) (Poison)
Zinc cyanide	UN1713, Zinc cyanide, 6.1, PG I (Marine Pollutant) (Poison)

CPSC banned the sale of products containing water-soluble metallic cyanides as constituents of consumer products intended for use in the United States. This regulatory ban is published at 16 C.F.R. §1500.17.

10.11-K TRANSPORTING METALLIC CYANIDES

When shippers transport a metallic cyanide in bulk, DOT requires them to identify the appropriate substance on an accompanying shipping paper. Table 10.11 provides some representative examples. DOT also requires shippers and carriers to comply with all applicable labeling, marking, and placards requirements.

When the name of a metallic cyanide or its solution is not listed in the Hazardous Materials Table at 49 C.F.R. §172.101, DOT requires shippers to provide the relevant shipping description generically and to provide the name of the specific compound parenthetically.

10.12 SULFUR DIOXIDE

Sulfur dioxide

Sulfur dioxide is a colorless, nonflammable, toxic gas having the sharp, pungent odor associated with burning matches or tires. Some other physical properties of this gas are noted in Table 10.12. These data indicate that sulfur dioxide may also be encountered commercially as a liquefied gas. When used as a commercial chemical product, sulfur dioxide is primarily produced by burning sulfur.

Today, sulfur dioxide is used mainly as a bleaching agent by the pulp and paper industry. It is also used by the agricultural industry to whiten refined sugar and to lengthen the shelf life of dried fruits. When grapes are dried and processed in an atmosphere of sulfur dioxide, raisins with a golden color are produced. When they are not exposed to sulfur dioxide, the raisins are dark brown. Foods that have been treated with sulfur dioxide do not ferment or support the growth of fungus and mold.

Sulfur dioxide is also used to fumigate storage compartments onboard ships, where it functions as an insecticide and rodenticide. In the chemical industry, it is used primarily for the production and manufacture of sulfuric acid (Section 8.7) and metallic sulfites,

TABLE 10.12	Physical Properties of Sulfur Dioxide
Melting point	−105°F (−76°C)
Boiling point	14°F (−10°C)
Specific gravity at 68°F (20°C)	1.436
Vapor density (air = 1)	2.22

TABLE 10.13	Ill Effects Caused by Inhaling Sulfur Dioxide[a]
SULFUR DIOXIDE (PPM)	**SIGNS AND SYMPTOMS**
<0.1	No observable effect
≥0.1	Bronchoconstriction in sensitive, exercising asthmatics
0.3–1	Generally detectable by taste and smell
1–2	Lung function changes in healthy, nonasthmatic individuals
<5	Repeated exposure may cause permanent pulmonary impairment
8	20-minute exposure produces reddening of the throat and mild nose and throat irritation
20–50	Objectionable irritation of the eyes and development of chronic respiratory symptoms including coughing, chest pains, shortness of breath, and constriction of the airways
50–100	Maximum tolerable exposure limit for 30–60 min
≥100	Immediately dangerous to life

[a]Adapted from multiple sources including *CHEMINFO,* Sulfur Dioxide, Canadian Centre for Occupational Health and Safety (May 1999).

especially sodium sulfite, which is used as a reducing agent at water treatment plants. Although it was once used as a refrigerant, the inherent toxicity of sulfur dioxide has resulted in its replacement with nontoxic alternatives.

10.12-A ILL EFFECTS CAUSED BY INHALING SULFUR DIOXIDE

Inhalation toxicity is the primary hazard associated with exposure to sulfur dioxide. As shown in Table 10.13, coughing, chest pains, shortness of breath, and constriction of the airways symptoms associated with an exposure to this gas. These toxic effects are derived wholly from the ability of sulfur dioxide to directly irritate the moist mucous membranes of the upper respiratory tract and the lungs. Although many individuals are able to tolerate very low concentrations of sulfur dioxide—such as those routinely found in polluted air—for short periods of time, an exposure to a concentration of 100 parts per million in air is usually fatal. The specific cause of respiratory failure from exposure to sulfur dioxide is most likely associated with rapid lowering of the blood's pH.

Asthmatics are especially susceptible to experiencing ill effects from inhaling sulfur dioxide. Typically, when exercising, they are unable to tolerate a short exposure to a concentration as low as 0.1 part per million without experiencing constriction of the bronchiole tubes.

10.12-B ENVIRONMENTAL EVENTS ASSOCIATED WITH THE GENERATION OF SULFUR DIOXIDE

In December 1952, coal was virtually the sole fuel used in London for heating residential dwellings and operating local factories. During the five-day period, December 5–9, as Londoners burned coal for these purposes, sulfur dioxide and particulate matter were emitted into the atmosphere until their concentrations became deadly. The situation was exacerbated by the absence of a wind that prevented the pollutants from dissipating, forcing Londoners to inhale them. Subsequently, 4000 Londoners died during the initial days and another 8000 died in the next two months. Today, the event is recalled as the **Lethal London Smog Episode**, or **Great Smog**. It drew attention to the fact that our well-being depends on the quality of the air we breathe.

Lethal London Smog Episode (Great Smog)
▪ The deadly event that occurred when London residents inhaled sulfur dioxide that could not dissipate due to meteorological conditions

Chapter 10 Chemistry of Some Toxic Substances   **377**

In contemporary times, sulfur dioxide is still a major air component. Almost two-thirds of the sulfur dioxide in the air is generated by the coal-fired power plants that operate in the United States, despite EPA regulations that require its curbing in stack emissions by the use of scrubbers and other means.

Sulfur dioxide is also produced in ways other than the burning of coal. It is produced naturally during the eruption of volcanoes. In Hawaii, for example, the Kilauea volcano has continuously spewed massive amounts of sulfur dioxide and particulate matter into the atmosphere since at least January 3, 1983, when scientists first began studying its eruption.[7]

Sulfur dioxide released into the air slowly oxidizes to sulfur trioxide, which in turn reacts with atmospheric moisture to form sulfuric acid.

$$2SO_2(g) \ + \ O_2(g) \ \longrightarrow \ 2SO_3(g)$$

Sulfur dioxide   Oxygen   Sulfur trioxide

$$SO_3(g) \ + \ H_2O(g) \ \longrightarrow \ H_2SO_4(aq)$$

Sulfur trioxide   Water   Sulfuric acid

acid rain ■ Precipitation having an approximate pH of 5.6 or less and caused by dissolved nitrogen oxides and sulfur dioxide

The presence of sulfuric acid in the atmosphere produces a type of **acid rain**, that is, precipitation having a pH of 5.6 or less. Acid rain has severely damaged building materials made of concrete, marble, mortar, and limestone. A startling example of the negative impact that acid rain has had on stone structures is evident from observing Cleopatra's Needle, an obelisk that was moved from Egypt to New York City in the late nineteenth century. Its surface has deteriorated more in 100 years from exposure to acid rain than it did during the 3000 years it stood in Egypt.

The release of sulfur dioxide to the atmosphere produces the environmental problem referred to as *atmospheric cooling*. When sulfur dioxide is ejected into the atmosphere by an erupting volcano, it oxidizes to sulfuric acid, which in turn seeds aerosol particles. These particles reflect the incoming sun's rays away from our planet back into space, thereby reducing the amount of radiation that reaches Earth's surface. This phenomenon can cause a severe drop in temperature that persists for eons. For example, scientists have proposed that the sulfuric acid aerosols produced during volcanic eruptions in Indonesia caused the Little Ice Age, the approximate period from 1550 to 1850 during which bitterly cold winters were experienced in many parts of the world.

10.12-C SULFUR DIOXIDE AT FIRE SCENES

Sulfur-containing compounds are constituents of many products including the following:

- Coal, natural gas, and crude oil
- Complex proteins present in wool, hair, and animal hides (Section 14.5-B)
- Several natural and synthetic polymers, including vulcanized rubber (Section 14.11-C)

When these products burn, the sulfur is converted into sulfur dioxide. Firefighters encounter it when they respond to fires involving their combustion or the burning of other products to which sulfur was added.

10.12-D WORKPLACE REGULATIONS INVOLVING SULFUR DIOXIDE

When the use of sulfur dioxide is needed in the workplace, OSHA requires employers to limit employee exposure to a maximum concentration of 5 parts per million (13 mg/m^3), averaged over an 8-hour workday.

---

[7]A.J. Sutton et al., "Implications for eruptive processes as indicated by sulfur dioxide emissions from Kilauea Volcano, Hawaii, 1979–1997," *J. Volcanol. Geotherm. Res.*,Vol. 108 (2001) pp. 283–302.

10.12-E ENVIRONMENTAL REGULATIONS INVOLVING SULFUR DIOXIDE

Using the authority of the Clean Air Act, EPA regulates the sulfur dioxide concentration in the ambient air as a criteria air pollutant. EPA set the daily and annual primary standard for sulfur dioxide at 0.14 parts per million (365 $\mu g/m^3$) and 0.03 parts per million (80 $\mu g/m^3$), respectively. EPA also set the secondary standard for sulfur dioxide at 0.50 parts per million (1300 $\mu g/m^3$) as a 3-hour average.

10.12-F TRANSPORTING SULFUR DIOXIDE

Sulfur dioxide is available commercially as a liquefied compressed gas. When shippers transport sulfur dioxide, DOT requires them to identify it on an accompanying shipping paper as follows:

UN1079, Sulfur dioxide, 2.3, (8), (Poison - Inhalation Hazard, Zone C)

DOT also requires shippers and carriers to comply with all applicable labeling, marking, and placarding requirements.

When first-on-the scene emergency responders are called to a transportation mishap involving the release of a bulk shipment of sulfur dioxide, they must acknowledge that the safety of the team, transportation personnel, and the general public is at risk. The *Emergency Response Guidebook* recommends isolation and evacuation distances when large spills of sulfur dioxide occur from multiple small cylinders, single ton cylinders, multiple ton cylinders, rail tankcars, and highway tank trucks and trailers under prevailing wind conditions.[8]

10.13 HYDROGEN SULFIDE

Hydrogen sulfide is a colorless, flammable, and toxic gas having the physical properties noted in Table 10.14. Perhaps its most immediately apparent feature is the disagreeable stench of rotten eggs. Humans detect this smell at just 2 parts per billion. The stench is truly offensive at concentrations as low as 3 to 5 parts per million. Sometimes individuals describe the odor of a material by saying that it smells like sulfur. Because elemental sulfur is an odorless solid, what they actually mean to say is that the material smells like hydrogen sulfide.

Hydrogen sulfide

TABLE 10.14	Physical Properties of Hydrogen Sulfide
Melting point	−117°F (−86°C)
Boiling point	−76°F (−60°C)
Specific gravity at 68°F (20°C)	1.54
Vapor density (air = 1)	1.18
Flashpoint	−116°F (−82°C)
Autoignition point	500°F (260°C)
Lower flammable limit	4.3% by volume
Upper flammable limit	46% by volume

---

[8]Table 3, *Emergency Response Guidebook* (Washington, DC: U.S. Department of Transportation, 2012), p. 355.

Although hydrogen sulfide is available as a liquefied compressed gas, it is not a popular commercial product. For its limited use in commerce, it is primarily derived from petroleum refineries and natural gas wells. In the chemical industry, hydrogen sulfide is used to produce elemental sulfur and sulfuric acid, to process mineral ores, to produce metallic sulfides such as nickel sulfide and molybdenum sulfide, and to manufacture phosphors used in television tubes.

Hydrogen sulfide is produced naturally by the decay of organisms in swamps, sewers, underground storm drains, and other anaerobic (nonoxygen) environments from which the gas seeps into the atmosphere. The decomposition of sewage gives rise to the term *sewer gas*, a term used by sanitary engineers who work in sewers and at sewage treatment facilities. The odor of hydrogen sulfide is also encountered at petroleum refineries, where the gas is recovered by the desulfurization of sour crude (Section 12.13-F), and as a constituent of mammalian flatulence. The body generates trace amounts that it uses to regulate metabolism rates.

Because hydrogen sulfide has a highly repulsive odor, most people become aware of its presence immediately upon exposure. Nonetheless, when they experience long-term exposure to low concentrations, their sense of smell is temporarily deadened. This phenomenon is called **olfactory fatigue**. Continued exposure causes them to become oblivious to the presence of hydrogen sulfide, increasing the possibility that they may unknowingly inhale a lethal concentration.

olfactory fatigue ■ The temporary inability to identify the odor of an airborne substance after prolonged exposure

10.13-A ILL EFFECTS CAUSED BY INHALING HYDROGEN SULFIDE

Table 10.15 indicates that inhalation toxicity is the primary hazard associated with exposure to hydrogen sulfide. Exposure initially gives rise to dizziness and the onset of a headache, but unconsciousness and respiratory paralysis can follow immediately. Inhalation of air having a hydrogen sulfide concentration of 1000 parts per million is regarded as fatal.

The mechanism by which hydrogen sulfide causes adverse ailments is similar to that previously noted for hydrogen cyanide poisoning: Hydrogen sulfide reacts with cytochrome c oxidase, thereby preventing cellular respiration. The treatment of individuals exposed to elevated concentrations of hydrogen sulfide with hyperbaric oxygen has been only minimally successful. There is no known antidote.

TABLE 10.15	Ill Effects Caused by Inhaling Hydrogen Sulfide[a]
HYDROGEN SULFIDE (PPM)	**SIGNS AND SYMPTOMS**
0.011	Odor threshold
2.8	Bronchial constriction in asthmatic individuals
5.0	Increased eye complaints
7 or 14	Decreased oxygen uptake
5–29	Eye irritation
28	Fatigue, loss of appetite, headache, irritability, poor memory, dizziness
>140	Olfactory paralysis
>560	Respiratory distress
≥700	Death

[a]Adapted in part from C.-H Selene J. Chou, "Hydrogen sulfide: Human health aspects," Concise International Chemical Assessment Document 53 (Geneva, Switzerland, World Health Organization, 2003), p. 14.

10.13-B HYDROGEN SULFIDE AT FIRE SCENES

Hydrogen sulfide is a flammable gas. When ignited in air, it readily burns as follows:

$$2H_2S(g) \ + \ 3O_2(g) \ \longrightarrow \ 2H_2O(g) \ + \ 2SO_2(g)$$

Hydrogen sulfide   Oxygen   Water   Sulfur dioxide

At fire scenes, the presence of hydrogen sulfide—although possible—is generally improbable. Although small concentrations may evolve during the thermal decomposition of certain materials manufactured from animal products (including leather items and wool carpeting), the gas is easily consumed by combustion.

10.13-C WORKPLACE REGULATIONS INVOLVING HYDROGEN SULFIDE

When the use of hydrogen sulfide is needed in the workplace, OSHA requires employers to limit employee exposure to a maximum concentration of 50 parts per million in a 10-minute maximum period. When an area may contain an atmosphere of hydrogen sulfide, a sign such as the following should be posted:

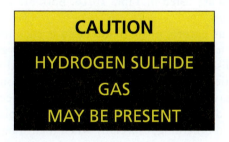

10.13-D TRANSPORTING HYDROGEN SULFIDE

Hydrogen sulfide is transported in steel cylinders and bulk transport vehicles. When shippers offer hydrogen sulfide for transportation, DOT requires them to provide its shipping description on an accompanying shipping paper as follows:

UN1053, Hydrogen sulfide, 2.3, (2.1), (Poison - Inhalation Hazard, Zone A)

DOT also requires shippers and carriers to comply with all applicable labeling, marking, and placarding requirements. When hydrogen sulfide is transported by rail, DOT requires carriers to post the POISON GAS placards on white squares with black borders on the bulk packaging used for shipment.

10.13-E RESPONDING TO INCIDENTS INVOLVING A RELEASE OF HYDROGEN SULFIDE

Most fatalities associated with hydrogen sulfide exposure have occurred when employees ignored safety practices at petroleum refineries and chemical manufacturing facilities, in sewer systems, and at other locations where hydrogen sulfide was stored or generated.

Special attention must be given to enclosures in which hydrogen sulfide may be generated unknowingly. For example, chambers are used for storing sewage onboard modern ships, until it can be treated on shore. Hydrogen sulfide can accumulate in these chambers at a lethal concentration. When it is necessary to repair fractures in the walls or their connecting pipes, workers must avoid exposure to this deadly gas by wearing

Nitric oxide

fully-encapsulated suits and breathing air from self-contained sources. Work should never be conducted in any area where hydrogen sulfide can potentially accumulate without first monitoring its concentration.

10.14 NITROGEN OXIDES

Eight oxides of nitrogen are known, but we are concerned here with only two, namely, nitric oxide (also known as nitrogen monoxide) and nitrogen dioxide. Their chemical formulas are NO and NO_2, respectively. The molecules of nitric oxide and nitrogen dioxide are stable free radicals.

$$:\!\overset{\cdot}{N}\!::\!\overset{\cdot\cdot}{O}\!: \qquad \cdot\overset{\cdot\cdot}{O}\!:\!N\!::\!\overset{\cdot\cdot}{O}\!:$$

Because they often are produced in combination, nitric oxide and nitrogen dioxide commonly are represented jointly as NO_x and are called the "nitrogen oxides." Nitric oxide is a colorless gas with an irritating odor, and nitrogen dioxide is a dark red-brown gas having a pungent, acrid odor. They are poisonous, nonflammable, and corrosive substances. Their physical properties are listed in Table 10.16. Both are available commercially as Hazard Class 2.3 gases.

Despite its poisonous nature, nitric oxide often is encountered in modern hospitals and clinics, where low doses are administered with oxygen to treat emphysema and other pulmonary diseases. Nitric oxide causes the blood vessels to dilate; consequently, respiratory patients experience reduced inflammation and lower blood pressure. Nitric oxide has saved the lives of premature babies, because it reduces the risk of fatalities from such diseases as bronchopulmonary dysplasia. It has also helped firefighters avoid labored breathing, especially when their lungs were coated with soot while combating fires.

Nitrogen dioxide is used by the aerospace industry to oxidize rocket fuels. For example, the Apollo astronauts used it with unsymmetrical dimethylhydrazine to leave the moon's surface.

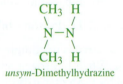

unsym-Dimethylhydrazine

A mixture of nitrogen dioxide and *unsym*-dimethylhydrazine is self-reactive, or **hypergolic**; that is, the substances react immediately on contact. The nitrogen dioxide and *unsym*-dimethylhydrazine act as an oxidizing agent and a reducing agent, respectively.

hypergolic ■ Pertaining to the ability of a substance to ignite spontaneously on contact with another substance

TABLE 10.16	Physical Properties of Nitric Oxide and Nitrogen Dioxide	
	NITRIC OXIDE	**NITROGEN DIOXIDE**
Melting point	−263°F (−164°C)	12°F (−11°C)
Boiling point	−243°F (−153°C)	68°F (20°C)
Specific gravity at 68°F (20°C)	1.34	1.49
Vapor density (air = 1)	1.04	1.59

Why do two nitrogen dioxide molecules combine to produce a single molecule of dinitrogen tetroxide?

Solution: The Lewis symbols for the nitrogen and oxygen atoms are $\cdot\ddot{N}\cdot$ and $\cdot\ddot{O}\cdot$, respectively. Using them, the following Lewis structure for nitrogen dioxide is written as follows:

$$\cdot\ddot{O}\!:\!N\!:\!:\!\ddot{O}\!: \qquad \text{or} \qquad O\!-\!N\!=\!O$$

Because this structure has an unpaired electron, it actually represents a free-radical molecule (Section 5.11). To complete the unfilled octet and achieve electronic stability, two nitrogen dioxide molecules combine to produce a single dinitrogen tetroxide molecule.

$$\begin{array}{c} O\!-\!N\!=\!O \\ | \\ O\!-\!N\!=\!O \end{array}$$

10.14-A PRODUCTION OF NO$_x$

Nitric oxide and nitrogen dioxide are produced for commercial use by the catalytic oxidation of ammonia as follows:

$$\underset{\text{Ammonia}}{4NH_3(g)} + \underset{\text{Oxygen}}{5O_2(g)} \longrightarrow \underset{\text{Nitric oxide}}{4NO(g)} + \underset{\text{Water}}{6H_2O(g)}$$

$$\underset{\text{Nitric oxide}}{2NO(g)} + \underset{\text{Oxygen}}{O_2(g)} \longrightarrow \underset{\text{Nitrogen dioxide}}{2NO_2(g)}$$

When confined in a cylinder at room temperature, the nitrogen dioxide combines with itself, mole for mole, to become dinitrogen tetroxide.

$$\underset{\text{Nitrogen dioxide}}{2NO_2(g)} \longrightarrow \underset{\text{Dinitrogen tetroxide}}{N_2O_4(l)}$$

Although dinitrogen tetroxide exists at room conditions, it is commonly still referred to as nitrogen dioxide.

10.14-B ENVIRONMENTAL ISSUES ASSOCIATED WITH THE NITROGEN OXIDES

In the lower atmosphere, nitrogen and oxygen combine very slowly to produce nitric oxide and nitrogen dioxide at ambient temperatures. However, the combination occurs rapidly at elevated temperatures.

$$\underset{\text{Nitrogen}}{N_2(g)} + \underset{\text{Oxygen}}{O_2(g)} \longrightarrow \underset{\text{Nitric oxide}}{2NO(g)}$$

$$\underset{\text{Nitric oxide}}{2NO(g)} + \underset{\text{Oxygen}}{O_2(g)} \longrightarrow \underset{\text{Nitrogen dioxide}}{2NO_2(g)}$$

At fossil-fuel-fired power plants and in the combustion chambers of motor vehicles, nitrogen dioxide is produced at the concentrations shown in Figure 10.9. For this reason, the gas is regarded as a major air pollutant. It is the principal culprit responsible for the brown atmospheric haze that hangs over many cities.

Nitrogen dioxide

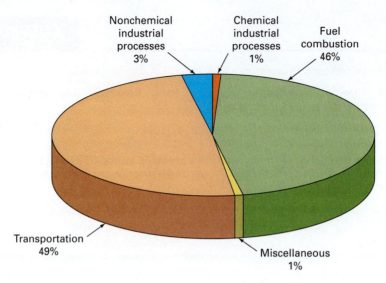

FIGURE 10.9 Nearly all the NO_x in the ambient air is generated during transportation and fuel-combustion activities. *(Courtesy of United States Environmental Protection Agency, Washington, DC.)*

Nonchemical industrial processes 3%

Chemical industrial processes 1%

Fuel combustion 46%

Transportation 49%

Miscellaneous 1%

The presence of NO_x in the lower atmosphere is directly linked with the production of ground-level ozone. Sunlight dissociates nitrogen dioxide into nitric oxide and oxygen atoms. The oxygen atoms rapidly combine with atmospheric oxygen to form ozone.

$$NO_2(g) \longrightarrow NO(g) + \cdot \ddot{O} \cdot (g)$$

Nitrogen dioxide   Nitric oxide   Oxygen atom

$$O_2(g) + \cdot \ddot{O} \cdot (g) \longrightarrow O_3(g)$$

Oxygen   Oxygen atom   Ozone

Nitrogen dioxide is also responsible for the formation in the lower atmosphere of a group of unstable compounds called *peroxyacyl nitrates*, or *PANs*. These compounds have the following general chemical formula, where R is an alkyl or aryl group (Sections 12.2-C and 12.11, respectively).

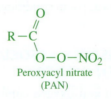

Peroxyacyl nitrate
(PAN)

These constituents of polluted air are especially powerful respiratory and eye irritants. The presence of ozone and PANs is characteristic of photochemical smog.

Concerned about the adverse impact on public health posed by NO_x, especially in urban atmospheres, EPA required automobile manufacturers to build cars that reduced the amount of NO_x emitted in vehicular exhaust. In response, catalytic converters were installed in new automobiles. One chemical reaction that occurs on the surface of the converter is the reduction of nitric oxide into environmentally friendly nitrogen and oxygen. The use of converters during the operation of modern-day motor vehicles has reduced the amount of NO_x that otherwise would be ejected into the atmosphere.

Although the nitrogen oxides can migrate to the ozone layer, their presence does not impact its quality. In the ozone layer, the nitrogen oxides react with ozone to produce nitrogen dioxide and normal oxygen. The nitrogen dioxide then reacts with oxygen atoms to reform nitrous oxide, which then proceeds again to react with ozone.

$$NO(g) \ + \ O_3(g) \ \longrightarrow \ NO_2(g) \ + \ O_2(g)$$

Nitrous oxide   Ozone   Nitrogen dioxide   Oxygen

$$NO_2(g) \ + \ \cdot\ddot{O}\cdot(g) \ \longrightarrow \ NO(g)$$

Nitrogen dioxide   Oxygen atom   Nitrous oxide

10.14-C ILL EFFECTS CAUSED BY INHALING NO_x

Inhalation toxicity is the primary hazard associated with exposure to both nitric oxide and nitrogen dioxide. Initial exposure produces irritation of the mucous membranes of the eyes, throat, nose, and lungs. As Table 10.17 shows, when an individual is exposed to nitrogen dioxide, these initial symptoms are followed by coughing, choking, headache, nausea, and fatigue. When the exposure is greater than 200 parts per million, the individual suffers from lung injury, bronchopneumonia, and possible fatal pulmonary edema.

Both nitric oxide and nitrogen dioxide are readily absorbed into the bloodstream, where they combine with hemoglobin to form **methemoglobin** [$(NO_2)Hb$]. For example, when nitrogen dioxide is inhaled, methemoglobin is produced as follows:

$$Hb(aq) \ + \ NO_2(g) \ \longrightarrow \ (NO_2)Hb(aq)$$

Hemoglobin   Nitrogen dioxide   Methemoglobin

methemoglobin
- Oxidized hemoglobin, produced when nitrogen dioxide reacts with the blood's hemoglobin

Methemoglobin is an oxidized form of hemoglobin. The iron in its molecules occurs as the ferric ion (Fe^{3+}), not the ferrous ion (Fe^{2+}). Like carboxyhemoglobin, methemoglobin cannot effectively transport oxygen to the body's tissues, so the host suffers from the ailment called **methemoglobinemia**. Individuals who inhale nitric oxide or nitrogen dioxide exhibit the initial symptoms of poisoning (dizziness, headache, diarrhea, and anemia), but their blood also becomes chocolate brown in color. It is this color that initially signals the severity of NO_x poisoning.

Hyperbaric oxygen therapy is ineffective when used to treat individuals afflicted with methemoglobinemia. Instead, blood exchange transfusions are usually needed to decrease the methemoglobin concentration in the bloodstream. Latent effects may also be experienced, and death can occur days after the initial exposure.

methemoglobinemia
- The health ailment resulting from an individual's exposure to nitrogen dioxide, metallic nitrites, metallic nitrates, and similar substances

TABLE 10.17	Ill Effects Caused by Inhaling Nitrogen Dioxide[a]
NITROGEN DIOXIDE (PPM)	**SIGNS AND SYMPTOMS**
5	Threshold limit for detection by smell
10–20	Mild irritation to the eyes, nose, and upper respiratory tract
25–40	No adverse effects to workers exposed over a period of years
50	Distinct irritation to the eyes, nose, and upper respiratory tract
80	Tightness in the chest after 3- to 5-min exposure
>200	Lung injury; bronchopneumonia; threat to life after 20- to 60-min exposure from pulmonary edema
250	Immediate danger of death following short-term exposure

[a]Adapted from multiple sources including Centers for Disease Control and Prevention, 1988 Permissible Exposure Limits Project Documentation, CDC 24/7.

The toxicity of nitrogen dioxide is exacerbated by the ability of the gas to react with water. A mixture of nitrous acid and nitric acid is produced as follows:

$$2NO_2(g) \; + \; H_2O(g) \; \longrightarrow \; HNO_2(aq) \; + \; HNO_3(aq)$$

Nitrogen dioxide   Water   Nitrous acid   Nitric acid

When it is inhaled, nitrogen dioxide reacts with atmospheric moisture to produce a mixture of these acids directly within the respiratory system. Because nitric acid is a strong acid, it damages the lining of the lungs and connective tissues, and even at low concentrations, its presence can cause pulmonary edema.

10.14-D NITROGEN OXIDES AT FIRE SCENES

Nitric oxide and nitrogen dioxide are the products of the incomplete and complete combustion, respectively, of nitrogen-containing materials. Examples of nitrogenous materials that burn to produce nitrogen dioxide include the polyurethane products used in building insulation, bedding, and furniture cushioning. Although its red-brown color is essentially unique, it may be impossible to detect the color in a predominantly black, smoky plume. Firefighters may determine whether they have been overly exposed to NO_x by use of a pulse CO-oximeter.

Certain commercial explosives also produce nitrogen dioxide when they detonate. Examples of these explosives include trinitrotoluene (Section 15.9), cyclonite (Section 15.10), tetryl (Section 15.11), and HMX (Section 15.13).

10.14-E WORKPLACE REGULATIONS INVOLVING THE NITROGEN OXIDES

When the use of nitric oxide and nitrogen dioxide is needed in the workplace, OSHA requires employers to limit employee exposure to maximum concentrations of 25 parts per million (30 mg/m^3) and 5 parts per million (9 mg/m^3), respectively.

10.14-F ENVIRONMENTAL REGULATIONS INVOLVING THE NITROGEN OXIDES

EPA regulates the concentration of NO_x in the ambient air as a criteria air pollutant. EPA has set both the primary and secondary national ambient air-quality standards for NO_x at 0.053 parts per million (100 μg/m^3) as an annual arithmetic mean.

10.14-G TRANSPORTING THE NITROGEN OXIDES

When shippers transport nitrogen dioxide (dinitrogen tetroxide) or nitric oxide, DOT requires them to provide the shipping description of the gas as shown in Table 10.18 on an accompanying shipping paper. DOT also requires shippers and carriers to comply with all applicable labeling, marking, and placarding requirements.

TABLE 10.18	Shipping Descriptions of the NO$_x$ Components
NITROGEN OXIDE	**SHIPPING DESCRIPTION**
Dinitrogen tetroxide	UN1067, Dinitrogen tetroxide, 2.3, (5.1), (8), (Poison - Inhalation Hazard, Zone A)
Nitric oxide	UN1660, Nitric oxide, compressed, 2.3, (5.1), (8), (Poison - Inhalation Hazard, Zone A)

TABLE 10.19	Physical Properties of Ammonia
Melting point	−108°F (−78°C)
Boiling point	−28°F (−33°C)
Specific gravity (liquid) at 68°F (20°C)	0.77
Specific gravity (gas) at 68°F (20°C)	0.68
Vapor density (air = 1)	0.589
Flashpoint	52°F (11°C)
Autoignition point	1204°F (651°C)
Lower flammable limit	16% by volume
Upper flammable limit	25% by volume

When these nitrogen oxides are transported by rail in bulk packaging, DOT requires the carriers to post the POISON GAS placards on white squares with black borders.

10.15 AMMONIA

Ammonia is a colorless gas that is easily detected and recognized by its familiar pungent odor. It has been known since the days of alchemy, when it was produced by heating either coal or animal hoofs and horns. Some of its important physical properties are noted in Table 10.19.

Over 80% of the ammonia produced in the United States is applied to farmlands as a fertilizer. Because the nitrogen content of ammonia is 82% by mass, it is an ideal agricultural fertilizer. When farmers apply it to farmland, they dispense the gas directly to the soil or into irrigation waters through a distribution pod from a tractor saddle tank or nurse tank like that shown in Figure 10.10. A **nurse tank** is a specific type of portable tank with a capacity of 3000 gallons (11 m^3) or less, is painted white or aluminum, and is never filled completely. It is mounted behind the tillage tool.

Less than 2% of the ammonia produced in the United States is used as a refrigerant in large refrigeration systems. It is the most economical coolant of choice for meatpacking plants, dairies, and similar facilities.

When ammonia is used as a refrigerant or fertilizer, it is encountered as the liquefied compressed gas called *anhydrous ammonia*. This product is transported by highway, railway, and watercraft and transferred by pipeline. In Alaska, anhydrous ammonia is the most prevalent hazardous substance throughout the state, where it is used as a refrigerant at seafood-processing plants and in radiators associated with the Trans-Alaska Pipeline System. In the latter, the ammonia disperses the heat that evolves from the hot crude oil as it travels through elevated parts of the 799-mile (1242-km) pipeline. The heat must be dispersed to prevent the melting of the underlying permafrost, the permanently frozen subsoil in which the supports for the elevated portions of the pipeline are embedded.

Anhydrous ammonia is also used in shuttlecraft, where it serves as the coolant in the air-conditioning system. It is also being used to cool the International Space Station.

10.15-A PRODUCTION OF AMMONIA

For commercial use, ammonia is produced from atmospheric nitrogen and hydrogen at elevated temperature and pressure. The production reaction is represented as follows:

$$N_2(g) + 3H_2(g) \longrightarrow 2NH_3(g)$$

Nitrogen   Hydrogen   Ammonia

Anhydrous ammonia

nurse tank ■ For purposes of DOT regulations, a cargo tank having a capacity of 3000 gallons (11 m^3) or less, painted white or aluminum, never filled to capacity, and used solely for the transportation of anhydrous ammonia

FIGURE 10.10 DOT requires a nurse tank used for the application of anhydrous ammonia to agricultural soils to be marked ANHYDROUS AMMONIA on all four sides and INHALATION HAZARD on two opposing sides. The DOT identification number 1005 must also be displayed across the center of green NON-FLAMMABLE GAS placards posted on all four sides of the tanks. (*Courtesy of The Fertilizer Institute, Washington, DC.*)

Chemical manufacturers generally produce ammonia at a pressure and temperature ranging from 3000 pounds per square inch (20,300 kPa) to 8900 pounds per square inch (10,800 kPa) and 842°F (450°C) to 1112°F (600°C), respectively.

10.15-B AMMONIA AT FIRE SCENES

Ammonia is a flammable gas within the range of 16% to 25% by volume. When it is released outdoors, ammonia dissipates with the wind, and a flammable mixture in air is not ordinarily attained. Under these circumstances, an ammonia fire cannot occur.

Nonetheless, ammonia concentrations within the flammable range are attained during some accident scenarios like the indoor release of ammonia into a confined area from which the ammonia cannot escape. In this circumstance, the occurrence of an ammonia fire is highly probable. When ammonia burns in air, the combustion evolves 9650 Btu/lb (22.5 kJ/g) and produces nitrogen and water vapor.

$$4NH_3(g) \ + \ 3O_2(g) \ \longrightarrow \ 2N_2(g) \ + \ 6H_2O(g)$$

Ammonia   Oxygen   Nitrogen   Water

This heat of combustion is low compared with the values listed in Table 5.2 for other substances. It is for this reason that ammonia fires are not sustainable unless they are exposed to an additional heat source.

TABLE 10.20	Ill Effects Caused by Inhaling Ammonia[a]
AMMONIA (PPM)	**SIGNS AND SYMPTOMS**
5–10	Detectable limit by odor
50	Odor detected, but no chronic effects
150–200	Very strong odor; general discomfort and irritation of nose, throat, and exposed moist skin; eye-tearing
400–700	Ordinarily, no serious results following short exposures; pronounced irritation, discomfort to the ears, nose, throat, bronchi, lungs, and moist skin, and severe eye irritation that may lead to loss of sight following long exposures
2000–3000	Barely tolerable for more than a few moments; convulsive coughing accompanied by severe eye irritation and blistering of the skin; danger of pulmonary edema, asphyxia, and death by suffocation even after a short exposure
5000–10,000	Respiratory spasms; rapid asphyxia; death by suffocation within minutes

[a]Adapted in part from OSHA Regulations Applicable to Ammonia (2007).

Ammonia may be generated at a fire scene in at least two ways:

- Materials manufactured from animal products—like leather items and wool carpeting—thermally decompose when they are exposed to intense heat. When these materials are present at fire scenes, the odor of ammonia often is perceptible, but the concentration generally is too low to cause harm.
- Fertilizers containing ammonium compounds decompose on exposure to heat. For example, when ammonium sulfate fertilizer is heated to high temperatures, it thermally decomposes to ammonia, sulfur dioxide, nitrogen, and water as follows:

$$3(NH_4)_2SO_4(s) \longrightarrow 4NH_3(g) + N_2(g) + 3SO_2(g) + 6H_2O(g)$$
Ammonium sulfate   Ammonia   Nitrogen   Sulfur dioxide   Water

10.15-C ILL EFFECTS CAUSED BY INHALING AMMONIA

When individuals are exposed to ammonia vapor, they experience the ill effects listed in Table 10.20. The eyes, skin, mouth, trachea, bronchi, and lungs are particularly susceptible to severe irritation from the exposure. The inhalation of ammonia at concentrations exceeding approximately 2000 parts per million causes death by suffocation.

10.15-D WORKPLACE REGULATIONS INVOLVING ANHYDROUS AMMONIA

When the use of anhydrous ammonia is needed in the workplace, OSHA requires employers to limit employee exposure to a concentration of 50 parts per million (35 mg/m^3), averaged over an 8-hour workday. OSHA also regulates how anhydrous ammonia is stored and handled in all workplaces other than ammonia manufacturing facilities and refrigeration plants using anhydrous ammonia solely as a refrigerant.

10.15-E TRANSPORTING ANHYDROUS AMMONIA

When shippers offer anhydrous ammonia for transportation, DOT requires them to provide the relevant shipping description shown in Table 10.21 on a shipping paper. DOT

TABLE 10.21	Shipping Descriptions of Anhydrous Ammonia
CHEMICAL COMMODITY	**SHIPPING DESCRIPTION**
Anhydrous Ammonia	UN1005, Ammonia anhydrous, 2.2 (Inhalation Hazard)[a]
Anhydrous Ammonia	UN1005, Ammonia, anhydrous, 2.3, (8) (Poison - Inhalation Hazard, Zone D)[b]

[a]Domestic transportation.
[b]International transportation.

also requires shippers and carriers to comply with all applicable labeling, marking, and placarding requirements.

10.15-F RESPONDING TO INCIDENTS INVOLVING A RELEASE OF ANHYDROUS AMMONIA

Ammonia injures the human organism through two different mechanisms:

- As a cold liquid at −28°F (−33°C), anhydrous ammonia freezes skin and other tissues at the points of contact.
- As previously noted, individuals who breathe ammonia experience the adverse health effects noted in Table 10.20.

Given the nature of the ill effects resulting from exposure to ammonia, the use of self-contained breathing apparatus is essential when emergency responders arrive at a scene involving a release of ammonia. When released directly from a liquid storage tank into the atmosphere, ammonia generally is visible as a white fog consisting of droplets of condensed atmospheric moisture. Although its presence severely limits visibility in the immediate environment, the initial formation of the fog helps workers locate the spot at which the ammonia is leaking from its storage tank or container.

The vapor density listed in Table 10.19 is an indication that ammonia is lighter than air. When it is released outdoors into dry air, ammonia quickly disperses into the atmosphere, especially under windy conditions. When anhydrous liquid ammonia is released into moist air, however, small liquid droplets of ammonia in water are produced. Under such conditions, these droplets behave as a dense gas, traveling along the surface of the ground instead of rising into the air and dispersing.

Water absorbs a very large volume of ammonia: At room conditions, one volume of water absorbs 1176 volumes of ammonia gas. This physical property is used to advantage when responding to incidents involving a release of ammonia. The recommended practice is to establish a water curtain downwind from the point at which the ammonia is being released. The water dissolves the ammonia and reduces the concentration that can travel farther.

SOLVED EXERCISE 10.6

To protect public health, safety, and the environment, what immediate action should be taken by first-on-the-scene responders when they encounter a strong odor of ammonia in an enclosed room at an ice-cream plant?

Solution: Ammonia cannot ignite in air unless a concentration within its flammable range is exposed to an ignition source. The flammable range of ammonia is 16% to 25% by volume. A concentration within this range is readily achieved when ammonia escapes from large refrigeration units into an enclosed room. When emergency responders first detect an especially strong odor of ammonia, their primary concern should be aimed at eliminating the risk of its ignition. They can do this by opening doors and windows to vent the ammonia from the enclosure while simultaneously taking precautions to avoid generating a spark.

Because an exposure to ammonia burns the eyes and nasal passageways, and the inhalation of elevated concentrations causes immediate asphyxiation, the implementation of any emergency response action involving ammonia should be undertaken only while wearing a full-body suit and using self-contained breathing apparatus. No activity should ever be conducted in an enclosure in which the ammonia concentration is within its flammable range, as survival is unlikely if the ammonia subsequently ignites. Rescue workers may be unable to successfully treat individuals who have inhaled ammonia. The treatment may be limited to administering oxygen until they can be moved to a medical facility.

When first-on-the scene emergency responders are called to a transportation mishap involving the release of a bulk shipment of anhydrous ammonia, they must acknowledge that the safety of the team, transportation personnel, and the general public is at risk. The *Emergency Response Guidebook* recommends isolation and evacuation distances when large spills of anhydrous ammonia occur from multiple small cylinders, single ton cylinders, multiple ton cylinders, rail tankcars, and highway tank trucks and trailers under prevailing wind conditions.[9]

The largest transportation spill of ammonia in the United States occurred in 2002, when 31 rail tankcars derailed near Minot, North Dakota, during the wreck of a Canadian Pacific Railway train. Five of the 31 tank cars catastrophically ruptured and instantly released 146,700 gallons (555 m^3) of ammonia into the atmosphere. Exposure to the ammonia resulted in the death of a local resident. More than 1400 others experienced some degree of respiratory impairment.[10]

10.15-G AMMONIA SOLUTIONS

Almost everyone has encountered ammonia as the gas that escapes from the commercial cleaning product called *household ammonia*. This is an aqueous solution of ammonia primarily used for cleaning glass, marble, and porcelain surfaces. Chemists refer to any aqueous solution of ammonia as *ammonium hydroxide*, *aqueous ammonia*, or *ammonia solution*. Its chemical formula is $NH_3(aq)$, or NH_4OH. The concentration of ammonia dissolved in household ammonia ranges from 2% to 5% by volume. The vapor is perceptible at concentrations as low as approximately 5 parts per million. Most individuals do not find this concentration irritating.

Commercial ammonia solutions contain dissolved ammonia that ranges in concentration from 10% to more than 50%. These aqueous solutions are hazardous primarily owing to the ammonia that rapidly escapes from them.

10.15-H TRANSPORTING AMMONIA SOLUTIONS

DOT regulates the transportation of ammonia solutions according to their ammonia concentrations. When shippers offer an aqueous ammonia solution for transportation, DOT requires them to provide the relevant shipping description shown in Table 10.22. DOT also requires shippers and carriers to comply with all applicable labeling, marking, and placarding requirements.

---

[9]Table 1, *Emergency Response Guidebook* (Washington, DC: U.S. Department of Transportation, 2012), p. 292.
[10]Railroad Accident Report, "Derailment of Canadian Pacific Railway Freight Train 292-16 and Subsequent Release of Anhydrous Ammonia Near Minot, North Dakota, January 18, 2002," NTSB/RAR-04/01, PB2004-916301 (Washington, DC: National Transportation Safety Board, 2004).

TABLE 10.22	Shipping Descriptions of Aqueous Ammonia Solutions
FORM OF AQUEOUS AMMONIA SOLUTION	**SHIPPING DESCRIPTION**
Aqueous ammonia solution, 10–35% ammonia, relative density between 0.880 and 0.957 at 15°C	UN2672, Ammonia solution, 8, PG III
Aqueous ammonia solution, 35–50% ammonia, relative density less than 0.880 at 15°C	UN2073, Ammonia solution, 2.2
Aqueous ammonia solution, >50% ammonia, relative density less than 0.880 at 15°C	UN3318, Ammonia solution, 2.2, (Inhalation Hazard)[a]
Aqueous ammonia solution, >50% ammonia, relative density less than 0.880 at 15°C	UN3318, Ammonia solution, 2.3, (8), (Poison - Inhalation Hazard, Zone D)[b]

[a]Domestic transportation.
[b]International transportation.

10.16 RESPONSE ACTIONS AT SCENES INVOLVING A RELEASE OF TOXIC SUBSTANCES

At the scenes involving transportation mishaps, emergency responders may establish that a toxic substance has been released to the environment by observing any of the following:

- The hazard class numbers 2.3 or 6.1 are included in the shipping description of a toxic substance on a shipping paper
- Either POISON or INHALATION HAZARD is marked on its packaging
- White POISON GAS, POISON INHALATION HAZARD, or POISON labels are affixed to the packaging
- A white POISON GAS placard is displayed on each side and each end of the bulk packaging or transport vehicle containing any amount of a poisonous gas.
- A white POISON INHALATION HAZARD placard is displayed on each side and each end of the bulk packaging or transport vehicle containing any amount of a toxic substance that poses a health hazard by inhalation and is classed in hazard Zone A or Zone B.
- A white POISON placard is displayed on each side and each end of the bulk packaging or transport vehicle used to transport 1001 pounds (454 kg) or more of a toxic substance in hazard class 6.1 other than those that pose a health hazard by inhalation in hazard Zone A or Zone B.

initial isolation distance
■ For purposes of DOT regulations, the distance from a transportation mishap involving the release of a hazardous material to which everyone should be quickly moved in a crosswind direction

initial isolation zone
■ For purposes of DOT regulations, the area surrounding a transportation mishap involving the release of a hazardous material in which persons may be exposed to dangerous, life-threatening concentrations of a gas or vapor that poses an inhalation hazard

When a toxic substance has spilled or leaked during a transportation mishap, DOT provides emergency responders with a table of initial isolation and protective-action distances in its *Emergency Response Guidebook* (Section 6.7). Excerpts of this table are provided in Table 10.23 for the toxic substances noted in this chapter. To use this table effectively, first-on-the-scene responders must first determine whether a "small" or a "large" spill of the toxic substance has occurred. This determination can ordinarily be made by speaking with the individual most immediately responsible for transportation of the hazardous material. A "small" spill is one that involves a single package such as a 55-gallon (208-L) drum, a small cylinder, or a small leak from a large package. A "large" spill involves a release from bulk packaging or multiple releases from several small packages.

Once a determination regarding the size of the release has been established, the guidebook is used to determine each of the following:

■ The **initial isolation distance** is the distance from an emergency scene to which everyone should be quickly moved in a *crosswind* direction to establish the **initial isolation zone** shown in Figure 10.11(a). The initial isolation zone is the area surrounding an

TABLE 10.23   Initial Isolation and Protective-Action Distances[a]

ID NO.	GUIDE	NAME OF MATERIAL	SMALL SPILLS (FROM A SMALL PACKAGE OR SMALL LEAK FROM A LARGE PACKAGE)							LARGE SPILLS (FROM A LARGE PACKAGE OR FROM MANY SMALL PACKAGES)							
			FIRST ISOLATE IN ALL DIRECTIONS		THEN PROTECT PERSONS DOWNWIND DURING				FIRST ISOLATE IN ALL DIRECTIONS		THEN PROTECT PERSONS DOWNWIND DURING					
					DAY		NIGHT				DAY		NIGHT			
			m	ft	km	mi	km	mi	m	ft	km	mi	km	mi		
1005	125	Ammonia, anhydrous	30	100	0.1	0.1	0.2	0.1	150	500	0.8	0.5	2.0	1.3		
1005	125	Anhydrous ammonia														
1016	119	Carbon monoxide	30	100	0.1	0.1	0.2	0.1	200	600	1.2	0.8	4.8	3.0		
1016	119	Carbon monoxide, compressed														
1017	124	Chlorine	60	200	0.4	0.2	1.5	1.0	500	1500	3.0	1.9	7.9	4.9		
1023	119	Coal gas	60	200	0.2	0.1	0.2	0.1	100	300	0.4	0.2	0.5	0.3		
1023		Coal gas, compressed														
1045	124	Fluorine	30	100	0.1	0.1	0.2	0.1	100	300	0.5	0.3	2.3	1.4		
1045	124	Fluorine, compressed														
1050	125	Hydrogen chloride, anhydrous	30	100	0.1	0.1	0.3	0.2	60	200	0.3	0.2	1.3	0.8		
1051	117	Hydrocyanic acid, aqueous solutions, with more than 20% hydrogen cyanide	60	200	0.2	0.1	0.6	0.4	400	1250	1.4	0.9	3.8	2.4		
1051	117	Hydrogen cyanide, anhydrous, stabilized														
1051	117	Hydrogen cyanide, stabilized														
1052[b]	125	Hydrogen fluoride, anhydrous	30	100	0.1	0.1	0.5	0.3	300	1000	1.5	0.9	3.2	2.0		
1053	117	Hydrogen sulfide	30	100	0.1	0.1	0.4	0.3	300	1000	1.7	1.0	5.6	3.5		
1062	123	Methyl bromide	30	100	0.1	0.1	0.2	0.2	100	300	0.6	0.4	1.9	1.2		
1067	124	Dinitrogen tetroxide	30	100	0.1	0.1	0.4	0.2	300	1000	1.1	0.7	2.7	1.7		

(continued)

TABLE 10.23   Initial Isolation and Protective-Action Distances (*continued*)

ID NO.	GUIDE	NAME OF MATERIAL	SMALL SPILLS (FROM A SMALL PACKAGE OR SMALL LEAK FROM A LARGE PACKAGE)						LARGE SPILLS (FROM A LARGE PACKAGE OR FROM MANY SMALL PACKAGES)					
			FIRST ISOLATE IN ALL DIRECTIONS		THEN PROTECT PERSONS DOWNWIND DURING				FIRST ISOLATE IN ALL DIRECTIONS		THEN PROTECT PERSONS DOWNWIND DURING			
					DAY		NIGHT				DAY		NIGHT	
			m	ft	km	mi	km	mi	m	ft	km	mi	km	mi
1067	124	Nitrogen dioxide												
1076	125	Phosgene	100	300	0.6	0.4	2.7	1.7	500	1500	3.1	1.9	10.8	6.7
1079[b]	125	Sulfur dioxide	100	300	0.7	0.4	2.8	1.7	1000	3000	5.6	3.5	11.0+[c]	7.0+
1242	139	Methyldichlorosilane (when spilled in water)	30	100	0.1	0.1	0.3	0.2	60	200	0.8	0.5	2.5	1.6
1250	155	Methyltrichloro-silane (when spilled in water)	30	100	0.1	0.1	0.3	0.2	100	300	0.9	0.6	2.6	1.7
1295	139	Trichlorosilane (when spilled in water)	30	100	0.1	0.1	0.3	0.2	60	200	0.7	0.4	2.2	1.4
1613	154	Hydrocyanic acid, aqueous solution, with not more than 20% hydrogen cyanide	60	200	0.2	0.1	0.2	0.1	150	500	0.5	0.3	1.3	0.8
1613	154	Hydrogen cyanide, aqueous solution, with not more than 20% hydrogen cyanide												
1614	152	Hydrogen cyanide, stabilized (absorbed)	60	200	0.2	0.1	0.7	0.4	150	500	0.5	0.4	1.7	1.1
1660	124	Nitric oxide	30	100	0.1	0.1	0.6	0.4	100	300	0.6	0.4	2.3	1.5
1660	124	Nitric oxide, compressed												
1680	157	Potassium cyanide (when spilled in water)	30	100	0.1	0.1	0.2	0.1	100	300	0.3	0.2	1.2	0.8

(continued)

TABLE 10.23 Initial Isolation and Protective-Action Distances (*continued*)

ID NO.	GUIDE	NAME OF MATERIAL	SMALL SPILLS (FROM A SMALL PACKAGE OR SMALL LEAK FROM A LARGE PACKAGE)								LARGE SPILLS (FROM A LARGE PACKAGE OR FROM MANY SMALL PACKAGES)							
			FIRST ISOLATE IN ALL DIRECTIONS		THEN PROTECT PERSONS DOWNWIND DURING						FIRST ISOLATE IN ALL DIRECTIONS		THEN PROTECT PERSONS DOWNWIND DURING					
					DAY		NIGHT						DAY		NIGHT			
			m	ft	km	mi	km	mi			m	ft	km	mi	km	mi		
1680	157	Potassium cyanide, solid (when spilled in water)																
1689	157	Sodium cyanide (when spilled in water)	30	100	0.1	0.1	0.2	0.1			100	300	0.4	0.2	1.4	0.9		
1689	157	Sodium cyanide, solid (when spilled in water)																
1717	155	Acetyl chloride (when spilled in water)	30	100	0.1	0.1	0.3	0.2			100	300	1.0	0.6	2.8	1.7		
1749	124	Chlorine trifluoride	60	200	0.3	0.2	1.2	0.8			300	1000	1.5	0.9	4.6	2.9		
1829	137	Sulfur trioxide, inhibited	100	300	0.4	0.2	0.9	0.5			400	1250	2.9	1.8	5.7	3.5		
1829	137	Sulfur trioxide, stabilized																
1831	137	Sulfuric acid, fuming	100	300	0.4	0.2	0.9	0.5			400	1250	2.9	1.8	5.7	3.5		
1831	137	Sulfuric acid, fuming, with not less than 30% free sulfur trioxide																
1953	119	Compressed gas, flammable, poisonous, n.o.s. (Inhalation Hazard Zone A)	100	300	0.5	0.3	2.2	1.4			600	2000	2.6	1.7	8.6	5.4		
1953	119	Compressed gas, flammable, poisonous, n.o.s. (Inhalation Hazard Zone B)	30	100	0.1	0.1	0.3	0.2			300	1000	1.3	0.8	3.5	2.2		

(continued)

TABLE 10.23 Initial Isolation and Protective-Action Distances (*continued*)

ID NO.	GUIDE	NAME OF MATERIAL	SMALL SPILLS (FROM A SMALL PACKAGE OR SMALL LEAK FROM A LARGE PACKAGE) FIRST ISOLATE IN ALL DIRECTIONS		THEN PROTECT PERSONS DOWNWIND DURING DAY		NIGHT		LARGE SPILLS (FROM A LARGE PACKAGE OR FROM MANY SMALL PACKAGES) FIRST ISOLATE IN ALL DIRECTIONS		THEN PROTECT PERSONS DOWNWIND DURING DAY		NIGHT	
			m	ft	km	mi	km	mi	m	ft	km	mi	km	mi
1953	119	Compressed gas, flammable, poisonous, n.o.s. (Inhalation Hazard Zone C)	30	100	0.1	0.1	0.3	0.2	200	600	1.0	0.7	3.2	2.0
1953	119	Compressed gas, flammable, poisonous, n.o.s. (Inhalation Hazard Zone D)	30	100	0.1	0.1	0.2	0.1	200	600	0.8	0.5	2.0	1.3
1953	119	Compressed gas, flammable, toxic, n.o.s. (Inhalation Hazard Zone A)	100	300	0.5	0.3	2.2	1.4	600	2000	2.6	1.7	8.6	5.4
1953	119	Compressed gas, flammable, toxic, n.o.s. (Inhalation Hazard Zone B)	30	100	0.1	0.1	0.3	0.2	300	1000	1.3	0.8	3.5	2.2
1953	119	Compressed gas, flammable, toxic, n.o.s. (Inhalation Hazard Zone C)	30	100	0.1	0.1	0.3	0.2	200	600	1.0	0.7	3.2	2.0
1953	119	Compressed gas, flammable, toxic, n.o.s. (Inhalation Hazard Zone D)	30	100	0.1	0.1	0.2	0.1	200	600	0.8	0.5	2.0	1.3
2199	119	Phosphine	60	200	0.2	0.2	1.0	0.7	400	1250	1.3	0.8	4.1	2.5

[a]Adapted from *Emergency Response Guidebook* (Washington, DC: U.S. Department of Transportation, 2012), pp. 292–343. (Courtesy of the U.S. Department of Transportation, Pipeline and Hazardous Materials Safety Administration, Washington, DC.)
[b]Consult Table 3 in the *Emergency Response Guidebook* for this material.
[c]"+" means the distance can be larger in certain atmospheric conditions.

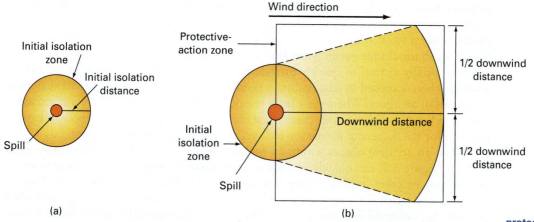

FIGURE 10.11 The initial isolation zone and the protective-action zone shown in (a) and (b), respectively, constitute regions in which individuals risk exposure to a hazardous material that poses a health hazard by inhalation. From a practical viewpoint, the initial isolation zone is the area of a circle whose radius equals the initial isolation distance, and the protective-action zone is the area of a square each of whose sides equals the protective-action (downwind) distance. The initial isolation distance and protective-action distance are published in the green section of the *Emergency Response Guidebook* (Washington, DC: U.S. Department of Transportation, 2012) (Section 6.7) for small and large spills of materials that pose a health hazard by inhalation.

protective-action distance ■ For purposes of DOT regulations, the minimum distance downwind from a transportation mishap involving the release of a hazardous material at which emergency responders and the public can reasonably anticipate that their health, safety, and welfare will be preserved

emergency scene in which persons may be exposed to dangerous, life-threatening concentrations of a toxic gas or vapor that poses an inhalation hazard. Emergency responders should first direct persons to areas outside the initial isolation zone.

■ The **protective-action distance** is the distance downwind from the scene that establishes the **protective-action zone** shown in Figure 10.11(b). The protective-action zone is the area in which persons may become incapacitated and unable to take protective action and/or incur serious, irreversible health effects. After they direct persons outside the initial isolation zone, emergency responders then should direct others outside the protective-action zone, beginning with those individuals nearest the right-hand semicircular component of the initial isolation zone.

protective-action zone ■ For purposes of DOT regulations, the area in which persons may become incapacitated, unable to take protective action, and/or incur serious, irreversible health effects

SOLVED EXERCISE 10.7

Show how an emergency responder uses Table 10.23 to identify the initial isolation and protective-action distances recommended by DOT for protection of public health at an 11:00 P.M. domestic transportation mishap involving a leaking 75-pound (34-kg) cylinder of hydrogen sulfide.

Solution: It is first necessary for the responder to identify whether this incident involves a "small" or a "large" spill of hydrogen sulfide. It is prudent to acknowledge that 75 pounds (34 kg) constitutes a "large" spill of a toxic substance. Furthermore, in most parts of the United States, it is dark at 11:00 P.M. Hence, we examine Table 10.23 under the heading "night" for the initial isolation and protective-action distances in relation to potential exposure to hydrogen sulfide. These distances are listed in Table 10.23 as 1000 feet (300 m) and 3.5 miles (5.6 km), respectively.

When emergency responders are required to encounter a toxic substance, they must first don appropriate protective clothing to prevent the possibility of skin contact. This protective clothing includes impermeable coveralls or similar encapsulating body suit, gloves, head coverings, and self-contained breathing apparatus. Because toxic substances enter the body not only by inhalation but also by skin absorption, the use of a fully encapsulating body suit minimizes or eliminates the risk of exposure to the substance by both routes. Once a toxic substance has come in contact with clothing, it is essential to thoroughly wash the clothing with water before removing it.

It is usually critical to quickly provide professional medical attention to individuals who have been exposed to toxic substances, even if they do not exhibit apparent signs or symptoms of an injury. Whenever a team responds to any accident involving the release of a toxic substance, information concerning the unique properties of the poison may be essential for the successful protection of lives, property, and the environment.

When an accident occurs within an enclosed area, the use of a confined-space entry monitor (Figure 3.2) can quickly provide an estimate of the vapor concentration to which the workers or other individuals are exposed. CHEMTREC and the American Association of Poison Control Centers may also be utilized as informational sources,[11] and when an injury has occurred in connection with the use of a commercial product, the U.S. Consumer Product Safety Commission should be contacted.[12]

Caution always needs to be exercised when emergency responders are called to a transportation mishap involving the release of a toxic substance. As a basis for making technical decisions, firefighters may compare the reading from a confined-space entry monitor with the STEL, PEL, and IDLH values for the toxic substance. If the decision to fight a fire involving a toxic substance is made, firefighters should also observe the following basic principles:

- Use impermeable coveralls or similar fully-encapsulated body suit and self-contained breathing apparatus.
- Avoid direct exposure to the smoke or fumes that evolve from these fires by attacking them from an upwind location.
- Use fog instead of direct streams of water to limit the generation of poisonous dust.
- Keep runoff water to a minimum, and channel runoff water into a temporary reservoir to prevent its entrance into local sewers or waterways.
- Notify the operators of the relevant storm and sanitary sewer system and water treatment plant of the ongoing fire.

10.17 CARCINOGENESIS

cancer ■ A large group of diseases characterized by the uncontrolled growth and spread of abnormal cells

Cancer is the name given to a relatively large group of diseases characterized by the uncontrolled growth and spread of abnormal cells. It is the most fearsome disease among populations living in developed countries. Despite decades of research work and billions of dollars spent to combat cancer, millions of individuals are still diagnosed with it annually.[13] Not only does cancer often cause their deaths directly, but the current cancer treatment processes—chemotherapy and radiation—compromise the immune systems of cancer patients so severely that individuals who survive treatment may ultimately succumb from pneumonia or other diseases.

carcinogen ■ A substance or agent that can contribute to causing the formation of a tumor in a living organism

In some instances, the onset of cancer may be traced to exposures to certain agents called **carcinogens**. Chemical carcinogens, physical carcinogens (like ultraviolet radiation), and viruses cause cancer in living tissues. In addition, many individuals inherit a

[11]The American Association of Poison Control Centers can be accessed by telephone at (800) 222–1222. It is essentially a library staffed by nonmedical personnel. Be prepared to give the name of the poisonous product involved in the emergency and any information provided on the label. The poison hotline may also be contacted when an individual overdoses on medication.

[12]The U.S. Consumer Product Commission may be contacted at (800) 638-CPSC.

[13]In 2010, the World Health Organization reported that cancer had become the leading cause of death worldwide, even surpassing heart disease. Over 12 million new cases are diagnosed annually.

genetic predisposition towards contracting cancer from their ancestors. In this text, our concern is primarily with an individual's potential exposure to chemical and physical carcinogens.

The two most common methods for determining whether exposure to a substance causes cancer are derived from studies conducted by toxicologists and epidemiologists. Toxicologists expose laboratory animals to varying doses of a substance to determine whether they develop cancer. Epidemiologists review historical exposures of humans to the same substance to determine whether the rate of development of cancer, if any, has increased in exposed populations.

Cancer may start as a disease of cells that malfunction because of an exposure to agents external to the body. These carcinogens enter or penetrate the cells (e.g., radiation) and corrupt them. They act with vastly different potencies, differing in their abilities to cause cancer by as much as a million-fold.

Although the mechanism of carcinogenesis—the production of cancer—is not completely understood, scientists now know that it consists of a multistepped cellular phenomenon that can be viewed in three stages: initiation, promotion, and progression.

■ During the **initiation** stage, the genetic material within a healthy cell or group of healthy cells becomes altered. Sometimes, this cellular alteration results from a single exposure to a carcinogen, whereas on other occasions it results from repeated exposures. Carcinogens that are initiators transform or mutate cellular DNA. Thus, the molecular pathway by which cancer can occur is established during the initiation stage. Overall, however, these initially damaged cells rarely cause cancer, because there are many cellular mechanisms for repairing the damaged DNA. It is when the cells are unable to repair the damage that they are more prone to becoming cancerous.

Cancer may also be initiated without exposure to any agent whatsoever. This happens when genetic errors occur naturally during the replication phase of cellular behavior.

■ During the subsequent **promotion** stage, the damaged cell divides into new but similarly damaged cells, each of which subsequently divides without restraint into new, similarly damaged cells. In most instances, cell division is accelerated when the damaged cells are exposed to carcinogens; that is, these carcinogens act as promoters rather than initiators. When cell division occurs over an extended period, the aggregates of damaged cells often create a mass of tissue called a **tumor**. (Leukemia is a cancer of the blood-forming parts of the body, particularly bone marrow and spleen, in which cancerous white blood cells form a diffuse liquid tumor throughout the circulatory system.) It is mind-boggling to realize that each tumor has descended from a single ancestral cell or group of cells that went awry. Some carcinogens are so powerful that they cause cancer without the need for promotion. Exposure to ionizing radiation (Section 16.2), for example, directly initiates a variety of cancers.

■ During the final **progression** stage, the damaged cells invade nearby tissues and migrate to other tissues and organs in the body, where they may provoke the emergence of tumors. Although tumors may be either benign or malignant, it is generally during the progression stage that malignant growths first develop. "Malignant" is a term applied to growths that penetrate the tissues in which they originated, spread further (metastasize) to other locations in the body, or cause death. Cancer is viewed, in part, as the transformation of previously healthy tissue cells into malignant tumors. The phenomenon associated with the formation of a malignant growth at one site in the body from cells derived from a malignancy located elsewhere is called **metastasis**. A tumor that is localized and noninvasive is benign, whereas a tumor that is invasive and metastatic is malignant.

Tobacco smokers represent a unique group of individuals for whom the incidence of cancer is especially high. More than 60 chemical carcinogens have been identified among the approximately 4000 substances known to be constituents of tobacco smoke. With each puff, cigarette smokers inhale these carcinogens and bathe the surfaces of their

initiation ■ The stage of cancer during which one or more healthy cells are triggered to subsequently respond to the actions of a promoter

promotion ■ The second stage of cancer, during which the formation of malfunctioning cells is accelerated to produce tumors

tumor ■ A mass of tissue that persists and grows independently of its surrounding structures and has no known physiological function

progression ■ The third stage of cancer, during which tumors evolve into malignant growths in the body

metastasis ■ The formation of a malignant growth at one site in the body from cells derived from a malignancy located elsewhere in the body

respiratory tracts with them. By so doing, heavy smokers assume a lifetime risk of acquiring lung cancer that is more than 20-fold higher than that of nonsmokers.

The emergence of tumors within the lungs and other organs of tobacco smokers may even be accelerated by exposure of the organs to carcinogens that are not constituents of tobacco smoke. For example, asbestos, ethanol, and radon are nontobacco carcinogens that increase the rate at which cancer develops in smokers compared to the rate of its development in nonsmokers. This phenomenon is an example of **synergism**; that is, the interaction of two or more carcinogens produces more profound effects than either carcinogen causes separately. The phenomenon may be evident by the onset of an adverse health effect other than cancer. For example, smokers exposed to asbestos are more vulnerable to contracting (noncancerous) asbestosis than are nonsmokers. Asbestos-exposed smokers are about 90% more likely to contract asbestosis than are nonsmokers with an equal exposure.

Firefighters represent another group that is more prone than the general population to develop cancer. Based on information derived from research studies,[14] firefighters are twice as likely to develop testicular cancer, multiple myeloma (a cancer of the bone marrow), non-Hodgkin's lymphoma, and prostate cancer compared to the general population. It is prudent to assume that at least in some instances, the firefighters initially contracted these cancers by exposure to carcinogens like certain asbestos fibers (Section 10.19); soot, polynuclear aromatic hydrocarbons, and other diesel fuel combustion products (Section 12.12-C); and polychlorinated dibenzofurans and polychlorinated dibenzo-*p*-dioxins (Section 13.4-B).

10.17-A CLASSIFICATION OF CHEMICAL CARCINOGENS

At least five organizations study whether exposure to a substance adversely impacts the human organism or experimental animals as a carcinogen: EPA, OSHA, NIOSH, the U.S. Department of Health and Human Services' National Toxicology Program (NTP), and the World Health Organization's International Agency for Research on Cancer (IARC). The NTP publishes biennially reports on carcinogens. The twelfth report, published in 2011, listed 240 substances as carcinogens.

Each of these agencies classifies the carcinogenic potential of a substance similarly, but yet, differently. For example, IARC provides the procedure of classification for given substances as shown in Table 10.24.

The various procedures employed for the study of carcinogenic potentiality have produced a scientific literature peppered with such terms as "known human carcinogen," "suspect carcinogen," "probable carcinogen," "potential occupational carcinogen," and "reasonably anticipated to be a human carcinogen." This classification sometimes is necessary, because the existing data do not always enable toxicologists to clearly ascertain

synergism ■ The phenomenon in which two or more substances interact in such a way that their combined effects are more severe than the sum of their individual effects

TABLE 10.24	IARC's Classification of Carcinogens
GROUP	**SHIPPING DESCRIPTION**
Group 1	Carcinogenic to humans
Group 2A	Probably carcinogenic to humans
Group 2B	Possibly carcinogenic to humans
Group 3	Not classifiable as to its carcinogenicity to humans
Group 4	Probably not carcinogenic to humans

---

[14]Grace LeMasters et al., "Cancer risk among firefighters: A review and meta-analysis of 32 studies," *J. Occup. Environ. Med.*, Vol. 48 (2006), pp. 1189–1202.

whether a substance causes cancer in the human organism. Research studies in carcinogenesis are complicated by the observation that certain substances appear incapable of directly causing cancer, but they can affect the rate of tumor formation when exposure to them occurs in combination with exposure to other substances.

Individuals are exposed to cancer-causing substances in a variety of ways. Because certain carcinogens occur naturally, exposure to them often is unavoidable. Carcinogens may be contained in commercial products, including gasoline and diesel fuel. When a carcinogen is known to be a constituent of a commercial product, some state laws require advising the consumer. Conveying this information is the responsibility of the seller, who is obligated to display the information at the locations where the product is sold.

10.17-B WORKPLACE REGULATIONS INVOLVING CARCINOGENS

When the use of carcinogens is required in the workplace, OSHA requires employers to provide employees with protective clothing, self-contained breathing apparatus, and a segregated area in which the work with carcinogens can be safely conducted. OSHA also requires employers to post in this area a sign that identifies the carcinogen and its physical hazards. For example, OSHA requires the area of a workplace where vinyl chloride is used to be posted as follows:

VINYL CHLORIDE
EXTREMELY
FLAMMABLE GAS
UNDER PRESSURE
CANCER-SUSPECT
AGENT

Vinyl chloride is a human carcinogen (Section 14.6).

To reduce or eliminate their exposure to carcinogens, emergency responders should heed this warning when they enter buildings in which carcinogens are stored or used. They also should heed the health hazard pictogram that OSHA requires the manufacturers, distributors, and importers of carcinogens to display on container labels.

In this and future chapters, we recognize the carcinogens of immediate concern to emergency responders as they are first encountered.[15]

10.18 COMPOUNDS OF TOXIC METALS

From a nutritional viewpoint, metallic compounds may be classified as follows:

■ *Essential elements.* Trace amounts of 14 metals and metalloids are essential in the diet for the proper functioning and survival of the human organism. For this reason, nutritionists refer to them as the **essential elements**. They are calcium, phosphorus, potassium, sulfur, sodium, chlorine, magnesium, iron, zinc, copper, molybdenum, cobalt, iodine, and selenium. Their proper balance and availability are essential for proper biological function. This is not meant to imply that the essential elements are beneficial in all doses. The ingestion of elevated concentrations may cause illness or death, often by damaging cellular components.

■ *Toxic metals.* Certain other metals and metalloids are not known to offer any dietary benefit to the human organism. These **toxic metals** include arsenic, barium, beryllium, lead, and thallium. The consumption of even minute amounts of their compounds can cause illness or death.

essential element ■ Any element in the human diet that is needed for the proper functioning and survival of the human organism

toxic metal ■ Any metals or metalloids whose consumption, even in minute amounts, can cause illness or death

---

[15]Unless otherwise indicated, the designations in this text are those denoted by the U.S. Department of Health and Human Services and the World Health Organization's International Agency for Research on Cancer (IARC).

Lead chloride

Certain other metals and metalloids may be either toxic or beneficial, depending on their concentration and other factors. They include the compounds of manganese, selenium, and tungsten. They can either exert a toxic effect or perform an essential biological function in the human body.

10.18-A LEAD AND ITS COMPOUNDS

Of all the toxic metals, the compounds of lead are most likely to be regularly encountered by emergency responders. For example, firefighters may be exposed to them as they battle fires in old structures.

Lead compounds formerly were used as pigments in surface coatings to provide color and prevent corrosion. Examples of these pigments are lead acetate, lead chromate, and lead oxide. The surface coatings are called **lead-based paints**. EPA banned the use of lead-based paint in buildings intended for residential use in 1978.

There remain in the United States an estimated 38 million homes built before 1978 containing some lead-based paint. When these structures collapse during ongoing building fires, lead-laden dust is dispersed into the atmosphere. Firefighters are then put at risk of inhaling lead particulates, although the use of respiratory protection equipment significantly reduces or eliminates their exposure.

Lead poisoning is the collection of adverse health problems caused by lead exposure. This affliction is most prominent in children. Lead may damage the nervous systems of children exposed to lead by causing them to experience behavioral and learning problems such as hyperactivity and reduced attention span, or to physically develop more slowly than nonexposed children.

Lead exposure in children also impairs their ability to learn and execute mental processes effectively. Furthermore, research studies demonstrate that IQ (Intelligence Quotient) measurements obtained for children during their school-age years relate similarly to their IQ measurements obtained as adults.[16] This finding suggests that the ill effects caused by lead exposure in children do not terminate, but continue into adulthood.

Children who live in older dwellings in economically disadvantaged communities are disproportionately affected compared to children who live elsewhere. The lead-based paint in these substandard dwellings is considered the greatest remaining source of childhood lead exposure. Formerly, experts concluded that the children forced to live in these dwellings were at risk of eating flaking paint or plaster containing lead. Today, however, they believe that the ingestion of household dust containing lead from deteriorating or abraded lead-based paint is the most likely cause of lead poisoning in children.

Adults may also be adversely affected by lead exposure. They experience reproductive problems, high blood pressure, nervous system disorders, and memory problems. When ingested or inhaled, lead primarily affects the human blood-forming, nervous, and kidney systems, but to a lesser extent, it also harms the reproductive, endocrine, hepatic, cardiovascular, immunologic, and gastrointestinal processes. Lead interferes with the body's mechanism for absorbing the essential elements calcium and iron. Elevated lead concentrations can cause severe health problems, including brain disease, colic, palsy, and anemia. Damage to the central nervous system in general, and to the brain in particular, is one of the most severe consequences of chronic lead poisoning. Exposure to certain lead compounds may also cause latent health effects. In particular, lead acetate and lead phosphate are regarded as probable human carcinogens.

lead-based paint ■ For the purposes of the TSCA regulations, paint and other surface coatings that contain lead in concentrations equal to or greater than 1.0 mg/cm^2, or 0.5% by weight

lead poisoning ■ The combination of health ailments caused by increased levels of lead in the body

---

[16]M. Mazumdar et al., "Low-level environmental lead exposure in children and adult intellectual function: a follow-up study," *Environmental Health*, Vol. 10 (2011), pp. 24–30.

10.18-B CONSUMER PRODUCT REGULATIONS INVOLVING LEAD

CPSC has published regulations to reduce or eliminate the risk of consumer exposure to lead. At 16 C.F.R. §1303.4, it bans lead in paint and other similar surface-coating materials; toys and other articles intended for use by children that bear lead-based paint; and furniture articles that bear lead-based paint. At 16 C.F.R. §1500.17, CPSC bans the sale of paint having a lead content above 0.06% by dry weight (other than artist's paint) in the United States.

Notwithstanding CPSC's efforts, lead-based paint has been used as a surface coating on toys, especially those manufactured outside the United States. Furthermore, the lead content in some children's products is sufficiently elevated that the lead may cause adverse health effects in the children who play with them. To minimize or eliminate the potential for lead exposure to children, CPSC set standards for both the lead in paint and the total lead content in toys and other products designed or intended primarily for use by children 12 years or younger. CPSC now requires the manufacturers and importers of children's products to test them for assurance that they do not contain more than 90 mg/kg of lead. CPSC also requires that the total lead content of any children's product be limited to less than 100 mg/kg.

CPSC also requires manufacturers and importers to affix permanent, distinguishing marks on children's products and their packaging to ascertain the sources of the products. These marks include the production location, date of production, name of manufacturer, and lot or batch number. The following is an example of the required marking:

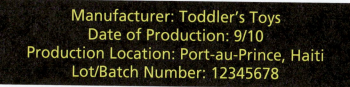

Manufacturer: Toddler's Toys
Date of Production: 9/10
Production Location: Port-au-Prince, Haiti
Lot/Batch Number: 12345678

10.18-C WORKPLACE REGULATIONS INVOLVING LEAD

When the use of lead compounds is needed in the workplace, OSHA requires employers to limit employee exposure to a concentration of 0.050 mg/m^3, averaged over an 8-hour workday. OSHA also requires the posting of the following warning sign in work areas where airborne lead is generated:

WARNING

POISON
LEAD WORK AREA
NO SMOKING,
EATING OR
DRINKING

OSHA also requires at 29 C.F.R. §1926.62 that employers include the following information on bags or containers of lead-contaminated protective clothing and equipment:

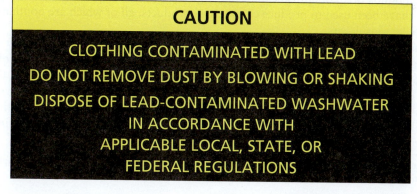

CAUTION

CLOTHING CONTAMINATED WITH LEAD

DO NOT REMOVE DUST BY BLOWING OR SHAKING

DISPOSE OF LEAD-CONTAMINATED WASHWATER

IN ACCORDANCE WITH

APPLICABLE LOCAL, STATE, OR

FEDERAL REGULATIONS

10.18-D ENVIRONMENTAL REGULATIONS INVOLVING LEAD

EPA regulates certain aspects of lead-based paint hazards pursuant to the Residential Lead-Based Paint Hazard Reduction Act of 1992, enacted using the legal authority of TSCA. In particular, the residential Lead-Based Paint Hazard Reduction Act influences the manner by which housing is now bought and sold. Some requirements include the following:

■ Sellers and landlords must disclose known lead-based paint hazards in the target housing and provide available reports to buyers and renters.

■ Sellers and landlords must provide an EPA-approved lead-hazard information pamphlet to potential homebuyers and renters.[17]

■ Sellers must grant homebuyers a 10-day period in which to conduct a lead-based paint inspection or risk assessment of the target home.

■ Sellers, lessors, and real estate agents share responsibility for ensuring compliance with all applicable aspects of the act

EPA certification is required for all firms who intend to conduct renovation, repair, and repainting projects that disturb lead-based paints in homes constructed before 1978.

Other environmental statutes, including the following, impact EPA's regulations concerning lead:

■ Using the legal authority of the Clean Air Act, EPA regulates the concentration of lead in the ambient air as a criteria air pollutant. In 2008, EPA set the primary and secondary national ambient air-quality standards for lead at $0.15 \ \mu g/m^3$ as a quarterly average.

■ Using the legal authority of RCRA, EPA regulates the treatment, storage, and disposal of lead-laden wastes that exhibit the RCRA characteristic of toxicity.

■ Using the legal authority of TSCA, EPA regulates the manner by which contractors become certified for removing lead-laden dust from housing as well as requiring certain procedures to be implemented during the actual removal actions. These regulations are published at 40 C.F.R. §§745.226 – 745.227.

10.18-E TRANSPORTING LEAD

When shippers offer a nonoxidizing lead compound for transportation, DOT requires them to identify it in the shipping description on an accompanying shipping paper. Three examples of these descriptions are "UN1616, Lead acetate, 6.1, PG III (Poison)," "UN1620, Lead cyanide, 6.1, PG II (Poison)," and "UN2291, Lead compounds, soluble, n.o.s., 6.1, PG III (Poison)." DOT also requires shippers and carriers to comply with all applicable labeling, marking, and placarding requirements.

We note the DOT requirements for transporting oxidizing lead compounds in Chapter 11.

10.19 ASBESTOS

Asbestos

Asbestos is the generic name of a complex naturally occurring mineral composed of several compounds that contain silicate (SiO_4^{4-}) groups linked into chains. Several types of asbestos have been used to produce commercial products, all of which are nonflammable and excellent insulators.

When mixed with magnesium oxide, asbestos was formerly used for a variety of fireproofing, insulating, soundproofing, and decorative purposes. It was a component of products like asbestos pipe coverings, flooring- and ceiling-tile products, paper products, thermal system insulation, antifriction materials (e.g., brake linings and clutch facings), roofing materials, fire walls or doors, and surface-coating and patching compounds. Their fibers could also be woven into strong, flexible threads that could be converted into fireproof

---

[17]An example is the EPA document, "Protecting Your Family From Lead in Your Home" (Washington, DC: U.S. Environmental Protection Agency, 2003), EPA747-K-99-001.

fabrics. Although these asbestos products once were prized for their desirable features, their availability in today's market has been sharply curtailed or entirely eliminated.

10.19-A ILL EFFECTS CAUSED BY EXPOSURE TO ASBESTOS

Attention to the adverse health effects caused by exposure to asbestos fibers typically has focused on six mineral types: the serpentine mineral *chrysotile* (commonly called *white asbestos* or *white serpentine*), and the amphibole minerals, *cummingtonite-grunerite asbestos* (*amosite*), *riebeckite asbestos* (*crocidolite*), *actinolite asbestos*, *anthophyllite asbestos*, and *tremolite asbestos*. Chrysotile was the most common form of asbestos formerly used in the United States. There is substantial scientific evidence supporting the observation that exposure to these fibers cause serious health problems, including at least two types of cancer.

Unless asbestos has been completely sealed into a product, as in asbestos floor tiles, it often breaks apart into tiny fibers, much smaller and more buoyant than ordinary dust. This occurs when the asbestos-containing material is damaged or disintegrates with age. It is said to be **friable**. It is friable asbestos that poses severe health hazards to those who inhale or swallow its fibers.

Figure 10.12 illustrates that asbestos fibers may range from 0.1 to 10 micrometers in length and have a diameter of only 30 nanometers. They float almost indefinitely in the air and may be readily inhaled or swallowed. When they are inhaled, asbestos fibers sometimes are removed by the cilia lining the respiratory tract. Then, they move into the throat, where they usually are swallowed. Once swallowed, they pose the threat of causing cancer of the esophagus, stomach, intestines, and rectum.

Exposure to friable asbestos potentially causes the onset of the following diseases, typically after a substantial latency period ranging from approximately 20 to 40 years:

- **Lung cancer**, the growth of malignant tumors in the lungs. Microscopic asbestos fibers have been observed in pieces of tumors surgically removed from the lungs. Such biopsies clearly link the onset of lung cancer to asbestos exposure.
- **Asbestosis**, a progressive, irreversible, and noncancerous scarring of lung tissue that occurs when sharp asbestos fibers deposit deep in the respiratory tract following prolonged exposure. There is no known cure or treatment for asbestosis.
- **Mesothelioma**, cancer of the **mesothelium**, the protective membrane or sac surrounding the lungs, abdomen, heart, and other internal organs. The most common asbestos-related form of this disease is **pleural mesothelioma**, cancer of the protective linings surrounding the lungs and chest. Mesothelioma is uniquely distinctive to asbestos exposure, because the disease essentially is never contracted by individuals who were not exposed to asbestos. It weakens the lungs so that they become more vulnerable to attack by other diseases. It is characterized by persistent coughing, weight loss, fever, rasping, and coughing up blood. Although mesothelioma may be treated by chemotherapy, radiation, and surgery, it is virtually always fatal.

friable ■ Brittle or capable of readily being crumbled, as in friable asbestos

lung cancer ■ Cancer characterized by the emergence of malignant tumors on the lungs

asbestosis ■ A scarring of lung tissue by asbestos fibers

mesothelioma ■ The cancer of the membranes lining the chest and abdominal cavities resulting from exposure to asbestos fibers

mesothelium ■ The protective linings and sacs surrounding certain of the body's internal organs and cavities

pleural mesothelioma ■ The cancer of the protective linings surrounding the lungs and chest

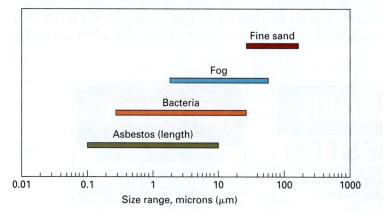

FIGURE 10.12 The typical size of an asbestos fiber varies from 0.1 to 10 microns, shown here relative to the size of some other materials. These sizes are too small to be visible to the human eye. (*Courtesy of the U.S. Environmental Protection Agency, "Asbestos-Containing Materials in School Buildings: A Guidance Document," Office of Toxic Substances, Washington, DC, EPA 450/2-78-014, March 1978.*)

The evidence associating mesothelioma in humans with asbestos exposure is overwhelming. Because of this association, asbestos is classified as a human carcinogen.

The likelihood that individuals will contract lung cancer, asbestosis, or mesothelioma from exposure to friable asbestos is significantly elevated when they are smokers. The impact caused by the simultaneous exposure to asbestos fibers and the carcinogens in tobacco smoke is greater than the impact that asbestos or the tobacco carcinogens cause separately. Two examples of this synergistic impact are the following:

■ Smokers exposed to asbestos fibers are about 50% more likely than nonsmokers to develop lung cancer.

■ Asbestos-exposed smokers are about 90% more likely than nonsmokers to contract asbestosis.

These examples suggest that tobacco smoking weakens the lungs and provides a vulnerability to contracting asbestos-related diseases that is not experienced by nonsmokers.

10.19-B CONSUMER PRODUCT REGULATIONS INVOLVING ASBESTOS

The CPSC has promulgated regulations to eliminate or reduce asbestos in consumer products sold for use in the United States. For example, at 16 C.F.R. §1304.4, the use of consumer patching compounds containing friable asbestos is banned; and at 16 C.F.R. §§1305.4 and 1500.17, the use of general-use garments, tape joint compounds, and artificial fireplace ash and embers containing friable asbestos is banned. The manufacturers of handheld hair dryers voluntarily ceased using asbestos in hair dryers.

10.19-C WORKPLACE REGULATIONS INVOLVING ASBESTOS

countable elongate mineral particle ■ Any mineral having the elemental composition and crystal structure of the forms of asbestos that cause cancer

When the use of asbestos is needed in the workplace, NIOSH requires employers to limit employee exposure to an airborne concentration of asbestos fibers denoted as "0.1 countable elongate mineral particles from one or more covered minerals per cubic centimeter averaged over 100 minutes." A **countable elongate mineral particle** is any mineral having the elemental composition and crystal structure of chrysotile, crocidolite, amosite, anthophyllite, tremolite, actinolite, or other forms of asbestos.

When asbestos is present in the workplace, OSHA requires employers to post the sign on the left below to identify the potential harm caused by asbestos exposure. OSHA also requires employers to post the sign on the right on the outside surfaces of bags or other containers of protective clothing and equipment, scrap, waste, and debris containing asbestos fibers.

10.19-D ENVIRONMENTAL REGULATIONS INVOLVING ASBESTOS

Using the authority of the Clean Air Act, EPA banned the use of asbestos in pipe and boiler coverings as well as all uses of asbestos materials that are applied by spraying. Using the authority of TSCA, EPA also banned the use of asbestos in corrugated paper, rollboard, commercial paper, specialty paper, and flooring felt.

To safeguard the health of school children and others who work in schools, EPA launched a program cited at 40 C.F.R. §763.91 in which each local educational agency is required to accomplish the following within their jurisdictions:

- Inspect for the presence of friable asbestos in its school buildings
- Determine whether friable asbestos fibers are being released into the air
- Remove and repair all damaged asbestos-containing material

EPA also regulates at 40 C.F.R. §61.145 how the asbestos is to be stripped from walls, ceilings, and other locations. The proper abatement procedures include the following:

- Sealing a work site before the asbestos-containing material is removed
- Preventing the asbestos fibers from becoming airborne by keeping them wet
- Removing, encapsulating, repairing, enclosing, or encasing the asbestos-containing material
- Placing the asbestos debris in leak-tight containers that subsequently are sealed
- Posting warning signs that contain relevant information noted earlier.

10.19-E TRANSPORTING ASBESTOS

When shippers offer asbestos or asbestos waste for transportation, DOT requires them to first comply with certain packaging requirements published at 49 C.F.R. §173.216. These regulations cite the use of rigid, leak-proof drums and portable tanks, and bags for shipment in closed freight containers, motor vehicles, or railcars. DOT requires shippers to affix CLASS 9 labels on two opposing sides of the chosen nonbulk packaging and to mark it with the notation "Asbestos, NA2212."

DOT also requires shippers to identify asbestos on an accompanying shipping paper as follows:

NA2212, Asbestos, 9, PG III

When bulk quantities of asbestos are transported, this shipping description is preceded by "RQ," because the reportable quantity of white asbestos is only 1 pounds (0.454 kg).

DOT requires carriers who transport bulk quantities of asbestos by highway or rail to display the identification number 2212 on each side and each end of the transport vehicle on orange panels or across the center of CLASS 9 placards or white square-on-point diamonds.

Although DOT generally requires carriers to post placards when using the square-on-point diamonds and orange panels to display the identification number of a hazardous material transported in bulk, it does not require carriers to post CLASS 9 placards on the vehicles used to transport asbestos.

The DOT regulation at 49 C.F.R. §172.101.156 notes that these transportation requirements do not apply to asbestos that has been immersed or fixed in a natural or synthetic binder material, such as cement, plastic, asphalt, resins, or mineral ore, or contained in manufactured products.

10.19-F RESPONDING TO INCIDENTS INVOLVING A RELEASE OF ASBESTOS

Despite efforts to eliminate asbestos in buildings, emergency responders are most likely to be exposed to airborne asbestos fibers as they combat fires in old buildings. The primary route of exposure route is inhalation of the dust generated when walls, ceilings, and roofs collapse. To reduce or eliminate this potential exposure, firefighters should combat these fires from upwind locations and use self-contained breathing apparatus.

Firefighters also put their families and colleagues at risk of contracting an asbestos-related disease. After firefighters combat fires in which asbestos exposure was probable, the fibers may lodge on their hair, skin, shoes, and clothing. Later, the fibers can be brought from the fire scene and into the home. To reduce or eliminate the likelihood that their family members will become exposed to asbestos fibers, firefighters should shower, change their clothes, and launder their uniforms separately from other clothing before leaving the firehouse.

10.20 PESTICIDES

As first noted in Section 1.4-B, EPA defines a *pesticide* in part at 40 C.F.R. §152.3 as any substance or mixture of substances intended for preventing, destroying, repelling, or mitigating any pest, or intended for use as a plant regulator, defoliant, or desiccant. In more general terms, a pesticide generally is regarded as a commercial product that has been intentionally designed to destroy or control the spread of insects, fungi, rodents, weeds, mold, mildew, algae, and other pests.

There are two chemical types of pesticides: inorganic pesticides and organic pesticides. Inorganic pesticides generally are identified either by the name of their active component or the use of a generic term like *arsenical pesticide*. Because their application leaves a hazardous substance in the environment to pose long-term damage, few inorganic pesticides are encountered in today's commercial markets.

Although organic pesticides may be classified in several ways, we divide them into the following four groups based on their chemical structures:

- Carbamate pesticides, derivatives of carbamic acid, $H_2N-\overset{\overset{O}{\|}}{C}\diagdown_{OH}$

- Organochlorine pesticides, chlorinated derivatives of certain complex hydrocarbons (Section 12.1)

- Organophosphorus pesticides, derivatives of phosphoric acid, $HO-\overset{\overset{O}{\|}}{\underset{\underset{OH}{|}}{P}}-OH$

- Thiocarbamate pesticides, derivatives of thiocarbamic acid, $H_2N-\overset{\overset{S}{\|}}{C}\diagdown_{OH}$

This chemical classification scheme is employed by DOT for generically describing organic pesticides.

10.20-A ENVIRONMENTAL REGULATIONS PERTAINING TO PESTICIDES

To advise consumers of the hazards posed by exposure to specific pesticides, EPA uses the authority of FIFRA (Section 1.4-B) to compel manufacturers to provide advisory information on pesticide product labels.

10.20-B TRANSPORTING PESTICIDES

DOT requires shippers and carriers to identify the shipping description of a pesticide on the shipping paper accompanying the shipment. The Hazardous Materials Table at 40 C.F.R. §172.101 lists specific chemical names as well as generic descriptions of pesticides.

When the shipping description is listed generically, DOT requires its technical name or the names of the principal constituents that cause the pesticide to meet the definition of a Division 6.1, Packing Group I or II, to be entered in parentheses in the shipping description. Some examples of these generic shipping descriptions are provided in Table 10.25. DOT also requires shippers and carriers to comply with all applicable labeling, marking, and placarding requirements.

10.20-C PESTICIDES AND FIREFIGHTING

Exposure to the airborne smoke and combustion products produced during pesticide fires could be fatal. Consequently, emergency responders who attempt to extinguish these fires should use self-contained breathing apparatus and wear fully encapsulating suits.

The common symptoms associated with pesticide exposure include blurred vision, headache, vomiting, salivation, and lack of coordination. These symptoms may be mistaken by responders as those associated with heat exhaustion.

Pesticide fires should always be fought from the upwind direction or at right angles. The use of fog is preferable to the use of direct streams of water because fog is less likely to raise toxic dusts. To protect the environment, the runoff water generated during the response action should be stored in a temporary reservoir, neutralized with soda ash, and absorbed into sawdust or clay following extinguishment of the fire. Thereafter, the residue may be disposed of in accordance with applicable environmental laws.

TABLE 10.25	Representative Generic Shipping Descriptions of Some Solid Pesticides
PESTICIDE	**SHIPPING DESCRIPTION**
Carbamate pesticides, solid, toxic	UN2757, Carbamate pesticides, solid, toxic (ethyl carbamate), 6.1, PG I (Poison) *or* UN2757, Carbamate pesticides, solid, toxic (ethyl carbamate), 6.1, PG II (Poison) *or* UN2757, Carbamate pesticides, solid, toxic, (ethyl carbamate), 6.1, PG III (Poison)
Organochlorine pesticides, solid, toxic	UN2761, Organochlorine pesticides, solid, toxic (Endosulfan), 6.1, PG I (Poison) *or* UN2761, Organochlorine pesticides, solid, toxic, (Endosulfan), 6.1, PG II (Poison)
Organophosphorus pesticides, solid, toxic	UN2783, Organophosphorus pesticides, solid, toxic (Chlorpyrifos), 6.1, PG I (Poison) *or* UN2783, Organophosphorus pesticides, solid, toxic (Chlorpyrifos), 6.1, PG II (Poison)
Pesticides, solid, toxic	UN2588, Pesticides, solid, toxic, n.o.s. (ammonium sulfamate), 6.1, PG I (Poison) *or* UN2588, Pesticides, solid, toxic, n.o.s. (ammonium sulfamate), 6.1, PG II (Poison) *or* UN2588, Pesticides, solid, toxic, n.o.s. (ammonium sulfamate), 6.1, PG III (Poison)
Thiocarbamate pesticides, solid, toxic	UN2771, Thiocarbamate pesticides, solid, toxic (ethylenebisdithiocarbamic acid), 6.1, PG I (Poison) *or* UN2772, Thiocarbamate pesticides, solid, toxic (ethylenebisdithiocarbamic acid), 6.1, PG I (Poison) *or* UN2772, Thiocarbamate pesticides, solid, toxic (ethylenebisdithiocarbamic acid), 6.1, PG I (Poison)

10.21 BIOLOGICAL WARFARE AGENTS

Certain microorganisms and toxins are so highly poisonous to humans, plants, and animals that during wartime, their dissemination could sicken, injure, incapacitate, or otherwise cause mass casualties. They are called **biological warfare agents**, or **germ warfare agents.** Their common forms are infectious pathogens and toxins, some examples of which are listed in Table 10.26.

biological warfare agent (germ warfare agent) ■ Any biocide that causes death or injury to humans, plants, or animals when intentionally disseminated throughout a population during warfare

infectious pathogen ■ Any bacterium, virus, or fungus that may cause a deadly disease to exposed individuals

toxin ■ A poisonous substance produced by a living organism or bacterium

- ■ An **infectious pathogen** consists of living bacteria, viruses, fungi, rickettsiae, or protozoa, exposure to which causes such deadly diseases as anthrax, botulism, bubonic plague, cholera, and smallpox.
- ■ A **toxin** is a poisonous substance biologically produced by certain plants and animals. Examples of organisms that produce toxins include certain mushrooms, spiders, frogs, toads, jellyfish, and snakes. On entry into the body, toxins typically act on either the central nervous system or the circulatory system, or both; that is, they act as neurotoxins and/or cytotoxins, respectively. They typically cause death by producing paralysis and respiratory failure.

Although the dissemination of infectious pathogens or toxins could cause large-scale impairment, disease, death, or biological malfunctions among the enemy, this warfare practice has been condemned by civilized countries as unethical and inhumane.

As noted in Section 7.3-A, the League of Nations introduced the Geneva Protocol in 1925. Despite its ratification by 130 nations, the mistrust and fear among nations has resulted in the production and stockpiling of biological warfare agents for defensive purposes.

TABLE 10.26	Some Infectious Pathogens and Toxins
INFECTIOUS PATHOGEN OR TOXIN	**ADVERSE HEALTH EFFECTS CAUSED BY INFECTION**
Bacillus anthracis, a spore-forming bacterium	Anthrax, usually fatal in the inhaled form (The symptoms of exposure are provided in Section 10.21-E.)
Bacillus cereus	Food-poisoning-like symptoms
Bacillus licheniformis	Produces a protease enzyme that interferes with protein metabolism
Bacillus megaterium, a spore-forming bacterium	Unknown
Bacillus subtilis	Unknown
Brucella abortus (multiple biotypes)	Brucellosis, an infectious disease that may cause an infected pregnant woman to abort her fetus
Brucella melitensis	Fatigue, loss of appetite, and other symptoms
Burkholderia mallei	Glanders, a disease typically acquired by humans from infected animals; characterized by fever, muscle aches, ulcers, chest pain, muscle tenderness, weakness, weight loss; seizure; light sensitivity, pus-forming skin infections, diarrhea, and headache
Burkholderia pseudomallei	Meliodosis, a disease similar to glanders and characterized by the same symptoms
Clostridium botulinum (seven biotypes)	Botulism, characterized by vomiting, paralysis of the muscles, constipation, thirst, headache, fever, dizziness, double or blurred vision, deadening of the nerves, and dilation of the pupils

(continued)

| TABLE 10.26 | Some Infectious Pathogens and Toxins (*continued*) |

INFECTIOUS PATHOGEN OR TOXIN	ADVERSE HEALTH EFFECTS CAUSED BY INFECTION
Clostridium perfringens	Enteritis, the inflammation of the small intestine; gas gangrene, respiratory distress, and failure of the body's internal organs
Clostridium tetani	Tetanus (lockjaw)
Corynebacterium diphtheriae	Diphtheria, an often fatal respiratory disease
Coxiella burnetii	Q fever, characterized by chills, fever, coughing, headache, fatigue, hallucination, and sometimes chest pain
Ebola virus (four types)	Ebola hemorrhagic fever, a highly contagious disease that has ended in death for 90% of the those exposed to the virus; characterized by bleeding from the eyes, ears, nose, mouth, and rectum, eye inflammation, seizures, coma, and delirium
Francisella tularensis	Tularemia, a disease often contacted by rodents and transmitted to humans; characterized by sudden chills, fever, coughing, severe weakness, and the development of large skin lesions; can lead to fatal pneumonia
Listeria monocytogenes	Fever; muscle ache, nausea, and diarrhea; can lead to meningitis, encephalitis, and intrauterine and cervical infections in infected women
Marburg virus	Marburg hemorrhagic fever, a disease similar to but distinct from Ebola hemorrhagic fever but characterized with the same symptoms
Plasmodium malaria	Malaria, a mosquito-borne infectious disease characterized in the most severe instances by hot-and-cold sweats, hemorrhaging, convulsions, coma, and death
Pneumocystis carinii	Pneumonia
Poxvirus variola	Smallpox, a highly infectious, often fatal disease characterized by high fever, headache, body ache, vomiting, and the appearance of a rash on the tongue and in the mouth that turns to blisters
Ricin	Acts as a hemotoxin (The symptoms of exposure are provided in Section 10.21-F.)
Rickettsia prowazekii	Typhus, a disease caused by louse-borne bacteria, characterized by severe headache, sustained high fever, severe muscle pain, rash, sensitivity to light, and delirium
Salmonella enteritidis	Salmonellosis, or "food poisoning," characterized by nausea, vomiting, abdominal cramps, diarrhea, intestinal inflammation, and ulceration
Salmonella typhimurium[a]	Salmonellosis, characterized in the same manner as exposure to Salmonella enteritidis; in severe cases, accompanied by typhoid fever and hemorrhaging in the intestines.
Staphylococcus aureus (multiple biotypes)	Staphylococcus food poisoning, characterized by sudden severe nausea, vomiting, abdominal cramps, muscle pain, severe diarrhea, headache, fever, and chills
Streptococcus pneumoniae	Pneumonia
Vibrio cholerae	Cholera, an acute intestinal disease characterized by a profuse watery diarrhea and vomiting
Yersinia pestis[b,c]	Bubonic plague, a disease associated with the production of ugly black welts and bulges called *buboes* in the groin or near the armpits

[a]The first known act of bioterrorism on American soil was carried out in 1984 by members of the Rajneeshees, an Indian religious cult. The bioterrorism act involved contaminating the foods displayed on salad bars in several Oregonian restaurants with *Salmonella typhimurium*. More than 750 people who consumed food from the salad bars were infected with salmonellosis. [Judith Miller, Stephen Engelberg, and William Broad, *Germs: Biological Weapons and America's Secret War* (New York: Simon & Schuster, 2001), pp. 15–33.]

[b]Bubonic plague has often been associated with the so-called Black Death that historians credited with killing 20 million Europeans during the fourteenth century. The conventional belief is that the loss of life was caused by deadly bacilli that were transmitted from fleas carried on rodents and then to humans. In contemporary times, however, scientists are seriously considering whether the Black Death was actually caused by the human consumption of beef tainted with an anthrax-like bacillus. [Norman F. Cantor, *In the Wake of the Plague* (New York: Free Press, 2001).]

[c]During World War II, thousands of Chinese and Manchurians died when they consumed rice and wheat that had been intentionally contaminated with fleas infected with *Yersinia pestis*. The foods were dropped by the Japanese from aircraft on Chinese and Manchurian towns and cities.

10.21-A SELECT AGENTS

Today, infectious pathogens and toxins are associated with the likelihood that bioterrorists will covertly obtain and deploy them to sicken and kill or create disruption and fear among segments of the American population. Two federal statutes seek to improve our ability to prevent, prepare for, and respond to acts of bioterrorism and other public health emergencies: The Public Health Security and Bioterrorism Preparedness and Response Act of 2002 and the Agricultural Bioterrorism Protection Act of 2002. They are referred to here as the **Bioterrorism Acts**.

Using the authorities of the Bioterrorism Acts, the U.S. Department of Health and Human Services (HHS) and the U.S. Department of Agriculture (DA) have prepared lists of infectious pathogens and toxins. Both departments refer to them as **select agents**. The appropriate department lists the following three types of select agents:

■ Those that potentially pose a threat to public health and safety. HHS publishes this list of select agents at 42 C.F.R. §73.3.

■ Those that potentially pose a threat to animal health and animal products. DA publishes this list of select agents at 7 C.F.R. §331.3.

■ Those that potentially pose a threat to plant health and plant products. DA publishes this list of select agents at 9 C.F.R. §121.3.

All examples of the infectious pathogens and toxins listed in Table 10.26 are select agents.

HHS's method for choosing a select agent is based on the following criteria: the impact on human health from an exposure, its degree of contagiousness, the methods by which the agent is transferred to humans, its immunization effectiveness, its potential for use as a weapon, and the threat posed by its use to vulnerable portions of the population such as children and the elderly. The USDA's criteria for choosing select agents are based largely on their adverse impact on animals and plants: the relative ease with which the agent may be disseminated or transmitted from animal to animal or into the environment where it could produce a deleterious effect upon animal or plant health, the potential impact for high animal or plant mortality rates, the potential impact for a major animal or plant health impact, and the likelihood for misuse resulting in mass panic or other forms of social or economic disruption.

The Bioterrorism Act regulations require research laboratories and other facilities that possess, use, or transfer select agents to register with the appropriate federal agency and to comply with regulations published at 7 C.F.R. Part 331, 9 C.F.R. Part 121, and 42 C.F.R. Part 73.

10.21-B DISSEMINATION OF INFECTIOUS PATHOGENS AND TOXINS

Infectious pathogens and toxins may be covertly disseminated by several means that include the following:

■ Containerization in canisters or cluster bombs that are then inserted into modified missile heads or artillery shells

■ Aerosolized dispersal from aircraft (e.g., crop dusters, helicopters, or low-flying robotic craft) in an enclosed space or over a city

■ Aerosolized dispersal using generators carried on trucks or boats in an enclosed space or over a city

■ Direct introduction into food or drinking water supplies

■ Inoculation of fabric, paper, and similar items

To maximize the adverse impact from exposure to a pathogen, a period of time must pass after it has been disseminated. The period may be hours, days, weeks, or even years. During this time the pathogen enters the bodies of its victims and multiplies exponentially until the symptoms and ailments of a disease are clearly evident. During this incubation

Bioterrorism Acts ■ The federal statutes that address the prevention, preparedness, and response to acts of bioterrorism and other public health emergencies from human, animal, or plant exposure to select agents

select agent ■ For purposes of Bioterrorism Acts' regulations, any infectious pathogen or toxin listed at 42 C.F.R. §73.3, 7 C.F.R. 331.3, or 9 C.F.R. 121.3

period, or gap between the exposure and the appearance of clinical symptoms, the disease-causing microorganism is insufficient in number to cause symptoms or infect other people, but it may be transmitted to others. An epidemic could exist before medical authorities are fully aware that the disease has spread.

Medical authorities categorically recommend the isolation of infected individuals from the general population until a bacterium is identified as contagious or noncontagious. Contagious infections may be transmitted from person to person by inhaling bacteria spewed into the air by coughing or sneezing, contacting body fluids, and simply touching something the infected person previously touched.

The isolation of persons infected with a contagious disease serves to prevent the spread of the disease to others. Following a bioterrorist event involving numerous individuals, state and local governments would most likely use their legal authorities to quarantine infected persons so they can be treated medically to prevent further spread of the disease.

10.21-C RESPONDING TO INCIDENTS INVOLVING A RELEASE OF INFECTIOUS PATHOGENS AND TOXINS

Hazardous material teams are now trained to identify infectious pathogens or toxins even more rapidly than medical personnel. They use detection equipment that first became available commercially during the late 2000s. Doctors and nurses stand ready to confirm their findings. In combination, the emergency response and medical personnel are the first individuals to clearly identify that an act of bioterrorism has occurred. Once the pathogen or toxin has been specifically identified, an appropriate antibiotic or vaccine can be administered to curb its impact and protect the exposed victims against death.

When a bioterrorism act has been committed, emergency responders need to cordon off the areas potentially contaminated with an infectious pathogen or toxin from the general public. While working in these areas, personnel need to wear fully encapsulating body suits and use self-contained breathing apparatus. They also should advise people who live in the communities surrounding the contaminated areas not only of the importance of seeking medical treatment but that failure to do so could well be fatal and adversely affect those with whom they live and work.

Areas contaminated with an infectious pathogen or toxin may be decontaminated by knowledgeable professionals, typically through use of a microbiocide such as chlorine dioxide (Section 11.8). For individual protection, these workers also need to wear fully encapsulating body suits and use self-contained breathing apparatus.

10.21-D TRANSPORTING INFECTIOUS SUBSTANCES

Although the general public fears exposure to biological warfare agents, microbiologists often study the nature of microorganisms as a routine component of their profession. Consequently, it is necessary to transport bacteria and viruses for legitimate purposes from their point of production to clinical laboratories and elsewhere.

DOT refers to infectious pathogens and toxins as **infectious substances**, and regulates their transportation as division 6.2 hazardous materials. Certain special rules apply that we note here.

When shippers and carriers intend to transport an infectious substance, they must first become aware of its risk group. A **risk group** is a ranking of a microorganism's ability to cause injury through disease as defined by criteria developed by the World Health Organization (WHO). These criteria are based on the severity of the disease caused by the microorganism, the mode and relative ease of transmission, the degree of risk to both an individual and a community, and the reversibility of the disease through the availability of known and effective preventative agents and treatment. The risk groups are identified by the numbers 1 through 4, and the criteria for each rank are provided in Table 10.27.

infectious substance
■ For purposes of DOT regulations, a material known to contain or suspected of containing an infectious pathogen or toxin

risk group ■ For purposes of DOT regulations, a ranking of a microorganism's ability to cause injury through development of a disease as defined by criteria developed by the World Health Organization (WHO) based on the severity of the disease caused by the organism, the mode and relative ease of transmission, the degree of risk to both an individual and a community, and the reversibility of the disease through the availability of known and effective preventive agents and treatment

TABLE 10.27 | Risk Groups for Pathogens

RISK GROUP	PATHOGEN	RISK TO INDIVIDUALS	RISK TO COMMUNITY
4	A pathogen that usually causes serious human or animal disease and that can be readily transmitted from one individual to another, directly or indirectly, and for which effective treatments and preventive measures usually are not available	High	High
3	A pathogen that usually causes serious human or animal disease but ordinarily does not spread from one infected individual to another, and for which effective treatments and preventive measures are available	High	Low
2	A pathogen that can cause human or animal disease but is unlikely to be a serious hazard, and although capable of causing serious infection on exposure, is one for which there are effective treatments and preventive measures available and the risk of spread of infection is limited	Moderate	Low
1	A microorganism that is unlikely to cause human or animal disease[a]	None or very low	Low

[a]Not subject to DOT regulations.

DOT regulates the transportation of hazardous materials in division 6.2 having the proper shipping names, "infectious substances, affecting humans," "infectious substances, affecting animals," and "diagnostic specimen." A **diagnostic specimen** is any human or animal material, including excreta, blood and its components, tissue, and tissue fluids being transported for diagnostic or investigational purposes (but excluding live infected humans or animals), when the source human or animal has or may have a serious human or animal disease from a Risk Group 4 pathogen.

When shippers prepare the applicable shipping paper, they are obligated to include the technical name of the infectious substance in its shipping description. For example, when shippers transport 1 gram of *Bacillus anthracis*, they enter the following shipping description on the accompanying shipping paper:

diagnostic specimen
■ For purposes of DOT regulations, a human or animal material, including excreta, blood and its components, tissue, and tissue fluids being transported for diagnostic or investigational purpose (but excluding live infected humans or animals), when the source human or animal has or may have a serious human or animal disease from a Risk Group 4 pathogen

UNITS	HM	SHIPPING DESCRIPTION (IDENTIFICATION NUMBER, PROPER SHIPPING NAME, PRIMARY HAZARD CLASS OR DIVISION, SUBSIDIARY HAZARD CLASS OR DIVISION, AND PACKING GROUP)	WEIGHT (g)
1 glass tube (UN6P)	X	UN2814, Infectious substances, affecting humans (*Bacillus anthracis*), 6.2	1

Although an infectious substance is not assigned a packing group, it is packaged to meet rigorous DOT performance tests. Figure 10.13 illustrates the triple packaging that generally is encountered when an infectious substance is transported. It consists of a primary receptacle, water-tight secondary packaging, and rigid outer packaging.

DOT requires shippers and carriers to mark the outer packaging with the following information:

■ Shipper or carrier's full name, address, and telephone number
■ Proper shipping name, including the name of the infectious substance
■ UN number

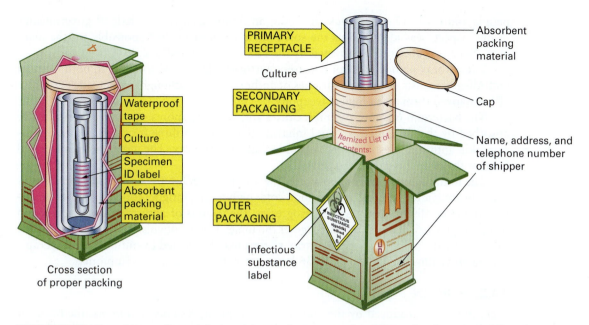

FIGURE 10.13 When shippers offer an infectious substance for transportation, DOT requires them to use triple packaging comprising a primary receptacle, water-tight secondary packaging, and rigid outer packaging.

- Precautionary statement
- 24-hour emergency response telephone number
- The net quantity of infectious substance

In addition, DOT requires shippers to display the orange-and-black BIOHAZARD marking shown to the right on two opposing sides or two ends other than the bottom of its packaging when it has a capacity less than 1000 gallons (3785 L) *or* on each side and each end when the packaging has a capacity equal to or greater than 1000 gallons (3785 L).

DOT also requires INFECTIOUS SUBSTANCE labels to be affixed to packaging containing an infectious substance. The INFECTIOUS SUBSTANCE label was previously illustrated in Figure 6.5. DOT also requires shippers to affix the applicable NON-FLAMMABLE GAS or CLASS 9 labels to the packaging *in addition to* the INFECTIOUS SUBSTANCE labels when the infectious substance is preserved in dry ice or nitrogen, respectively. When 50 milliliters or 50 grams of the infectious substance is to be transported by cargo aircraft, DOT requires them also to affix CARGO AIRCRAFT ONLY labels to the packaging.

Carriers are obligated to notify the consignee and either the National Response Center (Section 6.10) or the Centers for Disease Control and Prevention at (800) 232-0124 of an incident involving the release of an infectious substance. This telephone number is printed on the INFECTIOUS SUBSTANCE label.

DOT does not require carriers to placard the transport vehicle used to ship an infectious substance.

10.21-E ANTHRAX

Special concern about potential exposure to anthrax was first raised in the United States in 2001 just after the 9-11 attacks, following the appearance of anthrax-laden mailings in postal areas and throughout the Hart Senate Office Building in Washington, DC. Seven anthrax-tainted letters were mailed to U.S. senators and media outlets. These incidents constituted the first major acts of bioterrorism in our nation and caused the deaths of five people and made 17 others sick. In 2008, an Army scientist, the federal government's prime suspect, was identified as the sole individual most likely responsible for the mailings. As he was about to be arrested by the FBI, he committed suicide.

The anthrax pathogen is not new to bioterrorism. During World War II, both German and Allied scientists developed varieties of anthrax bacilli, although neither side ever actually employed them as biological weapons.

The bacterium *Bacillus anthracis* causes anthrax. The spores of the bacterium can be absorbed by the skin, ingested, and inhaled, giving rise to three forms of anthrax infection: skin anthrax, intestinal anthrax, and inhalation anthrax. The most potentially lethal form of the disease is inhalation anthrax.

The initial symptoms of exposure to the anthrax bacillus are similar to those experienced by individuals having the common cold: high fevers, fatigue, coughing, hard breathing, and chest discomfort. As time progresses, the victims of anthrax poisoning develop severe respiratory distress, hemorrhage of the lungs, heavy convulsions, and perspiration, followed by the onset of shock. Anthrax victims may be treated by antimicrobial therapy. Untreated victims usually die 24 to 36 hours after the first onset of symptoms.

10.21-F RICIN

Ricin is a toxin produced by the common castor plant, *Ricinus communis*, the beans of which also serve as a source of castor oil, the well-known laxative. Ricin is produced from the waste generated when castor beans are processed.

The chemical analysis of ricin reveals that it is a solid polymer (Chapter 14). In warfare jargon, it is known as *Agent W*. Its LD_{50} is estimated to be approximately 3 mg/kg. Ricin can be inhaled as a mist or powder, swallowed as a pellet or dissolved in a liquid, or injected into muscle. Its use is banned by chemical and biological weapons treaties.

The following incidents involving the use of ricin toxin as a biological warfare agent have been documented:

- In 1978, Georgi Markov, a Bulgarian exiled journalist living in London, was jabbed in the thigh with an umbrella by an individual suspected of being a KGB agent. Two days later, he died from cardiac arrest. An autopsy revealed the presence of a small metal ball with four holes in his thigh.
- In Paris, a similar platinum–iridium ball was removed from the back of a friend of the journalist. The wax used to seal the holes in the ball had not yet melted. Chemical analysis of the residue in the ball revealed the presence of ricin. The friend recalled being jabbed with an umbrella on the Paris Metro.
- In 1980, a Soviet citizen visiting Tyson Corners, Virginia, also was jabbed. He died en route to the local hospital. A similar platinum–iridium ball was removed from the deceased.
- In 2003, London antiterrorism police and security service agents arrested seven terrorists who were plotting to use ricin to kill and terrify the local population. The details of their methods were never made public.
- On several occasions in 2013, ricin was identified in envelopes addressed to certain public officials, including the President, a U.S. Senator, the New York Mayor, and judges in Mississippi and Washington.

When ricin is assimilated into the body, its victims experience fever, cough, shortness of breath, chest tightness, and low blood pressure within 8 hours. When inhaled, ricin can cause respiratory failure, and when ingested, it can cause organ dysfunction leading to death. Inhalation of ricin can cause death 36 to 72 hours after the initial exposure. There is no specific treatment for ricin exposure.

Basic Elements of Toxicology

1. When emergency responders are obliged to work in an atmosphere known to contain an elevated concentration of a toxic gas, why is it essential that they don fully encapsulating body suits?

Measuring Toxicity

2. One constituent of hardwood smoke is acrolein, exposure to which causes the eyes to tear. Acrolein is also a commercial chemical product. It is a toxic gas whose inhalation LD_{50} is 0.4 mg/kg. Using this measurement, determine whether the inhalation of 0.05 milligrams of acrolein by a 180-pound unprotected firefighter is likely to be fatal.

3. Governmental hygienists report the IDLH value for sulfur dioxide as 100 parts per million.
   (a) Is it safe for emergency responders to enter a room in which the concentration of sulfur dioxide is 750 parts per million if they are not using respiratory protection?
   (b) For what approximate period of time can a responder using respiratory protection remain in an environment containing sulfur dioxide at a concentration of 100 parts per million without experiencing any escape-impairing symptoms or irreversible health effects?

Toxicity of the Fire Scene

4. Why are active on-duty firefighters likely to experience the adverse health effects from inhaling 300 parts per million of carbon monoxide faster than nonactive off-duty firefighters who inhale the same concentration of the same substance?

Carbon Monoxide

5. In the United States, the number of carbon monoxide–poisoning incidents increases during cold weather. What factor is most likely associated with this rise in carbon monoxide–poisoning incidents?
6. Using the following partial sketch of a single-floor home, identify the locations at which carbon monoxide detectors should be installed to provide maximum protection for its occupants.
7. Why should carbon monoxide detectors be positioned at least 15 feet (4.6 m) from fuel-burning appliances?
8. How does oxygen act as an antidote when hyperbaric oxygen therapy is used to treat emergency responders who have been overexposed to carbon monoxide?

Hydrogen Cyanide

9. When causes the bluish coloration that develops in the fingernail beds and lips of firefighters who have been exposed to hydrogen cyanide?

Sulfur Dioxide

10. Why is it essential that volcanologists use self-contained breathing apparatus when conducting studies involving active volcanoes?

Hydrogen Sulfide

11. Is there a danger of fire and explosion within a segment of a sanitary sewer in which the hydrogen sulfide concentration measures 2.5% by volume in the surrounding air?

Nitrogen Oxides

12. Why do safety experts recommend separating nitric oxide cylinders from flammable gas cylinders during their storage by a minimum distance of 20 feet (6.1 m) or by a barrier of noncombustible material at least 5 feet (1.5 m) high, having a fire-resistance rating of at least ½ hour?

Ammonia

13. Although ammonia has a vapor density of 0.6 (air = 1), why does released ammonia sometimes linger near the surface of the ground rather than rising into the air and dispersing?

Metallic Cyanides

14. Why do experts advise emergency responders to remove contaminated clothing and wash their skin when they have been exposed to metallic cyanides?

Carcinogenesis

15. As early as 1761, John Hill, an English physician, noted that men who excessively sniffed tobacco snuff often developed nasal cancer. What does his observation indicate about the chemical nature of snuff?

Emergency Response Actions Involving Toxic Substances

16. On responding to a 2:00 A.M. domestic transportation mishap, a firefighting team determines that a gas is leaking from a 5-pound (2.3-kg) steel cylinder. The chemical commodity is described on the accompanying shipping paper as follows:

UNITS	HM	SHIPPING DESCRIPTION (IDENTIFICATION NUMBER, PROPER SHIPPING NAME, PRIMARY HAZARD CLASS OR DIVISION, SUBSIDIARY HAZARD CLASS OR DIVISION, AND PACKING GROUP)	WEIGHT (lb)
1 DOT-3A2000 gas cylinder	X	UN1953, Compressed gas, toxic, flammable, n.o.s. (carbon monoxide/butane mixture), 2.3, (2.1) (Poison - Inhalation Hazard, Zone D)	5

Use Table 10.23 to identify the initial isolation and protective-action distances recommended by DOT when responding to this mishap.

17. Use Table 10.23 to identify the initial isolation and protective-action distances recommended by DOT when responding to each of the following domestic transportation mishaps:

    (a) A rail tankcar leaking anhydrous ammonia at 9:30 A.M.

    (b) A truck that overturns at 7:45 P.M., causing a drum containing aqueous (25%) hydrocyanic acid to burst

Toxic Metallic Compounds

18. In 1992, Congress passed the Residential Lead-Based Paint Hazard Reduction Act, which requires housing owners to notify their tenants of the presence or suspicion of the presence of lead-based paint. What is the most likely reason Congress took this action?

Asbestos

19. The DOT regulation at 49 C.F.R. §173.216 requires that asbestos be transported in "rigid, leak-tight packaging" such as dust- and sift-proof bags. What is the most logical reason DOT requires the use of this special packaging?

Pesticides

20. Mecarbam is an insecticide and acaricide whose acute oral LD_{50} (rat) is 36 mg/kg. While combating a fire in a greenhouse, a 150-pound firefighter inadvertently swallows 0.5 gram of Mecarbam dust, despite using protective gear. Is the firefighter likely to survive?

Biological Warfare Agents

21. Why is it essential to seek medical attention as soon as possible after a suspected exposure to a biological warfare agent?

Storing and Transporting Toxic Substances

22. Chlorine trifluoride (ClF_3) is a greenish-yellow, liquefied compressed gas that typically is contained in DOT-3E1800 steel cylinders. It is a highly reactive chemical substance, even attacking water. Although it is nonflammable, chlorine trifluoride has the properties of a corrosive and toxic oxidizer. Even on short exposure, inhalation of the gas causes serious temporary or residual injury to humans. The permissible exposure limit for chlorine trifluoride as a time-weighted average is only 0.1 part per million.

    (a) Which integer or symbol should be inserted in each of the four quadrants of the NFPA hazard diamond displayed outside buildings where two cylinders of chlorine trifluoride are stored?

(b) When chlorine trifluoride is transported in two steel cylinders mounted within a rail boxcar, which DOT warning labels should be affixed to the exterior surfaces of the cylinders?

(c) Which DOT placards should be affixed to a railcar in which the two steel cylinders of chlorine trifluoride are transported to their use-destination?

(d) Identify the markings that DOT requires on the exterior surface of the cylinders used to transport chlorine trifluoride.

(e) When chlorine trifluoride is offered for transportation by rail, what shipping description does DOT require the shipper to enter on the accompanying waybill?

(f) Determine the initial isolation distance and protective-action distance recommended by DOT for use by emergency responders when they arrive at the scene of transportation mishaps involving a slow leak of chlorine trifluoride from a 5-pound (2.3-kg) steel cylinder during daytime hours.

11
Chemistry of Some Oxidizers

Courtesy of FEMA.

KEY TERMS

available chlorine, *p. 433*
chlorine bleach, *p. 433*
classes of oxidizers, *p. 427*
fireworks, *p. 438*
Fireworks 1.3G (special fireworks, display fireworks), *p. 439*
Fireworks 1.4G (common fireworks, consumer fireworks), *p. 439*
flare, *p. 443*

half-reaction, *p. 425*
hexavalent chromium compounds, *p. 452*
high-test hypochlorite (HTH), *p. 434*
ionic equation, *p. 424*
nonchlorine bleach, *p. 432*
oxidation number, *p. 423*
oxidizer (DOT), *p. 422*
oxidizer (NFPA), *p. 427*

oxidizer (OSHA), *p. 422*
oxidizing agent (oxidant), *p. 422*
safety matches, *p. 461*
signaling smoke, *p. 443*
smoke bomb, *p. 443*
spectator ion, *p. 424*
strike-anywhere matches, *p. 461*

OBJECTIVES

- Associate the physical and health hazards of the oxidizers noted in this chapter with the information provided by their hazard diamonds and GHS pictograms.
- Determine the oxidation numbers of the atoms and ions in a given substance.
- Describe how NFPA distinguishes the degree of hazard potentially posed by different oxidizers.
- Identify the primary industries that use the oxidizers noted in this chapter.
- Identify commercial products that contain oxidizers.
- Describe the OSHA and DHS regulations that pertain to the handling, storage, stowing, loading, unloading, or discharge of bulk quantities of ammonium nitrate.
- Identify the labels, markings, and placards that DOT requires on the packaging of oxidizers and the transport vehicles used for their shipment.
- Identify the response actions to be executed when oxidizers are released from their packaging into the environment.

Certain chemical reactions greatly benefit modern lifestyles. They include the combustion of fuels, the chlorination of water, the explosion of dynamite, the bleaching of fabrics, and the flaring of recreational fireworks. These specific phenomena are associated with oxidation–reduction reactions, also called *redox reactions*. These reactions and the hazardous materials associated with them constitute the subject matter of this chapter.

When oxidation–reduction reactions are conducted in a controlled fashion, the energy they release can be harnessed to advantage. For example, the reactions that occur in devices such as batteries, dry cells, and fuel cells provide a supply of portable electrical energy. However, when redox reactions occur in an uncontrolled fashion, the generated energy is released into the immediate environment, where it can initiate or intensify fire and explosion and cause the loss of life and property. This potential for destructiveness necessitates the examination of redox reactions as a component of the study of hazardous materials.

11.1 WHAT IS AN OXIDIZER?

It was noted in Chapter 7 that hydrogen burns in oxygen and chlorine to produce water and hydrogen chloride, respectively. These combustion reactions are represented by the following equations:

$$2H_2(g) \ + \ O_2(g) \ \longrightarrow \ 2H_2O(g)$$

Hydrogen   Oxygen   Water

$$H_2(g) \ + \ Cl_2(g) \ \longrightarrow \ 2HCl(g)$$

Hydrogen   Chlorine   Hydrogen chloride

oxidizing agent (oxidant) ■ An oxygen-rich substance capable of readily yielding some of its oxygen; the substance reduced during an oxidation–reduction reaction

The substances supporting these independent combustion processes are examples of oxidizers, **oxidants**, or **oxidizing agents**. The first two terms are used by emergency responders, but the third term is used by chemists. Hydrogen is an example of a reducing agent. Oxidizing agents always react with reducing agents in concert.

Because hydrogen burns in a chlorine atmosphere, it is evident that oxygen is not the sole substance that supports the combustion of other materials. Like oxygen, chlorine is an example of an oxidizer.

11.1-A DOT CLASSIFICATION OF OXIDIZERS

oxidizer ■ For the purposes of DOT regulations, any substance that may enhance or support the combustion of other materials, generally by yielding its oxygen

In the DOT regulations, an **oxidizer** is defined as a substance that may enhance or support the combustion of other materials, generally by yielding its oxygen. DOT also distinguishes between two classes of oxidizers: inorganic (or metallic) oxidizers, which are simply called oxidizers, and organic peroxides. Their properties are noted in this chapter and Section 13.9, respectively.

11.1-B OSHA/GHS IDENTIFICATION OF OXIDIZERS

oxidizer (OSHA) ■ Any gas, liquid, or solid that readily yields oxygen or other oxidizing gas, or that readily reacts to promote or initiate combustion of combustible materials and, under some circumstances, can undergo vigorous self-sustained decomposition due to contamination or heat exposure

Formerly, OSHA defined an oxidizer as a substance other than a blasting agent or explosive that initiates or promotes combustion in other materials, thereby causing fire either of itself or through the release of oxygen or other gases. Following the adoption of the GHS, OSHA now considers oxidizers as any of the three classes, oxidizing gases, oxidizing liquids, or oxidizing solids. As oxidizers, each of the latter states of matter cause or contribute to the combustion of other material, generally by providing oxygen.

Except for oxidizers that are organic compounds, OSHA minimally requires the manufacturers, distributors, and importers of oxidizers to post the GHS flame-over-the-letter-"O" pictogram to the labels of chemical products consisting of an oxidizing gas, liquid, or solid, in addition to appropriate signal words and hazard and precautionary statements.

11.2 OXIDATION NUMBERS

Chemists use the concept of an **oxidation number,** or **oxidation state,** to describe the combining capability of one ion for another ion or of one atom for another atom. In practice, an oxidation number is formally assigned to the atoms that make up a substance by using the following rules:

- The oxidation number of each atom in an element is zero; thus, the oxidation number of each atom in the elements H_2, O_2, Na, and Mg is 0.
- The algebraic sum of the oxidation numbers of the atoms in any substance is zero.
- The hydrogen atom in hydrogen-containing compounds other than metallic hydrides has an oxidation number of +1. The hydrogen in a hydride ion (H^-) has an oxidation number of −1.
- The oxygen atom in oxygen-containing compounds other than peroxides and superoxides has an oxidation number of −2. Each oxygen atom in a peroxide ion (O_2^{2-}) and a superoxide ion (O_2^-) has an oxidation number of −1 and −0.5, respectively.
- The oxidation number of a monatomic ion (i.e., having one atom) is the same as its net ionic charge; thus, the oxidation number of sodium is +1; of magnesium, +2; of the chloride ion, −1; and of the sulfide ion, −2.
- The algebraic sum of the oxidation numbers of the atoms in a polyatomic ion is equal to its ionic charge.

oxidation number (oxidation state) ■ A number assigned to an atom or ion by following established rules that aim to reflect its capacity for combining with other atoms or ions

The chemical formula of a substance must be known to use these rules and determine the oxidation numbers of its component atoms. As an example, let's use them to determine the oxidation number of each atom in sodium chlorate, an oxidizing agent used in fireworks and signaling flares. The chemical formula of sodium chlorate is $NaClO_3$. This formula indicates that the compound is composed of two ions, the sodium ion (Na+) and the chlorate ion (ClO_3^-). The oxidation number of each element in sodium chlorate is determined as follows:

CONSTITUENT	OXIDATION NUMBER	RELEVANT RULE
Sodium ion, Na^+	+1	The oxidation number of a monatomic ion is the same as its ionic charge.
Oxygen atom	−2	Other than in peroxides, the oxidation number of oxygen is −2.
Chlorate ion, ClO_3^-	−1	The algebraic sum of the oxidation numbers of the atoms in a polyatomic ion is equal to its ionic charge; that is, −1 = +5 + [3 × (−2)] = +5 + (−6).
Chlorine atom	+5	The algebraic sum of the oxidation numbers in any substance is zero; that is, 0 = +1 +5 + [3 × (−2)] = +6 + (−6).

Additional examples of determining the oxidation numbers of the atoms in some compounds are noted in Table 11.1.

11.3 OXIDATION–REDUCTION REACTIONS

Because a redox reaction occurs between an oxidizing agent and a reducing agent, the equation illustrating a redox reaction is always written in the following generalized form:

$$\text{Oxidizing agent} \ + \ \text{Reducing agent} \longrightarrow \text{Products}$$

A specific example of this is the reaction between iron(III) chloride and tin(II) chloride. We write the equation for this reaction as follows:

$$2FeCl_3(aq) \ + \ SnCl_2(aq) \longrightarrow 2FeCl_2(aq) \ + \ SnCl_4(aq)$$

Iron(III) chloride   Tin(II) chloride   Iron(II) chloride   Tin(IV) chloride

TABLE 11.1	Examples of Oxidation Numbers	
SYMBOL OF ELEMENT	**OXIDATION NUMBER**	**CHEMICAL FORMULA OF A REPRESENTATIVE COMPOUND**
F	−1	HF
O	−2	SO_2
	−1	H_2O_2
N	−3	NH_3
	−2	N_2H_4
	−1	NH_2OH
	+1	N_2O
	+2	NO
	+3	HNO_2
	+4	NO_2
	+5	HNO_3
Cl	−1	HCl
	+1	HClO
	+3	$KClO_2$
	+5	$KClO_3$
	+7	$HClO_4$
S	−2	H_2S
	+4	H_2SO_3
	+6	H_2SO_4
P	−3	AlP
	+3	H_3PO_3
	+5	H_3PO_4

Because the chloride ions do not participate in this redox reaction, they often are called **spectator ions**. We can eliminate them and write an **ionic equation** for the reaction as follows:

$$2Fe^{3+}(aq) \; + \; Sn^{2+}(aq) \; \longrightarrow \; 2Fe^{2+}(aq) \; + \; Sn^{4+}(aq)$$

In an ionic equation, only the symbols of the ions contributing to the reaction are written. In this representation, electrons have simply been transferred between the iron ions and tin ions. The information conveyed by this equation is summarized as follows:

spectator ion ▪ One or more of the ions that do not participate in an oxidation–reduction reaction

ionic equation ▪ An equation that depicts only the ions that participate in an oxidation–reduction reaction

- The iron(III) ions become iron(II) ions. The oxidation number of iron decreases from +3 to +2; thus, the iron(III) ions are *reduced*. Iron(III) chloride is called the *oxidizing agent*, because it oxidizes tin(II) chloride.
- The tin(II) ions become tin(IV) ions. The oxidation number of tin increases from +2 to +4; thus, the tin(II) ions are *oxidized*. Tin(II) chloride is called the *reducing agent*, because it reduces iron(III) chloride.

Because a simultaneous increase and decrease in oxidation numbers always accompanies oxidation and reduction, respectively, we can summarize the nature of this process in ionic systems as follows:

- *Oxidation* is the phenomenon associated with an *increase* in oxidation number and *loss* of electrons from an ion, atom, or group of atoms.
- *Reduction* is the phenomenon associated with a *decrease* in oxidation number and *gain* of electrons from an ion, atom, or group of atoms.

TABLE 11.2 — Some Common Oxidizing Agents

OXIDIZING AGENT	ELEMENT THAT CHANGES OXIDATION NUMBER	OXIDATION NUMBER IN REACTANT	OXIDATION NUMBER IN PRODUCT	EQUATION ILLUSTRATING HALF-REACTION
Sodium peroxide	Oxygen	−1	−2	$Na_2O_2(aq) + 2H_2O(l) + 2e^- \longrightarrow 2Na^+(aq) + 4OH^-(aq)$
Metallic hypochlorites	Chlorine	+1	−1	$ClO^-(aq) + 2H^+(aq) + 2e^- \longrightarrow Cl^-(aq) + H_2O(l)$
Metallic chlorates	Chlorine	+5	−1	$ClO_3^-(aq) + 6H^+(aq) + 6e^- \longrightarrow Cl^-(aq) + 3H_2O(l)$
Nitric acid (concentrated)	Nitrogen	+5	+4	$NO_3^-(aq) + 2H^+(aq) + e^- \longrightarrow NO_2(g) + H_2O(l)$
Nitric acid (dilute)	Nitrogen	+5	+2	$NO_3^-(aq) + 4H^+(aq) + 3e^- \longrightarrow NO(g) + 2H_2O(l)$
Metallic peroxydisulfates	Sulfur	+7	+6	$S_2O_8^{2-}(aq) + 2e^- \longrightarrow 2SO_4^{2-}(aq)$

- The substance that *accepts* electrons during a redox reaction is called the *oxidizing agent*.
- The substance that *loses* electrons during a redox reaction is called the *reducing agent*.

Oxidation and reduction phenomena can also be represented separately by equations like the following:

$$Sn^{2+}(aq) \longrightarrow Sn^{4+}(aq) + 2e^-$$
$$Fe^{3+}(aq) + e^- \longrightarrow Fe^{2+}(aq)$$

The phenomenon illustrated by each equation is called a **half-reaction**. One half-reaction represents oxidation, whereas the other represents reduction. The processes represented by half-reactions always occur simultaneously.

Several other examples of oxidation–reduction phenomena are provided in Table 11.2.

half-reaction ■
An equation that separately depicts either oxidation or reduction

SOLVED EXERCISE 11.1

Identify the element oxidized, the element reduced, the oxidizing agent, and the reducing agent in the redox reaction denoted by the following equation:

$$6FeSO_4(aq) + Na_2Cr_2O_7(aq) + 7H_2SO_4(aq) \longrightarrow$$

Iron(II) sulfate   Sodium dichromate   Sulfuric acid

$$3Fe_2(SO_4)_3(aq) + Na_2SO_4(aq) + Cr_2(SO_4)_3(aq) + 7H_2O(l)$$

Iron(III) sulfate   Sodium sulfate   Chromium(III) sulfate   Water

Solution: In this equation, the sodium and sulfate ions are identified as spectator ions, because the oxidation numbers of sodium, sulfur, and oxygen are the same on each side of the arrow. When the spectator ions are eliminated, the following ionic equation is written:

$$6Fe^{2+}(aq) + Cr_2O_7^{-2}(aq) + 14H^+(aq) \longrightarrow 6Fe^{3+}(aq) + 2Cr^{3+}(aq) + 7H_2O(l)$$

To determine the oxidation number of chromium in the dichromate ion, recall that the algebraic sum of the oxidation numbers of the atoms in a polyatomic ion is equal to its ionic charge. We thus determine that the oxidation number of chromium is +6.

$$-2 = [2 \times (+6)] + [7 \times (-2)]$$
$$-2 = -2$$

The oxidation number of the monatomic ions is the same as their ionic charge.

In the ionic equation, we readily observe that the iron(II) ions become iron(III) ions. The oxidation number of iron increases from +2 to +3, as each iron(II) ion loses an electron. Because an *increase* in oxidation number is associated with oxidation, the iron(II) ions are *oxidized*. Iron(II) sulfate is the *reducing agent*.

The ionic equation also notes that dichromate ions become chromium(III) ions. The oxidation number of chromium decreases from +6 in $Cr_2O_7^{-2}$ to +3 in Cr^{3+}, as each chromium ion gains three electrons. Because a *decrease* in oxidation number is associated with reduction, the dichromate ions are *reduced*. Thus, sodium dichromate is the *oxidizing agent*.

11.4 COMMON FEATURES OF OXIDIZERS

Oxidizers generally are perceived as relatively powerful chemical substances, because they often react rapidly, even at explosive rates, with other substances. These latter substances include fuels, lubricants, greases, oils, cotton, animal and vegetable fats, paper, coal, coke, straw, sawdust, and wood shavings. Because they are potentially powerful, oxidizers have been chosen by terrorists to intentionally cause mass destruction and chaos. In 1995, an American terrorist used an ammonium nitrate/fuel oil mixture as a weapon of mass destruction. Ignition of the mixture destroyed the Murrah Federal Building in Oklahoma City, shown in Figure 11.1. The incident killed 168 people and injured 850.

Different oxidizers can have dissimilar strengths. This variation in the strength of individual oxidizers can be approximated by examining the listing in Table 11.3. In this series, oxidizing agents are arranged according to their decreasing oxidizing power. Any substance whose name appears on this list is a stronger oxidizing agent than the substances whose names are listed below it. It is noteworthy that many substances are stronger oxidizing agents than oxygen itself.

FIGURE 11.1 The criminal act of detonating a weapon of mass destruction at the Oklahoma City federal building resulted in 168 fatalities and 850 injuries. The weapon was a 4000-pound (1800-kg) mixture of ammonium nitrate fertilizer and fuel oil (ANFO). The detonation occurred at the rate of approximately 13,000 ft/s (4000 m/s). (*Courtesy of FEMA.*)

TABLE 11.3	Relative Strength of Oxidizing Agents[a]
	Fluorine
	Ozone
	Hydrogen peroxide
	Hypochlorous acid
	Metallic chlorates[b]
	Lead dioxide
	Metallic permanganates[b]
	Metallic dichromates[b]
	Nitric acid (concentrated)
	Chlorine
	Sulfuric acid (concentrated)
	Oxygen
	Metallic iodates
	Bromine
	Iron(III) (Fe^{3+}) compounds
	Iodine
	Sulfur
	Tin(IV) (Sn^{4+}) compounds

[a]Listed in descending order of oxidizing power.
[b]In an acidic environment.

oxidizer (NFPA) ▪ As used in NFPA's standards and codes, any substance that can increase the rate of combustion of other material and, under some circumstances, can undergo vigorous self-sustained decomposition due to contamination or heat exposure

classes of oxidizers ▪ As used in NFPA's standards and codes, any of four classes into which individual oxidizers are assigned based on their ability to affect the burning rate of combustible materials or to undergo self-sustained decomposition

In its codes and standards, NFPA uses the term oxidizer to denote a substance that can increase the rate of combustion of other materials with which it contacts. NFPA distinguishes the degree of hazard posed by specific oxidizers by assigning them to **classes of oxidizers** as follows:[1]

▪ Class 1 oxidizer. This is a substance that does not moderately increase the burning rate of combustible materials with which it comes in contact.
▪ Class 2 oxidizer. This is a substance that causes a moderate increase in the burning rate of combustible materials with which it comes in contact.
▪ Class 3 oxidizer. This is a substance that causes a severe increase in the burning rate of combustible materials with which it comes in contact.
▪ Class 4 oxidizer. This is a substance that can undergo an explosive reaction due to contamination or exposure to thermal or physical shock, and that causes a severe increase in the burning rate of combustible materials with which it comes in contact.

Some specific oxidizers are denoted in Table 11.4 using this system of classification.

11.5 HYDROGEN PEROXIDE

Hydrogen peroxide is an important substance having the chemical formula and molecular structure H_2O_2 and H–O–O'–H, respectively. Specific concentrations of hydrogen peroxide solutions are designated by NFPA as members of each class of oxidizer. They are used primarily for the following purposes:

▪ Solutions having a concentration from 1% to 3% by mass are used as topical antiseptics on minor cuts and wounds, as well as sterilizing and disinfecting agents. They

Hydrogen peroxide, 35% solution

[1]NFPA Code No. 400, *Hazardous Materials Code* (Quincy, Massachusetts: National Fire Protection Association), 2009.

TABLE 11.4	Some Typical Oxidizers by NFPA Classification[a]

CLASS 1

All inorganic nitrates (unless otherwise classified)	Potassium dichromate
All inorganic nitrites (unless otherwise classified)	Potassium percarbonate
Ammonium persulfate	Potassium persulfate
Barium peroxide	Sodium carbonate peroxide
Calcium peroxide	Sodium dichloro-s-triazinetrione dihydrate
Calcium hypochlorite (66% or less available chlorine and a total water content of at least 17% by mass)	Sodium dichromate
	Sodium perborate (anhydrous)[b]
Hydrogen peroxide solutions (greater than 8% up to 27.5%)	Sodium perborate monohydrate
Lead dioxide	Sodium perborate tetrahydrate
Lithium hypochlorite (39% or less available chlorine)	Sodium percarbonate
Lithium peroxide	Sodium persulfate
Magnesium peroxide	Strontium peroxide
Manganese dioxide	Trichloro-s-triazinetrione
Nitric acid (40% concentration or less)	Zinc peroxide
Perchloric acid solutions (less than 50% by mass)	

CLASS 2

Barium bromate	Nitric acid (more than 40% but less than 86% by mass)
Barium chlorate	Perchloric acid solutions (more than 50% but less than 60% by mass)
Barium hypochlorite	
Barium perchlorate	Potassium perchlorate
Barium permanganate	Potassium permanganate
1-Bromo-3-chloro-5,5-dimethylhydantoin (BCDMH)	Potassium peroxide
Calcium chlorate	Potassium superoxide
Calcium chlorite	Silver peroxide
Calcium hypochlorite (less than 50% by mass)	Sodium chlorite (40% or less by mass)
Calcium permanganate	Sodium perchlorate
Chromium trioxide (chromic acid)	Sodium perchlorate monohydrate
Halane (1,3-dichloro-5,5-dimethyl hydantoin)	Sodium permanganate
Hydrogen peroxide (greater than 27.5% up to 52% by mass)	Sodium peroxide
Lead perchlorate	Strontium chlorate
Lithium chlorate	Strontium perchlorate
Lithium hypochlorite (more than 39% available chlorine)	Thallium chlorate
Lithium perchlorate	Urea hydrogen peroxide
Magnesium bromate	Zinc bromate
Magnesium chlorate	Zinc chlorate
Magnesium perchlorate	Zinc permanganate
Mercurous chlorate	

CLASS 3

Ammonium dichromate	Potassium bromate
Calcium hypochlorite (over 50% by mass)	Potassium chlorate
Chloric acid (10% maximum concentration)	Potassium dichloro-s-triazinetrione
	Sodium bromate
Hydrogen peroxide solutions (greater than 52% up to 91% by mass)	Sodium chlorate
	Sodium chlorite (over 40% by mass)
Mono-(trichloro)-tetra-(monopotassium dichloro)-penta-triazinetrione	Sodium dichloro-s-triazinetrione anhydrous
Perchloric acid solutions, 60% to 72.5% by mass	

CLASS 4

Ammonium perchlorate (particle size greater than 15 microns)	Guanidine nitrate
	Hydrogen peroxide solutions (more than 91% by mass)
Ammonium permanganate	

[a]Reprinted with permission from NFPA 400-2013, *Hazardous Materials Code*, Copyright © 2012, National Fire Protection Association, Quincy, MA. This reprinted material is not the complete and official position of the NFPA on the referenced subject, which is represented only by the standard in its entirety.
[b]Sodium perborate is the common name for the substance whose proper chemical name is sodium peroxoborate (Section 11.5-E).

are commonly available as over-the-counter products in drug stores. They are also widely used in hospitals and medical clinics to destroy germs like *Escherichia coli*, botulism, salmonella, and other infectious microorganisms that cause disease.

■ Solutions having a concentration of 6% by mass are used to bleach hair. In the process, hydrogen peroxide oxidizes the dark-colored pigment called melanin to colorless products.

■ Solutions having a concentration from 30% to 50% by mass are used in the chemical industry to synthesize peroxo-organic compounds (Section 13.9). The 30% solution is also used to bleach cotton, flour, wool, straw, leather, gelatin, and paper. The 30% solution can also be used instead of chlorine for treating drinking water. Contemporaneously, the solution has also achieved notoriety in connection with its misuse by terrorists for the production of the explosive triacetone triperoxide (Section 13.9-C).

■ Solutions having a concentration of 70% by mass are used by the chemical industry to execute oxidation-reduction reactions. They are often diluted before use and stored as 35% to 50% solutions.

■ Solutions having a concentration of 90% by volume are used by the aerospace industry as rocket-grade solutions. They have been used to oxidize fuels such as hydrazine, which launched the *Apollo* rockets and other payloads into space. Russia uses the solution to launch Soyuz rockets into space. The 90% solution has also been used by the U.S. military.

These solutions visibly resemble water in physical appearance, although they may have slightly pungent, irritating odors. Some other physical properties of hydrogen peroxide solutions are noted in Table 11.5. Although anhydrous hydrogen peroxide has been produced, it is unavailable commercially because it is likely to decompose violently.

The world's supply of hydrogen peroxide is manufactured by a number of methods. One involves the oxidation of isopropyl alcohol, during which acetone is coproduced.

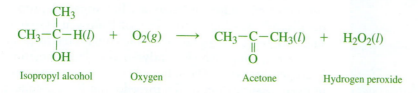

| Isopropyl alcohol | Oxygen | Acetone | Hydrogen peroxide |

The hydrogen peroxide and acetone are separated by distillation. The hydrogen peroxide solution may then be diluted to produce the desired concentration for commercial sale.

TABLE 11.5	Physical Properties of Some Hydrogen Peroxide Solutions		
	$H_2O_2(aq)$ (30%)	$H_2O_2(aq)$ (50%)	$H_2O_2(aq)$ (70%)
Melting point	−13°F (−25°C)	−62°F (−52°C)	−40°F (−40°C)
Boiling point	226°F (108°C)	239°F (115°C)	257°F (125°C)
Vapor density (air = 1)	1.17	1.0	
Specific gravity at 68°F (20°C)	1.11	1.20	1.29
Vapor pressure at 68°F (20°C)	25 mmHg	9 mmHg	
Solubility in water	Infinitely soluble	Infinitely soluble	Infinitely soluble

Hydrogen peroxide is an inherently unstable substance. It slowly decomposes as follows:

$$2H_2O_2(aq) \longrightarrow 2H_2O(l) + O_2(g)$$

Hydrogen peroxide   Water   Oxygen

The rate of the decomposition reaction is catalyzed by sunlight as well as certain metals, most notably, iron, copper, chromium, and silver. An aqueous solution of 8% hydrogen peroxide is completely decomposed following a 10-month exposure to light, whereas a similar solution kept in darkness for the same length of time remains virtually unaltered. In solutions of concentrations of less than 30%, the decomposition of hydrogen peroxide occurs so slowly when stored in dark glass bottles that it is virtually imperceptible.

Concentrated solutions of hydrogen peroxide ($>6\%$ H_2O_2) become intensely heated when they decompose. These hot solutions then vaporize. To prevent their decomposition from posing a hazard before the intended use of the oxidant, all commercial forms of hydrogen peroxide are stabilized by addition of a substance that retards its decomposition.

Concentrated hydrogen peroxide solutions have the following properties:

- When concentrated hydrogen peroxide solutions decompose, ample heat may be evolved to cause the spontaneous ignition of nearby combustible materials.
- Hydrogen peroxide solutions having a concentration in excess of 20% are highly corrosive. When exposed to skin, they cause severe irritation; and when exposed to the eyes, they can cause blindness.

FIGURE 11.2 In compliance with DOT regulations, the shipper affixes OXIDIZER and CORROSIVE labels to this intermediate bulk container holding a hydrogen peroxide solution consisting of 30% to 32% hydrogen peroxide. The shipper also posts OXIDIZER placards that display the DOT identification number 2014 across their center areas. (*Courtesy of Avantor Performance Materials, Inc., Center Valley, Pennsylvania.*)

11.5-A THE *KURSK*

Hydrogen peroxide has been linked with the naval disaster onboard the Russian submarine *Kursk*. In 2000, the *Kursk* exploded and sank in the Barents Sea. Aboard were a number of torpedoes, each of whose fuel systems consisted in part of highly concentrated hydrogen peroxide. Investigators proposed that the disaster was linked to a leak of hydrogen peroxide from a single torpedo. The oxidizer interacted with the torpedo's stainless steel casing, which catalyzed the decomposition of the hydrogen peroxide. The subsequent buildup of oxygen resulted in overpressurization of the torpedo and its subsequent explosion. This first explosion then initiated the detonation of other torpedoes within the storage compartment. The hull of the submarine burst, and the *Kursk* foundered and sank. There were no survivors.

11.5-B WORKPLACE REGULATIONS INVOLVING HYDROGEN PEROXIDE

When hydrogen peroxide is used in the workplace, OSHA requires employers to limit employee exposure to an inhalation concentration of 1 part per million, averaged over an 8-hour workday.

11.5-C TRANSPORTING HYDROGEN PEROXIDE

Hydrogen peroxide solutions may be transported by means of motor vans on public highways or by rail in boxcars, usually contained in nonbulk darkened glass or plastic bottles or intermediate bulk containers like the type shown in Figure 11.2. In addition, hydrogen peroxide may be transported in bulk in a tank truck or rail tankcar.

TABLE 11.6	Shipping Descriptions of Aqueous Hydrogen Peroxide Solutions
CONCENTRATION OF HYDROGEN PEROXIDE SOLUTION BY VOLUME	**SHIPPING DESCRIPTION[a]**
8% to 20%	UN2984, Hydrogen peroxide, aqueous solution (contains 8%–20% hydrogen peroxide), 5.1, PG III
20% to 40%	UN2014, Hydrogen peroxide, aqueous solution (stabilized) (contains 20%–40% hydrogen peroxide), 5.1, (8), PG II
40% to 60%	UN2014, Hydrogen peroxide, aqueous solution (stabilized) (contains 40%–60% hydrogen peroxide), 5.1, (8), PG II
>60%	UN2015, Hydrogen peroxide, stabilized) (contains >60% hydrogen peroxide), 5.1, (8), PG I

[a]Before shipment, DOT requires the addition of a substance to hydrogen peroxide to inhibit its decomposition.

When shippers offer a hydrogen peroxide solution for transportation, DOT requires them to provide the relevant shipping description shown in Table 11.6 on an accompanying shipping paper. DOT also requires shippers and carriers to comply with all applicable labeling, marking, and placarding requirements. When the solution contains more than 10% hydrogen peroxide, carriers must comply with a special marking requirement. DOT requires them to mark each side and each end of the tank HYDROGEN PEROXIDE, AQUEOUS SOLUTION *or* HYDROGEN PEROXIDE, STABILIZED.

SOLVED EXERCISE 11.2

When a shipper offers a carrier 2500 pounds (1135 kg) of an aqueous solution containing 30% to 32% hydrogen peroxide for transportation in a cargo tank by highway:

(a) What shipping description does DOT require the shipper to provide for this chemical commodity?
(b) Which DOT placards and markings does DOT require to be displayed on the cargo tank?

Solution:

(a) DOT requires the shipper to provide the following shipping description of an aqueous solution containing 30% to 32% hydrogen peroxide solution on the accompanying shipping paper:

UNITS	HM	SHIPPING DESCRIPTION (IDENTIFICATION NUMBER, PROPER SHIPPING NAME, PRIMARY HAZARD CLASS OR DIVISION, SUBSIDIARY HAZARD CLASS OR DIVISION, AND PACKING GROUP)	WEIGHT (lb)
1 cargo tank	X	UN2014, Hydrogen peroxide, aqueous solution (stabilized) (contains 30% to 32% hydrogen peroxide), 5.1, (8), PG II	2500

(b) Because more than 1001 pounds (454 kg) of the hydrogen peroxide solution is offered for transportation, DOT requires the carrier to display OXIDIZER and CORROSIVE placards on each side and each end of the cargo tank. Because the hydrogen peroxide solution is offered for transportation by highway, DOT also requires the carrier to display the identification number 2014 on an orange panel, across the center area of the OXIDIZER placards, or on white square-on-point diamonds posted on each side and each end of the tank. DOT prohibits the display of the identification number on the subsidiary CORROSIVE placards. Finally, as previously noted in Table 6.7, because the solution contains more than 10% hydrogen peroxide, DOT requires the carrier to mark each side and each end of the tank HYDROGEN PEROXIDE, AQUEOUS SOLUTION *or* HYDROGEN PEROXIDE, STABILIZED.

11.5-D RESPONDING TO INCIDENTS INVOLVING A RELEASE OF HYDROGEN PEROXIDE

When emergency-response teams encounter hydrogen peroxide solutions at accident scenes, they should attempt not only to extinguish ongoing fires but to dam or dike the water runoff generated while combating them. When a fire has not occurred, the solutions should be segregated from combustible materials.

11.5-E NONCHLORINE BLEACHES

A number of products are available commercially that remove undesirable stains during the laundering of garments and other fabrics. Those that dissolve in water to produce hydrogen peroxide are called **nonchlorine bleaches**. Their use during laundering operations provides the advantage that they do not destroy most fabrics or harm the appearance of their colors, which can occur when chlorine bleaches (Section 11.6) are used.

nonchlorine bleach ■ Any nonchlorine substance used as a bleaching agent during the laundering of fabrics

An example of a nonchlorine bleach is *OxiClean*. Its active component is an adduct of sodium carbonate and hydrogen peroxide whose chemical formula is $2Na_2CO_3 \cdot 3H_2O_2$. It is also known as sodium carbonate peroxyhydrate and sodium percarbonate. It is a white powder that dissociates into sodium carbonate and hydrogen peroxide when dissolved in warm or hot water.

$$2Na_2CO_3 \cdot 3H_2O_2(aq) \longrightarrow 2Na_2CO_3(aq) + 3H_2O_2(aq)$$

Sodium carbonate peroxyhydrate (Sodium percarbonate) — Sodium carbonate — Hydrogen peroxide

Sodium carbonate peroxyhydrate is also a component of *Tide Laundry Detergent* and several cleaning agents and laundry detergents manufactured by Arm & Hammer. Another example is *Clorox 2,* a liquid product whose active ingredient is hydrated sodium peroxoborate, commonly called sodium perborate.

$$2Na^+ \begin{bmatrix} HO & O-O & OH \\ & \diagdown / & & \diagdown / \\ & B & & B \\ & / \diagdown & & / \diagdown \\ HO & O-O & OH \end{bmatrix}^{2-} \cdot 4H_2O$$

Sodium peroxoborate (Sodium perborate)

When sodium peroxoborate is dissolved in cold water, it dissociates and forms hydrogen peroxide.

$$Na_2B_2(O_2)_2(OH)_4 \cdot 4H_2O(aq) \longrightarrow 2NaBO_2(aq) + 2H_2O_2(aq) + 4H_2O(l)$$

Sodium peroxoborate (Sodium perborate) — Sodium metaborate — Hydrogen peroxide — Water

When sodium peroxoborate is dissolved in hot water, the hydrogen peroxide decomposes to water and oxygen. Table 11.3 illustrates that hydrogen peroxide is a stronger oxidizing agent than oxygen; hence, a greater bleaching action is accomplished when fabrics are laundered with Clorox 2 in cold water.

11.6 METALLIC HYPOCHLORITES

Metallic hypochlorites are compounds composed of a metallic ion and a hypochlorite ion (ClO^-). The most common commercial examples are sodium hypochlorite and calcium hypochlorite, each of which is encountered as the active component of several household and commercial bleaching and sanitation products. Their chemical formulas are $NaClO$ and $Ca(ClO)_2$, respectively. When they are used as bleaching agents, these substances are distinguished from metallic peroxide bleaches by calling them **chlorine bleaches**.

During laundering, bleaching agents are mixed with detergents to remove undesirable stains from fabrics and other textiles. Chlorine bleaches can be used safely only on textiles made from cotton and linen, but not wool, silk, and nylon. Nonchemists generally contend that the bleaching action is accomplished by chlorine, but it is actually the oxidizing power of the hypochlorite ion.

When chlorine bleaches can be used safely during the laundering of fabrics, American textile manufacturers post on the fabric's label the open triangular symbol shown below on the left. When the use of chlorine bleaches should be entirely avoided, they post the middle symbol; and when neither a chlorine bleach nor a nonchlorine bleach can be used safely, they post the symbol shown below on the right.

chlorine bleach ■ Any metallic hypochlorite or other chlorine-containing substance used as a bleaching agent during the laundering of fabrics

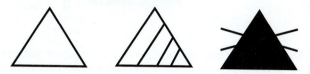

The effectiveness of the hypochlorite bleaching agents is commercially described by their **available chlorine**. This term is used to compare the effectiveness of bleaching agents with elemental chlorine, which has 100% available chlorine. A bleaching agent having an available chlorine value of 99.2% has the same bleaching power as a solution containing 99.2% chlorine by mass.

available chlorine ■ The amount of free chlorine that a substance yields when treated with an acid

The available chlorine in any chlorine bleach is determined by using the reaction between its active ingredient and an acid; for instance, the available chlorine in a bleaching agent containing calcium hypochlorite is determined by subjecting the agent to a reaction with hydrochloric acid. Chlorine is produced as follows:

$$Ca(ClO)_2(s) \;+\; 4HCl(aq) \;\longrightarrow\; CaCl_2(aq) \;+\; 2H_2O(l) \;+\; 2Cl_2(g)$$

Calcium hypochlorite   Hydrochloric acid   Calcium chloride   Water   Chlorine

The percentage of chlorine produced from the given amount of calcium hypochlorite is then determined.

Some hypochlorite bleaching agents are solids at room temperature, but they are also available commercially as aqueous solutions. When they are heated, metallic hypochlorites decompose to produce oxygen. For instance, the thermal decomposition of calcium hypochlorite is represented as follows:

$$Ca(ClO)_2(s) \;\longrightarrow\; CaCl_2(s) \;+\; O_2(g)$$

Calcium hypochlorite   Calcium chloride   Oxygen

When this decomposition reaction occurs at a fire scene, the oxygen supports combustion reactions and contributes to the degree of fire hazard.

Sodium hypochlorite

11.6-A SODIUM HYPOCHLORITE

Solid sodium hypochlorite is an unstable compound, but its aqueous solutions are sufficiently stable when their contact with light is averted. Even the solutions of the metallic hypochlorites decompose when exposed to the ultraviolet radiation in sunlight.

We are most familiar with sodium hypochlorite solutions that are used as bleaches and as components of other household laundry products. A well-known example is the household bleaching agent known as *Clorox*, an aqueous solution of sodium hypochlorite having 5.7% available chlorine.

Solutions that contain from 10% to 12.5% sodium hypochlorite are also available commercially. The solution known commercially as *Multi-chlor* is a sodium hypochlorite

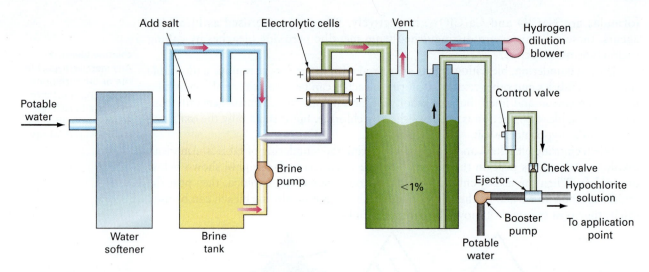

FIGURE 11.3 The process by which sodium hypochlorite may be electrolytically produced on-site from a brine (sodium chloride) solution. The oxidizer is used by hospital personnel to disinfect surfaces and equipment contaminated with microbial pathogens. It may also be used to control algae growth in large bodies of standing water.

Calcium hypochlorite

high-test hypochlorite (HTH) ▪ A commercial oxidizer composed of a metallic hypochlorite

solution having 12.5% available chlorine. It is used as a disinfectant and sanitizer for the treatment of water supplies, sewage effluents, and the water in swimming pools, spas, and hot tubs. It is also used to sanitize equipment and surfaces on dairy farms and poultry plants, food-processing plants, breweries, and beverage plants.

Sodium hypochlorite formerly was manufactured for commercial use by passing gaseous chlorine into an aqueous solution of sodium hydroxide. A mixture of sodium hypochlorite and sodium chloride was produced.

$$2NaOH(aq) \; + \; Cl_2(g) \; \longrightarrow \; NaClO(aq) \; + \; NaCl(aq) \; + \; H_2O(l)$$

Sodium hydroxide   Chlorine   Sodium hypochlorite   Sodium chloride   Water

Today, citing the desire to improve security, Clorox dilutes high-strength (15%) bleach with water to produce its household product. The new process eliminates the risks associated with storing, transporting, and using elemental chlorine (Section 7.3).

Sodium hypochlorite may also be generated for use on demand. In the system shown in Figure 11.3, for example, it is prepared by electrolyzing a portable brine solution.

$$NaCl(aq) \; + \; H_2O(l) \; \longrightarrow \; NaClO(aq) \; + \; H_2(g)$$

Sodium chloride   Water   Sodium hypochlorite   Hydrogen

11.6-B CALCIUM HYPOCHLORITE

Calcium hypochlorite is commonly used for large-scale bleaching operations, sanitizing municipal drinking water, disinfecting domestic and municipal swimming pools, and sewage treatment. The commercial grades of calcium hypochlorite are available as unique substances, solutions, and other mixtures with drying agents and other additives. It is frequently encountered in products known commercially by the trade name **high-test hypochlorite,** or **HTH.**

Calcium hypochlorite is manufactured by reacting chlorine with a lime slurry. A mixture of calcium hypochlorite and calcium chloride is produced.

$$2Ca(OH)_2(aq) \; + \; 2Cl_2(g) \; \longrightarrow \; Ca(ClO)_2(aq) \; + \; CaCl_2(aq) \; + \; 2H_2O(l)$$

Calcium hydroxide   Chlorine   Calcium hypochlorite   Calcium chloride   Water

Sodium chloride is then added to the solution. This causes the calcium hypochlorite to precipitate as the solid.

FIGURE 11.4 This test illustrates the enhanced rate at which cellulose powder burns when it is mixed with high-strength calcium hypochlorite (72.6% available chlorine and 5.6% water). (*By permission, Elizabeth Buc, Ph.D., P.E., Fire and Materials Research Laboratory, LLC, Final Report, Oxidizer Classification Research Project: Tests and Criteria, Fire Protection Research Foundation, Quincy, MA, November 2009.*)

Contact of calcium hypochlorite with flammable and combustible materials causes them to burn with an increased intensity and to generate a significant volume of gaseous products. This is illustrated by the test result shown in Figure 11.4.

11.6-C TRANSPORTING METALLIC HYPOCHLORITES

When shippers offer a metallic hypochlorite for transportation, DOT requires them to identify the appropriate substance as shown in Table 11.7 on an accompanying shipping paper. DOT also requires shippers and carriers to comply with all applicable labeling, marking, and placarding requirements.

TABLE 11.7	Shipping Descriptions of Some Representative Metallic Hypochlorites
METALLIC HYPOCHLORITE	**SHIPPING DESCRIPTION**
Barium hypochlorite (contains more than 22% available chlorine)	UN2741, Barium hypochlorite, 5.1, (6.1), PG II (Poison)
Calcium hypochlorite, dry, and calcium hypochlorite dry mixtures [contains more than 39% available chlorine (8.8% available oxygen)]	UN1748, Calcium hypochlorite, dry, 5.1, PG II *or* UN1748, Calcium hypochlorite, dry, 5.1, PG III
Calcium hypochlorite, hydrated and calcium hypochlorite hydrated mixtures (contains not less than 5.5% but not more than 16% water)	UN2880, Calcium hypochlorite, hydrated, 5.1, PG II *or* UN2880, Calcium hypochlorite, hydrated mixture, 5.1, PG II
Calcium hypochlorite, dry (contains more than 10% but not more than 39% available chlorine)	UN2208, Calcium hypochlorite, 5.1, PG III
Calcium hypochlorite mixtures, dry (contains more than 10% but not more than 39% available chlorine)	UN2208, Calcium hypochlorite mixtures, dry, 5.1, PG III
Lithium hypochlorite and lithium hypochlorite mixtures [contains more than 39% available chlorine (8.8% available oxygen)]	UN1471, Lithium hypochlorite, dry, 5.1, PG II *or* UN1471, Lithium hypochlorite mixture, 5.1, PG II

The shipping descriptions of all other metallic hypochlorites and their solutions are identified generically in the Hazardous Materials Table as "UN1791, Hypochlorites, inorganic, n.o.s., 8, PG II," "UN1791, Hypochlorite solutions (contains sodium hypochlorite), 8, PG II," *or* "UN1791, Hypochlorite solutions (contains sodium hypochlorite), 8, PG III." Shippers must include the name of the specific compounds parenthetically in the proper shipping name. For example, when shippers offer zinc hypochlorite for transportation, they enter the following shipping description on a shipping paper: "UN1791, Hypochlorites, inorganic, n.o.s. (contains zinc hypochlorite), 8, PG II."

Sodium trichloroisocyanurate

11.7 DI- AND TRICHLOROISOCYANURIC ACIDS AND THEIR SALTS

Although calcium hypochlorite continues to be a popular component of industrial sanitation products, it has been largely replaced in modern products with certain chlorinated derivatives of isocyanuric acid. These compounds include dichloroisocyanuric acid and trichloroisocyanuric acid. The molecular structures of these compounds are relatively complex.

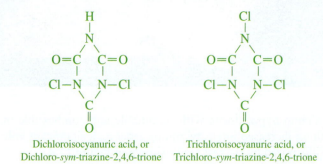

Dichloroisocyanuric acid, or
Dichloro-*sym*-triazine-2,4,6-trione

Trichloroisocyanuric acid, or
Trichloro-*sym*-triazine-2,4,6-trione

We study the nature of the bonding in these compounds more fully in Chapter 12.

Di- and trichloroisocyanuric acids are white tablets or powders having an available chlorine value that ranges from 90% to 99%. They are also available commercially as their sodium and potassium salts. These compounds are typically encountered in 50-pound (22.7-kg) pails or buckets.

11.7-A COMMERCIAL USES OF DI- AND TRICHLOROISOCYANURIC ACIDS AND THEIR SALTS

The major use of the di- and trichloroisocyanuric acid products is to chlorinate the water in swimming pools. For this purpose, these compounds are usually referred to as *dichlor* and *trichlor*, respectively. Di- and trichloroisocyanuric acids and their salts release their chlorine slowly into swimming pool water. They have much longer lifespans when dissolved in water than either free chlorine or the metallic hypochlorites, all of which rapidly decompose when exposed to the ultraviolet radiation in sunlight. Thus, di- and trichloroisocyanuric acids and their salts are considered more desirable for the purpose of chlorination, especially in areas where swimming pools are exposed to intense sunlight for extended periods.

Di- and trichloroisocyanuric acids are also components of dry laundry bleaches, detergents and other dishwashing compounds, scouring powders, and bactericides. They have also been used for the purification of drinking water and the nonshrinking treatment of wool.

11.7-B TRANSPORTING DI- AND TRICHLOROISOCYANURIC ACIDS AND THEIR SALTS

When shippers offer dichloroisocyanuric acid, trichloroisocyanuric acid, or the salts of these compounds for transportation, DOT requires them to provide the shipping description shown in Table 11.8 on an accompanying shipping paper. DOT also requires

TABLE 11.8	Shipping Descriptions of the Chlorinated Isocyanuric Acids
CHLORINATED ISOCYANURIC ACID	**SHIPPING DESCRIPTION**
Dichloroisocyanuric acid	UN2465, Dichloroisocyanuric acid, dry, 5.1, PG II
	or
	UN2465, Dichloroisocyanuric acid salts, dry, 5.1, PG II

shippers and carriers to comply with all applicable labeling, marking, and placarding requirements.

11.8 CHLORINE DIOXIDE

Although chemists recognize three oxides of chlorine, only one is commercially important: chlorine dioxide. At room temperature, this substance is a red-yellow gas having the chemical formula ClO_2. It is a highly unstable substance, decomposing into its elements at an explosive rate.

$$2ClO_2(g) \longrightarrow Cl_2(g) + 2O_2(g)$$

Chlorine dioxide   Chlorine Oxygen

Chlorine dioxide

Chlorine dioxide is routinely produced near the point of its intended use. In the simplified generator in Figure 11.5, it is produced in a two-step process, each of which is conducted under vacuum conditions. In the first step, sodium hypochlorite is reacted with hydrochloric acid to generate chlorine.

$$NaClO(aq) + 2HCl(aq) \longrightarrow NaCl(aq) + H_2O(l) + Cl_2(g)$$

Sodium hypochlorite Hydrochloric acid Sodium chloride Water Chlorine

In the second step, the chlorine is reacted with a sodium chlorite solution, which results in the generation of the chlorine dioxide.

$$2NaClO_2(aq) + Cl_2(g) \longrightarrow 2ClO_2(g) + 2NaCl(aq)$$

Sodium chlorite Chlorine Chlorine dioxide Sodium chloride

When chlorine dioxide is dissolved in water, an aqueous solution is produced known as *chlorine dioxide hydrate*. In commerce, this solution is frozen before it is shipped to a

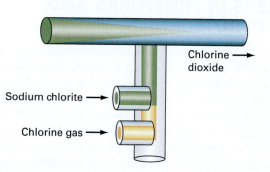

FIGURE 11.5 Chlorine dioxide usually is generated at the location where it is to be used. In this simplified schematic, two independent chemical reactions work in unison to generate the gas. Although its primary use is to bleach paper pulp, in modern times, chlorine dioxide frequently has been selected over other microbicides to disinfect areas contaminated with mold, bacteria, and viruses.

point of intended use. When frozen, the aqueous solution survives for a long time, but gentle heating causes it to decompose.

As with chlorine itself, the principal risk associated with exposure to chlorine dioxide is inhalation toxicity. Exposure to chlorine dioxide also causes severe respiratory and eye damage.

11.8-A COMMERCIAL USES OF CHLORINE DIOXIDE

Chlorine dioxide and elemental chlorine generally are used for the same purposes. By comparison, however, chlorine dioxide is approximately 2½ times more powerful than chlorine as an oxidizing agent. For this reason, chlorine dioxide is sometimes selected over chlorine for bleaching paper pulp and producing white wheat flour, even though chlorine dioxide is more costly.

Chlorine dioxide is sometimes also used as a microbicide. For example, it is used to disinfect water in the dairy, beverage, and other food industries. At water-treatment plants, its use is generally supplementary to the use of chlorine for killing bacteria. Because it is capable of killing *Cryptosporidium parvum*, the use of chlorine dioxide to disinfect drinking water compares favorably with ozone.

In 2002, when anthrax-laden mail was discovered in postal areas and in the Hart Senate Office Building, EPA contractors chose chlorine dioxide to destroy the anthrax. The mail packages and Senate offices were decontaminated by discharging the gas into the atmosphere of sealed-off areas, where it remained for several hours before neutralization. The hard surfaces within the Senate building were disinfected using liquid chlorine dioxide.

In 2005, following the Hurricane Katrina disaster along the Gulf coast, chlorine dioxide was also used to eradicate the mold that grew in homes that had been inundated with water. It has also been used to control the infestation of the pipe-clogging, thumbnail-sized quagga mussel in waterways.

Consumer Fireworks and Special Fireworks

11.8-B TRANSPORTING CHLORINE DIOXIDE

DOT allows chlorine dioxide to be transported only when it is frozen as its hydrate. When shippers offer the frozen hydrate for transportation, DOT requires them to maintain the gas in the solid state using dry ice or other means and to identify the gas on an accompanying shipping paper as follows:

NA9191, Chlorine dioxide, hydrate, frozen, 5.1, (6.1), PG II (Poison)

DOT also requires shippers and carriers to comply with all applicable labeling, marking, and placarding requirements.

11.9 OXIDIZERS IN FIREWORKS AND OTHER PYROTECHNICS

fireworks ■ Pyrotechnic devices intentionally designed for the purpose of producing a visible or audible effect by means of combustion, explosion, deflagration, or detonation reactions

Fireworks are items designed primarily for entertainment purposes with the intent of producing audible and visible pyrotechnic effects by combustion, explosion, deflagration, or detonation reactions. Some familiar examples include firecrackers, Roman candles, pinwheels, flares, serpents, sparklers, skyrockets, and toy torpedoes. Special-effects fireworks include comets, fountains, strobes, and kaleidoscope shells. To choreograph an aerial fireworks program, modern professionals use computers, flame generators, and firing systems that actuate a set of redox reactions, each of which is precisely timed to provide lavish aerial scenes such as chrysanthemum shells. In a modern program, lighting consoles synchronized with the activation of these reactions are integrated into the performance.

In the United States, the GHS system (Section 1.9) is used for the classification of fireworks. Two classes are especially important: Fireworks 1.3G and Fireworks 1.4G (formerly called special fireworks and common fireworks, respectively). "1.3G" and "1.4G" are the DOT division numbers of the relevant hazard class and its compatibility group (Section 15.4-B). **Fireworks 1.3G**, or **special fireworks**, are relatively large firework devices designed for use by professionally trained experts. **Fireworks 1.4G**, or **consumer fireworks**, are relatively small firework devices designed primarily as consumer products for use by the general public. The GHS symbols pertaining to fireworks in both hazard classes are provided in the marginal art, but the hazard diamond information is not noted since they vary according to their chemical composition.

At 16 C.F.R. §1500.17, CPSC has banned the use of certain fireworks devices intended to produce audible effects. In addition, child-protection laws have banned the possession, manufacture, and sale of cherry bombs, a type of fireworks physically resembling red cherries with green stems. Cherry bombs were activated by igniting fuses (the green stems) that contacted a pyrotechnic mixture of potassium perchlorate and metallic aluminum. The activation of this mixture caused the loss of hundreds of eyes and fingers each year; hence, CPSC instituted an outright ban against the sale and use of cherry bombs in the United States.

CPSC also compels manufacturers to provide special labeling information on the containers of certain consumer fireworks. Although there are far too many types and statements to list here, they may be accessed at 16 C.F.R. §1500.14(b)(7).

Although state and local laws attempt to prevent the widespread sale of fireworks to unauthorized individuals, incidents involving the misuse of fireworks annually cause numerous secondary fires, blindness, deafness, the mutilation of fingers, and death. In most states, to reduce or eliminate the likelihood of these incidents, only a licensed retail outlet that posts a permit is allowed to sell fireworks to the public. To further reduce or eliminate the risk of incidents caused by their misuse, state and local laws rigidly require that common fireworks be actuated only in accordance with the manufacturer's warnings and instructions. Display fireworks must be actuated by trained experts only. Even under the best of conditions, the actuation of all fireworks often constitutes a risk of fire and explosion.

One component of all pyrotechnics is a pyrotechnic mixture of an oxidizing agent and a reducing agent. The most commonly encountered oxidizers are sodium chlorite, sodium chlorate, and sodium perchlorate, all of which are also components of vehicular air bags, solid rocket fuels, and certain fertilizers. Sodium perchlorate is the oxidizing agent of choice for most pyrotechnic displays, and charcoal, sulfur, pulverized magnesium, and aluminum flakes are the reducing agents. To provide color when fireworks are actuated, one or more metallic compounds are added to fireworks formulations.

Sodium chlorite, sodium chlorate, and sodium perchlorate decompose to produce oxygen when they are heated.

$$NaClO_2(s) \longrightarrow NaCl(s) + O_2(g)$$

Sodium chlorite   Sodium chloride   Oxygen

$$2NaClO_3(s) \longrightarrow 2NaCl(s) + 3O_2(g)$$

Sodium chlorate   Sodium chloride   Oxygen

$$NaClO_4(s) \longrightarrow NaCl(s) + 2O_2(g)$$

Sodium perchlorate   Sodium chloride   Oxygen

Although the oxygen produced by these reactions aids in actuating the display, more complex redox reactions are responsible for the brilliant lighting and sound effects associated with fireworks displays. When the mixture of an oxidizer, charcoal, sulfur, and finely divided magnesium or aluminum is activated, the resulting redox reactions occur at explosive rates. The presence of magnesium powder enhances the brilliance of a fireworks display, whereas the addition of coarse aluminum flakes produces luminous tails. The sets of equations in Table 11.9 illustrate some of the redox reactions associated with the actuation of the reactive mixture.

The unique colors observed during the display of fireworks result as certain compounds vaporize or decompose. These compounds are almost exclusively metallic

Fireworks 1.3G (special fireworks, display fireworks) ■ Relatively large firework devices designed primarily for display by professional experts

Fireworks 1.4G (common fireworks, consumer fireworks) ■ Relatively small firework devices designed primarily for use as consumer products by the general public

TABLE 11.9		Examples of the Chemical Phenomena Associated with the Actuation of Fireworks

OXIDIZING AGENT	REDUCING AGENT	EQUATION REPRESENTING THE REDOX REACTION
Sodium chlorite	Charcoal	$C(s) + NaClO_2(s) \longrightarrow NaCl(s) + CO_2(g)$ Carbon Sodium chlorite Sodium chloride Carbon dioxide
	Sulfur	$S_8(s) + 8NaClO_2(s) \longrightarrow 8NaCl(s) + 8SO_2(g)$ Sulfur Sodium chlorite Sodium chloride Sulfur dioxide
	Magnesium	$2Mg(s) + NaClO_2(s) \longrightarrow 2MgO(s) + NaCl(s)$ Magnesium Sodium chlorite Magnesium oxide Sodium chloride
	Aluminum	$4Al(s) + 3NaClO_2(s) \longrightarrow 2Al_2O_3(s) + 3NaCl(s)$ Aluminum Sodium chlorite Aluminum oxide Sodium chloride
Sodium chlorate	Charcoal	$3C(s) + 2NaClO_3(s) \longrightarrow 2NaCl(s) + 3CO_2(g)$ Carbon Sodium chlorate Sodium chloride Carbon dioxide
	Sulfur	$3S_8(s) + 16NaClO_3(s) \longrightarrow 16NaCl(s) + 24SO_2(g)$ Sulfur Sodium chlorate Sodium chloride Sulfur dioxide
	Magnesium	$3Mg(s) + NaClO_3(s) \longrightarrow 3MgO(s) + NaCl(s)$ Magnesium Sodium chlorate Magnesium oxide Sodium chloride
	Aluminum	$2Al(s) + NaClO_3(s) \longrightarrow Al_2O_3(s) + NaCl(s)$ Aluminum Sodium chlorate Aluminum oxide Sodium chloride
Sodium perchlorate	Charcoal	$2C(s) + NaClO_4(s) \longrightarrow NaCl(s) + 2CO_2(g)$ Carbon Sodium perchlorate Sodium chloride Carbon dioxide
	Sulfur	$S_8(s) + 4NaClO_4(s) \longrightarrow 4NaCl(s) + 8SO_2(g)$ Sulfur Sodium perchlorate Sodium chloride Sulfur dioxide
	Magnesium	$4Mg(s) + NaClO_4(s) \longrightarrow 4MgO(s) + NaCl(s)$ Magnesium Sodium perchlorate Magnesium oxide Sodium chloride
	Aluminum	$8Al(s) + 3NaClO_4(s) \longrightarrow 4Al_2O_3(s) + 3NaCl(s)$ Aluminum Sodium perchlorate Aluminum oxide Sodium chloride

chlorides, which fluoresce strongly in the visible wavelengths. Some examples and the colors they exhibit on vaporization are noted here:

Barium chloride	yellow-green	Cupric chloride	blue-green
Lithium chloride	crimson red	Calcium chloride	orange red
Potassium chloride	lavender	Sodium chloride	golden yellow
Strontium chloride	carmine red		

Fireworks have also been used as weapons of mass destruction. The mixture of reactive substances in fireworks was the active explosive in the pressure-cooker bombs used during the 2013 Boston Marathon terrorist attack.

11.9-A TRANSPORTING FIREWORKS

When shippers offer fireworks for transportation, DOT requires them to describe the fireworks generically as shown in Table 11.10 on a shipping paper. DOT requires shippers to include the applicable EX-number (designated here as EX-xxxxx), Department of Defense

TABLE 11.10	Shipping Descriptions of Fireworks[a]
FIREWORKS	**SHIPPING DESCRIPTION**
Fireworks, division 1.1	UN0333, Fireworks, 1.1G, PG II (EX-xxxxx)
Fireworks, division 1.2	UN0334, Fireworks, 1.2G, PG II (EX-xxxxx)
Fireworks, division 1.3	UN0335, Fireworks, 1.3G, PG II (EX-xxxxx)
Fireworks, division 1.4	UN0336, Fireworks, 1.4G, PG II (EX-xxxxx) *or* UN0337, Fireworks, 1.4S, PG II (EX-xxxxx)

[a]Regulatory Guidelines for Shipping and Transporting Fireworks (Washington, DC: U.S. Department of Transportation, 2012).

(DOD) national stock number, product code, or other identifying information in the shipping description. The nature of these latter requirements is discussed in Section 15.4-C.

DOT requires shippers and carriers to comply with all applicable labeling, marking, and placarding requirements. It also requires carriers to block and brace packaging containing fireworks to restrict movement in a transport vehicle and protect them against exposure to ignition sources.

When emergency responders encounter fireworks during a transportation mishap, their packaging must identify the appropriate EX-number, EXPLOSIVE label, proper shipping name, UN identification number, and UN packaging specification marking. Each of these DOT requirements is illustrated in the example shown in Figure 11.6.

11.9-B DISPLAYING FIREWORKS

Local ordinances regulate how fireworks are displayed. Regardless of their nature, fireworks should never be actuated when climatic conditions are dangerously dry or when the wind speed exceeds 30 mi/hr (48 km/hr).

NFPA has also been concerned with the training of individuals who display fireworks. It requires them to comply with at least the following:[2]

- Professionals who publicly display fireworks must be experienced, responsible, and bonded.
- For both land and water displays, professionals may actuate aerial shell displays outdoors only by adhering to specified minimum separation distances between the actuation point and the spectators.

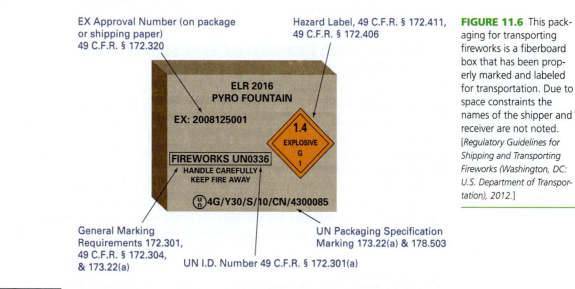

EX Approval Number (on package or shipping paper) 49 C.F.R. § 172.320

Hazard Label, 49 C.F.R. § 172.411, 49 C.F.R. § 172.406

General Marking Requirements 172.301, 49 C.F.R. § 172.304, & 173.22(a)

UN I.D. Number 49 C.F.R. § 172.301(a)

UN Packaging Specification Marking 173.22(a) & 178.503

FIGURE 11.6 This packaging for transporting fireworks is a fiberboard box that has been properly marked and labeled for transportation. Due to space constraints the names of the shipper and receiver are not noted. [*Regulatory Guidelines for Shipping and Transporting Fireworks (Washington, DC: U.S. Department of Transportation), 2012.*]

---

[2]NFPA Standard 1123, *Code for Fireworks Display* (Quincy, Massachusetts: National Fire Protection Agency), 2010.

METALLIC CHLORITE, CHLORATE, OR PERCHLORATE	SHIPPING DESCRIPTION
Barium chlorate	UN1445, Barium chlorate, 5.1, (6.1), PG II (Poison)
Barium perchlorate	UN1447, Barium perchlorate, 5.1, (6.1), PG II (Poison)
Calcium chlorate	UN1452, Calcium chlorate, 5.1, PG II
Calcium chlorite	UN1453, Calcium chlorite, 5.1, PG II
Calcium perchlorate	UN1455, Calcium perchlorate, 5.1, PG II
Magnesium chlorate	UN2723, Magnesium chlorate, 5.1, PG II
Magnesium perchlorate	UN1475, Magnesium perchlorate, 5.1, PG II
Potassium chlorate	UN1485, Potassium chlorate, 5.1, PG II
Potassium perchlorate	UN1489, Potassium perchlorate, solid, 5.1, PG II
Sodium chlorate	UN1495, Sodium chlorate, 5.1, PG II
Sodium chlorite	UN1496, Sodium chlorite, 5.1, PG II
Sodium perchlorate	UN1502, Sodium perchlorate, 5.1, PG II
Zinc chlorate	UN1513, Zinc chlorate, 5.1, PG II

TABLE 11.11 Shipping Descriptions of Some Representative Metallic Chlorites, Chlorates, and Perchlorates

11.9-C TRANSPORTING METALLIC CHLORITES, CHLORATES, AND PERCHLORATES

The oxidizers used in fireworks are available as commercial chemical products. When shippers offer these individual compounds for transportation, DOT requires them to provide the relevant shipping description on an accompanying shipping paper. Some examples for several representative metallic chlorites, chlorates, and perchlorates are listed in Table 11.11. When metallic chlorites, chlorates, and perchlorates are not specifically listed at 49 C.F.R. §172.101, DOT requires their shippers to provide a generic shipping description on shipping papers and name the specific compound parenthetically. It also requires shippers and carriers to comply with all applicable labeling, marking, and placarding requirements.

11.9-D ILL EFFECTS CAUSED BY EXPOSURE TO METALLIC PERCHLORATES

Considerable concern about the health effects likely to result from exposure to perchlorate ions has been publicly voiced, particularly because perchlorate has been identified in the water supplies of 35 states. This widespread groundwater contamination is associated with the former military use of metallic perchlorates as rocket fuels. Today, these compounds have been replaced in fireworks and roadside flares with compounds that are friendly to public health and the environment.

Does human exposure to perchlorates pose a health risk? Studies reveal a link between the consumption of perchlorate-contaminated water and damage to the thyroid gland, neurological problems, and the inability of humans to properly produce growth- and fetal-developmental hormones.[3] Perchlorate pollution potentially affects the health of

---

[3]National Research Council of the National Academy of Sciences, *Health Implications of Perchlorate Ingestion* (Washington, DC: National Academies Press, 2005).

tens of millions of American women and the children they conceive, especially because perchlorate prevents the proper production of the hormones needed for developing fetuses. In 2005, EPA set a daily dose of perchlorate that people can safely ingest at 0.7 µg/kg of body weight.

Perchlorate contamination is a hotly contested political issue in California and Nevada, two states in which perchlorate pollution is especially pronounced. Owing to public pressure, additional information on the health effects relating to perchlorate exposure is likely to be forthcoming.

11.10 OXIDIZERS IN FLARES, SIGNALING SMOKES, AND SMOKE BOMBS

Flares, **signaling smokes**, and **smoke bombs** are devices that contain a pyrotechnic mixture of an oxidizing agent and reducing agent. They are used by civilian police and the military to identify and obscure scenes of interest, cordon off accident sites, and coordinate activities during local assault operations. When they are used by the military, these devices are helpful in locating friendly units as well as enemy targets and controlling the laying and lifting of artillery.

In the civilian world, roadside flares are used to communicate distress, identify a transportation accident or mechanical failure, and control traffic during nighttime hours. For these purposes, the flares are composed of a mixture of strontium nitrate, sulfur, potassium perchlorate, and sawdust, the combination of which is inserted into cardboard tubes. When they are ignited, the sawdust burns at the approximate rate of 1 in./min. The redox reaction produces an intensely carmine-red glow. The chemical reaction is represented as follows:

$$2Sr(NO_3)_2(s) \quad + \quad S_8(s) \quad + \quad 4KClO_4(s) \longrightarrow$$
Strontium nitrate Sulfur Potassium perchlorate

$$2SrO(s) \quad + \quad 8SO_2(g) \quad + \quad 4KCl(s) \quad + \quad 4NO_2(g) \quad + \quad O_2(g)$$
Strontium oxide Sulfur dioxide Potassium chloride Nitrogen dioxide Oxygen

Military and civilian aviators and sailors use flares or flare/signaling smoke combinations to communicate distress when a crisis arises. The U.S. Coast Guard regulates the features of several types, including the handheld, red flare; floating orange-smoke flare; pistol-projected parachute red flare; self-contained, rocket-propelled parachute, red flare; handheld orange signaling smoke; floating orange signaling smoke, and red aerial pyrotechnic flare. A description of each device is provided at 46 C.F.R. Part 160. Each is composed of a unique mixture of substances.

11.10-A CHEMICAL ACTUATION OF FLARES, SIGNALING SMOKES, AND SMOKE BOMBS

In addition to an oxidizing agent and reducing agent, flares, signaling smokes, and smoke bombs often contain sodium bicarbonate and an organic dye. Most commonly, the oxidizing and reducing agents are potassium chlorate and elemental sulfur, respectively. The mixture of components is compressed into cartridges, hand grenades, and canisters.

When the chemical mixture in flares, signaling smokes, and smoke bombs is activated by ignition, at least two chemical reactions occur:

■ The oxidation–reduction reaction between potassium chlorate and sulfur

$$16KClO_3(s) \quad + \quad 3S_8(s) \longrightarrow 16KCl(s) \quad + \quad 24SO_2(g)$$
Potassium chlorate Sulfur Potassium chloride Sulfur dioxide

flare ■ Any article containing a pyrotechnic substance designed to illuminate, identify, signal, or warn

signaling smoke ■ Any type of smoke bomb whose actuation communicates prearranged information to troops or others who subsequently observe the signal from a distance

smoke bomb ■ Any device containing substances that produce smoke when actuated, generally for use in concealing military operations

- The thermal decomposition of sodium bicarbonate

$$2NaHCO_3(s) \longrightarrow Na_2CO_3(s) + CO_2(g) + H_2O(g)$$

Sodium bicarbonate   Sodium carbonate   Carbon dioxide   Water

As carbon dioxide and sulfur dioxide evolve into the air, they disperse the dye in the immediate area.

The use of smoke bombs and other pyrotechnics is especially vital to the success of military operations. Aviators routinely carry flares in their flight suits or on life rafts to use as distress signals if their aircraft fails. Flares are also used by military aviators as decoys to thwart heat-seeking missiles. When they are actuated, these flares emit an infrared signal that mimics the heat discharged by an aircraft engine. When a flare is ejected from the aircraft, the missile travels towards it rather than the aircraft.

Smoke bombs are used by the military to visually mask the movement of troops and vehicles during combat activities. A popular smoke bomb used by the U.S. Army during the Korean conflict was a pyrotechnic mixture containing 6.7% granulated aluminum, 46.7% zinc oxide, and 46.7% (C_2H_6, or CCl_3-CCl_3) hexachloroethane by mass. When the components of this mixture were actuated with a fuse, the products that appeared in the smoke were aluminum oxide, zinc chloride, and carbon (soot).

$$2Al(s) + CCl_3-CCl_3(s) + 3ZnO(s) \longrightarrow 3ZnCl_2(s) + Al_2O_3(s) + 2C(s)$$

Aluminum   Hexachloroethane   Zinc oxide   Zinc chloride   Aluminum oxide   Carbon

The zinc chloride attracted atmospheric moisture to form a fog, the aluminum oxide reflected light, and the carbon colored the smoke cloud gray.

11.10-B TRANSPORTING FLARES, SIGNALING SMOKES, AND SMOKE BOMBS

When shippers offer flares, signaling smokes, and smoke bombs for transportation, DOT requires them to identify the appropriate item generically as shown in Table 11.12. DOT also requires shippers and carriers to comply with all applicable labeling, marking, and placarding requirements.

11.11 THE THERMAL STABILITY OF AMMONIUM COMPOUNDS

Ammonium compounds are substances whose units contain an ammonium ion (NH_4^+) and a negative ion. Most are thermally unstable. When they are heated, ammonium compounds decompose in either of the following ways:

- Ammonium compounds may decompose into the compounds from which they were initially made. For instance, when ammonium chloride is heated, it thermally decomposes to form ammonia and hydrogen chloride.

$$NH_4Cl(s) \longrightarrow NH_3(g) + HCl(g)$$

Ammonium chloride   Ammonia   Hydrogen chloride

The ammonium compounds that decompose in this simple manner generally are not regarded as hazardous materials.
- Ammonium compounds may also decompose by oxidation–reduction. Examples of ammonium compounds that decompose in this manner are illustrated by the equations in Table 11.13. The majority of them decompose at explosive rates when heated. For this reason, DOT prohibits their transportation by any mode. Examples of these

FLARES, SMOKE BOMBS, OR SIGNALING SMOKES	SHIPPING DESCRIPTION
Aerial flares	UN0093, Flares, aerial, 1.3G, PG II (EX-xxxxx)
	or
	UN0403, Flares, aerial, 1.4G, PG II (EX-xxxxx)
	or
	UN0404, Flares, aerial, 1.4S, PG II (EX-xxxxx)
	or
	UN0420, Flares, aerial, 1.1G, PG II
	or
	UN0421, Flares, aerial, 1.2G, PG II (EX-xxxxx)
Smoke bombs (nonexplosive), with corrosive liquid and without initiating device	UN2028, Bombs, smoke, nonexplosive, 8, PG II
Smoke signals	UN0196, Signals, smoke, 1.1G, PG II (EX-xxxxx)
	or
	UN0197, Signals, smoke, 1.4G, PG II (EX-xxxxx)
	or
	UN0313, Signals, smoke, 1.2G, PG II (EX-xxxxx)
	or
	UN0487, Signals, smoke, 1.3G, PG II (EX-xxxxx)
Surface flares	UN0092, Flares, surface, 1.3G, PG II (EX-xxxxx)
	or
	UN0418, Flares, surface, 1.1G, PG II (EX-xxxxx)
	or
	UN0419, Flares, surface, 1.2G, PG II (EX-xxxxx)

TABLE 11.12 Shipping Descriptions of Flares, Signaling Smokes, and Smoke Bombs

ammonium compounds are ammonium azide, ammonium bromate, ammonium chlorate, ammonium fulminate, ammonium nitrite, and ammonium permanganate.

11.11-A COMMERCIAL USES OF AMMONIUM COMPOUNDS

Because few compounds containing the ammonium ion are thermally stable, the industrial uses of these compounds are fairly limited. Ammonium nitrate is by far the most important compound containing the ammonium ion. In Section 11.12, we observe that it is an important commercial fertilizer and explosive component.

Ammonium perchlorate is the only other ammonium compound used in significant amounts in the United States. It accounts for 70% of the solid propellants used by the aerospace industry to propel space shuttles. The ammonium perchlorate is mixed with a fuel like powdered aluminum and a binder. When ignited, an oxidation–reduction reaction occurs that generates sufficient gas to propel the shuttle into outer space.

$$6NH_4ClO_4(s) + 10Al(s) \longrightarrow 4Al_2O_3(s) + 2AlCl_3(s) + 12H_2O(g) + 3N_2(g)$$

Ammonium perchlorate Aluminum   Aluminum oxide Aluminum chloride   Water   Nitrogen

11.11-B TRANSPORTING AMMONIUM COMPOUNDS

When ammonium compounds are transported, DOT requires shippers to provide the relevant shipping description shown in Table 11.14 on an accompanying shipping paper.

TABLE 11.13   Examples of Ammonium Compounds That Are Hazardous Materials

AMMONIUM COMPOUND	EQUATION ILLUSTRATING THERMAL DECOMPOSITION
Ammonium azide[a]	$NH_4N_3(s) \longrightarrow 2N_2(g) + 2H_2(g)$ Ammonium azide Nitrogen Hydrogen
Ammonium bromate[a]	$2NH_4BrO_3(s) \longrightarrow 2NH_4Br(s) + 3O_2(g)$ Ammonium bromate Ammonium bromide Oxygen
Ammonium chlorate[a]	$2NH_4ClO_3(s) \longrightarrow 2NH_4Cl + 3O_2(g)$ Ammonium chlorate Ammonium chloride Oxygen
Ammonium dichromate	$(NH_4)_2Cr_2O_7(s) \longrightarrow Cr_2O_3(s) + N_2(g) + 4H_2O(g)$ Ammonium dichromate Chromium(III) oxide Nitrogen Water
Ammonium fulminate[a]	$2NH_4NCO(s) \longrightarrow 2N_2(g) + 4H_2(g) + 2CO(g)$ Ammonium fulminate Nitrogen Hydrogen Carbon monoxide
Ammonium nitrite[a]	$NH_4NO_2(s) \longrightarrow N_2(g) + 2H_2O(g)$ Ammonium nitrite Nitrogen Water
Ammonium perchlorate	$2NH_4ClO_4(s) \longrightarrow N_2(g) + Cl_2(g) + 2O_2(g) + 4H_2O(g)$ Ammonium perchlorate Nitrogen Chlorine Oxygen Water
Ammonium permanganate[a]	$2NH_4MnO_4(s) \longrightarrow 2MnO(s) + N_2(g) + O_2(g) + 4H_2O(g)$ Ammonium permanganate Manganese(II) oxide Nitrogen Oxygen Water
Ammonium peroxydisulfate[b]	$3(NH_4)_2S_2O_8(s) \longrightarrow 4NH_3(s) + N_2(g) + 3O_2(g) + 6SO_2(g) + 6H_2O(g)$ Ammonium peroxydisulfate Ammonia Nitrogen Oxygen Sulfur dioxide Water

[a]DOT prohibits the transportation of these materials.
[b]Also named ammonium persulfate.

TABLE 11.14   Shipping Descriptions of Some Representative Ammonium Compounds

AMMONIUM COMPOUND	SHIPPING DESCRIPTION
Ammonium dichromate	UN1439, Ammonium dichromate, 5.1, PG II
Ammonium nitrate fertilizer	UN2067, Ammonium nitrate based fertilizer, 5.1, PG III or UN2071, Ammonium nitrate based fertilizer, 9, PG III
Ammonium nitrate emulsion	UN3375, Ammonium nitrate emulsion, 5.1, PG II
Ammonium nitrate suspension	UN3375, Ammonium nitrate suspension, 5.1, PG II
Ammonium nitrate gel (intermediate for blasting explosives)	UN3375, Ammonium nitrate gel, 5.1, PG II
Ammonium nitrate/fuel oil mixtures (contains prilled ammonium nitrate and fuel oil)	NA0331, Ammonium nitrate—fuel oil mixtures, 1.5 D (EX-xxxxx)
Ammonium nitrate, liquid (hot concentrated solution)	UN2426, Ammonium nitrate, liquid, 5.1
Ammonium nitrate (contains more than 0.2% combustible substances)	UN0222, Ammonium nitrate, 1.1D, PG II (EX-xxxxx)
Ammonium perchlorate	UN0402, Ammonium perchlorate, 1.1D, PG II (EX-xxxxx) or UN1442, Ammonium perchlorate, 5.1, PG II
Ammonium persulfate	UN1444, Ammonium persulfate, 5.1, PG III

DOT also requires shippers and carriers to comply with all applicable labeling, marking, and placarding requirements.

11.12 AMMONIUM NITRATE

By far, ammonium nitrate is the most commercially important chemical product containing the ammonium ion. Its chemical formula is NH_4NO_3. Several grades are available including a fertilizer, dynamite, nitrous oxide, and technical grade, all of which are available commercially as crystals, flakes, grains, and prills (small, bead-like pellets). The characteristics of these grades are briefly noted next.

Ammonium nitrate

Dinitrogen oxide

- Fertilizer-grade ammonium nitrate is commonly encountered at farms and farm distribution centers. Fertilizer-grade ammonium nitrate is a formulation of ammonium nitrate with ammonium sulfate or calcium carbonate, each of which reduces the potential risk of its self-decomposition. As its name implies, it is intended for use as an agricultural fertilizer. It is commonly sold in 50-pound (23-kg) bags.
- Dynamite-grade ammonium nitrate is widely used during road construction and blasting operations in mines and quarries, where it is the active component of blasting agents. It is also a component of unique explosives like ammonal, a mixture of ammonium nitrate and powdered aluminum. Ammonium nitrate has also been combined with TNT in a bursting charge used in demolition bombs. It is also available commercially mixed with diesel fuel, called ammonium nitrate/fuel oil mixture, or ANFO. The presence of the diesel oil makes ANFO easier to detonate than ammonium nitrate alone. Although it can be safely handled and transported, ANFO detonates with a powerful force when it is ignited. ANFO has been misused in acts of terrorism to cause catastrophic damage to human health, safety, and national security.
- Nitrous oxide–grade ammonium nitrate is used to produce dinitrogen oxide, or nitrous oxide, by gentle heating. It is known commonly as *laughing gas*, an anesthetic used mainly by dentists to lessen the anxiety experienced by some patients.

$$NH_4NO_3(s) \longrightarrow N_2O(g) + 2H_2O(g)$$
Ammonium nitrate   Dinitrogen oxide   Water

- Technical-grade ammonium nitrate is produced for a variety of uses. For example, it is a component of cold packs used for treating minor athletic injuries. Because the dissolving of ammonium nitrate in water is an endothermic process, the solution is cold when first produced and provides relief from pain when applied to bruises.

11.12-A AMMONIUM NITRATE AT FIRE SCENES

Ammonium nitrate melts at 337°F (170°C) and decomposes between 350 and 410°F (177 and 210°C). When a bulk quantity of ammonium nitrate is present at a fire scene, it may undergo thermal decomposition at an explosive rate. The decomposition produces an array of products including those illustrated by the following two equations:

$$4NH_4NO_3(s) \longrightarrow 3N_2(g) + 2NO_2(g) + 8H_2O(g)$$
Ammonium nitrate   Nitrogen   Nitrogen dioxide   Water

$$2NH_4NO_3(s) \longrightarrow 2N_2(g) + O_2(g) + 4H_2O(g)$$
Ammonium nitrate   Nitrogen   Oxygen   Water

These decomposition products are gaseous.

When the gases produced by the decomposition of ammonium nitrate are vented from the fire scene, the threat of explosive decomposition is reduced or eliminated. However, when the decomposition products are confined or cannot be vented from the storage

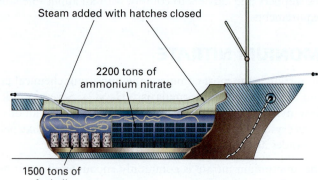

Steam added with hatches closed

2200 tons of
ammonium nitrate

1500 tons of
fuel oil

FIGURE 11.7 On April 15, 1947, fire was detected in two holds containing 2200 tons (1998 t) of ammonium nitrate onboard the SS *Grandcamp,* a French cargo ship docked in Texas City, Texas. An order was given to close the hatches to the holds and apply steam. As the fire continued to burn, the ammonium nitrate decomposed into a mixture of nitrogen, nitrogen dioxide, oxygen, and water vapor. This production of gases caused the buildup of internal pressure within the holds that was relieved only when the cargo ship subsequently exploded. Six hundred people were killed and another 3500 were injured. The property damage was comparable to that experienced during a major wartime bombing incident. The total property loss was estimated as $33 million based on 1947 costs.

area, a major explosion is likely to occur. This was the fate of the French cargo ship SS *Grandcamp,* a portion of whose features are illustrated in Figure 11.7.

In 1947, the *Grandcamp* was docked at the harbor in Texas City, Texas, when it caught fire with nearly 2280 tons (2070 t) of fertilizer-grade ammonium nitrate onboard stored within two holds. The heat generated by the fire caused the ammonium nitrate to melt and decompose.

Although fires involving oxidizing agents can effectively be extinguished through application of water, this attempt was never made onboard the *Grandcamp*. Instead, orders were issued to seal the hatches to the holds to eliminate the air supply to the fire and prevent damage to the cargo by water. Supported by oxidizing agents, however, the fire continued to burn in the absence of air. Internal pressure within the holds increased until a major explosion released the confined gases.

The explosion of the *Grandcamp* is now regarded as the worst maritime incident involving a chemical substance. It caused the deaths of nearly 600 people, including 27 of Galveston's 28 firefighters. It serves to illustrate dramatically the potentially hazardous nature of oxidizers in general and ammonium nitrate in particular.

In contemporary times, the worst industrial accident for emergency responders involved the explosion of ammonium nitrate at a fire scene. In 2013, 28 to 34 tons (25 to 31 t) of ammonium nitrate exploded at a fertilizer storage and distribution center in West, Texas. Fifteen people, including 11 firefighters, lost their lives and over 200 residents were injured.[4]

11.12-B WORKPLACE REGULATIONS INVOLVING THE BULK STORAGE OF AMMONIUM NITRATE

OSHA has promulgated regulations relating to the storage of bulk quantities of ammonium nitrate at 29 C.F.R. §1910.109. These regulations apply to farmers and other employers who store and use ammonium nitrate in workplaces, as well as the owner and lessee of any building, premise, or structure in which the ammonium nitrate is stored in quantities of 1000 pounds (454 kg) or more.

---

[4]Glenn Hess, "Fertilizer Blast Ignites Concerns," *Chem. Eng. News,* Vol. 91 (2013), p. 33.

The OSHA regulations pertain to three subject areas: the nature of the building in which storage of ammonium nitrate is intended; the storage of ammonium nitrate in containers or piles within an approved building; and the storage of ammonium nitrate with other substances. We note each subject area separately to appreciate the extent of the precautions that must be undertaken to prevent mishaps.

The Nature of the Building in Which Ammonium Nitrate Is Stored

- A building used to store ammonium nitrate may have a basement only when the basement has been constructed so it is open on at least one side.
- The building used to store ammonium nitrate is limited to one story in height.
- The building used for the storage of ammonium nitrate must be adequately ventilated or be of a construction that allows self-ventilating in the event of a fire.
- The wall on the exposed side of an ammonium nitrate storage building that is within 50 feet (15 m) of a combustible building, forest, piles of combustible materials and similar exposure hazards must be of fire-resistive construction. In lieu of a fire-resistive wall, other suitable means of exposure protection, such as a freestanding wall, may be used.
- All flooring in storage and handling areas within an ammonium nitrate storage building must be of noncombustible material or protected against impregnation of ammonium nitrate, and must be without open drains, traps, tunnels, pits, or pockets into which any molten ammonium nitrate could flow and be confined in the event of fire.
- The ammonium nitrate storage building and its structures must be dry and free from water seepage through the roof, walls, and floors.
- Unless constructed of noncombustible material, or unless adequate facilities for fighting a roof fire are available, bulk storage structures within an ammonium nitrate storage building cannot exceed a height of 40 feet (12 m).
- Suitable fire-control devices, such as small-hose or portable fire extinguishers, must be provided throughout an ammonium nitrate storage building and in the loading and unloading areas.
- Water supplies and fire hydrants must be located nearby an ammonium nitrate storage building in accordance with recognized good practices.

The Storage of Ammonium Nitrate in Containers

- Containers of ammonium nitrate cannot be accepted for storage when the temperature of the product exceeds 130°F (54°C).
- Bags of ammonium nitrate cannot be stored within 30 inches (76 cm) of the walls and partitions of the ammonium nitrate storage building.
- No more than 2500 tons (2270 t) of bagged ammonium nitrate may be stored in an ammonium nitrate storage building that is not equipped with an automatic sprinkler system.

The Storage of Ammonium Nitrate in Piles

- Ammonium nitrate cannot be accepted for storage in piles when the temperature of the product exceeds 130°F (54°C).
- Piles of ammonium nitrate ordinarily cannot exceed 20 feet (6 m) in height, 20 feet (6 m) in width, and 50 feet (15 m) in length. When the piles are stored within a building of noncombustible construction or protected with automatic sprinklers, the length of the piles is unlimited. Ammonium nitrate piles cannot be stacked closer than 36 inches (91 cm) below the roof or supporting and spreader beams.
- Aisles must be provided to separate ammonium nitrate piles by a clear space of not less than 3 feet (0.9 m) in width. At least one service or main aisle in the storage area that is not less than 4 feet (1.2 m) in width also must be provided.
- Bins must be kept clean and free of materials that can contaminate ammonium nitrate.

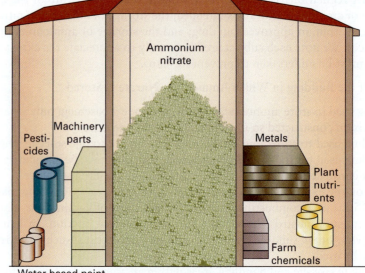

FIGURE 11.8 At 29 C.F.R. §1910.109, OSHA requires employers to store bulk quantities [>1000 pounds (454 kg)] of ammonium nitrate within a bin that is clean and free of contaminants. When the ammonium nitrate is stored as a pile, it must be sized in a manner that all material within the pile may be periodically moved. The bin must be separated by fire walls from flammable and combustible materials, corrosive materials, other oxidizers, and substances with which the ammonium could react. Employers may not store explosives or blasting agents within the same building.

- Owing to the corrosive and reactive properties of ammonium nitrate and to avoid contamination, galvanized iron, copper, lead, and zinc cannot be used in the construction of a bin unless they are suitably protected. Although aluminum and wooden bins protected against impregnation by ammonium nitrate are permissible, the partitions dividing the ammonium nitrate storage area from the storage of other products that could contaminate the ammonium nitrate must be of "tight" construction, as illustrated in Figure 11.8.

- Ammonium nitrate storage bins and piles must be clearly marked with signs reading as follows:

> ## AMMONIUM NITRATE

- The piles of ammonium nitrate must be sized and arranged so that all material in the piles is moved periodically to minimize caking of the ammonium nitrate during its storage.

- The height or depth of the ammonium nitrate piles is limited by the pressure-settling tendency of the product; however, in no case can the ammonium nitrate be piled higher at any point than 36 inches (91 cm) below the roof or the supporting and spreader beams overhead.

- Dynamite, other explosives, and blasting agents are prohibited for loosening ammonium nitrate that has caked.

The Storage of Ammonium Nitrate with Other Substances

- Ammonium nitrate must be stored in a separate building, separated by approved-type firewalls of not less than 1-hour fire-resistance rating, or separated by a space of at least 30 feet (9 m) from the storage of organic chemicals, acids, or other corrosive materials; materials that may require blasting during processing or handling; compressed flammable gases; and flammable and combustible materials or other contaminating substances, including but not limited to animal fats, baled cotton, baled rags, baled scrap paper, bleaching powder, burlap or cotton bags, caustic soda, coal, coke, charcoal, cork, camphor, excelsior, fibers of any kind, fish oils, fish meal, foam rubber, hay, lubricating oil, linseed oil, or other oxidizable or drying oils, naphthalene, oakum, oiled clothing, oiled paper, oiled textiles, paint, straw, sawdust, wood shavings, or vegetable oils.

■ Flammable liquids (such as gasoline, kerosene, solvents, and light fuels), liquefied petroleum gas, sulfur, and finely divided metals cannot be stored on the same premises where bulk ammonium nitrate is stored.

11.12-C HOMELAND SECURITY MEASURES INVOLVING AMMONIUM NITRATE FERTILIZER

Since 9-11, law enforcement agencies have taken measures to reduce the potential for the criminal misuse of ammonium nitrate fertilizer that is stored, handled, or transported within their jurisdictions. Given the unorthodox ways in which terrorists may obtain and use the fertilizer, law enforcement agencies must be provided with sufficient information to account for bulk quantities of ammonium nitrate potentially available to terrorists within their jurisdictions. Because ammonium nitrate can be used by terrorists, fertilizer companies must notify the Department of Homeland Security when they possess 1 ton (0.91 t) or more of agricultural material.

DHS and the U.S. Coast Guard (USCG) have been especially sensitive to the entry of bulk quantities of ammonium nitrate fertilizer at the nation's ports. At 49 C.F.R. §176.415 and 33 C.F.R. §126.28, respectively, DHS and USCG require the owner or operator of a watercraft loaded with bulk quantities of ammonium nitrate, or a watercraft on which bulk quantities are intended to be loaded, to receive written permission from the Captain of the Port before the loading or unloading operations occur. The permit stipulates that the owner or operator must load or unload the ammonium nitrate at a facility removed from congested areas or high-value or high-hazard industrial facilities and at which an abundance of water for firefighting purposes is available. DHS also requires the watercraft to be moored bow to seaward and maintained in a mobile status by the presence of tugs and the readiness of its engines so it may pass unrestricted to open water in the event of an emergency.

SOLVED EXERCISE 11.3

The OSHA regulation at 29 C.F.R. §1910.109 stipulates that ammonium nitrate fertilizer cannot ordinarily be stored in the same building with either sulfur or finely divided metals. What is the most likely reason OSHA requires ammonium nitrate to be segregated from these substances?

Solution: Ammonium nitrate is chemically incompatible with both elemental sulfur and finely divided metals. When exposed to an ignition source, the resulting combustion reactions are violent and occur at explosive rates. These reactions are denoted as follows, where metallic magnesium has been selected as the combustible metal:

$$S_8(s) + 32NH_4NO_3(s) \longrightarrow 8SO_2(g) + 8NO_2(g) + 64H_2O(g) + 28N_2(g)$$

Sulfur — Ammonium nitrate — Sulfur dioxide — Nitrogen dioxide — Water — Nitrogen

$$12Mg(s) + 16NH_4NO_3(s) \longrightarrow 12MgO(s) + 2NO_2(g) + 32H_2O(g) + 15N_2(g)$$

Magnesium — Ammonium nitrate — Magnesium oxide — Nitrogen dioxide — Water — Nitrogen

OSHA most likely requires the segregation of ammonium nitrate from sulfur and finely divided metals to reduce or eliminate the risk of fire and explosion.

At 33 C.F.R. §126.28, DHS has promulgated regulations relating to the handling, storage, stowing, loading, unloading, discharging, and transportation of bulk quantities of ammonium nitrate at a waterfront facility. These regulations are paraphrased as follows:

- The proper shipping name of the ammonium nitrate must be marked on the outside surface of all containers.
- The buildings on a waterfront facility used for the storage of ammonium nitrate must be constructed so as to afford good ventilation.
- The storage of ammonium nitrate must occur at a safe distance from electric wiring, steam pipes, radiators, or any other heating mechanism.
- The ammonium nitrate must be separated by a fire-resistive wall or by a distance of at least 30 feet (9 m) from organic materials or other substances that could cause contamination, such as flammable liquids, combustible liquids, corrosive liquids, metallic chlorates, metallic permanganates, finely divided metals, caustic soda, charcoal, sulfur, cotton, coal, fats, fish oils, or vegetable oils.
- Ammonium nitrate must be stored on a clean wooden or concrete floor or on pellets over a clean floor. When ammonium nitrate is stored on a concrete floor, the floor must first be covered with a moisture barrier such as a polyethylene sheet or asphaltic-laminated paper.
- Any spilled ammonium nitrate must be promptly and thoroughly cleaned up and removed from the waterfront facility. If ammonium nitrate has remained in contact with a wooden floor for any length of time, the floor must be scrubbed with water, and all spilled material must be thoroughly dissolved and flushed away.
- An abundance of water for firefighting must be readily available.
- Open drains, traps, pits, or pockets that could be filled with molten ammonium nitrate in the event of a fire must be eliminated or plugged.

11.13 OXIDIZING CHROMIUM COMPOUNDS

Chromium-containing compounds in which the chromium assumes the +6 oxidation state are called oxidizing chromium compounds. The +6 oxidation state of chromium is the hexavalent chromium ion. The oxidizing chromium compounds are also called **hexavalent chromium compounds**. They are metallic chromates, metallic dichromates, chromium trioxide, and chromium oxychloride.

hexavalent chromium compound ■ Any compound containing chromium in the +6 oxidation state

The metallic chromates and metallic dichromates are compounds composed of a positive ion and the chromate ion (CrO_4^{2-}) and dichromate ion ($Cr_2O_7^{2-}$), respectively. Potassium chromate and potassium dichromate are specific examples. Their chemical formulas are K_2CrO_4 and $K_2Cr_2O_7$, respectively.

The oxidizing chromium compounds are manufactured for commercial use from chromite ore, which we represent here as $FeCr_2O_4$. For example, to produce potassium dichromate the ore is first heated at kiln temperatures with potassium carbonate to produce potassium chromate as follows:

$$4FeCr_2O_4(s) \ + \ 8K_2CO_3(s) \ + \ 7O_2(g) \ \longrightarrow \ 8K_2CrO_4(s) \ + \ 2Fe_2O_3(s) \ + \ 8CO_2(g)$$

Chromite 　Potassium carbonate 　Oxygen 　　Potassium chromate 　Iron(III) oxide 　Carbon dioxide

Potassium dichromate is then produced by the reaction of sulfuric acid with potassium chromate.

$$2K_2CrO_4(aq) \ + \ H_2SO_4(aq) \ \longrightarrow \ K_2Cr_2O_7(aq) \ + \ H_2O(l) \ + \ K_2SO_4(aq)$$

Potassium chromate 　Sulfuric acid 　　Potassium dichromate 　Water 　Potassium sulfate

Potassium chromate may be produced from potassium dichromate, and vice versa, by altering the pH conditions as follows:

$$K_2Cr_2O_7(aq) \ + \ 2KOH(aq) \ \longrightarrow \ 2K_2CrO_4(aq) \ + \ H_2O(l)$$
Potassium dichromate   Potassium hydroxide   Potassium chromate   Water

$$2K_2CrO_4(aq) \ + \ 2HCl(aq) \ \longrightarrow \ K_2Cr_2O_7(aq) \ + \ H_2O(l) \ + \ 2KCl(aq)$$
Potassium chromate   Hydrochloric acid   Potassium dichromate   Water   Potassium chloride

The metallic chromates and metallic dichromates are solid yellow and orange compounds, respectively. Consequently, compounds such as lead chromate, strontium chromate, and zinc chromate have been useful as pigments in industrial paints. The oxidizing chromium compounds have also been used to manufacture textile dyes and preserve wood and leather. They are also the major components of chrome electroplating bath solutions.

Before the 1990s, hexavalent chromium compounds commonly were used as water-treatment agents to prevent the formation of a mineral scale that otherwise would form in the circulating waters used in air conditioners and industrial cooling towers. The production of this scale reduced the efficiency by which heat was transferred to the water. Cooling towers often are huge structures located at petroleum refineries, power plants, chemical plants, and elsewhere, where they are used to cool water by evaporation. The most readily recognizable type of cooling tower is probably the rounded hourglass structure used at nuclear power plants, but cooling towers are also located wherever large heating, ventilation, and air-conditioning or refrigeration systems are needed. The cooling tower shown in Figure 11.9 is a component of an air-conditioning and ventilation system that cools a major hotel.

In the past, the waste generated in connection with using hexavalent chromium-based water-treatment agents at petroleum refineries, power plants, and chemical plants was often discharged into onsite unlined surface impoundments. Rainwater caused the hexavalent chromium compound to seep into the subsurface and contaminate the underlying groundwater aquifer.

The absorption of a hexavalent chromium compound into the body can cause a number of health ailments. In particular, the inhalation of the dust or fines of hexavalent chromium compounds has been linked with the onset of lung cancer in humans. Thus, these compounds are ranked as human carcinogens by the inhalation route of exposure.

Erin Brockovich, the heroine in the blockbuster movie of the same name, alleged in the film that the long-term consumption of water contaminated with elevated levels of hexavalent chromium causes cancer and other health ailments. At the time of the movie's release in 2000, there was scant evidence to definitively link these health problems with the consumption of hexavalent chromium–contaminated water. Today, however, scientists have clear evidence that ingestion of this contaminant not only causes the onset of stomach ulcers and stomach and intestinal cancers but may also damage the kidneys and liver.[5]

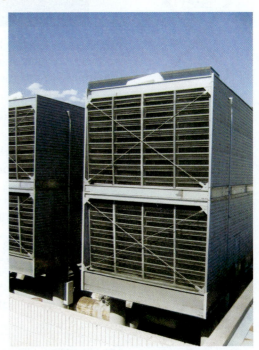

FIGURE 11.9 Cooling towers often are massive water-recirculation structures used to remove heat from the water by contacting the fluid with air. A commercial product containing a hexavalent chromium compound formerly was added to the water in industrial cooling towers to prevent the formation of a mineral scale on grates and other tower components. Because hexavalent chromium is a human carcinogen, EPA promulgated regulations at 40 C.F.R. §749.68 that now prohibit the use of a compound containing hexavalent chromium for this purpose. (*Courtesy of Eugene Meyer.*)

---

[5]Office of Environmental Health Hazard Assessment, *Evidence on the Developmental and Reproduction Toxicity of Chromium (Hexavalent Compounds)* (California Environmental Protection Agency, 2009).

Potassium dichromate

EPA has set a maximum contaminant level of 100 µg/L of total chromium in drinking water. EPA has also used the authority of the Toxic Substances Control Act to prohibit the use of water-treatment agents containing hexavalent chromium in heating and ventilation and air-conditioning and refrigeration systems. This ban aims to reduce or eliminate the potential of exposing individuals to their air emissions.

At 40 C.F.R. §749.68, EPA requires distributors to label containers of hexavalent chromium water-treatment agents distributed in commerce and intended for use in cooling towers as follows:

WARNING

This product contains hexavalent chromium. Inhalation of hexavalent chromium air emissions increases the risk of lung cancer. Federal law prohibits use of this substance in comfort cooling towers, which are towers that are open water-recirculation devices and that are dedicated exclusively to, and are an integral part of, heating, ventilation, air conditioning, or refrigeration systems.

EPA requires this label to be affixed to the containers with such conspicuousness that the warning statement is read and understood by the ordinary individual under customary conditions of purchase and use.

11.13-A POTASSIUM DICHROMATE

Potassium dichromate is frequently used commercially as a strong oxidizer; for example, it is a component of the bath solutions used for chromium plating. When heated, potassium dichromate decomposes to generate oxygen as follows:

$$2K_2Cr_2O_7(s) \longrightarrow 2Cr_2O_3(s) + 2K_2O(s) + 3O_2(g)$$

Potassium dichromate   Chromic oxide   Potassium oxide   Oxygen

Because it is a strong oxidizer, potassium dichromate is chemically incompatible with many other substances. For example, potassium dichromate and hydrochloric acid react to produce chlorine.

$$K_2Cr_2O_7(aq) + 14HCl(aq) \longrightarrow 2CrCl_3(aq) + 2KCl(aq) + 7H_2O(l) + 3Cl_2(g)$$

Potassium dichromate   Hydrochloric acid   Chromic chloride   Potassium chloride   Water   Chlorine

The emergency incidents involving this hazardous material generally are associated with its tendency to oxidize the substances with which it makes contact.

11.13-B CHROMIUM TRIOXIDE

Chromium trioxide

The acids associated with the metallic chromates and inorganic dichromates exist only in aqueous solutions, but when the water is evaporated from them, a compound formally called chromium(VI) oxide remains. This anhydrous compound is also known as chromium trioxide, chromium anhydride, and chromic acid. Its chemical formula is CrO_3.

Chromium trioxide is a red solid prepared industrially by adding concentrated sulfuric acid to a saturated solution of potassium dichromate.

$$K_2Cr_2O_7(aq) + 2H_2SO_4(conc) \longrightarrow 2KHSO_4(aq) + H_2O(l) + 2CrO_3(s)$$

Potassium dichromate   Sulfuric acid   Potassium bisulfate   Water   Chromium trioxide

Chromium trioxide is used for metal-treatment processes such as chrome plating, copper stripping, aluminum anodizing, and corrosion prevention.

Chromium oxychloride

11.13-C CHROMIUM OXYCHLORIDE

Chromium oxychloride is a dark red liquid that forms when a mixture of concentrated hydrochloric acid and sulfuric acid is added to a saturated solution of potassium dichromate. It is also called chromyl chloride, and its chemical formula is CrO_2Cl_2.

Chromium oxychloride is not only an oxidizer. It also reacts very violently with water, forming chromium(III) chloride, chromic acid, and chlorine.

$$6CrO_2Cl_2(l) \ + \ 4H_2O(l) \ \longrightarrow \ 2CrCl_3(aq) \ + \ 4H_2CrO_4(aq) \ + \ 3Cl_2(g)$$

Chromium oxychloride   Water   Chromium(III) chloride   Chromic acid   Chlorine

In this instance, the chromic acid is the "true" chromic acid (not chromium trioxide), which exists only in solution.

11.13-D AMMONIUM DICHROMATE

Ammonium dichromate is an orange solid used as a component of certain pyrotechnics. It is also used to dye fabrics and to alter the texture of animal hides during the production of leather. Its chemical formula is $(NH_4)_2Cr_2O_7$. When heated, ammonium dichromate decomposes to form nitrogen.

$$(NH_4)_2Cr_2O_7(s) \ \longrightarrow \ Cr_2O_3(s) \ + \ 4H_2O(g) \ + \ N_2(g)$$

Ammonium dichromate   Chromic oxide   Water   Nitrogen

11.13-E WORKPLACE REGULATIONS INVOLVING THE OXIDIZING CHROMIUM COMPOUNDS

When used in the workplace, OSHA requires employers to limit employee exposure to a hexavalent chromium inhalation concentration of 5 μg/m^3, averaged over an 8-hour workday.

11.13-F TRANSPORTING THE OXIDIZING CHROMIUM COMPOUNDS

When shippers offer an oxidizing chromium compound for transportation, DOT requires them to describe it as shown in Table 11.15 on an accompanying shipping paper. DOT also requires shippers and carriers to comply with all applicable labeling, marking, and placarding requirements.

TABLE 11.15	Shipping Descriptions of the Oxidizing Chromium Compounds
OXIDIZING CHROMIUM COMPOUND	**SHIPPING DESCRIPTION**
Chromic acid	UN1755, Chromic acid solution, 8, PG II *or* UN1755, Chromic acid solution, 8, PG III
Chromium nitrate	UN2720, Chromium nitrate, 5.1, PG III
Chromium oxychloride	UN1758, Chromium oxychloride, 8, PG I
Chromium trioxide	UN1463, Chromium trioxide, anhydrous, 5.1, (8), PG II
Chromosulfuric acid	UN2240, Chromosulfuric acid, 8, PG I

11.14 SODIUM PERMANGANATE AND POTASSIUM PERMANGANATE

Metallic permanganates are compounds in which manganese assumes the +7 oxidation state. They are generally encountered as dark violet, iridescent crystals. There are two commercially important members of this class of compounds: sodium permanganate and potassium permanganate. Their chemical formulas are $NaMnO_4$ and $KMnO_4$, respectively.

11.14-A PRODUCTION AND COMMERCIAL USES

The metallic permanganates are manufactured from manganese ore containing about 70% manganese dioxide. As the ore is heated with either sodium hydroxide or potassium hydroxide, the manganese undergoes oxidation to produce the corresponding alkali metal manganate. For example, potassium manganate is formed from the ore and potassium hydroxide.

$$2MnO_2(s) \;+\; 4KOH(l) \;+\; O_2(g) \longrightarrow 2K_2MnO_4(s) \;+\; 2H_2O(g)$$

Manganese dioxide   Potassium hydroxide   Oxygen   Potassium manganate   Water

A solution of the potassium manganate is then oxidized electrolytically to produce potassium permanganate.

$$2K_2MnO_4(aq) \;+\; 2H_2O(l) \longrightarrow 2KMnO_4(aq) \;+\; 2KOH(aq) \;+\; H_2(g)$$

Potassium manganate   Water   Potassium permanganate   Potassium hydroxide   Hydrogen

The potassium permanganate is then separated from the potassium hydroxide by crystallization.

You may be familiar with dilute solutions of sodium permanganate and potassium permanganate that are used pharmaceutically to cure dermatitis having a bacterial or fungal origin ("athlete's foot"). These same solutions are used to treat fish diseases. The concentrated solutions are used to remove objectionable matter from chemical and biological process wastes by oxidation. The effluent gases from many industrial sources are also often eliminated or reduced in concentration through redox reactions involving the use of sodium permanganate or potassium permanganate.

Potassium permanganate has also been used to control the infestation of the quagga mussel (Section 11.8-A) in waterways. Potassium permanganate solutions are injected into pipes where the mussels congregate and thereby disturb the flow of water in the intake and discharge systems. The oxidizing agent kills the pesky mussels.

11.14-B TRANSPORTING METALLIC PERMANGANATES

When shippers offer a metallic permanganate for transportation, DOT requires them to identify it as shown in Table 11.16 on an accompanying shipping paper. When shippers offer a metallic permanganate not listed in Table 11.16 for transportation, DOT requires them to identify the compound generically on the shipping paper as "UN1482, Permanganates, inorganic, n.o.s., 5.1, PG II," "UN1482, Permanganates, inorganic, n.o.s., 5.1, PG III," or "UN3214, Permanganates, inorganic, aqueous solution, n.o.s., 5.1, PG II." DOT also requires shippers to enter the name of the specific compound parenthetically in the shipping description. It also requires shippers and carriers to comply with all labeling, marking, and placarding requirements.

11.15 METALLIC NITRITES AND METALLIC NITRATES

Metallic nitrites and metallic nitrates are compounds composed of metallic ions and the nitrite ion (NO_2^-) and nitrate ion NO_3^-), respectively. In these compounds, the nitrogen atom assumes oxidation numbers of +3 and +5, respectively. Metallic nitrates

METALLIC PEMANGANATE	SHIPPING DESCRIPTION
Barium permanganate	UN1448, Barium permanganate, 5.1, (6.1), PG II (Poison)
Calcium permanganate	UN1456, Calcium permanganate, 5.1, PG II
Potassium permanganate	UN1490, Potassium permanganate, 5.1, PG II
Sodium permanganate	UN1503, Sodium permanganate, 5.1, PG II
Zinc permanganate	UN1515, Zinc permanganate, 5.1, PG II

are especially common oxidizers; for example, sodium nitrate and potassium nitrate are common components of blasting agents and other explosives. Several metallic nitrates are also used to produce color in activated signaling flares.

For decades, sodium nitrite and sodium nitrate have been added to raw meat (bacon, hot dogs, and luncheon meats), poultry, and fish to fix their color and inhibit the growth of *Clostridium botulinum,* the bacterial spores that cause botulism. This use was limited by the U.S. Food and Drug Administration when scientists demonstrated that nitrites can produce carcinogenic N-nitrosamines in the stomach. The nitrites react with stomach acid to form nitrous acid, which in turn reacts with the proteins in meat to form N-nitrosa-mines (Section 13.8-B), organic compounds containing the group of atoms $-N-N=O$. Low concentrations of N-nitrosamines have been identified in smoked fish, cured meat, beer, and cheese. Because test animals exposed to N-nitrosamines develop cancer, N-nitrosamines are regarded as probable human carcinogens.

When individuals ingest or inhale sodium nitrite or sodium nitrate in excess, they are likely to suffer from methemoglobinemia (Section 10.14-B). In rare instances, this ailment can also be contracted by absorption through the skin, e.g., from exposure to molten sodium nitrite or sodium nitrate.

11.15-A SOME PROPERTIES OF METALLIC NITRITES

Metallic nitrites can act as either oxidizing agents or reducing agents. They are oxidized to metallic nitrates and reduced to nitric oxide (NO) as follows:

$$NaNO_2(aq) \;+\; Na_2O_2(aq) \;+\; H_2O(l) \longrightarrow NaNO_3(aq) \;+\; 2NaOH(aq)$$

Sodium nitrite   Sodium peroxide   Water   Sodium nitrate   Sodium hydroxide

$$2NaNO_2(aq) \;+\; Na_2SO_3(aq) \;+\; 2HCl(aq) \longrightarrow$$

Sodium nitrite   Sodium sulfite   Hydrochloric acid

$$Na_2SO_4(aq) \;+\; 2NaCl(aq) \;+\; 2NO(g) \;+\; H_2O(l)$$

Sodium sulfate   Sodium chloride   Nitric oxide   Water

Although metallic nitrites generally are stable to heat, they decompose at elevated temperatures. For example, sodium nitrite decomposes as follows:

$$2NaNO_2(s) \longrightarrow Na_2O(s) \;+\; NO_2(g) \;+\; NO(g)$$

Sodium nitrite   Sodium oxide   Nitrogen dioxide   Nitric oxide

11.15-B SOME PROPERTIES OF METALLIC NITRATES

Metallic nitrates react only as oxidizing agents. In the presence of acids, they are converted to nitric oxide, nitrogen dioxide, or ammonia. This chemical behavior is exemplified by the chemical reactions of zinc with concentrated nitric acid noted in Section 8.8-A.

Sodium nitrite

Sodium nitrate

When heated, metallic nitrates decompose in the following ways:

- The alkali metal nitrates decompose to form the respective alkali metal nitrite and oxygen. For example, sodium nitrate thermally decomposes to form oxygen.

$$2NaNO_3(s) \longrightarrow 2NaNO_2(s) + O_2(g)$$

Sodium nitrate   Sodium nitrite   Oxygen

- Nitrates of the noble metals decompose to produce the metal, nitrogen dioxide, and oxygen. For example, silver nitrate thermally decomposes to produce silver, nitrogen dioxide, and oxygen.

$$2AgNO_3(s) \longrightarrow 2Ag(s) + 2NO_2(s) + O_2(g)$$

Silver nitrate   Silver   Nitrogen dioxide   Oxygen

- Metal nitrates other than the alkali metal nitrates and the noble metal nitrates decompose to form the respective metallic oxide, nitrogen dioxide, and oxygen. For example, lead(II) nitrate thermally decomposes to produce lead oxide, nitrogen dioxide, and oxygen.

$$2Pb(NO_3)_2(s) \longrightarrow 2PbO(s) + 4NO_2(g) + O_2(g)$$

Lead nitrate   Lead oxide   Nitrogen dioxide   Oxygen

11.15-C FDA REGULATIONS INVOLVING SODIUM NITRITE AND SODIUM NITRATE

At 21 C.F.R. §§172.170 and 172.175, FDA publishes maximum permissible concentrations of sodium nitrite and/or sodium nitrate in certain foods intended for use as follows:

- As a color fixative in smoked, cured tuna fish products so that the level of sodium nitrite does not exceed 10 parts per million in the finished product
- As a preservative and color fixative, with or without sodium nitrate, in smoked, cured sablefish, smoked, cured salmon, and smoked, cured shad so that the level of sodium nitrite does not exceed 200 parts per million and the level of sodium nitrate does not exceed 500 parts per million in the finished product
- As a preservative and color fixative, with sodium nitrate, in meat-curing preparations for the home-curing of meat and meat products (including poultry and wild game), with directions for use that limit the amount of sodium nitrite to not more than 200 parts per million in the finished meat product, and the amount of sodium nitrate to not more than 500 parts per million in the finished meat product

FDA also requires the finished product to be labeled with the name of the additive, its concentration, directions for use, and the statement "Keep Out of the Reach of Children."

11.15-D TRANSPORTING METALLIC NITRITES AND NITRATES

When shippers offer a metallic nitrite or metallic nitrate for transportation, DOT requires them to describe the appropriate substance on an accompanying shipping paper. Some examples for several representative metallic nitrites and nitrates are listed in Table 11.17. When shippers offer to transport a metallic nitrate or metallic nitrite that is not listed at 49 C.F.R. §172.101, DOT requires them to identify the compound generically and to enter the name of the specific compound parenthetically in the shipping description. DOT also requires shippers and carriers to comply with all applicable labeling, marking, and placarding requirements.

When carriers transport a potassium nitrate and sodium nitrate mixture by highway or rail, DOT requires them to display the name on two opposing sides of the packaging.

POTASSIUM NITRATE AND SODIUM NITRATE MIXTURE

METALLIC NITRITE OR NITRATE	SHIPPING DESCRIPTION
Aluminum nitrate	UN1438, Aluminum nitrate, 5.1, UN1438, PG III
Barium nitrate	UN1446, Barium nitrate, 5.1, (6.1), PG II (Poison)
Calcium nitrate	UN1454, Calcium nitrate, 5.1, PG III
Lead nitrate	UN1469, Lead nitrate, 5.1, (6.1), PG II (Poison)
Nickel nitrite	UN2726, Nickel nitrite, 5.1, PG III
Potassium nitrate	UN1486, Potassium nitrate, 5.1, PG III
Potassium nitrite	UN1488, Potassium nitrite, 5.1, PG II
Silver nitrate	UN1493, Silver nitrate, 5.1, PG II
Sodium nitrate	UN1498, Sodium nitrate, 5.1, PG III
Sodium nitrite	UN1500, Sodium nitrite, 5.1, PG II
Zinc nitrate	UN1514, Zinc nitrate, 5.1, PG II

Sodium peroxide

Barium peroxide

11.16 METALLIC PEROXIDES AND SUPEROXIDES

The compounds composed of metallic ions and peroxide ions [O_2^{2-} or $(-O-O-)^{2-}$] are called *metallic peroxides*, and the compounds composed of metallic ions and superoxide ions [O_2^- or $(-O-O-)^-$] are called *metallic superoxides*. The commercially important metallic peroxides are compounds that contain an alkali metal ion or an alkaline earth metal ion. Examples are sodium peroxide and barium peroxide, whose chemical formulas are Na_2O_2 and BaO_2, respectively. There is only one commercially important metallic superoxide: potassium superoxide. Its chemical formula is KO_2.

11.16-A PROPERTIES OF METALLIC PEROXIDES

The metallic peroxides are thermally unstable, highly reactive compounds. When heated, they decompose to the corresponding metallic oxide and oxygen.

$$2Na_2O_2(s) \longrightarrow 2Na_2O(s) + O_2(g)$$
Sodium peroxide ⟶ Sodium oxide + Oxygen

$$2BaO_2(s) \longrightarrow 2BaO(s) + O_2(g)$$
Barium peroxide ⟶ Barium oxide + Oxygen

They also oxidize finely divided combustible metals like powdered aluminum.

$$2Al(s) + 3Na_2O_2(s) + 3H_2O(l) \longrightarrow Al_2O_3(s) + 6NaOH(s)$$
Aluminum + Sodium peroxide + Water ⟶ Aluminum oxide + Sodium hydroxide

The metallic peroxides are also water-reactive substances.

$$2Na_2O_2(s) + 2H_2O(l) \longrightarrow 4NaOH(aq) + O_2(g)$$
Sodium peroxide + Water ⟶ Sodium hydroxide + Oxygen

To prevent metallic peroxides from absorbing atmospheric moisture, they should be stored in tightly closed, moisture-proof containers.

11.16-B PROPERTIES OF METALLIC SUPEROXIDES

The metallic superoxides are thermally unstable, water-reactive compounds. Their water reactivity is put to use in a type of chemical oxygen generator that provides breathable oxygen to

Potassium superoxide

TABLE 11.18	Shipping Descriptions of Some Representative Metallic Peroxides and Superoxides

METALLIC PEROXIDE OR METALLIC SUPEROXIDE	SHIPPING DESCRIPTION
Barium peroxide	UN1449, Barium peroxide, 5.1, (6.1), PG II (Poison)
Calcium peroxide	UN1457, Calcium peroxide, 5.1, PG II
Lithium peroxide	UN1472, Lithium peroxide, 5.1, PG II
Magnesium peroxide	UN1476, Magnesium peroxide, 5.1, PG II
Potassium peroxide	UN1491, Potassium peroxide, 5.1, PG I
Potassium superoxide	UN2466, Potassium superoxide, 5.1, PG III
Sodium peroxide	UN1504, Sodium peroxide, 5.1, PG I
Sodium superoxide	UN2547, Sodium superoxide, 5.1, PG III

its user for a period ranging from 5 minutes to several hours. The moisture and carbon dioxide in the user's breath are absorbed by potassium superoxide contained in a canister. The chemical action of the moisture and the superoxide generates oxygen. The oxygen enters a breathing bag for subsequent inhalation, and the exhaled breath repeats the cycle.

$$12KO_2(s) \ + \ 6H_2O(g) \ \longrightarrow \ 12KOH(s) \ + \ 9O_2(g)$$
Potassium superoxide   Water   Potassium hydroxide   Oxygen

$$K_2O(s) \ + \ CO_2(g) \ \longrightarrow \ K_2CO_3(s)$$
Potassium oxide   Carbon dioxide   Potassium carbonate

11.16-C TRANSPORTING METALLIC PEROXIDES AND SUPEROXIDES

DOT requires shippers who offer a metallic peroxide or superoxide for transportation to identify the relevant substance on an accompanying shipping paper. Some examples for several representative metallic peroxides and superoxides are listed in Table 11.18. When shippers offer a metallic peroxide or its solutions that are not listed at 49 C.F.R. 172.101, DOT requires them to identify the compounds or solutions generically and to list their specific names parenthetically. DOT requires shippers and carriers to comply with all applicable labeling, marking, and placarding requirements.

11.17 POTASSIUM PERSULFATE AND SODIUM PERSULFATE

Metallic persulfates are compounds composed of a metallic ion and the persulfate ion, also called the peroxydisulfate ion ($S_2O_8^{2-}$). The persulfate ion is also known as the peroxydisulfate ion. It has the following Lewis structure:

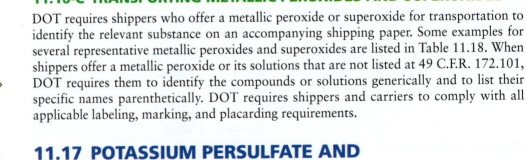

Two examples of metallic persulfates are potassium persulfate and sodium persulfate, whose chemical formulas are $K_2S_2O_8$ and $Na_2S_2O_8$, respectively. They are white solids.

Sodium persulfate

TABLE 11.19	Shipping Descriptions of Metallic Persulfates
METALLIC PERSULFATE	**SHIPPING DESCRIPTION**
Potassium persulfate	UN1491, Potassium persulfate, 5.1, PG II
Sodium persulfate	UN1505, Sodium persulfate, 5.1, PG III

11.17-A PROPERTIES AND COMMERCIAL USES

The metallic persulfates are the salts of peroxydisulfuric acid, whose chemical formula is $H_2S_2O_8$. The latter substance is a liquid prepared by mixing 30% hydrogen peroxide and 98% sulfuric acid in a 1:3 ratio by volume.

$$2H_2SO_4(l) \ + \ H_2O_2(l) \ \longrightarrow \ H_2S_2O_8(l) \ + \ 2H_2O(l)$$

Sulfuric acid   Hydrogen peroxide   Peroxydisulfuric acid   Water

This highly corrosive solution is popular for several technical applications. Most notably, in the computer-chip industry, where it is known as *piranha*, peroxydisulfuric acid is used to clean the wafers that are processed into chips.

The metallic persulfates are thermally unstable. For example, oxygen and sulfur dioxide are generated when sodium persulfate is heated.

$$2Na_2S_2O_8(s) \ \longrightarrow \ 2Na_2O(s) \ + \ 4SO_2(g) \ + \ 3O_2(g)$$

Sodium persulfate   Sodium oxide   Sulfur dioxide   Oxygen

11.17-B TRANSPORTING METALLIC PERSULFATES

When shippers offer potassium persulfate or sodium persulfate for transportation, DOT requires them to provide the relevant shipping description shown in Table 11.19 on an accompanying shipping paper.

When shippers offer metallic persulfates other than potassium persulfate or sodium persulfate and their solutions for transportation, DOT requires them to identify the substances generically as "UN3215, Persulfates, inorganic, n.o.s., 5.1, PG III" *or* "UN3216, Persulfates, inorganic, aqueous solution, n.o.s., 5.1, PG III." DOT also requires shippers to enter the names of the specific compounds parenthetically in the shipping description, and to affix an OXIDIZER label to their packaging. DOT also requires shippers and carriers to comply with all applicable labeling, marking, and placarding requirements.

11.18 MATCHES

Matches are items used to intentionally initiate fire when the mixture of substances contained in their heads is rubbed against a hard surface. They have long served as one of the earliest commercial products in which redox reactions were used to provide fire on demand. Even today, they are the most common items used to intentionally initiate fires.

The two types of matches shown in Figure 11.10 are commercially popular. They are called **strike-anywhere matches** and **safety matches**. It is appropriate to examine the chemistry associated with their burning, especially insofar as these processes involve the use of oxidizers.

11.18-A STRIKE-ANYWHERE MATCHES

The head of a strike-anywhere match consists of a mixture of tetraphosphorus trisulfide (also called phosphorus sesquisulfide), sulfur, lead(IV) oxide, powdered glass, and glue. This mixture is mounted on a small stick of wood and covered with paraffin wax.

strike-anywhere match ■ A match with a bulls-eye head consisting of a white tip containing tetraphosphorus trisulfide that is actuated by rubbing it against an abrasive strip

safety match ■ A match designed to ignite only when its head is rubbed on a prepared surface consisting of red phosphorus and powdered glass

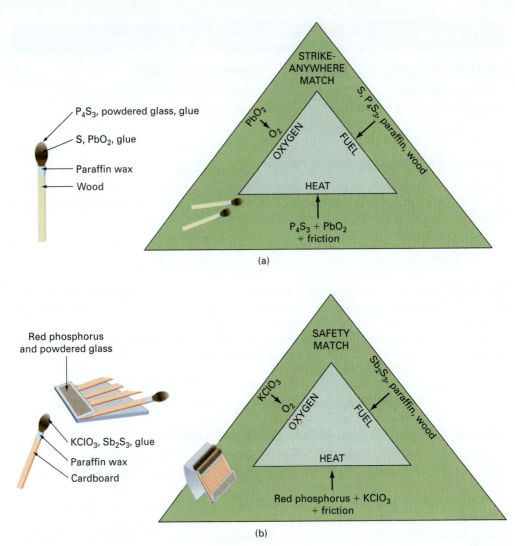

FIGURE 11.10 The strike-anywhere match in (a) ignites when the surface of the match head is rubbed against a hard surface. The safety match in (b) ignites when the surface of a match head is struck against a surface containing red phosphorus and powdered glass. The fire triangles to the right of each match illustrate the composition of the fuels, the substances that react to provide the activation energy, and the relevant oxidizer.

Tetraphosphorus trisulfide ignites as the match head is struck against a hard surface. The evolved heat of combustion initiates the combustion of the sulfur and wood.

$$P_4S_3(s) \quad + \quad 8O_2(g) \quad \longrightarrow \quad P_4O_{10}(s) \quad + \quad 3SO_2(g)$$

Tetraphosphorus trisulfide   Oxygen   Tetraphosphorus decoxide   Sulfur dioxide

11.18-B SAFETY MATCHES

The active components of a safety match consist of a mixture of antimony trisulfide, sulfur, and potassium chlorate, glued to a piece of cardboard. A safety match ignites by means of friction against a prepared surface consisting of red phosphorus and powdered glass. This surface is usually on the box, book, or card. When the match head is rubbed on this surface, the heat initiates a redox reaction as follows:

$$16KClO_3(s) \quad + \quad 3S_8(s) \quad \longrightarrow \quad 16KCl(s) \quad + \quad 24SO_2(g)$$

Potassium chlorate   Sulfur   Potassium chloride   Sulfur dioxide

Then, the energy evolved from the oxidation of the sulfur ignites the antimony trisulfide.

$$2Sb_2S_3(s) \quad + \quad 9O_2(g) \quad \longrightarrow \quad 2Sb_2O_3(g) \quad + \quad 6SO_2(g)$$

Antimony trisulfide   Oxygen   Antimony trioxide   Sulfur dioxide

TABLE 11.20	Shipping Descriptions of Matches
MATCHES	**SHIPPING DESCRIPTION**
Safety book matches	UN1944, Matches, safety book, 4.1, PG III
Strike-anywhere matches	UN1331, Matches, strike anywhere, 4.1, PG III

11.18-C TRANSPORTING MATCHES

When shippers offer matches for transportation, DOT requires them to provide the relevant shipping description as shown in Table 11.20 on the accompanying shipping paper. DOT also requires shippers and carriers to comply with all applicable labeling, marking, and placarding requirements.

11.19 RESPONDING TO INCIDENTS INVOLVING A RELEASE OF A METALLIC OXIDIZER

First-on-the-scene responders generally can identify the presence of a metallic oxidizer at a transportation mishap by observing at least one of the following:

■ The number 5.1 as a component of a shipping description of a hazardous material listed on a shipping paper
■ The word *OXIDIZER* and the number 5.1 printed on yellow labels affixed to packaging
■ The word *OXIDIZER* and the number 5.1 printed on yellow placards posted on each side and each end of a transport vehicle containing 1001 pounds (454 kg) or more of an oxidizer

The recommended method of extinguishing fires supported by most liquid or solid oxidizers is to deluge them with water. Metallic oxidizers generally are soluble in water. When they are diluted with water, their chemical reactivity is sharply reduced or eliminated.

Although water effectively extinguishes most fires supported by metallic oxidizers, emergency responders need to exercise caution when applying the water. Many solid oxidizers melt before they decompose. These hot molten materials flow to adjoining areas, where they may mix with combustible materials and support their ignition. Although water may be applied to fires supported by oxidizers, it should be applied only to bulk quantities of molten oxidizers as a fog to avoid the rapid generation of steam that may cause the molten material to splatter. Experts also recommend the use of sand on fires involving bulk quantities of molten oxidizers.

SOLVED EXERCISE 11.4

Why is it essential to dam or dike the water runoff generated during a response action involving a solid oxidizer?

Solution: When water is used during a response action involving a solid oxidizer, its primary function is to dilute the oxidizer appreciably so that the rate of an oxidation–reduction reaction is reduced. When the runoff is left unattended, however, the water evaporates and leaves the oxidizer in a dry state, whereupon it is again susceptible to supporting the combustion of matter. For this reason, the water runoff generated during a response action involving a solid oxidizer should always be dammed or diked so that the oxidizer may be neutralized before its ultimate disposition. Sodium sulfite is a typical neutralizing agent.

REVIEW EXERCISES

Basic Concepts Involving Oxidation–Reduction

1. What is the oxidation number of each underlined atom or ion in the following chemical formulas?
   (a) M<u>g</u>O and M<u>g</u>O$_2$
   (b) K<u>Mn</u>O$_4$ and <u>Mn</u>O$_2$
   (c) H$_2$<u>S</u>O$_3$ and H$_2$<u>S</u>O$_4$
   (d) <u>Ba</u>O and <u>Ba</u>O$_2$
   (e) <u>S</u>$_8$ and H$_2$<u>S</u>O$_4$
   (f) <u>O</u>$_2$ and K$_2$<u>O</u>
2. Identify the oxidizing agent, the reducing agent, the substance oxidized, and the substance reduced in each redox reaction noted by the following equations:
   (a) $MgO(s) + C(s) + Cl_2(g) \longrightarrow MgCl_2(l) + CO(g)$
   (b) $Br_2(g) + SO_2(g) + 2H_2O(l) \longrightarrow 2HBr(l) + H_2SO_4(l)$

Hydrogen Peroxide

3. Why is the use of water generally the best means of extinguishing fires in which hydrogen peroxide is involved?
4. Why are the bulk containers used to transport 30% to 50% hydrogen peroxide solutions vented to the atmosphere?

Hypochlorite Oxidizers

5. Before the advent of plastic bottles, why were sodium hypochlorite solutions stored in brown glass bottles?

Chlorination Agents Other Than Metallic Hypochlorites

6. Why must carriers take caution to segregate trichloroisocyanuric acid from combustible matter when loading it onboard railcars for shipment?

Oxidizers in Fireworks, Flares, and Signaling Smokes

7. Standard flash powder consists of a mixture of potassium perchlorate and powdered aluminum. It is the mixture of reactants in fireworks that provides a bright flash of light when ignited. What equation describes the phenomenon?

Oxidizing Ammonium Compounds

8. The OSHA regulation at 29 C.F.R. §1910.109(c)(4)(iii)(a) permits employers to store ammonium nitrate in a building having a basement but *only* when the basement is open on at least one side. What is the most likely reason that OSHA requires one side of the basement to be open?

9. The OSHA regulation at 29 C.F.R. §1910.109(c)(4)(iii)(*d*) requires employers to protect the flooring and handling areas in buildings against impregnation by ammonium nitrate and requires that the floors be constructed without open drains, traps, tunnels, pits, or pockets into which any molten ammonium nitrate could flow and be confined in the event of fire. What is the most likely reason OSHA mandates these structural features?

Metallic Nitrites and Metallic Nitrates

10. Why do nutritionists often advise individuals on low-sodium diets to avoid cured fish and meat products?
11. Write the equation that illustrates the thermal decomposition of lithium nitrate.

Metallic Peroxides

12. A chemical manufacturer in Denver, Colorado, intends to ship five 0.5-pound bottles of sodium peroxide to a customer in Dallas, Texas, by cargo aircraft. The five bottles are inserted into a fiberboard box and cushioned to prevent breakage.
    (a) What shipping description does DOT require the shipper to enter on the accompanying shipping paper? (DOT authorizes the transportation of sodium peroxide in the limited quantity of 15 kilograms by cargo aircraft.)
    (b) What labels and markings does DOT require the shipper to post on the box?

Fighting Fires Involving the Simple Oxidizers

13. When properly posted on the exterior wall of a building, what entry on an NFPA hazard diamond conveys that an oxidizer is stored therein to on-duty emergency responders?
14. Why is the use of carbon dioxide generally ineffective for extinguishing fires supported by oxidizers?

Transporting Oxidizers Other Than Organic Peroxides

15. A shipper offers a tighthead plastic drum containing 110 pounds (50 kg) of a sodium hypochlorite solution to a carrier as freight by rail.
    (a) Identify the manufacturer's markings required by DOT on the drum, assuming that it retained its integrity when 50 kilograms of material was contained therein and subjected to a hydrostatic test pressure of 200 kilopascals for Packing Group II. The drum was manufactured in 2009, and the manufacturer's registration number is SWA.
    (b) What markings does DOT require the shipper to enter on the drum?
16. A carrier contracts with a farmer to haul 2000 pounds (909 kg) of fertilizer-grade ammonium nitrate to a storage facility in a Class 2 light-duty truck. If the placard affixed to each side and each end of the truck resembles the following, has the carrier complied with DOT's placarding requirements?

Courtesy of Eugene Meyer.

KEY TERMS

466

OBJECTIVES

■ Associate the physical and health hazards of the organic compounds noted in this chapter with the information provided by their hazard diamonds and GHS pictograms.

■ Discuss the manner in which a carbon atom covalently bonds to nonmetallic atoms including other carbon atoms.

■ Describe the nature of carbon–carbon single bonds, carbon–carbon double bonds, and carbon–carbon triple bonds.

■ Describe the chemical bonds that exist in molecules of the hydrocarbons.

■ Illustrate that most hydrocarbons have structural isomers but only those with carbon–carbon double bonds have geometrical isomers.

■ Memorize and apply the rules for naming simple alkanes, cycloalkanes, alkenes, dienes, trienes, cycloalkenes, cyclodienes, cyclotrienes, and alkynes.

■ Describe the nature of the line markers required by DOT to identify the approximate locations of natural gas and petroleum transmission pipelines.

■ Identify the types of liquefied petroleum gas that are available for commercial use as bottled gas.

■ Identify the general nature of the labels required by the U.S. Federal Trade Commission on compressed natural gas and liquefied petroleum gas dispensers when these substances are provided to customers for potential use as alternative motor fuels.

■ Describe generally the manner in which acetylene is containerized in steel cylinders for storage and transportation.

■ Name the simple aromatic hydrocarbons and their derivatives when provided with their molecular structures and vice versa.

■ Identify the most common ways by which emergency responders are likely to be exposed to polynuclear aromatic hydrocarbons (PAHs).

■ Identify the petroleum products that are produced by the fractionation of crude petroleum.

■ Provide the molecular structures of the halogenated hydrocarbons having one and two carbon atoms per molecule and provide acceptable names for them.

■ Identify the primary risks associated with exposure to the halogenated hydrocarbons having one and two carbon atoms per molecule.

■ Describe the molecular nature of the most common chlorofluorocarbons.

■ Identify the principal ways in which chlorofluorocarbons and related substances were formerly used.

■ Describe the nature of the markings that EPA requires on containers and tanks of ozone-depleting substances.

- Describe the molecular structures of the polychlorinated biphenyls (PCBs).
- Identify the primary reason that PCBs are no longer manufactured in the United States.
- Identify the locations in a typical community where emergency responders are likely to encounter PCB transformers.
- Describe the nature of the markings that EPA requires on PCB-containing electrical equipment.
- Identify the labels, markings, and placards that DOT requires on packaging of the organic compounds noted in this chapter and the transport vehicles used for their shipment.

Organic compounds are constituents of many commercial products including heating and motor fuels, solvents, plastics, resins, fibers, surface coatings, refrigerants, textiles, and explosives. The fact that they are used so widely underscores the need to study them in some detail.

Many organic compounds of commercial interest are flammable gases or flammable liquids. It is hardly surprising to learn that many organic compounds are burning when they are encountered at transportation mishaps and other emergency scenes. Although the potential for fire and explosion is generally their primary hazard, exposure to organic compounds may also pose a health risk. Prolonged exposure to certain organic compounds causes a variety of acute and chronic health effects that includes damage to the liver, kidneys, and heart; depression of the central nervous system; and the onset of cancer.

This chapter is an introduction to the study of organic compounds and the properties of hydrocarbons and halogenated hydrocarbons that are commonly encountered commercially. Chapters 13, 14, and 15 discuss the properties of more complex organic compounds.

12.1 WHAT ARE ORGANIC COMPOUNDS?

The molecules of all organic compounds have one common feature: the presence of one or more carbon atoms. In most instances, the carbon atoms share electrons with other nonmetallic atoms. As shown by their molecular structures in Figure 12.1, methane, carbon tetrachloride, carbon monoxide, and carbon disulfide are examples of organic compounds having molecules in which the carbon atoms are bonded to other nonmetallic atoms.

Carbon atoms can also mutually share electrons with other carbon atoms. When the molecular structures of such compounds are examined, we find that two carbon atoms can share electrons to form any of the following: **carbon–carbon single bonds** (C—C); **carbon–carbon double bonds** (C=C); and **carbon–carbon triple bonds** (C≡C). Carbon–carbon single, double, and triple bonds consist of one, two, and three pairs of shared electrons, respectively, between two carbon atoms.

Figure 12.2 illustrates the bonding in molecules of ethane, ethylene, and acetylene, each of which has two carbon atoms. Molecules of ethane have carbon–carbon single bonds, molecules of ethylene have carbon–carbon double bonds, and molecules of acetylene have carbon–carbon triple bonds.

carbon–carbon single bond ■ A shared pair of electrons between two carbon atoms

carbon–carbon double bond ■ Two shared pairs of electrons between two carbon atoms

carbon–carbon triple bond ■ Three shared pairs of electrons between two carbon atoms

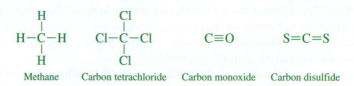

Methane   Carbon tetrachloride   Carbon monoxide   Carbon disulfide

FIGURE 12.1 A carbon atom may share its electrons with the electrons of hydrogen, chlorine, oxygen, and sulfur. The compounds that result from this electron sharing are methane, carbon tetrachloride, carbon monoxide, and carbon disulfide, respectively.

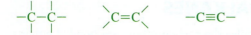

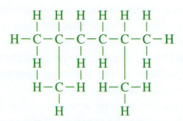

FIGURE 12.2 Two carbon atoms may share their electrons in one of three ways, resulting in the formation of carbon–carbon single bonds, carbon–carbon double bonds, and carbon–carbon triple bonds. When carbon atoms unite with hydrogen atoms, the compounds that result are ethane, ethene (ethylene), and ethyne (acetylene), respectively.

The covalent bonds between carbon atoms in the molecules of more complex organic compounds may be linked into chains, including branched chains, or into rings. Consider just the carbon skeleton of the following two molecules noted in the following illustration as (a) and (b):

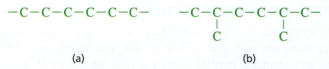

(a)   (b)

In (a), the carbon atoms are bonded to one another in a continuous chain, whereas in (b), the carbon atoms are bonded to one another in a branched pattern. This means that carbon atoms bond not only to other carbon atoms in long chains but also to groups of other carbon atoms as side chains attached to the main chain.

Aside from being bonded to other carbon atoms, each carbon atom displayed in either a straight- or a branched-chain manner is typically bonded to one or more atoms of hydrogen, oxygen, nitrogen, sulfur, or a halogen. For instance, when the carbon atoms in the compound having the carbon skeleton (a) are bonded to hydrogen atoms, the molecular structure of the compound is written as follows:

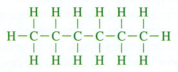

When the carbon atoms in (b) are bonded to hydrogen atoms, the molecular structure of the compound is written as follows:

$$
\begin{array}{c}
\ \ \ \ \ H\ \ H\ \ H\ \ H\ \ H\ \ H \\
\ \ \ \ \ |\ \ \ |\ \ \ |\ \ \ |\ \ \ |\ \ \ | \\
H-C-C-C-C-C-C-H \\
\ \ \ \ \ |\ \ \ \ \ \ \ |\ \ \ |\ \ \ \ \ \ \ | \\
\ \ \ \ \ H\ \ \ \ \ \ H\ \ H\ \ \ \ \ \ H \\
\ \ \ \ \ |\ \ \ \ \ \ \ |\ \ \ |\ \ \ \ \ \ \ | \\
\ \ \ \ H-C-H\ \ H-C-H \\
\ \ \ \ \ \ |\ \ \ \ \ \ \ \ \ \ \ | \\
\ \ \ \ \ \ H\ \ \ \ \ \ \ \ \ \ H
\end{array}
$$

There are many organic compounds that consist of continuous or branched chains of carbon atoms. The actual number of such compounds appears to be virtually unlimited. More than 6 million organic compounds are already known.

We begin the study of organic compounds by examining the simplest group, the **hydrocarbons**. These are compounds whose molecules are composed of only carbon and hydrogen atoms. All hydrocarbons are broadly divided into two groups: aliphatic and aromatic hydrocarbons. **Aliphatic hydrocarbons** are nonaromatic hydrocarbons. The **aromatic hydrocarbons** are a group of compounds characterized by a common feature of their molecular structures: They contain benzene rings. We examine this feature in more detail in Section 12.11.

hydrocarbon ■ Any compound whose molecules are composed solely of carbon and hydrogen atoms

aliphatic hydrocarbon ■ Any compound composed of molecules having only carbon and hydrogen atoms that are not arranged in benzene or a benzene-like structure

aromatic hydrocarbon ■ Any compound whose molecules are composed solely of carbon and hydrogen atoms, at least some of which are structurally similar to benzene

12.2 ALKANES AND CYCLOALKANES

alkane ■ Any hydrocarbon having molecules in which the carbon atoms are bonded solely as carbon–carbon single bonds (C—C) or (C—H) bonds

An **alkane** is a hydrocarbon in whose molecules the carbon atoms are bonded either to hydrogen atoms or to other carbon atoms solely by means of carbon–carbon single bonds. The general chemical formula of an alkane is C_nH_{2n+2}, where n is a nonzero integer. When the number of carbon atoms is only 1, the number of hydrogen atoms is 4. The corresponding compound is named methane. Its chemical formula is CH_4. When the number of carbon atoms is 2, the number of hydrogen atoms is 6. The corresponding compound is named ethane, and its chemical formula is C_2H_6.

saturated hydrocarbon ■ Any hydrocarbon having molecules that are composed solely of C—C and C—H bonds

The alkanes are called **saturated hydrocarbons**, because the four bonding electrons of each carbon atom are shared with the bonding electrons of four other atoms. They are said to be saturated with hydrogen. Several examples of the alkanes having from one to eight carbon atoms are noted in Table 12.1. Their names and formulas should be memorized.

12.2-A FORMULAS OF THE ALKANES

The molecular structures listed in Table 12.1 are straight-chain arrangements for the alkanes having from one to eight carbon atoms, but beginning with butane, several branched-chain arrangements may be written as well. Butane molecules have four carbon atoms per molecule, which can be represented by their Lewis structures as follows:

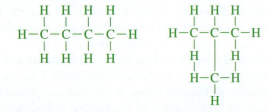

In the first structure, the four carbon atoms are bonded to one another in a continuous chain. In the second structure, only three carbon atoms are bonded in a continuous chain. The fourth carbon atom is bonded to the carbon atom in the middle of the chain.

structural isomerism ■ The phenomenon associated with compounds whose molecules have the same atoms but differ in the manner in which the atoms are structurally arranged

Because there are two correct ways to write molecular structures for the formula C_4H_{10}, they represent two distinct compounds. To distinguish them by name, the compound having the carbon atoms bonded in a continuous chain is called n-*butane*, whereas the compound with the branched structure is named *isobutane*.

The phenomenon associated with two or more compounds that have the same molecular formula but different structural arrangements of their atoms is called **structural isomerism**. One way by which structural isomers are identified is with an *n-* (for "normal") in front of the name of the compound that contains a continuous chain of carbon atoms, and *iso-* (meaning "the same") in front of the name of the compound that has a methyl group (CH_3-) bonded to a carbon atom next to the terminal (end) carbon atom. This system generally is referred to as the **common system of nomenclature**, and the names of the compounds are called **common names**.

common system of nomenclature ■ The system used historically for the naming of organic compounds

common name ■ The historical name for an organic compound

condensed formula ■ The formula of a compound in which symbols of the atoms are written next to the symbols of the atoms to which they are bonded and in which dashes are omitted or used only in a limited fashion

Although a molecular structure may be written for any alkane, it generally is convenient to condense it by simply writing the symbols of the atoms next to the symbol of the carbon atom to which they are bonded. When writing the formula for an alkane, the symbols of the hydrogen atoms are written next to the symbols of the carbon atoms to which they are bonded. The formulas derived in this manner are called **condensed formulas**. For example, the condensed formulas for *n*-butane and isobutane are noted here:

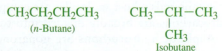

Condensed formulas convey the same bonding information as the more complete Lewis structures, but the dashes are either entirely omitted or used only in a limited fashion.

TABLE 12.1 | Simple Alkanes Having a Continuous Chain of Carbon Atoms

NAME	MOLECULAR FORMULA	LEWIS STRUCTURE	CONDENSED FORMULA
Methane	CH_4		CH_4
Ethane	C_2H_6		CH_3CH_3
Propane	C_3H_8		$CH_3CH_2CH_3$
Butane	C_4H_{10}		$CH_3CH_2CH_2CH_3$
Pentane	C_5H_{12}		$CH_3CH_2CH_2CH_2CH_3$
Hexane	C_6H_{14}		$CH_3CH_2CH_2CH_2CH_2CH_3$
Heptane	C_7H_{16}		$CH_3CH_2CH_2CH_2CH_2CH_2CH_3$
Octane	C_8H_{18}		$CH_3CH_2CH_2CH_2CH_2CH_2CH_2CH_3$
Nonane	C_9H_{20}		$CH_3CH_2CH_2CH_2CH_2CH_2CH_2CH_2CH_3$
Decane	$C_{10}H_{22}$		$CH_3CH_2CH_2CH_2CH_2CH_2CH_2CH_2CH_2CH_3$

12.2-B FORMULAS OF THE CYCLOALKANES

It is possible to write formulas for the alkanes in which the first and last carbon atoms in a continuous chain are chemically linked to each other in a cyclic arrangement. These compounds are called **cycloalkanes**, although in the petroleum industry, they are more commonly called *naphthenes*. Like the alkanes, the cycloalkanes are saturated hydrocarbons.

cycloalkane (naphthene)
■ Any hydrocarbon whose molecules are composed of a cyclic ring of carbon–carbon single bonds

The general chemical formula of the cycloalkanes is C_nH_{2n}. Chemists name them by placing the prefix *cyclo-* in front of the name of the parent hydrocarbon. Thus, the simplest cycloalkanes are cyclopropane, cyclobutane, cyclopentane, and cyclohexane.

Sometimes the structural formulas for cyclopropane, cyclobutane, and cyclopentane are represented by writing a triangle, square, and pentagon for C_3H_6, C_4H_8, and C_5H_{10}, respectively. The structural formula of cyclohexane (C_6H_{12}) is represented by two unique formulas called its *boat and chair conformations*.

| Cyclopropane | Cyclobutane | Cyclopentane | (boat) Cyclohexane (chair) |

In this text, the chair conformation solely is used to represent cyclohexane.

12.2-C THE IUPAC SYSTEM OF NOMENCLATURE

When a hydrogen atom is removed from an alkane, the resulting group is called an **alkyl group**, or **alkyl substituent**. Their general chemical formula is C_nH_{2n+1}. The most familiar examples are the **methyl group**, **ethyl group**, and ***n*-propyl group**.

$$-CH_3 \qquad -CH_2CH_3 \qquad -CH_2CH_2CH_3$$
Methyl group Ethyl group *n*-Propyl group

The names of the common alkyl substituents are provided in Table 12.2. These names and their molecular formulas should be memorized.

TABLE 12.2	Some Common Alkyl Substituents
NAME[a]	**CHEMICAL FORMULA**
Methyl	$-CH_3$
Ethyl	$-CH_2CH_3$, *or* C_2H_5-
n-Propyl	$-CH_2CH_2CH_3$, *or* C_3H_7-
Isopropyl	CH_3-CH-, *or* $(CH_3)_2CH-$ $\quad\quad\; |$ $\quad\quad CH_3$
n-Butyl	$-CH_2CH_2CH_2CH_3$, *or* C_4H_9-
Isobutyl	$\quad\quad\quad CH_3$ $\quad\quad\quad |$ $-CH_2- CH$, *or* $(CH_3)_2CHCH_2-$ $\quad\quad\quad |$ $\quad\quad\quad CH_3$
sec-Butyl[b]	$-CH-CH_2CH_3$, *or* CH_3CH_2-CH- $\quad |$ $\quad\quad\quad\quad\quad\quad\quad\quad |$ $\quad CH_3$ $\quad\quad\quad\quad\quad\quad\quad CH_3$
tert-Butyl[b]	$\quad\quad\quad CH_3$ $\quad\quad\quad |$ CH_3-C-, *or* $(CH_3)_3-C-$ $\quad\quad\quad |$ $\quad\quad\quad CH_3$
n-Pentyl[c]	$-CH_2CH_2CH_2CH_2CH_3$, *or* $C_5H_{11}-$

[a]The alkyl groups are named by replacing the *-ane* suffix of the parent hydrocarbon with *-yl*.
[b]The carbon atom to which a substituent is bonded may be identified as primary, secondary, or tertiary. If it is bonded to one other carbon atom, it is primary; if bonded to two carbon atoms, it is secondary (*sec-*); and if bonded to three carbon atoms, it is tertiary (*tert-*).
[c]Also called *n*-amyl.

alkyl group (alkyl substituent) ▪ Any group of atoms having the general chemical formula C_nH_{2n+1}, obtained by removing a hydrogen atom from the formula for an alkane

methyl group ▪ The designation for the group of atoms CH_3-

ethyl group ▪ The designation for the group of atoms CH_3CH_2-

***n*-propyl group** ▪ The designation for the group of atoms $CH_3CH_2CH_2-$

Although the simple hydrocarbons generally are referred to by their common names, the complex hydrocarbons are difficult to name using the common system of nomenclature. To overcome this hurdle, a second system of nomenclature was devised that is both simple and systematic. It was adopted by the International Union of Pure and Applied Chemistry (IUPAC) and is called the **IUPAC system of nomenclature**. The names of organic compounds derived from the use of this system are called **IUPAC names**.

In the IUPAC system, the following rules apply to the naming of alkanes from their molecular structures:

- Locate the longest chain of carbon atoms in the structure. This is called the "main chain." It is not necessary that the main chain be written horizontally to be continuous.
- Assign numbers consecutively to each carbon atom in the main chain starting from the end that gives the alkyl substituents attached to the chain the smaller numbers.
- Designate the position of each substituent by the number of the carbon atom along the main chain to which it is attached.
- Name the substituents alphabetically (ethyl before methyl, and so on) and place the names of the substituents as prefixes on the name of the main chain.
- If several substituents occur in the same compound, indicate the number of identical groups by the use of the following prefixes before the name of the substituent: *di-* for two identical groups, *tri-* for three identical groups, *tetra-* for four identical groups, and so on. The prefixes *di, tri, tetra, sec,* and *tert* are ignored when alphabetizing the substituents, but the prefixes *iso* and *cyclo* are not ignored.

The use of these IUPAC rules of nomenclature is illustrated in Table 12.3.

IUPAC system of nomenclature ■ The internationally recognized system used to name organic compounds

IUPAC name ■ The name of an organic compound derived from use of the IUPAC system of nomenclature

TABLE 12.3	Examples of Using the IUPAC Rules for Naming Alkanes When Their Chemical Formulas Are Known
ALKANE	**CHEMICAL FORMULA**
3-Methylheptane	$CH_3CH_2-CH-CH_3$ $\quad\quad\quad\quad\; \lvert$ $\quad\quad\quad\quad CH_2$ $\quad\quad\quad\quad\; \lvert$ $\quad\quad\quad\quad CH_2$ $\quad\quad\quad\quad\; \lvert$ $\quad\quad\quad\quad CH_2$ $\quad\quad\quad\quad\; \lvert$ $\quad\quad\quad\quad CH_3$
2,2-Dimethylbutane	$\quad\quad\; CH_3$ $\quad\quad\;\; \lvert$ $CH_3-C-CH_2CH_3$ $\quad\quad\;\; \lvert$ $\quad\quad\; CH_3$
2,4-Dimethylhexane	$CH_3-CH-CH_2-CH-CH_3$ $\quad\quad\;\; \lvert \quad\quad\quad\quad \lvert$ $\quad\quad\; CH_3 \quad\quad\; CH_2$ $\quad\quad\quad\quad\quad\quad\quad\; \lvert$ $\quad\quad\quad\quad\quad\quad\quad CH_3$
3-Ethyl-4-isopropylheptane	$\quad\quad\quad\; CH_3$ $\quad\quad\quad\;\; \lvert$ $CH_3-CH-CH-CH_2CH_2CH_3$ $\quad\quad\quad\quad\quad \lvert$ $\quad\; CH_3CH_2-CH-CH_2CH_3$
5-Ethyl-2,5-dimethylheptane	$\quad\quad\quad\quad CH_2CH_3$ $\quad\quad\quad\quad\;\; \lvert$ $CH_3CH_2-C-CH_2CH_2-CH-CH_3$ $\quad\quad\quad\quad \lvert \quad\quad\quad\quad\; \lvert$ $\quad\quad\quad\; CH_3 \quad\quad\quad\; CH_3$

Using the IUPAC system, name the alkane having the following condensed formula:

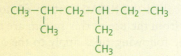

Solution: First, identify the longest continuous chain of carbon atoms. By examination, the longest chain contains six carbon atoms, which signifies that the compound is a derivative of hexane. Next, assign a number to each carbon atom of this longest chain.

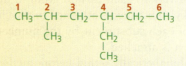

We see that the methyl group is bonded to the carbon atom numbered 2, and the ethyl group is bonded to the carbon atom numbered 4. Because the alkyl substituents are named in alphabetical order, the compound is correctly named 4-ethyl-2-methylhexane.

It is incorrect to number the carbon atoms from right to left along the horizontal chain of carbon atoms, because this numbering scheme gives larger numbers to the alkyl groups bonded to the main chain (5 for the methyl group and 3 for the ethyl group). However, it is correct to number the longest chain of carbon atoms in either of the following ways:

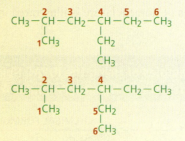

The use of either of these numbering schemes also gives 4-ethyl-2-methylhexane as the correct name of this compound.

The detailed analysis of a commercial rubber solvent provides the following approximate chemical composition:

(a) 50% by volume is a mixture of 2,3-dimethylbutane, 2,3-dimethylpentane, and 3,3-dimethylpentane;
(b) 30% is a mixture of 2,2-dimethylbutane, 3-methylpentane, 2,2-dimethylpentane, methylcyclopentane, methylcyclohexane, 1,2-dimethylcyclopentane, *n*-hexane, and *n*-heptane; and
(c) 20% is a mixture of *n*-butane, *n*-pentane, and 2,4-dimethylhexane.

Write the chemical formula for each substance.

Solution: The solvent's constituents have the following condensed formulas:

(a)

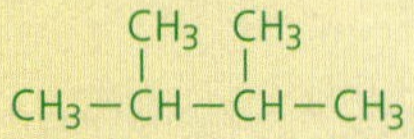

$$CH_3-CH-CH-CH_3$$ with CH_3 CH_3 branches
2,3-Dimethylbutane

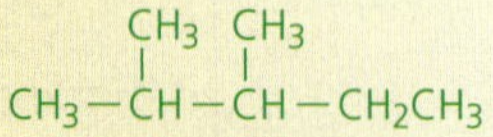

$$CH_3-CH-CH-CH_2CH_3$$ with CH_3 CH_3 branches
2,3-Dimethylpentane

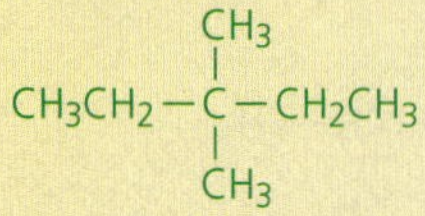

$$CH_3CH_2-C-CH_2CH_3$$ with CH_3 above and CH_3 below
3,3-Dimethylpentane

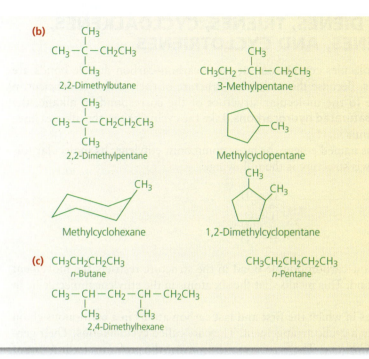

(b)

2,2-Dimethylbutane

3-Methylpentane

2,2-Dimethylpentane

Methylcyclopentane

Methylcyclohexane

1,2-Dimethylcyclopentane

(c) CH₃CH₂CH₂CH₃
n-Butane

CH₃CH₂CH₂CH₂CH₃
n-Pentane

2,4-Dimethylhexane

12.2-D TRANSPORTING ALKANES AND CYCLOALKANES

When shippers offer an alkane or cycloalkane for transportation, DOT requires them to enter the relevant shipping description on an accompanying shipping paper. Some examples for several representative alkanes and cycloalkanes are listed in Table 12.4. DOT also requires shippers and carriers to comply with all applicable labeling, marking, and placarding requirements.

TABLE 12.4	Shipping Descriptions of Some Representative Alkanes and Cycloalkanes
ALKANE OR CYCLOALKANE	**SHIPPING DESCRIPTION**
Butane	UN1011, Butane, 2.1
Cyclohexane	UN1145, Cyclohexane, 3, PG II
Ethane	UN1035, Ethane, 2.1
Methane, compressed	UN1971, Methane, compressed, 2.1
Methane, cryogenic	UN1972, Methane, refrigerated liquid, 2.1
Natural gas, compressed	UN1971, Natural gas, compressed, 2.1
Natural gas, cryogenic	UN1972, Natural gas, refrigerated liquid, 2.1
Petroleum gases, liquefied	UN1075, Petroleum gases liquefied, 2.1 *or* UN1075, Liquefied petroleum gas, 2.1
Propane	UN1978, Propane, 2.1

12.3 ALKENES, DIENES, TRIENES, CYCLOALKENES, CYCLODIENES, AND CYCLOTRIENES

alkene (olefin) ■ Any hydrocarbon having molecules that contain at least one carbon–carbon double bond (C=C)

unsaturated hydrocarbon ■ Any hydrocarbon whose molecules contain at least one carbon–carbon double bond or triple bond (C=C or C≡C)

Hydrocarbons whose molecules contain one or more carbon–carbon double bonds are called **alkenes**, or **olefins**. Because the molecular structure of each alkene is deficient in hydrogen atoms relative to the molecular structure of the corresponding alkane, the alkenes are said to be **unsaturated hydrocarbons**. Like the cycloalkanes, the alkenes have the general chemical formula C_nH_{2n}.

The simplest alkene is named ethene, or more commonly, ethylene. Its molecular formula is C_2H_4, and its Lewis structure is the following:

$$\begin{array}{c} H \quad\ H \\ \diagdown \quad \diagup \\ C=C \\ \diagup \quad \diagdown \\ H \quad\ H \end{array}$$

The presence of the carbon–carbon double bond in the structure restricts the movement of the atoms about the bond. This means that the six atoms in the ethylene molecule lie in the same plane.

cycloalkene ■ Any alkene whose carbon atoms are bonded to each other in a cyclic fashion

There are also alkenes in which the first and last carbon atoms in a continuous chain are joined to each other in a cyclic arrangement. They are called **cycloalkenes**. Their general chemical formula is C_nH_{2n-2}. They are named by placing the prefix *cyclo-* in front of the name of the parent alkene. The three simplest cycloalkenes are cyclobutene, cyclopentene, and cyclohexene. Their chemical formulas often are represented by using appropriate geometrical designs as shown here:

Cyclobutene   Cyclopentene   Cyclohexene

The molecules of alkenes can also have multiple carbon–carbon double bonds. When they have two and three double bonds, the compounds are called **dienes** and **trienes**, respectively.

diene ■ Any hydrocarbon whose molecules have two carbon–carbon double bonds

triene ■ Any hydrocarbon whose molecules have three carbon–carbon double bonds

12.3-A STRUCTURAL ISOMERISM IN ALKENES

Ethene and propene do not have structural isomers. For example, the condensed formula for propene can be correctly represented in either of the following ways:

$$CH_2=CH-CH_3 \qquad CH_3-CH=CH_2$$

Because either formula is the other one turned around end for end, it is a representation of the same substance.

However, we can write the following three condensed formulas for the structural isomers of the alkene having the formula C_4H_8.

$$CH_2=CHCH_2CH_3 \qquad CH_3CH=CHCH_3 \qquad \begin{array}{c} CH_3 \\ | \\ CH_2=C-CH_3 \end{array}$$

In the first two formulas, the carbon atoms are displayed in a continuous chain. The molecular structures differ only by the position of the carbon–carbon double bond. The third formula differs in that the carbon atoms are bonded to one another in a branched pattern.

12.3-B NAMING ALKENES, DIENES, TRIENES, CYCLOALKENES, CYCLODIENES, AND CYCLOTRIENES

Chemists use the *-ene* suffix to name alkenes. Consequently, in the IUPAC system, the first four members of the alkene series are named ethene, propene, butene, and pentene. Their common names are ethylene, propylene, butylene, and amylene, respectively.

In the IUPAC system, not only is the *-ene* suffix used to identify an alkene, but the position of the double bond in the main chain of its molecules also is indicated. The following rules are used to name alkenes:

- Consecutively number the carbon atoms in the main chain of continuous carbon atoms by beginning at the end that is *nearer* the double bond. The carbon–carbon double bond must always be included within the main chain of continuous carbon atoms.
- Indicate the position of the carbon–carbon double bond by the appropriate numerical prefix.
- Designate the position of each substituent by the number of the carbon atom along the main chain to which it is bonded.

For example, the compounds having the formulas $CH_2{=}CHCH_2CH_3$ and $CH_3CH{=}CHCH_3$ are named 1-butene and 2-butene, respectively. The compound having the formula $CH_2{=}\underset{\underset{CH_3}{|}}{C}{-}CH_3$

is named isobutene, isobutylene, or 2-methylpropene. The name used exclusively in commerce is isobutylene.

When dienes and trienes are named, the positions of the carbon–carbon double bonds are denoted with appropriate numbers, and *diene* or *triene* is used as the relevant suffix. The following examples illustrate the naming of a diene and a triene:

$$CH_2{=}CH{-}CH{=}CH_2 \qquad CH_3{-}CH{=}CH{-}CH{=}CH{-}CH{=}CH_2$$
1,3- Butadiene   1,3,5- Heptatriene

An alkyl derivative of a cycloalkene, cyclodiene, and cyclotriene is named by using one or more numbers to identify the location of the alkyl group relative to the location of the double bond. For example, the compounds designated by the following molecular structures are named 1-methylcyclopentene and 3-methylcyclopentene:

1-Methylcyclopentene   3-Methylcyclopentene

The atoms in the carbon–carbon double bond of a cycloalkene are always numbered 1 and 2. Consequently, as the name of a cycloalkane, methylcyclopentene may be preceded only by a 1-, 3-, or 4-.

Finally, just as there are alkyl groups, there are also **alkenyl groups**. The three most commonly encountered alkenyl groups are the **methylene group**, **vinyl group**, and **allyl group**.

$$-CH_2- \qquad CH_2{=}CH- \qquad CH_2{=}CH{-}CH_2-$$
Methylene group   Vinyl group   Allyl group

Although the vinyl and allyl groups are based on alkenes, the methylene group cannot contain a carbon–carbon double bond because it has only one carbon atom. The alkenyl groups are used in the common system to name simple organic compounds such as the chlorinated derivatives of methane, ethene, and propene.

$$CH_2Cl_2 \qquad CH_2{=}CH{-}Cl \qquad CH_2{=}CH{-}CH_2{-}Cl$$
Methylene chloride   Vinyl chloride   Allyl chloride
(Dichloromethane)   (Chloroethene)   (3-Chloropropene)

alkenyl group ■ Any noncyclic group of atoms having either the formula $-CH_2-$ or the general chemical formula C_nH_{2n-1} where $n \geqslant 2$

methylene group ■ The designation for the group of atoms $-CH_2-$

vinyl group ■ The designation for the group of atoms $CH_2{=}CH-$

allyl group ■ The designation for the group of atoms $CH_2{=}CH{-}CH_2-$

12.3-C GEOMETRICAL ISOMERISM IN ALKENES

geometrical isomerism
■ The phenomenon associated with compounds whose molecules have the same atoms but differ in the manner in which they are spatially bonded about a carbon–carbon double bond

The relative rigidity of the carbon–carbon double bond leads to a new kind of isomerism called **geometrical isomerism**. This phenomenon arises when the two substituents or atoms are different on each carbon atom that makes up the carbon–carbon double bond. For example, two Lewis structures can be written for 2-butene as follows:

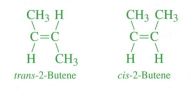

trans-2-Butene   *cis*-2-Butene

These structures illustrate that a hydrogen atom and a methyl group are bonded to each carbon atom in the carbon–carbon double bond. Because they are bonded to the same carbon atoms in each Lewis structure, they do not represent structural isomers, yet the arrangement of their atoms differs in space. When two compounds have the same structural formula but differ by the spatial arrangement of their atoms, they are called geometrical isomers.

Geometrical isomers are named by using either of the prefixes *trans*- or *cis*-, meaning on the opposite and same side of the carbon-carbon double bond, respectively. As previously shown, the two geometrical isomers of 2-butene are named *trans*-2-butene and *cis*-2-butene, respectively.

SOLVED EXERCISE 12.3

Using the IUPAC system, name the alkene having the following condensed formula:

$$CH_3CH_2-C=CH_2$$
$$|$$
$$CH_3CH_2$$

Solution: First, identify the longest continuous chain of carbon atoms that contains the carbon–carbon double bond. The longest chain contains four carbon atoms, which signifies that the compound is a derivative of butene. Assigning a number to each carbon atom from the right to the left of this longest chain, we see that the two carbon atoms in the carbon–carbon double bond are numbered 1 and 2, respectively.

$$\overset{4}{C}H_3\overset{3}{C}H_2-\overset{2}{C}=\overset{1}{C}H_2$$
$$|$$
$$CH_3CH_2$$

This means that the compound is a derivative of 1-butene. The ethyl group is bonded to the carbon atom numbered 2. The compound then is correctly named 2-ethyl-1-butene. (We do not assign a number to each carbon atom from the left to the right of the longest chain, because then the carbon atoms in the carbon–carbon double bond would be numbered 3 and 4. We always assign numbers to the carbon atoms so that the *smaller* numbers are assigned to the carbon atoms in the carbon–carbon double bond.)

12.3-D TRANSPORTING ALKENES, DIENES, TRIENES, CYCLOALKENES, CYCLODIENES, AND CYCLOTRIENES

When shippers offer an alkene, diene, triene, cycloalkene, cyclodiene, or cyclotriene for transportation, DOT requires them to enter the relevant shipping description on an accompanying shipping paper. Some examples for several representative compounds are listed in Table 12.5. DOT also requires shippers and carriers to comply with all other applicable labeling, marking, and placarding requirements.

TABLE 12.5	Shipping Descriptions of Some Representative Alkenes, Dienes, Trienes, and Cycloalkenes
ALKENE, DIENE, TRIENE, OR CYCLOALKENE	**SHIPPING DESCRIPTION**
Butadienes, stabilized[a]	UN1010, Butadienes, stabilized, 2.1
1-Butene[b]	UN1012, Butylene, 2.1
cis-2-Butene[b]	UN1012, Butylene, 2.1
trans-2-Butene[b]	UN1012, Butylene, 2.1
1,5,9-Cyclododecatriene	UN2518, 1,5,9-Cyclododecatriene, 6.1, PG III (Marine Pollutant)
Cycloheptatriene	UN2603, Cycloheptatriene, 3, PG II
Cyclohexene	UN2256, Cyclohexene, 3, PG II
Cyclopentene	UN2246, Cyclopentene, 3, PG II
Diisobutylene, isomeric compounds[c]	UN2050, Diisobutylene, isomeric compounds, 3, PG II
Ethylene	UN1962, Ethylene, 2.1
Hexadienes	UN2458, Hexadienes, 3, PG II
1-Hexene	UN2370, 1-Hexene, 3, PG II
1-Pentene	UN1108, 1-Pentene, 3, PG I
Propylene	UN1077, Propylene, 2.1

[a]Prior to their shipment, DOT requires the addition of a substance to 1,2- and 1,3-butadiene to inhibit their auto-polymerization and thereby promote their stabilization (Section 14.3).
[b]For purposes of DOT regulations, 1-butene, *cis*-2-butene, and *trans*-2-butene are designated "butylene."
[c]For purposes of DOT regulations, diisobutylene refers to a mixture of the compounds 2,4,4-trimethyl-1-pentene and 2,4,4-trimethyl-2-pentene.

12.4 ALKYNES

Hydrocarbons having one or more carbon–carbon triple bonds are called **alkynes**. Like the alkenes, they are unsaturated hydrocarbons. The general chemical formula of an alkyne is C_nH_{2n-2}.

alkyne ■ Any hydrocarbon whose molecules have at least one carbon–carbon triple bond ($C≡C$)

The simplest member of the alkyne series has the formula C_2H_2. It is named ethyne in the IUPAC system, but it is known more generally as acetylene, its common name. The Lewis structure of acetylene is denoted as follows:

$$H-C≡C-H$$

In the common system, alkynes are named as derivatives of acetylene. If an alkyne is represented by either of the general formulas $R-C≡C-H$ or $R-C≡C-R'$, the corresponding compound is named by identifying the alkyl groups, R and R', in their formulas. Thus, the derivatives of acetylene having one and two methyl groups are named methylacetylene and dimethylacetylene, respectively.

$$CH_3-C≡C-H \qquad CH_3-C≡C-CH_3$$
Methylacetylene   Dimethylacetylene

In the IUPAC system, the alkynes are named by replacing the *-ane* or *-ene* suffix on the associated alkane or alkene, respectively, with the suffix *-yne*. A numerical prefix is used to indicate the position of the carbon–carbon triple bond in the main chain of continuous carbon atoms. Consider the two alkyne isomers having the molecular formula

C_4H_6. The compound named 1-butyne is the isomer in which the carbon–carbon triple bond is located between a terminal carbon atom and the one immediately adjacent to it. The compound named 2-butyne is the C_4H_6 isomer in which the carbon–carbon triple bond is located between the two nonterminal carbon atoms.

$$H-C\equiv C-CH_2CH_3 \qquad CH_3-C\equiv C-CH_3$$

1-Butyne   2-Butyne

Although an alkyne may have structural isomers, it does not have geometrical isomers.

In the IUPAC system, the alkynes are named by using the following rules:

- Consecutively number the carbon atoms in the main chain of continuous carbon atoms by beginning at the end of the chain that is *nearer* the triple bond. The carbon–carbon triple bond must always be included within the main chain of continuous carbon atoms.
- Indicate the position of the triple bond by the appropriate numerical prefix.
- Designate the position of each substituent by the number of the carbon atom along the main chain to which it is bonded.

SOLVED EXERCISE 12.4

Using the IUPAC system, name the alkyne having the following condensed formula:

$$CH_3-CH-CH_2-C\equiv C-H$$
$$|$$
$$CH_2$$
$$|$$
$$CH_3$$

Solution: First, determine that the longest continuous chain of carbon atoms containing the carbon–carbon triple bond has six carbon atoms. This signifies that the compound is a derivative of 1-hexyne. Next, assign numbers to the carbon atoms from the right to the left and then downward. The two carbon atoms in the carbon–carbon triple bond are then numbered 1 and 2, respectively.

$$\overset{4}{C}H_3-\overset{3}{C}H-\overset{2}{C}H_2-\overset{}{C}\equiv\overset{1}{C}-H$$
$$|$$
$$5CH_2$$
$$|$$
$$6CH_3$$

Because the methyl group is bonded to the carbon atom numbered 4, the compound is correctly named 4-methyl-1-hexyne.

12.5 NATURAL GAS (METHANE)

The simple aliphatic hydrocarbons are flammable gases. They are most commonly encountered as domestic and industrial fuels. The simplest of them is methane, whose chemical formula is CH_4. Methane is the primary constituent of **natural gas**, roughly 70% by volume. The chemical industry uses natural gas as both a feedstock and an in-plant energy source.

Natural-gas-fired power plants are used to produce electricity. These plants now produce approximately 25% of the electricity used by Americans. Although the operation of natural-gas-fired power plants is not pollution-free, it is a more environmentally friendly practice when compared to the operation of either coal-fired or oil-fired power plants. Figure 12.3 illustrates that natural-gas fired power plants produce less pollution and fewer greenhouse gases.

The American electric utility industry's switch from the use of coal to natural gas substantially reduced the atmospheric carbon dioxide concentration during 2012.

natural gas ■ The flammable gas—primarily consisting of methane—formed in nature by the decomposition of animal and plant life

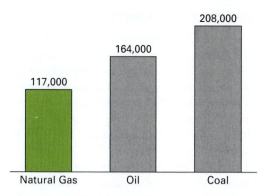

117,000 — Natural Gas
164,000 — Oil
208,000 — Coal

FIGURE 12.3 To generate the same amount of useful energy, natural-gas-fired power plants produce less pollution than either oil- or coal-fired plants. The emissions are noted as pounds of fossil fuel per billion Btu of energy produced. The contribution from natural gas, shown here in green, shows the environmental friendliness of natural gas when compared to oil and coal. [*Courtesy of U.S. Energy Information Administration and U.S. Department of Energy (Washington, DC), 2012.*]

Methane

Studies demonstrate that carbon dioxide emissions from energy generation in 2012 were the lowest since 1994.[1]

Methane constitutes 10% to 15% of the greenhouse gases linked to global warming. As noted in Table 5.3, its global warming potential over a 20-year time period is 72. This GWP reflects that methane is much more potent as a greenhouse gas than carbon dioxide.

Pure methane is an odorless, colorless, tasteless, and lighter-than-air gas. Some of its physical properties are provided in Table 12.6. Although odorless, when natural gas is encountered by emergency responders, it usually possesses a slightly offensive smell caused by the presence of sulfurous organic compounds that have been added intentionally to assist personnel attempting to detect gas leaks.

As shown by the data in Table 12.6, when methane is mixed with air, it burns at concentrations between approximately 5% and 15% by volume, producing a slightly luminous flame and releasing 958.1 Btu/ft^3 (39,820 kJ/m^3) to the surroundings. Fire and explosion are its primary hazards.

Methane can pose a health hazard as an asphyxiant (Section 10.3-A). Individuals who inhale the gas for a prolonged period lose consciousness because they are denied sufficient oxygen. Prolonged exposure constitutes a direct threat to life by suffocation.

Methane occurs primarily in nature as a result of the decay and alteration of animal and plant remains deep below Earth's surface. Consequently, it often accompanies nearby deposits of crude petroleum (Section 12.13-A), which forms by the same mechanism. Methane is also abundant in the atmosphere within coal mines, where it is called "fire-damp." When the mud at the bottom of stagnant pools, swamps, and rice paddies is disturbed, methane bubbles to the surface. Here, it is called "marsh gas."

TABLE 12.6	Physical Properties of Methane
Melting point	−297°F (−183°C)
Boiling point	−258°F (−161°C)
Specific gravity at 68°F (20°C)	0.42
Vapor density (air = 1)	0.553
Flashpoint	−366°F (−221°C)
Autoignition point	999°F (537°C)
Lower flammable limit	5% by volume
Upper flammable limit	15% by volume
Heat of combustion [32°F (0°C) and 1 atm]	958.1 Btu/ft^3 (39,820 kJ/m^3)

---

[1]U.S. Energy Information Administration (Washington, DC), 2013.

12.5-A METHANE IN THE ATMOSPHERE OF COAL MINES

As first noted in Section 7.6-C, methane is one of several constituents of coal gas. As coal is mined, the firedamp seeps from seams into the surrounding enclosed environment, where it accumulates and poses a serious risk of fire and explosion. The flammable range of methane is only 5% to 15% by volume in air. Even a small spark can cause it to ignite.

Worldwide, thousands of coal miners are killed annually, either directly when the flammable methane ignites within a mine, or by subsequently breathing the toxic atmosphere of carbon monoxide formed when methane and other gases burn incompletely. In the United States, using the legal authority of the Coal Act (Section 7.6-C), the Mine Safety and Health Administration (MSHA), a division of the U.S. Department of Interior, regulates health and safety conditions in coal mines by means of mandatory standards.

Safety standards within mines address the reduction of methane and other gases. To guard against the potentiality of hazardous incidents involving the presence of methane, carbon monoxide, and other gases within enclosed mines, the owners and operators of coal mines are required at 30 C.F.R. §75.330 "to provide ventilation [within mines] to dilute, render harmless, and to carry away flammable, explosive, noxious, and harmful gases, dusts, smoke, and fumes."

In 2010, the failure to comply with this regulation contributed to the deadliest coal mine accident experienced in the United States since 1970. An earth-shattering explosion occurred within several tunnels of a mine located in Beckley, West Virginia. Experts concluded that it was triggered by the ignition of accumulated methane and coal dust.[2] As a consequence of the explosion, 29 miners lost their lives, 19 from carbon monoxide poisoning.

12.5-B METHANE PRODUCTION BY THE DECOMPOSITION OF ORGANIC MATTER

methanogenesis ■ The biological production of methane by anaerobic bacteria

Methane is also produced when bacteria decompose organic wastes by means of the anaerobic process called **methanogenesis.** Hence, methane is the primary constituent of landfill gas and mammalian flatulence. It is also produced by cattle and other ruminants as they digest cellulose-like foods such as grass and straw. The culprit in these foods is actually lignin (Section 14.5-A), a group of complex compounds that exist in and between plant cell walls and provide the plant with its rigidity. The ruminants are unable to easily digest lignin. The process requires the help of microflora inhabiting their guts. Methane is produced as these microflora degrade the lignin, and the ruminants belch it into the atmosphere.

There are 1.8 billion ruminants worldwide. On average, a single cow exhales 634 quarts (600 L) of methane into the air daily. EPA has estimated that approximately 25% of the methane in the air originates with the belching of livestock. As amusing as this may first appear, the production of methane by livestock has become a serious matter. This fact has caused agricultural scientists to test ways of altering the volume of methane that livestock produce, primarily by providing ruminants with dietary adjustments.

The belching of livestock is not the only uncharacteristic way in which the atmospheric methane concentration increases. For example, methane has also been locked in the permafrost in Siberia since the last ice age, 10,000 years ago. The warming of planet Earth has caused the Siberian permafrost to thaw and slowly release its methane into the atmosphere.[3] Scientists estimate that approximately 9% of the methane emissions into the atmosphere originate in the Arctic tundra.

---

[2]J. Davitt McAteer, Upper Big Branch, Report to the Governor (May 2011).
[3]Natalia Shakhova et al., "Extensive methane venting to the atmosphere from sediments of the East Siberian Arctic Shelf," *Science*, Vol. 327 (2010), pp. 1246–1250.

12.5-C INDUSTRIAL SOURCES OF METHANE

The United States consumes approximately 25 trillion cubic feet (~0.71 m^3) of natural gas annually.[4] Approximately 85% now is obtained from domestic sources, with another 13% to 14% coming from Canada and Mexico by pipeline. It is also produced from crude petroleum and coal by a process called **cracking**. During cracking, methane is produced when more complex hydrocarbons are subjected to high temperatures under moderate pressure (**thermal cracking**) or relatively low temperatures in the presence of a catalyst (**catalytic cracking**).

Natural gas generally is retrieved from underground, porous reservoirs where it has accumulated for centuries. There is a vast wealth of natural gas worldwide, but it is locked in dense rock deep beneath Earth's surface where it often is difficult to retrieve. In the middle of the last decade, there was serious concern that America would experience a shortage in its supplies of natural gas. However, drillers perfected a method to extract natural gas from shale. The retrieval method is certain to boost U.S. gas production for decades to come, and is based on implementing a process called **hydraulic fracturing**, or **fracking**. During the process, drillers inject a mixture of water, sand, and chemical products under pressure deep into the ground to break up the shale rock formations and nearly impenetrable sands to create channels from which the entrained natural gas (and crude petroleum) may be recovered.

Shale is a fine-grained sedimentary rock composed mostly of clay and mud. Shale-rich regions exist in the Appalachian section of the northeastern United States, as well as in Texas and North Dakota. The largest known reservoir of natural gas in the United States is a geologic formation known as the Marcellus Shale. It underlies sizable portions of West Virginia, New York, Ohio, and Pennsylvania. The natural gas within this one shale may be as much as 490 trillion cubic feet (14,000 km^3).[5]

Natural gas is also available off U.S. shores beneath the seafloor. The U.S. Department of Interior estimates that the outer continental shelf holds 420 trillion cubic feet (12 trillion m^3) of unrecovered natural gas, which, if tapped, could heat American homes for the next 80 years. Although scientists perceive a benefit in harnessing this resource for future use by humankind, activists fear that environmental havoc would be wreaked by drilling operations that occur offshore in ultradeep waters.[6] Marine microorganisms produce approximately 4% of the methane that ultimately reaches the ocean surfaces and becomes a constituent of the atmosphere.

The prevailing pressure below the seafloor is approximately 30 times the atmospheric norm. Under these conditions, methane and water molecules coexist in a unique fashion: The methane molecules are trapped within "cages" composed of 5¾ water molecules. Many cages link together to form white crystals that physically resemble ice. Each unit of the substance, called **methane hydrate**, is represented by the formula $CH_4 \cdot 5.75H_2O(s)$. Methane hydrate is stable under pressure, but when brought to the ocean's surface, its units collapse and the methane wafts into the atmosphere.

cracking ■ The petroleum refining process associated with breaking down larger, heavier, and more complex hydrocarbon molecules into simpler and lighter molecules

thermal cracking ■ Any cracking process accomplished by the application of heat

catalytic cracking ■ Any cracking process accomplished in the presence of a catalyst

hydraulic fracturing (fracking) ■ A process that involves pumping water laced with sand and chemical products deep underground at high volumes and pressure to extract natural gas entrapped in shale

methane hydrate ■ Ice-encrusted natural gas located in porous rock in the arctic tundra under the permafrost and in the deep-ocean sediments of the continental slope and ocean floor

---

[4]U.S. Energy Information Administration (Washington, DC), 2013.

[5]For two major reasons, hydraulic fracturing has become a subject of intense environmental scrutiny. First, the chemical products used in the process include cancer-causing substances whose release into the environment has negatively impacted the quality of the groundwater in the regions where the processes are conducted. Second, hydraulic fracturing not only dislodges natural gas but also compounds of toxic metals like arsenic, barium, chromium, and zinc, all of which are natural components of shale. The fear is that these compounds could leach into the groundwater and taint it for future generations.

[6]Offshore drilling for natural gas and crude petroleum is also potentially replete with adverse environmental consequences. In 2010, the explosion of an offshore drilling rig in the Gulf of Mexico caused the deaths of 11 workers and the release of natural gas and crude petroleum through a massive pipeline leak. Approximately 57.1 million cubic feet (1.62 million m^3) of natural gas was flared off before the pipe could be capped. The hemorrhaging oil flowed toward the barrier islands, marshes, and beaches of the Gulf of Mexico, where it endangered regional wildlife, shut down large areas of commercial fishing, and threatened coastal tourism in four states.

Methane hydrate deposits are found not only below ocean floors, but also under the permafrost in areas where the climate is especially cold. Major discoveries of methane hydrate have been found in Alaska and Siberia.

12.5-D NATURAL GAS TRANSMISSION BY PIPELINE

Throughout many of the lower 48 states, large volumes of natural gas are directly transferred from petroleum fields or refineries by a network of pipelines until they ultimately are distributed to individual homes, apartment complexes, business buildings, chemical plants, and natural-gas-fired power plants. Approximately 60% of U.S. households use natural gas to provide the heat on the kitchen stovetop and to fire the home furnace, water heater, and clothes dryer.

As it moves through transmission pipelines, natural gas gradually loses its pressure and velocity. To overcome these problems, the gas is periodically passed through a number of compressor stations located along the pipeline. At each station, turbines recompress the natural gas to increase the rate at which it moves within the pipeline until reaching the next compressor station. The longest single network of natural-gas-transmission pipelines in the United States stretches a distance of over 13,000 miles (20,909 km), during which it passes through 100 compressor stations.

When utility and construction workers and emergency responders check for the possibility of pipeline damage or interference, they must quickly identify the location of natural gas pipelines. First, they generally spot signs like the following:

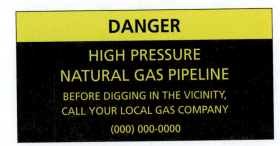

DANGER

HIGH PRESSURE
NATURAL GAS PIPELINE

BEFORE DIGGING IN THE VICINITY,
CALL YOUR LOCAL GAS COMPANY

(000) 000-0000

line marker ■ An aboveground sign indicating the approximate location of a natural-gas-transmission pipeline

Second, at 49 C.F.R. §195.410, DOT requires a **line marker** like that shown in (a) of Figure 12.4 to be posted at regular intervals along a pipeline, at road and railroad crossings, and at all aboveground facilities. The line marker identifies the approximate location of both aboveground and underground pipeline right-of-ways. Its posting is intended to alert

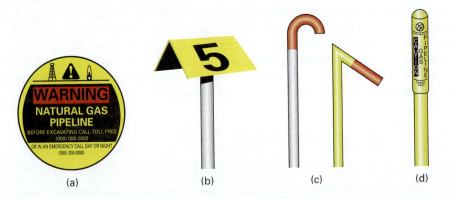

FIGURE 12.4 To alert workers and emergency responders of the presence of a natural-gas-transmission pipeline, DOT requires the posting of line markers along a pipeline route. (a) is a line marker posted near roads, railroads, and along pipeline right-of-ways. It contains the word WARNING followed by any of the words PIPELINE, GAS PIPELINE, or NATURAL GAS PIPELINE, the name of the operator, and the telephone number at which the operator may be reached at all times. The line marker colors are yellow, black, and red. (b) is an aerial line marker for use by pilots of pipeline patrol planes; (c) is a pipeline casing vent; and (d) is a painted metal or plastic post. The act of defacing, damaging, removing, or destroying any type of pipeline marker is a federal crime.

observers to exercise caution. Line markers do not indicate the exact location or depth of the pipeline, nor do they provide the pressure of the natural gas moving in the pipeline.

Line markers are positioned as follows:

■ The line marker for aboveground pipelines is placed and maintained along each section accessible to the public.

■ The line marker for underground pipelines is placed in right-of-way areas and at road, railroad, and waterway crossings.

Natural-gas-transmission pipelines themselves may represent hazards due to leaks caused by natural disasters, damage by digging and excavation, surface corrosion, and other troubles. In the United States, approximately half of them, 250,000 miles (402,000 km), are steel or cast-iron pipes that were installed before 1970. Prior to their installation, they were not generally coated or otherwise provided with anticorrosion protection, nor were they equipped with automatic shut-off valves. Consequently, when they rupture, the release of natural gas often must be stopped manually. Because many pipelines that were installed four and five decades ago now are badly corroded, they could rupture catastrophically at any time.

These pipelines resemble ticking time bombs, as noted during the past decade when ruptured natural-gas-transmission pipelines released massive fireballs resembling BLEVEs (Section 3.8-A). In 2010, for example, the rupture of a segment of a 30-inch-diameter pipeline in a residential area of San Bruno, California caused the release of approximately 47.6 million cubic feet (1.4 million m^3) of natural gas to the environment.[7] The gas ignited and burned for 95 minutes before a public utility employee could operate the manual valves that stopped the gas flow. The pipeline segment had been installed in 1956, and nearby segments had been installed in 1948. The fire caused eight fatalities and numerous injuries as well as the destruction of 38 homes and heavy damage to another 70 homes. Emergency responders could only attack the secondary fires ignited by the blaze.

The rupturing of a main transmission pipeline is not the sole hazard associated with delivering natural gas to the spot where it is needed. Before it is distributed to furnaces and appliances within buildings, natural gas first meanders within pipelines that were placed through shafts and ceilings, under floors, and around gas boilers. When the pipelines deteriorate, crack, or are otherwise damaged, natural gas leaks from them into the surrounding environment, where it may accumulate and pose the risk of fire and explosion. The detection of these leaks is the responsibility of the local gas utility company, but firefighters usually are asked to assist in accessing areas and preventing explosions.

Leaking natural gas may be initially identified by its odor, a hissing or roaring sound, discolored vegetation surrounding the pipeline, or water or dirt blowing into the air. However, when leaks of natural gas originate from buried pipes, the odorant may adsorb to particulates of the surrounding soils. Consequently, odor alone should never be used to determine the source or intensity of natural gas leaks.

A more practical method for detecting natural gas leaks involves the use of monitors that can be aimed at a buried pipeline to detect a natural gas leak or assess where a buried pipeline was accidentally damaged during an excavation operation. To detect the location of the leaks, emergency responders aim them from a safe distance within buildings at suspect areas like holes or vents, monitor the methane concentration, and assess whether the area may be safely accessed.

Natural-gas-distribution pipelines appear to be buried almost everywhere in residential areas, not just beneath the surface of streets and alleys. Firefighters often accompany gas company employees to scenes at which these pipelines have been inadvertently damaged during a routine landscaping or other soil-excavation operation. Damage to the pipeline may cause a natural gas leak that has the potential to produce service outages, evacuations, ignition, property damage, injury, or loss of life.

---

[7]Pipeline Accident Report, "San Bruno, California, Natural Gas Pipeline Explosion and Fire, September 9, 2010," NTSB No. PAR-11-01, PB2011-916501 (Washington, DC: National Transportation Safety Board, 2011).

To reduce or eliminate the occurrence of such incidents, local gas companies and local ordinances often require the following:

- Contact the local gas company before starting to excavate.
- Wait until a trained technician from the company has flagged the location of the pipeline (usually within less than 24 hours).
- Maintain the pipeline marks and avoid the flagged areas when excavating.
- Contact the company again if the pipeline is accidentally damaged during an excavation operation so that the damage can be professionally assessed. Even a minor dent or scraping of the pipeline may cause future leaks if it remains unrepaired.

A natural gas pipeline may be located near a sewer line. When the pipeline is damaged, a leak of natural gas may percolate through the subsurface soil and enter the sewer system. Then, bubbles of natural gas are observed rising through standing water on the surface or even in toilet bowls. The odor of natural gas may also be detected when using a rooter device to clear an obstruction or blockage in the sewer line.

When a natural gas leak is apparent or suspected anywhere, gas companies recommend that individuals comply with the following practices:

- Leave the area immediately by foot and leave the doors open.
- Do not light a match, start or stop an engine, use a telephone, turn light switches or appliances on or off, or perform any other activity that may generate a spark.
- Warn others to stay away from the scene.
- Contact 911 or the gas company from a distance of at least 1000 feet (300 m) from the leak.
- Do not attempt to extinguish a natural gas fire, but call the fire department.
- Do not attempt to operate pipeline valves.

Gas companies also recommend that residents and workers know the location of the incoming natural gas shutoff valves into their homes and other buildings so this information can be rapidly conveyed to first-on-the-scene responders when necessary.

12.5-E BULK SOURCES OF METHANE

compressed natural gas (CNG) ■ Methane as a compressed gas

liquefied natural gas (LNG) ■ Methane as a cryogenic liquid

Aside from its transference by pipeline, natural gas is also transported as **compressed natural gas**, or **CNG**, and the cryogenic fluid called **liquefied natural gas**, or **LNG**. Compressed natural gas is an alternative motor fuel whose use to power American motor vehicles has been strongly encouraged since at least the mid-2000s. When compared to motor gasoline, CNG is a desirable vehicular fuel, not only because its combustion provides a high heat value but because fewer volatile organic compounds (Figure 7.1-J) are components of the vehicle's exhaust. Given this positive environmental outlook, compressed natural gas has been chosen by several American municipalities for fueling an entire fleet of buses, taxi cabs, refuse trucks, and other city-owned vehicles. Businesses, too, like United Parcel Service, use CNG instead of diesel oil to power their surface vehicles. When vehicles are powered with CNG, the fuel is stored in cylinders located in the trunk area, under the side panel of a van or school bus, on the frame, or in the bed of a truck.

During the first decade of the twenty-first century, it first became popular to transport liquefied natural gas over long distances in large, cylindrical tanks on specially constructed refrigerated ships. Transporting LNG is more cost-effective compared to transporting CNG. As noted earlier in Table 2.10, when methane is cooled to −258°F (−161°C), the liquid volume is reduced to 1/650th of its gaseous volume. By cooling natural gas to a temperature equal to or less than this value, the gas is converted into LNG.

LNG is now transported from foreign plants in Nigeria, Trinidad, and Tobago to domestic ports in the United States. It then is transported on land by means of refrigerated tank trucks, delivered to regasification terminals where it is converted to CNG, or transferred as the gas by pipeline to its destination. It is also stored until needed in double-walled insulated tanks.

When shippers transfer CNG by pipeline, or transport CNG or LNG (or other flammable gases) by any means, DOT requires at 49 C.F.R. §192.625 that it be odorized by the addition of ethyl mercaptan, thiophane, or amyl mercaptan.

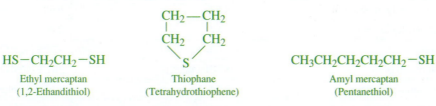

$$\text{HS}-\text{CH}_2\text{CH}_2-\text{SH}$$

Ethyl mercaptan
(1,2-Ethanedithiol)

Thiophane
(Tetrahydrothiophene)

$$\text{CH}_3\text{CH}_2\text{CH}_2\text{CH}_2\text{CH}_2-\text{SH}$$

Amyl mercaptan
(Pentanethiol)

These sulfurous organic compounds have such offensive odors that their detection is helpful when attempting to locate gas leaks.

At 16 C.F.R. §306.12, the U.S. Federal Trade Commission requires retail distributors to affix a black-and-orange label that resembles the following on CNG dispensers:

CNG

MINIMUM
90%
METHANE

This label identifies compressed natural gas as the alternative motor fuel and provides the minimum methane fuel rating as 90% by volume. The commission also requires new-vehicle manufacturers and used-vehicle dealers to affix the label on a visible surface of each vehicle powered by CNG.

12.6 LIQUEFIED PETROLEUM GAS

Most people are first introduced to propane or *n*-butane as the substances used to fuel a backyard barbeque grill. As noted earlier, propane and *n*-butane are alkanes having the chemical formulas C_3H_8 and C_4H_{10}, respectively. The liquid mixture of these simple hydrocarbons is called **liquefied petroleum gas**, or **LPG**. Although largely produced as a fuel, its constituents also are used as propellants in aerosol cans.

liquefied petroleum gas (LPG) ■ Any of three compressed commercial products consisting of a mixture of liquefied propane and butane

There are three types of LPG: a propane/butane mixture, commercial propane, and commercial butane. The propane/butane mixture consists of approximately 60% propane and 40% butane by volume; commercial propane is approximately 92% propane; and commercial butane is approximately 85% *n*-butane. Although the main constituents of LPG are propane and *n*-butane, small amounts of ethane, ethene, propene, butene, isobutane, isobutene, and isopentene also may be present. For commercial use, LPG is generally isolated as a by-product from the rectifiers used for treating natural gas.

At most ambient conditions, propane and *n*-butane are colorless, odorless, and tasteless gases, but DOT and OSHA require the addition of an odorant to each type so that leaks may be easily detected. Additional physical properties of propane and *n*-butane are noted in Table 12.7. These data show that when propane and *n*-butane burn, they evolve an amount of heat ranging from 2550 Btu/ft^3 (101,000 kJ/m^3) to 3200 Btu/ft^3 (133,000 kJ/m^3). Consequently, they are highly desirable as domestic and industrial fuels.

The data in Table 12.7 also show that propane and *n*-butane readily liquefy under moderate pressure. Hence, when pressurized in steel cylinders, each of the three types of LPG exists as a liquefied compressed gas. These data also show that *n*-butane liquefies near the freezing point of water. Consequently, its use as a fuel is limited when the temperature of the surroundings is colder than 31°F (−0.5°C).

LPG

TABLE 12.7	Physical Properties of Propane and *n*-Butane	
	PROPANE	***n*-BUTANE**
Melting point	−305°F (−187°C)	−216°F (−138°C)
Boiling point	−49°F (−45°C)	31°F (−0.5°C)
Specific gravity at 68°F (20°C)	0.58	0.60
Vapor density (air = 1)	1.52	2.04
Flashpoint	−156°F (−104°C)	−76°F (−60°C)
Autoignition point	874°F (468°C)	761°F (405°C)
Lower flammable limit	2.2% by volume	1.9% by volume
Upper flammable limit	9.5% by volume	8.5% by volume
Heat of combustion [32°F (0°C) and 1 atm]	2550 Btu/ft^3 (101,000 kJ/m^3)	3200 Btu/ft^3 (133,000 kJ/m^3)

LPG and its constituents pose a health hazard as an asphyxiant (Section 10.3-A). Individuals who inhale them for a prolonged period lose consciousness because they are denied sufficient oxygen. Prolonged exposure to LPG or either of its constituents poses a potential threat to life by suffocation.

SOLVED EXERCISE 12.5

Why is propane generally more suitable than butane for fueling mobile appliances and heaters during very cold weather?

Solution: It is the gaseous state of a fuel that ignites and burns. The data in Table 12.7 indicate that propane is a gas at temperatures equal to or greater than −49°F (−45°C), whereas butane is a gas at temperatures equal to or greater than 31°F (−0.5°C). These facts severely limit the use of butane as a fuel in cold climates because it remains liquid. There are few areas of the world where temperatures colder than −49°F (−45°C) are experienced; consequently, propane generally is more suitable than butane for fueling mobile appliances and heaters.

12.6-A BOTTLED GAS

The commercial types of LPG usually are shipped under pressure by motor or rail tankcar from petroleum gas wells and refineries to filling stations, where they are transferred into a bulk storage tank. Distributors then transfer them into horizontal or vertical dispensers from which they are further dispensed into far smaller portable, thick-walled cylinders and tanks. The contents then are said to be "bottled" and the gas is commonly referred to as **bottled gas**. The gas cylinders and tanks are convenient to deliver to rural areas and other locations where natural gas cannot be supplied economically by pipeline. They are always stored outdoors, because local codes and ordinances prohibit indoor storage.

Care must be exercised in storing portable LPG tanks. When it is rapidly subjected to intense heat, they can rupture. The tank may also be struck by lightning or be exposed to another ignition source. In these instances, the tank contents immediately ignite as a fireball as they expand into the atmosphere.

LPG is also transported by truck to fill permanently installed tanks. These tanks are always situated at prescribed distances from homes and other major buildings and connected by means of copper tubing to appliances and furnaces located indoors.

The largest volumes of LPG are found at LPG-filling stations, which are located in virtually every major city. At these filling stations, the LPG generally is contained within a bulk storage tank having a maximum capacity of less than 90,000 gallons (340 m^3).

bottled gas ■ Broadly, any gas stored under pressure in a portable gas cylinder, but more specifically, propane, butane, or a propane/butane mixture stored under pressure in a portable gas cylinder

12.6-B LPG AS AN ALTERNATIVE MOTOR FUEL

LPG can also be used as an alternative motor fuel. In 2008, Hyundai's *Elantra* was manufactured as the world's first LPG-fueled car.

At 16 C.F.R. §306.12, the U.S. Federal Trade Commission requires retail LPG distributors to affix orange-and-black labels on their LPG dispensers that resemble the two shown here:

These labels identify the alternative motor fuel as liquefied petroleum gas and provide the minimum percentage by volume as 90% propane (on the left) and 90% propane and 2% butane (on the right). The U.S. Federal Trade Commission also requires new-vehicle manufacturers and used-vehicle dealers to affix the label on a visible surface of each vehicle that is powered by LPG.

SOLVED EXERCISE 12.6

What special actions should firefighters implement when responding to the scene of a car crash involving a LPG-fueled vehicle?

Solution: Once the vehicle has been stabilized and secured against movement, firefighters should first ascertain that it is powered by LPG. The nature of the vehicular fuel may be readily identified by locating the label affixed to the vehicle. Having determined that the vehicular fuel is LPG, first-on-the-scene responders should then turn the gas cylinder valve handle to the "off" position and search for signs of fuel leaking from ruptured connection fittings and lines. Firefighters should extinguish all nearby fires and remove all potential sources of heat. The use of flares should be avoided although nonsparking markers or cones may be used to cordon off a hazard zone. Most importantly, firefighters should bear in mind that when LPG cylinders are exposed to intense heat, they will rupture and catastrophically release their contents to the environment as a BLEVE (Section 3.8-A). This situation should be avoided at all costs.

Ethylene   Propylene

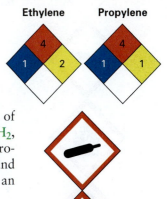

12.7 ETHYLENE AND PROPYLENE

Ethylene and propylene, or ethene and propene, respectively, are the common names of the simplest alkenes. As noted earlier, their chemical formulas are C_2H_4, or $CH_2{=}CH_2$, and C_3H_6, or $CH_3{-}CH{=}CH_2$, respectively. The physical properties of ethylene and propylene are provided in Table 12.8. The tabulated information indicates that ethylene and propylene pose the risk of fire and explosion when either gas is discharged within an enclosed area.

Ethylene and propylene are produced in the United States in larger amounts than all other substances. In the chemical industry, they are used primarily for the manufacture of polyethylene and polypropylene (Sections 14.6-A and 14.6-B), but they are also used as raw materials for the manufacture of a variety of other chemical products.

Ethylene is also used outside the chemical industry. For example, in the agricultural field, it is used as both a plant regulator and an herbicide:

■ Ethylene performs as a plant regulator by hastening the uniform ripening of apples, bananas, berries, figs, tomatoes, and citrus fruits; the stimulation of seed germination and

TABLE 12.8	Physical Properties of Ethylene and Propylene	
	ETHYLENE	**PROPYLENE**
Melting point	−272°F (−169°C)	−301°F (−185°C)
Boiling point	−155°F (−104°C)	−54°F (−48°C)
Specific gravity at 68°F (20°C)	0.001	0.51
Vapor density (air = 1)	1.0	1.5
Flashpoint	−213°F (−135°C)	−162°F (−108°C)
Autoignition point	842°F (450°C)	860°F (460°C)
Lower flammable limit	3.1% by volume	2% by volume
Upper flammable limit	32% by volume	11.1% by volume
Heat of combustion [32°F (0°C) and 1 atm]	21,600 Btu/lb (50,300 kJ/kg)	21,000 Btu/lb (48,900 kJ/kg)

sprouting; the curing of tobacco leaves; the production of flowers in pineapples; and the normal thinning of flowers, leaves, and fruit from trees as they mature.

■ Ethylene performs as an herbicide by controlling the spread of witchweed in corn, cotton, peanut, and soybean fields.

Ethylene and propylene can pose health hazards as asphyxiants. Individuals who inhale either gas for a prolonged period lose consciousness because they are denied sufficient oxygen. Prolonged exposure to these gases poses a potential threat to life by suffocation.

1,3-Butadiene

12.8 BUTADIENE

1,3-Butadiene is a colorless, highly flammable gas with a mild gasoline-like odor. Its physical properties are provided in Table 12.9. The gas often is manufactured at petroleum refineries by the dehydrogenation of 1-butene and 2-butene.

$$CH_3CH_2CH=CH_2(g) \longrightarrow CH_2=CH-CH=CH_2(g) + H_2(g)$$

      1-Butyne   1,3-Butadiene   Hydrogen

$$CH_3CH_2=CHCH_3(g) \longrightarrow CH_2=CH-CH=CH_2(g) + H_2(g)$$

      2-Butyne   1,3-Butadiene   Hydrogen

TABLE 12.9	Physical Properties of 1,3-Butadiene
Melting point	−164°F (−109°C)
Boiling point	23°F (−4.7°C)
Specific gravity at 68°F (20°C)	1.88
Vapor density (air = 1)	1.97
Flashpoint	−105°F (−76°C)
Autoignition point	804°F (429°C)
Lower flammable limit	2% by volume
Upper flammable limit	11.5% by volume

1,3-Butadiene is a very reactive substance. It even reacts spontaneously with itself, molecule for molecule, to form polybutadiene (Section 14.11-B). To prevent the reaction, the gas is mixed with an inhibitor during its production.

Commercially, the product is called butadiene, without use of the numerical prefixes. It often is transferred from production plants by means of pipelines to nearby petrochemical plants where the gas is used to manufacture various types of synthetic rubbers for automobile tires, appliance parts, pipes, and synthetic fibers.

1,3-Butadiene is produced as a product of the incomplete combustion of petroleum products, wood, and tobacco. Individuals who live or work in or near industrial areas in which butadiene is produced or used are more likely than the general population to breathe butadiene. Because butadiene is formed during the incomplete combustion of petroleum products, individuals who routinely breathe vehicular exhaust are also at risk of contracting cancer. In the workplace, OSHA requires employers to limit employee exposure to a maximum butadiene concentration of 1 part per million, averaged over an 8-hour workday.

When individuals breathe 1,3-butadiene, they experience nausea, eye, nose, and throat irritation, headache, and decreased blood pressure and pulse rate. Long-term exposure to 1,3-butadiene increases the risk of contracting stomach, blood, and lymphatic-system cancers.[8] Based on these studies, epidemiologists rank 1,3-butadiene as a human carcinogen.

Because 1,3-butadiene is one of several carcinogens formed during the incomplete combustion of tobacco,[9] smokers are exposed to 1,3-butadiene when they inhale tobacco smoke. Because it is highly reactive, scientists believe that the substance poses by far a more significant cancer risk to smokers and individuals exposed to secondhand smoke compared to the other carcinogenic components of tobacco smoke.

Acetylene

12.9 ACETYLENE

Acetylene is the simplest alkyne and the sole member of this class of compounds that has commercial importance. As noted earlier, its chemical formula is C_2H_2, or $H-C\equiv C-H$. The physical properties of acetylene are provided in Table 12.10.

TABLE 12.10	Physical Properties of Acetylene
Melting point	Sublimes
Boiling point	−118°F (−83°C)
Specific gravity at 68°F (20°C)	0.91
Vapor density (air = 1)	0.899
Flashpoint	0°F (−18°C)
Autoignition point	635°F (335°C)
Lower flammable limit	3% by volume
Upper flammable limit	82% by volume
Heat of combustion [32°F (0°C) and 1 atm]	1201 Btu/ft^3 (49,900 kJ/m^3)

---

[8]N. Sathiakumar *et al.*, "Mortality from cancer and other causes of death among synthetic rubber workers," *Occup. Environ. Med.*, Vol. 55 (1998), pp. 230–235.
[9]Report of the Surgeon General, *How Tobacco Smoke Causes Disease: The Biology and Behavioral Basis for Smoking-Attributable Disease*, U.S. Centers for Disease Control and Prevention, Atlanta, Georgia (2010) (ISBN-13: 978-0-16-084078-4).

In the United States, acetylene is produced mainly by cracking natural gas.

$$2CH_4(g) \longrightarrow C_2H_2(g) + 3H_2(g)$$

Methane   Acetylene   Hydrogen

dehydrogenation ▪ A chemical process typically conducted at a high temperature and pressure and during which molecular hydrogen is removed from a compound

In this instance, the cracking involves **dehydrogenation**. Although pure acetylene is a colorless gas with an ethereal odor, industrial-grade acetylene can be foul-smelling. As noted in Section 9.7-B, industrial-grade acetylene sometimes is produced by hydrolyzing calcium carbide. Because calcium carbide generally is contaminated with calcium phosphide, the foul-smelling phosphine is simultaneously produced.

Acetylene is an innately unstable substance, decomposing at an explosive rate into its elements when it has been compressed in excess of approximately 2 atmospheres (203 kPa).

$$C_2H_2(g) \longrightarrow 2C(s) + H_2(g)$$

Acetylene   Carbon   Hydrogen

The decomposition is especially likely when the compressed acetylene has been subjected to thermal or mechanical shock.

Not only is acetylene flammable and innately unstable, but it also reacts with compounds containing the copper(I) ion, especially in moist air. The product of this chemical reaction is copper(I) acetylide, which when dry, can decompose explosively.

$$H-C\equiv C-H(g) + Cu^+(s) \longrightarrow H-C\equiv C-Cu(s) + H^+(aq)$$

Acetylene   Copper(I) ion   Copper(I) acetylide   Hydrogen ion

To reduce or eliminate the potential for this explosive reaction, welders and other individuals who work with acetylene avoid using copper fittings or tubing on acetylene cylinders.

When sparked, a mixture of acetylene and oxygen burns with an intensely hot flame, reaching temperatures as high as 5400°F (2982°C). This combustion process yields 21,400 Btu/lb (49.9 kJ/g), which is put to use when welding and cutting steel and cladding metals.

Careless welding practices have caused many major fires. Heat that is sufficient to weld steel is capable of triggering the ignition of many other commonly encountered flammable materials. When acetylene is used to generate heat for welding, the risk that its heat of combustion will be transmitted to nearby flammable materials should always be acknowledged.

To counteract its potential decomposition, a special method is employed to safely compress and store acetylene within steel cylinders. This procedure takes advantage of the solubility of acetylene in acetone. One volume of acetone dissolves approximately 25 volumes of acetylene at 1 atmosphere (101.3 kPa) and 300 volumes at 12 atmospheres (1216 kPa). By dissolving acetylene in acetone, gas manufacturers increase the amount that may be safely compressed in cylinders by means of the three-step process shown in Figure 12.5. First, they pack the cylinders with a porous medium consisting of a mixture of monolithic filler and balsa wood. Then, they saturate this medium with liquid acetone. Finally, they charge acetylene at a prescribed pressure into the cylinders, where it dissolves in the acetone. As acetone solutions, acetylene typically is compressed in steel cylinders of various sizes whose volumes range from 9.9 cubic feet (0.28 m^3) to 390 cubic feet (11.04 m^3).

Acetylene poses a health hazard as an asphyxiant. Individuals who inhale acetylene for a prolonged period lose consciousness because they are denied sufficient oxygen for normal respiration. Prolonged exposure to the gas poses a potential threat to life by suffocation.

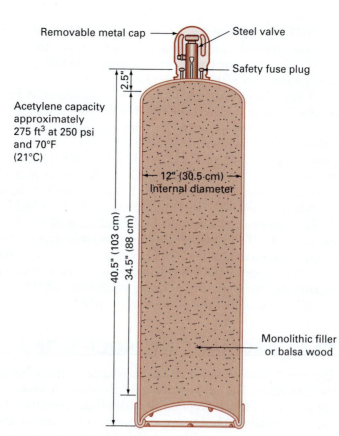

Removable metal cap — Steel valve

Safety fuse plug

2.5"

Acetylene capacity
approximately
275 ft³ at 250 psi
and 70°F
(21°C)

12" (30.5 cm)
Internal diameter

40.5" (103 cm)
34.5" (88 cm)

Monolithic filler
or balsa wood

FIGURE 12.5 This cutaway of an acetylene cylinder illustrates its unique features. The monolithic filler or balsa wood is a porous material that is charged with acetone, into which the acetylene subsequently dissolves.

12.10 TRANSPORTING THE SIMPLE GASEOUS HYDROCARBONS

When shippers offer acetylene, butadiene, ethylene, methane, or propylene for transportation, DOT requires them to identify the chemical commodity as shown in Table 12.11 on an accompanying shipping paper. All labeling, marking, and placarding requirements apply.

TABLE 12.11	Shipping Descriptions of the Simple Gaseous Hydrocarbons
HYDROCARBON	**SHIPPING DESCRIPTION**
Acetylene	UN1001, Acetylene, dissolved, 2.1
Butadiene	UN1010, Butadiene, stabilized, 2.1
Butane	UN1011, Butane, 2.1
Ethylene	UN1962, Ethylene, 2.1
Liquefied petroleum gas	UN1075, Petroleum gases liquefied, 2.1 *or* UN1075, Liquefied petroleum gas, 2.1
Methane, compressed	UN1971, Methane, compressed, 2.1
Methane, cryogenic	UN1972, Methane, refrigerated liquid, 2.1
Propane	UN1978, Propane, 2.1
Propylene	UN1077, Propylene, 2.1

DOT also requires the issuance of a hazardous materials safety permit (Section 6.10) to motor carriers before they transport methane as either a compressed gas or a cryogenic liquid with a methane content of at least 85% in bulk packaging having a capacity equal to or greater than 3500 gallons (13,248 L).

When LPG is transported in a DOT Specification M 330 or M 331 tank truck, DOT requires carriers to determine whether the commodity is corrosive and to indicate the appropriate information in the shipping description as one of the following: "RQ, UN1075, Liquefied petroleum gas, 2.1, (Non-corrosive)," "RQ, UN1075, Petroleum gases, liquefied, 2.1 (Noncor)," *or* "RQ, UN1075, Liquefied petroleum gas, 2.1 (Not for Q and T tanks)." Noncorrosive LPG may be transported in tank trucks that have been constructed from a special steel that is "quenched and tempered," or "Q and T."

When a flammable gas is transported in bulk by highway or rail, its name must be displayed on two opposing sides of the tankcar used for shipment. When it is transported by rail in a DOT-113 tankcar, DOT requires their carriers to display FLAMMABLE GAS placards on squares having a white background and black border.

When shippers offer nonbulk quantities of unodorized LPG for transportation, DOT requires them to mark the packaging NON-ODORIZED or NOT ODORIZED near the spot where the proper shipping name is marked.

12.11 AROMATIC HYDROCARBONS

The word *aromatic* suggests that aromatic hydrocarbons are compounds possessing fragrant odors. The odors of some simple aromatic hydrocarbons actually are fragrant, but otherwise, this perception is misleading. Aromatic hydrocarbons are regarded as compounds whose molecules are composed of one or more special rings of carbon atoms. These compounds are typified by the substance called *benzene*, the simplest aromatic hydrocarbon. Its chemical formula is C_6H_6. Although the molecular structure of benzene may be represented by either hexagon shown on the left in the following illustration, it is more commonly represented by a hexagon with a circle inside, as shown on the right:

Aromatic and aliphatic hydrocarbons thus are differentiated by the fact that the molecular structures of aromatic hydrocarbons have one or more hexagonal benzene rings, usually with side chains, while the molecular structures of aliphatic hydrocarbons do not. Hereafter, the molecular structure of benzene will be represented by the hexagonal structural formula with the inscribed circle. It is called the *benzene ring*. Although the symbols of the carbon atoms are not written as part of the hexagon, it is always understood that a carbon atom occupies each corner of the hexagon, where it is bonded to two other carbon atoms and a hydrogen atom.

When any one of the six hydrogen atoms in benzene is substituted with a methyl group, the compound that results is called methylbenzene, or more commonly, toluene. The chemical formula of toluene is $C_6H_5-CH_3$, and its molecular structure is represented as follows:

When two hydrogen atoms in the benzene molecule are substituted with methyl groups, the resulting compound has the chemical formula $C_6H_4-(CH_3)_2$. Because it is possible to position the methyl groups in three different ways on the benzene ring, three structural

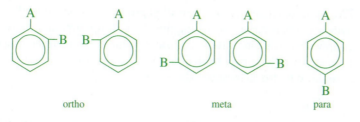

ortho   meta   para

FIGURE 12.6 The ortho-, meta-, and para- positions of B, an arbitrary substituent, relative to a fixed position of another substituent on the benzene ring. The chemical identity of A and B may be identical or different. The two configurations noted for the ortho- and meta- positions of B relative to A are equivalent ways of denoting the same compound.

isomers for the formula C_6H_4—$(CH_3)_2$ exist. These three compounds are the structural isomers of dimethylbenzene, or xylene. Their molecular structures are written as follows:

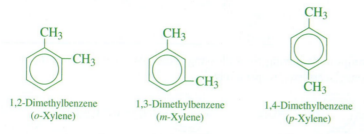

1,2-Dimethylbenzene   1,3-Dimethylbenzene   1,4-Dimethylbenzene
(*o*-Xylene)   (*m*-Xylene)   (*p*-Xylene)

In the common system of nomenclature, the prefixes **ortho- (o-)**, **meta- (m-)**, and **para- (p-)** are employed to identify disubstituted benzenes. *Ortho-* means "straight ahead," *meta-* means "beyond," and *para-* means "opposite." In commerce, the isomeric mixtures of xylene are sometimes referred to as either "xylene" or "xylene(s)" with total disregard for their isomeric distinctions.

The simplest derivatives of an aromatic hydrocarbon are compounds in which a single alkyl substituent has replaced a hydrogen atom on the benzene ring. These compounds are named by identifying the substituent followed by the word "benzene." For example, the compounds having the following molecular structures are named ethylbenzene and *n*-propylbenzene, respectively.

Ethylbenzene   *n*-Propylbenzene

When two substituents have replaced two hydrogen atoms, their locations on the benzene ring must be identified. The molecular structures of the three compounds in Figure 12.6 illustrate two arbitrary substituents, A and B, positioned on the benzene ring. B is located in the *ortho-*, *meta-*, and *para-* positions relative to the position of A. When naming the disubstituted benzenes, the italicized letters *o-*, *m-*, and *p-* are used as prefixes to identify the location of one substituent relative to the other one. When one of the two substituents is a methyl group, the compound often is named as a derivative of toluene as shown by the following examples:

ortho- (o-), meta- (m-), and para- (p-) ■ The prefixes used to name disubstituted benzenes in the following manner: *ortho-* refers to the substitution of the hydrogen atoms bonded to adjacent (the first and second) carbon atoms on the benzene ring; *meta-* refers to their substitution on the first and third carbon atoms; and *para-* refers to their substitution on the first and fourth carbon atoms

   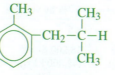

p-Ethyltoluene   *o*-Isobutyltoluene

In the IUPAC system, when two or more alkyl groups are bonded to the benzene ring, the resulting hydrocarbon is named by listing the alkyl groups alphabetically and numbering the first group with a 1. Each of the six carbon atoms in the benzene ring is then numbered from 1 to 6. Two examples that illustrate the use of this rule for naming the derivatives of aromatic hydrocarbons are shown next:

1-Ethyl-2-*n*-propylbenzene   1-Isobutyl-3-*n*-propylbenzene

Aromatic hydrocarbons sometimes are named by identifying the C_6H_5— and $C_6H_5CH_2$— groups of atoms. C_6H_5— and $C_6H_5CH_2$— are called the **phenyl group** and **benzyl group**, respectively. They are examples of **aryl groups**, or **aryl substituents**.

Phenyl group   Benzyl group

Thus, the compounds having the following molecular structures are named phenylethylene and dibenzylacetylene, respectively.

Phenylethylene   Dibenzylacetylene

12.11-A BENZENE, TOLUENE, AND XYLENE(S)

Benzene, toluene, and xylene are the three most commonly encountered aromatic hydrocarbons. They are sometimes referred to collectively by the acronym **BTX**. Benzene, toluene, and xylene(s) are colorless, water-insoluble, highly volatile liquids. Some important physical properties of benzene, toluene, and xylene(s) are provided in Table 12.12. Their

TABLE 12.12	Physical Properties of Benzene, Toluene, and the Isomeric Xylenes				
	BENZENE	**TOLUENE**	**o-XYLENE**	**m-XYLENE**	**p-XYLENE**
Melting point	41°F (5.4°C)	−139°F (−95°C)	−15°F (−26°C)	−54°F (−48°C)	55°F (13°C)
Boiling point	176°F (80°C)	231°F (111°C)	291°F (144°C)	282°F (139°C)	280°F (138°C)
Specific gravity at 68°F (20°C)	0.88	0.87	0.90	0.87	0.86
Vapor density (air = 1)	2.8	3.1	3.7	3.7	3.7
Vapor pressure at 68°F (20°C)	75 mmHg	22 mmHg	8 mmHg	8 mmHg	9 mmHg
Flashpoint	12°F (−11°C)	40°F (4.4°C)	90°F (32°C)	84°F (29°C)	81°F (27°C)
Autoignition point	1044°F (562°C)	997°F (536°C)	867°F (464°C)	982°F (528°C)	984°F (529°C)
Lower flammable limit	1.4% by volume	1.4% by volume	1.0% by volume	1.1% by volume	1.1% by volume
Upper flammable limit	8% by volume	6.7% by volume	6.0% by volume	7.0% by volume	7.0% by volume
Evaporation rate (ether = 1)	2.8	4.5	9.2	9.2	9.9
Heat of combustion [32°F (0°C) and 1 atm]	18,184 Btu/lb (42,300 kJ/kg)	18,200 Btu/lb (42,300 kJ/kg)	18,400 Btu/lb (42,800 kJ/kg)	18,400 Btu/lb (42,800 kJ/kg)	18,400 Btu/lb (42,800 kJ/kg)

flashpoints, flammable ranges, and heats of combustion indicate that the primary risk associated with them is fire and explosion. Their combustion is characterized by the production of smoky flames consisting of particulate matter to which the cancer-causing polynuclear aromatic hydrocarbons (Section 12.12-C) adsorb.

Benzene, toluene, and xylene are manufactured by several processes including the catalytic reformation of alkyl cycloalkanes at high temperature and pressure. **Reformation** is a chemical reaction involving the reforming of molecules into different molecules of other substances. It almost always occurs with the simultaneous loss of hydrogen.

When naphtha is subjected to the conditions of catalytic reformation, its constituent cycloalkanes are converted into compounds within the BTX group. The following equations illustrate reformation reactions that result in the production of benzene, toluene, and *m*-xylene:

reformation ■ Any chemical reaction in which certain substances are reformed into different substances, often accompanied by dehydrogenation

Benzene

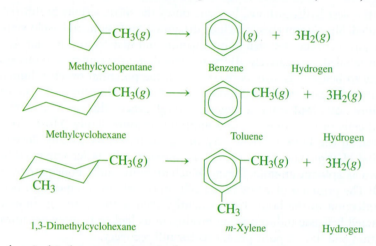

In the chemical industry, *p*-xylene is the most commercially important xylene isomer. It competes with naphthalene (Section 12.12-A) as the raw material needed to manufacture phthalic anhydride, a substance widely employed for the production of certain resins and polyesters including poly(ethylene terephthalate) (Section 14.2-B). Table 12.12 shows that the boiling points of *m*- and *p*-xylene are very similar. This similarity permits the *m*- and *p*- isomers to be separated from the *o*- isomer. The isomeric mixture is heated to vaporize its components within two temperature ranges; then the vapors are condensed and independently collected. Finally, the *p*- isomer crystallizes as a solid, which allows its isolation from *m*-xylene by filtration.

Although benzene once was a popular industrial solvent, it is no longer widely used for this purpose because workers' exposure to benzene vapor caused them to contract acute myelogenous leukemia and aplastic anemia (Section 12.11-B). Following this discovery, industrial employers turned to using toluene and xylene as replacement solvents.

Notwithstanding that worker exposure poses the risk of contracting cancer, the BTX group of compounds still is used by the chemical industry in substantial quantities as a raw material for the manufacture of other organic compounds. For example, benzene is used to produce ethylbenzene and isopropylbenzene.

Toluene

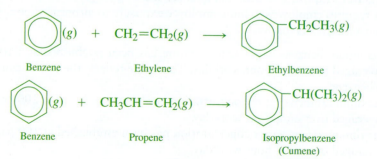

Xylene(s)

acute myelogenous leukemia ■ The cancer of the blood that develops in white blood cells, often associated with chronic exposure to benzene

aplastic anemia ■ A bone marrow injury, often associated with exposure to benzene

Each BTX compound is transported commercially in nonbulk and bulk containers. Considerable quantities of toluene and xylene formerly were used as components of solvent-based paints and other coatings, adhesives, inks, and cleaning agents, but in contemporary times, EPA regulations have reduced the use of these volatile organic compounds for these purposes.

Toluene is also the raw material needed for the production of trinitrotoluene and toluene diisocyanate. The xylene isomers are valuable constituents of motor and aviation gasolines because of their high octane numbers (Section 12.13). They are also used in the chemical industry as feedstocks for the production of compounds needed in the polymer industry.

12.11-B ILL EFFECTS CAUSED BY INHALING BTX VAPOR

Among the simplest hydrocarbons, benzene poses the most serious health risk when its vapor is inhaled. Short-term effects include dizziness, confusion, and asphyxiation. More-over, in humans, benzene is a well-known hemo- and neurotoxicant; that is, when benzene vapor is inhaled, it negatively affects the blood and the central nervous systems.

Exposure to benzene is also associated with the potential onset in humans of four types of leukemia (cancer of the blood-forming tissues and organs), one of which is **acute myelogenous leukemia**, a generally fatal cancer during which the immature white blood cells in the bone marrow multiply uncontrollably. Most blood cells are manufactured in the bone marrow. Myelogenous leukemia usually develops after a latency period of approximately 15 years. Its onset usually is accompanied by a second illness known as **aplastic anemia**, during which an individual's bone marrow is irreversibly injured. The presence of white blood cells in the blood is important, because these cells fight infection in the body. Consequently, even with treatment, individuals who have contracted benzene-induced acute myelogenous leukemia often are more susceptible to infection and have a poor prognosis for full recovery.

Because benzene exposure by humans is clearly linked with the onset of cancer, epidemiologists classify it as a human carcinogen. EPA estimates that a lifetime exposure to 4 parts per billion benzene in air results in one additional case of leukemia in a population of 10,000 exposed individuals.

Exposure to toluene and xylene is not linked with the onset of cancer. However, the inhalation of their vapors irritates the respiratory system and depresses the central nervous system. Initially, the inhalation of toluene and xylene vapors causes dizziness and nausea, and long-term inhalation can have a narcotic impact on the body. The long-term inhalation of toluene can also be addictive. This dangerous practice, colloquially known as *glue sniffing*, involves voluntarily inhaling the toluene vapor emitted by certain glue products like hobby glue. People who regularly practice glue sniffing risk injury to the liver, heart, and lungs.

12.11-C WORKPLACE REGULATIONS INVOLVING THE BTX HYDROCARBONS

OSHA publishes regulations at 29 C.F.R. §§1910.1028 and 1910.1000 that aim to protect workers from exposure to benzene, toluene, and xylene.

OSHA requires employers to limit employee exposure to airborne vapors of benzene, toluene, and xylene(s) as follows:

- The maximum benzene vapor concentration has been established at 1 part per million, averaged over an 8-hour workday. This value reflects the cancer-causing potential of benzene.
- The maximum toluene vapor concentration has been established at 200 parts per million, averaged over an 8-hour workday.
- The maximum xylene vapor concentration has been established at 100 parts per million, averaged over an 8-hour workday.

TABLE 12.13	Shipping Descriptions of Some Representative Aromatic Hydrocarbons
AROMATIC HYDROCARBON OR GROUPS THEREOF	**SHIPPING DESCRIPTION**
Benzene	UN1114, Benzene, 3, PG II
Diethylbenzene	UN2049, Diethylbenzene, 3, PG III (Marine Pollutant)
Ethylbenzene	UN1175, Ethylbenzene, 3, PG II
Isopropylbenzene	UN1918, Isopropylbenzene, 3, PG III (Marine Pollutant)
n-Propylbenzene	UN2364, *n*-Propylbenzene, 3, PG III (Marine Pollutant)
Toluene	UN1294, Toluene, 3, PG II
Xylenes	UN1307, Xylenes, 3, PG II *or* UN1307, Xylenes, 3, PG III

When benzene is present in the workplace, OSHA requires employers to establish regulated areas where the concentration of benzene vapor exceeds the permissible exposure limit of 1 part per million as a time-weighted average limit, or 5 parts per million as an average concentration over a 15-minute period. OSHA also requires employers to post the following warning sign at the entrances to these regulated areas:

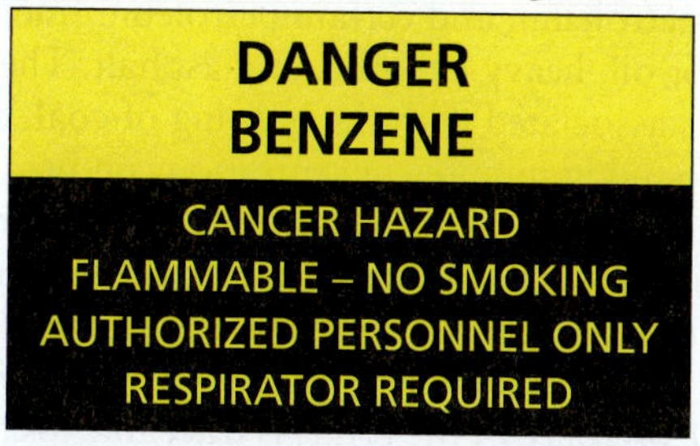

**DANGER
BENZENE**

CANCER HAZARD
FLAMMABLE – NO SMOKING
AUTHORIZED PERSONNEL ONLY
RESPIRATOR REQUIRED

12.11-D TRANSPORTING AROMATIC HYDROCARBONS

When shippers offer a simple aromatic hydrocarbon or any of its derivatives for transportation, DOT requires them to enter the relevant shipping description on an accompanying shipping paper. Some examples for several representative aromatic hydrocarbons are listed in Table 12.13. DOT also requires shippers and carriers to comply with all applicable labeling, marking, and placarding requirements.

12.12 POLYNUCLEAR AROMATIC HYDROCARBONS

There are more than 500 substances commonly called **polynuclear aromatic hydrocarbons**, or **PAHs**. The simplest members of this class of compounds are a group of structurally similar hydrocarbons whose molecules consist of two or more mutually fused benzene rings. "Mutually fused" means that the benzene rings share a pair of carbon atoms and the bond between them. The PAH having two mutually fused benzene rings is called naphthalene. The PAHs having three mutually fused benzene rings are called anthracene and phenanthrene.

polynuclear aromatic hydrocarbon (PAH)
■ Any aromatic hydrocarbon whose molecules have two or more mutually fused benzene or other rings

Naphthalene

Anthracene

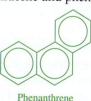
Phenanthrene

The molecules of the PAHs may also have four-, five-, six-, or seven-member rings. The molecules of some PAH derivatives consist of benzene rings fused to cyclopentenyl (C_5H_7-) and other rings. The molecules of some PAH derivatives may contain a ring nitrogen or oxygen atom (that is, a nitrogen or oxygen atom substitutes for a ring carbon atom). The substances whose molecules possess one or more ring atoms other than carbon are called **heterocyclic compounds**.

All PAHs are solid compounds at room temperature. Although individual PAHs have been isolated and characterized, their mixtures are of concern primarily as hazardous materials. They decompose only slowly to simpler substances; hence, once produced, they tend to remain in the environment.

heterocyclic compound
■ Any compound whose molecules contain one or more ring atoms other than carbon

Naphthalene

12.12-A NAPHTHALENE

Naphthalene is the only PAH of commercial importance. It is a white to colorless solid generally described as having the odor of mothballs. Although naphthalene once was the main constituent of a consumer product used as mothballs, this specific use is now discouraged. As noted in Section 12.11-A, it is used primarily in the chemical industry as a raw material for the manufacture of phthalic anhydride.

12.12-B SOURCES OF THE PAHs

Many processes associated with the incomplete combustion of organic materials generate mixtures of PAHs. Hence, many manufacturing and process industries often spew PAHs from their smokestacks into the air. PAHs are also dissolved constituents of coal tar and coal tar distillates, crude petroleum, and certain petroleum fractions including heavy diesel fuel, heating oils, motor oil, heavy naphtha, and asphalt. They are also components of the emissions and residues associated with the burning of coal, diesel oil, wood, tobacco, and peat. They are even found in minute concentrations on bread crusts, burnt toast, and the burnt surfaces of meats cooked on barbecue grills at high temperatures.

Because PAHs are widely scattered throughout the environment, they are frequently identified in samples of media collected from different sources. For example, they are identified as constituents of lake sediments and the particulate matter of tobacco smoke. In the former instance, the presence of the PAHs is traceable to the previous use of coal tar for sealing pavements.

The most common sources of the PAHs are diesel-engine exhaust and the smoke generated during the burning of wood, tobacco, and coal. Typically, they are adsorbed to the constituent soot particulates generated during combustion. Because many heavy-duty trucks, buses, and moving equipment are diesel-powered, the PAHs produced during their use become components of the environments into which their exhaust is discharged. This is a matter of significant concern when the exhaust is inadequately vented from enclosures.

During incomplete combustion processes, PAHs are produced simultaneously with soot by a multistep mechanism similar to that depicted in Section 5.11. When simple hydrocarbons are heated to high temperatures, molecular fragments—such as free radicals—are produced. These fragments undergo reactions that result in the production of new molecules. The original molecules and their fragments dehydrogenate while the new molecules similarly fragment, amalgamate, and continue to dehydrogenate. When dehydrogenation has occurred to its maximum extent, all that remains is soot or carbon black.

12.12-C ILL EFFECTS CAUSED BY INHALING PAHs

Diesel-engine exhaust is a complex mixture of particulate matter and gaseous compounds including carbon monoxide, carbon dioxide, nitric oxide, nitrogen dioxide, sulfur dioxide, methane, benzene, phenol, 1,3-butadiene, acrolein, and the vapors of several individual PAHs.

The particulate matter consists mainly of soot to which the PAHs and other compounds have adsorbed. When they are inhaled, the soot aggravates chronic respiratory problems.

Inhalation exposure to soot and certain individual PAHs has also been linked with the onset of lung and cardiovascular dysfunction. Moreover, massive numbers of mutated cells emerge in the bronchial passageways and bladders of individuals who regularly inhale diesel-engine exhaust, thereby causing lung and bladder cancer.[10,11] For this reason, soot is classified as a known human carcinogen.

The U.S. Department of Health and Human Services has concluded that exposure to at least 15 PAHs causes the emergence of malignant tumors at the site of contact. The molecular structures of these compounds are provided in Figure 12.7. These 15 compounds are among the substances that the scientific community now suspects of triggering the onset of cancer in humans.

As previously noted, the particulate matter of tobacco smoke is also a source of PAHs, especially the following: benz[*a*]anthracene, benzo[*b*]fluoranthene, benzo[*j*]fluoranthene, benz[*k*]fluoranthene, benzo[*a*]pyrene, dibenz[*a,h*]anthracene, dibenzo[*a,l*]pyrene, indeno[1,2,3-*c,d*]pyrene, 5-methylchrysene, and 7*H*-dibenzo[*c,g*]carbazole.[12] These compounds are not constituents of native tobacco leaves but form when the tobacco undergoes incomplete combustion. They are suspected of collectively contributing to the onset of the cancers experienced by tobacco smokers and individuals exposed to secondhand smoke.

12.12-D PAHs AND FIREFIGHTING

Among modern-day occupational settings, career-oriented firefighters are a major group of individuals most vulnerable to routinely inhaling PAHs. Firefighters are unwittingly exposed to PAHs in at least the following ways:

■ Firefighters are repeatedly exposed to airborne PAHs adsorbed to soot particulates at virtually every fire scene.

■ Firefighters are also exposed to PAHs adsorbed to soot particulates in enclosed areas where diesel engines operate. Such areas include firehouses, where diesel-power equipment is regularly serviced and maintained and fire trucks idle for immediate use.

These repeated exposures to soot and the cancer-causing PAHs could constitute the basis for long-term health concerns for firefighters. Scientists denote soot and diesel engine exhaust as human carcinogens.

To protect firefighters against inhaling diesel-engine exhaust, the enclosed areas of firehouses should be ventilated using a local exhaust ventilation system. As a general policy, the idling of vehicles inside the firehouse should be avoided.

SOLVED EXERCISE 12.7

When emitted from the exhaust pipe of an idling fire engine, does the exhaust tend to concentrate in the lower or upper regions of an enclosed firehouse?

Solution: Because diesel-engine exhaust is hotter than the temperature of the surrounding air, it rises to the ceiling or underside of the roof in an enclosed firehouse. Because the exhaust concentrates near the ceiling, exhaust fans should be located in the upper regions of an enclosed firehouse to vent the exhaust to the outside environment.

---

[10]Michael D. Attfield et al., "The diesel exhaust in miners study: A cohort mortality study with emphasis on lung cancer," *J. Nat. Can. Inst.*, Vol. 104 (2012), pp. 1–15.
[11]Neal D. Freeman et al., "Association between smoking and risk of bladder cancer among men and women," *J. Amer. Med. Assoc.*, Vol. 306 (2011), pp. 793–896.
[12]Report of the Surgeon General, *How Tobacco Smoke Causes Disease: The Biology and Behavioral Basis for Smoking-Attributable Disease*, U.S. Centers for Disease Control and Prevention, Atlanta, Georgia (2010) (ISBN-13: 978-0-16-084078-4).

FIGURE 12.7 Based on information from the U.S. Department of Health and Human Services, 15 PAHs are likely to cause cancer in exposed humans. They are produced during the incomplete combustion of wood, petroleum, coal, and tobacco, and are constituents of diesel-emission exhaust and coal tar. Their molecular structures show that they are complex compounds consisting of mutually fused benzene rings. Dibenz[a,h]acridine, dibenz[a,j]acridine, and 7H-dibenzo[c,g]carbazole are *heterocyclic PAHs*, meaning that their molecular structures have a nitrogen atom in place of a ring carbon atom. (*Report on Carcinogens,* 12th edition, U.S. Department of Health and Human Services, Public Health Service, National Toxicology Program: Research Triangle Park, North Carolina, 2011.)

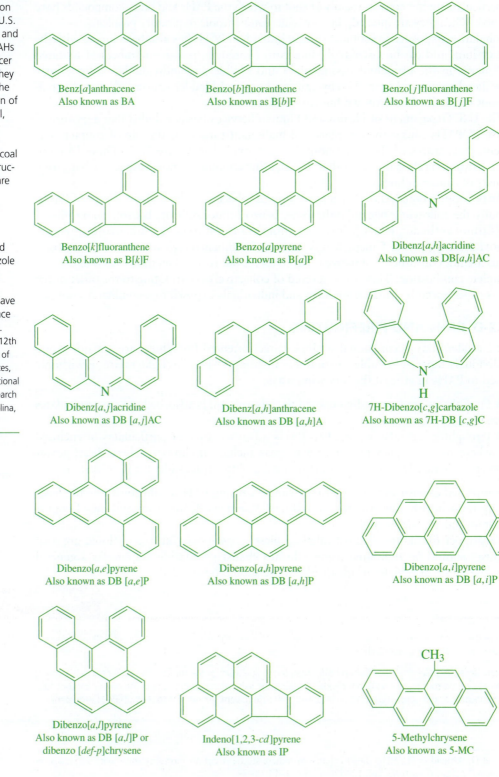

Benz[*a*]anthracene
Also known as BA

Benzo[*b*]fluoranthene
Also known as B[*b*]F

Benzo[*j*]fluoranthene
Also known as B[*j*]F

Benzo[*k*]fluoranthene
Also known as B[*k*]F

Benzo[*a*]pyrene
Also known as B[*a*]P

Dibenz[*a,h*]acridine
Also known as DB[*a,h*]AC

Dibenz[*a,j*]acridine
Also known as DB [*a,j*]AC

Dibenz[*a,h*]anthracene
Also known as DB [*a,h*]A

7H-Dibenzo[*c,g*]carbazole
Also known as 7H-DB [*c,g*]C

Dibenzo[*a,e*]pyrene
Also known as DB [*a,e*]P

Dibenzo[*a,h*]pyrene
Also known as DB [*a,h*]P

Dibenzo[*a,i*]pyrene
Also known as DB [*a,i*]P

Dibenzo[*a,l*]pyrene
Also known as DB [*a,l*]P or
dibenzo [*def-p*]chrysene

Indeno[1,2,3-*cd*]pyrene
Also known as IP

5-Methylchrysene
Also known as 5-MC

12.12-E WORKPLACE REGULATIONS INVOLVING THE PAHS

Although a permissible exposure limit for diesel-engine exhaust has not been published by OSHA or NIOSH, the limits for its gaseous constituents are available. In this text, the permissible exposure limits for carbon monoxide, nitric oxide, nitrogen dioxide, and sulfur dioxide are listed in Table 10.2. OSHA/NIOSH's publication of permissible exposure limits for individual PAHs is limited to naphthalene, anthracene, benzo[*a*]pyrene, chrysene, phenanthrene, and pyrene vapor. For naphthalene, the permissible exposure limit is 10 parts per million (50 mg/m^3), averaged over an 8-hour workday; for the other PAHs listed here, the value is 0.2 mg/m^3, averaged over an 8-hour workday.

12.13 PETROLEUM AND PETROLEUM PRODUCTS

The word *petroleum* is derived from the Greek *petra*, meaning "rock," and Latin *oleum*, meaning "oil." The name "rock oil" is reminiscent of humans' earliest contacts with this material: a liquid that oozed from fissures in rocks.

12.13-A THE NATURE AND NATURAL ORIGIN OF CRUDE OIL

Petroleum is a highly complex mixture consisting of many thousands of organic compounds, approximately 75% of which are hydrocarbons, each of whose molecules has from 3 to 60 carbon atoms. The actual number of individual hydrocarbons in petroleum has been estimated to be between 50,000 and 2,000,000. Petroleum occurs naturally deep below Earth's surface in certain areas around the world. Wells are drilled through the rock to the oil-bearing stratum, through which the petroleum is then pumped to the surface. In this form, it is called **crude oil, crude petroleum**, or **"crude."** In the United States, major oil fields are located in Texas, California, Louisiana, Oklahoma, and Alaska, from which crude oil is transported in tankers or transferred by pipeline, to plant sites known as **petroleum refineries**, where it is treated to produce **petroleum products**.

Although the mechanism by which crude oil formed in the earth is open to some debate, most scientists believe that it originated from the partial decomposition of animals and plants (zooplankton and algae) that lived millions of years ago. Because geological forces caused the position of Earth's crust to change over the passing millennia, these ancient organisms were buried at great depths. Their decomposition resulted from the enhanced temperature and pressure at these depths.

Both crude oil and natural gas formed when the remains of ancient organisms decomposed. Crude oil formed when the temperature was approximately 150°F (76°C), whereas natural gas formed when the temperature rose to approximately 200°F (93°C). Scientists estimate that approximately 13.0 pounds (5.9 kg) of crude oil resulted from the decomposition of 98 tons (89 t) of prehistoric matter.

A group of 14 Middle Eastern and South American countries now controls a large portion of the world's supply of crude oil through a cartel called the **Organization of Petroleum Exporting Countries**, or **OPEC**. This organization exerts control over the supply of crude oil by voluntarily restraining production in order to stabilize its price. Because the United States has an insatiable thirst for petroleum products, considerable efforts are now ongoing to shed U.S. dependence on foreign sources of crude oil by replacing them with alternatives.

12.13-B FIGHTING FIRES INVOLVING CRUDE PETROLEUM

Crude petroleum is a highly flammable liquid, because its flashpoint ranges from 20 to 90°F (−7 to 32°C). Fires involving this material have occurred at oil fields, in transit, during pipeline transfer, and during storage. Capping burning oil wells is a job for specially trained experts, but regular firefighters have also been called on to combat fires involving crude oil at storage facilities and during transfer and transit accidents.

crude oil (crude petroleum; crude) ■ The complex chemical mixture that forms from naturally occurring biomass in subsurface rock formations under high temperature and pressure conditions over geological time

petroleum refinery ■ A facility engaged primarily in producing on a commercial scale gasoline, kerosene, fuel oils, lubricants, and other products through fractionation, alkylation, cracking, blending, and other processes

petroleum product ■ Any petroleum-based fuel such as gasoline, jet fuel, kerosene, and heating oil, nonfuel like asphalt and lubricants, and petrochemicals like ethane, propane, and butane

Organization of Petroleum Exporting Countries (OPEC) ■ The intergovernmental organization of 14 nations whose objective is to coordinate and unify common oil-marketing policies among member countries

Although small crude oil fires can be extinguished using a deluging volume of water, most experts recommend the use of aqueous film-forming foam (AFFF) (Section 5.12-B) as the most practical means of extinguishing a crude oil fire inside a bulk storage tank. In these incidents, the use of water is recommended solely for cooling purposes. Water and petroleum are immiscible liquids, and petroleum is less dense than water. When discharged on a petroleum fire, the water sinks to the bottom of the storage tank, where it settles as a separate layer and serves no useful purpose.

Furthermore, the presence of a water layer at the bottom of a bulk storage tank poses a special concern for firefighters when combating a petroleum fire. Although the fire occurs at the tank's surface, where flammable vapor is emitted, the accompanying heat of combustion can be slowly transmitted by convection and radiation through the underlying petroleum to the water layer. Absorption of the intense heat causes the temperature of the petroleum and the underlying water to rise until ultimately, the water boils. The water vapor then forces its way upward, pushing the burning petroleum up and over the walls of the storage tank. This phenomenon is appropriately called a **boilover**.

When a crude petroleum fire occurs inside a storage tank, every attempt should be made to extinguish the fire before the boilover occurs. The burning, frothing petroleum that spills outside the storage tank has been known to flow substantial distances from the tank, triggering numerous secondary fires. The firefighters who combat these fires should be particularly wary, as the flowing, burning petroleum could overtake unsuspecting firefighters in its pathway. They must also attempt to confine water runoff to prevent an adverse impact on the environment and limit the department's liability.

12.13-C FRACTIONATION OF CRUDE PETROLEUM

One of the major operations occurring at petroleum refineries is the fractional isolation of the substances in crude petroleum. The process is called **fractionation**, or **fractional distillation**, and is accomplished by heating the crude within preestablished temperature ranges until its components vaporize. The vapors of these compounds then are condensed into liquids and collected in separate receivers. They are referred to as **petroleum distillates**.

A single petroleum fraction may be used directly as a commercial petroleum product, or it may be stored until it can be further processed. Bulk volumes of petroleum products are transported to major distribution centers, where they often are stored in multiple tanks in a tank farm.

The fractions most commonly isolated during the fractionation of crude petroleum are represented in Figure 12.8. They consist of the following:

■ *A gaseous mixture of methane (65% to 90%), ethane, propane, and butane.* This fraction, called **petroleum gas**, separates from crude petroleum at a temperature below 70°F (21°C). It is recovered and used directly as a *feedstock* for the production of other substances, or its components can be isolated to produce petroleum products such as bottled gas.

■ *A liquid mixture consisting mainly of alkanes and cycloalkanes having from 5 to 11 carbon atoms per molecule.* This fraction, called **light naphtha** or **ligroin**, generally is isolated within the boiling point range 158 to 284°F (70 to 140°C). At one time, light naphtha was used directly as the fuel formerly known as *straight-run gasoline*. However, the use of light naphtha is now considered environmentally undesirable for direct use, because it contains as much as 3% benzene by volume. (In 1995, using its authority under the Clean Air Act, EPA limited the benzene content in petroleum products to less than 1%.) Today, light naphtha is used as the feedstock either for the production of benzene, toluene, and the xylene isomers by catalytic reformation, or for the production of ethylene and propylene by catalytic cracking.

■ *A liquid mixture consisting mainly of alkanes and cycloalkanes having from 7 to 11 carbon atoms per molecule.* This fraction, called **heavy naphtha**, is generally isolated within

boilover ■ The phenomenon associated with the expulsion of crude oil and the production of steam during a fire within a crude oil storage tank

fractionation (fractional distillation) ■ The process of separating multiple components of a mixture based on the different boiling points of its constituents, during which the vapors are collected at specified temperatures or within specified temperature ranges and subsequently condensed to liquids and collected

petroleum distillate ■ A fraction of crude petroleum obtained by the vaporization of its components within a specific temperature range followed by its condensation

petroleum gas ■ A gaseous fraction of crude petroleum that does not condense at room temperature, consisting of methane, ethane, propane, and butane

light naphtha (ligroin) ■ The most volatile fraction of crude petroleum

heavy naphtha ■ A petroleum fraction of crude petroleum often used as the feedstock for the production of gasoline

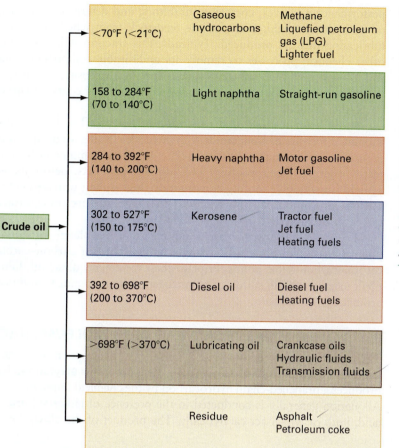

<70°F (<21°C)	Gaseous hydrocarbons	Methane Liquefied petroleum gas (LPG) Lighter fuel
158 to 284°F (70 to 140°C)	Light naphtha	Straight-run gasoline
284 to 392°F (140 to 200°C)	Heavy naphtha	Motor gasoline Jet fuel
302 to 527°F (150 to 175°C)	Kerosene	Tractor fuel Jet fuel Heating fuels
392 to 698°F (200 to 370°C)	Diesel oil	Diesel fuel Heating fuels
>698°F (>370°C)	Lubricating oil	Crankcase oils Hydraulic fluids Transmission fluids
	Residue	Asphalt Petroleum coke

FIGURE 12.8 Some products derived by the fractionation of crude petroleum. Liquefied petroleum gas (LPG) is marketed in the form that comes directly from the distillation tower, but the other fractions require further refinement to remove undesirable components, most notably, benzene and sulfurous and nitrogenous compounds. The refinement processes include vacuum distillation, catalytic reforming, hydrocracking, and catalytic cracking.

the boiling point range 284 to 392°F (140 to 200°C). It typically is used as the feedstock for the production of motor gasoline. The production of gasoline involves catalytic cracking, during which hydrocarbons having from 7 to 11 carbon atoms per molecule are produced. This fraction is also the feedstock for the production of toluene.

■ *A liquid petroleum product consisting of alkanes and cycloalkanes having from 9 to 16 carbon atoms per molecule.* This fraction, called **kerosene**, is often isolated within the boiling point range 302 to 527°F (150 to 275°C). Kerosene has been used as the fuel in space heaters, portable cooking stoves, and water heaters and is suitable as a light source when burned in wick-fed lamps. Today, kerosene is used primarily for the production of jet fuels.

■ *A liquid mixture of alkanes and cycloalkanes having from 10 to 15 carbon atoms per molecule.* This fraction, called **diesel oil**, **diesel fuel**, or **gas oil**, usually is isolated within the boiling point range 392 to 698°F (200 to 370°C). In addition to carbon and hydrogen atoms, the molecules of the component hydrocarbons may contain sulfur or nitrogen atoms. Diesel oil is blended with additives and used to heat buildings and to power passenger cars, pickup and heavy-duty trucks, buses, ships, and certain types of heavy equipment. More than 50% of the passenger cars now used by Europeans, and more than 90% of the cars in Italy alone are diesel-powered. In the United States over the past decades, the use of diesel-powered passenger vehicles has fluctuated.

■ *A liquid to semisolid fraction composed of polynuclear aromatic hydrocarbons having from 30 to 45 carbon atoms per molecule.* This mixture of compounds typically is isolated at temperatures above 698°F (370°C). The products produced from this fraction are broadly called **lubricating oils**, or **lubricants**. It typically is blended with various

kerosene ■ An oily fraction of crude petroleum often used as a tractor fuel, jet fuel, and heating fuel in space heaters

diesel oil (diesel fuel, gas oil) ■ A fraction of petroleum product commonly used as the fuel for operating diesel engines

lubricating oil (lubricant) ■ Any fraction of petroleum oil that is capable of producing a film for coating the surfaces of moving metal parts, so that when used, it reduces the friction generated between bearing surfaces

performance additives, corrosion inhibitors, and detergents and then used as crankcase oil or hydraulic fluid. It is also the feedstock from which other motor and nonvehicle lubricating oils are produced. Typically, this fraction is hydrotreated and subjected to other chemical processes to remove its undesirable components—metals, nitrogen, and sulfur—and then blended to achieve a specified viscosity, stability, and lubricity. When they are used, lubricating oils produce an oily film that coats the surfaces of moving metal parts, thereby reducing friction and wear.

■ *The solid residue of the distillation process.* The fractionation residue is called **asphalt**, a mixture of compounds with molecules having 40 or more carbon atoms. Asphalt ignites at approximately 400°F (204°C). It is used commercially as a component of protective coating, waterproofing, and adhesive products. Before its use for paving, road, airfield runway, and roofing construction, hot asphalt often is held in crucibles, **asphalt kettles**, and portable tank trucks. DOT regulates its transportation as an elevated-temperature material.

The hydrocarbons that are components of petroleum fractions are primarily compounds having carbon–carbon single bonds, but not carbon–carbon double bonds or carbon–carbon triple bonds. Prior to their treatment, diesel oil, lubricating oils, and asphalt usually contain sulfurous or nitrogenous compounds in addition to the polynuclear aromatic hydrocarbons.

12.13-D CHEMICAL TREATMENT OF PETROLEUM FRACTIONS

The nature of the processing and chemical treatment operations conducted at most petroleum refineries is noted in Table 12.14. One process is **alkylation.** This chemical reaction produces a branched-chain hydrocarbon by the chemical union of an alkane and an alkene. Alkylation generally is conducted in the presence of either sulfuric acid or hydrofluoric acid, both of which act catalytically. The product of the alkylation reaction is called an

TABLE 12.14	Major Treatment Processes Conducted at Petroleum Refineries
Thermal cracking and catalytic cracking of the constituents of petroleum fractions	The breaking down of large hydrocarbon molecules into smaller hydrocarbon molecules by the application of heat or the use of catalysts
Alkylation	The application of heat or pressure to produce a branched-chain hydrocarbon from an alkane and alkene in the presence of sulfuric acid or hydrofluoric acid
Isomerization	The rearrangement of a hydrocarbon molecule having a given number of carbon atoms to produce another molecule having the same number of carbon atoms
Catalytic reformation	The reforming of alkyl cycloalkanes into aromatic hydrocarbons (e.g., methylcyclohexane ⟶ toluene) by passing the vapor of a petroleum fraction over certain catalysts at high temperature and pressure
Steam cracking	The production of ethylene, propylene, and the butane isomers, from the ethane, propane, and butane, respectively, in natural gas by reacting the latter with steam
Catalytic hydrotreatment	The processing of a petroleum fraction with hydrogen to remove sulfur, nitrogen, heavy metals, and other impurities from its components
Hydrocracking	The processing of a petroleum fraction with hydrogen at high pressure and temperature to convert complex hydrocarbons into smaller ones for use as components in gasoline and other fuels

alkylate. Bulk quantities of alkylates often are present at petroleum refineries in bulk storage tanks. They are highly flammable substances and pose a fire and explosion hazard.

The most common alkylate produced by the petroleum industry is commonly called *isooctane*, but this name is actually a misnomer. The correct name is 2,2,4-trimethylpentane. It is produced industrially in massive amounts from the chemical union of isobutane and isobutene in the presence of a strong acid catalyst.

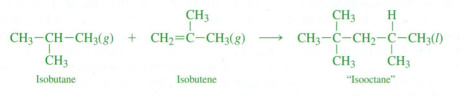

It is also produced by the catalytic conversion of isobutene to diisobutylene (footnote "c," Table 12.5), followed by hydrogenation.

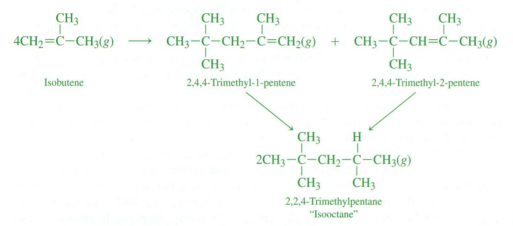

Aside from producing trimethylpentanes, alkylation reactions also produce dimethylhexanes and other branched-chain alkanes. These compounds are desirable additives in gasoline, because they improve the completeness of its combustion. In their absence, the phenomenon called **knocking** occurs. This term refers to the "putt-putt" noises, i.e., audible pings, sputtering, and rattling, that are generated as the fuel burns within the engine. Fuel additives that reduce the intensity of this knocking are called **antiknock agents**.

alkylate ■ Any branched-chain alkane resulting from the combination of an alkane and an alkene

knocking ■ The noises associated with the incomplete combustion of a petroleum fuel, especially in internal combustion engines

antiknock agent ■ Any petroleum fuel additive that reduces or eliminates the noises generated within a combustion chamber when the fuel burns

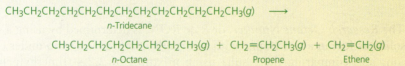

12.13-E GASOLINE

The following general types of gasoline are produced at petroleum refineries:

■ **Motor gasoline**, the fuel primarily produced by treating the heavy naphtha fraction and blending it with oxygenates (Section 13.2-F) and other performance additives. It is used mainly to fuel passenger cars and small trucks.

■ **Aviation gasoline**, the fuel primarily produced by treating the kerosene fraction and blending it with deicing inhibitors (to eliminate the formation of ice) and many other performance additives. It is used mainly to fuel aircraft. The terms "aviation gasoline" and "jet fuel" normally are differentiated in today's jargon by the manner in which they are intended for use. Aviation gasoline is used to power aircraft engines with spark plugs, whereas jet fuel is used to power jet turbine engines. Jet fuels are identified by names like JP-4, JP-7, and JP-8, where "JP" is the acronym for jet propulsion and the terminal number designates a specific composition.

The chemical composition of petroleum fuels varies, not only in different parts of the world but also at different times of the year. The chemical composition of a given fuel produced at a refinery can even change daily. There are numerous constituent components of both motor gasoline and aviation gasoline, but most are branched-chain alkanes. Hydrotreating and hydrocracking convert the unsaturated hydrocarbons in the petroleum fraction used for their production into branched-chain alkanes, thereby yielding alkanes having 7 to 11 carbon atoms per molecule in motor gasoline and 9 to 16 carbon atoms per molecule in aviation gasoline and jet fuel. These fuels are flammable liquids that pose the risk of fire and explosion.

The different commercial types of gasoline are distinguished by their inherent octane numbers. The **octane number**, or **octane rating**, is a representation of the antiknock properties of a fuel under laboratory or test conditions. Octane numbers of zero and 100 are arbitrarily assigned to *n*-heptane (a high "knocker") and 2,2,4-trimethylpentane (a low "knocker"), respectively. Heptane is an undesirable fuel, because it causes engine knocking as it burns in a combustion chamber, but 2,2,4-trimethylpentane ("isooctane") is a desirable fuel, because its combustion does not cause engine knocking.

$$CH_3CH_2CH_2CH_2CH_2CH_2CH_3$$

n-Heptane
Octane number = 0

$$CH_3-\overset{\overset{\displaystyle CH_3}{|}}{\underset{\underset{\displaystyle CH_3}{|}}{C}}-CH_2-\overset{\overset{\displaystyle CH_3}{|}}{CH}-CH_3$$

2,2,4-Trimethylpentane
Octane number = 100

The octane number of a given fuel is determined by comparing its knocking with the knocking of various mixtures of *n*-heptane and 2,2,4-trimethylpentane. When a sample of a fuel is found by testing to knock like a mixture of 85 parts 2,2,4-trimethylpentane and 15 parts *n*-heptane, the fuel is assigned an octane number of 85.

The octane rating of gasoline is commercially measured by the following two methods:

■ The *Research Octane Number*, or *R* or *RON*, is determined by using a sample of the gasoline as fuel in a test engine with a variable compression ratio under controlled conditions and comparing the results using mixtures of 2,2,4-trimethylpentane and *n*-heptane as fuels.

■ The *Motor Octane Number*, or *M* or *MON*, is determined in a similar manner using the same type of test engine, but with a preheated fuel mixture, higher engine speed, and variable ignition timing.

The expression *R* + *M*/2 is called the antiknock index.

There are three common grades of motor gasoline:

- **Regular-grade gasoline**, also called **regular unleaded**, which has a minimum octane rating of 82
- **Mid-grade gasoline**, which has an octane rating greater than or equal to 88 and less than or equal to 90
- **Premium-grade gasoline**, also called **supreme** and **super unleaded**, which has an antiknock index (R + M/2) greater than 90

By contrast, the individual types of aviation gasoline have varying octane numbers, but all are over 100.

At service stations, customers may dispense their choice of grade from fuel pumps like those shown in Figure 12.9 directly into a motor vehicle. A yellow label resembling the following is affixed to each pump to identify the minimum octane rating and the expression (R + M)/2 METHOD:

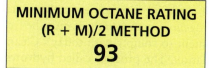

The latter is the average of the research octane number and the motor octane number.

regular-grade gasoline (regular unleaded)
- The grade of gasoline having a minimum octane rating of 82

mid-grade gasoline
- The grade of gasoline having an octane rating greater than or equal to 88 and less than or equal to 90

premium-grade gasoline (supreme, super unleaded) - The grade of gasoline having an antiknock index greater than 90

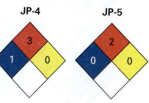

FIGURE 12.9 The different grades of gasoline fuels are differentiated by their octane numbers, which are established by subjecting a gasoline sample to test conditions that ascertain how well it knocks when ignited in a combustion chamber. At these fuel pumps, regular, mid-grade, and premium gasoline fuels having octane numbers of 87, 89, and 91, respectively, are provided for sale. As required by U.S. Federal Trade Commission regulations at 16 C.F.R. §306.12, the octane numbers are printed on labels in black on a yellow background. (*Courtesy of Eugene Meyer.*)

unleaded gasoline
■ Gasoline that does not contain a lead compound such as tetraethyl lead, formerly an additive that functioned as an antiknock agent

Diesel oil

All types of gasoline sold in the United States since 1975 are referred to as **unleaded gasoline**, meaning that they do not contain tetramethyllead, tetraethyllead, or other lead-based additives. Using the authority of the Clean Air Act, EPA banned the manufacture and use of lead compounds in gasoline to protect automotive catalytic converters from being inactivated by lead and to avoid the unnecessary dispersion of lead throughout the environment.

As previously noted in Section 5.8-A, to reduce the volume of carbon dioxide that operating motor vehicles emit to the atmosphere, the federal government is requiring that by 2016 and 2025, the fleets of auto companies must average 35.5 mi/gal (15 km/L) and 54.5 mi/gal (23 km/L), respectively. This almost doubles today's average of 27.5 mi/gal (11.7 km/L).

12.13-F DIESEL OIL

Diesel oil is the fuel used to power diesel engines in certain vehicles and machinery. Its flashpoint ranges from 110 to 190°F (43 to 88°C). In the United States, diesel oil is produced from the fraction of crude oil that refiners call **sour crude**. Because this fraction contains sulfurous and nitrogenous compounds, it must be treated to reduce or eliminate the sulfur and nitrogen content to comply with today's Clean Air Act standards. This is accomplished by hydrogenation. The process, called **hydrotreatment**, produces a diesel oil consisting of alkanes, cycloalkanes, and aromatic compounds. During hydrotreating, the sulfur and nitrogen atoms are converted into hydrogen sulfide and ammonia, respectively, and removed from the fuel.

Untreated and hydrotreated diesel oil are sometimes referred to as heavy diesel oil and light diesel oil, respectively. The combustion of untreated diesel oil yields a sooty smoke consisting of particulates to which polynuclear aromatic hydrocarbons adsorb. As noted in Section 12.12-C, the production of this plume poses a risk to public health and the environment, because inhalation exposure to PAHs is linked with the development of malignant tumors in the lungs. The combustion of light diesel oil yields an emission that contains fewer pollutants. To comply with the Clean Air Act standard established for fine particulate matter (Section 10.9-D), the American diesel industry began changing from heavy to light diesel oil in the mid-2000s. Today, light diesel oil is the main fuel used in diesel-powered vehicles that are driven on U.S. roadways.

The different commercial types of light diesel oil are rated by a system similar to the octane-number system used for rating different types of gasoline. For diesel oils, it is called the **cetane number**. Whereas the octane number measures the ability of a gasoline fuel to reduce engine knocking, the cetane number gauges the ease with which a diesel fuel autoignites when it is compressed in a cylinder of a diesel engine without a spark plug.

Cetane is the common name for *n*-hexadecane, an alkane having the formula $C_{16}H_{34}$. For test purposes, the cetane number of 2,2,4,4,6,8,8-heptamethylnonane and cetane are arbitrarily set at 15 and 100, respectively.

sour crude ■ Untreated crude oil or the fraction thereof that contains sulfurous and nitrogenous compounds

hydrotreatment ■ A petroleum refining process that upgrades the quality of petroleum fractions by using hydrogen to reduce or eliminate their nitrogenous and sulfurous content and to convert alkenes to alkanes

cetane number ■ A parameter that measures the combustion quality of diesel oil during its compression ignition

$$CH_3-\overset{\overset{\displaystyle CH_3}{|}}{C}-CH_2-\overset{\overset{\displaystyle CH_3}{|}}{C}-CH_2-\overset{\overset{\displaystyle CH_3}{|}}{CH}-CH_2-\overset{\overset{\displaystyle CH_3}{|}}{C}-CH_3$$
$$\underset{CH_3}{|}\qquad\underset{CH_3}{|}\qquad\qquad\underset{CH_3}{|}$$

2,2,4,4,6,8,8-Heptamethylnonane
Cetane number = 15

$$CH_3CH_2CH_2CH_2CH_2CH_2CH_2CH_2CH_2CH_2CH_2CH_2CH_2CH_2CH_2CH_3$$

n-Hexadecane (Cetane)
Cetane number = 100

The cetane number of a diesel fuel is determined in special variable-compression-ratio test engines.

Generally, diesel engines operate well when the diesel fuel has a cetane number between 40 and 55. The two diesel fuel types available commercially are referred to as **regular-grade diesel oil** and **premium-grade diesel oil**, which have cetane numbers ranging from 40 to 60 and 45 to 50, respectively.

12.13-G HEATING OILS

Some petroleum products are useful as **heating oils**. They are produced by fractionation, followed by blending with specified additives. For example, diesel fuel and heavy naphtha often are processed to produce heating oils, either by generating fractions within a narrower distillation range or by adding a specified amount of other petroleum fractions until the final blend possesses desirable ignition temperatures and heat (Btu) values. These products are used as fuels in home furnaces and boilers for factories, apartment and office buildings, schools, and steam-powered vessels, as well as for the generation of electricity.

Six grades of heating oils are commercially recognized in the United States, each designated by the words *Fuel Oil* followed by a number from 1 to 6. The grades are composed of hydrocarbons having 14 to 20 carbon atoms per molecule and are commercially distinguished by their ignition temperatures, each progressively increasing in the range from 444°F (229°C) to 765°F (407°C). In the United States, Fuel Oil No. 2 is the most commonly used heating oil.

12.13-H TRANSMISSION OF CRUDE PETROLEUM AND PETROLEUM PRODUCTS BY PIPELINE

Crude oil is frequently transferred by a network of pipelines from oil fields into storage tanks where it awaits processing. In the United States, approximately 55,000 miles (88,500 km) of transmission pipelines transfer crude oil from spot to spot. Similarly, all types of petroleum products are regularly transferred by pipeline from refineries into storage tanks awaiting an end-use.

A single pipeline used to transfer crude oil or a petroleum product may be either aboveground, underground, or both. Portions of the 799-mile (1242-km) Trans-Alaska Pipeline System previously noted in Section 10.15 are both located underground and elevated aboveground. The pipeline is used to transfer hot crude oil from the oil fields in Prudhoe Bay to Valdez, the northernmost ice-free American port. The crude oil then is transported in bulk by cargo ships to the west coast of the United States, from where it is again transferred by pipeline to refineries.

At 49 C.F.R. §195.410, DOT requires the posting of line markers like those shown in Figure 12.4 to indicate the approximate location of petroleum-transmission pipelines. Like the line markers used for locating natural-gas-transmission pipelines, the line markers for aboveground petroleum-transmission pipelines must be placed along each section located in an area accessible to the public and posted along right-of-ways and at road, railroad, and waterway crossings.

A major leak of crude petroleum or a petroleum product from a petroleum-transmission pipeline constitutes a major fire and explosion hazard, and may also cause a major environmental disaster. In 2010, a failed valve was the cause of a catastrophic leak from the trans-Alaska pipeline that resulted in the release of over 100,000 gallons (378 m^3) of crude oil to the environment. The oil was collected in a containment system from which it was thereafter retrieved. A major disaster associated with the leak was averted only because the prevailing temperature was so cold that the oil was unable to ignite.

Heating oils

regular-grade diesel oil ■ Diesel oil having a cetane number ranging from 40 to 60

premium-grade diesel oil ■ Diesel oil having a cetane number ranging from 45 to 50

heating oil ■ Any petroleum distillate used directly or as a blend to fuel furnaces and boilers in residences, large buildings, and factories

12.13-I TRANSPORTING CRUDE PETROLEUM AND PETROLEUM-BASED COMMODITIES

In the United States, crude petroleum and petroleum commodities are transported in large volumes, especially by public highway. When shippers offer them for transportation, DOT requires them to select the relevant shipping description from Table 12.15 for entry on an accompanying shipping paper. DOT also requires shippers and carriers to comply with all applicable labeling, marking, and placarding requirements.

When shippers offer hot asphalt for transportation in a crucible, kettle, or other appropriate bulk container as an elevated-temperature material, DOT requires them to display the HOT marking on both sides and ends of the container.

When shippers offer sour crude oil for transportation in bulk packaging, DOT requires them at 49 C.F.R. §172.327 to affix the following to the packaging, or to display a label, tag, or sign with the following precautionary wording:

DANGER
POSSIBLE HYDROGEN SULFIDE INHALATION HAZARD

12.13-J PETROCHEMICALS

During refinery operations, a petroleum fraction is subjected to further distillation to isolate its individual components, or it is appropriately treated to yield substances called **petrochemicals**. Broadly speaking, petrochemicals are substances produced in bulk directly from natural gas, crude petroleum, or a petroleum fraction. The most familiar petrochemicals are petroleum coke, ethylene, propylene, 1,3-butadiene, benzene, toluene, xylene(s), and methanol. They serve as indispensable raw materials for the plastics, synthetic rubber, and synthetic fiber industries.

When heavy petroleum fractions are strongly heated, volatile substances are produced that are condensed and used elsewhere in the refinery. A solid residue, called **petroleum coke**, also is produced. It is composed primarily of elemental carbon. When it has a low sulfur and metal content, petroleum coke may be used to manufacture electrodes for the manufacture of aluminum. It is also used to manufacture synthetic rubbers.

In the United States, by far the most important petrochemicals are ethylene and propylene. Their importance is associated with their use as the feedstocks for manufacturing polyethylene and polypropylene, respectively. They are routinely manufactured by cracking the ethane and propane extracted from natural gas.

$$C_2H_6(g) \longrightarrow CH_2{=}CH_2(g) + H_2(g)$$

Ethane   Ethylene   Hydrogen

$$2C_3H_8(g) \longrightarrow CH_4(g) + CH_2{=}CH_2(g) + CH_2{=}CHCH_3(g) + H_2(g)$$

Propane   Methane   Ethylene   Propylene   Hydrogen

12.14 SIMPLE HALOGENATED HYDROCARBONS

When one or more hydrogen atoms in the molecules of hydrocarbons are substituted with halogen atoms, the resulting compounds are called **halogenated hydrocarbons**, **alkyl halides**, or **aryl halides**. When one hydrogen atom is substituted with a halogen atom, the general chemical formula of the resulting halogenated hydrocarbon is $R{-}X$, where R is the formula of an arbitrary alkyl or aryl group, and X is the chemical symbol of a specific halogen. When the halogen is chlorine, the compounds are called *chlorinated hydrocarbons*, *alkyl chlorides*, or *aryl chlorides*.

---

petrochemical ■ Any substance obtained directly or indirectly from natural gas, crude petroleum, or a petroleum fraction

petroleum coke ■ The product manufactured at petroleum refineries by strongly heating heavy petroleum fractions

halogenated hydrocarbon (alkyl halide, aryl halide) ■ Any hydrocarbon in whose molecules at least one hydrogen atom has been replaced by halogen atoms

TABLE 12.15	Shipping Descriptions of Some Representative Crude-Oil-Based and Petroleum-Based Commodities
PETROLEUM-BASED COMMODITY	**SHIPPING DESCRIPTION**
Antiknock agents	UN1649, Motor fuel antiknock mixtures, 6.1, PG I
Asphalt (at or above its flashpoint) and other liquid tars including road oils and cutback bitumens	NA1999, Asphalt, 3, PG III *or* UN1999, Tars, liquid, 3, PG II
Aviation fuel, turbine engine	UN1863, Fuel, aviation, turbine engine, PG I *or* UN1863, Fuel, aviation, turbine engine, PG II *or* UN1863, Fuel, aviation, turbine engine, PG III
Crude oil	UN1267, Petroleum crude oil, 3, PG I *or* UN1267, Petroleum crude oil, 3, PG II *or* UN1267, Petroleum crude oil, 3, PG III
Diesel fuel, regular or premium	NA1993, Diesel fuel, 3, PG III
Fuel Oil (No. 1, 2, 4, 5, or 6) (domestic transportation)	NA1993, Fuel oil, 3, PG III
Gas oil	UN1202, Gas oil, 3, PG III
Kerosene	UN1223, Kerosene, 3, PG III
Liquefied petroleum gases	UN1075, Petroleum gases, liquefied, 2.1
Motor gasoline	UN1203, Gasoline, 3, PG II
Petroleum oil	NA1270, Petroleum oil, 3, PG I *or* NA1270, Petroleum oil, 3, PG II *or* NA1270, Petroleum oil, 3, PG III
Petroleum distillates	UN1268, Petroleum distillates, n.o.s., 3, PG I *or* UN1268, Petroleum distillates, n.o.s., 3, PG II *or* UN1268, Petroleum distillates, n.o.s., 3, PG III
Petroleum products	UN1268, Petroleum products, n.o.s., 3, PG I *or* UN1268, Petroleum products, n.o.s., 3, PG II *or* UN1268, Petroleum oil, n.o.s., 3, PG III

Consider the following Lewis structures of the chlorinated derivatives of methane:

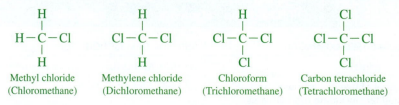

Methyl chloride (Chloromethane)

Methylene chloride (Dichloromethane)

Chloroform (Trichloromethane)

Carbon tetrachloride (Tetrachloromethane)

From left to right, each structure identifies an organic compound in which one, two, three, and four hydrogen atoms, respectively, have been substituted with chlorine atoms. These compounds are known mainly by their common names: methyl chloride, methylene chloride, chloroform, and carbon tetrachloride, respectively.

In the IUPAC system, the halogen atoms are named as substituents of the compound having the longest continuous chain of carbon atoms. As first noted in Section 5.14, the halogen atoms are named as substituents by replacing the *-ine* suffix on the name of the halogen with *-o*. The number of halogen atoms is indicated by the use of *mono-*, *di-*, *tri-*, *tetra-*, and so forth, for one, two, three, four, or more atoms of the same halogen, respectively. Hence, the IUPAC names of the chlorinated derivatives of methane are chloromethane (the mono- prefix is dropped), dichloromethane, trichloromethane, and tetrachloromethane.

SOLVED EXERCISE 12.9

The major component in the fire suppressant known as Pyro-Chem FM-200 is 1,1,1,2,3,3,3-heptafluoropropane. Show how this name is used to determine the molecular formula for this substance.

Solution: The name of this fire suppressant indicates that it is a fluorinated derivative of propane, whose chemical formula is C_3H_8. The use of *hepta* in the name of the fire suppressant means that seven of the eight hydrogen atoms in propane have been replaced with fluorine atoms. The notation "1,1,1,2,3,3,3" identifies precisely which hydrogen atoms along the three-carbon-atom chain have been replaced with fluorine atoms. Three fluorine atoms replace the three hydrogen atoms that are bonded to each of the first and third carbon atoms, and one fluorine atom replaces one of the hydrogen atoms bonded to the second carbon atom. Consequently, the molecular formula of Pyro-Chem FM-200 is the following:

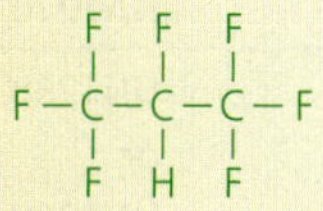

This formula may be condensed to $CF_3-CHF-CF_3$.

The molecular structures and names of several chlorinated derivatives of ethane are noted below:

CH_3-CH_2Cl

Chloroethane

1,1-Dichloroethane

$ClCH_2-CH_2Cl$

1,2-Dichloroethane

1,1,1-Trichloroethane

1,1,2-Trichloroethane

1,1,2,2-Tetrachloroethane

1,1,1,2-Tetrachloroethane

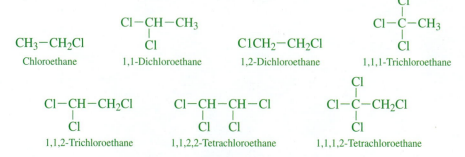

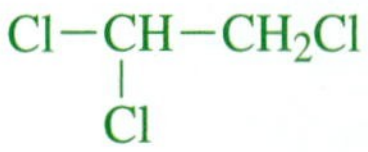

The molecular structures and names of the chlorinated derivatives of ethene are noted next:

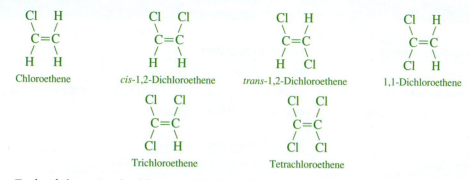

Chloroethene   cis-1,2-Dichloroethene   trans-1,2-Dichloroethene   1,1-Dichloroethene

Trichloroethene   Tetrachloroethene

Each of these simple chlorinated hydrocarbons is a colorless, water-insoluble, highly volatile liquid; most are nonflammable. For decades, this combination of physical properties was the technical basis for their popular use by many manufacturing and process industries. For example, trichloroethene, 1,1,1-trichloroethane, and 1,1,2,2-tetrachloroethane were once popular industrial **degreasers**. These liquids were used to effectively clean oil, grease, wax, and other undesirable material from metallic, textile, and glass surfaces. Carbon tetrachloride once was used as a fire-extinguishing agent. Methylene chloride formerly was widely used as an aerosol, and is still used as a solvent. Tetrachloroethene, also called perchloroethylene and *perc*, is still used as a popular dry-cleaning agent in many states, although in southern California, its use is illegal in new and upgraded facilities.

degreaser ■ Any organic compound used as a solvent to remove grease and oil from surfaces

When inhaled, the vapors of these simple chlorinated hydrocarbons can cause cancer. Furthermore, when released to the environment, they deplete stratospheric ozone. One by one, the simple chlorinated hydrocarbons have been replaced in the modern world with other substances that perform as well for a specific purpose but do not cause cancer or deplete stratospheric ozone. Chemical companies are still actively searching for substances to replace tetrachloroethene as an economical dry-cleaning agent because the days of its use are probably numbered. One nonhazardous agent that has been chosen for this purpose is a mixture of banana and orange extracts.

12.14-A ILL EFFECTS CAUSED BY INHALING THE VAPORS OF THE SIMPLE HALOGENATED HYDROCARBONS

When the vapors of the halogenated methanes and ethanes are inhaled at relatively low concentrations, they usually cause lightheadedness, dizziness, and fatigue. However, when they are inhaled at concentrations above their permissible exposure limits, they cause a lowering of consciousness that can be life-threatening.

Inhalation exposure to the vapors of the following halogenated hydrocarbons causes cancer in laboratory animals: bromodichloromethane, carbon tetrachloride, chloroform, 1,2-dichloroethane, hexachloroethane, methylene chloride, tetrachloroethene, tetrafluoroethene, and trichloroethene. EPA classifies trichloroethene as a human carcinogen,[13] but the other halogenated hydrocarbons in this list generally are acknowledged to be probable carcinogens.

12.14-B TRANSPORTING THE HALOGENATED HYDROCARBONS

When shippers offer a halogenated hydrocarbon for transportation, DOT requires them to enter the relevant shipping description on an accompanying shipping paper. Some examples for several representative halogenated hydrocarbons are listed in Table 12.16.

---

[13]"Toxicological Review of Trichloroethylene" (EPA/635/R-09/011F) (Washington, DC: U.S. Environmental Protection Agency, 2011).

TABLE 12.16	Shipping Descriptions of Some Representative Halogenated Hydrocarbons

HALOGENATED HYDROCARBON OR GROUPS THEREOF	SHIPPING DESCRIPTION
Allyl chloride	UN1100, Allyl chloride, 3, (6.1), PG I (Poison)
Carbon tetrachloride	UN1846, Carbon tetrachloride, 6.1, PG II (Marine Pollutant) (Poison)
Chlorodifluoromethane	UN1018, Chlorodifluoromethane, 2.2 or UN1018, Refrigerated gas R-22, 2.2
Chloroform	UN1888, Chloroform, 6.1, PG III (Poison)
Chlorotrifluoromethane	UN1022, Chlorotrifluoromethane, 2.2 or UN1022, Refrigerated gas R-13, 2.2
1,1-Dichloroethane	UN2362, 1,1-Dichloroethane, 3, PG II
Ethylene dichloride[a]	UN1184, Ethylene dichloride, 3, PG II
Methyl bromide	UN1062, Methyl bromide, 2.3 (Poison - Inhalation Hazard, Zone C)
Methyl chloride	UN1063, Methyl chloride, 2.1 or UN1063, Refrigerated gas R-40, 2.1
Methylene chloride	UN1593, Dichloromethane, 6.1, PG III (Poison)
1,1,2,2-Tetrachloroethane	UN1702, 1,1,2,2-Tetrachloroethane, 6.1, PG II (Poison) (Marine Pollutant)
Tetrachloroethylene	UN1897, Tetrachloroethylene, 6.1, PG III (Poison) (Marine Pollutant)
1,1,1-Trichloroethane	UN2831, 1,1,1-Trichloroethane, 6.1, PG III (Poison)
Trichloroethylene	UN1710, Trichloroethylene, 6.1, PG III (Poison)

[a]Ethylene dichloride and 1,2-dichloroethane are synonyms.

Methylene chloride

chlorofluorocarbon (CFC) ■ Any compound whose molecules are composed solely of carbon, chlorine, and fluorine atoms

DOT also requires shippers and carriers to comply with all applicable labeling, marking, and placarding requirements.

12.15 CHLOROFLUOROCARBONS AND THEIR RELATED COMPOUNDS

A special group of halogenated hydrocarbons are the **chlorofluorocarbons**, or **CFCs**. Their molecules contain only carbon, chlorine and fluorine atoms. They formerly were important commercial compounds, but for reasons soon to be discussed, their use has been sharply curtailed worldwide.

The commercially important chlorofluorocarbons used in the past were primarily members of the following classes of substances:

■ *Chlorofluoromethanes*. These compounds have the general chemical formula CF_nCl_{x-n}, where x and n are whole numbers less than 4.
■ *Chlorofluoroethanes*. These compounds have the general chemical formula $C_2F_nCl_{x-n}$, where x and n are whole numbers less than 6.

As a class of organic compounds, the chlorofluorocarbons are relatively inert substances. They readily vaporize at room temperature. Although many are nonflammable gases, some are flammable.

The chlorofluorocarbons typically were prepared by reacting hydrofluoric acid with a commercially available chlorinated hydrocarbon. For example, a mixture of chlorofluoromethanes was prepared from carbon tetrachloride as follows:

$$2CCl_4(g) \quad + \quad 3HF(l) \quad \longrightarrow \quad CCl_2F_2(g) \quad + \quad CCl_3F(g) \quad + \quad 3HCl(g)$$

Carbon tetrachloride   Hydrofluoric acid   Dichlorodifluoromethane   Trichlorofluoromethane   Hydrogen chloride

The two chlorofluorocarbons were then separated by distillation.

Some examples of the CFCs and their related compounds, hydrofluorocarbons and hydrochlorofluorocarbons, are noted in Table 12.17. Aside from their chemical names, they are also known by commercial names as *CFC-wxyz*, or *R-wxyz*, where *w*, *x*, *y*, and *z* are digits derived from their chemical formulas as follows:

> *w* = the number of carbon–carbon double bonds per molecule
> *x* = the number of carbon atoms per molecule minus one
> *y* = the number of hydrogen atoms per molecule plus one
> *z* = the number of fluorine atoms per molecule

When *w*, *x*, *y*, or *z* is zero, the digit is omitted. For example, *w*, *x*, *y*, and *z* for the chlorofluorocarbon having the formula $CFCl_3$ are 0, 0, 1, and 1, respectively. It is denoted commercially by the product names CFC-11 or R-11. Using the IUPAC system, it is named trichlorofluoromethane.

Sometimes, it is necessary to distinguish between two or more structural isomers of the chlorofluoroethanes in their commercial names. The compound within a group of these isomers having the smallest atomic mass difference on each of the two carbon atoms is denoted without a letter. A lowercase *a*, *b*, *c*, or *d* is appended to *wxyz* to differentiate the isomers as their masses diverge from this difference.

As individual compounds and blends thereof, the chlorofluorocarbons once were widely used for the following purposes:

- Refrigerants and coolants in residential and commercial refrigeration and air-conditioning equipment, including refrigerators, freezers, dehumidifiers, water coolers, ice machines, and air-conditioning units (including automotive air-conditioning units). The CFCs were first commercially introduced into the U.S. market in 1931 as safe alternatives to the flammable and toxic substances then in common use for cooling: methyl chloride, sulfur dioxide, and ammonia. In commerce, they are known collectively as **Freons**, or **Freon agents**.
- Cleaning fluids for electric, precision electronic, and photographic equipment and for maintaining aircraft
- Foam-blowing agents by extruded-polystyrene (Section 14.2-A) and rigid-polyurethane-foam manufacturers (Section 14.9). Because the CFCs have low boiling points, they readily vaporize when mixed with hot plastics. The bubbles of their vapors caused the plastics to expand until they resembled frothy mixtures that solidified as foams. These foams were then marketed commercially as insulation and packaging.
- Aerosols for dispensing consumer products containerized in metal cans. This use of CFCs is believed to have released into the atmosphere far greater quantities of CFCs than did refrigeration and air-conditioning units, which use closed systems to recycle fixed amounts.
- When mixed with ethylene oxide, sterilizers of heat-sensitive reusable hospital apparatus.
- A general inhalation anesthetic in hospitals and veterinary clinics, especially the compound known commercially as halothane, or fluothane.

$$
\begin{array}{ccc}
 & Cl & F \\
 & | & | \\
Br- & C---C & -F \\
 & | & | \\
 & H & F
\end{array}
$$

2-Bromo-2-chloro-1,1,1-trifluoroethane
(Halothane)

Trichloro-ethene

Tetrachloro-ethene

Carbon tetrachloride

Freon (Freon agent)
- The trademark of any chlorofluorocarbon, hydrochlorofluorocarbon, and hydrofluorocarbon used as a foam-blowing agent or refrigerant

TABLE 12.17	Some Commercially Important Chlorofluorocarbons and Hydrochlorofluorocarbons	
CFC OR HCFC DESIGNATION	**CHEMICAL NAME**	**MOLECULAR FORMULA**
CFC-11	Trichlorofluoromethane	$\begin{array}{c}Cl\\|\\Cl-C-Cl\\|\\F\end{array}$
CFC-12	Dichlorodifluoromethane	$\begin{array}{c}F\\|\\Cl-C-Cl\\|\\F\end{array}$
CFC-13	Chlorotrifluoromethane	$\begin{array}{c}F\\|\\Cl-C-F\\|\\F\end{array}$
CFC-21	Dichlorofluoromethane	$\begin{array}{c}H\\|\\Cl-C-Cl\\|\\F\end{array}$
CFC-113	1,1,2-Trichloro-1,2,2-trifluoroethane	$\begin{array}{c}F\\|\\Cl-C-H\\|\\F\end{array}$
CFC-114	1,2-Dichloro-1,1,2,2-tetrafluoroethane	$\begin{array}{c}F\quad F\\|\quad|\\Cl-C-C-F\\|\quad|\\Cl\quad Cl\end{array}$
HCFC-22	Chlorodifluoromethane	$\begin{array}{c}F\quad F\\|\quad|\\Cl-C-C-Cl\\|\quad|\\F\quad F\end{array}$
HCFC-123	2,2-Dichloro-1,1,1-trifluoroethane	$\begin{array}{c}Cl\quad F\\|\quad|\\Cl-C-C-F\\|\quad|\\H\quad F\end{array}$
HCFC-124	2-Chloro-1,1,1,2-tetrafluoroethane	$\begin{array}{c}F\quad H\\|\quad|\\F-C-C-F\\|\quad|\\F\quad Cl\end{array}$
HCFC-123a	1,2-Dichloro-1,1,2-trifluoroethane	$\begin{array}{c}F\quad F\\|\quad|\\Cl-C-C-Cl\\|\quad|\\F\quad H\end{array}$
HCFC-134a	1,1,1,2-Tetrafluoroethane	$\begin{array}{c}F\quad H\\|\quad|\\F-C-C-F\\|\quad|\\F\quad H\end{array}$

Although the CFCs served well for this combination of purposes, they also have less desirable features: They are among the most potent ozone-depleting substances, and they are greenhouse gases. The global warming potential of the commercialized CFCs ranges from 5310 to 11,000 over a 20-year time period. Because the presence of CFCs in our atmosphere seriously exacerbates stratospheric ozone depletion and global warming, environmentalists view positively the restriction in their worldwide manufacture.

12.15-A HOW THE CFCs DESTROY STRATOSPHERIC OZONE

The CFCs are extremely stable compounds with atmospheric lifetimes of hundreds of years—long enough to be transported to the stratospheric ozone layer. When their vapors enter the ozone layer, high levels of ultraviolet radiation cause them to decompose into simpler substances.[14] The CFC molecules absorb certain wavelengths of the ultraviolet radiation from the sun, which causes their carbon–chlorine bonds to rupture, resulting in the production of chlorine atoms. When a chlorofluoromethane molecule absorbs ultraviolet radiation, for example, chlorine atoms are produced.

$$CF_nCl_{x-n}(g) \longrightarrow CF_nCl_{x-n-1}\cdot(g) + :\overset{..}{\underset{..}{Cl}}\cdot(g)$$

A chlorofluorocarbon   A chlorofluorocarbon radical   Chlorine atom

When these chlorine atoms are exposed to ozone molecules in the ozone layer, they react by the following three-step process:

$$O_3(g) + :\overset{..}{\underset{..}{Cl}}\cdot(g) \longrightarrow Cl\overset{..}{\underset{..}{O}}\cdot(g) + O_2(g)$$

Ozone   Chlorine atom   Chlorine monoxide radical   Oxygen

$$Cl\overset{..}{\underset{..}{O}}\cdot(g) + O_3(g) \longrightarrow ClO_2(g) + O_2(g)$$

Chlorine monoxide radical   Ozone   Chlorine dioxide   Oxygen

$$ClO_2(g) \longrightarrow :\overset{..}{\underset{..}{Cl}}\cdot(g) + O_2(g)$$

Chlorine dioxide   Chlorine atom   Oxygen

The first two steps of this process result in the destruction of ozone. The actual ozone-destroying agents are the chlorine atom and chlorine monoxide radical. The cumulative impact of these reactions in the upper atmosphere caused ozone holes to form, through which ultraviolet radiation could penetrate and travel into the lower atmosphere.

To protect the integrity of the ozone layer, the member countries of the United Nations have banned the production of CFCs through implementation of the Montréal Protocol (Section 7.1-N). In the United States, EPA has used the authority of the Clean Air Act to independently ban the use of chlorofluorocarbons as refrigerants, foam-blowing agents, and aerosol propellants. Furthermore, the CPSC has used the authority of the Federal Hazardous Substances Act to ban the use of consumer products containing a CFC propellant. The impact of the combination of these actions has led to the reduction or disappearance of virtually all CFCs from the commercial marketplace.

12.15-B HYDROFLUOROCARBONS AND HYDROCHLOROFLUOROCARBONS

When the use of chlorofluorocarbons was banned, chemical manufacturers sought to find environmentally friendly replacements for use as refrigerants, coolants, and foam-blowing agents. Two groups of such transitional CFC replacements substances were identified:

---

[14]Mario J. Molina and F.S. Rowland, "Stratospheric sink for chlorofluoromethanes: Chlorine atom-catalyzed destruction of ozone," *Nature*, Vol. 249 (1974), pp. 810–812.

hydrofluorocarbons, or *HFCs*, and *hydrochlorofluorocarbons*, or *HCFCs*. HFC molecules have only C–H and C–F bonds, but HCFC molecules have C–H, C–F, and C–C1 bonds. Examples of substances in these two groups are difluoromethane and chlorodifluoromethane, respectively.

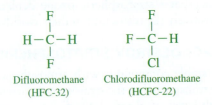

Difluoromethane
(HFC-32)

Chlorodifluoromethane
(HCFC-22)

Initial scientific research suggested that most HFCs and HCFCs were destroyed in the atmosphere at low altitudes and did not significantly diffuse upward to the ozone layer. When this initial information could not be corroborated, however, 190 countries of the United Nations agreed through the Montréal Protocol to phase out the production and use of all HCFCs other than those used as refrigerants by 2015. The protocol terminates the production of all HCFCs by 2030.

Several HCFCs have already been specifically banned by EPA for use as refrigerant gases, coolants, and foam-blowing agents. For example, HCFC-141b (1,1-dichloro-1-fluoroethane) was banned in 2003, and HCFC-142b (1-chloro-1,1-difluoroethane) and HCFC-22 (chlorodifluoromethane) were banned in 2010 although EPA permits their use in equipment manufactured before 2010. Several HCFCs may be used as EPA-approved halon substitutes (Section 12.15-C) for now. The production and use of the HFCs have not yet been restricted.

The HFCs and HCFCs are not only ozone-depleting substances; they are also greenhouse gases. The global warming potential of the common HFCs and HCFCs ranges from 43 to 12,000 and 273 to 5490, respectively, over a 20-year time period (Section 5.8).

Several HCFC replacements have been suggested by chemical manufacturers, including 1,1,1,3,3-pentafluoropropane and the hydrofluoroolefins (HFO), 2,3,3,3-tetrafluoropropene, *trans*-1,3,3,3-tetrafluoropropene, and *trans*-1-chloro-3,3,3-trifluoropropene. EPA approved the uses of 1,1,1,3,3-pentafluoropropane as a refrigerant, coolant, and foam-blowing agent and 2,3,3,3-tetraafluoropropene as an automotive air-conditioning refrigerant.

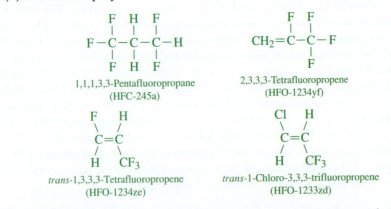

1,1,1,3,3-Pentafluoropropane
(HFC-245a)

2,3,3,3-Tetrafluoropropene
(HFO-1234yf)

trans-1,3,3,3-Tetrafluoropropene
(HFO-1234ze)

trans-1-Chloro-3,3,3-trifluoropropene
(HFO-1233zd)

In 2011, automakers first began to consider replacing HFC-134a with HFO-1234yf as the air-conditioning fluid in new-model cars. Although HFO-1234yf has a GWP of 4 over a 100-year time horizon, the automakers used it to replace HFC-134a, which has a GWP of 1430 over the same time horizon. However, the use of HFO-1234yf as an air-conditioning fluid is not problem-free: It is a flammable gas, causing automakers to be concerned that leaking HFO-1234yf may cause a fire during a head-on collision. Furthermore, when inhaled at elevated concentrations, HFO-1234yf causes asphyxiation and central nervous system disorders.

12.15-C SUBSTITUTES FOR THE HALON FIRE-EXTINGUISHING AGENTS

As first noted in Section 5.14, Halon 1301 (bromotrifluoromethane) and Halon 1211 (bromochlorodifluoromethane) once were popular fire extinguishing agents. They are examples of *bromofluorocarbons*, or *BFCs*, and *bromochlorofluorocarbons*, or *BCFCs*, respectively. As previously noted, Clean Air Act regulations banned their American production and manufacture in January 1994.

EPA has approved the use of a number of commercial fire extinguishing agents that may be substituted for Halon 1211 as a streaming agent for nonresidential uses only. Some substitute agents are subject to narrow-use limits. They neither deplete stratospheric ozone nor contribute as significantly to global warming as Halon 1211. They include the following: HCFC-123 (2,2-dichloro-1,1,1-trifluoroethane), HCFC-124 (2-chloro-1,1,1,2-tetrafluoroethane), Halotron I, HFC-227ea, HFC-236fa, and Novec 1230.

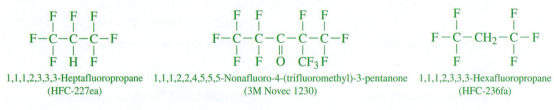

1,1,1,2,3,3,3-Heptafluoropropane (HFC-227ea)

1,1,1,2,2,4,5,5,5-Nonafluoro-4-(trifluoromethyl)-3-pentanone (3M Novec 1230)

1,1,1,2,3,3,3-Hexafluoropropane (HFC-236fa)

Halotron I consists of a three-component proprietary blend based on HCFC-123. It is most commonly discharged as a streaming agent. Following use, it does not leave a residue. Halotron I has been approved by the U.S. Coast Guard as a halon substitute for maritime application onboard watercraft.

The Federal Aviation Administration has approved HFC-227ea and HFC-236fa as fire-extinguishing agents to replace Halon 1211. The agents may be used not only onboard aircraft, but also on rescue and firefighting vehicles. Airbus uses only HFC-236fa, but Boeing uses both HFC-227ea and HFC-236fa.

12.15-D STORING OZONE-DEPLETING SUBSTANCES

Ozone-depleting substances, including the CFCs, HFCs, and HCFCs, are still likely to be encountered today within storage containers and tanks, especially at facilities that received an exemption from EPA or those that maintain, repair, or dispose of refrigeration and air-conditioning equipment. When an ozone-depleting substance is stored, EPA regulations require its owners to mark their containers and tanks with a warning statement that reads as follows: "WARNING: Contains [or manufactured with, if applicable] [insert name of substance], a substance which harms public health and environment by destroying ozone in the upper atmosphere." For example, at 40 C.F.R. §82.106, EPA requires owners of HCFC-22 to mark their containers and tanks of chlorodifluoromethane as follows:

WARNING:

CONTAINS CHLORODIFLUOROMETHANE (HCFC-22), A SUBSTANCE WHICH HARMS PUBLIC HEALTH AND THE ENVIRONMENT BY DESTROYING OZONE IN THE UPPER ATMOSPHERE

This statement serves to assist first-on-the-scene responders who encounter ozone-depleting substances in storage. To protect the environment, they must take precautionary action to prevent the rupture of these containers and tanks; for example, they may cool them with a prodigious volume of water during an ongoing fire. When rupture does occur, emergency responders should implement appropriate measures to confine the contaminated water for subsequent collection.

TABLE 12.18	Shipping Descriptions of Some Representative CFCs, HFCs, and HCFCs
CFC, HFC, OR HCFC	**SHIPPING DESCRIPTION**
Bromotrifluoromethane	UN1009, Bromotrifluoromethane, 2.2 *or* UN1009, Refrigerant gas R-13B1, 2.2
1-Chloro-1,1-difluoroethane	UN2517, 1-Chloro-1,1-difluoroethane, 2.1 *or* UN2517, Refrigerant gas R-142b, 2.1
1-Chloro-1,2,2,2-tetrafluoroethane	UN1021, 1-Chloro-1,2,2,2-tetrafluoroethane, 2.2 *or* UN1021, Refrigerant gas R-124, 2.2
Chlorotrifluoroethane	UN1022, Chlorotrifluoromethane, 2.2 *or* UN1022, Refrigerant gas R-13, 2.2
1,1,1,2-Tetrafluoroethane	UN3159, 1,1,1,2-Tetrafluoroethane, 2.2 *or* UN3159, Refrigerant gas R-134a, 2.2
1,1,1-Trifluoroethane	UN2035, 1,1,1-Trifluoroethane, compressed, 2.1 *or* UN2035, Refrigerant gas R-143a, 2.1

12.15-E TRANSPORTING CFCs, HFCs, AND HCFCs

When shippers offer a CFC, HFC, or HCFC for transportation, DOT requires them to identify the appropriate commodity on an accompanying shipping paper either by using its IUPAC name or "Refrigerant gas R-*wxyz*," where the meaning of the suffix *wxyz* was noted earlier. Some examples of the shipping names for several representative CFCs, HFCs, and HCFs are provided in Table 12.18.

When carriers transport CFCs, HFCs, or HCFCs by highway or rail, DOT requires them to display the words "Dispersant Gas" or "Refrigerated Gas" on two opposing sides of the transport vehicle.

12.16 POLYCHLORINATED BIPHENYLS

The *polychlorinated biphenyls*, or *PCBs*, are the chlorinated derivatives of the complex hydrocarbon *biphenyl*, which has the following molecular structure:

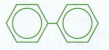

There are 209 structural isomers of the PCBs, each of which has the chemical formula $C_{12}H_xCl_y$, where x ranges from 0 to 9, and y equals $10 - x$. Approximately 130 isomers were components of the PCB products manufactured and marketed throughout the world from the 1930s through most of the 1970s.

At ordinary ambient conditions, products containing PCBs usually are viscous, sticky liquids. Even at elevated temperatures, their individual components are nonflammable, incredibly stable compounds that can be heated to their respective boiling points without undergoing decomposition or bursting into flame. This combination of properties formerly was considered ideal for implementing certain industrial practices that required high temperatures and fire resistance.

In the 1960s, however, the PCBs first became the focus of several scientific studies conducted on animals. These studies pointed out that the PCBs are extraordinarily stable substances that do not biodegrade in the environment; that is, they are *persistent*. The studies also showed that PCB exposure could cause cancer, birth defects, liver damage, chloracne (a disfiguring skin condition that resembles severe acne), impotence, and death. From these studies, we also learned that once ingested, the PCBs migrate into various receptor tissues of the thymus, lungs, spleen, kidneys, liver, brain, muscle, and testes, where they are primarily retained or *bioaccumulated* in the body's fat. Their periodic release from the fatty tissues has been linked with a variety of neurobehavioral disorders and birth abnormalities. For example, children who were exposed to high levels of PCBs during their mother's pregnancy often are born with reduced motor skills and short-term memories.

Today, we now know that PCBs cause physiological disorders by mimicking the action of the body's own **hormones** or by interfering with their normal functions. Hormones are substances secreted in minute quantities by the **endocrine glands**, the organs that control such activities as growth, development, metabolism, and reproduction. When absorbed into the body, the PCBs react like these hormones, sending false signals that interfere with the body's normal activities. They are examples of **endocrine disrupters**. Given this combination of adverse properties, PCBs are ranked by epidemiologists as human carcinogens.

12.16-A USE OF PCBS IN ELECTRICAL EQUIPMENT

From approximately 1930 to 1977, products containing PCBs were marketed for commercial utilization within certain types of electrical equipment such as transformers, voltage regulators, circuit breakers, capacitors, motors, electromagnets, reclosers, cables, and fluorescent light ballasts. They were also manufactured as heat-transfer, hydraulic, lubricating, and cutting fluids. PCBs were also used to manufacture carbonless copy paper, and they were incorporated as constituents in certain plastic, paint, wax, and rubber products. In all these instances, the purpose of the PCBs was primarily to provide these commercial products with a greater degree of fire resistance.

Because PCBs pose an unreasonable risk to public health and the environment, EPA banned their manufacture, sale, import, and distribution in 1978 by using the legal authority of TSCA. The relevant regulations are published at 40 C.F.R. §761 *et seq.* Although their production and manufacture is banned in the United States, EPA permits the indefinite use of PCBs in systems that were operating before 1978, albeit under specific regulatory conditions. Even today, PCBs can be legally used in a "totally enclosed manner," that is, within equipment that has been designed and constructed so as to ensure that no human or environmental exposure to them occurs. When this equipment is withdrawn from service or repaired, EPA requires replacement of the PCB fluid with a non–PCB-containing alternative product.

An electrical transformer is an example of a totally enclosed system. One type can often be observed near the tops of electric poles. It is a static device whose function is to transfer electrical energy between circuits of different voltage. Each transformer has a magnetic core around which there are several windings or coils of a metal such as copper. This assembly of core and current-carrying coils is then immersed in a fluid within a tank. The purpose of the fluid, called a **dielectric fluid**, is to increase the resistance of the unit to sustained arcing and to serve as a heat-transfer medium. Large numbers of such transformers are located at electrical substations operated by municipal utility companies.

Within the context of the TSCA regulations, EPA defines a **PCB transformer** at 40 C.F.R. §761.3 as a transformer that contains PCBs at a concentration of 500 parts per

hormone ■ A chemical substance produced in minute amounts by an endocrine gland and carried through the bloodstream to target organs or a group of cells of the body, where it regulates growth, reproduction, and other biological functions

endocrine gland ■ A ductless gland, such as the adrenal, thyroid, and pituitary glands, that secretes fluids directly into the general circulatory system

endocrine disrupter ■ Any compound that mimics a hormone or interferes with hormonal action when ingested into the body

dielectric fluid ■ The fluid used in certain types of electrical equipment to increase their resistance to arcing and serve as a heat-transfer medium

PCB transformer ■ For purposes of TSCA regulations, a transformer whose dielectric fluid contains polychlorinated biphenyls at a concentration equal to or exceeding 500 parts per million

mineral oil transformer
■ An electrical transformer containing mineral oil as its dielectric fluid

PCB-contaminated transformer ■ For purposes of TSCA regulations, a transformer whose dielectric fluid contains polychlorinated biphenyls at a concentration greater than 50 parts per million but less than 500 parts per million

non-PCB transformer
■ For purposes of TSCA regulations, a transformer whose dielectric fluid contains polychlorinated biphenyls at a concentration equal to or less than 50 parts per million

million or greater. A typical PCB transformer contains approximately 200 gallons (1500 L) of fluid with a PCB concentration of approximately 50% to 60% by volume, but a large PCB transformer may hold as much as 1000 gallons (7500 L) of PCB fluid. The remaining 40 to 50% consists of diluents, a mixture of trichlorobenzene and tetrachlorobenzene isomers.

The vast majority of transformers in service today were manufactured with the intent of using mineral oil as their dielectric fluid. Mineral oil is a petroleum-based oil, and these units are called **mineral oil transformers**. Although PCBs were never intentionally added to the mineral oil used in these transformers, the oil often became contaminated with PCBs when workers conducted routine equipment maintenance on both transformer types using the same hoses and pumping equipment. EPA defines a unit as a **PCB-contaminated transformer** when sampling and analysis indicates that its fluid contains PCBs at a concentration greater than 50 parts per million but less than 500 parts per million. A **non-PCB transformer** is a unit containing PCBs at a concentration equal to or less than 50 ppm.

12.16-B PCBs AND FIREFIGHTING

Although PCB fluids do not readily burn, they may ignite within an energized transformer when high-energy arcing, electrical overloading, or short-circuiting occurs. These fires are characterized by large amounts of oily, black soot in the accompanying smoke plume. When the energized transformer is located within a building, the soot and PCB vapors travel with the smoke plume through ventilation shafts, building ductwork, and other open-construction areas.

Today's firefighters respond rarely to PCB transformer fires, but because the service lifetime of a transformer is 40 to 85 years, such an action is not outside the realm of possibility. Firefighters are faced with an unwarranted health risk when they are exposed to the smoke from these class C fires. To reduce the magnitude of the risk, EPA requires the owners of the PCB transformers to prominently affix the M_L marking shown in Figure 12.10 so that it is easily read by firefighting personnel who respond to an electrical-equipment fire. EPA requires the M_s marking shown in Figure 12.10 to be affixed to PCB equipment that cannot accommodate the larger M_L marking. EPA also imposes mandatory reporting requirements at 40 C.F.R. §761.30 to ensure that the National Response Center (Section 1.13) is contacted immediately on the occurrence of an accidental fire involving PCBs.

EPA also imposes reporting requirements on the owners of facilities that still use electrical equipment containing PCBs. For example, EPA requires the registration of PCB transformers, including out-of-service transformers held in storage for reuse. Fire department personnel should identify the location of these transformers within their area of jurisdiction and plan a response action in case of an accidental PCB release.

When PCBs undergo incomplete combustion during transformer and other electrical-equipment fires, they produce the highly toxic compounds called polychlorinated dibenzofurans and polychlorinated dibenzo-*p*-dioxins (Section 13.4-B). During electrical-equipment fires involving PCBs, these substances also may be released into the environment. Firefighters should be especially wary when combating electrical equipment fires, because exposure to these toxic substances could cause long-term ill effects. Regardless of the magnitude of the fire, firefighters should always wear self-contained breathing apparatus to eliminate or reduce the potential for exposure to these substances. Public health can also be effectively protected by constructing suitable diking, diversion curbing, or grading to confine the spread of these contaminants and prevent their drainage into storm sewers.

A fire involving electrical equipment is not the only type during which PCBs, polychlorinated dibenzofurans, and polychlorinated dibenzo-*p*-dioxins may be released. Before 1977, PCBs were components of some protective and decorative paints, varnishes, lacquers, sealants, adhesives, caulking, and similar construction products. Some of them contained 5% to 30% PCBs. Although these products were used several decades ago, the

FIGURE 12.10 These two EPA markings are called the M_L and M_s markings, respectively. They differ only by their wording and size. The letters and striping on both markings are black on a yellow background and include the warning CAUTION CONTAINS PCBs and instructions to be implemented in the event of a PCB release. The M_L marking also includes the telephone number of the National Response Center. The TSCA regulation at 40 C.F.R. §761.45 requires a marking to be affixed to PCB transformers and PCB capacitors in service and PCB items containing or contaminated with PCBs in storage for disposal when the PCB concentration is equal to or greater than 50 parts per million. The regulation also requires the marking to be affixed to all doors or accesses leading from the outermost door to the door immediately accessing the PCB unit.

PCBs still remain in surface coatings, joint sealants around windows, caulking between masonry panels, and elsewhere. As fires occur in the buildings in which these PCB-containing products were once applied or used, the PCBs and their incomplete combustion products are released into the environment.

12.16-C THE ADVERSE IMPACT ON THE ENVIRONMENT CAUSED BY PCBs

Why are PCBs regarded as ultra-hazardous substances? The answer to this question is linked with the horrific impact they could have not only on public health but also on a wide segment of the environment. The worst PCB-contaminated site in the United States is a 200-mile stretch of the Hudson River in New York north of Albany. An estimated 1.3 million pounds (~0.6 million kg) of PCBs was allegedly dumped into the river by an electrical-transformer manufacturer from 1947 to 1977.

The adverse environmental impact from exposure to PCBs is linked with its passage from animal to animal. When PCBs are consumed by an animal that is then eaten by another animal, the PCBs pass progressively through the food chain: worms to fish and birds; fish to birds; and birds and fish to mammals. This process is called **biomagnification**, or **bioamplification**, because the PCBs increasingly concentrate as they pass up the food chain. The PCBs may thereby affect the well-being of many individual species, including humans who consume fish or meat contaminated with PCBs. In the United States alone, EPA has identified 703 bodies of water where health advisories have been posted to warn people against consuming PCB-ridden fish.

How do PCBs affect the health of birds and fish? The answer to this question centers on their immune systems. Fish who have consumed PCBs become sexually dysfunctional. The female fish are born with immature ovaries causing them to experience a reduced

biomagnification (bioamplification) ■ The retention of a substance in human and animal tissues at increasingly higher concentrations as the substance passes through successive levels of the food chain

ability to spawn. Both fish and birds who have consumed PCBs experience growth and reproductive impairment. In birds, these problems include eggshell thinning, lower egg production, decreased egg hatching, deformities in the juvenile birds that do emerge from eggs, and reduction in their growth and survival rates.

The environmental impact from exposure to PCBs is especially evident in aquatic organisms. Within the ocean hierarchy, for example, killer whales are contaminated with PCBs at concentrations higher than those observed in other marine animals because they are at the top of the food chain. Although herring may carry PCBs at an average concentration of only 1 part per million, the seals that eat them may contain 20 parts per million, and the killer whales that eat these contaminated seals may have PCB levels as high as 250 parts per million. The male killer whales carry the PCBs throughout their lives, but the females periodically rid these pollutants from their bodies by passing them to their calves. This could explain why the males live only half as long as the females, and why many calves die within several months of their birth.

12.16-D FDA REGULATIONS INVOLVING PCBs

Because PCBs have become an unavoidable contaminant in the environment, FDA publishes the following tolerance limits for PCB residues in foods, at 21 C.F.R. §109.30:

- 1.5 ppm in milk (fat basis)
- 1.5 ppm in manufactured dairy products (fat basis)
- 3 ppm in poultry (fat basis)
- 0.3 ppm in eggs
- 0.2 ppm in finished animal feed for food-producing animals (other than feed concentrates, feed supplements, and feed premixes)
- 2 ppm in animal feed components of animal origin, including fishmeal and other by-products of marine origin and in animal feed concentrates, feed supplements, and feed premixes intended for food-producing animals
- 2 ppm in the edible portions of fish and shellfish
- 0.2 ppm in infant and junior foods

12.16-E TRANSPORTING PCBs

When shippers offer polychlorinated biphenyls for transportation, DOT requires them to identify the PCBs on an accompanying shipping paper as one of the following:

> **UN2315, Polychlorinated biphenyls, liquid, 9, PG II (Marine Pollutant)**
> **UN2315, Polychlorinated biphenyls, solid, 9, PG II (Marine Pollutant)**

PCBs

DOT also requires them to affix CLASS 9 labels on its packaging.

When bulk quantities of PCBs are transported, DOT requires the identification number 2315 to be on displayed on orange panels or white square-on-point diamonds on the packaging. DOT does not require CLASS 9 placards to be displayed on the transport vehicle.

When shippers offer a PCB-containing electrical unit including a PCB-contaminated transformer for transportation, EPA requires them to affix the appropriate marking shown in Figure 12.10 on the unit when it is transported. EPA requires this marking to be affixed on an electrical unit even when the fluid has been drained from it, as the unit typically contains residual PCB fluid.

12.17 ORGANOCHLORINE PESTICIDES

The *organochlorine pesticides* are chlorinated organic compounds that farmers formerly used as insecticides. Most are chlorinated hydrocarbons like the PCBs. Also like the PCBs, they are extremely stable substances that do not biodegrade in the environment. Instead, their former use has resulted in their bioaccumulation through the food web in different animal species, ultimately causing their demise or decline in numbers. EPA now considers the organochlorine pesticides to pose a substantial risk of causing a negative impact on human health and the environment. Consequently, it does not issue FIFRA registrations (Section 1.4-B) to their potential manufacturers. Hence, these pesticides are either unavailable commercially, or their use is being phased out in the United States.

In 2001, the United Nations also took action on banning or severely limiting the use of organochlorine pesticides by implementation of the international treaty known as the **Stockholm Convention on Persistent Organic Pollutants**, or the **POPs Treaty.** The POPs are a group of the 13 most persistent, toxic substances that pose a threat to human health and the environment. The group is composed of aldrin, chlordane, dieldrin, endrin, heptachlor, hexachlorobenzene, Mirex^{TM}, Toxaphene^{TM}, DDT, PCBs, the polychlorinated dibenzofurans and dibenzo-*p*-dioxins, and hexabromocyclododecane (Section 14.2-A). The first nine members of this listing are organochlorine pesticides. Although it is doubtful that emergency responders would ever encounter them, this brief introduction is provided here to emphasize that some toxic substances are considered so ultra-hazardous that their production, manufacture, and use must be regulated.

Stockholm Convention on Persistent Organic Pollutants (POPs treaty) ■ The United Nations treaty that bans or limits the worldwide use of nine organochlorine pesticides, PCBs, the polychlorinated dibenzofurans and dibenzo-*p*-dioxins, and hexabromocyclododecane

REVIEW **EXERCISES**

Aliphatic Hydrocarbons

1. Determine whether each of the following chemical formulas represents an alkane, cycloalkane, alkene, diene, triene, cycloalkene, or alkyne:
   (a) C_3H_6
   (b) $CH_3-(CH_2)_5-CH_3$
   (c) $CH_3-CH=CH-CH_3$
   (d) $CH_3-(CH_2)_3-C\equiv C-CH_2CH_3$
   (e) $CH_3-CH=CH-CH=CH-CH=CH_2$
   (f) 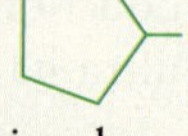 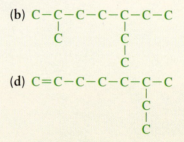 $-CH_2CH_3$

2. Write the condensed chemical formula of the aliphatic hydrocarbon represented by the following carbon skeletons:
   (a) C–C–C–C–C–C–C–C

   (b) C–C–C–C–C–C–C
         with C branch and C–C branch

   (c) C=C–C–C–C–C–C

   (d) C=C–C–C–C–C–C with C–C branch

3. Using the IUPAC system, name the alkanes having the following condensed formulas:
   (a) $CH_3CH_2-CH-CH_2CH_3$ with CH_3 branch
   (b) $CH_3-CH-CH_2-CH-CH_3$ with CH_2-CH_3 and CH_3 branches
   (c) $CH_3CH_2-CH-CH_2CH_3$ with $CH_2CH_2CH_3$ branch
   (d) $CH_3CH_2CH_2CH_2-C-CH_2CH_2CH_3$ with CH_3, CH_3-C-CH_3, CH_3 branches

4. Using the IUPAC system, name the alkenes having the following condensed formulas:
   (a) $CH_3CH_2CH_2-CH=CH_2$
   (b) H, CH_3 / C=C / H, CH_3
   (c) $CH_3CH_2CH_2-CH$ with CH_3, C=C, H, H
   (d) $-CH_2CH=CH_2$

5. Using the IUPAC system, name the alkynes having the following condensed formulas:
   (a) $CH_3CH_2CH_2-C\equiv C-CH_2CH_2CH_3$
   (b) $CH_3CH_2CH_2-CH-C\equiv C-H$ with $CH_2CH_2CH_3$ branch

6. Write the condensed formulas for the aliphatic hydrocarbons that are named as follows:
   (a) *n*-pentane
   (b) 2-pentene
   (c) *n*-heptane
   (d) acetylene
   (e) 3-methylheptane
   (f) 2-ethyl-1-butene
   (g) *cis*-2-butene
   (h) isobutane
   (i) cyclopentane
   (j) ethylcyclopentane
   (k) isopentane
   (l) 1-ethylcyclopentene
   (m) 3-ethyl-3-methylheptane
   (n) 3-methyloctane
   (o) 2-methyl-2-butene
   (p) 2,4-dimethyl-2-pentene
   (q) ethylcyclohexane
   (r) di-*n*-propylacetylene
   (s) 4-methyl-1,3-pentadiene
   (t) 3-hexene

Natural Gas (Methane)

7. At 30 C.F.R. §75.323(b)(1)(i), the Mine Safety and Health Administration (MSHA) of the U.S. Department of Labor requires mine owners and operators to accomplish the following: "When 1.0% or more methane is present in a working place or an intake air course . . . electrically powered equipment in the affected area must be deenergized, and other mechanized equipment shall be shut off." What is the most likely reason that MSHA promulgated this regulation?

8. Using the relevant data in Table 2.11, calculate the approximate volume in cubic feet of gaseous methane that results from the release of one ton of LNG to the environment during a transportation mishap.

Liquefied Petroleum Gas

9. The DOT regulation at 49 C.F.R. §173.315(b)(1) stipulates that liquefied petroleum gas must be odorized by the addition of a small amount of a substance such as ethyl mercaptan, thiophane, or amyl mercaptan. What is the most likely reason DOT requires liquefied petroleum gas to be odorized in this fashion?

10. When liquefied petroleum gas is transported in a portable tank by highway, what unique marking does DOT require on two opposing sides of the tank?

Ethylene and Propylene

11. Why is the U.S. production of ethylene and propylene far greater than the production of all other chemical substances?

Butadiene

12. When shippers offer butadiene for transportation, DOT requires them to enter the word "stabilized" with the shipping description. What is the most likely reason DOT requires the inclusion of this word in the shipping description?

Acetylene

13. Methyl acetylene and propadiene are minor constituents produced when acetylene is manufactured from natural gas.
    (a) Write the chemical formulas of these two substances.
    (b) Write the equations that illustrate their complete combustion.

Aromatic Hydrocarbons

14. Name each benzene derivative that is named as follows:

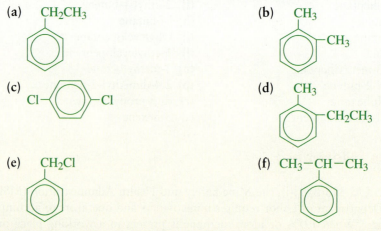

(a) CH$_2$CH$_3$

(b) CH$_3$ / CH$_3$

(c) Cl— —Cl

(d) CH$_3$ / CH$_2$CH$_3$

(e) CH$_2$Cl

(f) CH$_3$—CH—CH$_3$

15. Write the condensed formulas for the aromatic hydrocarbons that are named as follows:
    (a) *m*-xylene
    (b) *p*-diethylbenzene
    (c) 1,3,5-triethylbenzene
    (d) 1-phenylpropene
    (e) diphenylmethane
    (f) *tert*-butylbenzene

Polynuclear Aromatic Hydrocarbons

16. Which GHS pictogram(s) in Table 1.3 are manufacturers, distributors, or importers required to display on the labels of bottles containing benzo[*a*]pyrene?

Petroleum Products

17. Using standard methods of testing, *n*-butane is found to have an octane number of 99. What does this number signify?
18. When it burns, which produces the greater concentration of sulfur dioxide: light diesel fuel or heavy diesel fuel?
19. 2,3-Dimethylpentane is manufactured at some petroleum refineries by the alkylation of isobutane and propylene. Write the equation for this chemical reaction.
20. Write the condensed formulas for the following substances, each of which is a constituent of motor gasoline:
    (a) 3,3,5-trimethylheptane
    (b) isopropylbenzene
    (c) 2,2-dimethyloctane
    (d) 3-methylnonane
    (e) 3,3-dimethyloctane
    (f) 3-methylheptane
    (g) 1,1,2-trimethylcyclohexane
    (h) 2-methylpentane
    (i) 3,4-dimethylhexane
    (j) 2-ethylheptane
    (k) 2,4,4-trimethylhexane
    (l) *m*-ethyltoluene
    (m) 2-methyloctane
    (n) 3-ethylhexane
    (o) 2,3-dimethylhexane
    (p) 1,3,5-trimethylbenzene
    (q) *sec*-butylbenzene
    (r) 2-methyloctane

Simple Halogenated Hydrocarbons

21. Using the IUPAC system, name the halogenated hydrocarbons having each of the following condensed formulas:

   (a) $Br-CH_2CH_2-Br$

   (b) 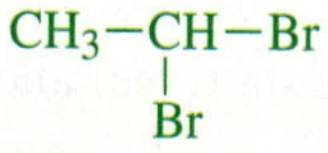 $CH_3-CH-Br$
                $|$
               Br

   (c) $CHCl_3$

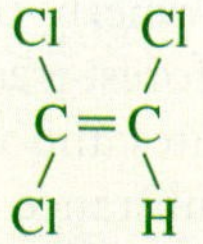

   (d) 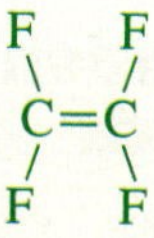 F   F
        \\ /
        C=C
        / \\
       F   F

   (e) Cl   Cl
        \\ /
        C=C
        / \\
      Cl H

   (f) CH_2CH_2Cl
                 $|$
       $CH_3-CH-CH_3$

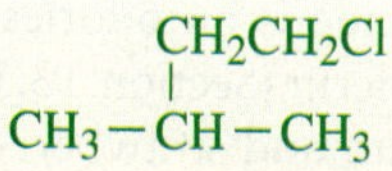

22. Write the condensed formulas for the halogenated hydrocarbons that are named as follows:
   (a) carbon tetrachloride
   (b) *cis*-2,3-dichloro-2-butene
   (c) 2-bromopropane
   (d) *o*-chlorotoluene
   (e) 1-phenyl-2-chloroethane
   (f) dichlorodifluoromethane

23. A motor carrier contracts to haul 1000 gallons (3785 L) of trichloroethylene in a tank truck from a chemical production facility to a customer's plant along a specified route. Does DOT require the carrier to prepare and implement a transportation security plan and obtain a hazardous materials safety permit before the shipment?

Chlorofluorocarbons, Hydrofluorocarbons, and Hydrochlorofluorocarbons

24. Chlorofluorocarbons formerly were used as refrigerants in soft-drink vending machines and ice-cream display cases. Why has their use been curtailed and, in most cases, eliminated?

25. Show that CFC-114 is the commercial name for the chlorofluorocarbon having the chemical name dichlorotetrafluoroethane.

26. What are the molecular structures and chemical names of the two structural isomers of dichlorotrifluoroethane?

Polychlorinated Biphenyls

27. The EPA regulation published at 40 C.F.R. §761.40(j)(ii) stipulates that owners of PCB electrical transformers must comply with the following: Mark the vault door, machinery room door, fence, hallway, or other means of access to these transformers with a marking shown in Figure 12.10; coordinate with the primary fire department; and document that the fire department knows, accepts, and recognizes the meaning of this marking. What is the most likely reason that EPA requires these steps to be taken?

28. What special action should be taken by firefighters when responding to a smoldering fire at a metal salvage yard where employees dismantled electrical transformers to retrieve their copper content?

Organochlorine Pesticides

29. When scientists measured the DDT concentration in samples of the tissues of various animals living in a Long Island salt marsh, they obtained the following data: shrimp, 0.16 ppm; eels, 0.28 ppm; needlefish, 2.07 ppm; and ring-billed gulls, 75.5 ppm. What do these data reveal about animals who have been exposed to DDT?

Responding to Incidents Involving a Release of Organic Compounds

30. Methyl bromide is an odorless, colorless, and poisonous gas. It formerly was used with chloropicrin (Section 13.11-D) as an agricultural fumigant by forest tree nurseries and tomato and strawberry growers to exterminate plant parasites and wipe out disease in soils. However, methyl bromide is an ozone-depleting substance. Its production and use have been largely phased out in the United States in compliance with conditions of the Montréal Protocol (Section 7.1-N).

    When first-on-the-scene responders arrive at the scene of a railway accident at 11:30 p.m. in San Diego, California, they note a ruptured, overturned tankcar and the following shipping description on the accompanying railroad waybill:

UNITS	HM	SHIPPING DESCRIPTION (IDENTIFICATION NUMBER, PROPER SHIPPING NAME, PRIMARY HAZARD CLASS OR DIVISION, SUBSIDIARY HAZARD CLASS OR DIVISION, AND PACKING GROUP)	WEIGHT (lb)
1 tankcar	X	UN1062, Methyl bromide, 2.3 (Inhalation Hazard, Zone C) (Placarded POISON GAS)	2500

    Use Table 10.15 to identify the isolation and protective-action distances recommended by DOT to protect public health, safety, and the environment.

31. When first-on-the-scene responders arrive at the scene of a highway accident, they note the HOT marking displayed on the exterior surface of a kettle that has been ruptured by impact with a motor van. The hot contents have spilled on the underlying highway. The responders note the following shipping description of the kettle contents:

UNITS	HM	SHIPPING DESCRIPTION (IDENTIFICATION NUMBER, PROPER SHIPPING NAME, PRIMARY HAZARD CLASS OR DIVISION, SUBSIDIARY HAZARD CLASS OR DIVISION, AND PACKING GROUP)	WEIGHT (lb)
1750-gal kettle	X	HOT, UN3257, Elevated-temperature liquid, n.o.s. (asphalt), 3, PG III	1200

    What immediate actions should the responders take to protect public health, safety, and the environment?

KEY TERMS

- Associate the physical and health hazards of the organic compounds noted in this chapter with the information provided by their hazard diamonds and GHS pictograms.
- Memorize the functional group that characterizes alcohols, ethers, aldehydes, ketones, organic acids, esters, amines, and peroxo-organic compounds.
- Memorize and apply the rules for naming simple alcohols, ethers, aldehydes, ketones, organic acids, esters, amines, and peroxo-organic compounds.
- Identify the adverse health effects that result from abusing the use of alcoholic beverages.
- Describe the most practical means of extinguishing bulk ethanol fires.
- Identify the risk associated with encountering an elevated concentration of peroxo-organic compounds within the containers used to store ethers.
- Identify the hazardous properties of the halogenated ethers, including the PCDFs, PCDDs, PBDFs, PBDDs, and PBDEs, and identify the most likely ways by which emergency responders are likely to be exposed to them.
- Identify the locations at which emergency responders are likely to encounter formaldehyde.
- Identify the general nature of the labels required by the U.S. Federal Trade Commission on biodiesel dispensers when types of this material are provided to customers for potential use as alternative motor fuels.
- Identify the labels, markings, and placards that DOT requires on packaging of the organic compounds noted in this chapter, especially organic peroxides, and the transport vehicles used for their shipment.

Chemists have learned from experience that organic compounds can be classified into families according to their common molecular features. For example, we noted in Chapter 12 that alkanes, alkenes, and alkynes are chemical families because each of their members has at least one carbon–carbon single, double, and triple bond, respectively. These structural similarities are primarily responsible for the reactions that are noted by the members of each family.

Certain organic compounds are composed of molecules in which one or more of their carbon atoms are bonded directly to an oxygen atom. Depending on the nature of the chemical bonding, these compounds are called alcohols, ethers, aldehydes, ketones, organic acids, esters, or peroxo-organic compounds. Other organic compounds, called amines, are composed of molecules in which the carbon atoms covalently bond to a nitrogen atom. In this chapter, we study the hazardous characteristics of some representative members of each family.

13.1 FUNCTIONAL GROUPS

One or more hydrogen atoms in the molecular structure of a hydrocarbon may be substituted with another atom or group of atoms. When this substitution occurs, a new organic compound is produced. The atom or group of atoms that substitutes for the hydrogen atom is an example of a **functional group**, because it represents the location on the molecule at which chemical reactions often occur. Multiple functional groups are components of complex molecules.

The functional groups that are components of the molecular structures of the most common organic compounds are listed in Table 13.1. The group of atoms associated with each class of organic compound should be memorized. The alkenes, alkynes, and halogenated hydrocarbons were introduced in Chapter 12; we study the other classes of organic compounds in this chapter.

functional group ■
The atom or group of atoms in the molecules of a substance that characterizes its chemical behavior

	Some Important Functional Groups in the Molecular Structures of Organic Compounds[a]	
TABLE 13.1		

CLASS OF ORGANIC COMPOUND	GENERAL FORMULA	FUNCTIONAL GROUP
Alkene	$\begin{array}{cc} R & R' \\ & \\ \diagdown C = C \diagup \\ \diagup & \diagdown \\ R'' & R''' \end{array}$	$C{=}C$, carbon-carbon double bond
Alkyne	$R-C{\equiv}C-R'$	$C{\equiv}C$, carbon-carbon triple bond
Halogenated hydrocarbon	$R-CH_2-X$ $\begin{array}{c} R' \\ | \\ R-CH-X \end{array}$ $\begin{array}{c} R' \\ | \\ R-C-X \\ | \\ R'' \end{array}$	$-X$ ($X{=}F$, Cl, Br, or I, referred to as fluoro, chloro, bromo, and iodo, respectively)
Alcohol	$R-CH_2-OH$ $\begin{array}{c} R' \\ | \\ R-CH-OH \end{array}$ $\begin{array}{c} R' \\ | \\ R-C-OH \\ | \\ R'' \end{array}$	$-OH$, hydroxyl
Ether	$R-O-R'$	$-O-$, oxy
Aldehyde	$\begin{array}{c} O \\ \| \\ R-C \\ | \\ H \end{array}$	$\begin{array}{c} -C- \\ \| \\ O \end{array}$, carbonyl
Ketone	$\begin{array}{c} R-C-R' \\ \| \\ O \end{array}$	$\begin{array}{c} -C- \\ \| \\ O \end{array}$, carbonyl
Carboxylic acid	$\begin{array}{c} O \\ \| \\ R-C \\ \diagdown \\ OH \end{array}$	$\begin{array}{c} O \\ \| \\ -C \\ \diagdown \\ O- \end{array}$, carboxyl
Ester	$\begin{array}{c} O \\ \| \\ R-C \\ \diagdown \\ O-R' \end{array}$	$\begin{array}{c} O \\ \| \\ -C \\ \diagdown \\ O- \end{array}$, carboxyl
Amine	$R-NH_2$ $R-NH-R'$ $\begin{array}{c} R-N-R' \\ | \\ R'' \end{array}$	$-NH_2$, amino
Hydroperoxide	$H-O-O-R$	$H-O-O-$, hydroperoxyl
Peroxide	$R-O-O-R'$	$-O-O-$, peroxyl

[a]R, R', R'', and R''' are arbitrary alkyl or aryl substituents.

OSHA regulates the storage of ethylene glycol monoethyl ether acetate as a category 2 flammable liquid. This substance has the following molecular formula:

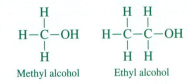

Use Table 13.1 to identify the functional groups in this molecule.

Solution: Table 13.1 indicates that the molecules of ethylene glycol monoethyl ether acetate have one carboxyl and one oxy group.

13.2 ALCOHOLS

alcohol ■ Any organic compound whose molecules contain at least one hydroxyl (—OH) group

Alcohols are organic compounds derived by substituting one or more hydrogen atoms in a hydrocarbon molecule with the *hydroxyl group* (—OH). Thus, the general chemical formula of the simplest alcohols is R—OH, where R is the formula of an arbitrary alkyl or aryl group.

When one hydrogen atom is substituted with the hydroxyl group in molecules of methane and ethane, the resulting compounds are methyl alcohol and ethyl alcohol, respectively.

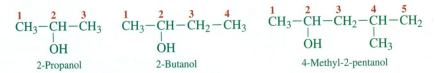

These are the common names of the two simplest alcohols.

In the IUPAC system, the simple alcohols are named by replacing the *-e* in the name of the corresponding alkane with *-ol*; hence, the compounds having the formulas CH_3OH and CH_3CH_2OH are also named methanol and ethanol, respectively.

To name more complex alcohols, it is necessary to indicate the position of the hydroxyl group by a number immediately preceding the name of the alcohol. We use the following rules:

■ Locate the longest chain of carbon atoms that contains the hydroxyl group.
■ Consecutively number them so that the lowest possible number is assigned to the carbon atom to which the hydroxyl group is bonded.

These rules are applied in the following examples:

$$
\begin{array}{ccc}
\overset{1}{C}H_3-\overset{2}{C}H-\overset{3}{C}H_3 & \overset{1}{C}H_3-\overset{2}{C}H-\overset{3}{C}H_2-\overset{4}{C}H_3 & \overset{1}{C}H_3-\overset{2}{C}H-\overset{3}{C}H_2-\overset{4}{C}H-\overset{5}{C}H_2 \\
| & | & | \qquad\qquad | \\
OH & OH & OH \qquad\quad CH_3 \\
\text{2-Propanol} & \text{2-Butanol} & \text{4-Methyl-2-pentanol}
\end{array}
$$

When one or more hydroxyl groups are present in a molecular structure, the corresponding compound sometimes is named as a hydroxy derivative of the parent compound. The number of hydroxyl groups in the structure is indicated by the use of *mono-*, *di-*, *tri-*, and *tetra-*, as relevant, for 1, 2, 3, and 4, respectively. When two or three hydroxyl groups are present in a molecular structure, the compound may also be named as a diol or triol,

TABLE 13.2	Physical Properties of Some Simple Alcohols		
	METHANOL	**ETHANOL**	**ISOPROPANOL**
Melting point	−144°F (−98°C)	−173°F (−114°C)	−128°F (−89°C)
Boiling point	149°F (65°C)	174°F (79°C)	180°F (82°C)
Specific gravity at 68°F (20°C)	0.79	0.79	0.79
Vapor density (air = 1)	1.11	1.59	2.07
Vapor pressure at 68°F (20°C)	98 mmHg	50 mmHg	33 mmHg
Flashpoint (without water) (with 15% water by volume) (with 24% water by volume)	54°F (12°C)	54°F (12°C) 68°F (20°C) 97°F (36°C)	53°F (12°C)
Autoignition point	867°F (464°C)	793°F (423°C)	750°F (399°C)
Lower flammable limit	6% by volume	3.3% by volume	2.3% by volume
Upper flammable limit	36.5% by volume	19% by volume	12.7% by volume
Evaporation rate (ether = 1)	5.2	7 (approximate)	7.7
Heat of combustion	9740 Btu/lb (22,700 kJ/kg)	12,800 Btu/lb (29,700 kJ/kg)	14,200 Btu/lb (33,000 kJ/kg)

respectively, of the parent compound. The following examples illustrate the use of these rules for naming alcohols with multiple hydroxyl groups:

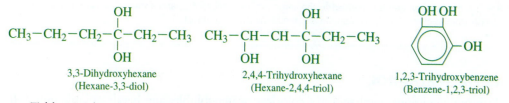

3,3-Dihydroxyhexane
(Hexane-3,3-diol)

2,4,4-Trihydroxyhexane
(Hexane-2,4,4-triol)

1,2,3-Trihydroxybenzene
(Benzene-1,2,3-triol)

Table 13.2 lists some physical properties of three simple aliphatic alcohols: methanol, ethanol, and isopropanol. These data indicate that when exposed to an ignition source, their vapors easily burn. Fire and explosion are considered their primary risks. Because oxygen is already a component of their molecular structures, these alcohols predominantly produce the products of complete combustion when they burn.

$$CH_3OH(g) \ + \ 2O_2(g) \ \longrightarrow \ CO_2(g) \ + \ 2H_2O(g)$$

Methanol   Oxygen   Carbon dioxide   Water

$$CH_3CH_2OH(g) \ + \ 3O_2(g) \ \longrightarrow \ 2CO_2(g) \ + \ 3H_2O(g)$$

Ethanol   Oxygen   Carbon dioxide   Water

$$2(CH_3)_2-CHOH(g) \ + \ 9O_2(g) \ \longrightarrow \ 6CO_2(g) \ + \ 8H_2O(g)$$

Isopropanol   Oxygen   Carbon dioxide   Water

When aliphatic alcohols burn, their combustion produces soot-less flames that are nearly imperceptible to the naked eye. When methanol, ethanol, and isopropanol burn, for example, the flames produced are pale blue and virtually invisible. This absence of visible flames and particulate matter also is characteristic of the burning of other aliphatic compounds that contain oxygen atoms in their molecular structures. Because alcohol flames cannot be detected with the naked eye, firefighters often encounter unique difficulties when combating fires involving them.

Why is little soot produced when methanol and ethanol burn in air? How does the absence of soot affect observation of the accompanying flames?

Solution: Methanol and ethanol are examples of organic compounds whose molecules contain oxygen atoms in addition to carbon and hydrogen atoms. This oxygen is used together with the available atmospheric oxygen when methanol and ethanol burn. Methanol and ethanol burn predominantly by complete combustion; i.e., they produce carbon dioxide instead of carbon monoxide and soot. Because flames are visibly manifested only when minute particulates of matter are heated to incandescence, the absence of soot causes the flames accompanying the combustion of these simple alcohols to be nearly imperceptible.

SOLVED EXERCISE 13.3

Name the alcohol whose molecular structure appears below:

$$
\begin{array}{ccccccc}
& \text{OH} & & & & \text{CH}_3 & \\
& | & & & & | & \\
\text{CH}_3 & - \text{CH} & - \text{CH}_2\text{CH}_2\text{CH}_2 & - \text{CH} & - \text{CH}_3 \\
\end{array}
$$

Solution: First, due to the presence of the hydroxyl group in the molecular structure, it is evident that the compound is an alcohol. Because the longest chain of carbon atoms containing the hydroxyl group has seven carbon atoms, this compound is a hydroxy derivative of heptane. To name the alcohol, the -e ending in heptane is replaced with -ol. Finally, the carbon atoms in the chain are consecutively numbered beginning at the end of the chain nearer to the hydroxyl group as follows:

$$
\begin{array}{ccccccc}
& \text{OH} & & & & \text{CH}_3 & \\
& | & & & & | & \\
\text{CH}_3 & - \text{CH} & - \text{CH}_2\text{CH}_2\text{CH}_2 & - \text{CH} & - \text{CH}_3 \\
1 & 2 & 3 \quad 4 \quad 5 & 6 & 7 \\
\end{array}
$$

Because the hydroxyl group is bonded to the carbon atom numbered 2, and a methyl group is bonded to the carbon atom numbered 6, the compound is named 6-methyl-2-heptanol.

13.2-A METHANOL

At room conditions, methanol is a colorless, water-soluble, and highly volatile liquid. It formerly was called **wood alcohol**, a term that acknowledges it as a constituent of the mixture that results when wood is strongly heated in the absence of air.

In the United States, methanol is manufactured mainly by the high-temperature, high-pressure hydrogenation of carbon monoxide in the presence of an appropriate catalyst.

$$\underset{\text{Carbon monoxide}}{CO(g)} + \underset{\text{Hydrogen}}{2H_2(g)} \longrightarrow \underset{\text{Methanol}}{CH_3OH(g)}$$

It is also manufactured by the incomplete combustion of natural gas.

$$\underset{\text{Methane}}{2CH_4(g)} + \underset{\text{Oxygen}}{O_2(g)} \longrightarrow \underset{\text{Methanol}}{2CH_3OH(g)}$$

Industrially, methanol is used as a polar solvent for shellac and gums; as a raw material for the manufacture of acetic acid and formaldehyde; and directly as an antifreeze, aircraft fuel-injection fluid, windshield washer fluid, automobile racing fuel, heat source for chafing dishes, and component of methanol fuel cells.

Fire and explosion are considered the primary risks associated with methanol. Its complete combustion is represented as follows:

$$\underset{\text{Methanol}}{2CH_3OH(g)} + \underset{\text{Oxygen}}{3O_2(g)} \longrightarrow \underset{\text{Carbon dioxide}}{2CO_2(g)} + \underset{\text{Water}}{4H_2O(g)}$$

wood alcohol ▪ The common name for methanol produced by heating wood in the absence of air

The secondary risk associated with methanol is posed by its ingestion. Death may occur when as little as 0.06 pint (30 mL) of methanol is ingested. Lesser amounts have been known to cause the onset of optic diseases, including retinal edema, that further lead to either partial or irreversible blindness. The toxicity of methanol is directly linked with its stepwise transformation into the toxic metabolic by-products formaldehyde and formic acid.

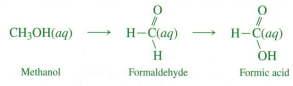

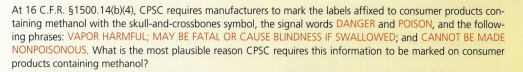

Methanol

Although the formaldehyde is produced by metabolism in the liver, it travels in the bloodstream throughout the body to various organs including the retina of the eyes, where the ultimate accumulation of formic acid causes the onset of optic diseases.

The production of formic acid during the metabolism of methanol also causes the pH of the blood, normally about 7.4, to decline. It is this condition, called *acidosis*, that causes victims of methanol poisoning initially to hyperventilate and ultimately to suffer disorders of the central nervous system. These symptoms are similar to those associated with ethanol intoxication. Thereafter, the victims of methanol poisoning experience blurred vision and decreased visual acuity. Absent treatment, these individuals may lose their sense of sight.

SOLVED EXERCISE 13.4

At 16 C.F.R. §1500.14(b)(4), CPSC requires manufacturers to mark the labels affixed to consumer products containing methanol with the skull-and-crossbones symbol, the signal words DANGER and POISON, and the following phrases: VAPOR HARMFUL; MAY BE FATAL OR CAUSE BLINDNESS IF SWALLOWED; and CANNOT BE MADE NONPOISONOUS. What is the most plausible reason CPSC requires this information to be marked on consumer products containing methanol?

Solution: As first noted in Section 1.2, to forewarn the public of the presence of hazardous substances in a consumer product, CPSC compels its manufacturers to label the product with certain information including advisory warnings and initial precautionary statements that identify the product's principal hazards. The most logical reason that CPSC takes this position with regard to consumer products containing methanol is that death and blindness could result from their ingestion. CPSC is congressionally mandated to provide special phrases on the labels of consumer products containing methanol so that the public is adequately alerted to avoid drinking them. These CPSC requirements are noted on the label displayed in Section 13.2-B.

13.2-B CONSUMER PRODUCT REGULATIONS INVOLVING METHANOL

Because death and blindness may result from the ingestion of methanol, the U.S. Consumer Product Safety Commission requires that the label affixed to methanol containers include the following statements:

13.2-C METHANOL AS AN ALTERNATIVE MOTOR FUEL

Methanol is potentially useful as an alternative motor fuel in either of the following ways:

■ *M85*, a blend of 85% methanol and 15% gasoline by volume, is used in specialty-built cars and trucks called flex-fuel vehicles. A **flex-fuel (flexible-fuel) vehicle**, or **FFV**, is an automobile specially designed to use blends of gasoline and methanol or ethanol such as the M85 and E85 fuels.

■ Pure methanol, called *M100*, can be used as the fuel in some heavy-duty trucks and transit buses.

At 16 C.F.R. §306.12, the U.S. Federal Trade Commission requires retail methanol distributors to affix the orange-and-black label shown below on methanol dispensers:

> **M-85**
>
> MINIMUM
> 85%
> METHANOL

In this instance, the label identifies the fuel as M85 and provides the minimum methanol concentration as 85% by volume. The U.S. Federal Trade Commission also requires new-vehicle manufacturers and used-vehicle dealers to affix this label on a visible surface of each M85-powered vehicle.

When compared to the combustion of motor gasoline, the combustion of M100 produces fewer toxic pollutants. Furthermore, the use of M100 in transit buses produces a far lesser concentration of particulate matter compared to the use of diesel oil. For these reasons, methanol is considered an environmentally friendly fuel.

When compared to motor gasoline, methanol is also safer to use as a fuel because it is more difficult to ignite. The information in Table 13.2 demonstrates that methanol possesses a substantially higher flashpoint than gasoline. [The average flashpoint of motor gasoline is −45°F (−43°C).]

Nonetheless, the heat of combustion of methanol is less than half that of gasoline: 9740 Btu/lb (22,700 kJ/kg) compared to 20,400 Btu/lb (47,300 kJ/kg). Consequently, vehicles powered by methanol require more frequent refueling than those powered by a petroleum fuel.

Because methanol is water-soluble, fires involving M100 can be effectively extinguished with water. However, fires involving M85 cannot be extinguished with water alone. Experts recommend the use of alcohol-resistant aqueous-film-forming foam (AR-AFFF) (Section 5.12-C) to extinguish fires that are fueled with M85.

13.2-D ETHANOL

Ethanol

Ethanol is also a colorless, water-soluble, highly volatile liquid. As the most commonly encountered active constituent of beer, wine, and the so-called "hard liquors," ethanol has been a staple of the human diet throughout recorded history. Some archeologists speculate that beverages were produced by **fermentation** as early as 100,000 years ago, when our human ancestors were first spreading out of Africa.

The production of ethanol in wine, champagne, and various brandies is accomplished by fermenting the sugars naturally present in ripe fruits. For example, the ethanol in wine is produced by fermenting the sugar in grapes; when the wine is charged with carbon dioxide, champagne is produced; and when the ethanol is distilled from fermented peaches, apricots, apples, and other fruits, brandies are produced.

Ethanol intended for consumption is also produced by fermenting the sugars generated by the enzymatic conversion of the starches in corn, potatoes, barley, rye, and wheat.

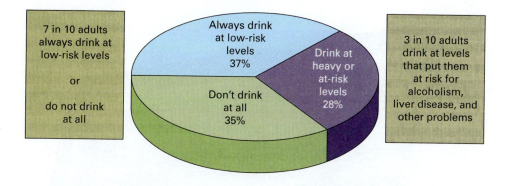

Always drink at low-risk levels 37%

Drink at heavy or at-risk levels 28%

Don't drink at all 35%

FIGURE 13.1 Alcohol use by American adults, aged 18 years or older. [*Courtesy of National Institutes of Health, Washington, DC: U.S. Department of Health and Human Services (2010).*]

The ethanol in beer is produced by fermenting malted barley; in whiskey by fermenting corn, barley, or wheat; in vodka by fermenting potato mash; and in gin is produced by fermenting rye. Because it is produced by fermenting cereal grains, the ethanol present in beers, whiskeys, gin, and vodka is sometimes called **grain alcohol**.

The concentration of ethanol in alcoholic beverages frequently is expressed by use of the term **alcoholic proof**. In the seventeenth century, people believed that spirits resided in beverages that caused the psychoactive effects experienced by individuals who consumed them. A procedure devised in England to test for the presence of these spirits consisted of preparing a mixture of gunpowder and a sample of the liquor being tested. If the gunpowder ignited after the alcohol had burned away, the event was regarded as "proof" or confirmation that spirits were actually present. If the gunpowder did not ignite, the action was taken to mean that spirits were absent. In reality, the liquor had been diluted with so much water it could not burn.

The percentage by volume of ethanol in an alcoholic beverage is defined as half its alcoholic proof. Thus, alcoholic beverages containing 90% and 95% ethanol by volume are 180-proof and 190-proof solutions, respectively.

Individuals often enjoy alcoholic beverages, either alone or when socializing with friends and family. When the National Institutes of Health surveyed 43,000 adults,[1] they determined that 28% of those studied placed themselves at a health risk by drinking too much alcohol. The result of their investigation is illustrated in Figure 13.1. Assuming that firefighters act similarly to the adults surveyed in this study, over 280,000 firefighters may be drinking too much alcohol.[2]

When individuals consume an alcoholic beverage, the ethanol is absorbed into the bloodstream, after which it is metabolized primarily in the liver and removed from the body. The rate of absorption is faster than the rate of metabolism. An estimate of the amount of alcohol that has been absorbed into the bloodstream can be approximated with an instrument like the one shown in Figure 13.2.

The metabolism of ethanol involves the stepwise oxidation to acetaldehyde and acetic acid.

grain alcohol ■ The common name of ethanol produced by the fermentation of grain

alcoholic proof ■ A measure of the volume of ethanol in an alcoholic beverage; in the United States, twice the percentage of ethanol by volume

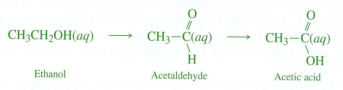

Ethanol   Acetaldehyde   Acetic acid

---

[1]"Rethinking Drinking: Alcohol and Your Health" (Washington, DC: U.S. Department of Health and Human Services, NIH Publication No. 10-3770, 2010), p. 1.

[2]In 2000, the number of career and active volunteer firefighters alone in the American emergency response community was 1,054,000. [Ari. N. Houser, Brian A. Jackson, James T. Bartis, and D. J. Peterson, "Emergency Responder Injuries and Fatalities" (Arlington, Virginia: Rand Science and Technology, 2004).]

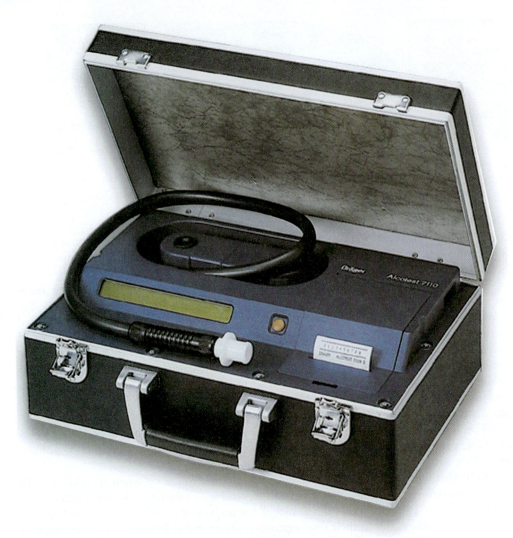

blood alcohol concentration (BAC) ■ The ratio of the volume of ethanol to total blood volume

When more alcohol is ingested than the liver is capable of metabolizing in a timely fashion, the **blood alcohol concentration,** or **BAC,** increases.

When ethanol is consumed, it acts simultaneously as a brain stimulant and a central nervous system depressant. The stimulation is responsible for the pleasurable, euphoric, and deinhibiting effects commonly associated with drinking alcohol, whereas the depression accounts for the anxiety, tension, and dulling of normal cognitive and motor processes. When alcoholic beverages are consumed, individuals experience physical, behavioral, and speech changes. The metabolic by-product acetaldehyde is responsible for these unpleasant physiological effects (including the dreaded hangover), but it is easier to measure the ethanol rather than the acetaldehyde in the consumer's blood. It is for this reason that an individual's behavior is routinely correlated with an individual's blood alcohol concentration.

At elevated altitudes, a person has to breathe harder to get ample oxygen into the blood; and when alcoholic beverages are drunk at high altitudes, the alcohol is absorbed more quickly into the bloodstream. These phenomena cause a person to experience the effects of consuming alcohol faster when compared to drinking the same amount at lower elevations.

Individuals with blood alcohol concentrations between 50 and 150 mg/dL, or between 0.05% and 0.15% by volume, typically experience a complete lack of coordination. [One deciliter (dL) is 100 milliliters, or 1/10 of a liter.] This condition prevents them from safely operating a motor vehicle. In most states, 80 mg/dL, or 0.08%, has been selected as the maximum legal limit above which a person is prohibited from driving on roadways, because at this BAC, braking, steering, lane changing, judgment, and ability to divide attention are affected significantly. Most countries other than the United States have selected 50 mg/dL, or 0.05%, as the threshold for establishing legally whether a driver is intoxicated.

How much alcohol is "too much"? Alcoholic beverages should be consumed only in moderation—between one and three drinks per day. The repeated consumption of alcohol that leads to drunkenness negatively impacts not only the lives of individual drinkers but the lives of their friends and family as well. *Binge drinking*, the indiscriminate consumption of multiple drinks in jest—one immediately after the other—can cause death within a few short hours of the drinker's attempt to discover amusement. This high-risk practice is clearly a dangerous form of recreation.

Because alcohol consumption dulls the senses and causes reflexes to be sluggish, emergency responders who consume alcohol in amounts beyond moderation cannot expect to perform to the best of their ability. Emergency responders serve their communities best by avoiding alcohol or drinking only in moderation. As a general practice, everyone should choose to drink responsibly at all times.

13.2-E ILL EFFECTS CAUSED BY ALCOHOL ABUSE

Individuals who consume alcohol in relatively large amounts over long periods of time often become addicted to alcohol. As alcoholics, they usually develop one or more alcohol-related liver diseases, of which there are three independent types:

■ **Fatty liver disease**, the build-up of extra fat in the liver. The drinkers exhibit no unique symptoms, but the disease ceases to exist when they abstain from the continued use of alcohol.

fatty liver disease ■ The disease associated with an increase of fat in the liver

■ **Alcoholic hepatitis**, the swelling of the liver. Drinkers do not always exhibit symptoms, but when they occur, the symptoms include nausea, vomiting, tenderness in the abdomen, and jaundice (a yellowishness of the skin). Again, the disease disappears when the drinkers cease to consume alcohol.

alcoholic hepatitis ■ The swelling of the liver caused by the misuse of alcohol

■ **Alcoholic cirrhosis**, a disease in which normal liver tissue is replaced by scar tissue. It is caused by the death of liver cells and is characterized by inflammation, pain, and jaundice. Because alcoholic cirrhosis causes serious impairment of the liver's biological function, it can be fatal.

alcoholic cirrhosis ■ A potentially fatal liver disease resulting from the misuse of alcohol

Studies[3] have also linked alcohol abuse with mouth, larynx, esophagus, liver, and breast cancers. Although it is unclear how ethanol causes the onset of these cancers, the prevailing theory is that metabolic acetaldehyde damages cellular DNA so severely that it is unable to repair itself.

Ethanol also enhances the cancer-causing effects of other carcinogens, particularly those found in tobacco smoke. The American Cancer Institute warns that frequent consumption of alcohol by a habitual smoker leads to as much as a 100-fold increase in the risk for contracting mouth, tracheal, or esophageal cancers compared with people who neither smoke nor drink.

Pregnant women and women who plan to become pregnant should be especially wary of consuming alcoholic beverages, because alcohol may cause their offspring to subsequently experience mental disabilities, physical deficiencies, and other birth disorders. Prenatal exposure to alcohol injures the neurological function of a developing fetus, ultimately causing a reduction in the child's intellectual potential for its entire life. This affliction is known as the **fetal alcohol syndrome**. The acetaldehyde formed during metabolism

fetal alcohol syndrome ■ The health disorder manifested by physical and mental impairment that potentially impacts the fetus due to the mother's consumption of alcohol during her pregnancy

---

[3]N.E. Allen *et al.*, "Moderate alcohol intake and cancer incidence in women," *J. Natl. Can. Inst.*, Volume 101 (2009), pp. 296–305.

is responsible for the onset of fetal alcohol syndrome, because it crosses the placental barrier and accumulates in the liver of the fetus.

To apprise the public of the potential dangers associated with alcohol consumption, the Alcohol and Tobacco Tax and Trade Bureau requires at 27 C.F.R. §16.21 labeling of all alcohol containers with the health-warning message shown here:

GOVERNMENT WARNING

According to the Surgeon General, women should not drink alcoholic beverages during pregnancy because of the risk of birth defects.

Consumption of alcoholic beverages impairs your ability to drive or operate machinery, and may cause health problems.

Posting of this label has been required on all alcohol containers sold in the United States since November 18, 1989.

Notwithstanding the combination of adverse effects associated with the consumption of alcohol, research studies also reveal that *moderate* drinking by mature adults may contribute to preventing heart attacks, regardless of the nature of the alcoholic beverage consumed. For example, one study[4] illustrates that individuals who drink alcoholic beverages three times a week experience approximately one-third fewer heart attacks than nondrinkers. This may mean that drinking a small amount of alcohol each day may be beneficial to one's health; nonetheless, the recommended guideline has always been to drink in moderation.

13.2-F INDUSTRIAL-GRADE ETHANOL

Aside from its use in alcoholic beverages, ethanol is also widely used for various industrial purposes. Manufacturing and process industries use large volumes of ethanol, most typically as a polar solvent in toiletries, cosmetics, pharmaceuticals, and surface coatings; a raw material in the manufacture of other substances; and as an oxygenate in vehicular fuels.

Ethanol is sometimes produced for industrial use by the acid-catalyzed vapor-phase reaction between ethylene and water.

$$CH_2{=}CH_2(g) \quad + \quad H_2O(g) \quad \longrightarrow \quad CH_3CH_2OH(g)$$

$$\text{Ethylene} \qquad\qquad \text{Water} \qquad\qquad\qquad \text{Ethanol}$$

Industrial-grade ethanol can also be produced from corn. The production of corn-based ethanol involves fermenting the sugar generated during the enzymatic conversion of the starch in corn. First, corn kernels are ground into flour and slurried with water; then, enzymes are added to the mixture to convert the starch into sugar; and finally, yeast is added to hasten the conversion of the sugar into ethanol, which subsequently is distilled from the mixture. Modifications of this process were used by moonshiners during Prohibition (1920 to 1933). In Brazil, the ethanol made from sugarcane is solely used as a **biofuel** (without any gasoline).

Industrial-grade ethanol is also produced from various cellulosic materials including switchgrass (a perennial warm-season grass native to North America), wood residues from the forest products industry, algae, and other nonedible parts of plants. When produced solely from these cellulosic sources for ultimate use as a biofuel, the ethanol is called **cellulosic ethanol**. When mixed with a petroleum-based fuel, cellulosic ethanol has been used successfully to fuel ground-based vehicles, as well as air- and watercraft.

biofuel ■ Any alternative motor fuel produced in whole or in part from domestic farm crops (such as corn kernels or soybeans) or crop residues, nonedible plant parts (such as wood, grass, switchgrass, or algae), or municipal or forest wastes

cellulosic ethanol ■ Ethanol produced from algae, corn waste, switchgrass, and other nonedible parts of plants

---

[4]Jennifer K. Pai et al., "Long-term alcohol consumption in relation to all-cause and cardiovascular mortality among survivors of myocardial infarction: the health professionals follow-up," *Eur. Heart. J.* (2012) (DOI:10.1093/eurheartj/ehs047).

Two forms of industrial-grade ethanol are available commercially:

- **Absolute alcohol** is ethanol that contains no more than 1% water.
- **Denatured alcohol** is an aqueous solution consisting of approximately 95% ethanol by volume to which a liquid called a **denaturant** has been added to impede its use as an alcoholic beverage or internal human medicine. Only certain substances may be added to ethanol as denaturants. They are published by the Alcohol and Tobacco Tax and Trade Bureau at 27 C.F.R. §19.1005. The presence of the denaturant may cause ethanol to be not only unpalatable but also poisonous. Denatured alcohol is intended for use primarily as a solvent.

13.2-G ETHANOL AS A BIOFUEL

To reduce foreign oil imports and greenhouse gas emissions, the U.S. Congress has overhauled the nation's energy policies through the introduction of a biofuels program that requires the blending of a biofuel into gasoline. Under the mandate of the Energy Independence and Security Act of 2007, refineries, blenders, and importers now are responsible for blending 36 billion gal/y (14×10^7 m^3/y) of biofuels into vehicular fuels by 2022. Given this mandate, it is apparent that the demand for biofuel use in U.S. vehicular fuels—including ethanol use—is certain to substantially increase.

In compliance with the intent of the Energy Independence and Security Act of 2007, ethanol is used as a biofuel in the United States when producing the following types of vehicular fuels:

■ *E10, a blend of 10% ethanol and 90% gasoline by volume.* E10 may be used as an additive the fuel in any motor vehicle manufactured in 2000 and earlier without engine modification. Because most petroleum fuels used in the United States now contain 10% ethanol as an additive, they are correctly denoted as E10 motor fuels.

■ *E15, a blend of 15% ethanol and 85% gasoline by volume.* E15 may be used as the fuel in light-duty motor vehicles, sport-utility vehicles, and flex-fuel vehicles manufactured in 2001 or later without engine modification. However, motorcycles, boats, snowmobiles, school buses, delivery trucks, off-road equipment such as lawnmowers, and chainsaws cannot use E15 without causing them to malfunction. To assure that customers are aware of the potential problems when purchasing E15, EPA requires E15 distributors to affix the following label to dispensers:

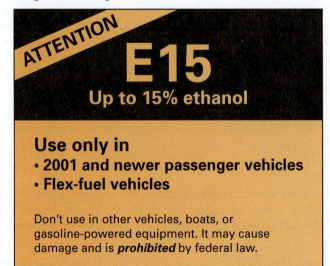

ATTENTION

E15
Up to 15% ethanol

Use only in
- **2001 and newer passenger vehicles**
- **Flex-fuel vehicles**

Don't use in other vehicles, boats, or gasoline-powered equipment. It may cause damage and is **prohibited** by federal law.

■ *E85, a blend of 85% ethanol and 15% gasoline by volume.* E85 may also be used as a biofuel in flex-fuel vehicles built in 2001 and later years. These cars and trucks have sensors to adjust the timing of spark plugs and fuel injectors so that the engine runs smoothly regardless of the fuel's ethanol concentration.

At 16 C.F.R. §306.12, the U.S. Federal Trade Commission requires retail distributors to affix the following orange-and-black label on the alcohol dispenser:

E-100

MINIMUM

95%

ETHANOL

This specific label identifies the fuel as E100 and provides the minimum ethanol concentration as 95% by volume. As with methanol, the commission also requires new-vehicle manufacturers and used-vehicle dealers to attach this label on a visible surface of each E85-powered vehicle. Ethanol with a denaturant only, is referred to as *E100*. Although it is available commercially, E100 has only limited use as an alternative motor fuel in the United States.

The sole use of ethanol as a biofuel is not free of problems, several of which are noted below:

■ Ethanol has only about 66% of the energy content of gasoline. Consequently, its use in vehicular fuels reduces mileage per gallon compared to the sole use of motor gasoline or diesel oil.

■ Ethanol absorbs moisture from the air, causing corrosion within pipelines. To avoid this problem, ethanol is now transported solely by rail tanks, barges, and tank trucks to regional fuel terminals or gasoline-blending racks.

■ The ethanol produced from corn diverts the use of corn throughout the agricultural and livestock industries and can contribute to the rise in certain food prices.

■ The ethanol produced from corn requires large amounts of pesticides and fertilizers, each of which requires the use of petroleum fuels for their production.

■ The ethanol produced from corn hampers the technological development of cellulosic ethanol and its commercialization.

The soaring demand for ethanol in the United States has resulted in an increase in transportation incidents involving ethanol. For example, in 2009, the derailment of a Canadian National Railway Company freight train in Cherry Valley, Illinois resulted in the explosion and burning of the ethanol contained in 13 tankcars.[5] The train consisted of 2 locomotives and 114 cars, 75 of which contained a total of 2,158,724 gallons (8169 m^3) of ethanol. Nineteen tankcars, all containing ethanol, derailed.

13.2-H EXTINGUISHING ETHANOL FIRES

Fire and explosion are considered the primary risks associated with ethanol. The combustion process, which at times can be nearly imperceptible, is represented as follows:

$$CH_3CH_2OH(g) \quad + \quad 3O_2(g) \quad \longrightarrow \quad 2CO_2(g) \quad + \quad 3H_2O(g)$$

Ethanol　　　　　Oxygen　　　　Carbon dioxide　　Water

Because ethanol is a water-soluble substance, the use of water on fires fueled by ethanol should bring them under control. However, the flashpoint data in Table 13.2 reveal that even when ethanol is diluted with water to produce a solution containing 24% water,

---

[5]Railway Accident Report, "Derailment of CN Freight Train U70691-18 with Subsequent Hazardous Materials Release and Fire, Cherry Valley, Illinois, June 19, 2009," NTSB/RAR-12/01, PB2012-916301 (Washington, DC: National Transportation Safety Board, 2012).

the solution is still flammable. From a practical viewpoint, the sole use of water may extinguish a nonbulk ethanol fire, but it is unlikely to extinguish a bulk ethanol fire. Testing reveals that when water alone is used on a bulk ethanol fire, a substantial volume (<5:1) is needed to extinguish the fire. In most instances, it is simply impractical to always have this volume of water available for use at a fire scene.

Experts concur that among the firefighting foams that are now commercially available, the use of alcohol-resistant aqueous-film-forming foam (AR-AFFF) (Section 5.12-C) serves best to extinguish a bulk ethanol fire. However, these experts are also quick to point out that the firefighting profession needs an economical foam that performs even more effectively than AR-AFFF.

13.2-I ISOPROPANOL

There are two propyl alcohols, *n*-propyl alcohol and isopropanol. Both are commercially important, but isopropanol is used in larger volume. Isopropanol is also known as isopropyl alcohol and 2-propanol. Most people recognize it as the most commonly encountered alcohol after ethanol. The formula of isopropanol is $CH_3-CH-OH$.
$$\underset{\displaystyle CH_3}{\vert}$$

In the chemical industry, isopropanol is manufactured by the acid-catalyzed vapor-phase reaction between propene and water.

$$CH_3CH{=}CH_2(g) \;+\; H_2O(g) \;\longrightarrow\; (CH_3)_2-CHOH(g)$$
Propene   Water   Isopropanol

Isopropanol is used in many ways. In the chemical industry, large volumes are needed for the production of hydrogen peroxide (Section 11.5), acetone (Section 13.5-C), and other substances. Most people recognize it as the main constituent of *rubbing alcohol*, a 60% to 70% solution (with methanol and ethanol) that is applied externally to the skin to relieve muscle and joint pains. As it evaporates, it cools and soothes the skin at the point of contact. Isopropanol is also used as a gasoline additive, where it serves as an oxygenate and "deicer." In the latter case, it dissolves water in the fuel line. Isopropanol is also used as a solvent for many lotions, oils, and other commercial products.

Although the major hazard associated with isopropanol is the risk of fire and explosion, the liquid is also a poison. The consumption of isopropanol causes permanent disabling illnesses, and when consumed in excess, it causes death. Ingested isopropanol metabolizes primarily to acetone.

glycol (diol) ■ Any compound whose molecules have two hydroxyl groups bonded to adjacent carbon atoms

13.2-J GLYCOLS

Glycols are diols whose molecules have two hydroxyl groups on adjacent carbon atoms. Glycols may be produced by oxidizing alkenes and hydrating the alkene oxide. For example, ethylene glycol is produced as follows:

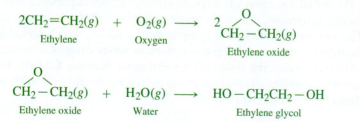

$$2CH_2{=}CH_2(g) \;+\; O_2(g) \longrightarrow 2\;\; \overset{O}{\overset{/\backslash}{CH_2-CH_2}}(g)$$
Ethylene   Oxygen   Ethylene oxide

$$\overset{O}{\overset{/\backslash}{CH_2-CH_2}}(g) \;+\; H_2O(g) \longrightarrow HO-CH_2CH_2-OH$$
Ethylene oxide   Water   Ethylene glycol

At room temperature, ethylene glycol and other simple glycols are slightly viscous liquids that are completely miscible in water.

The two most commonly encountered glycols are ethylene glycol and propylene glycol. These common names are derived by identifying the alkene portion of the alkene oxide used to synthesize them followed by the word *glycol*.

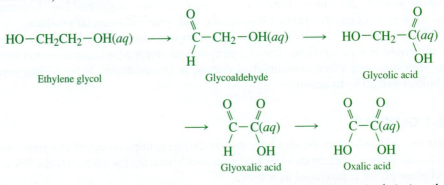

$$CH_2-CH_2(g) \quad + \quad H_2O(g) \quad \longrightarrow \quad HO-CH_2CH_2-OH(g)$$

Ethylene oxide Water Ethylene glycol
(1,2-Ethanediol)

$$CH_3-CH-CH_2(g) \quad + \quad H_2O(g) \quad \longrightarrow \quad CH_3-CH-CH_2-OH(g)$$

Propylene oxide Water Propylene glycol
(1,2-Propanediol)

Ethylene glycol and propylene oxide are used as liquid antifreeze agents in cooling and heating systems. An ultraviolet-light-sensitive dye is normally added to glycols intended for use as antifreeze agents to help locate hairline cracks in radiators. Ethylene glycol is also used in hydraulic brake fluids, deicing fluids for aircraft and airport runways, and as a solvent in paints and printer's inks. Propylene glycol is often the solvent used in certain medications that are intended to be inhaled or rubbed on the skin. It is also used as a raw material for the manufacture of certain polyester resins.

The human body responds initially in a similar fashion when increasing amounts of either ethanol or ethylene glycol is ingested; i.e., the consumers become intoxicated. Individuals may be tempted to drink ethylene glycol to experience the same "high" that they could obtain from drinking alcoholic beverages. Youngsters are attracted to ethylene glycol because it tastes sweet. However, when it is ingested, ethylene glycol acts as a toxic substance by destroying tissues and upsetting the normal pH of the blood. In adults, the ingestion of as little as 0.2 pint (100 mL) may cause death.

The toxicity of ethylene glycol is directly linked with its transformation into toxic metabolic by-products. The metabolism of ethylene glycol occurs within the liver and involves the stepwise oxidation to four toxic metabolites: glycoaldehyde, glycolic acid, glyoxalic acid, and oxalic acid.

$$HO-CH_2CH_2-OH(aq) \longrightarrow \underset{\text{Glycoaldehyde}}{\overset{\displaystyle O}{C}-CH_2-OH(aq)} \longrightarrow \underset{\text{Glycolic acid}}{HO-CH_2-\overset{\displaystyle O}{C}(aq)}$$

Ethylene glycol Glycoaldehyde Glycolic acid

$$\longrightarrow \underset{\text{Glyoxalic acid}}{\overset{\displaystyle O \quad O}{C-C}(aq)} \longrightarrow \underset{\text{Oxalic acid}}{\overset{\displaystyle O \quad O}{C-C}(aq)}$$

Glyoxalic acid Oxalic acid

Following the initial ingestion of ethylene glycol, victims experience inebriation, but as the metabolism progresses, the acids cause extensive cellular damage, especially in the kidneys. The victims experience the symptoms of acidosis, which includes hyperventilation and disorders of the central nervous system. In severe cases of ethylene glycol poisoning, the substance causes the onset of hypocalcemia (Section 8.11-B). Individuals who demonstrate the symptoms of ethylene glycol poisoning should be quickly rushed to emergency facilities for antidotal treatment.

13.2-K PHENOL

The **phenolic compounds** are regarded as derivatives of benzene in which one or more hydrogen atoms have been substituted with the hydroxyl group of atoms (—OH). The

phenolic compound ■
Any organic compound whose molecules have at least one hydroxyl group of atoms directly bonded to a benzene ring

Phenol

simplest phenolic compound is itself called **phenol**, or carbolic acid. Its chemical formula is C_6H_5OH.

phenol ■ An organic compound whose molecules have one hydroxyl group (—OH) bonded directly to the benzene ring

C_6H_5—O— is named the phenoxy group.

At room conditions, phenol is a colorless to white-pink crystalline solid that often darkens to red upon exposure to light. Because phenol readily absorbs atmospheric moisture, it is also encountered as a liquid. It has the sweet acrid odor characteristic of disinfectants.

Although phenol can be distilled from the middle coal-tar distillate (Section 7.6-C), it generally is produced in the chemical industry using other raw materials. The dominant production method involves the decomposition of cumene hydroperoxide by a two-step process. Cumene is first oxidized using air to cumene hydroperoxide, which then is decomposed into phenol and acetone.

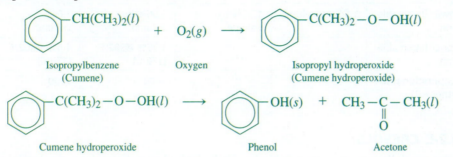

This production method is economically desirable, because the coproduct acetone is also a commercially important chemical product.

Phenol is an important industrial substance, because it is the raw material from which a number of derivatives and phenolic resins are manufactured. Phenolic derivatives are also used in surgical antiseptics and other germicidal solutions. Phenol itself was the first surgical antiseptic. A solution of one part phenol in 850 parts of water by mass prevents the multiplication of certain bacteria; for this reason, phenol derivatives are common constituents of mouthwashes, gargles, and sprays.

Some important physical properties of phenol are provided in Table 13.3. These data show that phenol burns, but its flashpoint is fairly elevated: 175°F (79°C). For this reason, phenol generally does not pose the risk of fire and explosion.

When dissolved in water, phenol dissociates into hydrogen ions and phenoxide ions, causing the phenol solution to be acidic.

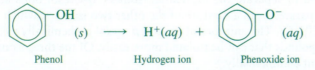

Thus, phenol is a corrosive material, especially when it contacts exposed skin, mucous membranes, and the eyes. Plastic surgeons use dilute solutions of phenol with soap and vegetable oil for chemical facial peels to produce a partial-thickness controlled burn of predictable depth when removing wrinkles and irregular facial pigmentation.

Phenol is also poisonous. Although it is absorbed into the body by all routes of exposure, the ingestion and absorption of elevated concentrations of phenol through the skin can be especially damaging. Phenol impairs liver and kidney function and profoundly disturbs the central nervous system. Exposed individuals often experience coma and may die from respiratory failure. Because its dermal LD_{50} is 630 mg/kg (rabbits), phenol is also considered a severe skin irritant.

Chapter 13 Chemistry of Some Hazardous Organic Compounds: Part II   **549**

TABLE 13.3 | Physical Properties of Phenol and the Isomeric Cresols

	PHENOL	o-CRESOL	m-CRESOL	p-CRESOL
Melting point	104°F (40°C)	88°F (31°C)	54°F (12°C)	95°F (35°C)
Boiling point	358°F (181°C)	376°F (191°C)	397°F (203°C)	396°F (202°C)
Specific gravity at 68°F (20°C)	1.07	1.05	1.03	1.04
Vapor density (air = 1)	3.24	3.7	3.7	3.7
Vapor pressure at 68°F (20°C)	0.357 mmHg	0.3 mmHg	<1 mmHg	<1 mmHg
Flashpoint	175°F (79°C)	178°F (81°C)	202°F (94°C)	202°F (94°C)
Autoignition point	1319°F (715°C)	1110°F (599°C)	1038°F (559°C)	1038°F (559°C)
Lower flammable limit	1.5% by volume	1.35% by volume	1.06% by volume	1.06% by volume
Upper flammable limit			1.35% @302°F (150°C)	1.4% @302°F (150°C)
Evaporation rate (ether = 1)			>400	>500

cresol ■ Any of the three methylated derivatives of phenol

13.2-L CRESOLS

A commercially important group of phenols is the hydroxy derivatives of toluene, called **cresols**. There are three isomeric cresols, named *o*-, *m*-, and *p*-cresol.

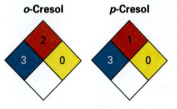

Their mixture is called *cresylic acid*. The physical properties of the cresol isomers are included in Table 13.3.

In the chemical industry, the cresol isomers generally are isolated as a mixture from the middle coal-tar distillate (Section 7.6-C), after which the mixture is subjected to fractional distillation to separate the individual isomers. *o*-Cresol boils at 376°F (191°C); thus, it easily separates from a mixture of the other two isomers, which boils at approximately 394°F (201°C). This mixture of *m*- and *p*-cresol is chemically treated with an acid to produce compounds that can be isolated more easily. Of the three isomers, the *p*-isomer is most important commercially.

When individuals are exposed to the cresol isomers, they are likely to experience the same adverse health effects as those noted for exposure to phenol. For example, both phenol and the cresols impair liver and kidney function and disturb the central nervous system when ingested or absorbed through the skin. The cresols are more corrosive to skin than phenol. The dermal LD_{50} (*o*-, *m*-, and *p*-) is 301 mg/kg (rabbits).

13.2-M WORKPLACE REGULATIONS INVOLVING ALCOHOLS

When methanol and ethanol are used in the workplace, OSHA requires employers to limit employee exposure to a maximum vapor concentration of 200 parts per million and 1000 parts per million, respectively, averaged over an 8-hour workday.

When phenol and the cresols are used in the workplace, OSHA requires employers to limit employee exposure by dermal contact to a maximum concentration of 5 parts per million, averaged over an 8-hour workday.

13.2-N TRANSPORTING ALCOHOLS

When shippers offer an alcohol for transportation, DOT requires them to enter the relevant shipping description on an accompanying shipping paper. Some examples for several representative alcohols are provided in Table 13.4. DOT also requires shippers and carriers to comply with all applicable labeling, marking, and placarding requirements.

When shippers offer for transportation an alcohol whose name is not listed at 49 C.F.R. §172.101, its shipping description is identified generically as "UN1987, Alcohols, n.o.s., 3, PG I," "UN1987, Alcohols, n.o.s., 3, PG II," "UN1987, Alcohols, n.o.s., 3, PG III," "UN1986, Alcohols, flammable, toxic, n.o.s., 3, PG I," "UN1986, Alcohols, flammable, toxic, n.o.s., 3, PG II," or "UN1986, Alcohols, flammable, toxic, n.o.s., 3, PG III." DOT requires the shipping description to include the name of the specific compound entered parenthetically. For instance, shippers describe a shipment of 1-octanol in PG III packaging as follows: "UN1987, Alcohols, n.o.s. (contains 1-octanol), 3, PG III (Marine Pollutant)."

m-Cresol

TABLE 13.4	Shipping Descriptions of Some Representative Alcohols
ALCOHOL OR GROUP THEREOF	**SHIPPING DESCRIPTION**
Alcoholic beverages	UN3065, Alcoholic beverages, 3, PG II *or* UN3065, Alcoholic beverages, 3, PG III
Allyl alcohol	UN1098, Allyl alcohol, 6.1, (3), PG I (Poison - Inhalation Hazard, Zone B)
Cresols, liquid	UN2076, Cresols, liquid, 6.1, (8), PGI I (Poison) (Marine Pollutants)
Cresols, solid	UN2076, Cresols, solid, 6.1, (8), PG II (Poison) (Marine Pollutants)
Ethanol	UN1170, Ethanol, 3, PG II *or* UN1170, Ethyl alcohol, 3, PG II *or* UN1170, Ethanol solutions, 3, PG II *or* UN1170, Ethyl alcohol solutions, 3, PG II
Isopropanol *or* Isopropyl alcohol	UN1219, Isopropanol, 3, PG II *or* UN1219, Isopropyl alcohol, 3, PG II
Methanol (international transportation)	UN1230, Methanol, 3, (6.1), PG II (Poison)
Methanol (domestic transportation)	UN1230, Methanol, 3, PG II
Phenol, molten	UN2312, Phenol, molten, 6.1, PG II (Poison)
Phenol, solid	UN1671, Phenol, solid, 6.1, PG II (Poison)
Phenol solutions	UN2821, Phenol solutions, 6.1, PG II (Poison)
n-Propanol	UN1274, *n*-Propanol, 3, PG II *or* UN1274, Propyl alcohol, 3, PG II

When shippers offer gasoline and ethanol for transportation within separate compartmented cargo tanks by highway:

(a) What shipping descriptions does DOT require them to enter on the accompanying shipping paper?
(b) What placards and markings does DOT require the carrier to display on the tanks?

Solution:

(a) DOT requires shippers to enter both of the following shipping descriptions on the accompanying shipping paper: UN1203, Gasoline, 3, PG II *and* UN1170, Ethanol, 3, PG II.
(b) DOT requires the carrier to display a FLAMMABLE placard on each side and each end of the cargo tank. Because more than one hazardous material is transported in separate compartments within a portable tank, DOT requires the carrier to display the identification numbers 1203 and 1170 on the ends of the tank *and* on the sides in the same sequence as the compartments containing the materials they identify. They may be displayed within either orange panels or white square-on-point diamonds on each side and each end of the tank near the FLAMMABLE placard, but it is improper to display the two numbers across the center area of a single placard.

When shippers offer for transportation E10 and E85 in separate cargo tanks, what shipping descriptions should be entered on the accompanying shipping papers?

Solution: E10 and E85 are mixtures of gasoline and not more than 10% and 85% ethanol by volume, respectively. We see in Appendix C that when these mixtures are offered for transportation, they should be properly described as "UN1203, gasoline, 3, PG II" and "UN3475, ethanol and gasoline mixture, 3, PG II," respectively.

13.2-O BISPHENOL A

BPA

When phenol reacts with acetone, the product predominantly produced is *p,p'*-isopropylidene diphenol, more commonly known by the product name *Bisphenol A*, or *BPA*. The production reaction is catalyzed by anhydrous hydrogen chloride.

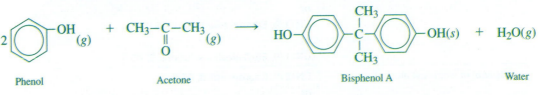

| Phenol | Acetone | Bisphenol A | Water |

Bisphenol A is used as an intermediate in the polymer industry to manufacture epoxy resins (Section 14.7) used in the production and manufacture of electrical, electronic, and sports-safety equipment. The epoxy resins are also used as protective coatings inside food, milk, and beverage containers, baby bottles, "sippy cups," and municipal and industrial water tanks. It is this latter combination of uses that has given rise to potential health concerns, because BPA residues leach from these coatings and become constituents of the contents. When the foods, beverages, and water are consumed, BPA is also ingested unknowingly.

BPA is also used as a color-developer that is coated on thermal-imaging paper used to print credit-card and cash-register receipts. The workers who handle these receipts risk exposure to the Bisphenol A via absorption through the skin.

The weight of scientific evidence indicates that BPA adversely affects human health. Studies[6] show that adult exposure to low concentrations of BPA can cause an increase in the rates of prostate and breast cancers, reproductive abnormalities, lowered sperm count, early onset of puberty in females, insulin-dependent diabetes, obesity, heart disease, and neurobehavioral abnormalities. Other studies[7] also indicate that BPA may contribute to adolescent obesity in exposed children.

When ingested, BPA behaves as an endocrine disrupter (Section 12.16) by preventing estrogen from acting in its customary way. Estrogen is a hormone that normally promotes and regulates the development of female characteristics.

In recognition of this combination of adverse information, chemical manufacturers voluntarily ceased the use of BPA in baby bottles and "sippy cups" in 2009, but FDA did not ban the practice until 2012. Furthermore, to avoid food poisoning and shorter shelf-lives, manufacturers in the food and beverage industry are eagerly seeking to find a replacement for the BPA now used to coat containers. Several states now prohibit the sale of products intended for use by infants and toddlers when the BPA concentration in their containers exceeds specified values. Canada also has banned the import, sale, and advertising of baby bottles that contain BPA. In the United States, however, FDA has rejected a plea to ban the use of BPA in food packaging.

13.3 ETHERS

An **ether** is any organic compound whose molecules have one or more oxygen atoms bridged between two alkyl or aryl groups. The simple ethers have the general chemical formula R—O—R′, where R and R′ are the formulas of arbitrary alkyl or aryl groups. In the chemical and petroleum industry, the simple ethers are used as solvents and oxygenates, respectively.

A special group of ethers is the **epoxides**. The molecules of these compounds have an oxygen atom bonded to two carbon atoms, each of which is also bonded to each other. Their general chemical formula follows, where R and R′ are arbitrary alkyl or aryl groups or hydrogen atoms:

$$R-\overset{\displaystyle O}{\overset{\displaystyle \diagup \diagdown}{CH-CH}}-R'$$

An epoxide

ether ■ Any organic compound whose general chemical formula is R—O—R′, where R and R′ represent arbitrary alkyl or aryl groups

epoxide ■ Any organic compound whose molecules have two carbon atoms bonded to an oxygen atom and to each other as a three-membered ring

In the chemical industry, epoxides typically are used as reactants for the preparation of other substances.

The common names of the simple ethers are determined by alphabetically noting the names of the alkyl or aryl groups bonded to the oxygen atom, followed by the word *ether*. For example, the compound having the chemical formula CH_3—O—CH_2CH_3 is named ethyl methyl ether. Here, R is the methyl group and R′ is the ethyl group. The simplest ether having a single aryl group is the substance whose formula is C_6H_5—O—CH_3. This compound is most frequently encountered by the name *anisole*.

In the IUPAC system, ethers are named by replacing the *-yl* suffix of the relevant alkyl group with *-oxy*, and naming the ether as a substituted alkane. Thus, ethyl methyl ether and anisole are named methoxyethane and methoxybenzene, respectively.

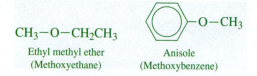

CH₃—O—CH₂CH₃

Ethyl methyl ether
(Methoxyethane)

Anisole
(Methoxybenzene)

[6]Iain A. Lang et al., "Association of urinary Bisphenol A concentration with medical disorders and laboratory abnormalities in adults," *J. Amer. Med. Assoc.*, Vol. 300 (2008), pp. 1303–1310.
[7]Leonardo Trasande, Teresa M. Attina, and Jan Blustein, "Association between urinary Bisphenol A concentration and obesity prevalence in children and adolescents," *J. Amer. Med. Assoc.*, Vol. 308 (2012), pp. 1113–1121.

An epoxide is usually named as an alkene oxide or epoxyalkane. In the latter instance, numbers are used to identify the carbon atoms to which the oxygen atom is bonded. These methods of nomenclature are illustrated in the examples that follow:

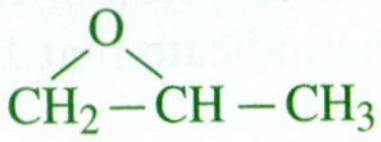

$$\underset{\substack{\text{Propylene oxide}\\ \text{(1,2-Epoxypropane)}}}{CH_2\overset{\displaystyle O}{\overbrace{-CH}}-CH_3} \qquad \underset{\substack{\text{Butene oxide}\\ \text{(1,2-Epoxybutane)}}}{CH_2\overset{\displaystyle O}{\overbrace{-CH}}-CH_2CH_3}$$

13.3-A REACTIONS OF ETHERS WITH ATMOSPHERIC OXYGEN

The primary hazard associated with most ethers is that they are highly volatile, flammable liquids; hence, they constitute dangerous fire and explosion hazards. Ethers are also hazardous substances because they produce potentially unstable peroxo-organic compounds (Section 13.9) by slowly reacting with atmospheric oxygen. For example, the following equations illustrate successive reactions that occur between diethyl ether and oxygen:

$$\underset{\text{Diethyl ether}}{CH_3CH_2-O-CH_2CH_3(l)} \longrightarrow \underset{\text{2-Ethoxyethyl hydroperoxide}}{CH_3CH_2-O-CH_2CH_2-O-OH(s)}$$

$$\longrightarrow \underset{\text{Diethyl peroxide}}{CH_3CH_2-O-O-CH_2CH_3(s)}$$

2-Ethoxyethyl hydroperoxide and diethyl peroxide are examples of peroxo-organic compounds that decompose at explosive rates.

The eight ethers having the following molecular formulas are the most vulnerable to forming peroxo-organic compounds:

$$\underset{\text{Diethyl ether}}{CH_3CH_2-O-CH_2CH_3} \qquad \underset{\substack{|\quad\quad|\\ \text{Diisopropyl ether}}}{CH_3-CH-O-CH-CH_3} \qquad \underset{\text{Tetrahydrofuran}}{} \qquad \underset{\text{1,4-Dioxane}}{}$$

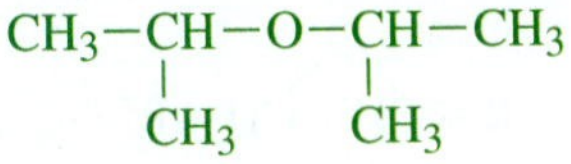

$$\underset{\substack{\text{Ethylene glycol dimethyl ether}\\ \text{(Glyme)}}}{CH_3-O-CH_2CH_2-O-CH_3} \qquad \underset{\substack{\text{Diethylene glycol dimethyl ether}\\ \text{(Diglyme)}}}{CH_3-O-CH_2CH_2-O-CH_2CH_2-O-CH_3}$$

$$\underset{\text{Butyl vinyl ether}}{C_4H_9-O-CH=CH_2} \qquad \underset{\text{Divinyl ether}}{CH_2=CH-O-CH=CH_2}$$

These ethers are popular polar solvents. To warn users of their potential reactivity with oxygen, ether manufacturers and distributors typically affix warning labels like the following on ether containers:

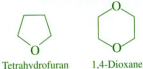

PROPERLY DISCARD 30 DAYS
AFTER OPENING OR
AFTER ONE YEAR IF UNOPENED

The chemical reactions between ethers and oxygen are catalyzed by light. Consequently, the rates at which peroxo-organic compounds are formed may be effectively altered by storing the ethers in metal cans or brown glass bottles that prevent the penetration of light. Because the peroxo-organic compounds are produced slowly, ethers that were purchased years ago are likely to be more susceptible to decomposition when compared to those that were recently purchased. Because ethers are also very volatile, the organic peroxides concentrate within the final residual volume of the liquid. Hence, the risk of an explosive reaction may be greater in ether containers containing small liquid residues than in full containers.

Safety engineers recommend storing ethers in a cooled, darkened room and marking the date received and the date first opened on a label affixed to their containers. An example of this label is shown below:

PEROXO-ORGANIC COMPOUND-FORMING ETHER

DATE RECEIVED: _____

DATE OPENED: _____

13.3-B DIETHYL ETHER

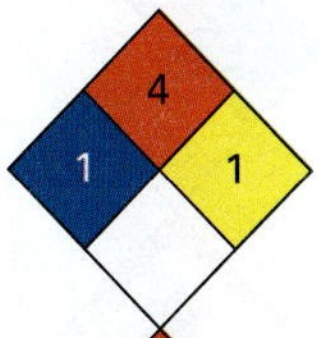

Diethyl ether is a colorless, water-soluble, flammable, and highly volatile liquid. In the chemical industry, it is produced by the dehydration of ethanol using concentrated sulfuric acid.

$$2CH_3CH_2OH(l) \longrightarrow CH_3CH_2-O-CH_2CH_3(l) + H_2O(l)$$

Ethanol   Diethyl ether   Water

On inhalation, the vapor of diethyl ether acts on the body as a short-lived muscle relaxant. Owing to this feature, it once was a well-known inhalation anesthetic. A generation ago, the odor of diethyl ether was commonplace in the surgical rooms of medical clinics and hospitals, where it was known simply as *ether*. The use of ether allowed surgeons to perform surgical operations while the patient was unconscious. However, the patient's recovery from exposure to ether was slow and unpleasant. Anesthesiologists now generally select alternatives to ether. Today, you are unlikely to detect the odor of ether in a clinic or hospital, but likely to detect its odor when using an automotive starting fluid during cold weather.

The physical properties of diethyl ether are provided in Table 13.5. The low flashpoint, wide flammable range, and relatively high vapor density attest to the flammable nature of diethyl ether. As demonstrated by the experiment in Figure 13.1, its vapor is heavier than air.

Diethyl ether

TABLE 13.5	Physical Properties of Diethyl Ether
Melting point	−189°F (−123°C)
Boiling point	94°F (34°C)
Specific gravity at 68°F (20°C)	0.71
Vapor density (air = 1)	2.55
Vapor pressure at 68°F (20°C)	442 mmHg
Flashpoint	−49°F (−45°C)
Autoignition point	320°F (160°C)
Lower flammable limit	1.85% by volume
Upper flammable limit	48% by volume
Evaporation rate (ether = 1)	1.0

FIGURE 13.3 In this laboratory illustration, a cloth is saturated with diethyl ether and placed at the top end of a trough that has been arranged at a 45° angle. Because the ether has a vapor density of 2.55 (air = 1), it moves *down* the trough, replacing the air, until it reaches the lighted candle. Then, the vapor ignites and flashes up the trough to the fuel source.

Methyl *tert*-butyl ether

Diethyl ether burns with the production of a virtually invisible, pale blue flame and no accompanying soot.

$$CH_3CH_2{-}O{-}CH_2CH_3(g) \quad + \quad 6O_2(g) \quad \longrightarrow \quad 4CO_2(g) \quad + \quad 5H_2O(g)$$

| Diethyl ether | Oxygen | Carbon dioxide | Water |

The presence of oxygen in the molecular structure of diethyl ether accounts for the virtually imperceptible flame associated with its combustion.

13.3-C METHYL *tert*-BUTYL ETHER

Methyl *tert*-butyl ether, or MTBE, was first introduced in the U.S. fuel market in 1995 to boost the oxygen content of gasoline, because it has an antiknock rating of 116.

$$CH_3{-}O{-}\underset{\underset{CH_3}{|}}{\overset{\overset{CH_3}{|}}{C}}{-}CH_3$$

Methyl *tert*-butyl ether
(*tert*-Butoxymethane)

oxygenate ■ A vehicular fuel additive that promotes complete combustion of the fuel and generates lesser amounts of carbon monoxide and other pollutants compared to the amounts produced when the fuel burns without the additive

MTBE formerly was a major organic compound manufactured in the United States. It was used as an **oxygenate** so that gasoline burned more cleanly and produced less carbon monoxide when compared to the petroleum fuels containing nonoxygenated antiknock agents.

In 1996, MTBE was identified as a low-level contaminant of a drinking water source in Santa Monica, California. Its origin was linked with leaking underground gasoline storage tanks. Santa Monica was obliged to close 7 of its 11 municipal groundwater wells.

Even at very low concentrations (40 µg/L), the presence of MTBE causes water to smell and taste foul. To ensure that drinking water supplies are simultaneously palatable and unlikely to have harmful constituents, EPA phased out its use as a gasoline additive. Today, MTBE is no longer a component of the existing pool of gasoline additives. Based on a scientific review of available data, EPA also concluded that MTBE causes cancer when it is consumed in high doses; hence, it is classified as a probable human carcinogen.

Eucalyptol

13.3-D EUCALYPTOL

Eucalyptol is an oily liquid produced by eucalyptus trees, where it concentrates primarily in the leaves. It provides the trees with their unique camphor-like fragrance.

Eucalyptol is a cyclic ether despite the use of the suffix *-ol* in its common name. Its preferred name is 1,3,3-trimethyl-2-oxabicyclo[2.2.2]octane.

1,3,3-Trimethyl-2-oxabicyclo[2.2.2]octane
(Eucalyptol)

It has a boiling point of 176.5°F (349.7°C) and a flashpoint of 120°F (49°C).

Although eucalyptus trees were not initially indigenous to the United States, they are nonetheless abundant in southern California and Hawaii. When exposed to an ignition source, their outer bark ignites readily, especially during dry, hot weather and droughts. The heat of combustion vaporizes the eucalyptol, and their leaves become ablaze with fire. Secondary fires are initiated readily in nearby homes and uncultivated lands. Fires involving eucalyptus trees are occasionally so difficult to extinguish that firefighters regard them as major hazards and have discouraged their use in landscaping.

13.3-E ETHYLENE GLYCOL ALKYL ETHERS

The **ethylene glycol alkyl ethers** are compounds having the following general chemical formula, in which R is an alkyl group, R′ is a hydrogen atom or alkyl group, and *n* is a nonzero integer:

$$R—(O—CH_2CH_2)_n—O—R'$$

These compounds are commercially known by the tradenames *Cellosolve* and *Carbitol*. Examples are noted in Table 13.6.

The simple ethylene glycol alkyl ethers generally are referred to by their common names, which are obtained by indicating the nature of R and R′ and the value of *n*. When *n* = 1, 2, and 3, *mono-*, *di-*, and *tri-* are used to respectively designate the number of —O—CH₂CH₂— chains, although *mono-* may be used only to avoid ambiguity. The

ethylene glycol alkyl ether ■ Any organic compound whose general chemical formula is R—(O—CH₂CH₂)ₙ—O—R′, where R is an alkyl group, R′ is a hydrogen atom or alkyl group, and *n* is a nonzero integer

Ethylene glycol monomethyl ether

TABLE 13.6	Some Ethylene Glycol Alkyl Ethers[a]	
COMMERCIAL NAME	**CHEMICAL NAME**	**CHEMICAL FORMULA**
Butyl Cellosolve	Ethylene glycol monobutyl ether, or 2-butoxyethanol	$C_4H_9—O—CH_2CH_2—OH$
Carbitol solvent	Diethylene glycol monoethyl ether	$HO—CH_2—CH_2—O—CH_2CH_2—O—C_2H_5$
Cellosolve solvent	Ethylene glycol monoethyl ether, or 2-ethoxyethanol	$C_2H_5—O—CH_2CH_2—OH$
Dibutyl Cellosolve	Ethylene glycol dibutyl ether, or 1,2-dibutoxyethane	$C_4H_9—O—CH_2CH_2—O—C_4H_9$
Diglyme	Diethylene glycol dimethyl ether, or 1-methoxy-2-(2-methoxy)-ethane	$CH_3—O—CH_2CH_2—O—CH_2CH_2—O—CH_3$
Dimethyl Cellosolve, or monoglyme	Ethylene glycol dimethyl ether, or 1,2-dimethoxyethane	$CH_3—O—CH_2CH_2—O—CH_3$
Methyl Cellosolve	Ethylene glycol monomethyl ether, or 2-methoxyethanol	$CH_3—O—CH_2CH_2—OH$
Triglyme	Triethylene glycol dimethyl ether, or 2,5,8,11-tetraoxadodecane	$CH_3—O—CH_2CH_2—O—CH_2CH_2—O—CH_2CH_2—O—CH_3$

prefixes *mono-* and *di-* are also used to designate the number (one or two) of the same alkyl group. For ethylene glycol dimethyl ether (or monoethylene glycol dimethyl ether), R and R′ are named "dimethyl" and *n* is 1; for diethylene glycol dimethyl ether, R and R′ are named "dimethyl" and *n* is 2; and for triethylene glycol dimethyl ether, R and R′ are again named "dimethyl" and *n* is 3.

In the IUPAC system, R or R′ are named with the oxygen atom from the ether as *alkoxy*; i.e., CH_3—O is named methoxy; CH_3CH_2—O— is named ethoxy; etc. Then, the ethylene glycol ethers are named as derivatives of alkanes, alcohols, or ethers, as appropriate.

The ethylene glycol alkyl ethers are widely used throughout a variety of commercial industries. For example, monoethylene glycol dimethyl ether is used as a solvent and as a component of electrolyte solutions for sealed lithium batteries; diethylene glycol dimethyl ether is used as a solvent for printing inks; and triethylene glycol dimethyl ether is a component of brake fluids. These three ethylene glycol alkyl ethers are also called *monoglyme*, *diglyme*, and *triglyme*, respectively, because —O—CH_2CH_2— is a component of their formulas one, two, and three times. Other ethylene glycol alkyl ethers are also ingredients in commercial products like surface coatings, adhesives, cleaning fluids, and consumer paint strippers. All are flammable liquids.

The ethylene glycol alkyl ethers are produced by reacting ethylene oxide with an appropriate alcohol. For example, ethylene glycol monoethyl ether is produced by reacting ethylene oxide and ethyl alcohol in the presence of an appropriate catalyst. It is produced as a mixture of mono-, di-, and triethylene glycol monoethyl ethers, each of which is isolated from the other two by fractional distillation.

$$CH_2-CH_2 \quad + \quad CH_3CH_2OH \longrightarrow CH_3CH_2-O-CH_2CH_2OH$$

Ethylene oxide   Ethanol   Ethylene glycol monoethyl ether

$$+ \quad CH_3CH_2-O-CH_2CH_2-O-CH_2CH_2OH$$

Diethylene glycol monoethyl ether

$$+ \quad CH_3CH_2-O-CH_2CH_2-O-CH_2CH_2-O-CH_2CH_2OH$$

Triethylene glycol monoethyl ether

Several ethylene glycol alkyl ethers are known reproductive toxins[8] via skin absorption and vapor inhalation. Pregnant women appear to be especially vulnerable to their adverse health risks, because they may experience miscarriages after inhalation exposure.

To emphasize the ill effects potentially posed to an exposed pregnant woman and her unborn child by the ethylene glycol alkyl ethers, manufacturers include the following warning statements on the labels of their commercial products:

> **May Impair Fertility**   **May Cause Harm to the Unborn Child**

13.3-F TRANSPORTING ETHERS

When shippers offer any ether for transportation, DOT requires them to enter the relevant shipping description on an accompanying shipping paper. Examples for several representative ethers are provided in Table 13.7. DOT also requires shippers and carriers to comply with all applicable labeling, marking, and placarding requirements.

When shippers offer for transportation a flammable ether other than those listed at 49 C.F.R. §172.101, DOT requires them to identify the commodity generically on a shipping paper as either "UN3271, Ethers, n.o.s., 3, PG II" *or* "UN3271, Ethers, n.o.s, 3, PG III." In both instances, the shipping description includes the name of the specific ether entered parenthetically.

---

[8]Bryan D. Hardin, "Reproduction toxicity of the glycol ethers," *Toxicology*, Vol. 27 (1983), pp. 91–102.

TABLE 13.7	Shipping Descriptions of Some Representative Ethers
ETHER	**SHIPPING DESCRIPTION**
Anisole	UN2222, Anisole, 3, PG III (Marine Pollutant)
Diethyl ether	UN1155, Diethyl ether, 3, PG I *or* UN1155, Ethyl ether, 3, PG I
Diisopropyl ether	UN1159, Diisopropyl ether, 3, PG I
1,1-Dimethoxyethane	UN2377, 1,1-Dimethoxyethane, 3, PG II
1,2-Dimethoxyethane	UN2252, 1,2-Dimethoxyethane, 3, PG II
Dimethyl ether	UN1033, Dimethyl ether, 2.1
Dioxane	UN1165, Dioxane, 3, PG II
Methyl *tert*-butyl ether	UN2398, Methyl *tert*-butyl ether, 3, PG II
Tetrahydrofuran	UN2056, Tetrahydrofuran, 3, PG II

TABLE 13.8	Shipping Descriptions of Some Representative Ethylene Glycol Methyl and Ethyl Ethers
ETHYLENE GLYCOL METHYL OR ETHYL ETHER	**SHIPPING DESCRIPTION**
Ethylene glycol diethyl ether	UN1153, Ethylene glycol diethyl ether, 3, PG II *or* UN1153, Ethylene glycol diethyl ether, 3, PG III
Ethylene glycol monoethyl ether	UN1171, Ethylene glycol monoethyl ether, 3, PG III
Ethylene glycol monomethyl ether	UN1188, Ethylene glycol monomethyl ether, 3, PG III

When shippers offer a glycol ether for transportation, DOT requires them to identify the substance on a shipping paper. Examples for the ethylene glycol methyl ethers and ethylene glycol ethyl ethers are listed in Table 13.8. Once again, all DOT labeling, marking, and placarding requirements apply.

13.4 HALOGENATED ETHERS

A **halogenated ether** is an organic compound whose molecules contain the —O— group and in which at least one hydrogen atom has been replaced with a halogen atom. The halogenated ethers of importance to emergency responders include the following: epichlorohydrin; the polychlorinated dibenzofurans and dibenzo-*p*-dioxins; the polybrominated dibenzofurans and dibenzo-*p*-dioxins; and the polybrominated diphenyl ethers.

halogenated ether ■ Any halogenated derivative of an ether

13.4-A EPICHLOROHYDRIN

In terms of chemical reactivity, the chlorinated epoxide known as epichlorohydrin is one of the more versatile halogenated ethers.

$$\overset{\displaystyle O}{\overset{\displaystyle \diagup\diagdown}{CH_2-CH-CH_2Cl}}$$

Epichlorohydrin
(1-Chloro-2,3-epoxypropane)

Epichlorohydrin

Methyl chloromethyl ether

TABLE 13.9	Physical Properties of Epichlorohydrin
Melting point	−54°F (48°C)
Boiling point	242°F (116°C)
Specific gravity at 68°F (20°C)	1.18
Vapor density (air = 1)	3.28
Vapor pressure at 68°F (20°C)	12.5 mmHg
Flashpoint	88°F (31°C)
Autoignition point	781°F (416°C)
Lower flammable limit	3.8% by volume
Upper flammable limit	21% by volume
Evaporation rate (ether = 1)	17.0

Although chemists use epichlorohydrin for a variety of purposes, its main industrial use is associated with the production of epoxy resins (Section 14.7).

Epichlorohydrin is a volatile colorless liquid. Some of its important physical properties are noted in Table 13.9. These data indicate that epichlorohydrin is a highly flammable liquid. In addition, the vapor of epichlorohydrin is highly poisonous when it is inhaled. Repeated exposures of moderate concentrations may cause pulmonary edema (Section 7.3-B). Upon contact, the liquid acts as a corrosive material that absorbs through the skin. IARC classifies it as a probable human carcinogen.

The combination of these hazardous properties is evident from its DOT shipping description: UN2023, Epichlorohydrin, 6.1, (3), PG II (Poison) (Marine Pollutant). When epichlorohydrin is transported, shippers and carriers must comply with all DOT labeling, marking, and placarding requirements.

13.4-B POLYCHLORINATED DIBENZOFURANS AND DIBENZO-*p*-DIOXINS

Bis(chloromethyl) ether

The **polychlorinated dibenzofurans** and **dibenzo-*p*-dioxins** are, respectively, the chlorinated derivatives of dibenzofuran and dibenzo-*p*-dioxin. These latter substances have molecules composed of two benzene rings linked to each other by one and two oxygen atoms, respectively.

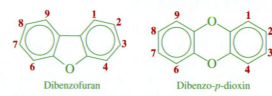

Dibenzofuran   Dibenzo-*p*-dioxin

The carbon atoms are numbered as shown from 1 to 10.

The polychlorinated dibenzofurans and dibenzo-*p*-dioxins are designated collectively as PCDFs and PCDDs, respectively. Their general molecular structures are as follows:

polychlorinated dibenzofuran (PCDF) ■ Any chlorinated derivative of dibenzofuran

polychlorinated dibenzo-*p*-dioxin (PCDD) ■ Any chlorinated derivative of dibenzo-*p*-dioxin

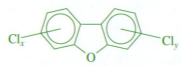

Polychlorinated dibenzofurans

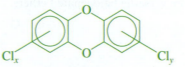

Polychlorinated dibenzo-*p*-dioxins

In these structures, x and y are integers, and the lines drawn from Cl_x and Cl_y to the benzene rings are chlorine atoms bonded at any available position. Using the indicated numbering system, the 17 compounds chlorinated at least at the 2, 3, 7, and 8 positions are of the greatest interest, because epidemiologists associate them with a relatively high degree of toxicity.

There are 135 polychlorinated dibenzofuran isomers and 75 polychlorinated dibenzo-p-dioxin isomers. Among them, the PCDF and PCDD having the following molecular structures have the highest degree of toxicity:

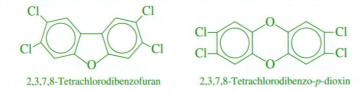

| 2,3,7,8-Tetrachlorodibenzofuran | 2,3,7,8-Tetrachlorodibenzo-p-dioxin |

2,3,7,8-Tetrachlorodibenzo-p-dioxin is commonly denoted as *dioxin*, or 2,3,7,8-TCDD, but the first name is chemically imprecise.

The PCDFs and PCDDs were never intentionally manufactured as commercial products, but they were generated as unwanted by-products during certain uncontrolled incineration, paper pulp bleaching, and chemical manufacturing operations. The latter included the production and manufacture of trichlorophenol, tetrachlorophenol, pentachlorophenol, and 2,4,5-trichlorophenoxyacetic acid.

Interest in dioxin was first stimulated by public health officials when the substance was identified as a trace contaminant in herbicides formerly used by the U.S. military during the Vietnam Conflict. Several color-coded herbicides were used as defoliants to clear jungle terrain, strip the Viet Cong of cover, and destroy enemy crops. Among them was *Agent Orange*, a 50:50 mixture of 2,4-dichlorophenoxyacetic acid and dioxin-contaminated 2,4,5-trichlorophenoxyacetic acid. 2,4-Dichlorophenoxyacetic acid and 2,4,5-trichlorophenoxyacetic acid are commonly known as 2,4-D and 2,4,5-TP, respectively.

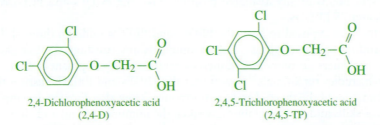

| 2,4-Dichlorophenoxyacetic acid (2,4-D) | 2,4,5-Trichlorophenoxyacetic acid (2,4,5-TP) |

Between 1962 and 1970, the U.S. Air Force sprayed approximately 18 million gallons ($68,100 \text{ m}^3$) of Agent Orange on vegetation in southern Vietnam.

The military personnel and local civilians who were exposed to the dioxin in Agent Orange subsequently contracted horrific diseases. The exposure in Vietnam veterans, for example, has been linked with the inception of four cancers: soft-tissue sarcoma; non-Hodgkin's lymphoma; Hodgkin's disease; and chronic lymphocytic leukemia. The exposure has also been associated with the onset of respiratory cancers (lung, bronchi, larynx, and trachea), prostate cancer, and multiple myeloma, as well as a number of noncancerous ailments including Parkinson's disease and type 2 diabetes.[9]

These adverse health effects do not appear to be limited solely to those personnel who were directly exposed to Agent Orange. Suggestive evidence associates the exposure with

---

[9]National Academy of Sciences, *Veterans and Agent Orange – Health Effects of Herbicides Used in Vietnam* (Washington, DC: National Academies Press, 1994).

serious birth defects in their offspring. In particular, this evidence links Agent Orange exposure with the inception of spina bifida, a congenital birth defect, in veterans' children who were born after a parent served on active duty in Vietnam.[10]

Although the Vietnam Conflict occurred during the 1960s and early 1970s, exposure to the PCDFs and PCDDs is possible even today, especially to firefighters. As first noted in Section 12.16-A, these substances are generated during certain electrical-equipment fires; and in Section 14.6, we will note that they are also generated during residential and other fires as products of the incomplete combustion of poly(vinyl chloride) (PVC) and similar plastics. Under these circumstances, exposure to the PCDFs and PCDDs may pose a pronounced health risk to the firefighters who respond to them.

IARC has ranked dioxin as a human carcinogen. It ranks the PCDDs and PCDFs whose molecules have chlorine atoms in the 2, 3, 7, and 8 positions as probable carcinogens. To avoid adverse health effects from dioxin exposure, EPA designates 6.4 femtograms per kilogram of body weight as the acceptable daily dosage for humans. (One femtogram is one quadrillionth (10^{-15}) of a gram.)

13.4-C POLYBROMINATED DIBENZOFURANS AND DIBENZO-*p*-DIOXINS

The polybrominated dibenzofurans and dibenzo-*p*-dioxins, or PBDFs and PBDDs, respectively, are the polybrominated derivatives of dibenzofuran and dibenzo-*p*-dioxin. They

fire retardant ■ An additive to commercial products that promotes resistance to burning

once were incorporated into certain plastic products to serve as **fire retardants**.

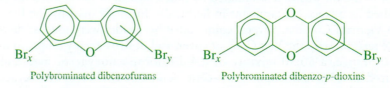

Polybrominated dibenzofurans   Polybrominated dibenzo-*p*-dioxins

Here again, x and y are integers, and the lines drawn from Br_x and Br_y to the benzene rings represent bonds at any available position. Like the PCDFs and PCDDs, there are 135 PBDFs and 75 PBDDs.

When they are exposed to heat, the PBDFs and PBDDs undergo thermal decomposition and produce bromine atoms. These bromine atoms consume the free radicals produced when polymers decompose and thereby retard the development of fire.

Research studies on animals[11] reveal that the PBDFs and PBDDs disrupt the normal biological function of the thyroid and sex hormones. They also damage developing brains, impair motor skills and mental abilities, weaken the immune system, and alter bone structure. Although these studies have not been performed on humans, scientists fear that exposure to the PBDFs and PBDDs may similarly affect the human organism.

Like the PCDFs and PCDDs, the compounds associated with elevated toxicities are those whose molecules are brominated in at least the 2, 3, 7, and 8 positions. These PBDFs and PBDDs are regarded as probable carcinogens.

It is assumed that emergency responders are at risk from exposure to polybrominated dibenzofurans and dibenzo-*p*-dioxins. During fires, plastics containing brominated fire retardants undergo incomplete combustion and release low concentrations of PBDFs and PBDDs to the surroundings. The inhalation of these substances can potentially subject firefighters and others to an unwarranted health risk.

---

[10]Jeff Johnson, *Environ. Sci. Technol.*, Vol. 30 (1996), p. 193A.
[11]John H. Mennear and Cheng-Chun Lee, "Polybrominated Dibenzo-*p*-dioxins and Dibenzofurans: Literature Review and Health Assessment," *Environ. Health Perspect.*, Vol. 102 (1994), pp. 265–274.

13.4-D POLYBROMINATED DIPHENYL ETHERS

The polybrominated diphenyl ethers, or PBDEs, are the polybrominated derivatives of diphenyl ether, whose chemical formula is $C_6H_5-O-C_6H_5$.

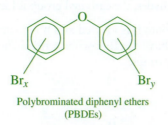

Polybrominated diphenyl ethers
(PBDEs)

In this general molecular structure, x and y are integers less than 5. There are 209 isomers of the PBDEs.

Like the polybrominated dibenzofurans and dibenzo-*p*-dioxins, the PBDEs formerly were manufactured for incorporation as fire retardants into the polymers used to make carpets and carpet padding and automobile and furniture cushioning. When used commercially, they often were manufactured as mixtures of the penta-, octa-, and decabromodiphenyl ethers. The pentabromodiphenyl ether isomers, collectively called the *penta- group*, once were formulated as fire retardants into the polyurethane foam used in upholstery. Also popular were the octabromodiphenyl ether isomers and decabromodiphenyl ether, all of which once were used in the housings for business machines and electrical appliances.

U.S. production and manufacture of the PBDEs ceased in 2004 when tests confirmed the presence of these compounds in human serum and breast milk. Nonetheless, despite their nonproduction, stockpiles of these compounds were formulated into U.S. products after 2004. Carpets, blankets, upholstery, fabrics, and other items manufactured before and after 2004 remain in U.S. homes. It is prudent to assume that firefighters are exposed to the PBDEs when they battle house fires. Decabromodiphenyl ether is the sole compound among the PBDEs that is still imported into the United States.

The concern about PBDEs is the risk posed by exposure to them. Although research studies linking adverse health effects with PBDE exposure in humans are ongoing, animal exposure has shown to cause the risks previously noted for exposure to PBDFs and PBDDs. The available information suggests that PBDEs may cause neurodevelopmental problems, endocrine disruption, and cancer. Epidemiologists rank them as probable human carcinogens.

EPA has shown that exposure to PBDE-laden dust was the key means by which household cats acquired high levels of certain PBDEs. The cats suffered from overactive thyroids. Although hyperthyroidism is treatable in both cats and humans, the health concern has focused on youngsters who are just as likely as cats to be exposed to the PBDEs in normal household dust. Unlike adults, they crawl on carpeted floors and chew blankets. Prudent parents may reduce the risk of child exposure to PBDEs by frequently vacuuming carpets, laundering blankets, and cleaning other items containing these flame retardants.

The PBDEs now are considered major environmental contaminants not only in the general U.S. population but also in animals. PBDEs enter the environment when the products in which they were incorporated are either incinerated as components of municipal waste or buried in landfills. The impact of their environmental presence is especially worrisome for the survival of the animals at the top of the food chain, like Arctic polar bears. Tests show that PBDE concentrations are more elevated in polar bears than in ringed seals or Arctic foxes, because these compounds biomagnify as they pass from prey to predator. Whether these animals continue to survive in the wild may well be connected with international efforts to ban the use of PBDEs and other brominated fire retardants.

13.5 ALDEHYDES AND KETONES

Aldehydes and **ketones** are organic compounds whose molecules contain the *carbonyl group* of atoms (C=O). In aldehydes, the carbonyl group is located at the end of a chain of carbon atoms, whereas in ketones it is located at a nonterminal position within the chain. Thus, aldehydes and ketones have the following general chemical formulas, where R and R' are arbitrary alkyl or aryl groups.

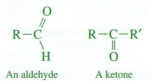

An aldehyde   A ketone

Naming Aldehydes

In the common system, an aldehyde is named by combining the word *aldehyde* with the prefix of the name of the acid to which it is oxidized. The aldehydes having one, two, three, and four carbon atoms per molecule are named formaldehyde, acetaldehyde, propionaldehyde, and *n*-butyraldehyde, respectively, because they oxidize to formic acid, acetic acid, propionic acid, and *n*-butyric acid, respectively.

In the IUPAC system, an aldehyde is named by replacing the *-e* in the name of the corresponding hydrocarbon with *-al*. The position of the carbonyl carbon atom does not have to be designated, because it is always located on a terminal carbon atom. The aldehydes having one, two, three, and four carbon atoms per molecule are then named methanal, ethanal, propanal, and butanal, respectively, by replacing the *-e* in methane, ethane, propane, and butane with *-al*.

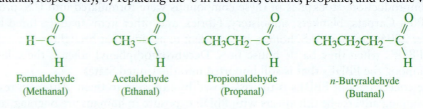

| Formaldehyde | Acetaldehyde | Propionaldehyde | *n*-Butyraldehyde |
| (Methanal) | (Ethanal) | (Propanal) | (Butanal) |

The simplest aromatic aldehyde is called benzaldehyde. Its chemical formula is C_6H_5—CHO.

Benzaldehyde

Naming Ketones

In the common system, a ketone is named by alphabetically identifying the alkyl or aryl groups of which the substance is composed. Thus, the compound having the formula CH_3—C—$CH_2CH_2CH_3$ (with O double bonded to C) is properly named methyl *n*-propyl ketone.

In the IUPAC system, a ketone is named by replacing the *-e* in the name of the corresponding hydrocarbon with *-one*; then, the chain of continuous carbon atoms is numbered so as to assign the *smallest possible number* to the carbonyl group. Hence, the IUPAC name for methyl *n*-propyl ketone is 2-pentanone.

$$\overset{1}{C}H_3 - \overset{2}{C} - \overset{3}{C}H_2\overset{4}{C}H_2\overset{5}{C}H_3$$
with O double bonded to C

Methyl *n*-propyl ketone
(2-Pentanone)

The ketone whose formula is $C_6H_5-\overset{\overset{\displaystyle O}{\|}}{C}-CH_3$ is named methyl phenyl ketone, but it is also known commonly as acetophenone.

Methyl phenyl ketone
(Acetophenone)

Using the IUPAC system, name the compound having the following condensed formula:

$$CH_3CH_2-\overset{\overset{\displaystyle O}{\|}}{C}-\overset{\overset{\displaystyle CH_3}{|}}{C}HCH_2CH_3$$

Solution: By examination of Table 13.1, we determine that the functional group represented as $C=O$ is the carbonyl group. When it is bonded to nonterminal carbon atoms, the carbonyl group characterizes the class of organic compounds called ketones. To name a ketone using the IUPAC system, we consecutively number each carbon atom in the longest continuous chain of carbon atoms that contains the carbonyl group, beginning at the end of the chain nearer to the carbonyl group. By so doing, we see that this particular ketone is the parent compound, 3-hexanone. This name is derived by replacing the -e with -one in hexane, the alkane having six carbon atoms per molecule.

$$\overset{1\quad 2\qquad 3\quad\ 4\quad\ 5\quad 6}{CH_3CH_2-\overset{\overset{\displaystyle O}{\|}}{C}-\overset{\overset{\displaystyle CH_3}{|}}{C}H-CH_2CH_3}$$

Then, we identify the methyl group on the carbon atom numbered 4, so the compound's correct IUPAC name is 4-methyl-3-hexanone.

It is incorrect to number the longest chain of continuous carbon atoms from right to left as follows, because a higher number (4) would then be assigned to the carbonyl group.

$$\overset{6\quad 5\qquad 4\quad\ 3\quad\ 2\quad 1}{CH_3CH_2-\overset{\overset{\displaystyle O}{\|}}{C}-\overset{\overset{\displaystyle CH_3}{|}}{C}H-CH_2CH_3}$$

13.5-A FORMALDEHYDE

The simplest aldehyde is formaldehyde, or methanal. Its chemical formula is HCHO.

$$H-\overset{\overset{\displaystyle O}{/\!/}}{\underset{\underset{\displaystyle H}{\diagdown}}{C}}$$

Formaldehyde is a colorless, flammable gas at room conditions that has a pungent, highly irritating odor detectable at concentrations as low as 1 part per million. Some of its important physical properties are provided in Table 13.10.

In the presence of water vapor, formaldehyde molecules react with each other and produce a solid substance called paraformaldehyde. This substance has an indefinite composition. Its chemical formula is $HO(CH_2O)_nH$, where n ranges from approximately 8 to 100.

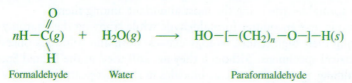

$$n\text{H}-\overset{\overset{\displaystyle O}{/\!/}}{\underset{\underset{\displaystyle H}{\diagdown}}{C}}(g)\ +\ H_2O(g)\ \longrightarrow\ HO-[-(CH_2)_n-O-]-H(s)$$

Formaldehyde   Water   Paraformaldehyde

Paraform-aldehyde

Formal-dehyde

Formalin (containing 20% methanol)

TABLE 13.10	Physical Properties of Formaldehyde
Melting point	−134°F (−92°C)
Boiling point	−3°F (−19°C)
Specific gravity at 68°F (20°C)	0.82
Vapor density (air = 1)	1.08
Vapor pressure at 68°F (20°C)	1.3 mmHg
Flashpoint	147°F (64°C)
Autoignition point	806°F (430°C)
Lower flammable limit	7.0% by volume
Upper flammable limit	73% by volume

Because formaldehyde is a **self-reactive material**, the gas is unavailable commercially. However, formaldehyde is soluble in water and alcohol, producing colorless, corrosive solutions whose properties vary with the formaldehyde concentration. These solutions are widely available commercially.

The liquid called **formalin** is an aqueous solution of formaldehyde containing approximately 40% water and 5% to 12% methanol. It is widely used as a disinfecting, sterilizing, and embalming agent. A formalin solution containing 37% water and 15% methanol flashes at 122°F (50°C), whereas an aqueous solution of formaldehyde without methanol flashes at 185°F (85°C) and is corrosive.

For industrial use, formaldehyde may be generated as the gas from paraformaldehyde and *sym*-trioxane:

■ Paraformaldehyde consists of several formaldehyde units bonded one to another in a short chain. Its formula is HO−[−(CH₂)ₙ−O−]−H, where *n* equals 8 or more. When it is mildly heated, paraformaldehyde decomposes into formaldehyde and water.

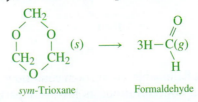

Paraformaldehyde   Formaldehyde   Water

■ *sym*-Trioxane is a combustible solid having a cyclic molecular structure. When mixed with a strong acid, *sym*-trioxane decomposes into formaldehyde.

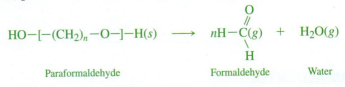

sym-Trioxane   Formaldehyde

Aldehydes often are produced as incomplete combustion products when organic substances burn within confined areas. Materials that produce formaldehyde upon their incomplete combustion include tobacco, wood, diesel oil, kerosene, natural gas, and ethanol. Aldehydes are the major class of organic compounds contained in diesel-engine exhaust, with formaldehyde being the most abundant among them.

Most persons first encounter formaldehyde while studying the anatomy of frogs and fetal pigs during their high school biology classes. Formalin solutions once were used to preserve biological specimens. Although they are still used in the United States by morticians as embalming agents, their use as biocides in the European Union now is banned.

The bulk of the formaldehyde produced in the United States is used as a raw material for production of formaldehyde-derived polymers, from which building-construction materials are manufactured. Prolonged exposure to formaldehyde may occur inside residential and commercial buildings constructed, insulated, or furnished with **composite wood products**, especially when the buildings are newly built and infrequently vented. Because mobile homes often are constructed from composite wood products, the presence of formaldehyde may be especially prominent inside them.

Formaldehyde leaches from composite wood products, especially particleboard in which urea-formaldehyde was used as the adhesive to bind the wood pieces. Some specific sources of these products are the following:

- Kitchen and bathroom cabinetry and furniture, shelving, countertops, flooring, and moldings
- Glues and adhesives that were used to bind wood pieces and fragments into particleboard, hardwood plywood, and fiberboard composites
- Foam insulation
- Synthetic padding and carpets

Figure 13.4 shows that low concentrations of formaldehyde often leach from these sources into the surrounding environment. These concentrations increase when the doors and windows of a mobile home are shut for extensive periods. The rate of release decreases as the age of these products increases and is dependent on the prevailing temperature and humidity. Formaldehyde is released into the air at its maximum rate when temperature and humidity are elevated.

Consumers may avoid their exposure to formaldehyde in pressed-wood products by buying only those labeled U.L.E.F., or ultra-low-emitting formaldehyde; N.A.F., or no-added formaldehyde; or C.A.R.B., or California Air Resources Board. C.A.R.B. refers to specific state of California standards that compel manufacturers to limit the formaldehyde emission levels in composite wood products.

self-reactive material ■ For purposes of DOT regulations, a material that is thermally unstable and can undergo a strongly exothermic decomposition even without the participation of atmospheric oxygen

formalin ■ A solution of formaldehyde in water and/or methanol

composite wood product ■ Any wood-based panel made from wood pieces, particles, or fibers bonded together with a phenol-formaldehyde or urea-formaldehyde resin

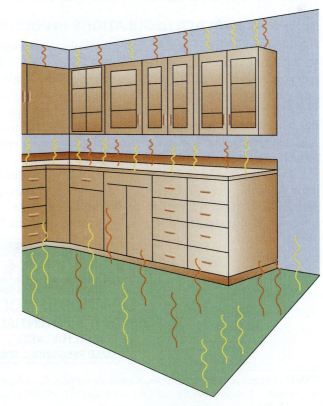

FIGURE 13.4 Formaldehyde gas escapes slowly from the following construction sources: kitchen and bathroom cabinetry and furniture; glue and adhesives used to bind wood particles in particleboard, hardwood plywood, and fiberboard; foam insulation; and synthetic padding and carpeting. When a mobile home is newly constructed, unvented, or heavily insulated, the interior air contains an elevated formaldehyde concentration, which, when inhaled by its occupants, may cause them to experience respiratory discomfort and other health problems.

TABLE 13.11	Adverse Health Effects Associated with Breathing Formaldehyde[a]

FORMALDEHYDE CONCENTRATION IN AIR (PPM)	SYMPTOMS
0.1–0.5	Nasal and eye irritation, neurological effects, increased risk of asthma and/or allergies
0.6–1.9	Nasal and eye irritation, eczema, change in pulmonary function
2.0–5.9	Nasal, eye, and throat irritation, eczema, or skin irritation, change in pulmonary function
6.0–10.9	Nasal, eye, throat and skin irritation, headache, nausea, discomfort in breathing, cough
11–50	Nasal, eye, and skin irritation, nasal ulceration, change in pulmonary function, possible liver dysfunction, testicular effects, nasal tumors, reduced survival (in animals)
>50.0	Possible pulmonary edema, pneumonitis, and possible death bloody nasal discharge, pulmonary edema (in animals)
>100.0	Death probable

[a]Agency For Toxic Substances & Disease Registry, Formaldehyde (Washington, DC: U.S. Department of Health and Human Services, 2008).

Exposure to formaldehyde can be worrisome, since inhalation causes the ill effects noted in Table 13.11. Furthermore, formaldehyde is a well-known human carcinogen. Toxicological studies have linked formaldehyde inhalation with the onset of upper-tract cancers, especially nasopharyngeal and sinonasal cancers. Embalmers comprise a group that is especially vulnerable, because they typically contract myeloid leukemia and rare cancers of the nasal passages and upper mouth.[12]

13.5-B WORKPLACE REGULATIONS INVOLVING FORMALDEHYDE

When formaldehyde is present in the workplace (such as anatomy and pathology laboratories), OSHA requires employers at 29 C.F.R. §1910.1048 to limit employee exposure to a maximum concentration of 0.75 part per million in air, averaged over an 8-hour workday, with a ceiling limit of 2 parts per million. The NIOSH recommended exposure limit is 0.016 part per million with a ceiling limit of 0.1 part per million, and the short-term exposure limit is 2 parts per million over a 15-minute period. These exposure limits are so low that those who routinely use formalin have been encouraged to select an alternative substance that accomplishes the same task.

At 29 C.F.R. §1910.1048(e)(1), OSHA also requires employers to establish regulated areas when the concentration of airborne formaldehyde exceeds the time-weighted average value (0.75 part per million) or the short-term exposure limit (2 parts per million), and to post all entrance and accessways with warning signs that bear the following information:

---

[12]M. Hauptmann et al., "Mortality from lymphohematopoietic malignancies and brain cancer among embalmers exposed to formaldehyde," *J. Natl. Cancer Instit.*, Vol. 95 (2003), pp. 1615–1623.

OSHA also requires formaldehyde-contaminated laundry to be placed in a plastic bag and labeled as follows:

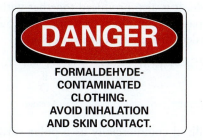

All formaldehyde-contaminated laundry must be cleaned through use of a specialized commercial laundry service.

13.5-C ACETONE

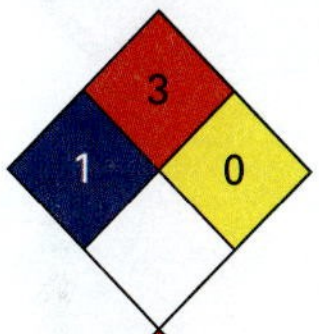

The simplest ketone is called acetone, propanone, or dimethyl ketone. Its chemical formula is $(CH_3)_2-C=O$.

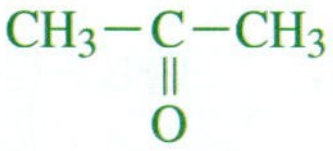

$$CH_3-\overset{\overset{\displaystyle }{\|}}{\underset{\displaystyle O}{C}}-CH_3$$

Acetone

It is a colorless, water-soluble, and highly volatile liquid with a sweet odor.

The data in Table 13.12 indicate that acetone and other simple ketones pose the risk of fire and explosion. When they burn, carbon dioxide and water vapor are the products of combustion.

$$(CH_3)_2-C=O(g) \ + \ 4O_2(g) \ \longrightarrow \ 3CO_2(g) \ + \ 3H_2O(g)$$

| Acetone | Oxygen | Carbon dioxide | Water |

Acetone

TABLE 13.12	Physical Properties of Acetone, Methyl Ethyl Ketone, and Methyl Isobutyl Ketone		
	ACETONE	**METHYL ETHYL KETONE (MEK)**	**METHYL ISOBUTYL KETONE (MIBK)**
Melting point	−137°F (−94°C)	−124°F (−87°C)	−121°F (−85°C)
Boiling point	133°F (56°C)	175°F (80°C)	243°F (117°C)
Specific gravity at 68°F (20°C)	0.79	0.81	0.80
Vapor density (air = 1)	2.0	2.5	3.5
Vapor pressure at 68°F (20°C)	181.7 mmHg	71 mmHg	15.7 mmHg
Flashpoint	0°F (−18°C)	20°F (−7°C)	73°F (23°C)
Autoignition point	1000°F (538°C)	960°F (515°C)	860°F (460°C)
Lower flammable limit	3% by volume	2% by volume	1.4% by volume
Upper flammable limit	13% by volume	10% by volume	7.5% by volume
Evaporation rate (ether = 1)	1.9	2.7	5.6

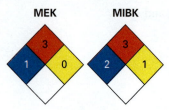

MEK **MIBK**

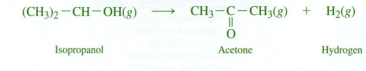

In the chemical industry, acetone is manufactured by several methods. As noted in Section 13.2-K, acetone is produced together with phenol by the decomposition of cumene hydroperoxide. It is also produced together with hydrogen peroxide by the oxidation of isopropanol. The most common method of production, however, involves the catalytic dehydrogenation of isopropanol.

$$(CH_3)_2-CH-OH(g) \longrightarrow CH_3-\underset{\substack{\|\\O}}{C}-CH_3(g) + H_2(g)$$

Isopropanol   Acetone   Hydrogen

Acetone is largely utilized commercially as a polar solvent in varnishes, lacquers, paints, fingernail polish remover, and acetylene. It is also used as a raw material for the manufacture of methyl methacrylate, methyl isobutyl ketone, and other substances.

13.5-D OTHER KETONES

Other commercially important ketones have physical properties similar to those of acetone. Two are ethyl methyl ketone (or, incorrectly, methyl ethyl ketone) and methyl isobutyl ketone.

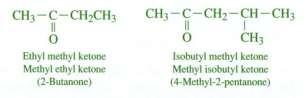

Ethyl methyl ketone   Isobutyl methyl ketone
Methyl ethyl ketone   Methyl isobutyl ketone
(2-Butanone)   (4-Methyl-2-pentanone)

In commerce, they are known more widely by their acronyms, MEK and MIBK, respectively. Both are flammable, water-soluble, highly volatile liquids at room conditions. They formerly were encountered as constituents of solvent mixtures, especially solvent-based paints.

Several commercially important ketones containing multiple carbonyl groups also are known. The compound called diacetyl is a volatile, highly flammable yellow liquid whose molecules have two carbonyl groups. They are bonded to the two nonterminal carbon atoms that provide a chain of four carbon atoms. Its IUPAC name is 2,3-butanedione. (Note that the *-e* in the name of the alkane is not dropped when naming a dione.)

Diacetyl

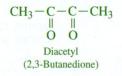

$$CH_3-\underset{\substack{\|\\O}}{C}-\underset{\substack{\|\\O}}{C}-CH_3$$

Diacetyl
(2,3-Butanedione)

Diacetyl once was used as a butter-flavoring agent and an aroma carrier in food products like microwave popcorn until it became apparent that the inhalation of elevated concentrations of its vapor could cause the rare lung disease *bronchiolitis obliterans*, known colloquially as *popcorn lung*. Workers exposed to relatively high concentrations of diacetyl experience shortness of breath, coughing, fatigue, and ultimately death. Popcorn manufacturers voluntarily eliminated it as a component of microwave popcorn in 2006, but its use in other butter-flavored foods continues.

13.5-E TRANSPORTING ALDEHYDES AND KETONES

When shippers offer an aldehyde or ketone for transportation, DOT requires them to enter the relevant shipping description on an accompanying shipping paper. Some examples for several representative aldehydes or ketones are listed in Table 13.13. When the name of an aldehyde or ketone is not listed in Table 13.13 or the Hazardous Materials

TABLE 13.13	Shipping Descriptions of Some Representative Aldehydes and Ketones
ALDEHYDE, KETONE, OR THEIR SOLUTIONS	**SHIPPING DESCRIPTION**
Acetaldehyde	UN1089, Acetaldehyde, 3, PG I
Acetone	UN1090, Acetone, 3, PG II
Benzaldehyde	UN1990, Benzaldehyde, 9, PG III
Butyraldehyde	UN1129, Butyraldehyde, 3, PG II (Marine Pollutant)
Cyclohexanone	UN1915, Cyclohexanone, 3, PG III
Ethyl methyl ketone	UN1193, Ethyl methyl ketone, 3, PG II *or* UN1193, Methyl ethyl ketone, 3, PG II
Formaldehyde solutions (flammable)	UN1198, Formaldehyde solutions, flammable, 3, PG III
Formaldehyde solutions (10–24.9% formaldehyde)[a]	UN3334, Aviation regulated liquid, n.o.s. (formaldehyde), 9 *or* NA3082, Other regulated substances, liquid, n.o.s. (formaldehyde), 9, PGIII
Formaldehyde solutions (≥25% formaldehyde)	UN2209, Formaldehyde solutions, 8, PG III
Methylcyclohexanone	UN2997, Methylcyclohexanone, 3, PG II
Methyl isobutyl ketone	UN1245, Methyl isobutyl ketone, 3, PG II
Paraformaldehyde	UN2213, Paraformaldehyde, 4.1, PG III
Propionaldehyde	UN1275, Propionaldehyde, 3, PG II (Marine Pollutant)

[a]For shipment by air.

Table at 49 C.F.R. §172.101, its shipping description is identified generically. DOT also requires shippers and carriers to comply with all applicable labeling, marking, and placarding requirements.

13.6 ORGANIC ACIDS

In Section 8.2-B, we learned that organic acids, or carboxylic acids, are compounds whose

molecules possess the group of atoms, $-\overset{\overset{\displaystyle O}{\parallel}}{\underset{\displaystyle OH}{C}}$, or $-COOH$. In molecules of the organic

acids, a hydrogen atom is bonded as shown to an oxygen atom in the *carboxyl group*. The simplest organic acids are monocarboxylic acids, compounds having one carboxyl group and the formula R—COOH, where R is an arbitrary alkyl or aryl group. Next are the di- and tricarboxylic acids, whose molecules have two and three carboxyl groups, respectively.

The simple organic acids are identified by their common historical names: formic acid, acetic acid, propionic acid, and so forth. In the IUPAC system, they are named by replacing the terminal -e in the name of the corresponding alkane with -*oic acid*. When necessary, the position of a substituent along a chain of carbon atoms is designated by a number. The carbon atom in the carboxyl group is assigned the number 1.

The organic acids having from one to four carbon atoms per molecule are most commonly encountered. These acids are named as follows:

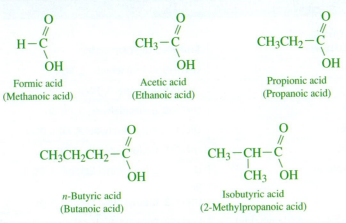

Formic acid
(Methanoic acid)

Acetic acid
(Ethanoic acid)

Propionic acid
(Propanoic acid)

n-Butyric acid
(Butanoic acid)

Isobutyric acid
(2-Methylpropanoic acid)

They are colorless, water-soluble liquids with characteristic odors. Formic acid, acetic acid, and propionic acid have pungent but not disagreeable odors, whereas n-butyric acid and isobutyric acid have highly disagreeable odors. Low concentrations of isobutyric acid are constituents of human perspiration and feces.

TABLE 13.14	Physical Properties of the Simple Organic Acids		
	FORMIC ACID	**ACETIC ACID**	**PROPIONIC ACID**
Melting point	47°F (8.2°C)	61°F (17°C)	−8°F (−22°C)
Boiling point	101°F (213°C)	244°F (118°C)	286°F (141°C)
Specific gravity at 68°F (20°C)	1.22	1.05	0.99
Vapor density (air = 1)	1.59	2.1	2.56
Vapor pressure at 68°F (20°C)	44.8 mmHg	11 mmHg	3 mmHg
Flashpoint	156°F (69°C)	109°F (43°C)	130°F (54°C)
Autoignition point	1114°F (601°C)	800°F (426°C)	955°F (513°C)
Lower flammable limit	18% by volume	4% by volume	2.9% by volume
Upper flammable limit	57% by volume	16% by volume	
Evaporation rate (ether = 1)	> 1	11.0	
	n-BUTYRIC ACID	**ISOBUTYRIC ACID**	
Melting point	18°F (−8°C)	−51°F (−47°C)	
Boiling point	327°F (164°C)	309°F (153°C)	
Specific gravity at 68°F (20°C)	0.96	0.949	
Vapor density (air = 1)	3.0	3.04	
Vapor pressure at 68°F (20°C)	0.84 mmHg	1.5 mmHg	
Flashpoint	151°F (66°C)	132°F (55°C)	
Autoignition point	846°F (452°C)	824°F (439°C)	
Lower flammable limit	2% by volume	2% by volume	
Upper flammable limit	10% by volume	10% by volume	

The simple organic acids are weak acids, meaning that they only partially ionize when dissolved in water. Although corrosiveness is the primary hazard of most aqueous solutions of formic acid, acetic acid, propionic acid, and *n*-butyric acid, flammability is regarded as the primary hazard of concentrated isobutyric acid. When they burn, these acids produce the products of complete combustion. For instance, it was noted in Section 8.13-C that carbon dioxide and water are produced when acetic acid burns. When they are diluted with water, the organic acids do not vaporize, and thus, are nonflammable.

The simplest aromatic acid is named *benzoic acid*. It is a white solid having the formula $C_6H_5—COOH$. Benzoic acid and its sodium salt have the structures shown:

Isobutyric acid

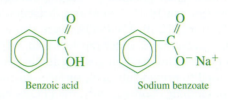

Benzoic acid   Sodium benzoate

Sodium benzoate is identified on the labels of many food containers to indicate its presence as a preservative and antimicrobial agent.

13.6-A THE FORMYL, ACETYL, AND BENZOYL GROUPS

The *formyl*, *acetyl*, and *benzoyl* groups are derived from formic acid, acetic acid, and benzoic acid, respectively. Their molecular structures are represented as follows:

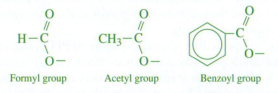

Formyl group   Acetyl group   Benzoyl group

These groups are encountered in many compounds. Acetyl chloride (Section 9.9-C) and benzoyl chloride, for instance, are the compounds having the formulas $CH_3—COCl$ and $C_6H_5—COCl$, respectively.

Acetyl chloride   Benzoyl chloride

Formyl chloride is unknown. When chemists attempt to prepare it, carbon monoxide and hydrogen chloride are produced instead. The acetyl and benzoyl groups are common components of peroxo-organic compounds (Section 13.9).

13.6-B PERFLUOROOCTANOIC ACID

Perfluorooctanoic acid is better known by its acronym PFOA. It is the totally fluorinated derivative of octanoic acid.

PFOA

Perfluorooctanoic acid

Two other compounds are often associated with perfluorooctanoic acid: ammonium perfluorooctanoate and perfluorooctane sulfonate, each of which is denoted as APFO and PFOS, respectively.

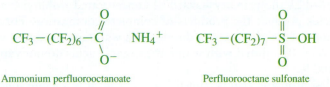

PFOA formerly was used as an essential processing aid during the manufacture of fluoropolymers like *Teflon* (Section 14.10-A) and as a surfactant in aqueous-film-forming foam fire extinguishers (Section 5.12-B). The PFOAs were also widely used to impart fire-resistance and oil-, stain-, grease-, and water-repellence to carpets, textiles, and paper. They were also used to make a low-friction coating for bearings, gears, and weapons components. PFOA was used to provide nonstick surfaces on cookware and waterproof, breathable membranes for clothing. Consequently, the PFOAs have been widely used for decades in dozens of commonly used commercial products.

Although there is no evidence that normal use of nonstick Teflon cookware is directly harmful, scientists note that when Teflon is heated to temperatures exceeding approximately 1000°F (600°C), the perfluorooctanoic acid is released. Exposure to it may be harmful when individuals heat Teflon cookware to these very high temperatures.

In the late 1990s, EPA first raised concern that PFOAs had been identified on a widespread basis in samples of blood withdrawn from virtually all Americans. The blood of approximately 95% of the American population is believed to contain at least a trace of PFOA. EPA also notes that these compounds persist in the environment and bioaccumulate within the fatty tissues of animals.

Research studies[13] indicate that animal exposure to PFOA can cause several types of tumors and neonatal death and may have toxic effects on the immune, liver, and endocrine systems. This finding was so startling that many major American PFOA manufacturers voluntarily ceased their production of this substance. In 2005, EPA[14] reported that additional studies performed on laboratory animals linked PFOA exposure with the onset of breast, testes, pancreas, and liver cancer. Based on the results from these studies, PFOA is now regarded as a probable human carcinogen. EPA has asked PFOA manufacturers to phase out the production of this substance.

In 2011, studies[15] conducted on humans revealed that children with high blood levels of PFOA and PFOS attain puberty later in life when compared to the norm. Elevated levels of PFOA were associated with the onset of delayed puberty in girls only, whereas elevated levels of PFOS were linked to the onset of delays in both boys and girls.

13.6-C TRANSPORTING ORGANIC ACIDS

When shippers offer an organic acid for transportation, DOT requires them to enter the relevant description on a shipping paper. Some examples of several representative organic acids are listed in Table 13.15. DOT also requires shippers and carriers to comply with all applicable labeling, marking, and placarding requirements.

[13]K. Steenland and Fletcher T. Savitz, "Epidemiologic evidence on the health effect of perfluorooctanoic acid (PFOA)," *Environ. Health Perspect.*, Vol. 118 (2010), pp. 1100–1108.

[14]Draft Risk Assessment of the Potential Health Effects Associated with Exposure to Perfluorooctanoic Acid and Its Salts (Washington, DC: U.S. Environmental Protection Agency, 2005).

[15]M. J. Lopez-Espinosa et al., "Association of perfluorooctanoic acid (PFOA) and perfluorooctane sulfonate (PFOS) with age of puberty of children living near a chemical plant," *Environ. Sci. Technol.*, Vol. 45 (2011), pp. 8160–8166.

TABLE 13.15	Shipping Descriptions of Some Representative Organic Acids
FORM OF ORGANIC ACID OR ITS SOLUTION	**SHIPPING DESCRIPTION[a]**
Acetic acid, glacial (contains >80% acid by mass)	UN2789, Acetic acid, glacial, 8, (3), PG II *or* UN2789, Acetic acid solutions, 8, (3), PG II
Acetic acid solutions, containing 50–80% acid by mass	UN2790, Acetic acid solution, 8, PG II
Acetic acid solutions containing 10–50% acid by mass	UN2790, Acetic acid solution, 8, PG III
Formic acid with more than 85% acid by mass	UN1779, Formic acid, 8, (3), PG II
Isobutyric acid	UN2529, Isobutyric acid, 3, (8), PG III

[a]See also Table 8.18.

When shippers offer for transportation a liquid organic acid whose name is not listed at 49 C.F.R. §172.101 for transportation, DOT requires them to identify the commodity generically as "UN3265, Corrosive liquid, acidic, organic, n.o.s., 8, PG I," "UN3265, Corrosive liquid, acidic, organic, n.o.s., 8, PG II," or "UN3265, Corrosive liquid, acidic, organic, n.o.s., 8, PG III." The shipping description includes the name of the specific organic acid entered parenthetically.

13.7 ESTERS

An **ester** is an organic compound having the general chemical formula $R-\overset{\displaystyle O}{\underset{\displaystyle O-R'}{C}}$. It is the substance that results when an organic acid reacts with an alcohol.

$$R-\overset{O}{\underset{OH}{C}} \;+\; R'-OH \;\longrightarrow\; R-\overset{O}{\underset{O-R'}{C}} \;+\; H_2O$$

An organic acid   An alcohol   An ester   Water

ester ■ Any organic compound whose general chemical formula is $R-\overset{O}{\underset{O-R'}{C}}$ where R and R' represent arbitrary alkyl or aryl groups

This type of chemical reaction is called **esterification**.

Esters are most commonly denoted by their IUPAC names. In the IUPAC system, an ester is named by changing the *-ic* suffix of the organic acid to *-ate*, preceded by the name of the alkyl or aryl group of the alcohol used to produce it. In other words, the R' group is named first followed by the name of the acid with *-ic acid* replaced by *-ate*.

esterification ■ The process of producing an ester by the reaction of an organic acid with an alcohol

The simple esters are colorless, highly volatile, and flammable liquids that are only partially soluble in water. Many possess pleasant, fruity odors. For example, the compounds responsible for the odors of pineapples, bananas, and apples are ethyl butyrate, isoamyl acetate, and ethyl-2-methylbutyrate, respectively. Several synthetic esters are added to foods as flavoring agents.

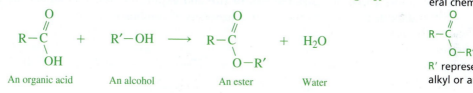

Ethyl butyrate   Isoamyl acetate   Ethyl-3-methylbutyrate

TABLE 13.16	Physical Properties of Ethyl Acetate
Melting point	−119°F (−84°C)
Boiling point	171°F (77°C)
Specific gravity at 68°F (20°C)	0.90
Vapor density (air = 1)	3.04
Vapor pressure at 68°F (20°C)	76 mmHg
Flashpoint	24°F (−4.4°C)
Autoignition point	800°F (427°C)
Lower flammable limit	2.18% by volume
Upper flammable limit	9% by volume
Evaporation rate (ether = 1)	2.7

Ethyl acetate

13.7-A ETHYL ACETATE

Ethyl acetate is a commonly encountered constituent of nitrocellulose-based lacquers and other protective coatings. It is produced by reacting ethanol and acetic acid.

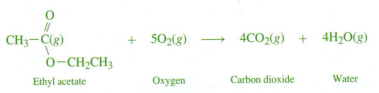

$$CH_3CH_2OH(aq) \ + \ CH_3-\overset{O}{\overset{\|}{C}}-OH(aq) \ \longrightarrow \ CH_3-\overset{O}{\overset{\|}{C}}-O-CH_2CH_3(aq) \ + \ H_2O(l)$$

Ethanol   Acetic acid   Ethyl acetate   Water

Ethyl acetate is a colorless liquid with a pleasing, fragrant odor. As reflected by the data in Table 13.16, it poses the risk of fire and explosion. When ethyl acetate burns, carbon dioxide and water vapor are the products of combustion.

$$CH_3-\overset{O}{\overset{\|}{C}}-O-CH_2CH_3(g) \ + \ 5O_2(g) \ \longrightarrow \ 4CO_2(g) \ + \ 4H_2O(g)$$

Ethyl acetate   Oxygen   Carbon dioxide   Water

13.7-B DI(2-ETHYLHEXYL) PHTHALATE

phthalate ■ Any ester of o-, m-, or p-phthalic acid

The **phthalates** are esters of 1,2-, 1,3-, and 1,4-benzenedicarboxylic acid, commonly called o-, m-, and p-phthalic acid. The esters of o-phthalic acid are commercially important. Their chemical formula generally is represented as follows, where R and R′ are arbitrary alkyl or aryl groups:

DEHP

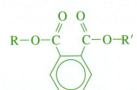

The individual phthalates are identified by naming R and R′, followed by the word *phthalate*. The commercially important phthalates include di-*n*-butyl phthalate (DBP), benzyl butyl phthalate (BBzP), di(2-ethylhexyl) phthalate (DEHP), diisononyl phthalate (DINP), diisodecyl phthalate (DIDP), and diisooctyl phthalate (DIOP). They are produced by reacting o-phthalic acid with appropriate alcohols.

DEHP is representative of the commercial phthalates. It is produced by reacting o-phthalic acid with 2-ethylhexanol.

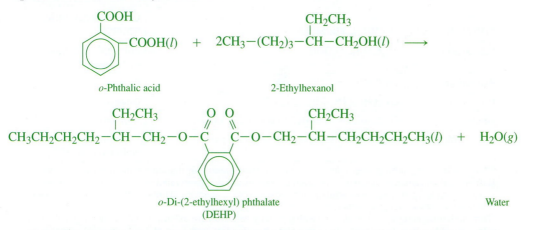

o-Phthalic acid   2-Ethylhexanol

o-Di-(2-ethylhexyl) phthalate
(DEHP)   Water

DEHP is a component of various rubber and other polymeric formulations in which its function is to increase the flexibility and toughness of the related products. When used in this fashion, the DEHP is called a **plasticizer**. DEHP is the most widely used plasticizer for the manufacture of products made from the polymer poly(vinyl chloride).

DEHP is also used as a plasticizer with certain explosives, where its presence provides the malleability that allows explosives to be molded into shapes and safely handled without detonating prematurely.

DEHP is also a component of certain cosmetics, where its function is to facilitate the addition of pleasant-smelling substances to shampoos, soaps, lotions, and deodorants. The presence of a phthalate ester in the composition of perfumes and nail polishes also provides them with an adhering or clinging quality.

When DEHP is used to manufacture plasticized products, it is mixed with the polymers but only loosely bonds with them. Hence, it is capable of leaching from the products. Persons often encounter DEHP unknowingly while searching to buy a new car. DEHP slowly vaporizes from the vinyl upholstery and provides the "new-car smell" in vehicles whose windows and doors have been kept closed.

In animal and human studies,[16] health concerns have been raised over the impact of the DEHP that leaches from plasticized medical products such as intravenous bags, dialysis tubing, and syringes. When blood is stored in DEHP-plasticized bags, for example, the phthalate ester leaches from the plastic and dissolves in the blood. Then, patients unknowingly receive DEHP during blood infusions. The prevailing fear is that phthalates are endocrine disrupters that may interfere with development. Animal exposure to DEHP has been shown to interfere with normal kidney and liver function and cause physiological abnormalities in the reproductive systems of the animals, especially their male offspring.

Whether phthalates can interfere with the hormones generated by humans is a controversial topic. FDA initially concluded that patient exposures to DEHP generally are well below the levels expected to cause adverse effects, although it does concede that children, especially infants and toddlers, may represent a special population at increased risk from exposure. In recognition of this guarded view, the FDA proposed voluntary restrictions on the use of DEHP-plasticized devices by pregnant women and newborn infants. The FDA also proposed labeling medical devices to indicate the absence of DEHP.

In 2008, the CPSC permanently banned DEHP, DBP, and DBP at concentrations exceeding 0.1% in any children's toy or childcare article intended for use by children 12 years of age and younger. Three additional phthalates, DIDP, DINP, and DIOP, also have

plasticizer ■ A substance, such as a phthalate ester, which when added in a prescribed amount to a polymeric formulation, facilitates processing and provides flexibility and toughness to the end product

---

[16]"Toxicological Profile of Diethylhexyl Phthalate" (Washington, DC: U.S. Department of Health and Human Services), 2002.

COMMON NAME	CHEMICAL FORMULA[a,b]	COMMON NATURAL SOURCES
TABLE 13.17	**Some Representative Examples of Fatty Acids**	
Stearic acid	$CH_3(CH_2)_{16}$—COOH	Butterfat, lard, beef fat, cottonseed oil
Palmitic acid	$CH_3(CH_2)_{14}$—COOH	Palm oil
Oleic acid	$CH_3(CH_2)_7CH$=$CH(CH_2)_7$—COOH	Olive oil
Linoleic acid	$CH_3(CH_2)_4CH$=$CHCH_2CH$=$CH(CH_2)_7$—COOH	Soybean oil
α-Linolenic acid[c]	CH_3CH_2CH=$CHCH_2CH$=$CHCH_2CH$=$CH(CH_2)_7$—COOH	Soybean oil; canola oil; walnuts; kale

[a]The saturated fatty acids possess only carbon-carbon single bonds, but the unsaturated fatty acids contain at least one carbon-carbon double bond. Although the formulas of the unsaturated fatty acids are represented here in a linear fashion, the compounds actually exist as *cis-* and *trans-* geometrical isomers. Nature produces predominantly, but not exclusively, *cis-* isomers of the unsaturated fatty acids.
[b]Fatty acids occur naturally in combination with glycerol as fats. They are colloquially called saturated and unsaturated fats. When they are consumed in one's diet, saturated fats contribute to the build-up of plaque in the arterial walls causing arteriosclerosis—narrowing of the arteries. The triglycerides of the *trans* isomers of unsaturated fatty acids are colloquially called *trans fats*. Their presence in the diet has been implicated with the onset of cardiovascular disorders.
[c]Nutritionists assert that linoleic acid and α-linolenic acid are desirable components of one's diet because they are important for growth and brain function and help to lower triglycerides and cholesterol.

been temporarily prohibited, pending further study and review. The European Parliament also banned the use of certain phthalate esters, including DEHP, in children's toys and child-care items like pacifiers and teething rings.

13.7-C LINSEED OIL

Linseed oil is a clear to yellowish oil obtained from the seeds of the flax plant *Linum usitatissimum* (Section 14.5-A). The oil is a mixture of the triesters of 1,2,3-propanetriol, an alcohol commonly known as glycerol. The triesters of glycerol are produced naturally when glycerol reacts with long-chain saturated and unsaturated **fatty acids**. Some representative examples of fatty acids are provided in Table 13.17.

The triester of glycerol is called a **triglyceride.** It has the following general chemical formula:

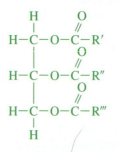

Here, R′, R″, and R‴ are alkyl or alkenyl groups that possess from 12 to 22 carbon atoms (usually an even number).

Linseed oil is a mixture of the triglycerides of the following fatty acids:

- The saturated fatty acids, palmitic acid (~7%) and stearic acid (3.4%–4.6%)
- The monounsaturated fatty acid, oleic acid (18.5%–22.6%)
- The doubly unsaturated fatty acid, linoleic acid (14.2%–17%)
- The triply unsaturated α-linolenic acid (51.9%–55.2%)

fatty acid ▪ Any carboxylic acid derived from the triglycerides present in animal and vegetable fats and oils

triglyceride ▪ Any triester of glycerol, usually produced by reacting glycerol with fatty acids

Linseed oil

Although it was formerly used extensively as a carrier in paints and varnishes, this use of linseed oil has declined seriously since the 1950s. Today, linseed oil is used mainly to polish the surface of furniture and other commercial products made from wood. It is also used during the production and manufacture of linoleum flooring.

Linseed oil is commercially available in two forms: raw and boiled. However, the expression *boiled linseed oil* is a misnomer because linseed oil does not boil. The term refers to raw linseed oil containing additives that accelerate drying.

As linseed oil dries, its constituents evaporate. Unless the heat of vaporization dissipates into the atmosphere, items moist with linseed oil are likely to burn. Cotton rags and other porous materials containing linseed oil are prone to undergo spontaneous combustion. Hobbyists have caused many domestic fires by failing to follow precautions when discarding these items.

13.7-D BIODIESEL FUEL, BIODIESEL BLENDS, AND BIOMASS-BASED DIESEL FUELS

Biofuels have become popular substitutes for petroleum-based diesel oil for the following reasons:

- Their comparatively lower cost
- Chemical compatibility with hoses and other accessories needed for storage and delivery
- A comparatively lower risk of fire and explosion
- Lesser amounts of smoke, VOCs, carbon monoxide, and sulfur dioxide (none) produced on combustion. (The NOx concentration remains unchanged in the emissions.)

Three types of biofuels are used to power diesel-operated vehicles and equipment:

■ **Biodiesel fuel** is an alternative motor fuel produced from naturally occurring fats and oils derived from plants and animals including olive oil, soybean oil, coconut oil, cottonseed oil, peanut oil, tallow, rapeseed oil, and canola oil. Recycled restaurant grease, poultry fats, tallow, and other rendered fats also are used to produce biodiesel fuel. The biodiesel fuel produced from soybean oil is most commonly encountered in the United States. It has a cetane number of 47. The biodiesel fuels produced from rapeseed oil and canola oil are more commonly encountered in Europe. They have cetane numbers of approximately 54.

The manufacturers of biodiesel fuels treat the fat-and-oil feedstocks with methanol or ethanol. During their treatment, the fatty acid triglycerides are converted into mixtures of methyl or ethyl fatty acid esters.

When biodiesel fuel has not been blended with other fuels, it is called B100. It generally is encountered only during its production, transportation, or storage. B100 possesses a variable flashpoint that depends upon the chemical composition of the biofuel. Nonetheless, the flashpoint is always greater than 200°F (93.3°C); hence, B100 is an OSHA category 4 flammable liquid, or an NFPA class IIIB combustible liquid.

■ **Biodiesel blend** is an alternative motor fuel consisting of a mixture of biodiesel and biomass-based diesel fuels.

■ **Biomass-based diesel** is an alternative motor fuel produced from the remains of recently living plants or animals (i.e., grease and rendered tallow and other animal fats), but they are not chemically treated.

Biodiesel fuel is also encountered as a mixture with petroleum-based fuels, but it is not then regarded as either a biodiesel blend or a biomass-based diesel. For example, the blend known as B20 is a mixture of 20% biodiesel fuel and 80% petroleum-based diesel oil. It is used by the city of San Francisco to fuel its entire fleet of diesel-operated vehicles from ambulances to street-sweepers. A mixture of Diesel #1, Diesel #2, or JP-8 with 7.7%

biodiesel fuel ■ Any alternative motor fuel produced by biologically or chemically converting the triglycerides in plant and animal fats and oils into their fatty acid methyl or ethyl esters

biodiesel blend ■ Any alternative motor fuel derived by blending biodiesel and biomass-based diesel fuels

biomass-based diesel ■ Any biofuel produced from nonpetroleum renewable resources, but that does not consist of a mixture of the esters of fatty acids

to 15% ethanol and other additives is another popular blend. Biomass-based diesel also may be mixed with petroleum-based diesel oil.

At 16 C.F.R. §306.12, the U.S. Federal Trade Commission requires biodiesel distributors to affix on their dispensers, as relevant, a blue-and-black or orange-and-black label resembling those shown here:

B-100 Biodiesel	Biodiesel Blend	100% Biomass-Based Diesel	Biomass-Based Diesel Blend
contains 100 percent biodiesel	contains biomass-based diesel or biodiesel in quantities between 5 percent and 20 percent	contains 100 percent biomass-based diesel	contains biomass-based diesel or biodiesel in quantities between 5 percent and 20 percent

The blue-and-black labels characterize either a biodiesel fuel or a biomass-based diesel blend, and the orange-and-black labels characterize either a biomass-based diesel fuel or a biomass-based diesel blend. The commission requires these distinctively colored labels to highlight and distinguish their chemical natures.

The blue-and-black biodiesel blend label shown here is affixed specifically to dispensers of B20 biodiesel blend having a biodiesel concentration that ranges from 5% to 20% by volume. Biodiesel blends with 5% or less biodiesel or biomass-based diesel are exempt from this labeling requirement. The commission also requires new-vehicle manufacturers and used-vehicle dealers to affix the relevant label on a visible surface of each vehicle to identify the manner by which it is powered.

13.7-E TRANSPORTING ESTERS

When shippers offer an ester for transportation, DOT requires them to provide the relevant shipping description on an accompanying shipping paper. Some examples of several representative esters are listed in Table 13.18.

When they transport a flammable ester whose name is not provided in the Hazardous Materials Table at 49 C.F.R. §172.101, DOT requires them to identify the commodity generically on a shipping paper as either "UN3272, Esters, n.o.s., 3, PG II" *or* "UN3272, Esters, n.o.s., 3, PG III." The generic shipping description includes the name of the specific ester entered parenthetically. DOT also requires shippers and carriers to comply with all applicable labeling, marking, and placarding requirements.

When carriers transport biodiesel fuels, they must identify the commodity on the shipping paper in the following manner, as relevant:

■ For biodiesel blends equal to or less than B5: NA1993, Diesel fuel, 3 PG III, UN1202, Diesel fuel, 3, PG III, *or* UN1202, Gas oil, 3, PG III.

■ For biodiesel blends over B5:NA1993, Diesel fuel solution, 3, PG III, UN1202, Diesel fuel solution, 3, PG III, *or* UN1202, Gas oil solution, 3, PG III.

TABLE 13.18	Shipping Descriptions of Some Representative Esters
ESTER	**SHIPPING DESCRIPTION**
Ethyl acetate	UN1173, Ethyl acetate, 3, PG II
Isopropyl propionate	UN2409, Isopropyl propionate, 3, PG II
Methyl formate	UN1243, Methyl formate, 3, PG I
Methyl propionate	UN1248, Methyl propionate, 3, PG II
n-Propyl acetate	UN1276, *n*-Propyl acetate, 3, UN1276, PG II
Propyl formates	UN1281, Propyl formates, 3, PG II

13.8 AMINES

The **amines** are organic derivatives of ammonia, that is, compounds whose molecules contain one, two, or three alkyl or aryl groups bonded to a nitrogen atom. Their general formulas are $R-NH_2$, R_2NH, and R_3N, where R is an arbitrary alkyl or aryl group.

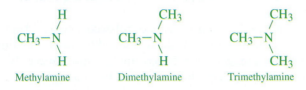

For example, the amines whose molecules have one, two, and three methyl groups bonded to the nitrogen atom are the compounds whose structures follow:

$$
\begin{array}{ccc}
& H & CH_3 & CH_3 \\
& / & / & / \\
CH_3-N & CH_3-N & CH_3-N \\
& \backslash & \backslash & \backslash \\
& H & H & CH_3 \\
\text{Methylamine} & \text{Dimethylamine} & \text{Trimethylamine}
\end{array}
$$

They are synthesized industrially from ammonia and methanol in the presence of an aluminum oxide catalyst at 842°F (450°C).

$$CH_3OH(g) + NH_3(g) \longrightarrow CH_3NH_2(g) \longrightarrow (CH_3)_2NH \longrightarrow (CH_3)_3N$$

Methanol   Ammonia   Methylamine   Dimethylamine   Trimethylamine

Table 13.19 provides their physical properties, which reflect the fact that the primary risk associated with them is fire and explosion. The incomplete combustion of amines produces nitric oxide, carbon monoxide, and water vapor, and complete combustion produces nitrogen dioxide, carbon dioxide, and water vapor.

The simple amines generally are used by manufacturing and process industries as coagulants, flocculating agents, corrosion and rust inhibitors, bactericides, and fungicides. In the pharmaceutical industry, they are used to manufacture drugs, and in the chemical industry, they are used to manufacture dyes and other substances.

The simplest amines can pose a health risk because their inhalation causes serious eye, nose, and throat irritation. The more complex amines usually are flammable and corrosive liquids. The nature of their corrosivity is linked with the similarity in chemical behavior that the amines share with ammonia.

amine ■ Any organic compound whose general chemical formula is $R-NH_2$, R_2NH, or R_3N, where R is an arbitrary alkyl or aryl group

Methylamine and Dimethylamine

TABLE 13.19	Physical Properties of the Simple Amines		
	METHYLAMINE	**DIMETHYLAMINE**	**TRIMETHYLAMINE**
Melting point	−137°F (−94°C)	−134°F (−92°C)	−179°F (−117°C)
Boiling point	21°F (−6°C)	45°F (7°C)	39°F (4°C)
Specific gravity at 68°F (20°C)	0.69	0.68	0.66
Vapor density (air = 1)	1.08	1.65	2.0
Flashpoint	34°F (1°C)	−58°F (−50°C)	8 to 18°F (−13 to −8°C)
Autoignition point	806°F (430°C)	755°F (402°C)	374°F (190°C)
Lower flammable limit	4.9% by volume	2.8% by volume	2% by volume
Upper flammable limit	20.7% by volume	14.4% by volume	11.6% by volume

Trimethylamine

Some amines are biologically important compounds. Some, like dopamine and adrenaline, are among the substances that transmit nerve impulses.

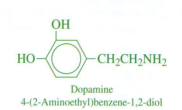

Dopamine
4-(2-Aminoethyl)benzene-1,2-diol

Adrenaline, or Epinephrine
4-[1-Hydroxy-2-(methylamino)ethyl]
benzene-1,2-diol

Dopamine is produced naturally in the brain and is responsible for the feeling of desire, loss of appetite, and inability to sleep. These symptoms are commonly associated with "falling in love."

Several aromatic amines are constituents of tobacco smoke.[17] Undoubtedly the best known is nicotine, the primary stimulant that causes addiction to smoking and accelerates the rates at which the carcinogens in tobacco smoke cause cancer. Two other aromatic amines in tobacco smoke are β-naphthylamine and 4-aminobiphenyl.

Nicotine
3-(1-Methylpyrrolidine-2-yl)pyridine

β-Naphthylamine
(2-Naphthylamine)

4-Aminobiphenyl

Nicotine

Nicotine is the only constituent of tobacco that has commercial value. Its poisonous nature (LD_{50} = 55 mg/kg in rats) is the basis for its use as a potent insecticide. However, the insecticidal use of nicotine in the United States will cease in January 2014. Using the authority of TSCA, EPA then will terminate the legal production and use of nicotine, due in large part to the negative impact that its release has on the environment.

The inhalation of β-naphthylamine and 4-aminobiphenyl in tobacco smoke is regarded as the primary cause of human bladder cancer. Research studies[18] reveal that smokers are four times more likely than nonsmokers to develop bladder cancer. Because they cause cancer of the urinary bladder in humans, these aromatic amines are ranked as human carcinogens.

Diamines are compounds having two amino groups per molecule. The diamines putrescine and cadaverine have especially foul odors.

$$H_2N-CH_2CH_2CH_2CH_2-NH_2 \qquad H_2N-CH_2CH_2CH_2CH_2CH_2-NH_2$$

Putrescine
(1,4-Butanediamine)

Cadaverine
(1,5-Pentanediamine)

Their combined stench is evident in decaying flesh. Cadaver dogs are trained to detect the odor when searching for human remains. Both are poisonous liquids without commercial value.

---

[17]Report of the Surgeon General, "How Tobacco Smoke Causes Disease: The Biology and Behavioral Basis for Smoking-Attributable Disease," U.S. Centers for Disease Control and Prevention, Atlanta, Georgia (ISBN-13: 978-0-16-084078-4).

[18]Neal D. Freeman et al., "Association between smoking and risk of bladder cancer among men and women," *J. Amer. Med. Assoc.,* Vol. 306 (2011), pp. 793–896.

Another diamine is benzidine, which formerly was used by homicide detectives to verify the presence of blood at crime scenes.

Benzidine
4,4′-Diaminobiphenyl

An enzyme in blood causes benzidine to oxidize, thereby forming a blue-colored derivative. Like several other aromatic amines, benzidine is a human carcinogen. Although formerly used widely in the chemical industry to produce certain dyes, its commercial use now is severely limited.

Several means are used for naming amines. The simplest amines are known by their common names. They are identified by naming the alkyl or aryl group bonded to the nitrogen atom, followed by the suffix *-amine*. The names of the simplest amines—methylamine, dimethylamine, and trimethylamine—are examples. Although *di-* and *tri-* are used to name two and three identical groups bonded to a nitrogen atom, respectively, the prefixes *bis* and *tris*, instead of *di* and *tri*, are used when complex amino groups are components of a formula.

Amines may also be named as aminoalkanes, alkylaminoalkanes, and dialkylaminoalkanes. A number is used to show the position of the amino group along the carbon–carbon chain. For example, the compounds whose formulas are $CH_3CH_2CH_2NH_2$, $(CH_3)CH-NH_2$, and $(CH_3)_2CH-NHCH_3$ are then named as follows:

$$CH_3CH_2CH_2-NH_2 \qquad CH_3-\underset{\underset{NH_2}{|}}{C}-CH_3 \qquad CH_2-\underset{\underset{CH_3}{|}}{CH}-NHCH_3$$

1-Aminopropane   2-Aminopropane   2-Methylaminopropane

In the IUPAC system, amines are named by replacing the *-e* in the name of the alkane with *-amine*. A number is again assigned to show the position of the amino group along the chain. The prefix *N-* (in italics) is used to designate the bonding of an alkyl or aryl substituent on the nitrogen atom. Then, the substituents are listed alphabetically regardless of whether they are bonded to the nitrogen atom or along the chain of carbon atoms.

$$CH_3CH_2-\underset{\underset{NHCH_2CH_3}{|}}{CH}-CH_2CH_2CH_3 \qquad CH_3-\underset{\underset{CH_3}{|}}{N}-CH_2CH_2CH_3$$

N-Ethyl-3-hexanamine   *N,N*-Dimethylpropanamine

The simplest aromatic amine is named phenylamine, or more commonly, aniline; its chemical formula is $C_6H_5-NH_2$. The substances having the chemical formulas $C_6H_5-NHCH_3$ and $C_6H_5-N(CH_3)_2$ are named *N*-methylaniline and *N,N*-dimethylaniline, respectively.

13.8-A WORKPLACE REGULATIONS INVOLVING THE SIMPLE AMINES

When the simplest amines are present in the workplace, OSHA requires employers to limit employee exposure to them at the following maximum concentrations in air, averaged over an 8-hour workday: methylamine, 10 parts per million; dimethylamine, 10 parts per million; trimethylamine, 10 parts per million; ethylamine, 10 parts per million; and aniline, 2 parts per million (skin).

13.8-B *N*-NITROSAMINES

N-nitrosamine ■ Any organic compound whose molecules have a nitroso group bonded to the nitrogen atom of an amine

The ***N*-nitrosamines** are compounds whose molecules have a nitroso group $(-N{=}O)$ bonded to an amine nitrogen atom. The aliphatic *N*-nitrosamines have the following general chemical formula, where R and R′ are arbitrary alkyl or aryl groups:

$$\underset{R'}{\overset{R}{\diagdown}}N-N{=}O$$

These compounds are produced when nitrous acid reacts with those amines whose molecules have only one hydrogen atom bonded to the nitrogen atom. As previously noted in Section 11.15, when cured meat products are consumed, the sodium nitrite used during the curing process reacts with stomach acid to produce nitrous acid, which in turn reacts with the proteins contained in all meat products to produce *N*-nitrosamines. This issue is potentially problematic because these compounds are probable carcinogens.

Several *N*-nitrosamines have been detected in tobacco smoke at concentrations much higher than those detected in cooked bacon and other cured meat products.[19] Included among them are the following:

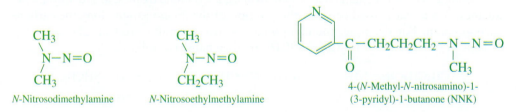

N-Nitrosodimethylamine   *N*-Nitrosoethylmethylamine   4-(*N*-Methyl-*N*-nitrosamino)-1-(3-pyridyl)-1-butanone (NNK)

These nitrosamines are produced during the fermentation, curing, and burning of tobacco. Because they may contribute to the cancers experienced by tobacco smokers and individuals exposed to secondhand smoke, *N*-nitrosamines are considered probable human carcinogens.

EPA has noted that *N*-nitrosamines may also form when metalworking fluids are used for cooling, lubricating, and corrosion or rust inhibition during grinding, polishing, and other machining operations. The *N*-nitrosamines are produced when metallic nitrites or similar substances mix with these fluids. The complex amine diethanolamine is a representative compound used in metalworking fluids. When it reacts with nitrous acid, *N*-nitrosodiethanolamine is produced.

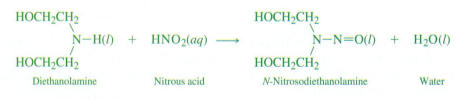

Diethanolamine   Nitrous acid   *N*-Nitrosodiethanolamine   Water

EPA used its authority under TSCA to reduce or eliminate the potential for human exposure to *N*-nitrosamines when metal-machining work required the use of fluids containing

---

[19]Report of the Surgeon General, "How Tobacco Smoke Causes Disease: The Biology and Behavioral Basis for Smoking-Attributable Disease," U.S. Centers for Disease Control and Prevention, Atlanta, Georgia (2010) (ISBN-13:978-0-16-084078-4).

amines. At 40 C.F.R. §747.115, EPA prohibits the mixing of metallic nitrites or similar substances with metalworking fluids and requires all containers of metalworking fluids distributed in commerce to be labeled as follows:

<table>
<tr><td>

WARNING!

Do Not Add Nitrites to This Metal-working Fluid Under Penalty of Federal Law

Addition of Nitrites Leads to Formation of a Substance Known to Cause Cancer

This Product is Designed to be Used without Nitrites

</td></tr>
</table>

EPA requires this label to be affixed to the containers with such conspicuousness that the warning statement is read and understood by the ordinary individual under customary conditions of purchase and use.

13.8-C METHAMPHETAMINE

Methamphetamine is a controlled substance, i.e., an illegal drug that can stimulate or dull an individual's senses and become addictive when used repeatedly over time. As a legitimate drug, it is prepared in pharmaceutical laboratories by reacting methylamine with phenyl-2-propanone in the presence of a reducing agent.

Methamphetamine

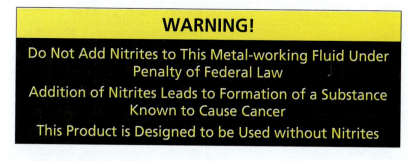

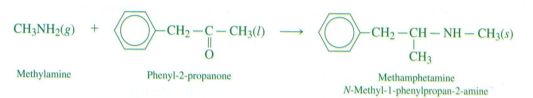

Methylamine   Phenyl-2-propanone   Methamphetamine
N-Methyl-1-phenylpropan-2-amine

Physicians generally prescribe methamphetamine for use as a stimulant.

Aside from its legitimate production, methamphetamine is produced for use by drug abusers, who refer to it as *speed*, *splash*, *crystal*, or *meth*. The unlawful production usually involves the reduction of the legal drugs, ephedrine or pseudoephedrine (both decongestants), with flammable solvents and other hazardous materials.

Clandestine laboratories used to produce methamphetamine can represent extremely dangerous settings. Emergency response personnel who encounter them must take special preservation measures because the illicit production of any controlled substance is a criminal activity. These measures are accomplished in major cities by Methamphetamine Response Teams whose responsibilities include the proper collection of relevant evidence from the crime scenes for use when the drug producers are prosecuted.

13.8-D TRANSPORTING AMINES

When shippers offer an amine for transportation, DOT requires them to provide the relevant shipping description on a shipping paper. Some examples for several representative amines are listed in Table 13.20. When shippers offer for transportation an amine whose name is not listed at 49 C.F.R. §172.101, DOT requires them to determine whether the commodity is solely corrosive, or corrosive *and* flammable. Then, the shipping description of the commodity is provided generically on the shipping paper that accompanies the shipment. DOT also requires shippers and carriers to comply with all applicable labeling, marking, and placarding requirements.

TABLE 13.20	Shipping Descriptions of Some Representative Amines
AMINE	**SHIPPING DESCRIPTION**
Aniline	UN1547, Aniline, 6.1, PG II
Diethylamine	UN1154, Diethylamine, 3, (8), PG II
N,N-Diethylaniline	UN2432, *N,N*-Diethylaniline, 6.1, PG III
Dimethylamine (anhydrous)	UN1032, Dimethylamine, anhydrous, 2.1
Ethylamine	UN1036, Ethylamine, 2.1
Hexamethylenetetramine	UN1328, Hexamethylenetetramine, 4.1, PG III
Methylamine, anhydrous	UN1061, Methylamine, anhydrous, 2.1
Trimethylamine, anhydrous	UN1083, Trimethylamine, anhydrous, 2.1

13.9 PEROXO-ORGANIC COMPOUNDS

organic hydroperoxide ■ Any derivative of hydrogen peroxide in which one hydrogen atom in the H_2O_2 molecule has been substituted with an alkyl or aryl group; any member of a class of organic compounds having the general chemical formula H—O—O—R, where R is an arbitrary alkyl or aryl group

organic peroxide ■ A derivative of hydrogen peroxide in which both hydrogen atoms in the H_2O_2 molecule have been substituted with alkyl or aryl groups; any organic compound whose general chemical formula is R—O—O—R', where R and R' are arbitrary alkyl or aryl groups; for purposes of DOT regulations, an organic compound containing oxygen in the bivalent —O—O— structure and that may be considered a derivative of hydrogen peroxide in which one or more of the hydrogen atoms have been replaced by organic radicals

Peroxo-organic compounds are either **organic hydroperoxides** or **organic peroxides**. They respectively differ by the substitution of one or both of the hydrogen atoms in the H_2O_2 molecule with alkyl or aryl groups (R and R'). Consequently, their chemical formulas are R—O—O—H and R—O—O—R', respectively.

The simplest organic hydroperoxides are identified by the name of the alkyl or aryl group bonded to the H—O—O— group followed by the word *hydroperoxide*. Examples are *tert*-butyl hydroperoxide and isopropylbenzene hydroperoxide, whose chemical formulas are $(CH_3)_3$—C—O—OH and C_6H_5—C(CH$_3$)$_2$—O—OH, respectively.

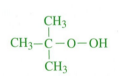

tert-Butyl hydroperoxide

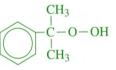

Isopropylbenzene hydroperoxide
(Cumene hydroperoxide)

The simplest organic peroxides are identified by the names of the alkyl or aryl groups bonded to the —O—O— group followed by the word *peroxide*. When the alkyl or aryl group is the same (R=R'), the name of the group is preceded by the prefix *di*. Examples are diacetyl peroxide and di-*tert*-butyl peroxide, whose chemical formulas are $(CH_3CO_2)_2$ and $[(CH_3)_3$—C—$]_2O_2$, respectively.

Diacetyl peroxide

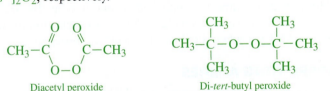

Di-*tert*-butyl peroxide

Di-*tert*-butyl peroxide is a substance sometimes added in low concentration to diesel oil to improve its cetane number (Section 12.13-F).

Some commercially important peroxo-organic compounds are perketones, peracids, and peresters. These substances are oxidized acids, ketones, and esters, respectively. The peracids are usually named by inserting the prefix *per*, or *peroxy-*, before the name of the oxidized acid. The common peracids have from 1 to 4 carbon atoms per molecule. For example, perpropionic acid, or peroxypropionic acid, is the compound with the following molecular structure:

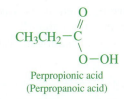

Perpropionic acid
(Perpropanoic acid)

The perketones commonly are named by inserting the word *peroxide* after the name of the ketone from which it was produced. These substances generally contain multiple functional groups, one of which is the peroxide group, as illustrated by the following molecular structures of ethyl methyl ketone peroxide and cyclohexanone peroxide.

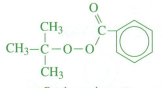

Bis(2-hydroperoxy-*sec*-butyl)peroxide
(Ethyl methyl ketone peroxide)
(Methyl ethyl ketone peroxide)

1-Hydroxy-1′-hydroperoxydicyclohexyl peroxide
(Cyclohexanone peroxide)

A perester often is identified by inserting the term *peroxy* before the name of alkyl or aryl group that is bonded to the carboxyl group, as in *tert*-butylperoxybenzoate.

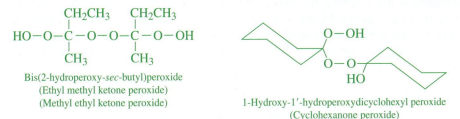

tert-Butylperoxybenzoate

Sometimes, however, the perester is named as a dialkylperoxydicarbonate, whose general chemical formula is the following:

For example, when R is the isopropyl group, $(CH_3)_2CH-$, the peroxyester is named diisopropylperoxydicarbonate.

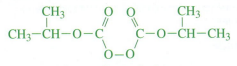

Diisopropylperoxydicarbonate

The peroxo-organic compounds are commonly employed in the chemical industry as raw materials in synthetic processes. Peroxo-organic compounds are also used widely to induce polymerization, a process essential to the production and manufacture of plastics. Although they are either liquids or solids at room conditions, peroxo-organic compounds often are encountered dissolved in water or an appropriate organic solvent. These solvents serve to stabilize peroxo-organic compounds against premature thermal decomposition.

Although the peroxo-organic compounds are flammable substances, they are also powerful oxidizers containing active oxygen within their molecular structures. When they burn, peroxo-organic compounds support their own combustion. This combination of properties poses a particularly pronounced risk of fire and explosion. When ignited, the peroxo-organic compounds often burn furiously and more intensely than other

deflagration ■ The chemical process during which a substance burns intensely instead of detonating

combustible substances. This burning phenomenon is called **deflagration**. Sufficient energy is generated during deflagration to enable the burning to proceed and accelerate without the input of additional energy from another source. An example is illustrated by the following equation:

$$[(CH_3)_3-C-]_2O_2(s) \quad + \quad 9O_2(g) \longrightarrow 3CO_2(g) \quad + \quad 5CO(g) \quad + \quad 9H_2O(g)$$

Di-*tert*-butyl peroxide   Oxygen   Carbon dioxide   Carbon monoxide   Water

Most peroxo-organic compounds are unstable. When exposed to an ignition source, they are just as likely to undergo rapid, autoaccelerated decomposition as they are to burn. Some are so sensitive to friction, heat, or shock that they cannot be safely handled unless maintained at low temperatures and diluted within an inert solid material or solvent. Furthermore, organic peroxides are skin and eye irritants and are very likely to cause gastrointestinal problems when swallowed. These potentially hazardous features are illustrated on the label affixed to containers of dibenzoyl peroxide and provided in Figure 13.5.

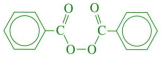

Dibenzoyl peroxide

FIGURE 13.5 This label affixed to nonbulk containers of dibenzoyl peroxide complies with OSHA's regulations published at 29 C.F.R. §1910.1200(f).

13.9-A TRANSPORTING PEROXO-ORGANIC COMPOUNDS

Before a peroxo-organic compound may be offered for transportation, DOT requires its manufacturer or other responsible party to test a sample of the substance using prescribed procedures to determine whether it decomposes under the temperature conditions likely to be encountered during transportation. At 49 C.F.R. §173.21(f), DOT prohibits the transportation of a package when the testing data reveal that the substance is likely to decompose at a **self-accelerating decomposition temperature**, or **SADT**, of 122°F (50°C) or less, or polymerize at a temperature of 130°F (54°C) or less, with an evolution of a dangerous quantity of heat or gas.

When shippers prepare the shipping description of a peroxo-organic compound, they consult the Hazardous Materials Table and the Organic Peroxides Table published at 49 C.F.R. §§172.101 and 173.225, respectively. DOT does not distinguish between organic hydroperoxides and organic peroxides in the hazardous materials regulations. For DOT's purposes, each is referenced as "organic peroxide."

The shipping descriptions of organic peroxides are listed generically in the Hazardous Materials Table. Several examples are provided in Appendix B. Each listing denotes a type of organic peroxide and its state of matter (liquid or solid). There are seven organic peroxide types, each of which is designated by a capital letter A through G, which collectively form a continuum of decreasing hazard. Their features are noted in Table 13.21.

The types of the generic peroxo-organic compounds noted in the Hazardous Materials Table refer only to types A through F. As noted earlier, DOT prohibits the transportation of type A organic peroxides. Some table entries include the words "temperature-controlled."

The letter G appears in column 1 of the Hazardous Materials Table for each proper shipping name of an organic peroxide. As noted in Section 6.2-A, this signifies that the technical name of the organic peroxide must be entered in parentheses in its shipping description. This name is included in the shipping information provided in the Organic Peroxides Table, an excerpt of which is provided in Table 13.22.

self-accelerating decomposition temperature (SADT) ■ The lowest temperature at which a peroxo-organic compound in a typical package undergoes self-accelerating decomposition that causes the package to rupture

TABLE 13.21	Generic Types of Organic Peroxides[a]
TYPE	**IDENTIFYING FEATURES**
Type A	Can detonate or rapidly deflagrate as packaged for transport. The transportation of type A organic peroxides is prohibited by DOT.
Type B	Neither detonates nor deflagrates rapidly when correctly packaged for transportation, but can undergo a thermal explosion
Type C	Neither detonates nor deflagrates rapidly when correctly packaged for transportation and cannot undergo a thermal explosion
Type D	Detonates only partially, but does not deflagrate rapidly, and is not affected by heat when confined; or does not detonate, deflagrates slowly, and manifests no violent effect if heated when confined; or does not detonate or deflagrate and manifests a medium effect when heated under confinement
Type E	Neither detonates nor deflagrates and manifests either a low or no effect when heated under confinement
Type F	Will not detonate in a cavitated state, does not deflagrate, manifests only a low or no effect if heated when confined, and possesses either a low or no explosive power
Type G[b]	Will not detonate in a cavitated state, will not deflagrate at all, shows no effect when heated under confinement, and manifests no explosive power

[a]49 C.F.R. §173.128.
[b]A type G organic peroxide is not subject to DOT's transportation requirements for organic peroxides if it is "thermally stable," i.e., its self-accelerating decomposition temperature is 122°F (50°C) or higher for a 110-pound (50-kg) package. An organic peroxide that possesses all characteristics of type G other than thermal stability and requires temperature control is classed as a type F, temperature-controlled organic peroxide.

TABLE 13.22 Organic Peroxides Table[a,b]

TECHNICAL NAME (1)	IDENTIFICATION NUMBER[c] (2)	CONCENTRATION (MASS %) (3)	DILUENT (MASS %)			CONCENTRATION OF WATER (MASS %) (5)	TEMPERATURE (°C)	
			(4A)	(4B)	(4C)		CONTROL (7A)	EMERGENCY (7B)
Acetyl acetone peroxide	UN3105	≤42	≥48			≥8		
tert-Butyl hydroperoxide	UN3103	<79 to 90				≥10		
Cumyl hydroperoxide	UN3107	<90 to 98	≤10					
Cyclohexanone peroxide(s)	Exempt	≤32			≥68			
Diacetyl peroxide	UN3115	≤27		≥73			20	25
Dibenzoyl peroxide	UN3102	<77 to 94				≥6		
Dicumyl peroxide	UN3110	<52 to 100	≤48					
Di-(2-methylbenzoyl) peroxide	UN3112	≤87				≥13	30	35
Methylcyclohexanone peroxide	UN3115	≤67		≥33			35	40
Peroxyacetic acid, Type D, stabilized	UN3105	≤43						
1,1,3-Tetramethylbutyl hydroperoxide	UN3105	≤100						

[a]Excerpted from the DOT Organic Peroxides Table, 49 C.F.R. §173.225. The information tabulated in columns 6 and 8 of original reference has been redacted.
[b]The columnar entries noted in this table signify the following:

Column 1   The technical name of the compound
Column 2   The DOT identification number. When the word "Exempt" appears in column 2, the transportation of the applicable substance is not regulated as an organic peroxide.
Column 3   The concentration limits, if any.
Column 4   The type and concentration of a diluent or inert solid, when required, that must be mixed with the peroxo-organic compound for effective desensitization.

■ A *Diluent type A* is an organic liquid that does not detrimentally affect the thermal stability or increase the hazard of the peroxo-organic compound and possesses a boiling point of not less than 302°F (150°C).
■ A *Diluent type B* is an organic liquid that is compatible with the peroxo-organic compound and possesses a boiling point of less than 302°F (150°C), but at least 140°F (60°C), and a flashpoint greater than 41°F (5°C).
■ An *inert solid* is a solid substance that does not detrimentally affect the thermal stability or increase the hazard of the peroxo-organic compound.

Column 5   The minimum concentration of water in mass percent that must be present in the formulation of the peroxo-organic compound.
Column 7   The control temperature and emergency temperature, when temperature controls are required

■ The *control temperature* is the temperature above which DOT prohibits carriers from transporting the peroxo-organic compound.
■ The *emergency temperature* is the temperature at which DOT prohibits carriers who are transporting a peroxo-organic compound to implement emergency measures owing to the imminent danger that the compound may undergo thermal decomposition. The emergency temperature is set by DOT at 10°C below the SADT.

[c]These identification numbers are assigned to the generic listings of organic peroxides at 49 C.F.R. §172.101, not to individual peroxy-organic compounds.

When shippers prepare the shipping description of a peroxo-organic compound, DOT requires them to provide the relevant generic shipping description, parenthetically identify the name of the specific compound offered for transportation, and include the concentration or the concentration range of the substance. For example, shippers use the following shipping description when they transport 50 pounds (110 kg) of 80% dibenzoyl peroxide within a plastic-lined cardboard box:

UNITS	HM	SHIPPING DESCRIPTION (IDENTIFICATION NUMBER, PROPER SHIPPING NAME, PRIMARY HAZARD CLASS OR DIVISION, SUBSIDIARY HAZARD CLASS OR DIVISION, AND PACKING GROUP)	WEIGHT (lb)
1 box (UN4G)	X	UN3012, Organic peroxide type B, solid, 5.2, (1), PG II (dibenzoyl peroxide, paste, 80%)	50

When shippers offer for transportation a peroxo-organic compound whose control temperature and emergency temperature are listed in column 7 of the Organic Peroxides Table, DOT requires them to include these temperatures on the shipping paper as shown in the following example:

UNITS	HM	SHIPPING DESCRIPTION (IDENTIFICATION NUMBER, PROPER SHIPPING NAME, PRIMARY HAZARD CLASS OR DIVISION, SUBSIDIARY HAZARD CLASS OR DIVISION, AND PACKING GROUP)	WEIGHT (lb)
2 boxes (UN4G)	X	UN3115, Organic peroxide type D, liquid, temperature-controlled, 5.2, PG II (diacetyl peroxide, 27%, and dimethyl phthalate, 73%) [Control temperature 68°F (20°C)] [Emergency temperature, 77°F (25°C)]	60

When shippers offer a peroxo-organic compound for transportation, DOT requires them to affix an ORGANIC PEROXIDE label to its packaging. In certain shipments, an EXPLOSIVE label is also required. DOT also requires shippers to mark each package containing a peroxo-organic compound with the following information:

- Name and address of the shipper and its receiver
- The proper shipping name
- Identification number of the commodity
- Applicable specifications, instructions, and precautions

In addition, when shippers offer the packaging for transportation by aircraft, DOT requires them to affix the KEEP AWAY FROM HEAT handling mark shown in Figure 13.6 on the packaging.

When warranted, DOT requires carriers to post ORGANIC PEROXIDE placards on the bulk packaging or transport vehicle used for shipment. Carriers display ORGANIC PEROXIDE placards when they ship by highway or rail any amount of a liquid or solid type B, temperature-controlled organic peroxide or 1001 pounds (454 kg) or more of the organic peroxides other than type B.

FIGURE 13.6 When packages containing self-reactive substances of Division 4.1 or organic peroxides of Division 5.2 are transported by aircraft, DOT requires carriers to post this KEEP AWAY FROM HEAT handling mark. The color of the starburst is red, but the other symbols, letters, and border are black on a white background.

Keep away from heat

13.9-B STORING PEROXO-ORGANIC COMPOUNDS

As a general policy, manufacturers of peroxo-organic compounds advise their customers to store limited quantities of these compounds in a segregated area at a temperature less than their emergency temperature.

13.9-C IDENTIFYING CONTAINERIZED PEROXO-ORGANIC COMPOUNDS

The hazards of peroxo-organic compounds may be readily determined by observing the GHS pictograms that OSHA requires manufacturers, distributors, and importers to affix on container labels. There are two types:

■ The exploding-bomb pictogram is posted on product labels when the containers hold any type A and certain type B organic peroxides.

■ The flame pictogram is posted on product labels when the containers hold certain type B organic peroxides and all type C, D, E, and F organic peroxides.

OSHA does not require organic peroxide manufacturers, distributors, and importers to affix a GHS pictogram on the labels of containers holding type G organic peroxides.

13.9-D RESPONDING TO INCIDENTS INVOLVING A RELEASE OF PEROXO-ORGANIC COMPOUNDS

Emergency responders identify the presence of peroxo-organic compounds at transportation mishaps by observing any of the following:

■ The number 5.2 as a component of a proper shipping description of a hazardous material on a shipping paper
■ The words *ORGANIC PEROXIDE* and the number 5.2 on yellow and red labels affixed to their packaging
■ The words *ORGANIC PEROXIDE* and the number 5.2 on yellow and red placards posted on the vehicle used to transport the packages

Because many organic compounds are flammable gases or flammable liquids, their fires are fought using the general techniques noted in Sections 3.5 and 3.8. However, this statement does not ordinarily apply to fires involving the peroxo-organic compounds, because they are likely to undergo thermal decomposition before the fires are extinguished. The generally recommended practice for fighting fires involving peroxo-organic compounds is to implement either of the following procedures:

- Use unmanned monitors to cool the area where these reactive materials are stored and assure that fire does not reach them.
- When a fire has engulfed the immediate area where peroxo-organic compounds are located, evacuate all personnel and do not combat the fire.

13.9-E TERRORISTS' MISUSE OF PEROXO-ORGANIC COMPOUNDS

At least two peroxo-organic compounds have been activated by terrorists to achieve their evil acts: triacetone peroxide, or TATP, and hexamethylene triperoxide diamine, or HMTD. These compounds are not used commercially due to their sensitivity to shock, friction, and heat. Nonetheless, terrorists produce them by the acid-catalyzed union of hydrogen peroxide with acetone and hexamethylenetetramine, respectively.

The molecular formulas of TATP and HMTD contain three peroxo groups, as shown below:

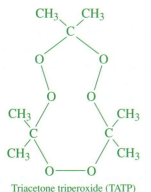

Triacetone triperoxide (TATP)
3,3,6,6,9,9-Hexamethyl-1,2,4,5,7,8-hexaoxocyclononane

Hexamethylene triperoxide diamine (HMTD)
3,4,8,9,12,13-Hexaoxa-1,6-diazabicyclo[4.4.4]tetradecane

Terrorists have clandestinely produced and used, or attempted to use, TATP and HMTD. The following terrorist events are well acknowledged:

- In December 2001, while traveling to the United States onboard an aircraft carrying 197 people from Paris to Miami, a terrorist called the "shoe-bomber" attempted to use TATP to accelerate the detonation of the explosive pentaerythritol tetranitrate (Section 15.12). The terrorist had packed a blend of TATP and PETN within the soles of his shoes but was overtaken by an alert flight attendant and passengers before he could successfully activate the mixture.

- TATP and (possibly) HMTD were used by four terrorists responsible for the London transit bombings on July 7, 2005 (p. 2). Packing the unstable substance and chemical explosives within backpacks, the terrorists detonated the mixture inside London's bus and underground subway system. Quantities of TATP were subsequently identified in the apartment used by one of the terrorists.

- TATP and HMTD probably were connected with an attempted terrorist plot in August 2006, during which terrorists allegedly sought to bomb multiple trans-Atlantic passenger jets bound for the United States from England. The terrorists intended to bring liquids onboard the planes in water bottles stored within their hand luggage. The nature of these liquids has never been clearly identified. British authorities were forewarned of the terrorists' plans and captured the conspirators before they were able to execute their plans.

- In September 2006, TATP was identified during the arrest of seven suspected terrorists in Vollsmose, Denmark.
- During September 2007, TATP was also identified during the arrest of eight suspected terrorists in Copenhagen, Denmark.
- Also during September 2007, German antiterrorist forces foiled plans initiated by three German terrorists who aimed to bomb the U.S. airbase at Ramstein as well as the U.S. and Uzbek consulates in Germany. The German forces uncovered more than 1600 pounds (727 kg) of a 35% hydrogen peroxide solution, which the terrorists presumably intended to use for production of TATP.
- On Christmas Day 2009, a Nigerian citizen with connections to Muslim extremists attempted to detonate a mixture of TATP and pentaerythritol tetranitrate while traveling on a flight to Detroit from Amsterdam. Flight attendants and passengers thwarted his activity. Because a packet of TATP was sewn into his underwear, the Nigerian sometimes is referred to as the "Underwear Bomber."
- HMTD was identified as an explosive component of a bomb that an al-Qaeda terrorist intended to activate at Los Angeles International Airport on New Year's Eve 1999/2000. The terrorist's activities were thwarted. He is known as the Millennium Bomber.

13.10 CARBON DISULFIDE

Carbon disulfide is a colorless, water-insoluble, and highly volatile liquid whose chemical formula is CS_2. Whereas pure carbon disulfide has a pleasant odor, the industrial grades of carbon disulfide generally are yellow and exhibit cabbagelike odors. Some other physical properties of carbon disulfide are listed in Table 13.23.

Carbon disulfide usually is prepared by reacting methane with vaporized sulfur at 1200°F (650°C).

$$2CH_4(g) \ + \ S_8(g) \ \longrightarrow \ 2CS_2(g) \ + \ 4H_2S(g)$$

Methane   Sulfur   Carbon disulfide   Hydrogen sulfide

The compound is used commercially as a solvent and as a raw material for the manufacture of viscose rayon, cellophane, and other textiles.

Because its flashpoint is −22°F (−30°C), carbon disulfide is a highly flammable liquid. Its flammable range extends from 1% to 44% by volume, and its vapor is 2.6 times heavier than air. The autoignition point of carbon disulfide is extremely low, 212°F (100°C), and the ignition of its vapor may be initiated by exposure to a hot steam pipe or

TABLE 13.23	Physical Properties of Carbon Disulfide
Melting point	−169°F (−112°C)
Boiling point	115°F (46°C)
Specific gravity at 68°F (20°C)	1.26
Vapor density (air = 1)	2.6
Vapor pressure at 68°F (20°C)	300 mmHg
Flashpoint	−22°F (−30°C)
Autoignition point	212°F (100°C)
Lower flammable limit	1% by volume
Upper flammable limit	44% by volume
Evaporation rate (ether = 1)	1.6

an electric lightbulb located a considerable distance away. This combination of properties constitutes cause for grave concern to firefighters who identify carbon disulfide at an emergency response scene. When bulk quantities of carbon disulfide are encountered, they usually are stored under water to reduce their potential to ignite.

Sulfur dioxide and carbon monoxide form as products of incomplete combustion when carbon disulfide burns.

$$2CS_2(g) \ + \ 5O_2(g) \ \longrightarrow \ 4SO_2(g) \ + \ 2CO(g)$$

Carbon disulfide   Oxygen   Sulfur dioxide Carbon monoxide

To protect against inhaling these toxic substances, the use of self-contained breathing apparatus is essential when firefighters respond to major fires involving carbon disulfide.

The inhalation of carbon disulfide vapor also is harmful. The repeated inhalation of the vapor damages the liver and kidneys and permanently affects the central nervous system. The prolonged contact of carbon disulfide with the skin also is harmful, as the liquid is absorbed through the skin. In the worst case, the absorption is fatal.

13.10-A WORKPLACE REGULATIONS INVOLVING CARBON DISULFIDE

When carbon disulfide is present in the workplace, OSHA requires workers to limit employee exposure to a maximum vapor concentration of 20 parts per million, averaged over an 8-hour workday.

13.10-B TRANSPORTING CARBON DISULFIDE

When shippers offer carbon disulfide for transportation, DOT requires them to describe the substance on a shipping paper as follows: UN1131, Carbon disulfide, 3, (6.1), PG I. DOT also requires shippers and carriers to comply with all applicable labeling, marking, and placarding requirements.

13.11 CHEMICAL WARFARE AGENTS

Some substances are so highly toxic to humans that their use during wartime can injure, incapacitate, or cause mass casualties among the enemy. They are called **chemical warfare agents.** Their use has been scorned by many countries; nonetheless, their production, use, and stockpiling have occurred due to mutual mistrust and fear.

Chemical warfare agents are primarily liquids that can be sealed into canisters or charged into missile warheads or other explosive devices. Because they vaporize when released into the atmosphere, these substances readily provide a lethal concentration.

In 1997, the United Nations spearheaded attempts to create the international treaty known as the **Chemical Weapons Convention.** Known formally as the Convention on the Prohibition of the Development, Production, Stockpiling, and Use of Chemical Weapons and on their Destruction, it prohibits the development, production, acquisition, stockpiling, retention, transfer, or use of chemical warfare agents. There are 188 signatories to the treaty, including the United States. Its watchdog agency is the Organization for the Prohibition of Chemical Weapons, or OPCW, but the U.S. Department of Commerce regulates its activities in the United States at 15 C.F.R. §§710 - 722.

Aside from the directives noted in its formal title, the Chemical Weapons Convention also requires its signatories to destroy their stockpiles of chemical weapons and declare in writing the quantity remaining in them. Of the member countries that have acknowledged possessing chemical weapons, Albania, India, and South Korea have totally eliminated their arsenals. Russia and the United States have destroyed 62.5% and 89.8% of their declared stockpiles, respectively, and plan complete disposal by 2015 and 2023. In the United States, stockpiles remain in storage for ultimate disposal at sites in Pine Bluff, Arkansas; Tooele, Utah; and Umatilla, Oregon.

chemical warfare agent ■ A nerve agent, vesicant, blood agent, or other substance that can inflict harm or cause mass casualties among exposed individuals

Chemical Weapons Convention ■ A United Nations' treaty that prohibits the development, production, acquisition, stockpiling, retention, transfer, or use of chemical warfare agents

13.11-A NERVE AGENTS

Nerve agents are volatile organofluorophosphorus compounds of which the following molecular structures are representative:

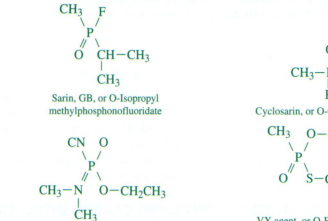

Sarin, GB, or O-Isopropyl methylphosphonofluoridate

Tabun, or Ethyl N,N-dimethylphosphoroamidocyanidate

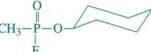

Cyclosarin, or O-Cyclohexylmethylfluorophosphonate

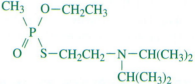

VX agent, or O-Ethyl-S-(2-diisopropylaminoethyl)-methyl phosphonothiolate

These formulas are similar to those of organophosphorus pesticides (Section 10.20), although the latter substances are not fluorinated.

Nerve agents are fairly simple to synthesize from readily available raw materials. For this reason, law enforcement authorities are sensitive to the fact that they could easily be obtained and used by terrorists as weapons of mass destruction.

The physical properties of sarin, cyclosarin, CF, tabun, GA, and VX are noted in Table 13.24. These data illustrate that the vapors of nerve agents are much heavier than air. Consequently, when the liquid agents are initially released from their containers, their vapors seek out low-lying areas. Sarin is the most volatile of them, and VX is the most lethal. Aside from their potential use as chemical weapons of mass destruction, the nerve agents may also be used by terrorists to contaminate water and food supplies.

Nerve agents generally cause their ill effects by two mechanisms: inhalation and absorption through the skin. The initial exposure often occurs by penetration of the agent's vapor through the eyes. Once in the body, they interact with substances that cause the nerves to transmit impulses to nearby muscles. This action paralyzes the muscles, which in turn, causes convulsions, respiratory failure, and immediate death.

Nerve gases were produced by Nazi Germany but never used against its enemies. Toward the end of World War II, U.S. troops seized at least 4100 pounds (1900 kg) of the nerve gas tabun and brought it to the United States for safekeeping. Pursuant to the terms of the Chemical Weapons Convention, the stockpile was finally incinerated in 2011—six decades after the end of the war.

TABLE 13.24	Physical Properties of Some Nerve Agents			
	SARIN	**CYCLOSARIN**	**TABUN**	**VX**
Melting point	−69°F (−57°C)	−22°F (−30°C)	−58°F (−50°C)	−58°F (−50°C)
Boiling point	316°F (158°C)	462°F (239°C)	464°F (240°C)	568°F (298°C)
Specific gravity at 68°F (20°C)	1.11	1.12	1.07	1.01
Vapor density (air = 1)	4.86	6.2	5.63	9.2
Vapor pressure at 68°F (20°C)	1.48 mmHg	0.044 mmHg	0.037 mmHg	0.00044 mmHg

During the Cold War, the U.S. military produced and stockpiled sarin, but never used it. By contrast, the use of chemical warfare agents in weapons of mass destruction has been used in contemporary times as noted by the following:

■ Iraq used the nerve agent sarin in its war against Iran during 1984–1988, as well as against its own citizens in 1988. An estimated 5000 people were killed and another 65,000 sickened. Iraq ratified the articles of the Chemical Weapons Convention in 1995.

■ In 1995, sarin was released in the Tokyo subway system by members of Aum Shinrikyo, a Japanese religious movement. The release caused the deaths of 13 persons and sickened at least 1000 other individuals.

■ In 2013, France and England reported forensic evidence that confirmed the use of sarin gas as a weapon of mass destruction by Syria against its own citizens in the ongoing civil war. An attack near the capital city of Damascus killed approximately 1400 civilians.

13.11-B VESICANTS

Vesicants are chemical warfare agents that blister skin and damage the eyes, mucous membranes, and respiratory tracts of the exposed enemy. The best known is *mustard gas*, a yellowish brown, oily liquid having a faint odor of garlic or mustard. Its chemical name is 2,2′-dichloroethyl sulfide.

vesicant ■ A chemical warfare agent that can blister the skin and other body tissues of wartime enemies

$$ClCH_2CH_2-S-CH_2CH_2Cl$$

2,2′-Dichloroethyl sulfide
(Mustard gas)

Mustard gas may be poured on the ground, sprayed into the air, or loaded into artillery shells and dropped from planes to serve as a chemical weapon of mass destruction.

The grisly success of mustard gas as a vesicant is due in part to its high vapor density of 5.4 (air = 1). The vapor hovers for a relatively long period at ground level. The vapor subsequently contacts the moisture on the skin and in the eyes or lungs and produces corrosive hydrochloric acid.

$$ClCH_2CH_2-S-CH_2CH_2Cl(g) + 2H_2O(l) \longrightarrow HOCH_2CH_2-S-CH_2CH_2OH(g) + 2HCl(g)$$

Prolonged exposure of mustard gas to the eyes causes temporary blindness. When it is inhaled, mustard gas may cause cancer of the respiratory tract.

The use of mustard gas by Germany against enemy troops during World War I is well documented. Discharged as a liquid, its vapor moved in the direction of the wind along the surface of the ground and into the trenches where soldiers were seeking a level of protection against live artillery. It was within the trenches that the deadly vapor inflicted the maximum harm on unsuspecting troops. Approximately 400,000 British and French soldiers lost their lives during World War I from exposure to mustard gas.

More than 6000 tons (5455 t) of mustard gas produced for use by the U.S. military still remains in storage, mainly in Utah. Destruction of this vesicant began in 2006.

Second-generation vesicants include the compounds known as the *nitrogen mustards*. These compounds are similar in molecular structure to 2,2′-dichloroethyl sulfide in that an amino nitrogen atom replaces the sulfur atom. There are three well-known nitrogen mustards: methyl bis(2-chloroethyl)amine, ethyl bis(2-chloroethyl)amine, and tris(2-chloroethyl)amine.

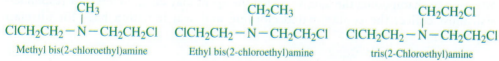

Methyl bis(2-chloroethyl)amine   Ethyl bis(2-chloroethyl)amine   tris(2-Chloroethyl)amine

These compounds are liquids whose vapors are especially heavy [vapor densities = 5.9, 5.4, and 7.1 (air = 1), respectively.] Although the nitrogen mustards had been produced and were available for use by Germany and the Allies, neither side used them as vesicants during World War II.

In 2011, shells filled with mustard gas were identified by rebel fighters at two sites in central Libya. The shells allegedly were supplied to Libya by Iran. Whether the vesicant was ordinary mustard gas or a nitrogen mustard is unknown.

13.11-C BLOOD AGENTS

blood agent ▪ A chemical warfare agent that can interfere with a wartime enemy's utilization of oxygen

Blood agents are highly volatile chemical warfare agents that can cause seizures, respiratory failure, and cardiac arrest upon inhalation exposure. An example of a blood agent is hydrocyanic acid. In Section 10.11-F, we noted that hydrogen cyanide inhibits the effective utilization of oxygen at the cellular level.

13.11-D CHOKING AGENTS

choking agent ▪ A chemical warfare agent that can damage the respiratory passageways of wartime enemies so severely that they choke

Choking agents are chemical warfare agents that cause the bronchial passageways of exposed enemies to constrict upon inhalation. Involuntary choking and suffocation then follow, and prolonged exposure may cause the onset of pulmonary edema. The most common examples of choking agents are elemental chlorine, phosgene, diphosgene, and chloropicrin.

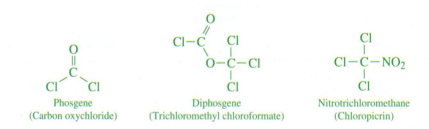

During World War I, chlorine and phosgene were responsible for numerous casualties on both sides.

13.11-E RESPONDING TO INCIDENTS INVOLVING THE RELEASE OF A CHEMICAL WARFARE AGENT

Imagine how gruesome it must be to encounter a scene at which a chemical warfare agent has been dispersed! Mass casualties are likely to appear everywhere. The survivors anxiously seek medical attention from health care personnel, who also are anxious that they will be exposed to the causative agent.

The major job of the first-on-the-scene responders may be restricted to providing calm and order to the prevailing pandemonium and delivering immediate help to those people who were fortunate to survive the ordeal. Thereafter, these first-on-the-scene responders should quickly move the exposed individuals to an agent-free environment while wearing fully encapsulated suits and breathing air from self-contained sources. They must also encourage exposed individuals to quickly remove their contaminated clothing and physically wash any exposed areas with soap and water. To ensure complete removal of the agent, experts recommend washing three times, giving special attention to shampooing the hair, to which the agent may cling. Emergency responders also must collect the contaminated clothing and seal it within bags for ultimate disposition.

It is only after the victims have been completely decontaminated that appropriate medical attention is given by health-care providers. The victims should have access to respirators, given an antidote designed to counteract the impact of the specific agent to which they were exposed, and monitored at a health-care facility for at least 24 hours.

Emergency responders must also tend to the deceased. The local coroner can provide appropriate directions for removing clothing and decontaminating corpses before their transferral to the local morgue.

13.12 LACRIMATORS

As noted in Section 10.9-A, lacrimators are substances that cause the eyes to involuntarily tear and close upon exposure. Their use by the military during warfare usually is limited to activities such as clearing enemies from tunnels. More commonly, lacrimators are dispersed in the environment by law enforcement personnel during civilian disputes to control mobs and discourage unlawful acts without resorting to the use of firearms. During their use, lacrimators render the recipients temporarily incapable of resistance or flight. They are commonly known as **riot-control agents**, or **tear gases**.

Riot-control agents are used defensively to control the action of activists and troops by temporarily impairing their vision and causing them to choke and breathe painfully. It is the *temporary* aspect that causes the use of lacrimators by authorized individuals to be considered humane. Exposure to riot-control agents can be a particularly irritating experience for recipients, especially within confined spaces.

Some representative molecular formulas of lacrimators are shown here:

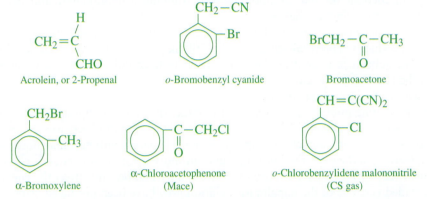

Acrolein, or 2-Propenal

o-Bromobenzyl cyanide

Bromoacetone

α-Bromoxylene

α-Chloroacetophenone
(Mace)

o-Chlorobenzylidene malononitrile
(CS gas)

Lacrimators have been extensively used during warfare since at least World War I, when the German army dispensed benzyl bromide at Russian and French troops. Since the 1960s, CS gas also has been used worldwide by police to incapacitate aggressive assailants. The gas is charged into cans as a 5% solution in methyl isobutyl ketone.

The lacrimators used by the general public are **personal self-defense agents**. The most common of them is *pepper spray*, which is an extract derived from *Capsicum* pepper plants. The principal component is *N*-[(4-hydroxy-3-methoxyphenyl)methyl]nonanamide, known more commonly as capsaicin.

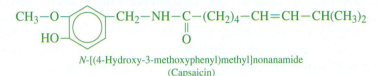

N-[(4-Hydroxy-3-methoxyphenyl)methyl]nonanamide
(Capsaicin)

It is the compound mainly responsible for the pungent, hot taste of jalapeño, habañero, cayenne, and chili peppers, especially of their ribs and seeds, as well as the discomforting

Isobutyric acid

riot-control agent (tear gas) ■ Any substance that rapidly can produce sensory irritation or a disabling physical effect among exposed individuals

personal self-defense agent ■ Any temporarily incapacitating substance

feeling that results from its use as a personal self-defense agent. Capsaicin is also the active ingredient of certain analgesic ointments including *Capzasin-HP*, a medication capable of reducing or eliminating mild arthritic and other pains.

For use as a personal self-defense agent, oleoresin capsicum is generally dissolved in a solvent to which a dye has been added and charged into an aerosol can. To affect incapacitation, the user dispenses it by aiming the discharge directly into an attacker's face. The dye helps police identify the attacker. The topical application of capsaicin to skin surfaces causes irritation, pain, prolonged sneezing, and coughing. When sprayed into the eyes, the pain and inflammation are especially severe.

Pepper spray can be used effectively in several ways. Women can use it to incapacitate aggressive or violent assailants; postal workers can use it against attacking dogs; police can use it to disperse unwieldy crowds; and park rangers can use it to ward off bears and other wild animals. Although the purchase of pepper spray by the general public is legal, most states have enacted laws that restrict the manner of its use.

13.13 NAPALM

incendiary agent ▪ Any substance used during warfare to intentionally set fire to objects or cause burn injury to exposed enemies through the action of flames or heat

napalm ▪ An aluminum triglyceride that produces a jellylike mixture with gasoline for use as an incendiary agent during warfare

Substances called **incendiary agents** are used during warfare to intentionally initiate fires. The use of white phosphorus in incendiary bombs, and triethylaluminum in flamethrowers, was previously noted in Sections 7.4 and 9.4, respectively.

Several materials have been developed to thicken petroleum products for their potential use as incendiary agents. The first incendiary agent of this type was called **napalm**. It consisted of a mixture of aluminum compounds made from several triglycerides found naturally in coconut oil. The compounds were aluminum naphthenate and palmitate, which denotes the origin of the name.

Napalm thickens gasoline until a mixture containing approximately 4% napalm by volume is produced. The resulting material is jellylike in consistency. When napalm or a napalm-like substance is used as the active agent in firebombs, an explosive substance triggers the burning of the gasoline, jet fuel, or kerosene. When these firebombs are activated, massive fireballs often are produced.

For conducting military activities, the use of fuel thickened with napalm is associated with results that are provided by few other materials. Napalm increases the range of flamethrowers, imparts slower-burning properties compared with gasoline alone, adds a clinging feature, and causes the burning flames to move around corners and rebound off walls and other surfaces. Its use during warfare can psychologically affect the enemy.

Upgraded versions of the napalm formulation now have been produced. Modern versions consist of jet fuel, kerosene, or benzene thickened with a polystyrene-based gel. During World War II and the Korean and Vietnam conflicts, napalm formulations were used as incendiary agents in firebombs and portable and mechanical flamethrowers. The burning of napalm controlled the extent of unwanted vegetation and routed the enemy from hiding places within jungles and other densely vegetated terrains.

Today, however, most countries have agreed to limit the use of all incendiary weapons if it is likely to adversely affect civilian populations. The Convention on Certain Conventional Weapons (Section 7.5) effectively limits the future use of napalm and napalm-like incendiary agents. Because the United States did not ratify this protocol, the U.S. military continues to use incendiary agents in warfare. For example, during the initial advance into Baghdad in 2003, U.S. pilots dropped napalm-like incendiary bombs called *Mark 77 firebombs* on Iraqi troops; later, these firebombs were again used during an attack on an observation post at Safwan Hill, a location near the Iraq–Kuwait border.

Classes of Organic Compounds

1. Use Table 13.1 to identify the class of organic compound represented by each of the following molecular formulas:

   (a) $CH_3CH_2-O-CH_2CH_3$

   (b) 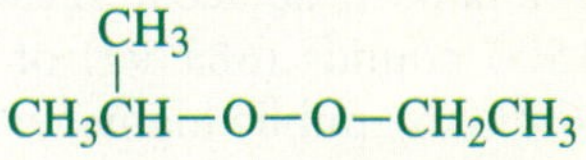 $CH_3\overset{\underset{|}{CH_3}}{CH}-O-O-CH_2CH_3$

   (c) 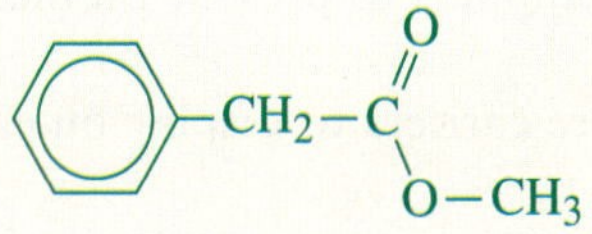 (benzene ring)$-CH_2-\overset{\overset{O}{\|}}{C}\overset{|}{\underset{O-CH_3}{}}$

   (d) 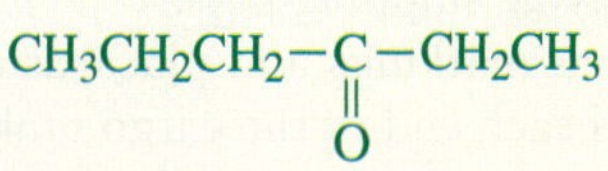 $CH_3CH_2CH_2-\overset{\overset{}{\underset{\|}{C}}}{\underset{O}{}}-CH_2CH_3$

2. Ethylene oxide is a flammable gas having the following molecular structure:

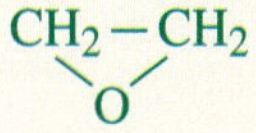

   $$CH_2-CH_2$$
   $$\diagdown O \diagup$$

   Use Table 13.1 to identify its class of organic compound.

Alcohols

3. Provide the IUPAC name of the alcohols having each of the following condensed formulas:

   (a) $CH_3CH_2CH_2CH_2OH$

   (b) 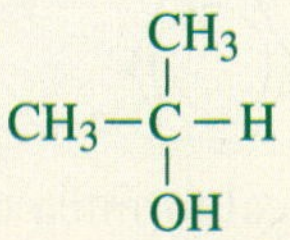 $CH_3-\overset{\overset{CH_3}{|}}{\underset{\underset{OH}{|}}{C}}-H$

   (c) $H-\overset{\overset{CH_3}{|}}{\underset{\underset{CH_3}{|}}{C}}-CH_2OH$

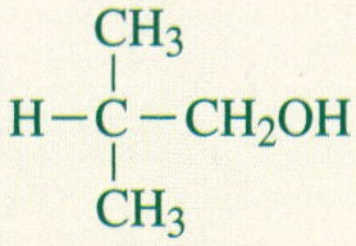

   (d) 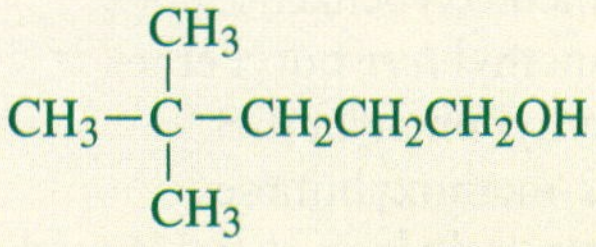 $CH_3-\overset{\overset{CH_3}{|}}{\underset{\underset{CH_3}{|}}{C}}-CH_2CH_2CH_2OH$

4. Provide the chemical formula of the alcohols that are named as follows:
   (a) isopropyl alcohol
   (b) *p*-cresol
   (c) *tert*-butyl alcohol
   (d) 3-ethylphenol
   (e) 3-heptanol
   (f) 1-phenyl-2-propanol

5. Propofol is a potent general anesthetic whose misuse was linked with his untimely death in 2009 by "King-of-Pop" idol Michael Jackson. Its molecular structure is shown here:

   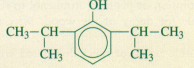

   What is the IUPAC name of this drug?

6. Between 1964 and 2006, M100 was used as the fuel in racecars competing in the Indianapolis 500 Race, but in 2007, it was replaced with race-grade E100. What

advantages does the use of these alternative motor fuels have compared to the use of gasoline or other petroleum-based fuels for racetrack competition?

7. Considering only their relative flashpoints, which alcoholic beverage poses the greater risk of fire and explosion: 90- or 100-proof whiskey?

8. When E85 is transported in bulk by rail tankcar:
   (a) Which identification number does DOT require carriers to display on orange panels affixed to each side and each end of the tankcar?
   (b) Which placard does DOT requires carriers to post on each side and each end of the rail tankcar adjacent to each orange panel?

9. When 1500 pounds (681 kg) of molten phenol is transported in a cargo tank at 140°F (60°C) by public highway:
   (a) What shipping description does DOT require shippers to provide on the accompanying shipping paper?
   (b) What markings and placards does DOT require carriers to display on each side and each end of the cargo tank?

Ethers

10. Provide an acceptable name of the ethers having each of the following condensed formulas:

   (a) $CH_3CH_2-O-CH_2CH_2CH_3$

   (b) $CH_3-\underset{\underset{CH_3}{|}}{\overset{\overset{CH_3}{|}}{C}}-O-CH_2CH_3$

   (c) 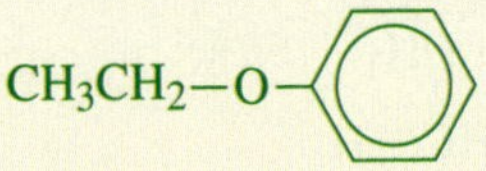 $CH_3CH_2-O-\hexagon$

   (d) $CH_3-\underset{\underset{}{\overset{CH_3}{|}}}{CH}-O-\underset{}{\overset{CH_3}{|}}{CH}-CH_3$

11. Provide the chemical formula of the ethers that are named as follows:
   (a) methoxybenzene
   (b) methyl *tert*-butyl ether
   (c) ethyl phenyl ether
   (d) 2-methoxybutane
   (e) tetraethylene glycol dimethyl ether

12. Why should ethers be purchased in small quantities, kept in tightly sealed containers, and used promptly?

13. In 2011, EPA first used the authority of TSCA to require chemical manufacturers to notify it in advance of any intended new uses of certain ethylene glycol alkyl ethers. What is the most logical reason EPA took this action?

Halogenated Ethers

14. In the past, before being mounted along a railroad track, wooden railroad ties were first soaked in a solution of pentachlorophenol to retard the action of the microorganisms responsible for decay of the wood. Why is this treatment procedure no longer practiced?

15. Provide an acceptable name of the halogenated ethers having the following condensed formulas:

   (a) 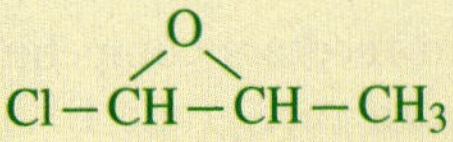 $Cl-\overset{\overset{O}{\triangle}}{CH}-CH-CH_3$

   (b) $ClCH_2-O-CHCl_2$

Aldehydes and Ketones

16. Provide an acceptable name of the aldehydes or ketones having each of the following condensed formulas:

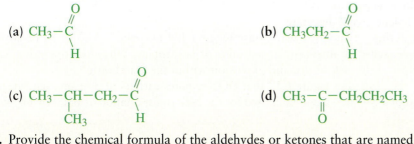

17. Provide the chemical formula of the aldehydes or ketones that are named as follows:
(a) 2-methylpropanal
(b) pentanal
(c) methyl *n*-propyl ketone
(d) 3-heptanone

18. Air-monitoring exercises conducted by the Centers for Disease Control and Prevention during 2008 established that the average concentration of formaldehyde was 77 parts per billion in the air inside trailers provided by the Federal Emergency Management Agency (FEMA) to the victims of Hurricane Katrina in Mississippi, Louisiana, and Alabama. What is the most logical technical explanation for the presence of formaldehyde in these trailers?

Organic Acids

19. Provide an acceptable name of the organic acids having each of the following condensed formulas:

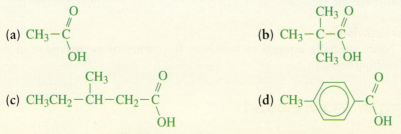

20. Provide the chemical formula of the organic acids that are named as follows:
(a) acetic acid
(b) propionic acid
(c) methanoic acid
(d) phenoxyacetic acid

21. Use the data in Table 13.14 to determine for each listed acid its OSHA category of flammable liquid.

Esters

22. Provide an acceptable name of the esters having each of the following condensed formulas:

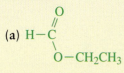

23. Provide the chemical formulas of the esters that are named as follows:
    (a) *n*-propyl formate
    (b) isobutyl benzoate
    (c) phenyl acetate
    (d) cyclohexyl propanoate
    (e) diethylene glycol diacetate

24. Upon entering an enclosed warehouse where ethyl butyrate is known to be stored, firefighters sense an intensely strong odor of pineapples. Why does this odor warn firefighters of the risk of fire and explosion within the warehouse?

25. Identify the placard in Figure 6.13 that DOT requires carriers to post on each side and each end of a MC-306 cargo tank used for the bulk transportation of B100 by highway.

Amines

26. Provide an acceptable name of the amines having each of the following condensed formulas:

    (a) $CH_3CH_2CH_2NH_2$

    (b) $CH_3-CH-CH_3$ with NH_2

    (c) ⬡—CH_2NH_2

    (d) CH_3CH_2-N with H and CH_3

27. Provide the chemical formula of the amines that are named as follows:
    (a) *N*-ethylaniline
    (b) 2-aminopentane
    (c) *n*-propylmethylamine
    (d) diisopropylamine
    (e) 2-butanamine
    (f) dimethylethylamine

28. Why are police specially attuned to searching for caches of methylamine in drug-infested neighborhoods?

Peroxo-Organic Compounds

29. Provide an acceptable name for the peroxo-organic compounds having each of the following condensed formulas:

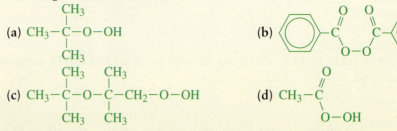

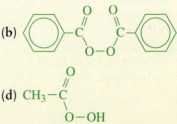

30. Provide the chemical formulas of the peroxo-organic compounds that are named as follows:
    (a) di-*n*-hexyl peroxide
    (b) bis-*p*-chlorobenzoyl peroxide
    (c) diisopropylbenzene hydroperoxide
    (d) acetyl benzoyl peroxide

31. How do firefighters determine rapidly that peroxo-organic compounds are stored within an isolated building from the integers displayed on a hazard diamond affixed to the exterior wall of the building?

32. A chemical manufacturer desires to transport 400 kilograms of liquid 1,1,3,3-tetra-methylbutyl hydroperoxide by motor vehicle within 10 plastic drums. Testing of this compound establishes that it neither detonates nor deflagrates rapidly, and when confined and heated, it manifests no violent effect.

    (a) What shipping description does DOT require the manufacturer to enter on the accompanying shipping paper?

    (b) Which labels does DOT require the manufacturer to affix on the outside surface of each package containing 1,1,3,3-tetramethylbutyl hydroperoxide?

    (c) Which markings does DOT require the manufacturer to print on the outside surface of each package?

    (d) Which placards, if any, does DOT require the carrier to post on the motor vehicle used to transport the packages?

Carbon Disulfide

33. Why is the use of self-contained breathing apparatus essential when firefighters respond to fires involving carbon disulfide?

Chemical Warfare Agents

34. Why should individuals who are likely to be exposed to a nerve agent seek out refuge above the point of its discharge?

Responding to Incidents Involving a Release of Organic Compounds

35. When first-on-the-scene responders arrive at the scene of a rail freight accident, they note that at least one side of an overturned boxcar is posted with a yellow ORGANIC PEROXIDE placard. There is not an ongoing fire. A member of the train crew provides the responders with the waybill, the relevant portion of which reads as follows:

UNITS	HM	SHIPPING DESCRIPTION (IDENTIFICATION NUMBER, PROPER SHIPPING NAME, PRIMARY HAZARD CLASS OR DIVISION, SUBSIDIARY HAZARD CLASS OR DIVISION, AND PACKING GROUP)	WEIGHT (lb)
75 plastic drums (1H2)	X	UN3115, Organic peroxide type D, liquid, temperature-controlled, 5.2, PG II (methylcyclohexanone peroxide, 67%) (Placarded ORGANIC PEROXIDE) (Control temperature, 35°C; Emergency temperature, 40°C)	1050

What immediate actions should the emergency responders execute to protect public health, safety, and the environment?

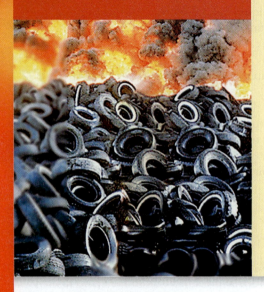

CHAPTER

14
Chemistry of Some Polymeric Materials

Courtesy of Pyrocool Technologies, Inc., Monroe, Virginia.

KEY TERMS

addition polymer, *p. 609*

amide, *p. 620*

amino acid, *p. 619*

atactic polymer, *p. 628*

autopolymerization, *p. 616*

cellulose, *p. 623*

condensation polymer, *p. 609*

copolymer, *p. 609*

cotton, *p. 621*

cross-linking, *p. 609*

depolymerization, *p. 617*

distillate-aromatic-extract oil, *p. 641*

ebonite, *p. 640*

elastomer, *p. 608*

fiber, *p. 608*

Flammable Fabrics Act, *p. 621*

flashover, *p. 618*

foam rubber, *p. 641*

inhibitor, *p. 616*

isocyanate, *p. 633*

isotactic polymer, *p. 628*

linen, *p. 621*

macromolecule, *p. 607*

monomer, *p. 608*

natural rubber, *p. 637*

neoprene, *p. 639*

nitrile, *p. 620*

nitro compound, *p. 620*

plastics, *p. 607*

polyamide, *p. 633*

polyester, *p. 613*

polymerization, *p. 608*

polymer, *p. 607*

polyurethane, *p. 633*

rubberized asphalt, *p. 641*

silk, *p. 625*

styrene–butadiene rubber, *p. 639*

syndiotactic polymer, *p. 628*

synthetic rubber, *p. 638*

textile, *p. 608*

thermoplastic polymer, *p. 607*

thermosetting polymer, *p. 607*

urethane (carbamate), *p. 633*

vinyl polymer, *p. 626*

vulcanization, *p. 638*

vulcanized rubber, *p. 638*

wool, *p. 625*

OBJECTIVES

- Associate the physical and health hazards of the monomers noted in this chapter with the information provided by their hazard diamonds and GHS pictograms.
- Describe the general nature of the polymerization reaction.
- Distinguish between addition and condensation polymerization reactions.
- Describe how cross-linking and the use of plasticizers alters the physical features of polymers.

- Discuss the general phenomena that occur when polymers burn on exposure to heat.
- Identify the toxic gases produced when polymers thermally decompose or burn.
- Describe and compare the macromolecular structures of the common vegetable and animal fibers.
- Identify the common products made from polyethylene, polypropylene, poly(vinyl chloride), polyacrylonitrile, poly(methyl methacrylate), polyacrylamide, phenol-formaldehyde, urea-formaldehyde, melamine-formaldehyde polymers, and polyurethane, and identify the toxic gases produced when they burn or smolder.
- Describe how natural rubber is vulcanized.
- Identify the labels, markings, and placards that DOT requires on packaging of the monomers noted in this chapter and the transport vehicles used for their shipment.

Over the past 75 years, the polymer industry has dramatically altered our way of life. It is unlikely that a modern civilized society could long survive without the wares it provides. In today's world, we regularly use products manufactured from both natural and synthetic polymers. Our clothing is made from polymeric fibers, including cotton, polyesters, nylon, and polyacrylics. Our homes are constructed from wood, insulated with polystyrene, carpeted with polypropylene, coated with polyacrylic paints, and decorated with polyacrylonitrile and other polymeric fabrics.

An appreciable portion of our automobiles also have been manufactured from polymers. Often, the bumpers are made of an acrylonitrile–butadiene–styrene copolymer, the roofs are made of poly(vinyl chloride), the upholstery is cushioned with polyurethane, and the rubber tires are manufactured from a styrene–butadiene copolymer.

Because they are stable at ambient conditions and do not routinely pose a health risk, polymers are not ordinarily considered hazardous materials. However, most polymeric products burn and generate toxic gases on combustion. Not only do most polymeric products burn, but their combustion is involved in virtually all common fires. For these reasons, the burning of polymers is a topic of great concern to firefighters.

In this chapter, to understand why they pose special hazards during fires, we examine the features and structural characteristics of several commonly encountered polymers.

14.1 WHAT ARE POLYMERS?

Polymers are substances that are best characterized by the relatively sizable nature of their molecules. Because these molecules are substantially larger than those otherwise encountered, chemists call them **macromolecules**. Each polymer macromolecule comprises a number of repeating smaller units; a polymer is a compound typically composed of hundreds or thousands of repeating units.

Polymers sometimes are described by their response to heat. Some soften when exposed to heat and may be physically manipulated to produce new shapes. Upon cooling, they retain these shapes until they are heated again. These polymers are called **thermoplastic polymers**. The following are examples of thermoplastic polymers: polyethylene, polypropylene, polystyrene, poly(ethylene terephthalate), and poly(vinyl chloride).

Other polymers solidify or set irreversibly when they are heated. They are called **thermosetting polymers**. Polyurethane is an example of a thermosetting polymer.

Polymers are also classified according to the ways their manufactured products are used. For instance, some polymers are used to produce **plastics**—items that can be shaped by means of molding, casting, extrusion, calendering, laminating, foaming, and blowing. Examples of some common plastics are poly(vinyl chloride), polystyrene, and poly(methyl methacrylate). The polymers in plastic products usually are combined with other materials including fillers, reinforcing agents, and fire retardants.

polymer ■ A high-molecular-weight substance produced by the linkage and cross-linkage of its multiple subunits (monomers)

macromolecule ■ The giant molecule of which polymers are composed, comprising an aggregation of hundreds or thousands of atoms and typically consisting of repeating chemical units linked together into chains and cross-linked into complex three-dimensional networks

thermoplastic polymer ■ Any polymer that softens when heated but returns to its original condition on cooling to ambient temperature

thermosetting polymer ■ Any polymer that cannot be remolded once it has solidified

plastics ■ Any of a variety of synthetic polymeric substances that can be molded and shaped

Some natural and synthetic polymers are used as threads or yarns collectively called **fibers**. These polymers are characterized by a high tenacity and a high ratio of length to diameter (several hundred to one). They differ widely in form, flexibility, durability, and porosity.

Although some fibers occur naturally, others are produced synthetically. *Natural fibers* are a class of polymers derived from vegetable and animal sources. They include cotton and wool, respectively. Synthetic fibers are a class of polymers made by some nonnatural means. Two examples of synthetic fibers are nylon and poly(acrylonitrile). Synthetic fibers also include polymers that have been chemically modified in some way. They are used to manufacture numerous goods including rope, woven cloth, matted fabrics, brushes, shingles and other building and insulating materials, as well as the stuffing in pillows and upholstery.

Natural and synthetic fibers are woven or knitted to produce **textiles**. These are items such as garments, carpets, carpet padding, towels, curtains, blankets, mattresses, and upholstery fabrics.

A polymer may also be an **elastomer**. This is a type of synthetic polymer that is characterized by the ability of its molecules to elongate when strained and to reversibly assume their original shape when the tension has been released. Examples of elastomers are neoprene and *cis*-1,4-polybutadiene. Products made from them include rubber bands, belts, footwear, vehicular tires, and the inner tubes of tires.

Various means have been used to name polymers. Although several common names are used, chemists often place the prefix *poly-* before the name of the substances used to produce specific polymers. Polyethylene, poly(vinyl chloride), and polystyrene are so named because they are produced from ethylene, vinyl chloride, and styrene, respectively. When a polymer is manufactured from a number of substances, all their names may be included in the name of the polymer. The acrylonitrile–butadiene–styrene copolymer commonly used in plastic sewer pipes is so named because it is made from acrylonitrile, butadiene, and styrene. In this instance, the name has been abbreviated for simplicity to *ABS* copolymer, where each letter represents a monomer used in its production.

fiber ■ A polymeric substance that can be separated into threads or thread-like structures

textile ■ A fabric produced by weaving fibers

elastomer ■ A synthetic polymer that elongates or stretches under strain but is incapable of retaining the deformation when the strain is released

14.2 POLYMERIZATION

Polymerization is a unique type of chemical reaction involving the union of certain substances called **monomers**. The equations for such reactions generally are denoted as follows, where A and B are arbitrary monomers and *a* and *b* are relatively large numbers:

$$a\text{A} + b\text{B} \longrightarrow \text{Polymer}$$

Only a select number of compounds undergo polymerization.

A polymerization reaction is characterized by the nature of the macromolecules it produces. These macromolecules can be described as follows:

polymerization ■ The chemical reaction during which monomer molecules are linked and cross-linked into macromolecules

monomer ■ One or more of the single substances that combine to produce a polymer

■ Figure 14.1 illustrates one type of polymeric macromolecule. Its structure resembles the midsection of a freight train having identical boxcars. Like a boxcar, the repeating unit in this portion of the macromolecule has couplings at its front and rear ends, here symbolized by chemical bonds.

■ Figure 14.2 illustrates a second type of polymeric macromolecule. Its structure resembles the midsection of a freight train that has alternating dissimilar railcars.

$$-\text{CH}_2-\text{CH}_2-\text{CH}_2-\text{CH}_2-\text{CH}_2-\text{CH}_2-$$

FIGURE 14.1 This midsection of a freight train consists of identical interconnected railcars. It resembles the partial structure of polyethylene, whose macromolecules have the repeating unit ∿CH$_2$—CH$_2$∿.

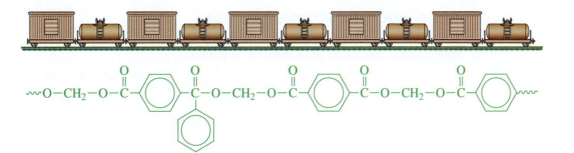

FIGURE 14.2 This midsection of a freight train consists of alternating interconnected railcars and tankcars. It resembles the macromolecular structure of polyethylene terphthalate, whose dissimilar units may alternate regularly or irregularly.

The substance having such macromolecules is called a **copolymer**. Although the portion of the macromolecule shown in Figure 14.2 has regularly repeating units, a copolymer may also be composed of units that alternate irregularly.

When chemists examine the three-dimensional structures of polymers, they find that these chains of repeating units are invariably cross-linked. Figure 14.3 illustrates a macromolecule in which one chain has attached itself to another chain by means of a chemical bond. Plastics manufacturers often attempt to intentionally increase the degree of **cross-linking** in thermosetting plastics. The resulting polymers are denser, and thus stronger and more durable, than those whose macromolecules have unlinked chains of atoms. Cross-linking the chains within macromolecules also potentially makes the products manufactured from them more elastic.

Macromolecular chains within polymers can also be folded, coiled, stacked, looped, or intertwined into definite three-dimensional shapes. Although these complex configurations give polymers their unique properties, we require only the information conveyed by their one-dimensional patterns.

Polymer manufacturers sometimes discover that their products are too stiff and brittle for their intended use. These undesirable features often can be overcome by adding a *plasticizer* to the polymer. This is usually a liquid that manufacturers dissolve within the polymer. It causes the polymer to become flexible by lowering the attraction between the polymer chains. The most common plasticizers are phthalates (Section 13.7-B).

The production of polymers is a major activity in the chemical industry. These processes are accomplished primarily by unique chemical reactions called *addition* and *condensation*. The polymers resulting from these reactions are called **addition polymers** and **condensation polymers**, respectively. We review them independently.

copolymer ▪ A polymer produced from two or more different monomers

cross-linking ▪ The production of chemical bonds in multiple dimensions within a polymer's macromolecules, typically associated with its strength, durability, and elasticity

addition polymer ▪ A polymer resulting from the addition of a molecule, one by one, to a growing polymer chain

condensation polymer ▪ A polymer produced by a chemical reaction along with a small molecule like water or ammonia

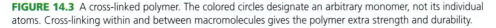

FIGURE 14.3 A cross-linked polymer. The colored circles designate an arbitrary monomer, not its individual atoms. Cross-linking within and between macromolecules gives the polymer extra strength and durability.

14.2-A ADDITION POLYMERIZATION

Polystyrene is an example of an addition polymer. It is produced by the polymerization of the monomer styrene, or phenylethene, as follows:

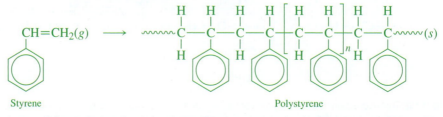

Styrene   Polystyrene

The chemical formula on the left of the arrow represents styrene, the monomer. The formula on the right of the arrow represents a section of the polystyrene macromolecule. The portion of this formula represented within the brackets is repeated over and over again (n times, where n is a very large integer). The symbol ～～～ denotes that the unit is repeated.

In the chemical industry, polymerization is initiated in a controlled fashion. One way to initiate polymerization is to use substances capable of forming free radicals when they are exposed to heat or light. Examples of such chemical initiators are peroxo-organic compounds (Section 13.9) and the Ziegler–Natta catalysts (Section 9.4). Once free radicals are produced by the thermal decomposition of a peroxo-organic compound, they combine with neutral monomer molecules to form more complex free radicals that react with other monomer molecules until the supply of the monomer has been exhausted.

For illustrative purposes, consider the polymerization of styrene induced by free radicals resulting from the dissociation of dibenzoyl peroxide. This polymerization occurs by means of a number of independent steps, some of which are represented as follows:

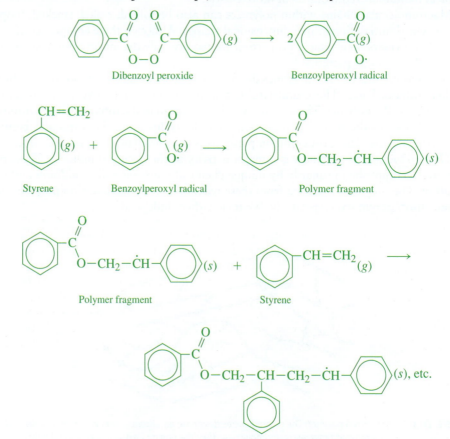

In the first equation, the oxygen–oxygen bond in dibenzoyl peroxide is ruptured, resulting in the production of benzoylperoxyl free radicals. In the second equation, a benzoylperoxyl free radical reacts with a styrene molecule, forming a more complex free radical. In the third equation, this free radical reacts with another molecule of styrene to form a still more complex free radical. Additional steps beyond those illustrated add successively more units to the chain in a self-propagating fashion until a long chain of the polymer has been produced and the monomer source has been consumed.

The substance produced by the polymerization of styrene is called *polystyrene*. The repeating unit in this polymer is the following:

Polystyrene is used commercially to manufacture products such as brushes, combs, disposable coffee cups, thermally insulated equipment, building and electrical insulation, coaxial television cable, compact disk cases, yogurt cups, refrigerator interiors, and the peanuts used in packaging. For several of these purposes, polystyrene is mixed with a foam-blowing compound during processing and manufacturing. The foam is produced by blowing and entrapping a vapor such as 1,1,1,3,3-pentafluoropropane within the polystyrene until it hardens.

When intended for use as building insulation, polystyrene sometimes was mixed with the fire retardant 1,2,5,6,9,10-hexabromocyclododecane, or HBCD, and produced as rigid panels of foam board. During construction projects, these panels were positioned against concrete or drywall.

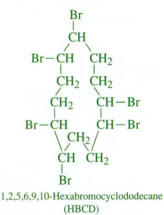

1,2,5,6,9,10-Hexabromocyclododecane
(HBCD)

HBCD was also used as a fire retardant when treating consumer textiles like upholstered furniture and automobile cushions. However, the continued use of HBCD in the United States in any material is questionable because health concerns about its safety have been raised. HBCD has been found to accumulate in fatty tissues and human breast milk, persist in the environment, and kill aquatic organisms at low concentrations. Nations that are parties to the POPs treaty (Section 12.17) agreed in 2013 to ban its future manufacture and use. However, while fighting fires, emergency responders may inhale HBCD since products containing HBCD still remains in residential dwellings.

Polystyrene is recyclable. Its recycling symbol is the familiar arrowed triangle enclosing the number 6, beneath which appear the letters *PS*.

PS

Several other examples of addition polymers are provided in Table 14.1.

TABLE 14.1 — Examples of Addition Polymers

MONOMER	REPEATING UNIT
Ethylene 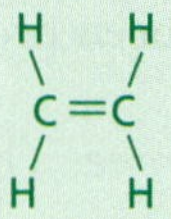 $H \quad\; H$ $\;\; \backslash \;\;\;\; /$ $\;\;\; C=C$ $\;\; / \;\;\;\; \backslash$ $H \quad\; H$	**Polyethylene** A chain of carbon atoms, each bearing two H atoms: $-CH_2-$ repeated along a backbone of 14 carbons. 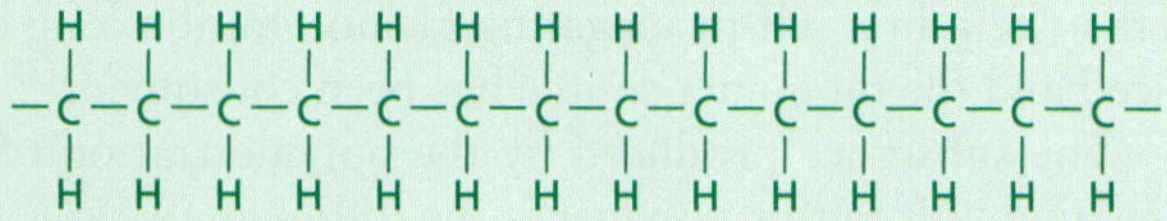
Vinyl chloride $H \quad\; H$ $\;\; \backslash \;\;\;\; /$ $\;\;\; C=C$ $\;\; / \;\;\;\; \backslash$ $H \quad\; H$ 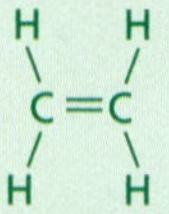	**Poly(vinyl chloride)** 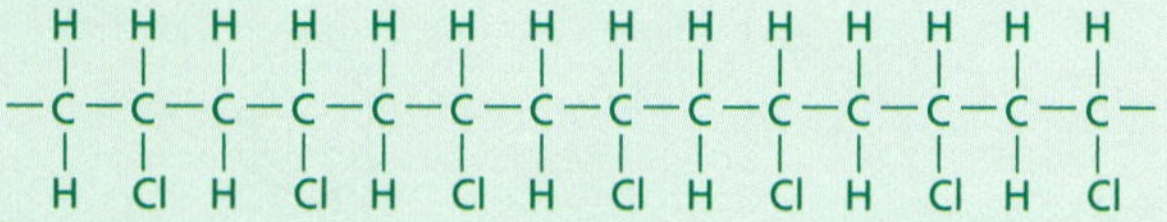 Carbon backbone alternating $-CH_2-CHCl-$ units.
Acrylonitrile $H \quad\; H$ $\;\; \backslash \;\;\;\; /$ $\;\;\; C=C$ $\;\; / \;\;\;\; \backslash$ $H \quad\; CN$ 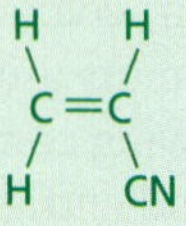	**Polyacrylonitrile** Carbon backbone alternating $-CH_2-CH(CN)-$ units. 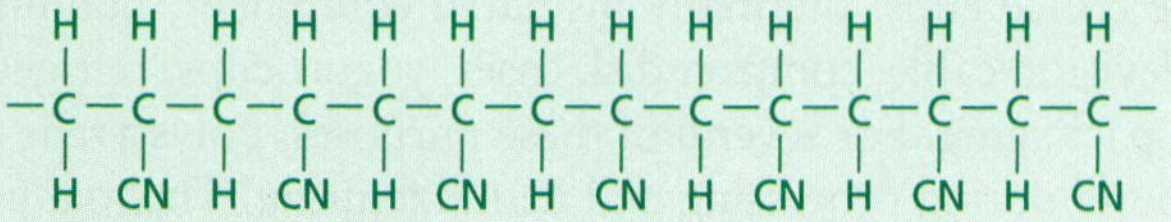
Tetrafluoroethylene $F \quad\; F$ $\;\; \backslash \;\;\;\; /$ $\;\;\; C=C$ $\;\; / \;\;\;\; \backslash$ $F \quad\; F$ 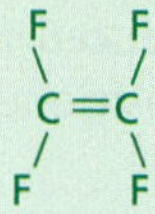	**Poly(tetrafluoroethylene)** Carbon backbone of $-CF_2-$ repeating units. 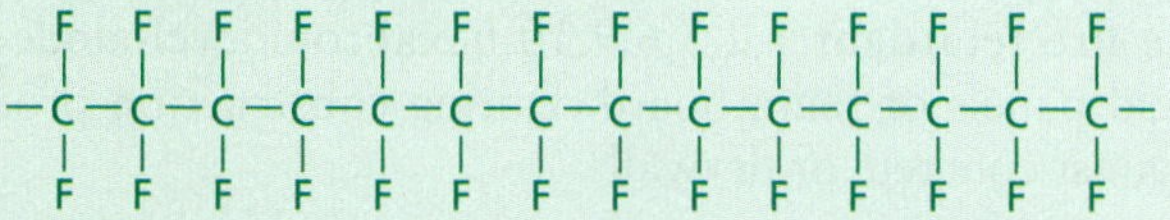
Styrene $CH=CH_2$ (attached to benzene ring) 	**Polystyrene** Carbon backbone with phenyl groups on alternating carbons. 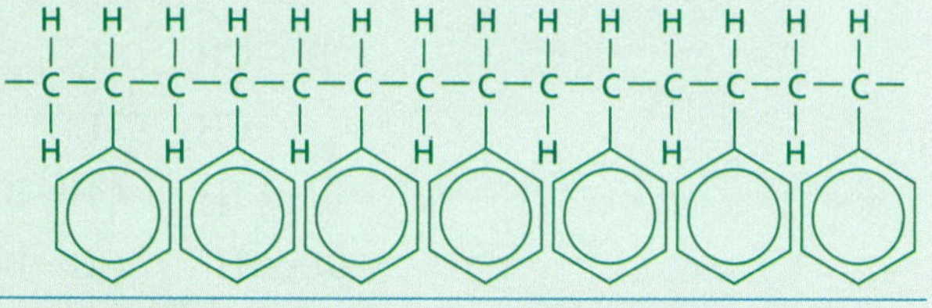
Methyl methacrylate $\qquad CH_3$ $\qquad /$ $CH_2=C$ $\qquad \backslash$ $\qquad C=O$ $\qquad /$ $\qquad O$ $\qquad \backslash$ $\qquad CH_3$ 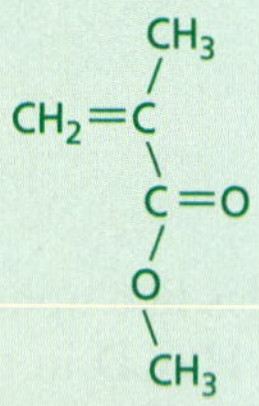	**Poly(methyl methacrylate)** 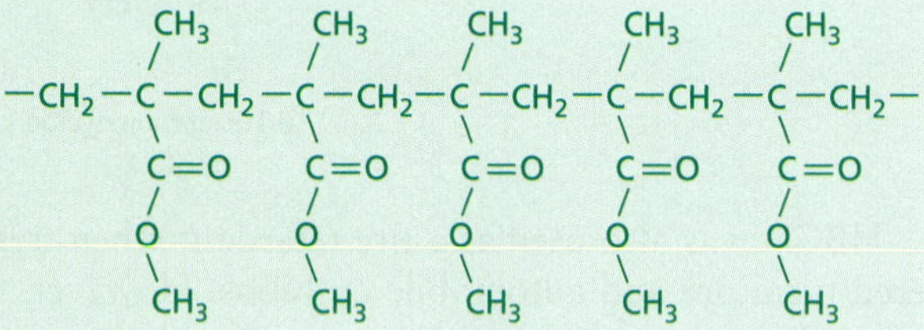 $-CH_2-C(CH_3)(COOCH_3)-$ repeating units.

SOLVED EXERCISE 14.1

Poly(methyl methylacrylate) is produced by the addition polymerization of methyl methacrylate, a substance having the following molecular formula:

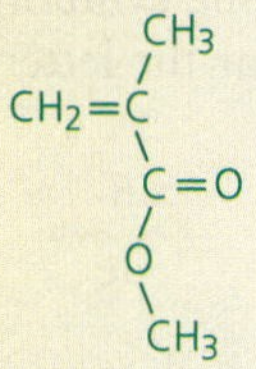

$$CH_2=C\begin{matrix} CH_3 \\ \\ C=O \\ \\ O \\ \\ CH_3 \end{matrix}$$

Dibenzoyl peroxide is used to initiate polymerization. Write the stepwise equations for this polymerization reaction.

Solution: Addition polymerization reactions occur by a mechanism that includes the production of free radicals. The free radicals are first produced when a peroxo-organic compound such as dibenzoyl peroxide is heated. The heat causes the dibenzoyl peroxide to melt and decompose.

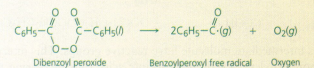

Dibenzoyl peroxide   Benzoylperoxyl free radical   Oxygen

Then, by adding to a molecule of methyl methacrylate, a benzoylperoxyl radical initiates polymerization.

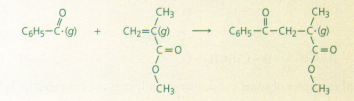

The latter radical reacts with another molecule of methyl methacrylate to produce an even larger free radical.

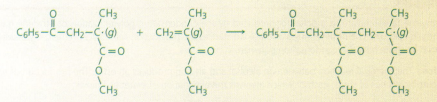

This process repeats until hundreds or thousands of the repeating units are produced within the macromolecule.

14.2-B CONDENSATION POLYMERIZATION

Condensation polymers are produced by a chemical reaction called *condensation polymerization*. A common example of their production involves the reaction between certain alcohols and organic acids. For example, when a glycol is heated with a dicarboxylic acid (Section 13.6), the substances combine with the simultaneous elimination of water. The polymer that results is called a **polyester**.

Consider the chemical reaction between the monomers ethylene glycol and succinic acid. The first step in this reaction is illustrated by the following equation:

polyester ▪ A polymer consisting of recurring ester groups

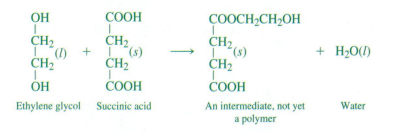

Ethylene glycol   Succinic acid   An intermediate, not yet a polymer   Water

The intermediate compound resulting from this reaction has potentially reactive groups at both ends of the carbon–carbon chain. It can react with two molecules of ethylene glycol, eliminating two molecules of water and producing an even more complex intermediate as follows:

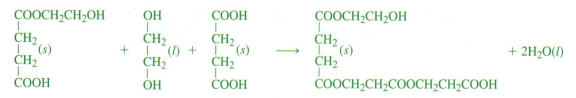

Both ends of this intermediate molecule have reactive groups. This intermediate can also combine with more ethylene glycol and succinic acid, thus increasing the length of the macromolecular chain. Finally, when the amount of either monomer is exhausted, the polyester that remains is described by the following formula:

$$H\text{www}[\text{www}O-CH_2CH_2-O\underset{\text{C}-CH_2CH_2-\text{C}}{\overset{O\quad\quad O}{}}O\text{www}]_n\text{www}H$$

Additional examples of some condensation polymers are provided in Table 14.2.

SOLVED EXERCISE 14.2

Poly(ethylene terephthalate) is a polyester produced by the condensation polymerization of ethylene glycol and terephthalic acid. The chemical formulas of these substances are provided in Table 14.2. Write the stepwise equations that illustrate how this condensation polymerization occurs.

Solution: The chemical reaction between an alcohol and an acid produces an ester. The production of a polyester occurs by a series of steps, each of which involves the elimination of water as follows:

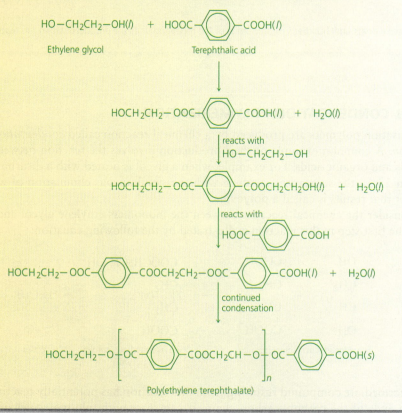

TABLE 14.2	Examples of Condensation Polymers

REACTANTS	REPEATING UNIT

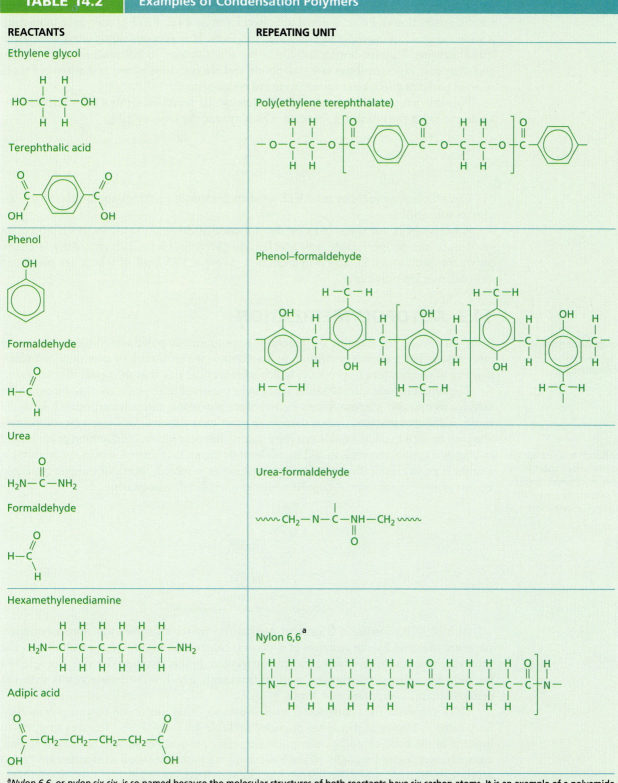

Ethylene glycol

Terephthalic acid

Poly(ethylene terephthalate)

Phenol

Phenol–formaldehyde

Formaldehyde

Urea

Urea-formaldehyde

Formaldehyde

Hexamethylenediamine

Nylon 6,6 [a]

Adipic acid

[a]*Nylon 6,6*, or *nylon six-six*, is so named because the molecular structures of both reactants have six carbon atoms. It is an example of a polyamide (Section 14.9).

A commonly encountered polyester is the tongue-twister poly(ethylene terephthalate), more commonly known as PETE, PET, or the DuPont trademark Mylar. The mechanism for the production of PETE is noted in Solved Exercise 14.2. PETE is used to manufacture a wide range of industrial and consumer products. It is predominantly used during the manufacturing of plastic bottles and fibers, but it is also used to manufacture audio, video, and computer tapes; sailboat sails; telephone and electric cable wires; and dozens of types of food packaging like ketchup and mustard containers.

Poly(ethylene terephthalate) is also recyclable. Its recycling symbol is an arrowed triangle enclosing the number 1, beneath which appear the letters *PETE*.

Coca-Cola now recycles the PETE used to manufacture the plastic bottles that contain its soft drinks.

A polyester is not the sole example of a condensation polymer. In this chapter, we shall note the properties of polyacrylamide (Section 14.6-C), formaldehyde-based polymers (Section 14.8), and polyurethanes (Section 14.9), all of which are produced as condensation polymers.

14.3 AUTOPOLYMERIZATION

autopolymerization
■ Spontaneous polymerization

Some monomers are capable of undergoing spontaneous polymerization, or **autopolymerization**. The autopolymerization of a substance within a tank or container can result in a rise in temperature or pressure, which increases the risk that the vessel may rupture.

The autopolymerization of a substance can be prevented so the substance can ultimately be used for its intended purpose. When the substance is gaseous, autopolymerization is hindered by the addition of nitrogen, carbon dioxide, or other inert diluents. Although inert compounds also may be used to dilute liquids that are prone to autopolymerize, the common practice used by chemical manufacturers is to add an inhibitor to them. As the name implies, an **inhibitor** is a substance that chemically reacts in such a fashion as to inhibit, retard, or stabilize the liquid against autopolymerization. A popular inhibitor is *tert*-butylhydroquinone.

inhibitor ■ A substance or mixture of substances used to prevent, arrest, or retard the speed of autopolymerization

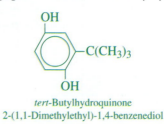

tert-Butylhydroquinone
2-(1,1-Dimethylethyl)-1,4-benzenediol

Although there is not a consistent mechanism by which substances autopolymerize, some are initiated by the reaction of a peroxo-organic compound that results when the substance reacts with dissolved atmospheric oxygen. In this instance, the inhibitor selected for retarding autopolymerization is an antioxidant, which immediately reacts with the peroxo-organic compound as it forms.

DOT regulates the transportation of substances that are prone to autopolymerization, some examples of which are provided in Table 14.3. DOT also requires shippers to indicate in the basic shipping description that their products have been stabilized against autopolymerization. Examples of the basic shipping names for such substances are "ethyl acrylate, stabilized" and "vinylidene chloride, stabilized."

To forewarn emergency responders of the presence of a substance that is susceptible to autopolymerization, chemical manufacturers and users sometimes insert a *P* in the bottom quadrant of the hazard diamond posted in the relevant storage area. In the *Emergency Response*

TABLE 14.3	Examples of Monomers That Autopolymerize Unless Effectively Stabilized	
MONOMER	**SHIPPING DESCRIPTION**	
Acrylonitrile	UN1093, Acrylonitrile, stabilized, 3, (6.1), PG I (Poison - Inhalation Hazard)	
Chloroprene	UN1991, Chloroprene, stabilized, 3, (6.1), PG I (Poison - Inhalation Hazard)	
Ethyl acrylate	UN1917, Ethyl acrylate, stabilized, 3, PG II (Marine Pollutant)	
Isoprene	UN1218, Isoprene, stabilized, 3, PG I	
Methyl acrylate	UN1919, Methyl acrylate, stabilized, 3, PG II	
Methyl methacrylate	UN1247, Methyl methacrylate monomer, stabilized, 3, PG II	
Styrene	UN2055, Styrene, stabilized, 3, PG II (Marine Pollutant)	
Vinyl acetate	UN1301, Vinyl acetate, stabilized, 3, PG II	
Vinyl bromide	UN1085, Vinyl bromide, stabilized, 2.1	
Vinyl butyrate	UN2838, Vinyl butyrate, stabilized, 3, PG II	
Vinyl chloride	UN1086, Vinyl chloride, stabilized, 2.1	
Vinylidene chloride	UN1303, Vinylidene chloride, stabilized, 3, PG I	

Guidebook, a substance susceptible to autopolymerization is identified by the insertion of the letter *P* following the listing of a guide number in the yellow- and blue-bordered pages.

14.4 POLYMER DECOMPOSITION AND COMBUSTION

Most products produced from natural and synthetic polymers are combustible when exposed to an ignition source. Their combustion is associated with the following general features:

- Polymeric products often melt and thermally decompose into the monomers from which they were made or a mixture of simpler substances.
- The surfaces of some polymeric products tend to char as they burn.
- Burning polymeric products release considerable heat.
- Burning polymeric products can evolve voluminous amounts of smoke, carbon monoxide, and other hazardous gases, vapors, and fumes.

The melting of polymeric products at a typical fire scene is associated with both beneficial and detrimental effects. The melting often causes the polymer to drip from its source, as from ceiling tile to an underlying floor. Dripping molten polymer closely resembles dripping hot candle wax. The dripping serves as a cooling mechanism, removing heat from the immediate site of combustion and hindering the continued combustion of the polymer at that site. However, when polymers remain in the fire zone, they begin to thermally decompose in their molten state. Then, their decomposition products ignite and the fire spreads.

Before their ignition, polymers frequently undergo thermal degradation into simpler chemical species; that is, when they are exposed to heat, polymers decompose into relatively simpler substances. The decomposition of unique organic polymers occurs by different mechanisms, several of which involve scission of the macromolecular chains. The following two types of thermal decomposition are characteristic of how polymers decompose:

depolymerization
- The thermal decomposition process during which a polymer produces the monomers from which it was made

- When heated, some polymers predominantly produce the monomers from which they were initially produced. They are said to *depolymerize*, and the decomposition process is called **depolymerization**. For example, when poly(methyl methacrylate)

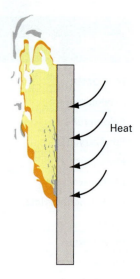

FIGURE 14.4 Heat radiated through this section of a wall gives rise to the phenomenon known as a flashover. It also causes the thermal decomposition of the polymeric paneling affixed to its opposite side. The decomposition of a polymer produces simple organic substances that slowly migrate from their point of origin and mix with the surrounding air. When a concentration within the flammable range is reached, exposure to an ignition source causes the mixture to ignite.

Heat

thermally decomposes, its monomer, methyl methacrylate, accounts for 91 to 98% by mass of the substances produced. The same is true of poly(ethylene terephthalate). When heated, it reverts into the substances from which it was produced: terephthalic acid and ethylene glycol. Polymers that depolymerize are desirable candidates for recycling.

■ When certain other polymers are heated, they produce an array of gaseous decomposition products. For example, when polypropylene is heated, it decomposes to form methane, ethane, propane, butane, pentane, propene, 2-methylpropene, 1- and 2-pentene, 2-methyl-1-butene, and other hydrocarbons.

The vapors produced by thermal decomposition initially diffuse to the surface of the polymer, where they mix with atmospheric oxygen and ignite. As the heat increases in intensity, these vapors migrate from the immediate burning area and accumulate elsewhere, such as near the ceiling of a room, where they mix with air and burn.

Heat may also be conducted or radiated through a polymeric material, thereby causing the polymer to decompose at a location isolated from the heat source. Consider the section of a wall shown in Figure 14.4. It is constructed of wooden support beams to which polymeric paneling has been affixed. Although the heat from a fire impinges on only one side of the wall, it can conduct or radiate *through* the wall, causing the polymer in the paneling to decompose. The mixture of combustible gases subsequently produced by thermal decomposition readily ignites.

This generation of flammable vapors at a location isolated from the source of heat is associated with a **flashover**, the phenomenon largely responsible for the spread of fire from one room into an adjacent room. At flashover, the entire contents of a room are ignited simultaneously by radiant heat. This situation makes living conditions within the room untenable. Safe exit for residents is impossible. At flashover, room temperatures typically range from approximately 1100 to 1470°F (~600 to 800°C).

Fires pose special problems in large public buildings that have been constructed in part from plastic products, since the polymers in these products usually burn differently from burning wood or other natural materials. The thermal characteristics of some common polymers in air are provided in Table 14.4. By comparison, hardwoods self-ignite in air at an average temperature of 781°F (416°C) and have an average heat of combustion of 8500 Btu/lb. Although hardwoods and synthetic polymers begin to burn at approximately the same temperature, polymers like polyethylene, polypropylene, and polystyrene emit more than twice as much heat as the same amount of burning wood.

flashover ■ The phenomenon associated with the spread of fire from the burning area to other areas physically isolated from the initial fire source

TABLE 14.4 Thermal Characteristics of Some Common Polymers

POLYMER	SPECIFIC GRAVITY	DECOMPOSITION RANGE	SELF-IGNITION TEMPERATURE	HEAT OF COMBUSTION
Polyethylene (high-density)	0.965	644–824°F (340–440°C)	662°F (350°C)	20,050 Btu/lb (46,500 kJ/kg)
Polypropylene	0.91	626–770°F (330–410°C)	734–770°F (390–410°C)	19,800 Btu/lb (46,000 kJ/kg)
Polystyrene	1.05	572–752°F (300–400°C)	914°F (490°C)	18,100 Btu/lb (42,000 kJ/kg)
Poly(methyl methacrylate)	1.18	338–572°F (170–300°C)	842°F (450°C)	11,210 Btu/lb (26,000 kJ/kg)
Poly(vinyl chloride)	1.40	392–572°F (200–300°C)	851°F (455°C)	8620 Btu/lb (20,000 kJ/kg)

14.4-A THE CHEMICAL NATURE OF THE GASES AND VAPORS PRODUCED DURING POLYMERIC FIRES

The fatalities that occur at fire scenes often result when individuals are exposed to the gases and vapors produced during the fires. Burning polymers produce massive concentrations of carbon monoxide. Within an enclosure, carbon monoxide and other gases soar to life-threatening concentrations within a matter of seconds.

Fires involving burning polymers can affect areas far removed from where the fire originated when hot gases and thermal degradation products travel by convection through ventilation systems, trash chutes, and similar openings. Although this movement spreads the fire, it also causes people to be unsuspectingly exposed to toxic fumes generated elsewhere within a building.

The burning of products made from synthetic polymers often produces a mixture of gases and fumes different from that generated by the burning of nonplastic products. Carbon monoxide is still the most prevalent gas at a fire scene, but other gases associated with burning plastics are also produced. These include hydrogen chloride, ammonia, hydrogen cyanide, sulfur dioxide, and nitrogen dioxide.

At a fire scene, the origin of the nitrogenous and sulfurous gases can be traced to the chemical nature of the polymeric materials that have burned or undergone thermal decomposition. Thermal decomposition occurs primarily when the polymers are exposed to the heat emitted during slow-burning processes. The polymers that produce hydrogen cyanide, ammonia, nitric oxide, and nitrogen dioxide are nitrogenous organic compounds. They are natural and synthetic polymers that have one or more of the functional groups listed in Table 14.5. When they thermally decompose, hydrogen cyanide and ammonia are produced; and when they burn incompletely and completely, nitric oxide and nitrogen dioxide are produced, respectively.

Polymers containing sulfur atoms usually are natural polymers having an animal origin. Their nonwater matter is composed of proteins, which in turn are composed of **amino acids**. Although there are 21 amino acids that make up the structures of nearly all proteins, of concern here are only the two whose molecules contain sulfur atoms in their composition: methionine and cysteine.

amino acid ■ Any carboxylic acid having the —NH$_2$ group of atoms

$$CH_3-S-CH_2CH_2-\underset{\underset{NH_2}{|}}{CH}-\overset{\overset{O}{\|}}{C}\underset{OH}{\diagdown}$$

Methionine

$$HS-CH_2-\underset{\underset{NH_2}{|}}{CH}-\overset{\overset{O}{\|}}{C}\underset{OH}{\diagdown}$$

Cysteine

Chapter 14 Chemistry of Some Polymeric Materials **619**

TABLE 14.5   Some Nitrogenous Organic Compounds

CLASS OF ORGANIC COMPOUND	FUNCTIONAL GROUP	EXAMPLE
Amine	$-NH_2$	$CH_3CH_2CH_2-NH_2$ *n*-Propylamine, or 1-Aminopropane
Amide	$\overset{\displaystyle O}{\underset{\displaystyle NH_2}{-\overset{\|\|}{C}}}$	$CH_3-\overset{\displaystyle O}{\underset{\displaystyle NH_2}{\overset{\|\|}{C}}}$ Acetamide
Amino acid	$\underset{\displaystyle NH_2}{-CH}-\overset{\displaystyle O}{\underset{\displaystyle OH}{\overset{\|\|}{C}}}$	$CH_3-\underset{\displaystyle NH_2}{CH}-\overset{\displaystyle O}{\underset{\displaystyle OH}{\overset{\|\|}{C}}}$ 2-Aminopropanoic acid
Isocyanate	$-N=C=O$	$CH_3CH_2-N=C=O$ Ethyl isocyanate
Nitrate	$-O-NO_2$	$CH_3CH_2CH_2CH_2-O-NO_2$ *n*-Butyl nitrate
Nitrile, or cyanide	$-C\equiv N$	$CH_2=\underset{\displaystyle C\equiv N}{\overset{\displaystyle H}{C}}$ Acrylonitrile, or Propenenitrile, or Vinyl cyanide
Nitro compounds	$-NO_2$	⬡—NO_2 Nitrobenzene

Acrolein

Because sulfur atoms are components of these amino acids, they are also constituents of the proteins biologically produced from them in leather, wool, and animal hair. Their thermal decomposition produces hydrogen sulfide and ammonia, and their combustion produces carbon monoxide, carbon dioxide, sulfur dioxide, nitric oxide, nitrogen dioxide, and water.

Another toxic substance whose presence has been detected in smoke is the unsaturated aldehyde known as acrolein, or 2-propenal. Its chemical formula is $CH_2=CHCHO$.

$$CH_2=CH-\overset{\displaystyle O}{\underset{\displaystyle H}{\overset{\|\|}{C}}}$$

Acrolein (2-Propenal)
(Acrylic aldehyde)

This is a pungent-smelling, intensely irritating lacrimator. A 1-minute exposure to an air concentration of 1 part per million causes nasal and eye irritation. A concentration of acrolein in air ranging from 2500 to 5000 parts per million has been established as the lethal dose to laboratory animals. During World War I, acrolein was used offensively as a lacrimator (Section 10.9-A).

14.4-B SMOKE PRODUCED DURING POLYMERIC FIRES

The character of the smoke produced during fires involving polymeric materials varies with the chemical nature of the polymer. In particular, the amount of smoke produced in fires involving polymers made from aromatic monomers is typically far greater than the amount produced during the burning of polymers made from aliphatic monomers. A fire involving materials produced from polystyrene, for instance, produces considerably more soot than one involving materials produced from polyethylene.

As was first noted in Section 10.9-C, the carbon particulates in smoke adsorb toxic gases on their surfaces. When smoke is inhaled, these particulates serve as the vehicles that draw toxic gases into the bronchi and lungs. Because considerable smoke is produced during the burning of materials produced from aromatic monomers, they pose a greater risk to one's health compared to materials produced from aliphatic monomers.

14.4-C CONSUMER PRODUCT REGULATIONS PERTAINING TO TEXTILES

Polymers have been used to manufacture hundreds of different consumer products. Because most polymers burn, the consumer products manufactured from them are now evaluated to determine whether the likelihood of their ignition is delayed or eliminated when their fibers have been treated with fire retardants or flameproofing agents.

Congress has directed the CPSC to reduce injuries and deaths caused by consumer products in a variety of settings. In response to this mandate, CPSC published a standard at 16 C.F.R. §§1633.1–1633.13, which aims to minimize or delay flashover during typical mattress/bedding fires that occur within the home. In Figure 14.5, the comparison between igniting two mattresses, one manufactured conventionally and the other manufactured in accordance with the CPSC standard, is demonstrated. It is apparent that the mattress manufactured in accordance with the CPSC standard is more desirable as a consumer product, because exposure to an open flame provides the occupants with time to discover the fire and escape from the home.

Figure 14.5 also illustrates that although treatment procedures may improve the safety of products for consumer use, they do not totally eliminate their flammability. After polymeric fibers have been treated, the products made from them still burn, especially when they are exposed to the intense heat experienced during major fires. However, treated consumer products are more resistant to ignition and more prone to self-extinguish once fire has been initiated.

State and federal laws require manufacturers to provide consumers with materials that are unlikely to easily ignite. Their combined use has helped reduce injuries when plastics and textiles are involved in fires. The CPSC has used the legal authority of the **Flammable Fabrics Act** and other statutes to respond to public concern over accidents involving the use of products like brushed rayon in high-pile sweaters, and children's cowboy chaps that flash-burn when ignited. The CPSC now requires manufacturers to submit samples of apparel fabrics like children's sleepwear to an ignition test and a rate-of-burn test. The use of these tests also applies to the manufacture of carpets, rugs, and other home furnishings.

14.5 VEGETABLE AND ANIMAL FIBERS

Many common textiles are produced from naturally occurring vegetable and animal fibers. **Cotton** and **linen** are examples of vegetable fibers, whereas wool and silk are examples of animal fibers. These naturally occurring fibers may be used to produce textiles, or they can be chemically altered to produce synthetic fibers from which the textiles are produced.

Flammable Fabrics Act
■ The federal statute that empowers the Consumer Products Safety Commission to establish flammability standards for clothing textiles as well as interior furnishings, including paper, plastic, foam and other materials used in wearing apparel and interior furnishings

cotton ■ A naturally occurring fiber of vegetable origin

linen ■ The naturally occurring fiber of the flax plant

FIGURE 14.5 In this experiment, two mattresses are simultaneously exposed to an open-flame ignition source to simulate the burning of mattresses during a bedroom fire. Open-flame ignition sources are candles, matches, lighters, cigarettes, and similar items. The top mattress was manufactured by following the standard published at 16 C.F.R. §§1633.1–1633.13. It burns or smolders at a noticeably slower growth rate compared to the burning rate of the untreated mattress at the bottom of the figure. (*Courtesy of the U.S. Consumer Product Safety Commission, Washington, DC*)

TABLE 14.6	Some Animal and Vegetable Products
PRODUCT	**SHIPPING DESCRIPTION**
Cotton	NA1365, Cotton, 9[a]
Cotton, waste, oily	UN1364, Cotton waste, oily, 4.2, PG III[b]
Cotton, wet	UN1365, Cotton, wet, 4.2, PG III[c]
Fibers, animal *or* Fibers, vegetable *burnt, wet, or damp*	UN1372, Fibers, animal, 4.2, PG III *or* UN1372, Fibers, vegetable, 4.2, PG III
Fibers, vegetable, dry	UN3360, Fibers, vegetable, dry, 4.1, PG III
Fibers *or* Fabrics, animal or vegetable *or* Synthetic, n.o.s. *with animal or vegetable oil*	UN1373, Fibers, animal, 4.2, PG III *or* UN1373, Fibers, vegetable, 4.2, PG III *or* UN1373, Fibers, synthetic, n.o.s. (containing animal oil), 4.2, PG III *or* UN1373, Fibers, synthetic, n.o.s. (containing vegetable oil), 4.2, PG III
Fibers *or* fabrics impregnated with weakly nitrated nitrocellulose	UN1353, Fibers impregnated with weakly nitrated nitrocellulose, n.o.s., 4.1, PG III *or* UN1353, Fabrics impregnated with weakly nitrated nitrocellulose, n.o.s., 4.1, PG III
Paper, unsaturated oil-treated *incompletely dried (including carbon paper)*	UN1379, Paper, unsaturated oil-treated, 4.2, PG III
Wool waste	UN1387, Wool waste, wet, 4.2, PG III[c]

[a]For domestic transportation only on water
[b]For transportation by air or on water
[c]For international transportation only by air or on water

As they occur naturally, vegetable and animal fibers often are mixed with combustible oils; for example, cotton contains cottonseed oil, and wool contains lanolin. DOT regulates the transportation of the vegetable and animal fibers listed in Table 14.6 as hazardous materials. We note the properties of some common fibers and pay special attention to their combustible nature.

14.5-A CELLULOSE AND ITS DERIVATIVES

The polymer that remains when the natural binding agent is removed from wood and other plants is called **cellulose**. It serves as the primary structural component of the cell walls in plants, and thus is generally regarded as nature's most important polymer. Wood is approximately 50% cellulose by mass, whereas cotton and linen are nearly 100% cellulose.

cellulose ■ The substance that forms the cell walls of all plants

Cellulose fibers are derived from cotton and linen. Cotton comes from any of four species of *Gossypium* that grow in warm climates throughout the world. To produce it as a yarn, raw cotton is first boiled in a dilute solution of sodium hydroxide to remove any wax that naturally adheres to the fibers. Then, it is bleached with chlorine, sodium hypochlorite, or a similar substance and passed through a vat of diluted sulfuric acid to neutralize any remaining alkaline materials. Finally, the cotton is washed with water. At this point, it is ready to be spun into a yarn that may be woven into cloth.

Linen is nearly pure cellulose. Its fibers are derived from the stalk of the flax plant *Linum usitatissimum*. Linen fibers are among the strongest naturally occurring fibers. Linen fabric absorbs moisture faster than any other fabric; however, linen lacks resiliency, the ability to spring back when stretched. This lack of resiliency causes it to wrinkle easily.

The difference between the physical properties of cotton and linen is associated with the nature of their fibers. When examined under a microscope, individual cotton fibers resemble short, twisted, flattened tubes; linen fibers resemble long transparent tubes that have periodic junctions for cross-linking.

Cotton is commercially available not only in its natural form but also as mercerized cotton. Cotton fibers swell when they are immersed in a concentrated solution of sodium hydroxide, a process called *mercerizing*, after the discoverer John Mercer. When each flattened carbon fiber is mercerized, it becomes rounded and more lustrous and acquires additional strength. Dyes penetrate fabrics made from mercerized cotton more easily than those made from untreated cotton. Mercerized cotton can be chemically treated to produce fabrics that are wrinkleproof, as well as fabrics that can be washed and drip-dried with few wrinkles. Cotton may also be physically combined with synthetic fibers to produce fabrics that are commercially available as blends with names such as *50% cotton, 50% acrylics*.

The chemical formula of cellulose often is abbreviated as $(C_6H_{10}O_5)_n$ or $[(C_6H_7O_2(OH)_3]_n$, where *n* ranges from 7000 to 12,000. The following structure represents the more complete chemical formula, in which carbon atoms are located at the intersecting lines:

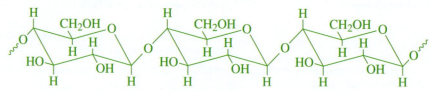

The repeating unit shown in this structure is a substance called *β-glucose*.

When triggered by an ignition source, cellulose and cellulosic products burn without melting, but they do not burn readily. Because burning cellulosic materials are class A fires, they routinely are extinguished with water.

There are two alternative pathways by which cellulose responds to the application of heat:

■ On exposure to temperatures less than approximately 570°F (<300°C), cellulose depolymerizes and eliminates water. As the dehydration occurs, a slow-burning or smoldering char forms that may ultimately convert into ash.

■ On exposure to temperatures greater than approximately 570°F (>300°C), cellulose thermally degrades into a black gooey tar. When this goo is further heated, it converts into a mixture of hydrocarbons, alcohols, aldehydes, and ketones. It is only when the individual compounds in this mixture vaporize and mix with atmospheric oxygen that fire occurs.

Synthetic Fibers Obtained from Cellulose

Synthetic fibers may be derived from the chemical treatment of cellulose. For example, cellulose acetate is the fiber produced when wood cellulose is reacted with acetic acid or acetic anhydride. In commerce, the fiber is called *acetate rayon*. Although it has a pronounced strength, acetate rayon can be ignited with a hot iron and destroyed by some dry-cleaning solvents. Given these adverse features, acetate rayon did not retain popularity as a fabric of choice, but it is still used today in motion-picture film, airplane wings, and safety glasses.

Another synthetic product derived from cellulose is cellulose xanthate, commonly called rayon. The production of rayon involves reacting highly purified wood pulp with sodium hydroxide, followed by chemical treatment with carbon disulfide. The

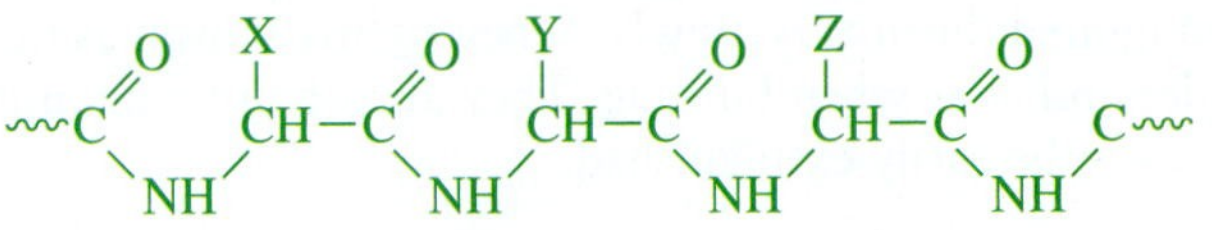

The structure at top:

$$\text{~~C} \overset{O}{\underset{NH}{\|}} \text{---} \overset{X}{\underset{NH}{|}} CH\text{---}C \overset{O}{\|} \text{---} \overset{Y}{\underset{NH}{|}} CH\text{---}C \overset{O}{\|} \text{---} \overset{Z}{\underset{NH}{|}} CH\text{---}C \overset{O}{\|} \text{---} C\text{~~}$$

FIGURE 14.6 In this segment of the macromolecular structure of wool, X, Y, and Z represent molecules of amino acids, compounds that have the following chemical formula:

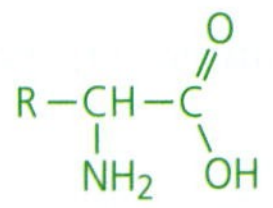

$$R\text{---}\overset{}{\underset{NH_2}{|}}CH\text{---}C\overset{O}{\underset{OH}{\|}}$$

Only 21 amino acids are commonly found in naturally occurring proteins. The interested reader may consult more advanced chemistry textbooks for additional information concerning proteins.

resulting solution is extruded through a spinneret, a metal disk having numerous tiny holes, and then into an acid solution that regenerates the cellulose as tiny transparent fibers.

Another example of a synthetic product derived from cellulose is nitrocellulose. This substance is manufactured by reacting cellulose with nitric acid. Although highly nitrated cellulose is a chemical explosive, the lesser grades of nitrated cellulose are dissolved in a solvent and applied to cloth to produce the fabric known as patent leather. Products produced from this artificial leather are strong and flexible. They include vehicle seat covers and convertible tops. Nitrocellulose is also used as a film-producing agent in lacquers, printing inks, and enamel nail polishes. The nitrocellulose formerly used to make film, called nitrate film, was highly flammable. A strip burned at a rate three times faster than an identically sized piece of paper.

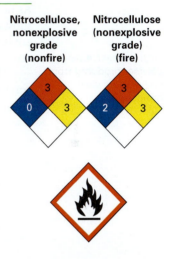

Nitrocellulose, nonexplosive grade (nonfire)

Nitrocellulose (nonexplosive grade) (fire)

14.5-B WOOL AND SILK

The most common animal fabrics employed in textiles are **wool** and **silk**. Wool is the curly hair of sheep, goats, and llamas. Under a microscope, wool fibers resemble tiny, overlapping scales, much like those of fish. These fibers bend and conform to a variety of physical shapes. They also possess resiliency and tend to hold their shape.

Silk is the soft, shiny fiber produced by silkworms to form their cocoons. Silk fibers are very strong, elastic, and smooth. Under a microscope, silk fibers appear semitransparent, which accounts for their lustrous sheen. Silk is resilient; its fibers readily spring back to their original position when stretched or folded. These qualities have made silk one of the most useful fibers for the textile market. Unwinding the long, delicate silk threads of a cocoon is a tedious process; hence, the fabric woven from silk fibers is relatively expensive.

Like all other forms of animal hair, wool and silk are composed of proteins, which are biological substances whose molecules have recurring amino groups. The condensed formulas of wool and silk are $C_{42}H_{157}N_5SO_{15}$ and $C_{15}H_{23}N_5O_6$, respectively. A portion of the macromolecular structure of wool is shown in Figure 14.6.

Silk consists of a mixture of two relatively simple proteins called *silk fibroin* and *sericin*. The macromolecular structure of silk fibroin resembles the structure of wool in which X, Y, and Z are invariably $-CH_3$, $HO-C_6H_4-CH_2-$, or $HO-CH_2-$. Sericin has a similar macromolecular structure, but the primary recurring group in the protein is the following:

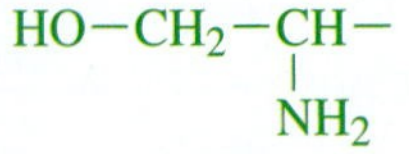

$$HO-CH_2-\overset{}{\underset{NH_2}{|}}CH-$$

Because wool and silk are proteins, both have similar properties. Each has an ignition temperature in excess of approximately 1058°F (570°C); hence, wool and silk are difficult

wool ▪ The naturally occurring fiber obtained from the coats of sheep, llamas, goats, and several other animals

silk ▪ A naturally occurring protein produced by the action of silkworms

to ignite and when ignited, burn very slowly. When tightly woven as in rugs, woolen textiles tend to smolder and char when burning. They absorb great quantities of water, thus permitting their fires to be easily extinguished.

Because the macromolecules of animal fibers contain bonding groups such as $-\overset{\overset{\textstyle O}{\|}}{C}\diagdown_{NH_2}$,

$-NH-$, and $-S-S-$, the presence of ammonia, hydrogen cyanide, and sulfur dioxide at fire scenes involving textiles is possible.

14.6 VINYL POLYMERS

As noted in Section 12.3-B, the vinyl group of atoms is $CH_2=CH-$. For our purposes, a substance containing the vinyl group is a vinyl compound. Examples of vinyl compounds are listed in Table 14.1 under the heading "Monomer." A polymer produced by the addition polymerization of one or more vinyl compounds is called a **vinyl polymer**.

vinyl polymer ■ Any polymer produced from one or more vinyl compounds

The following equations illustrate the production of several commercially important vinyl polymers from their respective vinyl compounds:

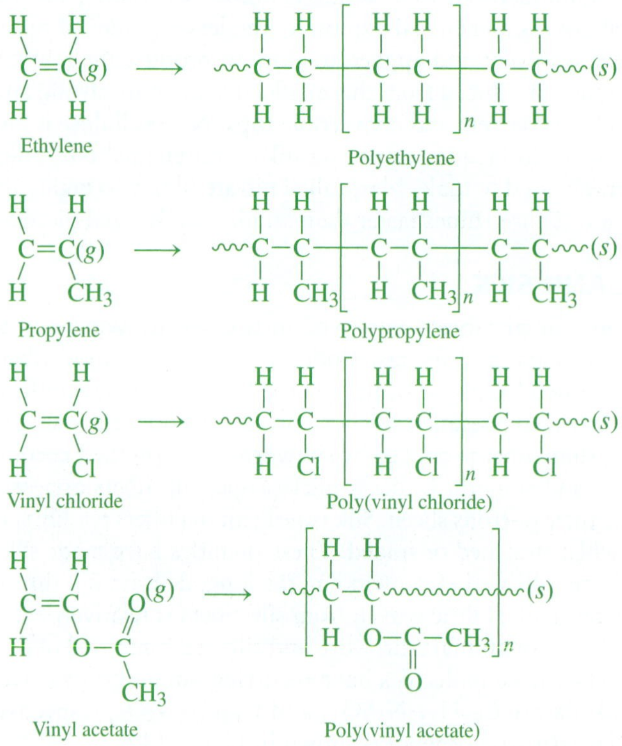

Ethylene → Polyethylene

Propylene → Polypropylene

Vinylidene chloride

Vinyl chloride → Poly(vinyl chloride)

Vinyl acetate → Poly(vinyl acetate)

A vinyl polymer may also be produced from multiple vinyl compounds. For example, the vinyl chloride–vinylidene chloride copolymer is produced from vinyl chloride and vinylidene chloride. Vinyl chloride and vinylidene chloride are synonyms for chloroethene and 1,1-dichloroethene, respectively.

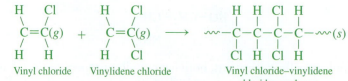

Vinyl chloride + Vinylidene chloride → Vinyl chloride–vinylidene chloride copolymer

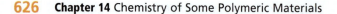

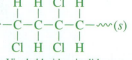

Commercially, the vinyl chloride–vinylidene chloride copolymer is known as *Saran*, or poly (vinylidene dichloride). Manufacturers form it into sheets, tubes, rods, fibers, and other molded items. Its fibers are used to produce a number of textiles, including carpets, curtains, and upholstery fabrics.

14.6-A POLYETHYLENE

Ethylene is prepared in the petrochemical industry, primarily by cracking ethane and propane. When it is polymerized, the resulting product is called *polyethylene* and is the most widely manufactured polymer in the United States. Polyethylene macromolecules have the repeating unit $\sim\sim CH_2 - CH_2 \sim\sim$.

Ethylene

Polyethylene is encountered primarily in two forms: *low-density* (cross-linked) (LDPE) and *high-density* (linear) (HDPE). A third form, low-molecular-weight polyethylene, is uniquely used in coatings and polishes. Both LDPE and HDPE are white solids. LDPE is a thermosetting polymer, whereas HDPE is a thermoplastic polymer. The difference between the densities of the two polymers is attained by their methods of production: LDPE and HDPE are produced by polymerizing ethylene in the presence of an organic peroxide and a Ziegler-Natta catalyst, respectively. The density of LDPE may be as low as 0.915 g/mL, and of HDPE as high as 0.965 g/mL.

Low-density polyethylene is used largely for making baby diapers and molded products such as toys. For example, the "noodles" that children use in swimming pools are manufactured from low-density polyethylene. High-density polyethylene is used mainly to manufacture filmed products and hardy storage containers. The polyethylene films are used in the building industry as vapor and moisture barriers and in the agricultural industry for mulching, silage covers, greenhouse glazings, pond liners, and animal shelters. Included among the containers are plastic milk jugs, detergent and bleach containers, sandwich bags, and liners for trash cans, drums, and other containers.

Alarm has been raised by consumers over the use of polyethylene film to manufacture garbage bags, dry-cleaning bags, and other containers that could inadvertently cause the suffocation of infants and small children who contact them. To minimize the number of deaths resulting from contact with the film, the following voluntary statement is often embossed on bags in English, Spanish, and French:

> **WARNING**
>
> KEEP THIS BAG AWAY FROM BABIES AND CHILDREN. DO NOT USE IN CRIBS, CARRIAGES, OR PLAYPENS. THE THIN FILM MAY CLING TO NOSE AND MOUTH AND PREVENT BREATHING.

Both low- and high-density polyethylene may be recycled. Their recycling symbols are arrowed triangles enclosing the numbers 4 and 2, respectively, and beneath which appear the letters *LDPE* and *HDPE*, as follows:

Products made from both LDPE and HDPE burn when they are exposed to fire. The data in Table 14.4 show that polyethylene is distinctive among burning polymers insofar as it releases more heat per unit mass than any other commercially popular polymer. During combustion, polyethylene products tend to disintegrate into numerous burning, molten globules or liquid pools. Because the macromolecules of polyethylene are composed of only carbon and hydrogen atoms, carbon monoxide, carbon dioxide, and water vapor are produced as combustion products.

Propylene

isotactic polymer
■ Any polymer whose macromolecules have side chains positioned on the same side of a chain of carbon atoms

syndiotactic polymer
■ Any polymer whose macromolecules have alternately positioned side chains along a chain of carbon atoms

atactic polymer ■ Any polymer whose macromolecules have randomly positioned side chains along a chain of carbon atoms

Vinyl chloride

14.6-B POLYPROPYLENE

The cracking of propane produces methane, ethylene, and propylene. In the petrochemical industry, the propylene is separated from this mixture of gases and polymerized to manufacture the polymer called *polypropylene*. Its macromolecules have the repeating unit $\sim\!\!\sim\!CH_2-CH\!\sim\!\!\sim$.
$$\overset{|}{CH_3}$$

Polypropylene is used to manufacture commercial products ranging from silky fibers to rigid containers. Examples are outdoor tables and chairs, shatterproof glasses, artificial grass and turf, pipes, ropes, nets, twines, carpets, and carpet padding. The plastic items in motor vehicles are largely made from polypropylene. They include the dashboards, bumpers, carpeting, and occasionally, the upholstery.

By using a Ziegler–Natta catalyst to initiate polymerization, manufacturers have been able to produce polypropylene so that the branching methyl group is arrayed regimentally along the carbon–carbon backbone, giving rise to the following three types of polypropylene macromolecules:

■ **Isotactic polymer**, in which the methyl groups are all pointing in the same direction:

$$\sim\!\!\sim\!CH_2-CH-CH_2-CH-CH_2-CH-CH_2-CH-CH_2-CH-CH_2-CH\!\sim\!\!\sim$$
$$\quad\quad CH_3 \quad\quad CH_3 \quad\quad CH_3 \quad\quad CH_3 \quad\quad CH_3 \quad\quad CH_3$$

■ **Syndiotactic polymer**, in which alternate methyl groups point in opposite directions:

$$\quad CH_3 \quad\quad\quad\quad\quad CH_3 \quad\quad\quad\quad\quad CH_3$$
$$\sim\!\!\sim\!CH_2-CH-CH_2-CH-CH_2-CH-CH_2-CH-CH_2-CH-CH_2-CH\!\sim\!\!\sim$$
$$\quad\quad\quad\quad CH_3 \quad\quad\quad\quad\quad CH_3 \quad\quad\quad\quad\quad CH_3$$

■ **Atactic polymer**, in which the methyl groups are randomly oriented:

$$\quad CH_3 \quad\quad\quad\quad\quad CH_3$$
$$\sim\!\!\sim\!CH_2-CH-CH_2-CH-CH_2-CH-CH_2-CH-CH_2-CH-CH_2-CH\!\sim\!\!\sim$$
$$\quad\quad\quad\quad CH_3 \quad\quad\quad\quad\quad CH_3 \quad\quad CH_3 \quad\quad CH_3$$

These individual polypropylene types may be cast into shapes, drawn into sheets, or extruded into fibers, thus producing a range of diversified products.

Polypropylene is a recyclable polymer. The recycling symbol for polypropylene is an arrowed triangle enclosing the number 5, beneath which appear the letters *PP*.

PP

Polypropylene does not ignite easily. However, on exposure to intense heat, polypropylene thermally decomposes into a mixture of hydrocarbon vapors. This mixture catches fire and burns.

14.6-C POLY(VINYL CHLORIDE)

Vinyl chloride, or chloroethene, is a gas primarily manufactured by the catalytic dehydrochlorination of 1,2-dichloroethane.

$$ClH_2C-CH_2Cl(g) \longrightarrow CH_2=\overset{\displaystyle H}{\underset{\displaystyle Cl}{C}}(g) \;+\; HCl(g)$$

$$\text{1,2-Dichloroethane} \quad\quad\quad \text{Vinyl chloride} \quad \text{Hydrogen chloride}$$
$$\text{(Chloroethene)}$$

628   **Chapter 14** Chemistry of Some Polymeric Materials

It is a flammable gas that is subject to autopolymerization.

Short-term exposure to high levels of vinyl chloride in the air affects the central nervous system and can cause dizziness, drowsiness, and headaches. The gas is regarded as a known human carcinogen, because human exposure by inhalation has been affirmatively linked with the onset of liver angiosarcoma. The latency period for the onset of this cancer ranges from 15 to 40 years. By acknowledging its potential to cause cancer, the CPSC prohibits the use of vinyl chloride in self-pressurized consumer products. OSHA requires employers to limit the vinyl chloride concentration to which employees are exposed in the workplace at 1 part per million, averaged over an 8-hour workday.

Poly(vinyl chloride) is produced when vinyl chloride polymerizes. It is commonly recognized by its acronym, *PVC*. The unit that recurs in its macromolecules is the following:

$$\sim\!\!\sim\!\!CH_2CH\!\!\sim\!\!\sim$$
$$|$$
$$Cl$$

The presence of the chlorine atoms gives PVC some unique properties. For example, the data in Table 14.4 show that when PVC burns, less heat is emitted per mass compared with other common polymers that burn.

PVC is the polymer used in many home-construction products like floor tiles, lighting fixtures, vinyl panels and siding, conduits, and wall coverings. Figure 14.7 shows that PVC is also the polymer used to manufacture pipe and pipe fittings. PVC is the plastic insulation sheath used in virtually all modern electrical wiring. PVC is also encountered in imitation leather, shower curtains, upholstery material, vinyl raincoats, plastic packaging

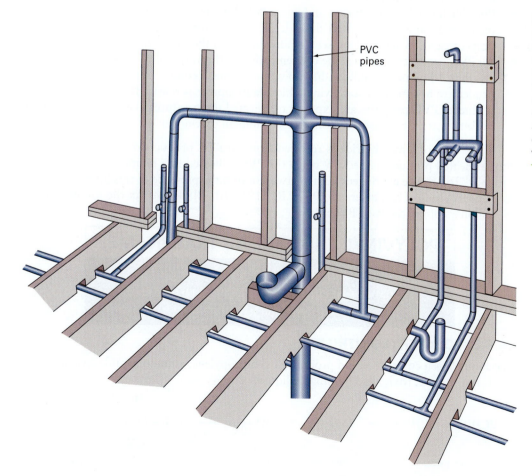

PVC pipes

FIGURE 14.7 PVC plumbing pipes often are used within bathrooms and kitchens for hot- and cold-water delivery to fixtures, to carry drainage and waste, and to vent odors. The PVC pipes usually are joined to copper or galvanized steel pipes with transition fittings.

materials, garden hoses, and medical products. Approximately 70% of the poly(vinyl chloride) produced in the United States is used in building construction materials. PVC copolymers are also commercially available in products such as films, fibers, sheeting, and moldings. Their mechanical properties vary from rigid to elastomeric.

PVC is recyclable. Its recycling symbol is an arrowed triangle enclosing the number 3, beneath which appears the letter *V*.

PVC is highly susceptible to degradation, which results in unsightly discoloring and loss of mechanical properties. For example, the vinyl upholstery in an automobile is initially soft and supple, but it becomes brittle and cracks as the automobile ages. As time passes, the plasticizer vaporizes, and the upholstery takes on an entirely new character.

Although PVC does not easily ignite, the same is not true of the plasticizers added to PVC products. The plasticizer most commonly added to PVC products is di(2-ethylhexyl) phthalate (DEHP). The ease with which PVC products ignite increases as more and more plasticizer is incorporated into them.

At elevated temperatures, PVC thermally decomposes and its decomposition products burn. Given the abundance of PVC products that firefighters are likely to encounter during a normal fire, the decomposition of poly(vinyl chloride) may constitute a potential health concern for the following reasons:

- Hydrogen chloride is produced when PVC burns. Hydrogen chloride poses the hazard of inhalation toxicity.
- Polychlorinated dibenzofurans and dibenzo-*p*-dioxins are produced during the incomplete combustion of poly(vinyl chloride). Exposure to minutely low dioxin concentrations causes a variety of illnesses in individuals and their offspring.
- All PVC products contain some concentration of the vinyl chloride monomer, usually less than 10 parts per million, that does not become a component of the polymer. It is prudent to assume that some vinyl chloride monomer is released to the environment whenever PVC products are exposed to elevated temperatures. This may represent a health concern, because the monomer is a human carcinogen.

Some level of vinyl chloride is released from all PVC products during its processing into fabricated products, especially when the PVC is heated to a temperature that causes it to melt. To provide employees with an awareness of the presence of this human carcinogen, OSHA requires employers at 29 C.F.R. §1910.1017(l)(4) to affix the following sign to containers of unprocessed PVC:

Acrylonitrile

POLYVINYL CHLORIDE (or TRADENAME)

CONTAINS VINYL CHLORIDE
VINYL CHLORIDE IS A KNOWN HUMAN CARCINOGEN

14.6-D POLYACRYLONITRILE

Acrylonitrile is a colorless liquid produced from propylene, ammonia, and oxygen. It is a volatile, toxic, and flammable liquid.

$$2CH_3CHC{=}CH_2(g) \ + \ 2NH_3(g) \ + \ 3O_2(g) \ \longrightarrow \ CH_2{=}\underset{CN}{\overset{H}{C}}(g) \ + \ 6H_2O(g)$$

Propylene   Ammonia   Oxygen   Acrylonitrile   Water

Polyacrylonitrile results when acrylonitrile is polymerized. The unit that recurs in its macromolecules is the following:

$$\sim\!\!\sim CH_2CH \sim\!\!\sim$$
$$|$$
$$CN$$

This synthetic polymer was the first to become commercially popular in the form of *acrylic fibers*, a term referring to those fibers composed of at least 85% by mass of acrylonitrile units. Today, acrylic fibers are used primarily in carpets and other textiles, because they are resistant to sunlight, quick-drying, and easy to launder. They are popularly known by the tradenames *Orlon* and *Acrilan*. Polyacrylonitrile is also used to manufacture plastics and nitrile rubber.

At fire scenes, hydrogen cyanide is produced when materials made of polyacrylonitrile undergo thermal decomposition. Life-threatening concentrations of hydrogen cyanide could be generated. Whether this is a significant factor in causing the deaths of firefighters is unknown. Nonetheless, it is prudent to assume that exposure to this toxic gas occurred, especially while fighting residential fires.

14.6-E POLY(METHYL METHACRYLATE)

Methyl methacrylate is a colorless liquid that is manufactured from acetone, hydrogen cyanide, sulfuric acid, and methanol. It is very volatile and highly flammable.

Several polyvinyl polymers are produced from the esters of acrylic acid and methacrylic acid. An example is the commercially popular poly(methyl methacrylate), which is frequently designated as *PMMA*. This polymer is produced by polymerizing methyl methacrylate. The repeating unit in PMMA is the following:

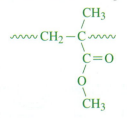

PMMA is a component of the plastic products known commercially as Plexiglas and Lucite. It is as clear as glass but can be manufactured as a transparent, translucent, or opaque material. Because PMMA is virtually unbreakable under normal conditions of stress and tension, it is useful in windshields, windows, and other products simultaneously requiring weather resistance, strength, and transparency. The largest use of PMMA is associated with the manufacture of displays and advertising signs, but the polymer is also used to make lighting fixtures, building panels, and plumbing and bathroom fixtures. The military uses PMMA in cockpit canopies, windows, gun turrets, and bombardier enclosures. It is also used as a component of latex and enamel paints, a drying oil for varnishes, and a finishing compound on leather.

14.6-F POLYACRYLAMIDE

Polyacrylamide is a water-soluble polymer produced by the polymerization of the monomer acrylamide, as follows:

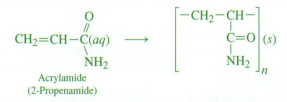

It is used primarily during paper manufacturing and water-treatment processes to coagulate the suspended solids in water. It is also used as a component of a sealant or grout used

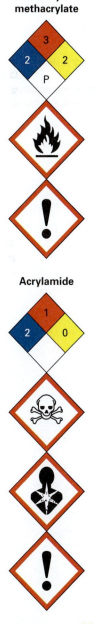

to repair leaking sewers. In 1991, EPA proposed to ban the manufacture, distribution, importation, and use of the acrylamide grout using the authority of TSCA. However, EPA withdrew the proposal and instead urged the adoption of management practices to abate the potential risks caused by exposure to acrylamide.

Polyacrylamide is nontoxic, but there are health concerns associated with exposure to its monomer. Acrylamide is both a neurotoxin and a probable carcinogen. Worker exposure may occur inadvertently because residual acrylamide remains in the polymer.

Health concerns related to acrylamide exposure have even been raised by nutritionists, because the substance forms at low concentrations during the high-temperature frying, roasting, and baking of foods made from plants, especially French fries and potato chips. FDA's position is that the acrylamide concentration in fried foods does not pose a health risk for most people.

As noted earlier, the monomer methyl methacrylate is produced when PMMA is exposed to heat. Within a fire environment, the monomer readily burns.

14.7 EPOXY RESINS

epoxy resin ▪ Any polymer produced by the condensation of a diol and an epoxide

In Section 13.2-N, we noted that Bisphenol-A is used in the polymer industry to manufacture **epoxy resins**. These are thermosetting polymers produced by the condensation of a diol and an epoxide. For example, Bisphenol-A and epichlorohydrin react as follows:

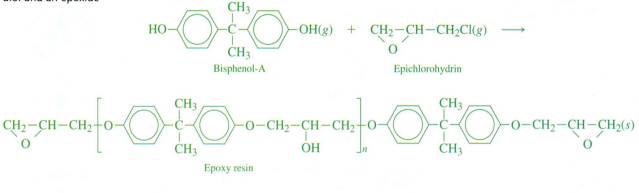

When the resin is cross-linked, the resulting polymers are especially hard, chemically resistant, and noncorrosive. They are the strongest adhesives known.

Epoxy resins are used in many commercial applications, including protective coatings for sports equipment, the hulls of ships, metal containers, and kitchen appliances.

14.8 FORMALDEHYDE-DERIVED POLYMERS

There are several formaldehyde-derived polymers used commercially. All are thermosetting polymers, and all are produced by condensation polymerization. The following three are noted briefly:

▪ *Phenol-formaldehyde.* This was the first formaldehyde-derived polymer to be discovered. It is best known by the trademark Bakelite, and has been used in molded electrical cases, adhesives, laminates, and varnishes. The molecular structure of its repeating unit is provided in Table 14.2.

▪ *Urea-formaldehyde.* As noted in Section 13.5-A, the major use of urea-formaldehyde is as the binder in particle board. The molecular structure of its repeating unit also is provided in Table 14.2.

■ *Melamine-formaldehyde.* This polymer is used to produce the hardest plastic items in commerce: laminated countertops like Formica; laminated flooring; kitchen cabinets; automotive topcoats; finishes for appliances and metal furniture; molded plastic tableware and kitchen utensils; and food containers. The molecular structure of its repeating unit is the following:

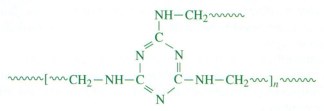

14.9 POLYURETHANE

A **urethane**, or **carbamate**, is an organic compound produced by the catalyzed reaction between an alcohol and an organic isocyanate. The chemical formula of a simple organic **isocyanate** is $R-N=C=O$. The production of a urethane may be illustrated by the following equation, where R and R' are arbitrary alkyl or aryl groups:

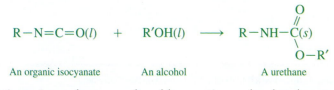

An organic isocyanate	An alcohol	A urethane

urethane (carbamate)
■ Any organic compound having the general chemical formula

$$\begin{matrix} & O \\ & \parallel \\ R-NH-C \\ & \diagdown \\ & O-R' \end{matrix}$$

isocyanate ■ An organic compound whose general chemical formula is $R-N=C=O$

polyurethane ■ Any polymer produced by the condensation of a glycol and an organic diisocyanate

A **polyurethane** is a polymer produced by reacting a glycol and an organic diisocyanate. Organic diisocyanate molecules have two isocyanate groups. A popular polyurethane product is a solid foam produced through the application of a foam-blowing agent to the reaction mixture of toluene 2,4-diisocyanate and ethylene glycol. The equation for this polymerization reaction is written as follows:

$$O=C=N-\text{(ring)}-N=C=O(l) \quad + \quad HO-CH_2CH_2-OH(l) \longrightarrow$$

Toluene 2,4-diisocyanate	Ethylene glycol

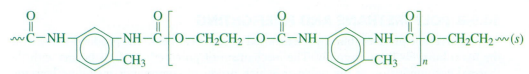

An example of a polyurethane

The heat released during the polymerization reaction causes the blowing agent to vaporize. Bubbles of vapor subsequently are captured within the viscous liquid as it polymerizes and expands. A polyurethane foam results when the frothy mixture solidifies.

Several commercially available polymers are composed of macromolecules in which the following group of atoms is repeated along their molecular chains:

$$\begin{matrix} & O \\ & \parallel \\ -C \\ & \diagdown \\ & NH- \end{matrix}$$

polyamide ■ Any polymer composed of repeating amide groups along a molecular chain in the following manner

$$\begin{matrix} & O \\ & \parallel \\ -C \\ & \diagdown \\ & NH- \end{matrix}$$

These are called **polyamides**. Nylon 6,6 and the polyurethanes are examples of polyamides.

14.9-A COMMERCIAL USES OF THE POLYURETHANES

Emergency responders are usually familiar with at least one use of polyurethane: to temporarily seal storm and sewer drains. The polyurethane products that are available commercially for this purpose are either sheets or sealants that can provide a barrier over entrance grates. The sheets are simply laid over the grates, whereas the foam is discharged under pressure from a cylinder into the grates and allowed to harden. Both products prevent contaminated water run-off and spills and leaks of liquid hazardous materials from entering the drainage systems.

Far more commonly, polyurethane foams are encountered as rigid or flexible plastics that have been manufactured with a wide range of additives like stabilizers, dyes, fire retardants, and fungicides. The rigid and flexible types of polyurethane are used primarily for the following purposes:

- The rigid foam is primarily used as insulation, sound-deadening boards, and wall panels. The insulation is conveniently sprayed on the surface of concrete, between wall and roof joists, and around window and door frames. It has also been used to insulate external fuel tanks on space shuttles.
- The flexible polyurethane foams are used in carpet padding and bedding, pillows, furniture, shoe soles, medical splints, and automobile-seat cushioning. In these foam products, the softness and resiliency of polyurethane is the desirable feature.

SOLVED EXERCISE 14.3

Which toxic gases are produced when piles of polyurethane jackets smolder?

Solution: Polyurethane macromolecules have hundreds or thousands of a unit such as the following:

Because there is an abundance of nitrogen atoms in the unit, when the polyurethane jackets smolder, the toxic gases produced are hydrogen cyanide, carbon monoxide, nitric oxide, and nitrogen dioxide.

14.9-B POLYURETHANE AND FIREFIGHTING

All polyurethane products burn when exposed to sufficient heat. The products made from the foam burn easily and rapidly. The mechanism of polyurethane fires begins with the thermal decomposition of the polymer, which produces compounds including benzene, toluene, acetaldehyde, acetone, propane, alkenes, and hydrogen cyanide, all of which readily ignite upon exposure to an ignition source. The formation of hydrogen cyanide during polyurethane fires may pose an inhalation hazard. One polyurethane combustion study showed that the maximum yield of hydrogen cyanide per gram of polyurethane foam ranged from 0.37 to 0.93 milligrams under nonflaming conditions and from 0.5 to 1.02 milligrams under flaming combustion.[1]

Polyurethane fires may be extinguished effectively with the application of water, but because the burning products retain considerable heat, it is essential to check for total fire extinguishment to prevent their reignition.

---

[1] D. S. Sklarew and D. J. Hayes, "Trace nitrogen-containing species in the offgas from 2 oil shale retorting processes," *Environ. Sci. Technol.*, Vol 18 (1984), pp. 600–603.

It has been well acknowledged that caution must be exercised when fighting fires involving polyurethane products due to the potential for inhaling nitrogen dioxide. Prodigious quantities of this gas are produced during polyurethane fires. Its presence poses an increased risk of inhalation toxicity to building occupants and firefighters. This risk is considered so great that safety engineers recommend the use of a warning label such as the following on polyurethane building products:

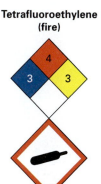

Tetrafluoroethylene (fire)

FIRE HAZARD WHEN USED INSIDE BUILDINGS

USE ONLY ON EXTERIOR APPLICATIONS OR, WHEN CONFINED, BETWEEN WALLS OF MATERIALS CONSIDERED ACCEPTABLE FOR CONSTRUCTION

The smoldering and open-flame ignition of consumer products containing polyurethane foam is a primary cause of death by fire within the home. In particular, the polyurethane foam in upholstered furniture constitutes a serious fire hazard. In just 4 minutes, a sofa fire may engulf a living room in flames and produce smoke and toxic gases. In this short time, the temperature of the surroundings may elevate to a staggering 1400°F (760°C). Polyurethane foam distributors generally provide warning notices similar to the following to upholstered furniture manufacturers.

WARNING

All Polyurethane Foam Can Burn!

In case of fire, serious injury or death can result from extreme heat, rapid oxygen depletion, and the production of toxic gases. When ignited, polyurethane foam, like other organic materials, may burn rapidly and generate thick dark smoke and toxic gases leading to confusion, incapacitation, and even death.

Do not expose polyurethane foam to any intense radiant heat or open flames, such as space heaters, open-burning operations, cigarettes, welding operations, naked lights, matches, electric sparks, or other intense heat sources.

Depending upon the intended use of the polyurethane foam, suitable warnings should be passed on to the ultimate product users.

To reduce their potential for ignition, the polyurethane products encountered in the contemporary market have been formulated with fire retardants like the polybrominated diphenyl ethers (Section 13.4-D) and chlorinated organophosphates, or otherwise treated to reduce their ease of combustion. CPSC has expressed especial concern about the use of tris(1,3-dichloroisopropyl) phosphate, or TDCPP, as a plasticizer and fire retardant in foam padding and other polyurethane products to which infants and toddlers often are exposed. Infants and toddlers are especially vulnerable to TDCPP exposure because they spend significant time in contact with the treated foam in mattresses, car seats, and pillows. Exposure to TDCPP is considered a health risk because studies link exposure with the development of cancer. In response to this potential health concern, New York prohibited the sale of products containing TDCPP intended for use by infants and toddlers after December 1, 2013.[2]

---

[2]Cancer Hazard Assessment Branch, *Evidence on the carcinogenicity of tris(1,3-dichloroisopropyl phosphate)*, California Environmental Protection Agency (2011).

TDCPP has the following molecular structure:

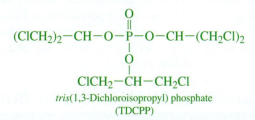

$$(ClCH_2)_2-CH-O-\overset{\overset{\displaystyle O}{\|}}{\underset{\underset{\displaystyle O}{|}}{P}}-O-CH-(CH_2Cl)_2$$

$$ClCH_2-CH-CH_2Cl$$

tris(1,3-Dichloroisopropyl) phosphate
(TDCPP)

Its acceptable daily intake concentration was established by CPSC as 0.005 mg/kg of body weight. Yet, polyurethane products often contain TDCPP at concentrations that exceed this value.

14.10 HEAT- AND FIRE-RESISTANT POLYMERS

Research chemists have devoted considerable time to learning how to produce and manufacture polymers that are heat and fire resistant. Their research efforts have resulted in the discovery of several synthetic polymers with these sought-after qualities. Heat- and fire-resistant polymers are very desirable for potential use in certain types of commercial products, especially building construction materials and products for home use.

14.10-A POLY(TETRAFLUOROETHYLENE)

Tetrafluoroethylene (nonfire)

Tetrafluoroethylene is a gas produced from chloroform and hydrofluoric acid. It is used to manufacture poly(tetrafluoroethylene) (PTFE), whose commercial name is *Teflon*. The recurring unit in its macromolecules is $\sim\sim CF_2-CF_2\sim\sim$.

Teflon is well known as the polymer used to produce nonstick cookware. It also is used to produce virtually indestructible tubing, gaskets, valves, and cable insulation. Chemists often use Teflon tubing in laboratories, and cardiologists use it as artificial veins and arteries. It is also used in pacemakers, dentures, and many other manufactured products. Teflon is also the polymer used in the waterproof fabric called *GORE-TEX*, which is used to manufacture the sportswear worn by campers, skiers, and golfers.

Teflon is extraordinarily heat resistant. Teflon products perform well when exposed to temperatures ranging from −400 to 482°F (−240 to 250°C). Nonstick cookware coated with Teflon can be heated to temperatures in excess of 932°F (500°C) without burning. Teflon is also extraordinarily unreactive when exposed to hot corrosive acids. For this reason, Teflon is coated on the interior walls of tanks intended for storing acids.

14.10-B NOMEX

Nomex is the tradename of a polyamide produced by the condensation of *m*-diaminobenzene and isophthaloyl chloride as follows:

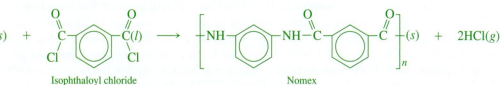

m-Diaminobenzene   Isophthaloyl chloride   Nomex

From the fibers of this polymer, a heat-resistant fabric is produced. Emergency responders are familiar with this fabric because it often is used in the uniforms worn by firefighters and racecar drivers. The fabric provides an extra element of thermal protection to its users, because Nomex carbonizes and thickens when exposed to intense heat. This increases the protective barrier between the heat source and the skin, thus minimizing the potential for burn injuries.

14.10-C KEVLAR

Kevlar is the commercial name of the polyamide produced by the condensation of *p*-diaminobenzene and terephthaloyl chloride.

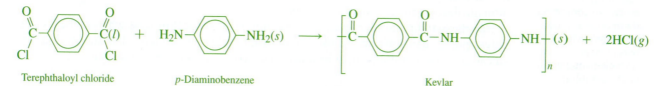

Terephthaloyl chloride   *p*-Diaminobenzene   Kevlar

Kevlar is flameproof and may be drawn into fibers having a strength five times stronger than steel. Given this astounding resiliency, fabrics made from Kevlar fibers are used as the reinforcement in bullet-resistant vests and helmets. A 16.4-pound (7.5-kg) vest lined with Kevlar and ceramic plates can stop armor-piercing bullets shot from high-power rifles. A 4-pound (1.8-kg) helmet lined with up to 24 layers of Kevlar is approximately 40% more resistant to shrapnel than the steel helmets formerly used by the military. Kevlar has also been reinforced with plastic resins to provide protection against multiple hazards.

No single product compares with Kevlar in terms of the number of lives saved through its use. Thousands of active police officers are alive today because they wore Kevlar vests and helmets during the line of duty.

Aside from its use in military and police body armor, Kevlar is also used in rip-resistant jeans, protective gloves, boots, skis, hockey sticks, golf balls, ropes, cables, boat hulls, aircraft-structural parts, reinforced-suspension-bridge structures, and flame-resistant mattresses. Some tire manufacturers now are using Kevlar to reinforce the tread of radial automobile tires and make them resistant to torsion, tension, and heat. By using Kevlar, tire manufacturers have produced off-road tires that are nearly indestructible when driven over rough terrains. Because Kevlar is stable at high temperatures, it is also used in some brands of the protective clothing worn by firefighters.

Styrene

14.11 RUBBER AND RUBBER PRODUCTS

The term **rubber** refers to any of the natural or synthetic polymers having two main properties: deformation under strain and elastic recovery after vulcanization (described in Section 14.10-A). These rubberlike polymers are called elastomers.

rubber ■ Any natural or synthetic polymer that is simultaneously elastic, airtight, water-resistant, and long-wearing

14.11-A NATURAL RUBBER

Natural rubber is produced from natural latex, a white fluid that exudes from cuts in the bark of the South American rubber tree *Hevea brasiliensis*. The natural latex consists of approximately 30% to 35% by mass of *cis*-1,4-polyisoprene. Its general chemical formula is $[\sim\sim CH_2-C=CH-CH_2\sim\sim]_n$ where n is a large integer. When geometrical isomerism is
$\hspace{3.3cm}|$
$\hspace{3.3cm}CH_3$
considered, the macromolecular structure is represented as follows:

natural rubber ■ The polymer produced from the latex that exudes from cuts in the bark of the South American rubber tree

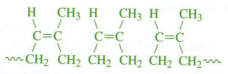

This structure illustrates that the methyl groups are oriented in one direction about the carbon–carbon double bonds. Nature selectively produces only the *cis*-isomer of 1,4-polyisoprene.

By itself, natural rubber is not entirely desirable for production of commercial products. It is soft and sticky, especially in warm climates. In 1839, however, Charles Goodyear accidentally discovered that these undesirable features could be eliminated by heating the natural rubber with elemental sulfur. This discovery, called **vulcanization**, led to the ultimate development of synthetic rubbers.

At the macromolecular level, vulcanization is a chemical process wherein the *cis*-1,4-polyisoprene chains are cross-linked with disulfide ($-S-S-$) bonds. The structure of the vulcanized macromolecule may be illustrated two-dimensionally as follows:

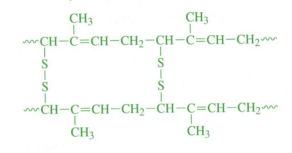

vulcanization ■ The process of converting thermoplastic rubbers to thermosetting rubbers by heating them with elemental sulfur or certain sulfur-bearing compounds to produce disulfide cross-linking bonds within segments of their macromolecules

---

SOLVED EXERCISE 14.4

Why do foam rubber products catch fire easily inside a hot clothes dryer or under a heating blanket, whereas vehicular rubber tires are comparatively difficult to ignite even on exposure to intense heat?

Solution: Foam rubber products are produced by blowing and entrapping an inert gas within the complex structure of rubber until it hardens. This process causes an expansion of the rubber and an increase in its surface area. By contrast, an inert gas is not entrapped within the rubber used for the production of rubber tires. According to the principles outlined in Section 5.5-B, an increase in the surface area of the reactants increases the reaction rate. Because the rubber in foam rubber products has a greater surface area than the rubber in rubber tires, foam rubber products ignite and burn at a comparatively faster rate than rubber tires.

Rubber manufacturers usually accelerate the rate at which they vulcanize rubber through the addition of substances like tetramethylthiuramdisulfide, zinc diethyldithiocarbamate, diphenylthiourea, or peroxo-organic compounds.

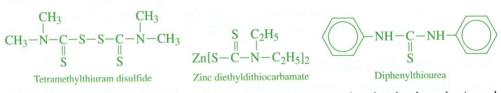

Tetramethylthiuram disulfide   Zinc diethyldithiocarbamate   Diphenylthiourea

vulcanized rubber ■ Any natural or synthetic rubber that has been vulcanized

Each resulting product is called a **vulcanized rubber**. It is tougher, harder, less plastic and sticky, and more elastic compared to unvulcanized rubber. Vulcanized rubber is also capable of retaining a firm shape over a relatively wide temperature range. Vulcanization also permits the casting of natural rubber into unique shapes such as automobile tires. A mixture of rubber, carbon, and sulfur is inserted into a mold that, when sealed and heated, produces a tire carcass.

14.11-B SYNTHETIC RUBBERS

synthetic rubber ■ Any polymer produced from substances other than, or in addition to, the latex from the South American rubber tree and having properties similar to those of natural rubber

Natural rubber still accounts for approximately 35% of the demand for rubber in the United States. Today, however, the world no longer relies solely on the latex from rubber trees for its rubber. Chemists have synthesized elastomers called **synthetic rubbers**. These have physical properties similar to or better than those of natural rubber. We briefly note the properties of five synthetic rubbers: *cis*-1,4-polybutadiene rubber, styrene–butadiene rubber, acrylonitrile–butadiene rubber, neoprene, and polysulfide rubber.

■ cis-*1,4-Polybutadiene* is the simplest synthetic rubber. It is produced by the polymerization of 1,4-butadiene using a Ziegler–Natta catalyst. In the two-dimensional segment of the macromolecule shown below, the hydrogen atoms are located on the same side of the carbon–carbon double bonds.

$$\text{CH=CH}\quad\text{CH=CH}\quad\text{CH=CH}\quad\text{CH=CH}\quad\text{CH=CH}$$
$$\text{\textasciitilde\textasciitilde CH}\quad\text{CH}_2\quad\text{CH}_2\quad\text{CH}_2\quad\text{CH}_2\quad\text{CH}_2\text{\textasciitilde\textasciitilde}$$

cis-1,4-Polybutadiene has many of the properties of natural rubber and may be vulcanized with elemental sulfur. It is used in the production and manufacture of vehicular tires.

■ **Styrene–butadiene rubber** (SBR) was first known as *GRS* (government rubber styrene), because it was produced on behalf of the U.S. military during World War II. Because it is a copolymer of 1,3-butadiene and styrene, it became known outside government circles as *Buna S* (Bu for butadiene, Na for sodium, and S for styrene). (The production reaction is catalyzed by sodium.) The repeating unit of SBR is represented as follows:

styrene–butadiene rubber ■ The synthetic rubber produced by the polymerization of styrene and butadiene

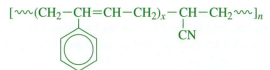

$$[\text{\textasciitilde\textasciitilde}(\text{CH}_2-\text{CH=CH}-\text{CH}_2)_x-\text{CH}-\text{CH}_2\text{\textasciitilde\textasciitilde}]_n$$

SBR is vulcanized using elemental sulfur. Because it resists wear more than any other synthetic rubber, SBR is used to manufacture most tire treads. It also is the major component of many adhesives.

■ Buna N rubber is a copolymer of butadiene and acrylonitrile (*N* stands for *nitrile*). It is also known as acrylonitrile–butadiene rubber. The repeating unit in Buna N rubber is the following:

$$[\text{\textasciitilde\textasciitilde}(\text{CH}_2-\text{CH=CH}-\text{CH}_2)_x-\text{CH}-\text{CH}_2\text{\textasciitilde\textasciitilde}]_n$$
$$\text{CN}$$

neoprene ■ A synthetic elastomer composed of cis-1,4-polychloroprene

Buna N rubber has the unique feature of withstanding heat up to 350°F (177°C); other synthetic rubbers soften, melt, or burn at lower temperatures.

■ Polychloroprene is more commonly known as **neoprene**. It is a chlorinated rubber produced by heating chloroprene, or 2-chloro-1,3-butadiene, a known human carcinogen. The simplest structure of the repeating unit in neoprene is the following:

2-Chloro-1,3-butadiene

$$[\text{\textasciitilde\textasciitilde CH}_2-\text{C=CH}-\text{CH}_2\text{\textasciitilde\textasciitilde}]_n$$
$$\text{Cl}$$

Neoprene is vulcanized with zinc oxide, during which single neoprene strands cross-link with oxygen atoms and replace some of the chlorine atoms. Macromolecular segments of the vulcanized neoprene have two-dimensional structures such as the following:

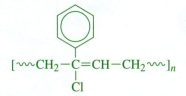

$$\text{Cl}$$
$$\text{\textasciitilde\textasciitilde CH}_2-\text{C=C}-\text{CH}_2\text{\textasciitilde\textasciitilde}$$
$$\text{O}$$
$$\text{\textasciitilde\textasciitilde CH}_2-\text{C=C}-\text{CH}_2\text{\textasciitilde\textasciitilde}$$
$$\text{Cl}$$

Neoprene rubber does not possess the resilience necessary for use in tires, but it is stretchable, waterproof, and chemically resistant to petroleum products and ozone. This feature makes it suitable for specialized uses such as roofing tile, flexible hose, wetsuits, running shoes, computer-mouse pads, iPod holders, and other products.

■ Thiokol is also known as polysulfide rubber. Although there are several types, the most common polysulfide rubber has repeating units like the following:

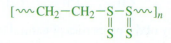

It is used primarily to manufacture hoses for inserting gasoline and oil into containers.

14.11-C CONSUMER PRODUCTS PRODUCED FROM SYNTHETIC RUBBERS

Synthetic rubbers vulcanized with sulfur or sulfur compounds are used to manufacture a variety of products, some of which are illustrated in Figure 14.8. Their manufacture requires controlling the amount of sulfur used for vulcanization, as demonstrated by the following:

■ Natural rubber mixed with only 3% sulfur by mass is soft and elastic. It is used for manufacturing inner tubes and rubber bands.

■ With additional sulfur, vulcanized rubber becomes hard and loses some of its elasticity. Natural rubber having up to 10% sulfur by mass is used to manufacture automobile tires.

■ When rubber has been vulcanized so that it contains 68% sulfur by mass, it becomes a black solid called **ebonite**, or *hard rubber*. It is used to make black piano keys and the casings for lead–acid storage batteries.

Other substances that are components of rubber formulations include the following:

■ Carbon black is often added as a reinforcing filler to make rubber stronger, abrasive, and easier to elongate, and to prevent it from breaking and tearing.

■ Silica fillers are added to provide surface abrasion to rubber intended for use in vehicular tires.

ebonite ■ A form of hard rubber used primarily to make the casings for automobile storage batteries

FIGURE 14.8 Some common products made either directly from rubber or from other polymers containing rubber additives.

■ **Distillate-aromatic-extract oils** (DAE oils) are added to soften rubber before its use, i.e., they act as plasticizers during processing. Their addition to vehicular tire formulations also provides improved performance characteristics like "wet grip" to road surfaces. DAE oils are products of the petroleum refining industry. They consist of mixtures of high-boiling compounds including polynuclear aromatic hydrocarbons. In 2010, DAE oils were banned in tires intended for use in the European Union; hence, alternative products having similar properties but lacking the cancer-causing potential most likely will be substituted for them.

■ A specified amount of previously used rubber (reclaimable rubber) often is recycled by adding it to a new batch of rubber.

■ Depending on the projected use of the rubber product, air or ammonium carbonate also may be mixed into the rubber latex. When this mixture subsequently is heated, the ammonium carbonate decomposes to ammonia and carbon dioxide. These gases become entrapped within the complex rubber structure as it hardens. The product, called **foam rubber**, is used to manufacture cushions and mattresses.

■ In the warm climates of the southwestern states, previously used rubber is pulverized and mixed with hot oil, sodium hydroxide, rock, and sand to produce **rubberized asphalt**. It is used for resurfacing roadways in much the same way as traditional asphalt is used elsewhere. Rubberized asphalt cannot be used successfully for the same purpose in areas of the United States that experience cold climates during the winter months, because it does not survive the impact of chains and snowplows.

14.11-D RESPONDING TO INCIDENTS INVOLVING THE BURNING OF RUBBER

To understand the nature of the substances produced when rubber products burn, it is necessary to recollect the general features of their chemical formulations. During the combustion process, the constituents of these formulations combine with atmospheric oxygen in the following ways:

- As they burn, rubber products vulcanized with sulfur or sulfur-bearing compounds produce carbon monoxide, sulfur dioxide, and water vapor.
- The smoke associated with rubber fires is extraordinarily dense and black, as in the scene depicted in Figure 14.9. Both features of these fires are linked with the presence of carbon black in the rubber formulation.

FIGURE 14.9 A fire within a massive pile of vehicle tires may be initiated by a lightning strike. As the tires burn, the smoke that evolves is intensely dense and black. To bring this type of fire under control, the use of special fire extinguishers generally is required. (*Courtesy of Pyrocool Technologies, Inc., Monroe, Virginia.*)

- The smoke associated with rubber tires is highly toxic, not only because of the presence of toxic gases like carbon monoxide and sulfur dioxide, but also due to the presence of the PAHs added in DAE oils.
- As neoprene burns, carbon monoxide, water vapor, and hydrogen chloride are produced.
- As polysulfide rubber burns, carbon monoxide, sulfur dioxide, and water vapor are produced.

Because the atmosphere near burning rubber consists of a mixture of these gases, the toxic environment of the scene must be acknowledged when selecting a proper firefighting strategy. The dense, black smoke associated with rubber fires is composed largely of airborne carbon particles. Because soot is a known human carcinogen, breathing it could initiate long-term adverse health effects. To prevent or reduce fatalities among firefighters, the use of self-contained breathing apparatus always is warranted.

Nature of Polymerization

1. Write the chemical formula for macromolecular segments of each of the following polymers:
   (a) polyethylene having eight recurring units
   (b) polypropylene having five recurring units
   (c) poly(methyl methacrylate) having three recurring units
2. Poly(vinylidene chloride) is produced by the addition polymerization of vinylidene chloride, or 1,1-dichloroethene. Write the stepwise equations that illustrate how this addition polymerization occurs.

Autopolymerization

3. DOT prohibits shippers from offering acrolein for transportation unless it has been "stabilized." DOT then denotes its proper shipping name as "acrolein, stabilized."
   (a) What specific property of acrolein must be stabilized?
   (b) Use Appendix C to determine the shipping description that DOT requires a shipper to enter on the accompanying waybill when offering 2500 pounds of acrolein for transportation by rail tankcar.
   (c) Determine the markings that DOT requires the carrier to post on both sides and ends of the tankcar.
   (d) Determine the placards that DOT requires carriers to post on the tankcar.

Combustion of Polymers

4. Identify the combustion products that result when polymers having each of the following recurring units burn in air:

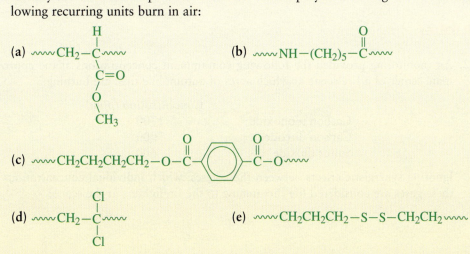

Vegetable and Animal Fibers

5. What thermal decomposition products are produced when a woolen sweater is exposed to intense heat?

6. Which symbol in Section 11.6 is typically affixed to the label on linen garments manufactured in the United States?

Vinyl Polymers

7. Use the information in Table 14.4 to determine which of the following requires the application of a larger volume of water for effective total extinguishment: 1000 pounds (454 kg) of burning polyethylene or the same mass of burning polystyrene?

8. *Dacron* fabrics are produced from poly(ethylene terephthalate) fibers. The recurring unit in the macromolecules of this polyester is the following:

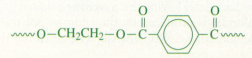

Identify the combustion products that are produced when Dacron fabrics burn.

9. Poly(vinylidene dichloride) decomposes when heated to a temperature ranging from 437 to 527°F (225 to 275°C). The recurring unit in poly(vinylidene dichloride) macromolecules is $\sim\text{CH}_2-\text{CH}-\text{CH}_2-\text{CCl}_2\sim$.
   $\quad\quad\quad\quad\quad\quad\quad\quad\quad\quad\quad\quad\quad\quad\;\;\mid$
   $\quad\quad\quad\quad\quad\quad\quad\quad\quad\quad\quad\quad\quad\;\;\text{Cl}$

   (a) Identify the toxic gases that are most likely to be produced when the polymer decomposes.
   (b) In what ways could the presence of these gases at a fire scene adversely affect the health of emergency responders?

10. The chemical analysis of an ash collected from a municipal waste incinerator revealed a 2,3,7,8-tetrachloro-*p*-dioxin concentration of 90 µg/g. How is the presence of 2,3,7,8-TCDD in the ash most logically explained?

Heat- and Fire-Resistant Polymers

11. Why are Kevlar fibers incorporated into the body armor used by emergency responders?

Rubber

12. Air monitoring provided the following contaminant concentrations at the approximate center of a fire scene at which acres of automobile tires are burning:

	Concentration (ppm)
Carbon monoxide	1700
Carbon dioxide	7000
Sulfur dioxide	600

Ignoring synergistic effects between these gases, which individual concentrations of these gases are considered life threatening to the firefighters at the scene?

CHAPTER

15

Chemistry of Some Explosives

Courtesy of Eugene Meyer.

KEY TERMS

black powder, *p. 660*

blasting agent, *p. 652*

blasting cap, *p. 651*

booster, *p. 654*

brisance, *p. 649*

bursting charge, *p. 654*

chemical explosive, *p. 646*

classification code, *p. 657*

compatibility group, *p. 656*

detonation, *p. 649*

detonation velocity, *p. 649*

detonator, *p. 654*

detonation temperature, *p. 649*

dynamite, *p. 663*

EX-number, *p. 656*

explosive, *p. 646*

explosive article, *p. 648*

explosive substance, *p. 648*

explosive train, *p. 654*

explosophore, *p. 652*

forbidden explosive, *p. 656*

guncotton, *p. 665*

high explosive (detonating explosive), *p. 650*

igniter, *p. 653*

improvised explosive device, *p. 647*

initiator (initiating explosive; primer), *p. 676*

low explosive (deflagrating explosive), *p. 650*

magazine, *p. 654*

nitramine, *p. 669*

nitrate ester, *p. 661*

plastic explosive, *p. 670*

preferred route (preferred highway), *p. 660*

primary explosive, *p. 651*

propellant, *p. 654*

secondary explosive, *p. 651*

sensitivity, *p. 650*

small-arms ammunition, *p. 666*

smokeless powder, *p. 666*

TNT equivalent, *p. 668*

vasodilator, *p. 662*

OBJECTIVES

- Associate the physical and health hazards of the explosives noted in this chapter with the information provided by their hazard diamonds and GHS pictograms.
- Distinguish between the combustion and the detonation of an explosive.
- Distinguish between an explosive substance and an explosive article.
- Describe the chemical nature of a blasting agent.

- Identify the function of each component of a round of artillery ammunition.
- Identify the general nature of ATF's regulations that apply to parties who are licensed to store commercial explosives.
- Identify the chemical composition of black powder and note the general nature of the oxidation-reduction reaction that occurs when they are activated.
- Identify the composition of the three commercial types of dynamite.
- Compare the detonation and deflagration of nitrocellulose.
- Identify why explosive articles containing either TNT or cyclonite are chosen by the military for use during warfare more frequently than other explosives.
- Provide the general nature of a plastic explosive.
- Identify the labels, markings, and placards that DOT requires on packaging of explosives and the transport vehicles used for their shipment.
- Identify the response actions to be executed when explosives are released from their packaging into the environment.

explosive ■ For purposes of DOT regulations, a substance or article, including a device, that is designed to function by explosion, or that, by chemical reaction with itself, is able to function in a similar manner even if not designed to function by explosion

chemical explosive
■ Any compound or mixture of compounds that undergoes rapid, self-propagating, heat-producing decomposition when subjected to heat, impact, friction, or shock

An **explosive** is a chemical or nuclear material that can be initiated to undergo a rapid, self-propagating decomposition that results in the formation of more stable substances, the liberation of heat, and the development of a sudden-pressure effect or blast by the action of heat on the gases produced during the decomposition. Our interest in this chapter is **chemical explosives**. We examine nuclear explosives in Chapter 16.

Chemical explosives generally are encountered as components of each of the following:

- The ammunition used for hunting and other sporting activities
- The charges implanted during the mining of coal and ores, drilling of oil wells, tunneling through mountains, clearing land, loosening of underground rock formations, creating firebreaks, and imploding unwanted structures. A step in the mining of coal, for example, includes the use of explosives to blast walls of rock and rubble into manageably sized pieces in order to expose a vein of coal.
- The artillery and munitions used during wartime to destroy cities, sink ships, and disable the enemy

Unwanted incidents involving explosives once were associated primarily with events in which the procedures designed to safely transport, store, and handle these materials had not been rigorously followed. These incidents formerly were rare compared with those linked with other hazardous materials.

Explosives have been the active components of weapons of mass destruction most commonly chosen by terrorists for their misdeeds. Some examples are the following:

■ We previously noted in Figure 11.1 that an explosive mixture of ammonium nitrate and fuel oil (ANFO) was used in 1995 during an act of domestic terrorism to destroy the Oklahoma City Federal Building.

■ In 1988, terrorists downed Pan Am Flight 103 over Lockerbie, Scotland, by detonating an explosive embedded within a suitcase. In this incident, 259 passengers and crew were killed along with another 11 people on the ground.

■ In the same year, nearly simultaneous car bombings attributed to al-Qaida destroyed the U.S. embassies in Nairobi, Kenya, and Dar es Salaam, Tanzania, killing 231 people, including 12 Americans, and wounding more than 5,000.

■ In 2000, al-Qaida homicide terrorists rammed an explosives-laden boat into the destroyer USS *Cole* in Yemen, killing 17 American sailors.

■ Since 2000, international terrorists have also used explosives to carry out acts of violence in Afghanistan, Britain, Colombia, Djerba, India, Indonesia, Iraq, Israel, Kenya, Lebanon, Morocco, Pakistan, Philippines, Russia, Saudi Arabia, Spain, Turkey, and elsewhere.

■ During the Iraqi War, we first heard news reports concerning homicide terrorists who were willing to commit suicide by detonating improvised explosive devices strapped to their bodies. An **improvised explosive device, or IED,** is any explosive article that has been fabricated and designed to destroy or incapacitate personnel or vehicles during unconventional warfare or terrorist events.

improvised explosive device (IED) ■ Any explosive device that has been fabricated and designed to destroy or incapacitate personnel or vehicles

■ Following the U.S. involvement in Iraq, IEDs continue to be used against Iraqi civilians by insurgents. Their use is also evident against the American military by Afghan insurgents. The IEDs used by the Afghans are homemade bombs primarily containing ANFO (Section 11.12). The Afghans obtain fertilizer-grade calcium ammonium nitrate that has been smuggled into Afghanistan from Pakistan. The calcium ammonium nitrate is a mixture of ammonium nitrate and limestone. They remove the limestone, mix the ammonium nitrate with fuel oil, and pack the mixture into jugs or other containers for use as roadside bombs that can be detonated remotely.

■ Two ethnic Chechens detonated explosives near the finish line of the Boston Marathon, killing three people, injuring over 260 other individuals, and putting the nation into panic mode.

In the United States, the Bureau of Alcohol, Tobacco, Firearms and Explosives (ATF) is charged with the responsibility of preventing terrorism, reducing violent crime, and protecting the nation. It requires persons and facilities producing, selling, transferring, or purchasing explosives to obtain a license or permit that grants them permission to perform these activities. The bureau also provides periodically a list of chemical explosives to which the licensing and permitting apply.[1]

Explosive production and manufacturing are legal practices only when the party has been issued a license by ATF; otherwise, they are federal crimes. Emergency responders who encounter illicit explosive production and manufacturing should immediately report the activity to law enforcement officials and DHS, and treat the pertinent area as a crime scene.

The Department of Commerce regulates the export and re-export of explosives. A "re-export" is a shipment or transmission subject to the certain regulations from one foreign county other than the United States to another foreign county. Specifically, licensees are authorized to export or re-export explosives only to certain countries identified by the department, and consistent with the terms of the issued export license.

The Department of Homeland Security (DHS) has authority to issue regulations governing the security of high-risk chemical facilities, including any facility that possesses explosives. DHS also registers persons that operate facilities producing, selling, transferring, and purchasing explosives.

OSHA minimally requires the manufacturers, distributors, and importers of explosives to post the GHS explosive pictogram on the labels of explosives in hazard class divisions 1.1, 1.2, 1.3, and 1.4, in addition to a signal word and appropriate hazard and precautionary statements.

OSHA does not require the posting of any pictogram on the labels of packaging containing an explosive in hazard class divisions 1.5 or 1.6.

Individuals who use explosives for peaceful or wartime purposes must receive specialized training in their handling. Although such training requires more rigorous instruction than may be provided here, we learn in this chapter about the chemical composition of some explosives, their different types, and the procedures that are recommended for responding to mishaps involving them.

---

[1]See, *e.g., Federal Register*, Vol. 75 (January 8, 2010), pp. 1085–1087.

15.1 GENERAL CHARACTERISTICS OF EXPLOSIVES

There are two general types of explosives:

explosive substance
■ For purposes of DOT regulations, any substance such as black powder or smokeless powder contained in packaging other than an article

explosive article ■ For purposes of DOT regulations, any commercially available item such as a detonator or flare containing an explosive substance as a component

■ **Explosive substances** are unique compounds that readily detonate or can be made to detonate.
■ **Explosive articles** are manufactured items that contain explosive substances as components, like cartridges and military bombs. Some specific examples are provided in Table 15.1. Under ordinary circumstances, the chemical substance in an explosive article is intentionally initiated to detonate by mechanical, electrical, chemical, or hydrostatic means as needed.

An explosive substance may be either an inorganic or organic compound. In both instances, it simultaneously represents the oxidizing agent *and* the reducing agent. Parts of its molecules are oxidized while other parts are reduced. An explosive substance undergoes a chemical transformation suddenly when compared with the rates at which other chemical reactions occur. Large volumes of gases and vapors are produced as well as an extraordinary amount of energy.

TABLE 15.1	Some Representative Explosive Articles[a]
Ammunition, illuminating	Cord, detonating
Ammunition, incendiary	Cord igniter
Ammunition, practice	Fireworks
Ammunition, proof	Flares
Ammunition, smoke	Flash powder
Ammunition, tear-producing	Fuse[b]
Ammunition, toxic	Fuze[b]
Black powder	Igniters
Bombs, military	Lighters, fuse
Boosters	Mines
Cartridges, blank	Powder cake (powder paste)
Cartridges, flash	Powder, smokeless
Cartridges for weapons	Primers, cap-type
Cartridges, oil well	Primers, tubular
Cartridges, power device	Projectiles
Cartridges, signal	Propellants, liquid
Cartridges, small-arms	Propellants, solid
Charges, bursting	Rocket motors
Charges, demolition	Signals
Charges, depth	Sounding devices
Charges, expelling	Torpedoes
	Warheads

[a]49 C.F.R. §173.59.
[b]*Fuse* refers to a cordlike igniting device, whereas *fuze* refers to a device used in ammunition that incorporates mechanical, electrical, chemical, or hydrostatic components to initiate an explosive train by deflagration or detonation.

The decomposition of an explosive is called a **detonation**, which may be expressed by an equation similar to the following:

$$A(s,l) \longrightarrow B(g) + C(g)$$

Here, A is the chemical formula of an explosive, and B and C are the chemical formulas of its decomposition products.

A detonation is associated with the rapid passing of an energy wave through the body of an explosive. The speed of this wave is called the **detonation velocity**. Most explosive substances are exceptionally powerful not solely because of the great amount of energy released when they detonate but also because of the exceptionally high rate at which this energy propagates. On average, the detonation velocity of an explosive is a staggering 23,000 ft/s (700 m/s).

An explosive also detonates at a unique temperature. This property is appropriately called the **detonation temperature**. The magnitude of a detonation temperature is difficult to accurately measure, because the value is heavily dependent on the explosive's density, geometrical shape, and other factors. It generally approximates an explosive's melting point.

When an explosive detonates, its decomposition products are routinely carbon monoxide, carbon dioxide, nitrogen, oxygen, and/or water vapor. Carbon particulates (soot) and nitrogen oxides also may be produced. These gases and vapors absorb some of the energy that is generated. They are then heated to extremely high temperatures, typically around 5400°F (3000°C). This increase in temperature causes the substances simultaneously to expand so rapidly that the passage of the accompanying shock wave exerts an exceedingly high pressure on the nearby surroundings.

Consider the detonation of the explosive substance nitroglycerin. When it detonates, nitroglycerin transforms in less than one-millionth of a second into carbon dioxide, nitrogen, oxygen, and water vapor as follows:

$$
\begin{array}{c}
CH_2-O-NO_2 \\
| \\
4CH-O-NO_2\,(l) \longrightarrow \quad 12CO_2(g) \;+\; 6N_2(g) \;+\; O_2(g) \;+\; 10H_2O(g) \\
| \\
CH_2-O-NO_2
\end{array}
$$

Nitroglycerin   Carbon dioxide   Nitrogen   Oxygen   Water

The volume increase is comparatively massive: Twenty-nine gaseous molecules are produced for every four nitroglycerin molecules that detonate. The energy of detonation travels at the rate of 22,600 ft/s (6900 m/s).

Unlike combustion, a detonation does not require the interplay of atmospheric oxygen or any other oxidizer with the explosive. Instead, a detonation involves the rapid decomposition of a single substance that simultaneously acts as both an oxidizing agent and a reducing agent. At the elevated temperatures accompanying the detonation of any explosive, the heated gases expand so rapidly they can occupy approximately 10,000 times their initial volume. This prodigious expansion is accompanied not only by the simultaneous production of a shock wave but also by a luminescence and an extraordinarily loud sound.

The shock waves produced during the detonation of an explosive substance are linked with their potential shattering power. This property is called the **brisance** of an explosive. The brisance is an important factor when an explosive substance is selected for a specific purpose. For example, when constructing a roadway, an engineer may consider whether the brisance generated by a detonation is likely to clear a given mass of rock. Some explosives are chosen for shattering hard rock, while others are chosen if the intent is to move rock but retain large pieces of it. Thus, when the engineer considers an explosive's brisance, he or she is actually evaluating whether the brisance is likely to shatter or fracture the rock.

detonation ■ The chemical transformation of an explosive, characterized by the essentially instantaneous passage of an energy wave through its mass, the production of large volumes of gases and vapors, the evolution of energy, and a rise in pressure

detonation velocity ■ The speed at which an energy wave moves through an explosive

detonation temperature ■ The temperature at which an explosive detonates

brisance ■ The shattering power of an explosive substance

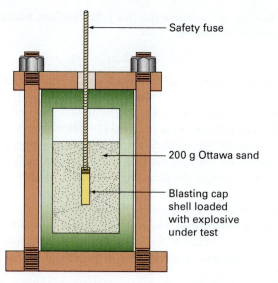

Safety fuse

200 g Ottawa sand

Blasting cap shell loaded with explosive under test

FIGURE 15.1 The type of equipment used for conducting a sand test, from which the brisance of an explosive is determined. The sand test measures the quantity of Ottawa sand that a chemical explosive crushes inside a heavy, thick-walled confining bomb. Ottawa sand is characterized by its inability to pass through a 30-mesh screen. After the chemical explosive is detonated, the sand is rescreened and the amount that then passes through a 30-mesh screen is weighed. The result is measured in grams of sand. Of the explosives noted in this text, PETN possesses the highest brisance, 62.7 grams of sand.

Scientists measure the brisance of an explosive by determining the weight of graded sand that is crushed when a given amount of the explosive is detonated within the device shown in Figure 15.1.

Emergency responders should always acknowledge that explosives possess a high degree of hazard. Exposure to the shock waves that develop during the detonation of an explosive can cause eardrums to rupture and lungs to collapse. Death may result when responders encounter a shockwave of high intensity. They may also suffer physically if they encounter the shrapnel or flying debris that is generated when an explosive detonates.

Although the detonation of an explosive substance does not require the presence of an oxidizer, most commercial explosives are mixed with oxidizers that provide additional oxygen and chemical energy. Some so-called explosives are actually pyrotechnic mixtures of multiple substances, at least one of which is an oxidizer like ammonium nitrate. These mixtures undergo oxidation–reduction reactions so vigorously that they *appear* to detonate like true explosives. They include the formulations used in blasting agents and gunpowder. Mixtures of pyrotechnic substances are also used as the active components of certain illuminating, incendiary, tear-producing, and smoke-producing devices.

15.2 CLASSIFICATION OF EXPLOSIVES

Individual explosives usually are classified into the following two groups, depending on the rapidity and sensitivity with which they detonate or ignite:

- **High explosives, or detonating explosives**
- **Low explosives, or deflagrating explosives**

Sensitivity refers to the amount and intensity of shock, friction, or heat needed to cause an explosive to detonate or ignite. Test methods have been developed to measure the sensitivity to impact, friction, and heat. When noted in this chapter, the sensitivity of a specific explosive is cited by considering all three characteristics in a general sense as either "high" or "low," and in the case of nitroglycerin, "very high."

high explosive (detonating explosive) ■ Any explosive that is capable of sustaining a detonation shock wave to produce a powerful blast

low explosive (deflagrating explosive) ■ Any substance that usually deflagrates rather than detonates, such as black powder and igniter cord

sensitivity ■ The ease with which an explosive detonates or ignites

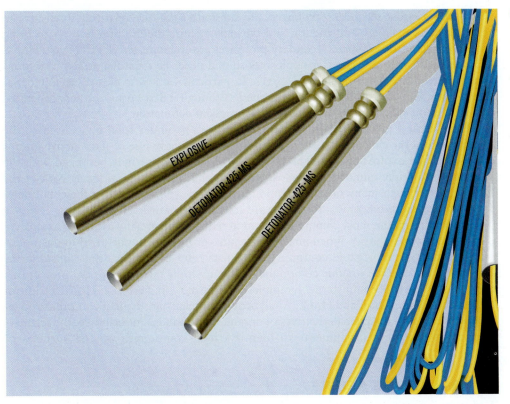

FIGURE 15.2 A common type of blasting cap that when activated, initiates the explosion of another explosive.

High and low explosives are distinguishable in a broad sense by the differences in their detonation velocities. High explosives are substances that detonate with an accompanying blast. Examples are triacetone triperoxide, nitroglycerin, and dynamite.

By contrast, low explosives are substances that deflagrate when unconfined; that is, they propagate energy comparatively slowly by burning. Examples are black powder, fireworks, and safety fuses.

A low explosive is also a component of a **blasting cap**, a small charge of a detonator designed to be embedded in a high explosive and ignited by a spark or burning fuse. One type is shown in Figure 15.2. The activation of a few milligrams of the low explosive within the blasting cap generally is sufficient to initiate the detonation of a high explosive.

The classification of explosives as high or low is not an entirely satisfactory one. Some explosives may be categorized into either class, depending on their state of subdivision, density, and degree of confinement. Smokeless powder (Section 15.8), for instance, typically is considered a low explosive, but under appropriate conditions, its detonation produces the physical effects of a detonating high explosive.

Although blasting agents have the features of low explosives, sometimes they are regarded as a separate class of explosives. For example, regulations promulgated by ATF at 27 C.F.R. §55.202 cite three classes of explosives: high explosives, low explosives, and blasting agents.

Explosives are also classified into one of the following two groups based on their susceptibility to detonate: **primary explosives** and **secondary explosives**. Both are dangerous hazardous materials, because a considerable brisance is associated with their detonations. Primary explosives are substances that are extremely sensitive to heat, mechanical shock, and friction; hence, they develop shock waves within an extremely short time. Typical primary explosives are mercury fulminate, lead azide, and lead styphnate.

blasting cap ■ A small charge of a detonator designed to be embedded in dynamite and ignited by means of a spark or a burning fuse

primary explosive ■ Any explosive that is highly susceptible to detonation by exposure to heat, mechanical shock, and friction

secondary explosive ■ Any explosive whose susceptibility to detonation occurs only when subjected to circumstances such as ignition by a primary explosive

By comparison, secondary explosives are substances that are relatively insensitive to heat, mechanical shock, and friction and typically require the use of a booster to bring about their detonation. Typical secondary explosives are cyclonite, tetryl, PETN, and HMX.

Low explosives or secondary explosives should never be regarded as less hazardous than high explosives or primary explosives. In particular, a low explosive usually deflagrates, but it *may* also detonate when confined. Although it first burns, the heat generated during its combustion serves to initiate the subsequent detonation.

The chemical instability of explosives often may be understood by examining the composition of their molecules. Some of these molecules have one or more active groups of atoms called **explosophores**. The instability of an explosive may be linked with the presence of multiple numbers of explosophores within the structural framework of the substance. Although many organic compounds having explosophores in their molecules are chemically unstable, only a select few are suitable as commercial explosives.

The explosophore most commonly encountered in commercially available explosives is the *nitro group* ($-NO_2$). When one or more nitro groups are bonded within the molecular structure of an organic compound, chemists refer to the resulting substance as a nitro derivative of an organic compound. When multiple nitro groups are present in a substance, it generally is unstable. Examples include the commercial explosives trinitrotoluene, cyclonite, tetryl, and HMX. Analytical instruments are available to rapidly identify the presence of the nitro derivatives of organic compounds. For example, the scanners at airports use these instruments to identify these compounds on the surfaces of luggage.

15.2-A BLASTING AGENTS

A **blasting agent** is any single material or mixture that has been designated for blasting, subjected to prescribed tests, and shown to have little probability of initiating an explosion or of burning and then exploding. The term generally refers to ammonium nitrate mixtures that have been sensitized with other substances. Figure 15.3 illustrates a commercially available blasting agent.

explosophore ■ The group of atoms bonded within the molecular or unit structure of a substance that provides its explosive nature (e.g., $-N_3$ or $-NO_2$)

blasting agent ■ For purposes of DOT regulations, a material designated for blasting that has been subjected to prescribed tests and has been shown to have little probability of initiating an explosion or of burning and then exploding

FIGURE 15.3 A DOT type B blasting agent normally is stored and transported in 50-pound (23-kg) bags. A DOT type B blasting agent is a mixture of ammonium nitrate and a nonexplosive, combustible substance that is intended exclusively for blasting. ANFO is the acronym for an ammonium nitrate/fuel oil mixture. Based on a review of data provided to it by the manufacturer, DOT determined that this blasting agent has the properties of a division 1.5 explosive in compatibility group D. In addition to posting a division 1.5 EXPLOSIVE label indicating the compatibility group D, DOT requires shippers who offer this blasting agent for transportation to mark the name of the contents, its identification number, the relevant EX-number, and special permit number (Section 6.1-E) on the outside surface of the bags. (*Courtesy of Dyno Nobel Inc., Salt Lake City, Utah.*)

DOT regulates the transportation of the following five types of blasting agents:

- *Type A blasting agents* are liquid nitrate esters, such as nitroglycerin, or a mixture of ingredients with one or more of the following: nitrocellulose, ammonium nitrate, other metallic nitrates, nitro derivatives of organic compounds, or combustible materials such as wood meal and aluminum powder. Type A blasting agents exist in powdery, gelatinous, plastic, and elastic forms. They include dynamite, blasting gelatin, and gelatin dynamite.
- *Type B blasting agents* are mixtures of ammonium nitrate or other metallic nitrate with an explosive such as trinitrotoluene, with or without other substances such as wood meal or aluminum powder, or a mixture of ammonium nitrate or other metallic nitrates with other nonexplosive, combustible substances. Type B blasting agents do not contain nitroglycerin, similar liquid nitrate esters, or metallic chlorates.
- *Type C blasting agents* are mixtures of either potassium chlorate or sodium chlorate, and potassium perchlorate, sodium perchlorate, or ammonium perchlorate, with a nitro derivative of an organic compound or combustible material such as wood meal, aluminum powder, or a hydrocarbon. Such explosives do not contain nitroglycerin or any similar liquid organic nitrate.
- *Type D blasting agents* are mixtures of an organic nitrate and combustible materials such as hydrocarbons and aluminum powder. They do not contain nitroglycerin, any similar liquid organic nitrate, metallic chlorate, or ammonium nitrate. This type generally includes plastic explosives.
- *Type E blasting agents* are mixtures of nitro derivatives of organic compounds, hydrocarbons, or aluminum powder with a high proportion of ammonium nitrate or other oxidizer, some or all of which is dissolved in an aqueous solution. The term includes explosives that have been formulated as emulsions, slurries, and water gels.

15.2-B ARTILLERY AMMUNITION

The sensitivity of individual explosives may be appreciated by examining the role performed by each component of a round of artillery ammunition. The structure of a typical round is shown in Figure 15.4. Each of its six components performs the following functions:

- *Primer.* In a round of artillery ammunition, the primer generally is a mixture of an oxidizer and some other arbitrary substance that readily burns. This mixture usually is confined within a cap or tube at one end of the round and is activated by a percussion action, such as a firing pin.
- *Igniter.* In a round of ammunition, the **igniter** is almost always black powder (Section 15.5). The oxidation–reduction reaction between the components of black powder is initiated by the heat released from burning the primer.

igniter ■ Any of several nitrocellulose explosives

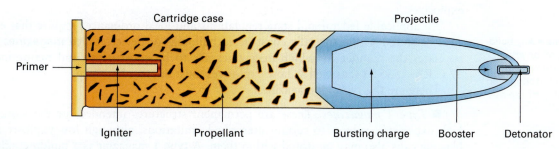

FIGURE 15.4 The component parts of a typical round of artillery ammunition. Each part serves a distinct function in the detonation of the round. An explosive sometimes is identified by its function (e.g., primer or booster).

■ *Propellant.* The **propellant** is a material such as smokeless powder that deflagrates and produces a relatively large volume of gases and vapors. In a round of artillery ammunition, the quantity is intentionally limited to prevent a major explosion but is sufficiently sizable to generate an adequate volume of gases and vapors to propel the projectile forward. The explosive that performs the role of the propellant is physically separated from the other explosives within the round by means of a thin wall.

■ *Detonator.* The **detonator** is the component of the round that explodes from the mechanical shock it experiences when it hits a target. In a round of artillery, the detonator is present in a limited quantity.

■ *Booster.* The **booster** is the component of the round that is induced to explode by the heat generated from the explosion of the detonator. As with the detonator itself, the quantity of the booster is limited so as to generate a prescribed brisance. The function of the booster is to intensify the brisance resulting from the explosion of the detonator.

■ *Bursting charge, or main charge.* The **bursting charge** is typically a high explosive whose detonation is initiated by the shock resulting from the booster explosion. It is the main charge in a round of artillery ammunition.

The chain of events from initiation of the primer to detonation of the bursting charge is called an **explosive train.** Each component of the train must be activated in the proper order for a round of ammunition to activate effectively.

The components of all types of ammunition available today may be regarded as modifications of those found in a typical round of artillery ammunition. A rifle cartridge, for instance, contains only a primer and propellant, or a primer, igniter, and propellant; the projectile generally is a mass of metal such as lead or steel pellets. A bomb that is dropped from a plane is also a modification of a round of artillery ammunition; generally, however, such a bomb lacks a propellant, because the detonator explodes from the force achieved by dropping the bomb from a high altitude.

When explosives are used for nonmilitary purposes such as blasting, the presence of a propellant is ordinarily unnecessary. In these instances, a blasting cap holds a prescribed amount of a high explosive that generally is detonated by means of a booster. The assembly containing the bursting charge and booster is connected to several hundred yards of electric wire and is activated from a safe distance by using an electrical detonator or other detonating device.

15.3 STORING EXPLOSIVES

ATF is responsible for the investigation and prevention of federal offenses involving the unlawful use, manufacture, and possession of explosives. ATF regulates, through the issuance of licenses and permits, the purchase, possession, and storage of explosives. Our concern in this section is directed at the ATF regulations that address the storage of explosives.

All applicable federal and state regulations and local ordinances require that explosives be stored within specially constructed buildings or devices called **magazines.** ATF regulates the manner by which magazines are to be constructed and maintained at 27 C.F.R. §555.213, as well as how explosives must be stored in them.

There are five types of magazines:

■ *Type 1 magazines.* These are permanent structures intended for the storage of high explosives, subject to certain prescribed limitations, although low explosives and blasting caps also may be stored within them. A type 1 magazine is a building, igloo or Army-type structure, tunnel, or dugout that is bullet-, fire-, weather-, and theft-resistant and ventilated.

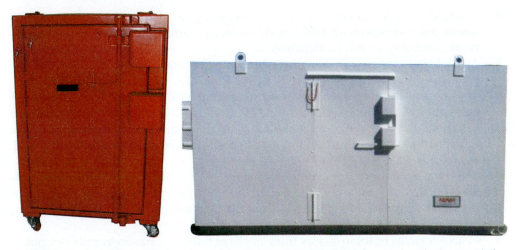

FIGURE 15.5 These type 2 magazines are used for storing certain explosives indoors and outdoors, respectively. (*Courtesy of ARMAG Corporation, Bradstown, Kentucky.*)

■ *Type 2 magazines.* These are mobile and portable, indoors- and outdoors-structures intended for the storage of high explosives, subject to certain prescribed limitations, although low explosives and blasting agents also may be legally stored within them. Figure 15.5 illustrates a type 2 magazine for use indoors and outdoors, but it may also be a trailer, semitrailer, or other mobile facility.

■ *Type 3 magazines.* These are portable outdoor structures for the temporary storage of high explosives while attended, subject to certain prescribed limitations, although low explosives and blasting agents also may be stored within them.

■ *Type 4 magazines.* These are structures intended for the storage of low explosives, subject to certain prescribed limitations. They are also used to store detonators that do not mass detonate. A type 4 magazine is a building, igloo, tunnel, dugout, trailer, semi-trailer, or other mobile structure.

■ *Type 5 magazines.* These are structures intended solely for the storage of blasting agents, subject to certain prescribed limitations. A type 5 magazine is a building, igloo, tunnel, dugout, bin, trailer, semitrailer, or other mobile structure.

The following list of some specific regulations promulgated by ATF concerns how commercial explosives must be stored. Their implementation constitutes a combination of measures that aim to prevent explosives from accidentally detonating or being directly involved in a fire.

■ Explosives may be stored only within the appropriate type of magazine that has been constructed and maintained according to specifications published by ATF.

■ To maintain the integrity of the explosives stored within them, magazines must be kept clean, dry, and free of grit, paper, empty packages and containers, and rubbish at all times.

■ The land surrounding a magazine must be kept clear of brush, dried grass, leaves, and dead trees at a distance of at least 25 feet (7.6 m) in all directions.

■ The relevant placard shown in Figure 6.15 that DOT requires carriers to post on transport vehicles and freight containers holding explosives must also be displayed on an outer surface of 5 magazines. Similar placarding is not required on type 1, type 2, type 3, and type 4 magazines.

■ The lighting within magazines must be either battery-activated safety lights or lanterns or electric lighting meeting the standards prescribed by the National Electrical Code. All electrical switches must be located outside the magazines.

- Smoking or the use of matches, open flames, and spark-producing devices is prohibited in any magazine, within 50 feet (15 m) of any outdoor magazine, or within any room containing an indoor magazine.
- An explosive may be stored in a magazine only in a quantity that conforms with the requirements published by ATF at 27 C.F.R. §555.213.
- The doors of the magazine must be constructed of solid wood or metal. Hinges and hasps must be attached to them by welding, riveting, or bolting and must be installed so they cannot be removed when the doors are closed or locked. Doors must also be equipped with locks or padlocks of a type specified in regulations published by ATF at 27 C.F.R. §555.207–§555.211.
- The contents of a magazine must be inventoried daily. This inventory must include the manufacturer's name or brand name, the total quantity received into and removed from each magazine during each day, and the total remaining on hand at the end of each day.
- Magazines must be located away from inhabited buildings, passenger railways, public highways, and other magazines in conformance with prescribed distances established by the Institute of Makers of Explosives[2] and published at 27 C.F.R. §555.218.

15.4 TRANSPORTING EXPLOSIVES

DOT regulates the transportation of the specific types of explosive articles *and* explosive substances listed in the Hazardous Materials Table. For purposes of DOT regulations, these explosives are divided into six divisions that form a continuum of decreasing hazard from 1.1 to 1.6.

When the word "forbidden" appears in column 3 of the Hazardous Materials Table, the designation signifies a **forbidden explosive**, one whose transportation is prohibited by any mode. Nonetheless, DOT may allow the same explosive substance to be transported when it has been suitably "desensitized" or reduced in its ability to detonate. Manufacturers desensitize an explosive by mixing it with an inert material called a stabilizer.

15.4-A EX-NUMBERS

Before an explosive may be offered for transportation, DOT requires its manufacturer or other party to test and evaluate the explosive following prescribed criteria and procedures to determine the division in which it should be classified. Then, the manufacturer or other party must formally apply for an **EX-number** by submitting information regarding the physical and chemical composition of the explosive, thermal-stability-test data, relevant background data, and other technical information for evaluation. When DOT approves this application, it assigns an EX-number corresponding to the specific explosive at issue. This assignment signifies that DOT approves the explosive for transportation when the shippers and carrier follow all applicable regulations. DOT also uses the information to assign a shipping description, division number, and compatibility group for the explosive.

15.4-B COMPATIBILITY GROUPS

The **compatibility group** of an explosive is one of the 12 uppercase letters *A, B, C, D, E, F, G, H, J, K, L,* or *S.* The assignment of the compatibility group is based on the relevant description of the explosive provided in Table 15.2. DOT enters the compatibility group

forbidden explosive
■ For purposes of DOT regulations, an explosive that does not have DOT's authorization for transport by any mode

EX-number ■ For purposes of DOT regulations, a number preceded by the prefix EX assigned by DOT to an explosive that has been evaluated by the provisions prescribed at 49 C.F.R. §173.56 and approved for transportation

compatibility group
■ For purposes of DOT regulations, a designated alphabetical letter used to classify an explosive for control during its storage and transportation

---

[2]The Institute of Makers of Explosives is the safety association of the commercial explosives industry in the United States and Canada. It promotes the safety and protection of employees, users, the public, and the environment throughout all aspects of the manufacture and use of explosives in industrial blasting and other essential operations.

TABLE 15.2	Compatibility Group Assignments[a]		
EXPLOSIVE		COMPATIBILITY GROUP	CLASSIFICATION CODE
Primary explosive substance.		A	1.1A
Article containing a primary explosive substance and not containing two or more effective protective features. Some articles, such as detonators for blasting/detonator assemblies for blasting and cap-type primers, are included, even though they do not contain primary explosive substances.		B	1.1B, 1.2B, or 1.4B
Propellant explosive substance or other deflagrating explosive substance or article containing such explosive substance.		C	1.1C, 1.2C, 1.3C, or 1.4C
Secondary detonating explosive substance or black powder or article containing a secondary detonating explosive substance, in each case without means of initiation and without a propelling charge; or article containing a primary explosive substance and containing two or more effective protective features.		D	1.1D, 1.2D, 1.4D, or 1.5D
Article containing a secondary detonating explosive substance, without means of initiation, with a propelling charge (other than one containing flammable liquid gel or hypergolic liquid).		E	1.1E, 1.2E, or 1.4E
Article containing a secondary detonating explosive substance with its means of initiation, with a propelling charge (other than one containing flammable liquid gel or hypergolic liquid) or without a propelling charge.		F	1.1F, 1.2F, 1.3F, or 1.4F
Pyrotechnic substance or article containing a pyrotechnic substance, or article containing both an explosive substance and an illuminating, incendiary, tear-producing, or smoke-producing substance (other than a water-activated article or one containing white phosphorus, phosphide, or flammable liquid or gel or hypergolic liquid).		G	1.1G, 1.2G, 1.3G, or 1.4G
Article containing both an explosive substance and white phosphorus.		H	1.2H or 1.3H
Article containing both an explosive substance and flammable liquid or gel.		J	1.1J, 1.2J, or 1.3J
Article containing both an explosive substance and a toxic chemical agent.		K	1.2K or 1.3K
Explosive substance or article-containing substance and presenting a special risk (e.g., owing to water activation or presence of hypergolic liquids, phosphides, or pyrophoric substances) needing isolation of each type.		L	1.1L, 1.2L, or 1.3L
Articles containing only extremely insensitive detonating substances.		N	1.6N
Substance or article so packed or designed that any hazardous effects arising from accidental functioning are limited to the extent that they do not significantly hinder or prohibit firefighting or other emergency response efforts in the immediate vicinity of the package.		S	1.4S

[a]49 C.F.R. §173.52.

after the hazard class and division number for a specific explosive in column 3 of the Hazardous Materials Table. Each division number followed by a compatibility group is called the **classification code** of the explosive.

DOT requires shippers to display the compatibility group on the applicable EXPLOSIVE 1.1, 1.2, 1.3, 1.4, 1.5, or 1.6 label that they affix to packaging containing the explosive. For explosives in hazard class divisions 1.1, 1.2, and 1.3, the compatibility group is designated as a component of the classification code in the lower quadrant of

classification code
■ The division number of an explosive followed by its compatibility group

the label and above hazard class 1. For explosives in hazard class divisions 1.4, 1.5, and 1.6, the compatibility group is written separately above hazard class 1 in the lower quadrant of the label, and the hazard class and division are written in the upper quadrant. DOT also requires the carrier to display the compatibility group on the EXPLOSIVES 1.1, 1.2, 1.3, 1.4, 1.5, or 1.6 placards posted on a transport vessel or freight container. These placards and labels display the compatibility group in an identical fashion.

SOLVED EXERCISE 15.1

What is the most likely incendiary agent contained within firebombs ("ammunition, incendiary") whose DOT hazard class is each of the following?

(a) 1.3J
(b) 1.2H

Solution:

(a) By referring to Table 15.2, we see that a flammable liquid or gel is contained within explosive articles that are assigned the compatibility group J. Consequently, gasoline, jet fuel, or kerosene thickened with either napalm or a napalm-like gel is the most likely incendiary agent contained within a firebomb that has been assigned the DOT hazard class 1.3J.

(b) By again referring to Table 15.2, we see that white phosphorus is contained within explosive articles that are assigned the compatibility group H. Consequently, white phosphorus is the most likely incendiary agent that is contained within a firebomb that has been assigned the DOT hazard class 1.2H.

15.4-C SHIPPING DESCRIPTIONS

When shippers offer an explosive for transportation, DOT requires them to identify the explosive in its shipping description and to mark the identification on the packaging used for shipment. The explosive may be identified through use of its assigned EX-number or, in lieu of it, the national stock number issued by the Department of Defense, product code, or other identifying information.

The national stock number, or NSN, is a 13-digit number used to identify a variety of commercial products including the types of explosive. NSNs are used primarily by the U.S. Department of Defense (DOD) for purposes of procurement. The NSN corresponding to individual explosives may be obtained from DOD publications and other sources.

When shippers offer an explosive for transportation, they also include the net explosive mass in the shipping description. When the explosive is an article, the net explosive mass is expressed in terms of the net mass of either the article or the explosive contained within the article.

15.4-D LABELING, MARKING, AND PLACARDING REQUIREMENTS

When an explosive is transported, DOT requires shippers and carriers to comply with its labeling, marking, and placarding requirements. In particular, DOT requires shippers to affix the relevant label shown in Figure 6.5 to its packaging. DOT also requires shippers to mark the packaging with the information specified in Section 6.6-A, including the following: proper shipping name, identification number, instructions, precautions, and the applicable EX-number, national stock number, product code, or other identifying information.

When carriers transport by highway or rail any quantity of an explosive in divisions 1.1, 1.2, or 1.3, DOT requires them to post the applicable EXPLOSIVES 1.1, EXPLOSIVES 1.2, or EXPLOSIVES 1.3 placards shown in Figure 6.15 on the transport vehicle or freight container and to display on the placard the applicable compatibility group. When carriers transport 1001 pounds (454 kg) or more of an explosive in divisions 1.4, 1.5, or 1.6, DOT requires them to post EXPLOSIVES 1.4, EXPLOSIVES 1.5, and EXPLOSIVES 1.6 placards with the applicable compatibility group on the transport vehicle or freight container. However, DOT prohibits them from displaying the UN/NA identification number of an explosive material across the center face of the EXPLOSIVES 1.1, 1.2, 1.3, 1.4, 1.5, or 1.6 placards, or an EXPLOSIVES SUBSIDIARY placard.

When carriers transport more than 55 pounds (25 kg) of a division 1.1, 1.2, or 1.3 explosive in a motor vehicle, railcar, or freight container, DOT requires them to prepare and implement a transportation security plan (Section 6.9) whose components comply with the requirements of 49 C.F.R. §172.802. Motor carriers are also required to obtain a hazardous materials safety permit (Section 6.10) before transporting more than 55 pounds (25 kg) of a division 1.1, 1.2, or 1.3 explosive or more than 1001 pounds (454 kg) of a division 1.5 explosive. Issuance of the safety permit requires motor carriers to prepare a written route plan that complies with the requirements of 49 C.F.R. §397.67 when transporting a division 1.1, 1.2, or 1.3 explosive. Other special controls may also apply.

When more than one explosive article or explosive substance having *different* compatibility groups is transported, DOT requires carriers to display only one placard based on the following scheme:

- Explosive articles of compatibility groups C, D, or E are placarded displaying compatibility group E.
- Explosive articles of compatibility groups C, D, or E, when transported with those in compatibility group N, are placarded displaying compatibility group D.
- Explosive substances of compatibility groups C and D are placarded displaying compatibility group D.
- Explosive articles of compatibility groups C, D, E, or G, except for fireworks, are placarded displaying compatibility group E.

SOLVED EXERCISE 15.2

A consignment consists solely of explosive articles having the following shipping descriptions, where EX-xxxxx are four different EX-numbers:

UN0028, Black powder, compressed, 1.1D, PG II (EX-xxxxx)
UN0006, Cartridges for weapons, with bursting charge, 1.1E, PG II (EX-xxxxx)
UN0271, Charges, propelling, 1.1C, PG II (EX-xxxxx)
UN0408, Fuzes, detonating, with protective features, 1.1D, PG II (EX-xxxxx)

Which placards, if any, does DOT require the carrier to display when these items are transported by motor carrier from Lexington, Kentucky, to Baltimore, Maryland?

Solution: When carriers simultaneously transport explosive articles having compatibility groups C, D, and E, DOT requires them to display the compatibility group E on the appropriate placards that are posted on the motor vehicle. Because the division number for each of the referenced explosive articles is 1.1, DOT requires carriers to display EXPLOSIVES 1.1 placards bearing the compatibility group E on each side and each end of the motor vehicle in which the explosive articles are transported.

15.4-E EXPLOSIVE WASTES

It was noted in Section 5.9 that an explosive waste may exhibit the RCRA characteristic of reactivity, and Table 6.2 noted that DOT requires shippers to identify the nature of this waste when it is offered for domestic transportation. For example, shippers who offer for transportation waste dynamite that exhibits the reactivity characteristic identify its shipping description on an accompanying waste manifest by using either of the following, where EX-xxxxx is the appropriate EX-number:

NA3077, Hazardous waste, solid, n.o.s. (dynamite), 9, PG III (EPA reactivity) (EX-xxxxx)
NA3077, Hazardous waste, solid, n.o.s. (dynamite), 9, PG III (D003) (EX-xxxxx)

15.4-F OTHER DOT REQUIREMENTS

When an explosive is moved by public highway, DOT may designate the routing or impose curfews, time-of-travel restrictions, lane restrictions, or route/weight restrictions on the shipper or carrier. The identity of such routing designations must be made available to the public in the form of a map, list, and/or a road sign. The route or highway that DOT designates for the domestic transportation is called a **preferred route**, or **preferred highway**. Its selection complies with the regulations published at 49 C.F.R. §397.71.

DOT also requires special precautions to be implemented when explosives are loaded, unloaded, or handled. Onboard watercraft in particular, a fire hose of sufficient length to reach every part of the loading area with an effective stream of water must be laid and connected to the water main, ready for use. In the event of fire, firefighting personnel proceed immediately to prevent a premature cargo explosion.

15.5 BLACK POWDER

Black powder is the oldest known explosive.[3] It was discovered by the Chinese before 1000 A.D. Although its chemical composition varies, black powder generally consists of an intimate mixture of charcoal, sulfur, and either potassium nitrate or sodium nitrate. It is regarded as a low explosive. Black powder formerly was employed as a blasting agent, but today, it is more often encountered as a component of fireworks, signal flares, and certain forms of ammunition.

When activated, the components of black powder undergo an oxidation–reduction reaction. Although the nature of the reaction is open to some debate, the following equation most likely expresses the phenomenon:

$$3S_8(s) + 16C(s) + 32KNO_3(s) \longrightarrow 16K_2O(s) + 16CO_2(g) + 16N_2(g) + 24SO_2(g)$$

| Sulfur | Carbon | Potassium nitrate | Potassium oxide | Carbon dioxide | Nitrogen | Sulfur dioxide |

Because it is composed of such a reactive mixture of substances, black powder may be activated by a spark or static electricity. It autoignites at 867°F (464°C) and detonates at a relatively low velocity of 1300 ft/s (400 m/s).

15.5-A RESPONDING TO INCIDENTS INVOLVING THE RELEASE OF BLACK POWDER

Under routine conditions, responding to an emergency incident involving explosives should be undertaken only by competent, experienced individuals who have received special training in the handling of explosives. Perhaps the single exception to this general rule involves encountering black powder. Because black powder contains an oxidizer, the mixture is easily desensitized by diluting it with water. Black powder cannot detonate when it is soaked with water.

15.5-B TRANSPORTING BLACK POWDER

When shippers offer black powder for transportation, DOT requires them to identify it as shown in Table 15.3 on an accompanying shipping paper. DOT also requires shippers and carriers to comply with all applicable labeling, marking, and placarding requirements.

---

[3]Black powder formulations were known to the ancient Chinese, who used them in warfare and pyrotechnic displays. Although the color of black powder is indeed black, the material is not named for its color but for the direct translation of the German word *Schwarzpulver*, named for Berthold Schwarz, who experimented with black powder formulations in the fourteenth century.

---

preferred route (preferred highway)
■ For purposes of DOT regulations, the designated route or highway on which a shipper or carrier is required to transport an explosive or radioactive material

black powder ■ A mixture of charcoal and sulfur with potassium nitrate or sodium nitrate in prescribed amounts

TABLE 15.3	Shipping Descriptions of Black Powder
FORM OF BLACK POWDER	**SHIPPING DESCRIPTION**
Black powder, compressed or in pellets	UN0028, Black powder, compressed, 1.1D, PG II (EX-xxxxx) *or* UN0028, Black powder, in pellets, 1.1D, PG II (EX-xxxxx) *or* UN0028, Gunpowder, 1,1D, PG II (EX-xxxxx) *or* UN0028, Gunpowder, in pellets, 1.1D, PG II (EX-xxxxx)
Black powder, granular	UN0027, Black powder, 1.1D, PG II (EX-xxxxx) *or* UN0027, Gunpowder, 1.1D, PG II (EX-xxxxx)
Black powder for small arms	NA0027, Black powder for small arms, 4.1, PG I (EX-xxxxx)

15.6 NITROGLYCERIN

When alcohols react with nitric acid, the compounds called **nitrate esters** are produced. The relevant reactions may be denoted by the following general equation, where R is an arbitrary alkyl or aryl group:

$$R-OH(l) + HNO_3(l) \longrightarrow R-O-NO_2(l) + H_2O(l)$$

Nitroglycerin is the best known nitrate ester. Its proper chemical name is glyceryl trinitrate.

$$
\begin{array}{l}
CH_2-O-NO_2 \\
| \\
CH-O-NO_2 \\
| \\
CH_2-O-NO_2
\end{array}
$$

It is painstakingly prepared by drizzling glycerol, or 1,2,3-trihydroxypropane, into a cooled mixture of nitric acid and sulfuric acid. The sulfuric acid functions by extracting the elements of water from glycerol and nitric acid.

Pure nitroglycerin is a thick, oily liquid whose physical appearance otherwise resembles water. Other physical properties are noted in Table 15.4. When it is encountered, however, nitroglycerin generally is a viscous, pale yellow liquid that is extremely sensitive to spontaneous decomposition. Even a slight jarring or dropping it on a hard surface may trigger its premature detonation. Given its sensitivity to decomposition, pure nitroglycerin cannot be safely transported and is impractical for general use as an explosive. Notwithstanding this fact, nitroglycerin often is mixed with other explosives as a component of ammunition.

Nitroglycerin hydrolyzes when it is exposed to atmospheric moisture. The hydrolysis produces a mixture of glycerol and nitric acid. When it has absorbed moisture, nitroglycerin becomes extremely sensitive to spontaneous decomposition. For this reason, the repeated exposure of nitroglycerin to a humid atmosphere is likely to produce a highly dangerous mixture that could detonate with the slightest provocation.

Aside from its potentially explosive nature, nitroglycerin is also a toxic substance by ingestion, inhalation, and skin absorption. When absorbed into the blood system, nitroglycerin causes the small veins, capillaries, and coronary vessels to dilate. These effects are actually caused by nitric oxide (Section 10.14), which is slowly released from the

nitrate ester ■ Any organic compound whose general chemical formula is $R-O-NO_2$, where R is an arbitrary alkyl or aryl group

Nitroglycerin

TABLE 15.4	Physical Properties of Nitroglycerin
Melting point	55°F (13.1°C)
Detonation temperature	424°F (218°C)
Specific gravity at 68°F (20°C)	1.60
Vapor density (air = 1)	7.84
Vapor pressure at 68°F (20°C)	0.00026 mmHg
Brisance (g of sand)	51.5
Detonation velocity	25,200 ft/s (7700 m/s)
Sensitivity	Very high; almost a primary explosive

nitroglycerin. The nitric oxide causes the development of severe headaches, flushing of the face, and a drop in blood pressure. In severe instances, exposure to nitroglycerin may also cause death.

15.6-A COMMERCIAL USE OF NITROGLYCERIN OTHER THAN AS AN EXPLOSIVE

vasodilator ■ A substance, like nitroglycerin, exposure to which is capable of causing the blood vessels to widen

Notwithstanding its toxic nature, nontoxic doses of nitroglycerin are used medicinally for the treatment of heart and blood-circulation diseases. A compound like nitroglycerin that causes dilation of the blood vessels is called a **vasodilator**. Individuals having coronary problems may receive their medication by one of three routes: They can put a tiny nitroglycerin tablet under the tongue when they experience chest pain, consume a medication called *spirits of nitroglycerin* (which consists of nitroglycerin dissolved in a solvent), or use a transdermal patch. The nitroglycerin slowly releases nitric oxide within the bloodstream as it decomposes. The nitric oxide relaxes nearby muscle cells and temporarily lowers blood pressure.

15.6-B WORKPLACE REGULATIONS INVOLVING NITROGLYCERIN

When nitroglycerin is present in the workplace, OSHA requires employers to limit employee exposure by both inhalation and skin contact. The permissible exposure limit is 0.2 mg/m^3 of air, averaged over an 8-hour workday.

15.6-C TRANSPORTING NITROGLYCERIN

DOT prohibits the domestic transportation of nitroglycerin unless the shipper successfully demonstrates that it has been desensitized against unwanted decomposition. Nitroglycerin is usually desensitized by dissolving it in a simple alcohol such as methanol, ethanol, or isopropanol. Because these alcohols are flammable liquids, DOT may require shippers to assign these commercially available forms of nitroglycerin to hazard class 3 or 4.1.

When nitroglycerin is transported, DOT requires its shipper to provide the relevant shipping description shown in Table 15.5. DOT also requires shippers and carriers to comply with all applicable marking, labeling, and placarding requirements.

DOT also requires shippers to affix EXPLOSIVE 1.1 labels bearing the compatibility group D on packages containing desensitized nitroglycerin. When the nitroglycerin poses an inhalation hazard, DOT also requires them to display POISON INHALATION HAZARD labels on the packages.

TABLE 15.5	Shipping Descriptions of Nitroglycerin
FORM OF NITROGLYCERIN	**SHIPPING DESCRIPTION**
Nitroglycerin, desensitized	UN0143, Nitroglycerin, desensitized, 1.1D, (6.1), PG II (EX-xxxxx) (Poison)
Nitroglycerin, solution in alcohol	UN1044, Nitroglycerin, solution in alcohol, 1.1D, PG II (EX-xxxxx) (Poison)

When carriers transport nitroglycerin in any amount, DOT requires them to post EXPLOSIVES 1.1 placards bearing the compatibility group D on the bulk packaging or transport vehicle used for shipment. When carriers transport nitroglycerin that poses an inhalation hazard, DOT also requires them to display POISON INHALATION HAZARD placards on each side and each end of the bulk packaging or transport vehicle.

15.6-D TERRORISTS' MISUSE OF NITROGLYCERIN

It appears that most terrorists have avoided the use of nitroglycerin owing to its sensitivity to detonation. Nonetheless, during the early 2000s, Palestinian forces routinely used nitroglycerin in suicide bombs directed against the Israelis. Furthermore, a Muslim extremist planned to use nitroglycerin to blow up several U.S. planes, but the plan was foiled when police located the explosive in the terrorist's apartment.

15.7 DYNAMITE

dynamite ■ A detonating explosive composed of nitroglycerin, similar organic nitrate esters, and one or more oxidizers, all of which are mixed into a stabilizing absorbent

In 1867, Swedish engineer Alfred Nobel discovered that nitroglycerin could be absorbed into a porous material such as siliceous earth. The resulting mixture became known as **dynamite**. Nobel acquired a substantial fortune from his discovery and the subsequent manufacture of dynamite. He used the fortune to establish a monetary fund for the world-famous Nobel prizes.

Dynamite may be handled more safely than nitroglycerin, a property that allows it to be transported and used with less risk of spontaneous decomposition. In practice, dynamite requires a detonating cap to activate its detonation. Notwithstanding this fact, dynamite is a high explosive that is sensitive to heat, shock, and friction. The equation denoting the detonation of nitroglycerin was previously noted in Section 15.1.

Today, dynamite is produced by absorbing a mixture of nitroglycerin and diethylene glycol dinitrate into wood pulp, sawdust, flour, starch, or similar carbonaceous materials. Diethyleneglycol dinitrate is also an explosive, but its function in the production of dynamite is to depress the solution's freezing point. Calcium carbonate often is added to dynamite to neutralize the nitric acid produced by hydrolysis. Oxidizers are routinely added as a source of additional internal energy. This mixture of substances is then packed into cylindrical cartridges made of waxed paper that resemble those shown in Figure 15.6. They vary in size from 7/8 to 8 inches (2 to 20 cm) in diameter and from 4 to 30 inches (10 to 76 cm) in length. They routinely weigh approximately 0.5 pounds (230 g).

When they are involved in fires, small quantities of dynamite usually burn with a bluish flame without detonating. Nonetheless, the heat generated during these fires may activate the detonation of the remaining nitroglycerin. Acknowledging this possibility, experts recommend that firefighters avoid fighting fires involving dynamite.

From the 1920s through the 1930s, dynamite was the most popular explosive used for peacetime purposes, but accidental detonations involving dynamite occurred frequently. Since the 1930s, new ways to containerize dynamite that are far safer to transport, store,

Dynamite

and use have been developed. Even though these products are commercially available, some explosives experts still prefer to use dynamite sticks for unique demolition assignments.

15.7-A FORMS OF DYNAMITE

Three commercial forms of dynamite are commonly encountered: ammonia dynamite, straight dynamite, and gelatin dynamite. Information about their individual chemical compositions is provided in Table 15.6. Although nitroglycerin and diethyleneglycol dinitrate are contained in each form, ammonia dynamite and straight dynamite also contain ammonium nitrate and sodium nitrate, respectively. Gelatin dynamite contains about 1% nitrocellulose, which serves to thicken the nitroglycerin.

When their constituent nitroglycerin detonates, the three forms of dynamite produce more brisance than an equivalent amount of nitroglycerin, because additional energy is provided through the chemical action of ammonium nitrate, sodium nitrate, and nitrocellulose.

TABLE 15.6	**Physical Properties of Some Commercial Forms of Dynamite**		
	STRAIGHT DYNAMITE	**AMMONIA DYNAMITE**	**GELATIN DYNAMITE**
Specific gravity at 68°F (20°C)	1.3	0.8–1.2	1.3–1.6
Chemical composition	20–60% nitroglycerin depending on grade, sodium nitrate, carbonaceous materials, antacid, and moisture	20–60% nitroglycerin depending on grade, ammonium nitrate, carbonaceous materials, sulfur, antacid, and moisture	20–90% nitroglycerin depending on grade, sodium nitrate, and moisture, gelatinized in nitrocellulose
Detonation velocity	10,000–20,000 ft/s (3050–6100 m/s)	8000–13,000 ft/s (2440–3960 m/s)	15,400–17,800 ft/s (4700–5400 m/s)
Sensitivity	High	High	High

TABLE 15.7	Shipping Descriptions of Dynamite
TYPE OF DYNAMITE	**SHIPPING DESCRIPTION**
Type A	UN0081, Explosive, blasting, type A, 1.1D, PG II (dynamite) (EX-xxxxx)
Type B	UN0082, Explosive, blasting, type B, 1.1D, PG II (dynamite) (EX-xxxxx)
	or
	UN0331, Explosive, blasting, type B, 1.5D, PG II (dynamite) (EX-xxxxx)
	or
	UN0331, Agent blasting, Type B, 1.5D, PG II (dynamite) (EX-xxxxx)
Type C	UN0083, Explosive, blasting, type C, 1.1D, PG II (dynamite) (EX-xxxxx)
Type D	UN0084, Explosive, blasting, type D, 1.1D, PG II (dynamite) (EX-xxxxx)
Type E	UN0241, Explosive, blasting, type E, 1.1D, PG II (dynamite) (EX-xxxxx)
	or
	UN0332, Agent blasting, Type E, 1.5D, PG II (dynamite) (EX-xxxxx)

15.7-B TRANSPORTING DYNAMITE

When shippers offer dynamite for transportation, DOT requires them to provide the relevant shipping description shown in Table 15.7 on the accompanying shipping paper. The shipping description of dynamite is a generic one that cites the type of blasting agent.

DOT also requires shippers and carriers to comply with all applicable marking, labeling, and placarding requirements.

15.8 NITROCELLULOSE

As first noted in Section 14.5-A, nitrocellulose is a polymer produced by reacting the cellulose in cotton with nitric acid. Because cellulose may be nitrated to different degrees, varying forms of nitrocellulose may be produced. One molecular structure of nitrocellulose is noted here:

Nitrocellulose (explosive grade)

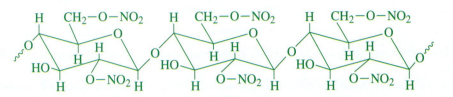

In this structure, the cellulose is almost completely nitrated. Although highly nitrated cellulose has been used as a rocket propellant, the lesser grades of nitrocellulose are used for other purposes. In the United States, the explosive is now produced almost exclusively by the federal government. When the nitration produces a substance having a nitrogen content exceeding 13.2% by mass, the resulting material is called **guncotton**.

Nitrocellulose is a white solid that resembles cotton in physical appearance. When intended for use as an explosive, it may be blocked, gelled, flaked, granulated, or powdered. To desensitize it for domestic transportation and storage, nitrocellulose usually is wetted with either water or an aqueous solution of ethanol. Notwithstanding the presence of a desensitizing agent, every nitrocellulose-based explosive is inherently unstable. The nitrocellulose slowly decomposes, producing traces of nitrogen oxides.

guncotton ▪ Any of several nitrocellulose explosives produced by the chemical reaction between cotton and nitric and sulfuric acids

When nitrocellulose is activated, it usually deflagrates. Because its flashpoint is only 55°F (13°C), nitrocellulose burns readily and produces carbon dioxide, water vapor, and nitrogen, but not nitrogen dioxide. The deflagration of nitrocellulose is a furiously burning phenomenon represented as follows, where n is a nonzero integer:

$$2[C_5H_5(CH_2ONO_2)O_2(ONO_2)_2]_n(s) + 10nO_2(g) \longrightarrow 12nCO_2(g) + 7nH_2O(g) + 3nN_2(g)$$

| Nitrocellulose | Oxygen | Carbon dioxide | Water | Nitrogen |

Nitrocellulose burns at a faster rate than virtually any other flammable solid. Tons of nitrocellulose may be consumed by fire within minutes.

Because the products of its combustion are colorless, nitrocellulose often is called **smokeless powder**. The absence of a visible smoke provides a tactical advantage when it is used during warfare. It prevents the enemy from easily identifying the location from which artillery was fired.

Nitrocellulose detonates when it is confined or accumulated in large quantities. The detonation reaction may be expressed as follows:

$$2[C_5H_5(CH_2ONO_2)O_2(ONO_2)_2]_n(s) \longrightarrow nCO_2(g) + 11nCO(g) + 7nH_2O(g) + 3nN_2(g)$$

| Nitrocellulose | Carbon dioxide | Carbon monoxide | Water | Nitrogen |

smokeless powder ■ A form of nitrocellulose that deflagrates to produce colorless gases and negligible smoke

Nitrocellulose is most commonly encountered as a component of **small-arms ammunition**, although it is also used as the propellant in artillery ammunition. It is also sometimes selected for demolition work, wherein it is usually combined with a second explosive. This combination of explosives is called a *double-base formulation*.

small-arms ammunition ■ Shotgun, rifle, pistol, or revolver cartridges

15.8-A FORMULATIONS OF SMOKELESS POWDER

Three formulations of nitrocellulose are called the single-, double-, and triple-base smokeless powders. Together with nonexplosive ingredients, these formulations contain one, two, or three explosives, respectively. The single-base formulation contains only nitrocellulose. The most popular variation is the double-base formulation, which consists of a mixture of nitrocellulose and nitroglycerin. It is used as a component of handgun and rifle ammunition. This formulation has been encountered in illegally improvised devices such as pipe bombs.

The triple-base smokeless powder usually contains the three explosives nitrocellulose, nitroglycerin, and nitroguanidine.

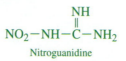

Nitroguanidine

It is most commonly encountered as a component of the ammunition used in artillery guns.

A stabilizer is added to all nitrocellulose-based formulations. It is usually diphenylamine, which functions by reacting with the nitrogen oxides to form innocuous compounds.

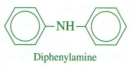

Diphenylamine

15.8-B TRANSPORTING NITROCELLULOSE

When shippers offer desensitized nitrocellulose for transportation, DOT requires them to provide the shipping description shown in Table 15.8 on an accompanying shipping paper. DOT also requires shippers and carriers to comply with all applicable labeling, marking, and placarding requirements.

TABLE 15.8	Shipping Descriptions of Nitrocellulose
FORM OF NITROCELLULOSE	**SHIPPING DESCRIPTION**
Nitrocellulose, dry or wetted with less than 25% water or alcohol by mass	UN0340, Nitrocellulose, dry, 1.1D, PG II (EX-xxxxx)
Nitrocellulose, plasticized with not less than 18% plasticizing substance by mass	UN0343, Nitrocellulose, plasticized, 1.3C, PG II (EX-xxxxx)
Nitrocellulose, unmodified, or plasticized with less than 18% plasticizing substance by mass	UN0341, Nitrocellulose, 1.1D, PG II (EX-xxxxx)
Nitrocellulose, wetted with not less than 25% alcohol by mass	UN0342, Nitrocellulose, 1.3C, PG II (EX-xxxxx)
Smokeless powder	UN0160, Powder, smokeless, 1.4C (EX-xxxxx) *or* UN0161, Powder, smokeless, 1.3C (EX-xxxxx)
Smokeless powder for small arms	NA3178, Smokeless powder for small arms, 4.1, PG I

15.9 TRINITROTOLUENE

2,4,6-Trinitrotoluene, or TNT, is a pale yellow solid, although its commercial grade generally is yellow to dark brown.

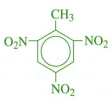

TNT was used on a very large scale as a military explosive during World War II. Movie buffs are especially familiar with its use in depth charges catapulted from destroyers to disable enemy submarines. During peacetime, it is still used for demolition during construction and mining projects.

The equation denoting the detonation of TNT is provided in Solved Exercise 15.3. Some physical properties of TNT are provided in Table 15.9.

Trinitrotoluene

TABLE 15.9	Physical Properties of TNT
Melting point	178°F (81°C)
Boiling point	563°F (295°C) with decomposition
Detonation temperature	450°F (232°C)
Specific gravity at 68°F (20°C)	1.65
Vapor density (air = 1)	7.85
Vapor pressure	0.046 mmHg
Brisance (g of sand)	48
Detonation velocity	22,600 ft/s (6900 m/s)
Sensitivity	Low

Write the balanced equation for the detonation of TNT by using its condensed formula, $C_7H_5(NO_2)_3$.

Solution: Before we can write the equation representing the detonation of TNT, we need first to identify the decomposition products produced. During any detonation, the hydrogen and oxygen atoms combine to form water, and the carbon and oxygen atoms combine to form either carbon monoxide or carbon dioxide, or both. If carbon and oxygen remain after production of carbon monoxide and/or carbon dioxide, they are released during the detonation as their respective elements. Because there are seven carbon atoms in each TNT molecule, some elemental carbon inevitably is produced during the detonation of TNT. The nitrogen atoms unite to form elemental nitrogen, or they react with oxygen atoms to produce either nitrogen monoxide or nitrogen dioxide, or both.

After some consideration, it becomes apparent that the TNT detonation products are nitrogen, water vapor, carbon monoxide, and elemental carbon. The presence of carbon indicates that the detonation is sooty in appearance. The unbalanced equation is then written as follows:

$$C_7H_5(NO_2)_3(s) \longrightarrow CO(g) + H_2O(g) + N_2(g) + C(s)$$

Trinitrotoluene   Carbon monoxide Water   Nitrogen   Carbon

Finally, we balance this equation.

$$2C_7H_5(NO_2)_3(s) \longrightarrow 7CO(g) + 5H_2O(g) + 3N_2(g) + 7C(s)$$

Trinitrotoluene   Carbon monoxide Water   Nitrogen   Carbon

TNT is prepared by nitrating toluene using a mixture of nitric acid and sulfuric acid. It has been used primarily as a military explosive, because it has the following features:

- TNT is relatively safe to handle. Although it is a high explosive, TNT is unusually insensitive to heat, shock, and friction. When it is engulfed in fire, relatively small quantities often burn but do not detonate. Ordinarily TNT detonates only when confined or when relatively large amounts are intentionally activated.
- TNT does not react with atmospheric moisture.
- TNT is not susceptible to spontaneous decomposition, even after it has been kept in storage for years.
- TNT may be melted using steam with little fear of explosive decomposition. This feature is used when commercial explosive mixtures containing TNT are prepared. The molten TNT may be safely mixed with other explosives or oxidizers, cast into the shape of blocks, or poured into ammunition shells.

Commercial explosives often are prepared by mixing TNT with other substances. A popular explosive containing TNT is amatol, a mixture consisting of 80% ammonium nitrate and 20% TNT by mass. It has been used widely as a blasting agent and a military and industrial explosive. Two other commercial explosives containing TNT are cyclonite (Composition B) and tetrytol.

Caution must be exercised when handling and using TNT, because it is toxic by ingestion, inhalation, and skin absorption. Table 15.9 indicates that even as a solid, TNT has a significant vapor density. Inhalation of TNT vapor can be fatal because the TNT reacts with hemoglobin to form methemoglobin in the bloodstream. Because animals exposed to TNT have contracted cancer, the explosive is ranked as a probable human carcinogen.

15.9-A TNT EQUIVALENTS

TNT equivalent ■ A comparison of the mass of an explosive substance with the mass of TNT that produces the same explosive power

It is customary to compare the mass of an explosive substance to the mass of TNT that produces the same explosive power. This comparison is called a **TNT equivalent**. For example, when 1 pound (454 g) of amatol detonates, the same explosive power is produced as when 0.59 pound (268 g) of TNT detonates. When 1 pound (454 g) of

TABLE 15.10	Shipping Descriptions of Trinitrotoluene
TRINITROTOLUENE	**SHIPPING DESCRIPTION**
Trinitrotoluene, dry or wetted with less than 30% water by mass	UN0209, Trinitrotoluene, 1.1D, PG II (EX-xxxxx) (Poison) *or* UN0209, TNT, 1.1D, PG II (EX-xxxxx)
Trinitrotoluene/trinitrobenzene mixtures	UN0388, Trinitrotoluene and trinitrobenzene mixtures, 1.1D, PG II (EX-xxxxx) (Poison) *or* UN0388, TNT and trinitrobenzene mixtures, 1.1D, PG II (EX-xxxxx) (Poison)
Trinitrotoluene/trinitrobenzene/ hexanitrostilbene mixtures	UN0389, Trinitrotoluene mixtures containing trinitrobenzene and hexanitrostilbene, 1.1D, PG II (EX-xxxxx) (Poison) *or* UN0389, TNT mixtures containing trinitrobenzene and hexanitrostilbene, 1.1D, PG II (EX-xxxxx) (Poison)

nitroglycerin detonates, it produces the same explosive power as when 1.49 pounds (676 g) of TNT detonates.

TNT equivalents are used to compare the relative effectiveness of different explosives to accomplish a specified task.

15.9-B WORKPLACE REGULATIONS INVOLVING TNT

When TNT is present in the workplace, OSHA requires employers to limit employee exposure by both inhalation and skin contact. The permissible exposure limit is 1.5 mg/m^3, averaged over an 8-hour workday.

15.9-C TRANSPORTING TNT

When shippers offer desensitized TNT for transportation, DOT requires them to provide the relevant shipping description shown in Table 15.10 on an accompanying shipping paper. DOT also requires shippers and carriers to comply with all applicable labeling, marking, and placarding requirements.

15.10 CYCLONITE

Cyclonite is the common name for the explosive whose chemical name is cyclotrimethylene-trinitramine, or hexahydro-1,3,5-trinitro-1,3,5-triazine.

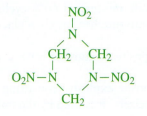

Cyclonite

Another name for cyclonite is *RDX*, the acronym for Research Department Explosive or Royal Demolition Explosive.

Cyclonite is an important member of a class of explosives called **nitramines**, organic compounds having the following group of atoms:

$$-N-NO_2$$

nitramine ■ Any organic compound whose general chemical formula is R—N—NO₂, where R is an arbitrary alkyl or aryl group

TABLE 15.11	Physical Properties of Cyclonite
Melting point	396°F (204°C)
Detonation temperature	Detonates as it melts
Specific gravity at 68°F (20°C)	1.82
Brisance (g of sand)	60.2
Detonation velocity	27,400 ft/s (8350 m/s)
Sensitivity	High

Other nitramines noted in this chapter include tetryl and cyclotetramethylenetetranitramine.

Cyclonite is prepared from hexamethylenetetramine, nitric acid, ammonium nitrate, and acetic anhydride. It is a white crystalline solid. Other physical properties of cyclonite are listed in Table 15.11.

When the chemical formula for cyclonite is condensed to $(CH_2)_3N_3(NO_2)_3$, its detonation is represented as follows:

$$(CH_2)_3N_3(NO_2)_3(s) \longrightarrow 3CO(g) + 3N_2(g) + 3H_2O(g)$$

Cyclonite   Carbon monoxide   Nitrogen   Water

One pound of cyclonite produces the same explosive power as 1.2 pounds (545 g) of TNT. Although it is used by itself as an explosive, it is also mixed with other explosives for use. For instance, the mixture of cyclonite, TNT, and aluminum fines called *torpex* has been used in warfare as the explosive component of mines, depth charges, and torpedo warheads.

Although cyclonite is extremely sensitive to explosive decomposition, it may be desensitized by mixing it with beeswax, which stabilizes it even when exposed to high temperatures. The fact that cyclonite is thermally stable when desensitized makes it potentially useful when the need arises to use explosives during firefighting. Mixed with beeswax in varying proportions, cyclonite was also used widely through World War II as the bursting charge in aerial bombs, mines, and torpedoes.

The following formulations of cyclonite are encountered:

- Composition A-3 is a composite mixture of 91% cyclonite and 9% beeswax by mass. This explosive substance often is selected by explosive experts because it detonates so rapidly. A charge of Composition A-3 in a 1-ton bomb detonates in approximately 0.25 millisecond. This relatively high rate of detonation yields an extraordinary brisance.
- *Composition B* is a composite mixture of 60% cyclonite, 39% trinitrotoluene, and 1% beeswax by mass. This mixture has largely replaced Composition A-3 in artillery shells.
- Composition B-4 is a composite mixture of 60% cyclonite, 39.5% trinitrotoluene, and 0.5% calcium silicate.
- Composition C-4 is a composite explosive consisting of a mixture of 91% cyclonite and 9% nonexplosive plasticizers. It is used by the military in buster charges.

plastic explosive ■ A high explosive that has been mixed with a gummy binder and manufactured in a flexible, hand-malleable form for an intended use at 77°F (25°C) such as demolition

Composition C-4 sometimes is mixed with a gummy binder and hand-molded into a putty-like shape for easy use. This flexible or malleable mixture is an example of a **plastic explosive**. When mixed with the binder, the explosive substance retains its brisance when detonated but is more suitable for use when a specific shape is desired. The main disadvantage in using a plastic explosive is that it may become brittle in cold weather, thereby losing its plasticity.

Caution must be exercised when using cyclonite, because it is toxic when ingested or inhaled. Exposure to high concentrations of the explosive has been linked with the onset of seizures. Animal exposure to cyclonite has been shown to cause cancer; hence, the explosive is ranked as a probable human carcinogen.

15.10-A WORKPLACE REGULATIONS INVOLVING CYCLONITE

When cyclonite is present in the workplace, OSHA requires workers to limit employee exposure to a concentration of 1.5 mg/m^3, averaged over an 8-hour workday. NIOSH recommends a short-term exposure limit of 1.5 mg/m^3 over a 40-hour workweek.

15.10-B TRANSPORTING CYCLONITE

When shippers offer desensitized cyclonite for transportation, DOT requires them to identify it on an accompanying shipping paper as one of the following, as relevant: UN0483, Cyclotrimethylenetrinitramine, desensitized, 1.1D, PG II (EX-xxxxx) (Poison) *or* UN0072, Cyclotrimethylenetrinitramine, wetted, 1.1D, PG II (EX-xxxxx) (Poison). DOT also requires shippers and carriers to comply with all applicable labeling, marking, and placarding requirements.

15.10-C TERRORISTS' MISUSE OF CYCLONITE

Terrorists have used cyclonite as a weapon of mass destruction. Cyclonite and PETN were identified as the active ingredients in the plastic explosive used to destroy Pan Am Flight 103 in 1988. Residues of these explosives were also identified following a terrorist event in India in 2003. Furthermore, al-Qaida terrorists used the C-4 formulation of cyclonite to cripple the USS Cole in 2000.

15.11 TETRYL

Tetryl is the commercial explosive whose proper chemical name is 2,4,6-trinitrophenyl-*N*-methylnitramine.

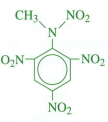

It is a yellow solid; other physical properties are provided in Table 15.12.

Tetryl

TABLE 15.12	Physical Properties of Tetryl
Melting point	266°F (130°C)
Detonation temperature	369°F (187°C)
Specific gravity (pressed) at 68°F (20°C)	1.60
Brisance (g of sand)	54.2
Detonation velocity	23,300 ft/s (7100 m/s)
Sensitivity	High

When the chemical formula for tetryl is condensed to $(NO_2)_3C_6H_2N(NO_2)(CH_3)$, its detonation is represented as follows:

$$2(NO_2)_3C_6H_2N(NO_2)(CH_3)(s) \longrightarrow 3C(s) + 11CO(g) + 5N_2(g) + 5H_2O(g)$$

Tetryl   Carbon   Carbon monoxide   Nitrogen   Water

The detonation of tetryl produces approximately the same explosive power as the same mass of TNT.

Tetryl is well known for its exceptionally high brisance. Since World War II it has been the standard explosive used by the military as the booster in artillery ammunition.

When tetryl is mixed with molten trinitrotoluene and a small amount of graphite, the popular explosive substance called tetrytol is produced. Tetrytol sometimes is used by the military as the bursting charge in artillery ammunition.

Caution must be exercised when handling and using explosives containing tetryl, because tetryl is toxic by inhalation, ingestion, and skin absorption. Exposed individuals experience a host of symptoms, including coughing, fatigue, headache, nosebleed, nausea, vomiting, and skin rashes. Research studies suggest that exposure to tetryl may also affect kidney, liver, and spleen function.

15.11-A WORKPLACE REGULATIONS INVOLVING TETRYL

When tetryl is present in the workplace, OSHA requires employers to limit employee exposure to a concentration of 1.5 mg/m³, averaged over an 8-hour workday. NIOSH recommends a maximum exposure limit of 1.5 mg/m³ over a 40-hour workweek.

15.11-B TRANSPORTING TETRYL

When shippers offer tetryl for transportation, DOT requires them to identify it on an accompanying shipping paper as follows:

UN0208, Trinitrophenylmethylnitramine, 1.1D, PG II (EX-xxxxx) (Poison)

DOT also requires shippers and carriers to comply with all applicable labeling, marking, and placarding requirements.

Pentaerythritol

15.12 PETN

PETN (pronounced *pettin*) is the active component of the commercial explosive substance whose proper chemical name is pentaerythritol tetranitrate, or pentaerythrite tetranitrate.

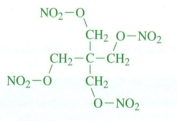

The condensed formula for this substance is $C(CH_2-O-NO_2)_4$. Like nitroglycerin, PETN is an example of a nitrate ester. It is a white solid; some of its other physical properties are provided in Table 15.13.

Although PETN is sometimes used as the booster in artillery ammunition, it is more commonly encountered in the detonation fuses, *Primacord* and *Cordtex*. These product trademarks refer to thin, flexible tubes containing PETN and wrapped in a fabric sheath. Both Primacord and Cordtex are used to initiate the detonation of other high explosives. PETN is also a key ingredient in the explosive called Semtex.

TABLE 15.13	Physical Properties of PETN
Melting point	282°F (139°C)
Boiling point	437°F (225°C)
Detonation temperature	374–410°F (190–210°C)
Specific gravity at 68°F (20°C)	1.75
Brisance (g of sand)	62.7
Detonation velocity	27,200 ft/s (8,300 m/s)
Sensitivity	High

When the chemical formula for PETN is condensed to $C(CH_2-O-NO_2)_4$, its detonation is represented as follows:

$$C(CH_2-O-NO_2)_4(s) \longrightarrow 3CO_2(g) + 2CO(g) + 2N_2(g) + 4H_2O(g)$$

PETN   Carbon dioxide   Carbon monoxide   Nitrogen   Water

When 1 pound (454 g) of PETN detonates, the same explosive power is produced as when 1.29 pounds (586 g) of TNT detonates.

15.12-A USE OF PETN OTHER THAN AS AN EXPLOSIVE

PETN acts as a vasodilator when consumed internally. Thus, like nitroglycerin, it is useful medicinally for the treatment of heart and circulatory diseases.

15.12-B TRANSPORTING PETN

When shippers offer PETN for transportation, DOT requires them to identify the relevant form as shown in Table 15.14 on an accompanying shipping paper. DOT also requires shippers and carriers to comply with all applicable labeling, marking, and placarding requirements.

When carriers transport PETN in hazard class 4.1 in an amount exceeding 1001 pounds (454 kg), DOT requires them to post FLAMMABLE SOLID placards on the bulk packaging or transport vehicle used for shipment. When carriers transport PETN in hazard class 1.1 in any amount, DOT requires them to post EXPLOSIVES 1.1 placards bearing the compatibility group D on the bulk packaging used for shipment.

15.12-C TERRORISTS' MISUSE OF PETN

al-Qaida operatives have chosen PETN on multiple occasions for executing terrorist events. The explosive is the terrorist's choice of weapons, presumably due to the wide availability of Primacord and Cordtex.

The nature of the terrorist events involving PETN is summarized below:

■ As noted in Section 13.9-D, in 2001, the "Shoe-Bomber" terrorist intended to detonate PETN while traveling to the United States onboard an international aircraft. A mixture of PETN and triacetone triperoxide (TATP) had been packed within the soles of his shoes, but passengers overtook him before the mixture was activated.

■ On Christmas Day 2009, a Nigerian citizen with connections to al-Qaida attempted to detonate 80 grams (0.2 lb) of PETN with TATP while traveling on a Detroit-bound international flight. Because a packet of the explosives was sewn into his underwear, he is now known by both of the names "Underwear Bomber" and "Christmas Day Bomber." His activities were thwarted by passengers and the flight crew.

TABLE 15.14	Shipping Descriptions of Pentaerythritol Tetranitrate
FORM OF PENTAERYTHRITOL TETRANITRATE	**SHIPPING DESCRIPTION**
Pentaerythritol tetranitrate, with not less than 7 percent wax by mass	UN0411, Pentaerythrite tetranitrate, 1.1D, PG II (EX-xxxxx) *or* UN0411, Pentaerythritol tetranitrate, 1.1D, PG II (EX-xxxxx) *or* UN0411, PETN, 1.1D, PG II (EX-xxxxx)
Pentaerythritol tetranitrate mixture, with more than 10% but not more than 20% PETN by mass	UN3344, Pentaerythrite tetranitrate mixture, desensitized, solid, n.o.s., 4.1, PG II *or* UN3344, Pentaerythritol tetranitrate mixture, desensitized, solid, n.o.s., 4.1, PG II *or* UN3344, PETN mixture, desensitized, solid, n.o.s., 4.1, PG II
Pentaerythritol tetranitrate, wetted, with not less than 25% water by mass	UN0150, Pentaerythrite tetranitrate, wetted, 1.1D, PG II (EX-xxxxx) *or* UN0150, Pentaerythritol tetranitrate, wetted, 1.1D, PG II (EX-xxxxx) *or* UN0150, PETN, wetted, 1.1D, PG II (EX-xxxxx) *or* UN0150, Pentaerythrite tetranitrate, desensitized, 1.1D, PGII (EX-xxxxx) *or* UN0150, Pentaerythritol tetranitrate, desensitized, 1.1D, PGII (EX-xxxxx) *or* UN0150, PETN, desensitized, 1.1D, PGII (EX-xxxxx)

■ In 2009, PETN was used in a failed suicide attempt against a top Saudi counterterrorism official. Although the official survived, the terrorist died when the PETN detonated.

■ In November 2010, al-Qaida terrorists in Yemen hid 300 grams (0.7 lb) and 400 grams (0.9 lb), respectively, of PETN in the toner cartridges of two computer printers that were being mailed from Yemen as packages addressed to synagogues in Chicago, Illinois. The packages traveled on two passenger planes that transited through Yemen and Britain, where they were intercepted and seized by authorities before the PETN detonated. The terrorists had timed for at least one explosive package to detonate while the plane was flying in Canadian airspace or over the eastern seaboard of the United States.

15.13 HMX

HMX is the secondary high explosive whose proper chemical name is cyclotetramethylenetetra-nitramine, or 1,3,5,7-tetranitro-1,3,5,7-tetrazocine.

Cyclotetramethylene-tetramine

TABLE 15.15	Physical Properties of HMX
Melting point	527°F (275°C)
Detonation temperature	319°F (159°C)
Specific gravity at 68°F (20°C)	1.89
Brisance (g of sand)	60.4
Detonation velocity	29,890 ft/s (9110 m/s)
Sensitivity	High

The origin of the acronym HMX is most likely linked with the names "*high-molecular-weight explosive*, *high-melting explosive*, or *Her Majesty's eXplosive*". Other commercial names for this explosive are *Octogen* and *Rowanex 2000*.

The molecular structure of HMX may be condensed to $(CH_2)_4(N-NO_2)_4$. Some of its important physical properties are provided in Table 15.15.

The detonation of HMX can be represented as follows:

$$(CH_2)_4(N-NO_2)_4(s) \longrightarrow 4CO(g) + 4N_2(g) + 4H_2O(g)$$

HMX   Carbon monoxide   Nitrogen   Water

When 1 pound (454 g) of HMX detonates, the same explosive power is produced as when 1.26 pounds (572 g) of TNT detonates. HMX is regularly used in the military warfare actions of today, generally as shaped-charge warhead explosives and rocket propellants.

Like cyclonite, HMX is produced from hexamethylenetetramine, nitric acid, ammonium nitrate, and acetic anhydride. HMX and cyclonite are chemically similar. Their molecules are eight- and six-membered-ring-shaped nitramines, respectively.

Caution must be exercised when using HMX, because studies conducted on rats, mice, and rabbits reveal that HMX may be harmful to the hepatic (liver) and central nervous systems when ingested or absorbed through the skin.

15.13-A USE OF HMX WITH OTHER EXPLOSIVES

An explosive called Octol 70/30 is a mixture of 70% HMX and 30% TNT by mass. It is used as a rocket propellant by the military. The detonation of HMX produces approximately the same explosive power as the same mass of TNT. The American military engaged in the war in Afghanistan has used a mixture of Octol 75/25 (75% HMX/25% TNT) and potassium chlorate as the explosive in ammunition.

15.13-B TRANSPORTING HMX

When shippers offer HMX for transportation, DOT requires them to identify it on an accompanying shipping paper as either of the following, as relevant: UN0484, Cyclotetramethylenetetra-nitramine, desensitized, 1.1D, PG II (EX-xxxxx) *or* UN0226, Cyclotetramethylenetetranitramine, wetted, 1.1D, PG II (EX-xxxxx). DOT also requires shippers and carriers to comply with all applicable labeling, marking, and placarding requirements.

TABLE 15.16	Physical Properties of the Primary Explosives		
	MERCURY FULMINATE	**LEAD AZIDE**	**LEAD STYPHNATE**
Melting point	Detonates	Detonates	Detonates
Detonation temperature	>302°F (150°C)	660°F (350°C)	500°F (260°C)
Specific gravity at 68°F (20°C)	4.42		2.9 (anhydrous)
Brisance (g of sand)	23.4	19	24
Detonation velocity	13,940 ft/s (4250 m/s)	17,500 ft/s (5330 m/s)	17,000 ft/s (5180 m/s)
Sensitivity	High	High	High

15.14 PRIMARY EXPLOSIVES

As first noted in Section 15.2, primary explosives are substances whose detonation is sensitive to a stimulus like heat, shock, or friction. For this reason, they are extremely dangerous to handle. In practice, primary explosives are used in small quantities to initiate the detonation of a larger charge of a main explosive. When used for this purpose, they are called **initiators, initiating explosives**, or **primers**.

initiator (initiating explosive; primer)
■ Any explosive material used to detonate the main charge

Three primary explosives frequently are used as initiators in percussion caps, shell cartridges, detonators, and fuses. When activated, they produce the detonation wave that initiates the booster, or bursting charge. These primary explosives are mercuric fulminate, lead azide, and lead styphnate. As compounds of mercury and lead they are highly toxic. For this reason, their use by explosives experts is declining. Their physical properties are provided in Table 15.16.

15.14-A MERCURY FULMINATE

Mercury fulminate

The name *mercury fulminate* is synonymous with mercury(II) cyanate, a substance whose chemical formula is $Hg(CNO)_2$. The word *fulminate* is derived from a Latin term meaning to strike with lightning.

$$O-N\equiv C-Hg-C\equiv N-O$$

It is a white-to-gray solid produced by pouring a nitric acid solution of mercury(II) nitrate into ethanol, but the mechanism of the chemical reaction is not entirely understood.

Mercury fulminate is extremely sensitive to heat, shock, and friction. When it detonates, mercury fulminate decomposes violently into mercury, carbon monoxide, and nitrogen.

$$Hg(CNO)_2(s) \longrightarrow Hg(g) + 2CO(g) + N_2(g)$$
Mercury fulminate   Mercury   Carbon monoxide   Nitrogen

It has been used for the past 300 years to manufacture blasting caps and detonators for military, industrial, and sporting purposes.

15.14-B LEAD AZIDE

This substance is unique among commercial and military explosives in that it is the only one whose chemical composition does not include oxygen. Its chemical formula is $Pb(N_3)_2$.

It is a colorless solid prepared by reacting aqueous solutions of sodium azide and lead acetate.

Lead azide

$$2NaN_3(aq) \quad + \quad Pb(C_2H_3O_2)_2(aq) \quad \longrightarrow \quad Pb(N_3)_2(s) \quad + \quad 2NaC_2H_3O_2(aq)$$

Sodium azide   Lead acetate   Lead azide   Sodium acetate

Lead azide often is packed into aluminum detonator capsules. When lead azide detonates, it decomposes into lead and nitrogen.

$$Pb(N_3)_2(s) \quad \longrightarrow \quad Pb(s) \quad + \quad 3N_2(g)$$

Lead azide   Lead   Nitrogen

15.14-C LEAD STYPHNATE

Lead styphnate and lead trinitroresorcinate are synonyms for the substance having the following formula:

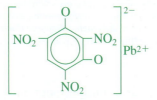

It is a yellow-orange solid made from styphnic acid, or trinitroresorcinol.

When the formula for lead styphnate is condensed to $Pb[C_6HO_2(NO_2)_3]$, its detonation is represented as follows. The products are a mixture of gases and vapors.

$$2Pb[C_6HO_2(NO_2)_3](s) \quad \longrightarrow \quad Pb(s) \quad + \quad 3N_2(g) \quad + \quad H_2O(g) \quad + \quad 5CO(g) \quad + \quad 5CO_2(g) \quad + \quad 2C(s)$$

Lead styphnate   Lead   Nitrogen   Water   Carbon monoxide   Carbon dioxide   Carbon

15.14-D TRANSPORTING PRIMARY EXPLOSIVES

When shippers offer a primary explosive for transportation, DOT requires them to identify it on an accompanying shipping paper as one of the following, as relevant:

Lead styphnate

> UN0135, Mercury fulminate, wetted, 1.1A, PG II (EX-xxxxx) (Poison)
>
> UN0129, Lead azide, wetted, 1.1A, PG II (EX-xxxxx) (Poison)
>
> UN0130, Lead styphnate, wetted, 1.1A, PG II (EX-xxxxx) (Poison)

When shipping mercury fulminate, DOT requires shippers and carriers to comply with all applicable labeling, marking, and placarding requirements. DOT classifies lead azide and lead styphnate as forbidden explosives.

15.15 RESPONDING TO INCIDENTS INVOLVING A RELEASE OF EXPLOSIVES

There are five potential types of emergency incidents affecting a release of explosives: a transportation mishap involving explosives; a fire within a magazine; disowned or abandoned ordnance; the detection of an improvised explosive device; and the treatment or disposal of a waste explosive. *Responding to any of these emergency incidents should be undertaken only by competent, experienced individuals who have received special training in the handling of explosives.* To save lives at a scene where an explosive may have been detonated with felonious intent, untrained responders often must limit their activities to evacuating unnecessary individuals to the distances specified in Table 15.17.

TABLE 15.17		Surface Explosion Information[a]	
TNT EQUIVALENT		**RECOMMENDED EVACUATION DISTANCES**	
lb	kg	ft	m
5	2.3	11–17	4–6
50	23	24–37	8–12
500	227	51–79	16–24
1000	454	64–99	20–31
4000	1814	101–157	31–48
10,000	4536	137–213	42–65
30,000	13,608	198–307	61–94
60,000	27,216	249–387	76–118

[a]The evacuation distances from the scene of a potential explosion were generated through the use of Sandia National Laboratory's BLAST computer model. Excerpted with permission from Michael B. Dillon, Ronald L. Baskett, Kevin T. Foster, and Connee S. Foster, *The NARC Emergency Response Guide to Initial Airborne Hazard Estimates*, UCRL-TM-202990 (Livermore, CA: U.S. Department of Energy, 2004), p. 8.

The presence of an explosive at a transportation mishap usually may be verified by observing the following:

- The number 1 as a component of a shipping description of a hazardous material on a shipping paper
- Either of the expressions *EXPLOSIVE* or *BLASTING AGENT* and the number 1 on orange labels affixed to packaging
- The expression *EXPLOSIVES* and the number 1 on orange placards posted on each side and each end of a transport vehicle containing an explosive in divisions 1.1, 1.2, or 1.3; the expression *1.4 EXPLOSIVES* or *1.6 EXPLOSIVES* and the number 1 on orange placards posted on each side and each end of a transport vehicle containing 1001 pounds (454 kg) or more of an explosive in divisions 1.4 or 1.6; or the expression *1.5 BLASTING AGENTS* and the number 1 on orange placards posted on each side and each end of a transport vehicle containing 1001 pounds (454 kg) or more of a blasting agent in division 1.5.

When explosives are involved in a transportation mishap, DOT advises first-on-the-scene responders to evacuate all unnecessary individuals to a distance of 0.5 miles (800 m).

When called to an emergency incident involving an explosive magazine, the Institute of Makers of Explosives advises first-on-the-scene responders to implement the procedures in Figure 15.7. Fires outside a magazine should be controlled to ensure that they do not reach the exterior perimeter of the magazine, but fires inside the magazine should not be fought.

SOLVED EXERCISE 15.4

The first-on-the-scene responders to a transportation mishap note the presence of an overturned motor van along a route specifically designated for the transportation of explosives. From a distance of approximately 300 feet (91 m), they also observe a small fire that appears to be spreading upward from a tire on the vehicle toward the overlying cargo van. Orange placards have been posted on the visible sides of the vehicle, although the DOT division number is not readily discernible. What immediate actions should these responders execute to protect public health, safety, and the environment?

Solution: DOT requires the posting of orange placards on a transport vehicle to warn that packages of explosives are onboard for transit. Given that the transport vehicle is on fire, the crew should promptly recognize that the highest degree of hazard is associated with potential detonation of the explosives. Because the DOT division of the explosive cannot be discerned, it is prudent to assume that it is 1.1.

The responding crew should do the following:

- Evacuate all individuals to 5000 feet (1500 m) in all directions from the transportation incident. This includes stopping traffic in all directions.
- Do not attempt to extinguish or control the spread of the fire *until* the driver, shipper, or carrier accurately confirms the identity of the division number of the explosive onboard.
- Without knowledge of the division number of the explosive, acknowledge that the risk to the lives of the responders is so great that combating the fire is absolutely unwarranted.

When emergency responders are called to incidents involving material that has been disowned or abandoned, perhaps by activists or terrorists, it is prudent to suspect that this material consists of unexploded ordnance. In this type of incident, the suspect material generally is removed remotely into a total-explosive-containment vessel such as that

FIGURE 15.7 The "Emergency Procedures" poster published by the Institute of Makers of Explosives. (*Courtesy of Institute of Makers of Explosives, Washington, DC.*)

FIGURE 15.8 This total containment device allows explosives experts and hazmat technicians to safely containerize, transport, sample, and dispose of explosives and chemical/biological devices having up to 15 lb (6.8 kg) of C-4 explosives. (*Courtesy of Nabco, Inc., Cononsburg, Pennsylvania.*)

shown in Figure 15.8. Once the suspect material has been transferred, the vessel is driven to a remote location at which the chemical nature of the material may be determined by experts.

Disowned or abandoned waste explosives are RCRA-regulated hazardous wastes that exhibit the characteristic of reactivity. When a decision is made to treat or destroy them, EPA should be contacted for guidance relating to the manner of treatment or disposal that best protects public health and the environment.

General Characteristics of Explosives

1. Why is a shock wave generated when an explosive detonates?
2. Following the detonation of an explosive, a plume often appears, much like a smoke plume during a fire. Why is this detonation plume often red?
3. The OSHA regulation at 29 C.F.R. §1910.109(e)(2) stipulates that empty boxes, paper, and fiber packing materials that previously contained high explosives cannot be used again for any purpose; instead, they must be destroyed by burning them at an approved, isolated location outdoors. No person may be nearer than 100 feet (31 m) after the burning begins. What is the most likely reason that OSHA requires the destruction of the packaging materials in this fashion?

Transporting Explosives

4. Which explosive article in each of the pairs labeled as follows potentially poses the greater detonation hazard?
   (a) EXPLOSIVE 1.4S or EXPLOSIVE 1.2G
   (b) EXPLOSIVE 1.1D or EXPLOSIVE 1.4S
5. A shipper intends to offer fourteen 1-pound (0.5-kg) wooden boxes of trinitrotoluene for transportation by a private carrier in a sole-use truck.
   (a) Use Appendix C to determine the shipping description that the shipper must enter on the accompanying shipping paper if the material is wetted with less than 30% water by mass.
   (b) Use Appendix C to determine the shipping description that the shipper must enter on accompanying shipping paper if the material is wetted with more than 30% water by mass.
   (c) For each situation noted in parts (a) and (b) of this question, identify the labels, if any, that DOT requires the shipper to affix to the boxes.
   (d) For each situation noted in parts (a) and (b) of this question, identify the placards, if any, that DOT requires the carrier to post on the truck.
6. An ammunition manufacturer in Massachusetts desires to offer incendiary ammunition for transportation by motor vehicle to a military facility in Virginia. The ammunition consists of an explosive encased within a flammable gel. The manufacturer packs the ammunition into 10 cylindrical cartridges made of waxed cardboard, which then are packed into two wooden boxes. The ammunition explosive has a net explosive mass of 24 pounds (11 kg). DOT has assigned EX-1325867 to the manufacturer.
   (a) What shipping description of the ammunition does DOT require the manufacturer to enter on the accompanying shipping paper?
   (b) Which labels does DOT require the manufacturer to affix to the outside surface of each wooden box?
   (c) Which markings does DOT require the manufacturer to display on the outside surface of each wooden box?
   (d) Which placards, if any, does DOT require the carrier to post on the motor vehicle used to transport the wooden boxes to the military facility?

Black Powder

7. Black powder consists of a mixture of substances whose components, when ignited, undergo an oxidation–reduction reaction. One of the components is an oxidizer. What is the most likely reason DOT regulates the transportation of black powder as an explosive rather than as an oxidizer?

8. When emergency responders encounter black powder while combating a fire at a sporting goods store, how may they substantially reduce its explosive potential?

Nitroglycerin and Dynamite

9. Why should dynamite sticks be handled remotely when they are discolored, excessively soft, crumbly, or show visible signs of exterior crystallization?

10. Although nitroglycerin is an explosive substance, cardiologists often prescribe low doses of this substance to their patients. How does the nitroglycerin aid the patients?

Nitrocellulose

11. Why do explosive experts strongly advise against storing nitrocellulose in bulk quantities?

Trinitrotoluene

12. During World War II, TNT often was used as a military explosive. What combination of properties caused the military to select TNT over other explosives?

Cyclonite, Tetryl, PETN, and HMX

13. Which explosive is a nitrate ester: cyclonite, tetryl, PETN, or HMX?

14. Military explosives containing cyclonite, tetryl, PETN, or HMX sometimes are *aluminized*; that is, they are mixed with aluminum dust and an oxidizing agent. What is the role of these latter substances in the explosive articles?

Primary Explosives

15. Why is lead azide selected for use in virtually *all* blasting caps and other hot-wired initiated detonators?

Responding to Emergencies Involving a Release of Explosives

16. When first-on-the-scene responders arrive at the scene of an accident on a heavily trafficked highway, they note that orange EXPLOSIVES 1.4 placards bearing the compatibility group D are posted on each side and each end of a vehicle that has been

struck by another vehicle. Fuel escaping from the second vehicle has ignited. A representative of the carrier provides the accompanying shipping paper, which reads in part as follows:

UNITS	HM	SHIPPING DESCRIPTION (IDENTIFICATION NUMBER, PROPER SHIPPING NAME, PRIMARY HAZARD CLASS OR DIVISION, SUBSIDIARY HAZARD CLASS OR DIVISION, AND PACKING GROUP)	WEIGHT (lb)
30 plywood boxes (UN4D)	X	UN0104, Cord, detonating, mild effect (contains pentaerythritol tetranitrate), 1.4D, PG II (EX-9806054) (DOT-SP 4850)	1500 lb

What immediate actions should the responders execute to protect public health, safety, and the environment?

INTERNATIONAL
FACILITY   5698A

01/07-03/07

Dr. Susan Winters

BRIAN
SUMMERS
RADIOLOGY
UNIVERSITY
MEDICAL CENTER

Pa

luxel+
LANDAUER

InLight Systems
LANDAUER

103702

qs1501

Courtesy of Landauer, Inc., Glenwood, Illinois.

KEY TERMS

OBJECTIVES

- Describe the phenomenon of radioactivity and the concept of a half-life for a given radioisotope.
- Describe the nature of each mode by which a radioisotope may decay.
- Describe the differentiating features of alpha, beta, and gamma radiation.
- Describe the nature of a sealed radiation source and the features of the ionizing radiation symbol used to identify its presence.
- Identify the units used for the measurement of activity and radiation dose.
- Identify the general aspects of OSHA regulations requiring employers to limit radiation exposure in the workplace, including the posting of signs in areas where radioactive materials are stored or used.
- Describe the phenomenon of nuclear fission, including spontaneous fission.
- Identify the primary health concern posed by the presence of radon within residential dwellings.
- Identify the labels, markings, and placards that DOT requires on packaging of radioactive materials and the transport vehicles used for their shipment.
- Identify the response actions to be executed when radioactive materials are released from their packaging into the environment.
- Describe the response action to be executed when a radiological dirty bomb has been activated and discharged into the environment.

When we hear the term *radioactive*, two fearsome incidents generally come to mind: the accidental release of material from a nuclear power plant and the deployment of nuclear weapons.[1] We discover in this chapter that both events are associated with the occurrence of one or more nuclear processes. The forms of matter that display them are called **radioactive materials**.

We also learn in this chapter that serious health risks are linked with exposure to radioactive materials. To eliminate or minimize the impact of these risks, the U.S. Congress delegated the following responsibilities to the regulatory bodies listed:

- The U.S. Nuclear Regulatory Commission (NRC) regulates the civilian nuclear energy industry by licensing the construction and operation of the nation's nuclear power plants.
- The U.S. Department of Energy (DOE) oversees the research and development of new and creative means for reducing the burgeoning supply of nuclear waste. In addition, it oversees the construction and operation of nuclear waste disposal sites and responds to releases of radioactive material from any source. DOE also certifies the integrity of the unique packages in which radioactive material is transported, provided they are for the purpose of national security.

radioactive material
- A material containing an isotope that spontaneously emits ionizing radiation; for DOT purposes, a material whose specific activity and total activity in a consignment exceed values published for listed radioisotopes at 49 C.F.R. §173.436

---

[1]Nuclear weapons, or nuclear bombs, initially were called *atomic bombs*. The term *nuclear* is preferable to *atomic*, because the phenomenon responsible for their detonation is a nuclear one.

- EPA establishes radiation-exposure limits to protect public health. These limits apply to radiation arising in the environment, including natural radiation and the radiation from spent radioactive materials in storage. EPA also monitors the levels of radioactivity in air, precipitation, drinking water, and milk at 164 monitoring stations spread throughout the 50 states.

- OSHA establishes radiation-exposure limits that protect employees who use radioactive materials within those workplaces that are not regulated by NRC or DOE.

- DOT ensures that shippers and carriers adopt procedures to eliminate or minimize the risks associated with transporting radioactive materials.

In combination, these agencies serve to provide a degree of protection against the hazards associated with inadvertent exposure to radioactive materials.

16.1 FEATURES OF ATOMIC NUCLEI

In Section 4.4, we noted that the atomic nucleus has two primary constituents: protons and neutrons. Although the nuclei of all atoms of the same element have the same number of protons, they may have different numbers of neutrons. These nuclei are called isotopes.

The number of protons found in the nucleus of an atom is its atomic number. The number of protons equals the number of electrons in a neutral atom. Because the atomic numbers of the elements are compiled in the periodic table, we may use a periodic table to readily identify the number of protons in a given nucleus. The total number of protons and neutrons in a particular isotope is called its **mass number**.

Hydrogen has an atomic number of 1; this means that every hydrogen atom has one proton. Each hydrogen atom has one electron, but when a hydrogen atom is ionized, it is stripped of its electron, and only its nucleus remains.

Hydrogen atoms exist in any of three isotopic forms having the following unique names and compositions:

- **Protium** is the simplest of the hydrogen isotopes and, in fact, the simplest of all nuclei. Protium has a single proton as its nucleus. When a protium atom is ionized, only a proton remains.

- **Deuterium** is the second hydrogen isotope. Deuterium has a nucleus consisting of a proton and a neutron. When a deuterium atom is ionized, the remaining nucleus is composed of one proton and one neutron and is called a **deuteron**.

- **Tritium** is the third hydrogen isotope. Tritium has a nucleus consisting of a proton and two neutrons. The nucleus of an ionized tritium atom is called a **triton**.

Each isotope is designated by the symbol $^A_Z X$, where Z and A are the atomic number and mass number, respectively; X is the element's chemical symbol. Using this format, the three hydrogen isotopes are designated by the symbols $^1_1 H$, $^2_1 H$, and $^3_1 H$, respectively. For each, the symbol Z equals 1, the number of protons. The number of neutrons is the difference between the superscript and the subscript: for protium, the number is 0; for deuterium, the number is 1; and for tritium, the number is 2. Deuterium and tritium are also represented as D and T, respectively.

Only the isotopes of hydrogen have unique names. The isotopes of other elements are named by identifying the name or symbol of the element and the mass number of the isotope at issue. Thus, nuclei designated as $^{12}_6 C$, $^{235}_{92} U$ and $^{40}_{19} K$ are named carbon-12, uranium-235, and potassium-40, or C-12, U-235, and K-40, respectively.

All nuclei can have three or more isotopes, many of which are stable; that is, they retain their compositions and do not undergo spontaneous changes. However, many other nuclei are subject to spontaneous transformations. They are said to "transform," "decay," or "disintegrate." The phenomenon is called **radioactivity**, and the intrinsically unstable nuclei are

mass number ■ The total number of protons and neutrons in a given nucleus

protium ■ The hydrogen isotope composed of one proton and no neutrons

deuterium ■ The hydrogen isotope composed of one proton and one neutron

deuteron ■ The nucleus of the deuterium atom

tritium ■ The hydrogen isotope composed of one proton and two neutrons

triton ■ The nucleus of the tritium atom

radioactivity ■ The property associated with the spontaneous emission of alpha, beta, and/or gamma radiation from radioisotopes, their capture of extranuclear electrons, or their spontaneous fission

said to be radioactive. They are called **radioisotopes**, or **radionuclides**. Protium and deuterium are stable, nonradioactive isotopes of hydrogen, but tritium is a radioisotope.

Radioactivity generally is not affected by any physical or chemical change in a substance. Hence, when radioactive materials are subjected to changes in pressure, temperature, or chemical nature, the spontaneous disintegration of the relevant radioisotope usually is not altered.

When a radioisotope undergoes a change, it usually emits a particle; less commonly, it absorbs an electron. Both processes frequently are accompanied by the simultaneous emission of energy. When the transformation occurs, the radioisotope is converted into a new nucleus, which is either stable or radioactive itself. Radioisotopes often undergo several transformations before they are converted into stable nuclei.

Each radioactive transformation is associated with a specific period of time. The time during which an arbitrary number of nuclei are reduced to half the number is called the **half-life** of that radioisotope. For instance, suppose a volume of tritium gas containing 500,000 molecules is set aside for 12.32 years. Because the tritium molecule is diatomic, there are 1 million atoms of tritium in this volume. After a 12.32-year lapse, only 500,000 tritium atoms remain; and after another 12.32-year lapse, only 250,000 tritium atoms remain. Hence, 12.32 years is the half-life of tritium.

A half-life is typically symbolized as $t_{1/2}$ and is one of the characteristic properties of a radioisotope. Its value may range from nanoseconds to billions of years. Each element has at least one radioisotope. The hundred or so elements collectively have nearly 3100 known radioisotopes. Table 16.1 shows that few radioisotopes are present in nature. If they were present when Earth was initially formed, most disappeared long ago, because the planet is 3.6 billion years old.

Some commercially available radioisotopes are listed in Table 16.2. Although they have many potential purposes, most commercial radioisotopes are used to image the

radioisotope (radionuclide) ■ An atomic nucleus that undergoes a spontaneous change by emitting a particle, absorbing an extranuclear electron, or undergoing spontaneous fission

half-life ■ The time period during which an arbitrary amount of a radioisotope is transformed into half that amount

TABLE 16.1	Some Naturally Occurring Radioisotopes[a]		
RADIOISOTOPE	**HALF-LIFE (y)**	**RELATIVE ISOTOPIC ABUNDANCE (%)**	**MODE OF DECAY[b]**
Tritium	12.32	0.00013	β^-
Carbon-14	5700		β^-
Potassium-40	1.248×10^9	0.012	β^-, EC, β^+
Rubidium-87	4.81×10^{10}	27.8	β^-
Indium-115	4.41×10^{14}	95.8	β^-
Lanthanum-138	1.02×10^{11}	0.089	β^-, EC
Neodymium-144	2.29×10^{15}	23.9	α
Samarium-147	1.06×10^{11}	15.1	α
Lutetium-176	3.76×10^{10}	2.60	β^-
Rhenium-187	4.33×10^{10}	62.9	β^-
Platinum-190	6.5×10^{12}	0.012	α
Radium-226	1600		α
Thorium-232	1.40×10^{10}	100	α
Uranium-235	7.04×10^8	0.72	α
Uranium-238	4.468×10^9	99.28	α

[a]Chart of Nuclides, National Nuclear Data Center, Brookhaven National Laboratory, Long Island, New York (2012).
[b]Section 16.2.

TABLE 16.2	Some Commercially Available Radioisotopes	

RADIOISOTOPE	HALF-LIFE[a]	APPLICATION OR USE
Americium-241	432.6 y	Smoke detectors
Cesium-137	30.08 y	Radiation source for treatment of cancer; sealed radiation source for irradiation of foods
Chromium-51	27.7025 d	Radiation source for determination of red blood cell volume and total blood volume
Cobalt-57	271.74 d	Radiation source for instrument calibration and determination of the effectiveness of the body's uptake of vitamin B_{12}
Cobalt-60	5.27 y	Sealed source for industrial radiography; sealed source for treatment of cancer, irradiation of foods, and inducing cross-linking within polyethylene and rubber macromolecules; determination of the effectiveness of the body's uptake of vitamins; sterilization of medical devices
Fluorine-18	109.77 min	Brain- and bone-imaging drug; radiation source for tumor imaging using positron emission tomography
Gadolinium-153	240.4 d	Radiation source for determining bone density and the extent of bone mineralization
Gallium-67	3.2617 d	Diagnostic drug for tumor detection
Indium-111	2.8047 d	Diagnostic drug for tumor detection; radiation source for imaging the gastric and cardiac systems
Iodine-123	13.2235 hr	Radiation source for imaging the brain, thyroid, and renal systems
Iodine-125	59.407 d	Cancer therapeutic drug; brain, blood, and metabolic-function diagnostic drug; surgically implanted as a component of "seeds" for the internal treatment of prostate cancer
Iodine-131	8.0252 d	Brain, pulmonary, and thyroid diagnostic drug; thyroid-cancer chemotherapeutic drug
Iridium-192	73.829 d	Sealed source for industrial radiography; surgically implanted as brachytherapy needles into organ for internal cancer treatment
Iron-59	44.495 d	Radiation source for measuring the rate of formation and lifetime of red blood cells
Palladium-103	16.991 d	Cancer therapeutic drug; surgically implanted as a component of "seeds" for the internal treatment of prostate cancer
Phosphorus-32	14.262 d	Radiation source for detection of skin cancer
Plutonium-238	87.7 y	Power source for the thermoelectric generators used in equipment for planetary studies, including the Martian rover, *Curiosity*
Potassium-40	1.248×10^9 y	Dating geological formations
Radium-223	11.43 d	Radioimmunotherapeutic drug
Radium-226	1600 y	Radiation source for treatment of cancer
Rhenium-187	4.33×10^{10} y	Dating meteorites
Selenium-75	119.79 d	Pancreatic cancer diagnostic drug
Sodium-24	15.997 hr	Radiation source for detection of obstructions within the circulatory system
Strontium-90	28.90 y	Industrial gauges; thermoelectric generators
Technetium-99*m*	6.0067 hr	Radiation source for imaging the brain, thyroid, liver, kidney, lung, and cardiovascular systems
Thallium-201	3.0421 d	Cardiac diagnostic drug
Tritium, or H-3	12.32 y	Radiation source for determination of total body water
Yttrium-90	64.053 hr	Radioimmunotherapeutic drug

[a]Chart of Nuclides, National Nuclear Data Center, Brookhaven National Laboratory, Long Island, New York (2012).

organs in patients having diseases in the heart, bones, and elsewhere. Generally, to produce a graphic image of the structure or metabolic activity of an organ, radiologists inject a radioisotope into the bloodstream and film the activity of the radioisotope as it moves or concentrates.

Some radioisotopes are used to treat cancer patients. The radiation source delivers a high dose for consecutive daily treatments to a specific location in the body. Less commonly, a radioisotope is surgically implanted to deliberately radiate an organ over a long time period. During the treatment of prostate cancer, for example, a surgeon may implant iodine-125 or palladium-103 encased in tiny pellets or "seeds" within the prostate. The radioisotope kills cells in the nearby region, including those that are malignant. Once treatment is completed, the seeds are surgically removed.

SOLVED EXERCISE 16.1

Cesium-137 is a radioisotope often used as a sealed source (Section 16.3) of gamma radiation for the treatment of cancer. If a clinic purchases a source containing 7.60×10^{15} atoms of cesium-137 today, how many atoms will remain in 121 years?

Solution: The half-life of cesium-137 is listed in Table 16.2 as 30.08 years. The number of half-lives in 121 years is then determined by division:

$$\text{Number of half-lives} = 121\,\text{y}/30.08\,\text{y} = 4.0$$

This means that after the passage of 121 years, cesium-137 will have disintegrated through four half-lives; hence, one-sixteenth of the atoms will remain.

$$\tfrac{1}{2} \times \tfrac{1}{2} \times \tfrac{1}{2} \times \tfrac{1}{2} = (\tfrac{1}{2})^4 = 1/16$$

Multiplying 7.60×10^{15} by 1/16 gives 4.75×10^{14} atoms.

$$7.60 \times 10^{15}\,\text{atoms} \times 1/16 = 4.75 \times 10^{14}\,\text{atoms}$$

Thus, after the passage of 121 years, 4.75×10^{14} atoms of cesium-137 remain in the sealed source.

16.2 MODES OF RADIOISOTOPIC DECAY

Nuclear transformations of radioisotopes principally occur by means of one or more of the modes illustrated in Figure 16.1: alpha decay, beta decay (negatron emission, positron emission, and electron capture), gamma decay, and spontaneous fission. Alpha, beta, and gamma decay gives rise to a specific type of radiation. Because the radiation ionizes the matter through which it passes, each is an example of **ionizing radiation**. We shall examine them in Sections 16.2-A, 16.2-B, and 16.2-C.

The unit used to express the energy associated with alpha, beta, and gamma decay typically is the **electron volt**, or **eV**. One electron volt is the amount of energy acquired by an electron when it is accelerated by an electric potential of one volt. It is equivalent to 1.602×10^{-19} J. The energy typically emitted by radioisotopes is expressed in million electron volts. One million electron volts, or 1 MeV, is equivalent to 1.602×10^{-13} J.

16.2-A ALPHA DECAY

Many radioisotopes, especially those having atomic numbers greater than 83, disintegrate by spontaneously emitting particles composed of two protons and two neutrons. These particles are called **alpha particles,** and the process is called **alpha decay**. Alpha particles are the nuclei of doubly ionized helium atoms and are symbolized as either ^4_2He or the

ionizing radiation
■ Types of radiation that ionize matter upon impact

electron volt (eV) ■ The amount of energy acquired by an electron when it is accelerated by an electric potential of one volt

alpha particle ■ A particle emitted from certain radioisotopes and having the properties of a doubly ionized helium atom

alpha decay ■ A mode of radioisotopic decay associated with the emission of an alpha particle by a nucleus

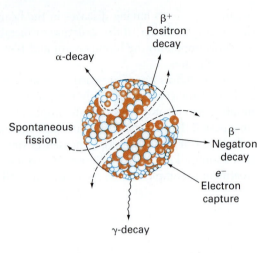

FIGURE 16.1 The modes by which radioisotopes decay, wherein the open circles represent protons, and the solid red circles represent neutrons. The most common modes of decay are those associated with the production of beta and gamma radiation. The rarest mode of decay is spontaneous fission.

α-decay

β⁺ Positron decay

Spontaneous fission

β⁻ Negatron decay

e⁻ Electron capture

γ-decay

Greek letter α. When many nuclei decay by alpha emission, the combination of alpha particles is called **alpha radiation.**

Alpha radiation is associated with a relatively large amount of energy that ranges from 4 to 8 MeV; but because alpha particles are doubly ionized, this energy is readily dissipated by its passage through a few centimeters of air or by absorption in a thin piece of matter. For instance, alpha radiation is absorbed by the thickness of this page.

When alpha decay occurs, the atomic number and mass number of the associated nuclei decrease by 2 and 4, respectively. An example of a radioisotope that disintegrates by alpha decay is uranium-238. This phenomenon is represented by either of the following equations:

$$^{238}_{92}\text{U} \longrightarrow {}^{234}_{90}\text{Th} + {}^{4}_{2}\text{He}$$
$$^{238}_{92}\text{U} \longrightarrow {}^{234}_{90}\text{Th} + \alpha$$

Both equations represent the change that the uranium-238 nucleus undergoes by emitting an alpha particle. The symbol of the particle is written to the right of the arrow to designate that the alpha particle has been emitted from the uranium-238 nucleus.

Equations denoting nuclear phenomena are not balanced in the chemical sense. Instead, a nuclear equation is balanced when each of the following is fulfilled: The sums of the atomic numbers and the sums of the mass numbers are the same on each side of the arrow.

16.2-B BETA DECAY

The second mode of radioactive disintegration is called **beta decay.** This process occurs when a radioisotope either emits a negatron or positron or combines with an extranuclear electron. These individual types of beta decay may occur individually or in combination.

Negatron Emission

The first type of beta decay is equivalent in result to the emission of an electron from the nucleus. When electrons are discussed in nuclear phenomena, they ordinarily are called **negatrons** and are designated as β^-, or $^{0}_{-1}e$. When beta decay occurs by the emission of negatrons, the mass numbers of the associated radioisotopes remain unchanged, but their atomic numbers increase by 1.[2]

An example of a radioisotope that disintegrates by negatron emission is thorium-234. This is the nucleus produced when uranium-238 disintegrates. On emitting a negatron,

---

[2]The types of beta-decay are also associated with the production of neutral subatomic particles called *neutrinos* and *antineutrinos*. These particles are of no interest here.

thorium-234 becomes protoactinium-234. This phenomenon is represented by either of the following equations:

$$^{234}_{90}\text{Th} \longrightarrow {}^{234}_{91}\text{Pa} + {}^{0}_{-1}e$$

$$^{234}_{90}\text{Th} \longrightarrow {}^{234}_{91}\text{Pa} + \beta^{-}$$

The emission of a negatron from a nucleus raises a basic question: How can it be emitted *from* the nucleus when the electron is not a component *of* the nucleus? The process is apparently more involved than a single equation represents. Nuclear scientists have determined that during negatron emission, each neutron within the unstable nucleus transforms into a proton and an electron. The proton then becomes part of the new nucleus, and the electron is simultaneously emitted. This conversion of the neutron ($^{1}_{0}n$) into a proton and an electron is designated as follows:

$$^{1}_{0}n \longrightarrow {}^{1}_{1}\text{H} + {}^{0}_{-1}e$$

Negatrons possess a range of energies, but these energies generally are no greater than 4 MeV. They usually are absorbed by a 1-centimeter-thick sheet of aluminum. The combination of multiple negatron-decay processes represents one type of **beta radiation**.

Positron Emission

The second type of beta decay involves the emission of a **positron** from a nucleus. A positron is a particle like an electron in most features, but it is positively charged. It is symbolized as $^{0}_{+1}e$, or β^{+}. When a radioisotope emits positrons, the mass numbers of the associated nuclei remain unchanged, but their atomic numbers decrease by 1. Each nuclear event involves the conversion of a proton into a neutron and positron, expressed as follows:

$$^{1}_{1}\text{H} \longrightarrow {}^{1}_{0}n + {}^{0}_{+1}e$$

Sodium-22 is an example of a radioisotope that decays by emitting positrons. This nucleus spontaneously transforms into neon-22. The event is denoted by either of the following equations:

$$^{22}_{11}\text{Na} \longrightarrow {}^{22}_{10}\text{Ne} + {}^{0}_{+1}e$$

$$^{22}_{11}\text{Na} \longrightarrow {}^{22}_{10}\text{Ne} + \beta^{+}$$

Like negatrons, positrons possess a range of energies, but generally they are no greater than 3 MeV. Like negatrons, they usually are absorbed by a 1-centimeter-thick sheet of aluminum. The combination of multiple positron-decay processes is also representative of beta radiation.

Electron Capture

The phenomenon associated with the third type of beta decay involves the capture of an unstable nucleus and an orbital electron. A radioisotope that undergoes electron capture decreases in atomic number by 1 but its mass number remains unchanged. Each electron captured by the nucleus reacts with a proton, thereby forming a neutron, which then becomes part of the structure of the new nucleus. This nuclear event is represented by the following equation:

$$^{1}_{1}\text{H} + {}^{0}_{-1}e \longrightarrow {}^{1}_{0}n$$

Argon-37 is an example of a radioisotope that undergoes electron capture. The phenomenon is represented as follows:

$$^{37}_{18}\text{Ar} + {}^{0}_{-1}e \longrightarrow {}^{37}_{17}\text{Cl}$$

In this instance, the symbol of the electron is written to the left of the arrow to designate that the electron has combined with the argon-37 nucleus.

Indium-111 is a radioisotope used in medicine as a tumoral diagnostic drug and for gastric and cardiac imaging. It undergoes electron capture and has a half-life of 2.8 d.

(a) How many protons and neutrons are present in the indium-111 nucleus?

(b) Identify the product of its radioactive transformation.

(c) What percentage of indium-111 remains in the bloodstream 8.4 days after administration of the drug?

Solution: Referring to either Figure 4.3 or Appendix B at the back of this text, we see that the chemical symbol and atomic number of indium are determined to be In and 49, respectively. Thus, the symbol for the indium-111 nucleus is $^{111}_{49}$In.

(a) Because the atomic number is the number of protons in a nucleus, there are 49 protons in the indium-111 nucleus. Because the atomic mass of $^{111}_{49}$In is 111, the total number of protons and neutrons in this nucleus is 111. The number of neutrons in $^{111}_{49}$In is 111 − 49, or 62.

(b) Indium-111 captures an orbital electron, a process represented as follows:

$$^{111}_{49}\text{In} + {}^{0}_{-1}\text{e} \longrightarrow {}^{111}_{48}\text{Cd}$$

The product of this transformation is cadmium-111.

(c) For indium-111, a period of 8.4 days represents three half-lives. After that time, the percentage of the administered drug that remains in the bloodstream is 12.5%.

$$\text{Final percentage} = 100\% \times \frac{1}{2} \times \frac{1}{2} \times \frac{1}{2} = 12.5\%$$

16.2-C GAMMA DECAY

electromagnetic radiation ■ The entire range of energy that travels as waves through space

gamma radiation ■ The collective combination of gamma rays emitted from a nucleus

gamma ray (photon) ■ A massless packet of electromagnetic energy emitted by certain radioisotopes

Alpha and beta decay frequently are accompanied by the simultaneous emission of the form of **electromagnetic radiation** called **gamma radiation**. Like X-ray, infrared, and ultraviolet radiation, gamma radiation has neither mass nor charge.

In the portion of the electromagnetic spectrum shown in Figure 16.2, the various forms of radiant energy are characterized by their wavelengths. Ultraviolet, infrared, and radio waves are said to have "long" wavelengths, whereas gamma radiation and X-ray radiation have "short" wavelengths. Visible light is the balance point between long and short wavelengths, and the only form that is detected by our eyes. The components of the electromagnetic spectrum with short wavelengths are very energetic, so much so that they ionize the matter through which they pass, but the components with long wavelengths are relatively nonenergetic and do not ionize matter.

Each individual component of gamma radiation is called a **gamma ray**, or **photon**, and is represented by the symbol γ. Because it does not possess a charge, a gamma ray is extremely penetrating and absorbed only by dense forms of matter such as thick blocks of lead.

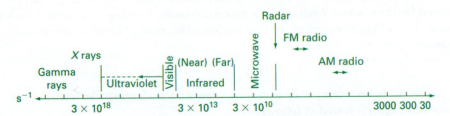
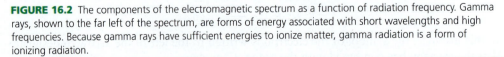

FIGURE 16.2 The components of the electromagnetic spectrum as a function of radiation frequency. Gamma rays, shown to the far left of the spectrum, are forms of energy associated with short wavelengths and high frequencies. Because gamma rays have sufficient energies to ionize matter, gamma radiation is a form of ionizing radiation.

The process involving the emission of gamma rays from a nucleus is called **gamma decay**. When a radioisotope undergoes gamma decay, no change in either its atomic number or mass number occurs. With the emission of each gamma ray, some fraction of the energy of excitation that causes the nucleus to be unstable is removed.

Imagine a radioisotope that exists in only two energy states. The more energetic form—the excited state—may emit one or more gamma rays from the nucleus. The phenomenon may be illustrated by the following equation, where the excited state is represented by an asterisk:

$$({}_Z^A X)^* \longrightarrow {}_Z^A X + \gamma$$

In this process, the radioisotope gives up a fraction of its excitation energy to become a more stable form of the *same* radioisotope.

Excited states that decay by emitting gamma rays generally have especially short half-lives ($<<10^{-8}$ s), but some excited states have half-lives in the range of $\sim 10^{-8}$ second to several years. In the latter case, the long-lived excited state is referred to as a **metastable state**, or **isomeric state**, of the radioisotope. The metastable state is designated by adding an *m* following the isotope's mass number. For example, technetium-99*m* is an excited state of technetium-99 that has a half-life of 6.0067 hours. It decays by gamma-ray emission as follows:

$$^{99m}_{43}\mathrm{Tc} \longrightarrow {}^{99}_{43}\mathrm{Tc} + \gamma$$

The decay of a metastable state of a radioisotope like technetium-99*m* is called an **isomeric transition**, or **IT**. Table 16.2 notes that, the gamma rays emitted from technetium-99*m* can be specifically used by radiologists to image the organs of the body.

The phenomenon of gamma-ray emission is not always represented by means of an equation. It also is represented by the following general diagram, where each horizontal line designates a discrete energy state of the atomic nucleus:

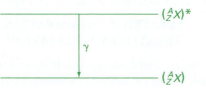

16.3 SEALED RADIATION SOURCES

Several radioisotopes listed in Table 16.2 are encountered as **sealed sources**. Typically, a manufacturer seals high levels of these radioisotopes within double-skinned steel tubes, which then are housed within a medical or other device. The radioisotopes remain sealed within these tubes throughout the period of their use.

Cobalt-60 is an example of a radioisotope that is used as a sealed source. As it decays, negatrons and gamma rays are emitted to the environment. First, each nucleus emits a negatron and transforms into an excited state of nickel-60, which then emits two gamma rays having energies of 1.173 MeV and 1.332 MeV. We represent this phenomenon as follows:

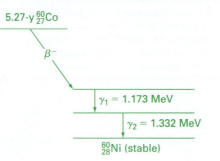

gamma decay ■ A spontaneous mode of radioisotopic decay associated with the emission of gamma rays from a nucleus, often accompanying alpha and beta decay

metastable state (isomeric state) ■ An excited state of a radioisotope that has a half-life in the range of $<10^{-8}$ second to several years

isomeric transition (IT) ■ The decay of a radioisotope by the emission of a photon from an excited state to a less energetic state

sealed source ■ An encapsulated radioisotope used in irradiation equipment and elsewhere

When cobalt-60 is used for a specific purpose, the sealed source is positioned so that the gamma rays emitted by the radioisotope pass through its steel container and intentionally penetrate a material. This process is referred to as **irradiation**. Gamma rays are used to irradiate polyethylene or rubber to induce cross-linking within their macromolecules. They are also used to irradiate spices and other harvested foods and to sterilize medical devices. Since 2008, when anthrax-tainted letters were mailed to U.S. senators (Section 10.21-E), the government has rerouted and irradiated mail to the White House, Congressional offices, and other federal government offices in certain zip code areas before its delivery.

When gamma rays or other forms of ionizing radiation are used to irradiate foods, they disrupt the fast-growing cells of insects, molds, and microbes on perishable meat, poultry, and produce. Irradiation also destroys the microorganisms that cause food spoilage, but it does not render the food radioactive or cause it to lose its nutritive value.

A primary advantage associated with irradiating foods is its success at killing the bacteria within raw meat and produce. It is the only known method of eliminating the potentially deadly bacteria *Escherichia coli*, *Salmonella*, and *Campylobacter* from raw fruits and vegetables. Typically, fruits and vegetables tolerate an irradiation of 1.0 kGy (Section 16.5-B), which inactivates 99.999% of the bacteria.

Although gamma-ray irradiation eliminates pests in fresh foods and extends their shelf life, irradiation also creates free radicals, the presence of which could negatively affect the inherent quality of foods by producing small amounts of undesirable substances. Although there is no technical basis for concluding that these irradiated foods are unsafe to consume, the widespread use of gamma rays for irradiating foods remains a controversial subject.

FDA's approval is required to sell irradiated foods in American stores. As a component of its approval, FDA requires manufacturers and distributors to affix the Radura symbol shown in Figure 16.3 on packages of irradiated foods. FDA also requires food distributors to mark packages of irradiated foods with either of the following statements:

TREATED WITH RADIATION
TREATED BY IRRADIATION

Sealed radiation sources are highly dangerous if they become unsealed. Acknowledging the need to warn people of the presence of radioisotopes in sealed sources, the United Nations' International Atomic Energy Agency, or IAEA, and the International Organization for Standardization, or ISO, introduced the ionizing radiation symbol shown in Figure 16.4. The symbol shows waves radiating from a three-bladed propeller called a trefoil, a skull-and-crossbones symbol, and a running person.

IAEA and ISO recommend affixing this symbol to devices that house a high-level radioisotope in a sealed source, exposure to which could cause death or serious injury. The sealed source usually is a sealed capsule that contains the radioisotope. It is further sealed between layers of nonradioactive material or firmly fixed to a nonradioactive surface by electroplating or other means to prevent leakage or escape of the radioisotope.

FIGURE 16.3 At 21 C.F.R. §179.26(c), FDA requires this international Radura symbol to be posted on irradiated packaged foods, bulk containers of unpackaged foods, on placards at the point of purchase for fresh produce, and on invoices for irradiated ingredients and products sold to food processors. The logo is dark green and displayed on a white background. (*Courtesy of FDA—Food and Drug Administration.*)

FIGURE 16.4 The IAEA/ISO ionizing radiation symbol used to warn individuals that a dangerous level of ionizing radiation in a sealed radioactive source is nearby. The triangular symbol is red with a black border and has black and white waves radiating from a trefoil, a skull-and-crossbones symbol, and a running person.

Normally the radioisotope is visible only when attempts are made to disassemble the equipment in which it is maintained. The intent of the ionizing radiation symbol is to warn people to distance themselves from the radiation source. Because it is not affixed to building-access doors, containers, or transport vehicles, this symbol supplements the warning trefoil required by NRC, OSHA and DOT on signs, labels, and placards.

16.4 DETECTION OF RADIOACTIVITY

Several radiation detection instruments are commercially available. The type shown in Figure 16.5 often is used by emergency responders for detecting the presence of a radiation source. The operator may specifically determine whether a radiation source is nearby, how close it is, its identity, and its intensity.

The total radiation to which an individual has been exposed, including the occupational radiation dose, is determined through the use of personal-monitoring equipment like the optically stimulated luminescent dosimeters shown in Figure 16.6(a) and the pocket or pen dosimeters illustrated in Figure 16.6(b).

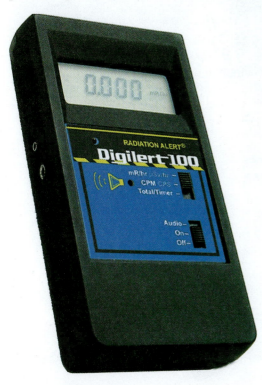

FIGURE 16.5 A portable handheld radiation detector is commonly used by first-on-the-scene responders to measure the intensity of alpha, beta, gamma, and X-ray radiation. This model is known as the Inspector. Its digital display provides readings in milliroentgens per hour (mR/h), counts per minute (c/min), or microsieverts per hour (µSv/h). (Courtesy of S. E. International, Inc., Summertown, Tennessee.)

FIGURE 16.6 (a) Optically stimulated luminescent dosimeters and (b) pocket or pen dosimeters. The use of these dosimeters provides a measurement of the total amount of radiation to which an individual has been exposed. (*Courtesy of Landauer, Inc., Glenwood, Illinois, and S.E. International, Inc., Summertown, Tennessee.*)

(a)

(b)

16.5 UNITS OF RADIATION AND RADIATION DOSE

The intensity of a radioisotope is called its **activity**, and the activity per unit mass is called its **specific activity**. As scientists developed radiation detection instruments, they simultaneously defined units of radiation measurement to serve as a means of accounting for the activity of a specific radioisotope. We cite them in the following sections.

activity ▪ The intensity of a radiation source

specific activity ▪ The activity of a radioisotope per unit mass

16.5-A UNITS OF ACTIVITY

Each type of radiation detector provides the intensity of a radioisotope by counting the number of nuclei that disintegrate during a time period. The intensity of the radioisotope may be provided directly in disintegrations per second, or it may be converted into multiples or fractions of units called the curie and becquerel.

▪ The **curie (Ci)** is a measure of the number of radioactive disintegrations occurring each second in a sample. One curie is the amount of radiation equal to 3.7×10^{10} disintegrations per second. Because the curie is an extremely large unit, the intensity of a sample of a radioactive material usually is measured in millicuries (mCi), microcuries (μCi), and picocuries (pCi).

curie (Ci) ▪ The amount of a radioisotope that decays at the rate of 3.7×10^{10} disintegrations per second

$$1\text{mCi} = 3.7 \times 10^7 \text{ disintegrations/s}$$
$$1\mu\text{Ci} = 3.7 \times 10^4 \text{ disintegrations/s}$$
$$1\text{pCi} = 3.7 \times 10^{-2} \text{ disintegrations/s}$$

▪ The **becquerel (Bq)** is the SI unit used to measure radioactive disintegrations per second. One Bq is equivalent to one nuclear disintegration per second, which can be written as follows:

becquerel (Bq) ▪ The SI unit of radioactivity equivalent to 1 disintegration per second

$$1\text{Ci} = 3.7 \times 10^{10} \text{ Bq}$$

The becquerel is a small amount of activity; hence, it generally is used with a prefix like *tera-*. EPA, DOE, and NRC regulations usually list the activities of radioisotopes in terabecquerels (TBq) and curies. One TBq equals a trillion becquerels or a trillion disintegrations per second.

$$1\text{TBq} = 10^{12} \text{ Bq}$$

SOLVED EXERCISE 16.3

Technetium-99*m* is used during diagnostic imaging of the body's internal organs. When a radiological technician injects a patient intravenously with 24.5 mCi of technetium-99*m*, how many becquerels of the radioisotope were received by the patient?

Solution: One curie equals 3.7×10^{10} becquerels; hence, by calculation, 24.5 mCi equals 9×10^8 Bq.

$$24.5 \text{ mCi} \times \frac{\text{Ci}}{10^3 \text{ mCi}} \times \frac{3.7 \times 10^{10} \text{ Bq}}{\text{Ci}}$$
$$= 9 \times 10^8 \text{ Bq}$$

16.5-B UNITS OF RADIATION DOSE

When human tissue is exposed to ionizing radiation, we are concerned with the quantity of radiation absorbed per unit of mass. The amount of radiation absorbed per body weight is called the *DOSE*. It is measured by using units like the roentgen, rad, rem, gray, and sievert.

roentgen (R) ■ The amount of radiation that produces 2.1×10^9 units of charge in 1 milliliter of dry air at standard atmospheric pressure

radiation absorbed dose (rad) ■ The amount of radiation absorbed per gram of body tissue

roentgen equivalent man (rem) ■ The amount of ionizing radiation that produces the same biological effect as 1 roentgen of X rays or gamma rays

gray (Gy) ■ The unit of an absorbed dose of radiation equivalent to 1 joule per kilogram of living tissue

sievert (Sv) ■ The SI unit of dose-equivalent, equal to 100 rem

■ The **roentgen** is symbolized with a capital letter "R." Once used as the international unit of radiation quantity for X-ray and gamma radiation, 1 R was defined as the amount of X-ray or gamma radiation that produces 2.58×10^{-4} coulombs of charge in 1 kilogram of dry air at 0°C and standard atmospheric pressure. Although its use is encountered in the older literature, the roentgen now is essentially obsolete.

■ The term **radiation absorbed dose**, or **rad**, is the amount of radiation absorbed per gram of body tissue. The roentgen and the rad are so close in magnitude that they are considered identical. For practical purposes, 1 R = 1 rad.

■ **Rem** is the acronym for **roentgen equivalent man**. It is a measure of the dose of any ionizing radiation to body tissue in terms of its estimated biological effect relative to a dose of 1 R of X rays or gamma rays. The relation of the rem to other dose units depends on the biological effect under consideration. One rem is the amount of radiation that produces the same damage as 1 R of X rays or gamma rays. The millirem (mrem), or one-thousandth of a rem, is a commonly encountered fraction of the rem. The average American receives around 620 mrem/y of radiation from natural and artificial radiation sources.

■ The **gray (Gy)** is the amount of radiation equivalent to the transfer of 1 Joule of energy to 1 kilogram of living tissue. One gray is equal to 100 rad.

■ The **sievert (Sv)** is the preferred unit for measuring exposure to ionizing radiation. One sievert equals 100 rem. It is the SI unit of radiation-dose-equivalent.

16.6 ILL EFFECTS CAUSED BY RADIATION EXPOSURE

All persons are constantly exposed to cosmic radiation and other inescapable low-level ionizing radiation emitted from the naturally occurring radioisotopes listed in Table 16.1. Everyone is also exposed to artificially produced radioisotopes in certain consumer products and when radioactive pharmaceuticals are used to detect the presence of tumors or to therapeutically destroy the cells within diseased tissue.

background radiation ■ Cosmic radiation and inescapable low-level ionizing radiation that is emitted from naturally occurring and artificially produced radioisotopes; 300 mrem/y, or 3 mSv/y

radiation sickness ■ The health ailment following short- or long-term exposure to radiation

The combination of ionizing radiation from natural and artificial sources in and around Earth is called **background radiation**, the sources of which are noted in Figure 16.7. The most significant sources are X rays, the naturally occurring radioisotopes, and the radioisotopes used to image the organs and tissues of the body.

The acute adverse health effects resulting from exposure of the human organism to different doses of radiation are provided in Table 16.3. They are manifested as the illness called **radiation sickness**.

The consequences to specific body organs, especially the reproductive and active blood-forming organs, are likely to be experienced as chronic effects. Radiation scientists assert that all types of ionizing radiation are human carcinogens. The onset of cancer is likely when an individual is exposed to ionizing radiation, because the radiation damages cellular DNA. In 2006, the National Academy of Sciences estimated that one out of 100 individuals is likely to develop cancer from an acute dose of 100 millisieverts of radiation.[3] This one additional cancer is in addition to the 42 that normally would be expected in the same U.S. population from all other causes of cancer. The cancerous cells develop as the radiation-damaged cells incorrectly repair themselves. Cancers of the skin, blood (leukemia), lung, stomach, esophagus, bone, thyroid, and brain are common diseases caused by exposure to ionizing radiation. There is no threshold of exposure below which a dose of ionizing radiation is entirely harmless—although the likelihood of acquiring cancer from radiation exposure diminishes as the dosage declines or when the total dosage is broken into smaller increments.

---

[3]National Research Council of the National Academies, *Health Risks from Exposure to Low Levels of Ionizing Radiation*, BEIR BII Phase 2 (Washington, DC, National Academies Press, 2006).

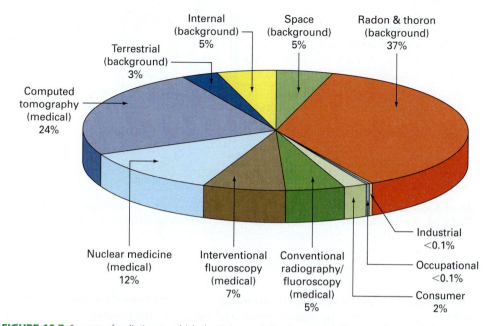

FIGURE 16.7 Sources of radiation to which the U.S. population is exposed. Computed tomography, interventional fluoroscopy, and conventional radiography are medical procedures that use X rays to image parts of the body. Thoron is a synonym for radon-220, formed during the stepwise decay of the naturally occurring thorium radioisotope, thorium-232. (*Reprinted with permission of the National Council on Radiation Protection and Measurements, Bethesda, Maryland, Ionizing Radiation Exposure of the Population of the United States, NCRP Report No. 160.*)

The radiation dose-equivalent rate is called the **radiation level**, which is measured in millisieverts per hour (mSv/hr). In most countries, the current maximum permissible radiation level to exposed workers is 20 mSv/y when it is averaged over 5 years, with a maximum radiation level of 50 millisieverts in any one year.

The official position of all federal agencies is that any exposure to radiation produces some risk of cancer; however, for exposures at doses below legal limits, the risk is within the range of other risks commonly accepted.

radiation level ■ The dose-equivalent to a person, expressed in millisieverts per hour (mSv/hr)

TABLE 16.3	Acute Radiation Effects Caused by a Single-Dose Exposure of Radiation to the Whole Body[a]
DOSE (Gy)	**SIGNS AND SYMPTOMS**
0–0.75	No detectable signs or symptoms in most exposed individuals although vomiting (retching) may occur in a few
0.75–1.5	Nausea, vomiting (retching), anorexia, fever, and infection
1.5–3.0	Nausea, vomiting (retching), anorexia, fatigue, weakness, bleeding, fever, and infection; death rate <5%
3.0–5.3	Nausea, vomiting (retching), anorexia, fatigue, weakness, bleeding, fever, infection, and ulceration; death rate <5–50%
8.3–11.0	Nausea, vomiting (retching), anorexia, fatigue, weakness, bleeding, fever, infection, hypotension, dizziness, disorientation, fluid loss/electrolyte imbalance, headache, and fainting; death rate 100%

[a]Adapted from *NATO Handbook on the Medical Aspects of NBC Defensive Operations AMedP-6(b)*, Chapter 6, General Medical Effects of Nuclear Weapons: Diagnosis, Treatment, and Prognosis (February 1996).

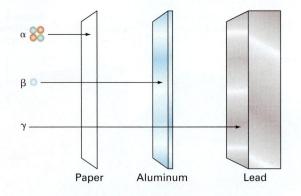

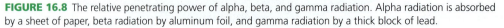

Paper   Aluminum   Lead

FIGURE 16.8 The relative penetrating power of alpha, beta, and gamma radiation. Alpha radiation is absorbed by a sheet of paper, beta radiation by aluminum foil, and gamma radiation by a thick block of lead.

The biological impact of radiation exposure depends on the type of radiation absorbed, its energy, the nature of specific radioisotopes, and the age of the exposed individual. We note the impact of these factors separately in the sections that immediately follow.

16.6-A GENERAL EXPOSURE TO IONIZING RADIATION

As demonstrated in Figure 16.8, alpha, beta, and gamma radiation penetrate matter differently. Thus, when evaluating the potential health effects resulting from exposure to ionizing radiation, it is important to know the specific type of the radiation to which an individual has been exposed. These ill effects are noted below:

■ Although alpha radiation is very energetic, it is easily absorbed externally by the epidermis, the outermost layer of the skin. Because the epidermis serves as a protective covering of the underlying living skin cells, external exposure to alpha radiation does not cause an adverse health effect. If it is ingested, however, the energetic alpha radiation may localize in a tissue or bone, where the subsequent biological damage may be severe.

■ Beta radiation penetrates deeper into tissue than alpha radiation and ionizes the substances it encounters. A 2-MeV negatron, for instance, travels through 0.4 inch (10 mm) of tissue before being absorbed. Beta radiation particles equal to or greater than 2 million electron volts may cause biological damage in exposed individuals.

■ Gamma radiation also passes readily through tissue and ionizes the substances it encounters. Internal and external exposure to gamma radiation usually causes more severe biological damage than exposure to either alpha or beta radiation.

16.6-B EXPOSURE TO SPECIFIC RADIOISOTOPES

Aside from the general effects caused by radiation exposure, several radioisotopes can damage the human organism at specific locations where they concentrate and mimic the role of essential nutrients. For example, the radioisotopes of strontium and radium may replace calcium in bone structures. This is potentially problematic because they may become deposited in spaces ordinarily occupied by marrow, a tissue containing the blood-forming cells. The irradiation of these cells can then cause an increase in the host victim's chances of developing leukemia, myelofibrosis, or other diseases.

Localized damage may also occur within the human organism from exposure to iodine radioisotopes. This exposure specifically damages the thyroid, a butterfly-shaped gland located near the base of the neck. The thyroid manufactures and secretes the hormones responsible for regulating the rates of body growth and metabolism. Two such hormones are thyroxine and triiodothyronine, whose molecular formulas are noted here:

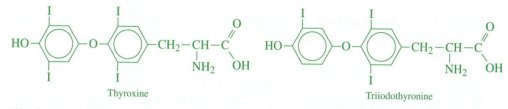

Thyroxine   Triiodothyronine

The human body acquires iodine mainly from the diet (e.g., in seafood). Iodized salt is used to prevent iodine deficiency. A supply of thyroxine and triiodothyronine remains stored in the thyroid for use as the body demands.

When the body is exposed to iodine radioisotopes, the thyroid gland captures and stores the amount that is not secreted. The concentration of iodine radioisotopes in the thyroid and their low removal rates increase an individual's chance of developing thyroid cancer or other disorders. Approximately 40,000 people develop thyroid cancer each year.

16.6-C IMPACT OF AN INDIVIDUAL'S AGE ON RADIATION EXPOSURE

When human cells divide, their nuclear membranes dissolve. This then allows DNA to spread across the entire volume of the cells and increases their likelihood of exposure to the free radicals produced by radiation. Under normal conditions, the nuclear membrane protects the DNA from exposure to free radicals; but, when cells frequently divide, their DNA is likely to be damaged. Because the cells in unborn children, infants, and youngsters are dividing frequently to support growth, these individuals are more susceptible to the effects of radiation exposure compared to adults.

16.7 WORKPLACE REGULATIONS INVOLVING RADIATION EXPOSURE

For activities not addressed by NRC or DOE, OSHA regulates employee exposure to radiation in the workplace at 29 C.F.R. §1910.96 by requiring the following:

■ No employer shall possess, use, or transfer sources of ionizing radiation in such a manner as to cause any individual to receive during one calendar quarter (qtr) a dose in excess of the limits quoted in Table 16.4 from sources in the employer's possession or control. These radiation doses are permissible exposure limits (Section 10.6-D).

TABLE 16.4	Permissible Exposure Limits in the Workplace[a]
PERMISSIBLE (RADIATION) EXPOSURE LIMIT (mSv/qtr)	**AREA OF THE BODY**
12.5	Whole body; head and trunk; active blood-forming organs; lens of the eyes; or gonads
18.75	Hands and forearms; feet and ankles
75	Skin of the whole body

[a]Adapted from OSHA 29 C.F.R. §1910.96(b).

■ An employer may permit an employee to receive a dose greater than those provided in Table 16.4 as long as:

    a. The dose to the whole body during any calendar quarter does not exceed 3 rem, *and*

    b. The dose to the whole body, when added to the accumulated occupational dose to the whole body, does not exceed $5 \times (N - 18)$ rem, where N is the individual's current age in years.

The expression "dose to the whole body" includes any dose to the entire body or a specific organ (gonads, active blood-forming organs, head and trunk, or lens of the eye).

■ No employer may permit any employee who is under 18 years of age to receive in any one calendar quarter a dose in excess of 10% of the limits in Table 16.4.

At 10 C.F.R. §835, the NRC establishes occupational radiation protection standards for DOE employees, and OSHA subsequently adopted them. These standards can also protect on-duty emergency responders. One common requirement is the posting of signs in areas where individuals work or gather or where a radiation hazard could potentially exist. At 10 C.F.R. §835.603, NRC requires the conspicuous posting of the signs like those illustrated in Figure 16.9 in given areas, each accessible to individuals. The nature of each area is defined at 29 C.F.R. §1910.1096 and 10 C.F.R. §20.1003. These definitions are summarized below:

■ A **radioactive materials area** is any area within a controlled area, accessible to individuals, in which items or containers of radioactive material exist and the total activity of the radioactive material exceeds the applicable values published at 10 C.F.R. §20, Appendix E. A "controlled area" is an area to which access is managed by or for DOE to protect individuals from exposure to radiation and/or radioactive material.

(a)   (b)   (c)

(d)   (e)

FIGURE 16.9 At 10 C.F.R. §20.192, 29 C.F.R. §§1910.1096(e)(2), 1910.1096(e)(3)(i), 1910.1096(e)(4)(ii), 1910.1096(e)(5)(i), 1910.1096(e)(6), and 10 C.F.R. §835.603, OSHA and NRC respectively require the conspicuous posting of these caution signs in areas where workers and other individuals could receive a dose of ionizing radiation to the whole body. The conventional color of the trefoil is magenta or purple imposed on a yellow background.

- A **radiation area** is any area, accessible to individuals, in which radiation levels could result in an individual receiving an equivalent dose to the whole body in excess of 0.05 millisievert in 1 hour at 11.8 inches (30 cm) from the source or from any surface that the radiation penetrates.

- A **high-radiation area** is any area, accessible to individuals, in which radiation levels could result in an individual receiving an equivalent dose to the whole body in excess of 0.1 millisievert in any 1 hour at 11.8 inches (30 cm) from the radiation source or from any surface that the radiation penetrates.

- An **airborne radioactivity area** is any room, enclosure, or area in which airborne radioactive material, composed wholly or in part of licensed material, exist in either of the following concentrations:
  - **(a)** in excess of the derived air concentrations (DACs) specified at 10 C.F.R. §20.1001, Appendix B; or
  - **(b)** to such a degree that an individual present in the area without respiratory protection could exceed, during the hours an individual is present in a week, an intake of 0.6% of the annual limit or 12 DAC-hours.

  "Licensed material" means certain radioactive materials received, possessed, used, transferred, or disposed of under the conditions of a license issued by the NRC.

- A **very high radiation area** is an area, accessible to individuals, in which radiation levels could result in an individual receiving an absorbed dose in excess of 5 gray in any 1 hour at 39.37 inches (1 m) from a radiation source or from any surface that the radiation penetrates.

Each sign in Figure 16.9 bears a warning trefoil and one of the following expressions:

CAUTION, RADIOACTIVE MATERIALS

CAUTION, RADIATION AREA

CAUTION, HIGH RADIATION AREA

CAUTION, AIRBORNE RADIOACTIVITY AREA, or

DANGER, AIRBORNE RADIOACTIVITY AREA

GRAVE DANGER, VERY HIGH RADIATION AREA

DOE and OSHA also require the conspicuous posting of labels affixed to the containers used to store certain quantities of radioactive materials. The labels bear the standard warning trefoil and the words *CAUTION, RADIOACTIVE MATERIAL* or *DANGER RADIOACTIVE MATERIAL*.

Although the intention of these signs and labels is to warn workers and other individuals that radioactive materials are present at the locations where they are posted, the signs and labels also serve to inform emergency responders of the presence of radioactive materials in the specified areas.

16.8 EFFECTS OF IONIZING RADIATION ON MATTER

When ionizing radiation passes through matter, its energy is dispersed among the atoms or molecules of which the matter is composed. The process of irradiating the matter is called **radiolysis**.

The initial action resulting from irradiating matter is the ionization of the substances through which the radiation passes. The ionization is represented for a molecule of an arbitrary substance A by the following equation:

$$A \rightsquigarrow A^+ + e^2$$

The wiggly arrow indicates that the relevant reaction was induced by ionizing radiation, e^- symbolizes an electron, and A^+ is the symbol for a molecule–ion.

radiation area ■ For purposes of NRC and OSHA regulations, any area, accessible to individuals, in which radiation levels could result in an individual receiving an equivalent dose to the whole body in excess of 0.05 mSv in 1 hour at 11.8 inches (30 cm) from the source or from any surface that the radiation penetrates

high-radiation area ■ For purposes of NRC and OSHA regulations, any area, accessible to individuals, in which radiation levels could result in an individual receiving an equivalent dose to the whole body in excess of 0.1 mSv in any 1 hour at 11.8 inches (30 cm) from the radiation source or from any surface that the radiation penetrates

airborne radioactivity area ■ For purposes of NRC and OSHA regulations, any room, enclosure, or area in which certain airborne radioactive material exists at either concentration cited at 29 C.F.R. §1910.1096 and 10 C.F.R. §20.1003.

very high radiation area ■ For purposes of NRC and OSHA regulations, any area, accessible to individuals, in which radiation levels could result in an individual receiving an absorbed dose in excess of 5 grays in any 1 hour at 39.37 inches (1 m) from a radiation source or from any surface that the radiation penetrates

radiolysis ■ The combination of chemical reactions resulting from the exposure of a substance to ionizing radiation

When such events occur repeatedly, the number of ions and electrons increases. Then, several secondary phenomena usually occur. In particular, the ions may combine with any electron to form an excited state of A as follows:

$$A^+ + e^- \longrightarrow A^*$$

Here, A^* refers to a molecule of A that has excess energy. A^* is an unstable molecule. Because it possesses excessive energy, it may dissociate entirely into molecules of new substances as follows:

$$A^* \longrightarrow B + C$$

The ions that form when radiation initially strikes a substance may also react with neutral molecules. Such reactions produce new ions and free radicals. The direct exposure of a substance to radiation also may result in the production of free radicals. As noted before, free radicals are highly reactive chemical species that may trigger a variety of chemical reactions.

Let's consider specifically the phenomena associated with exposing water to gamma radiation. Water constitutes three-fourths of all the body's tissues; hence, considerable research efforts have been devoted to examining the nature of the chemical species produced when water is exposed to radiation. When foreign substances form by means of the radiolysis of water, the cellular biochemistry may be significantly altered, thereby causing damage or death to the affected cells.

When water is exposed to gamma radiation, primary ionization of the water first occurs, as follows:

$$H_2O(l) \rightsquigarrow (H_2O)^+(l) + e^-$$

Then, the water molecule–ion reacts with a neutral water molecule to form a hydroxyl free radical.

$$(H_2O)^+(l) + H_2O(l) \longrightarrow H^+(aq) + \cdot OH(aq)$$

The parenthetical expression "aq" on the right of the arrow accounts for the molecules of water listed on the left. The exposure of water to gamma radiation may also produce hydrogen atoms and hydroxyl free radicals.

$$H_2O(l) \longrightarrow H\cdot(aq) + \cdot OH(aq)$$

The repeated occurrence of this latter event causes the concentration of hydroxyl radicals to increase. Ultimately two hydroxyl radicals unite to form hydrogen peroxide, a substance known to damage living tissues.

$$2 \cdot OH(aq) \longrightarrow H_2O_2(l)$$

16.9 NUCLEAR FISSION

nuclear reactor ■ A device that initiates, maintains, and controls a chain reaction involving fission to produce energy, generally for use as electricity

nuclear fission ■ The nuclear process during which certain nuclei are split into two lighter-mass nuclei with the simultaneous emission of one or more neutrons and a substantial amount of energy

fission product ■ Any isotope produced when a fission event occurs

Scientists have been able to synthetically produce numerous radioisotopes that are not found naturally. One method of production involves radiating stable or long-lived isotopes in nuclear accelerators such as cyclotrons and bevatrons. Samples of such isotopes generally are radiated with alpha particles, protons, deuterons, or other particles that have been accelerated to relatively high velocities. These high-velocity particles possess substantial energies. When they are used to bombard nuclei, nuclear reactions often occur. This phenomenon artificially transforms one nucleus into another. Many of the artificially produced radioisotopes used in medicine are prepared in this fashion.

A second method of production involves using the nuclear reactions that occur within a **nuclear reactor**, the heart of a nuclear power plant. The reactor functions because of the phenomenon called **nuclear fission**. As illustrated in Figure 16.10, nuclear fission is a process whereby a nucleus splits into two lighter-mass nuclei, accompanied by the release of a relatively large amount of energy (~200 MeV) and several neutrons (2.5 on the average). The fragments that are produced by individual fission events are called **fission products.**

In the nuclear power industry, engineers harness this energy and convert it into electrical energy in the manner illustrated in Figure 16.11. The heat generated during fission

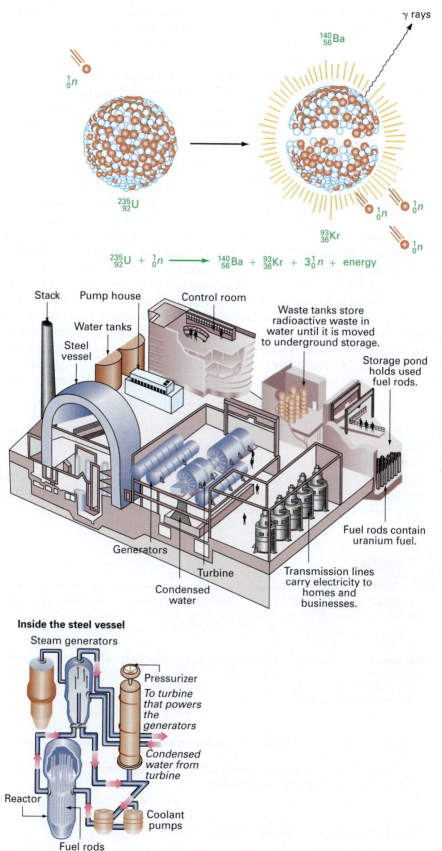

FIGURE 16.10 The fission of a uranium-235 nucleus induced by a low-velocity neutron shown at the upper left. In this instance, barium-140, krypton-93, and three neutrons are produced.

γ rays

$^{1}_{0}n$

$^{140}_{56}Ba$

$^{235}_{92}U$

$^{93}_{36}Kr$

$^{1}_{0}n$

$^{1}_{0}n$

$^{1}_{0}n$

$$^{235}_{92}U + ^{1}_{0}n \longrightarrow ^{140}_{56}Ba + ^{93}_{36}Kr + 3^{1}_{0}n + \text{energy}$$

FIGURE 16.11 In a nuclear power plant, the energy generated by fission is used to produce electricity. A nuclear power plant does not generate air pollutants like a fossil-fuel-fired power plant, but it does generate spent fuel and high-level radioactive waste, both of which require disposal.

Stack

Pump house

Control room

Water tanks

Steel vessel

Waste tanks store radioactive waste in water until it is moved to underground storage.

Storage pond holds used fuel rods.

Fuel rods contain uranium fuel.

Generators

Turbine

Condensed water

Transmission lines carry electricity to homes and businesses.

Inside the steel vessel

Steam generators

Pressurizer

To turbine that powers the generators

Condensed water from turbine

Reactor

Coolant pumps

Fuel rods

is absorbed by the water that is piped through the reactor. Next, the heated water is piped from the reactor to a steam generator, where the water vaporizes and is sent to a turbine. The steam spins the turbine blades, which generate electricity much like they do during the operation of a fossil-fuel-fired power plant.

The neutrons generated by fission may also be used to bombard other nuclei. This process serves as a means of artificially producing radioisotopes that do not exist naturally. Many radioisotopes listed in Table 16.2 are produced by neutron bombardment in nuclear reactors. Cobalt-60, for example, may be produced artificially by radiating naturally occurring cobalt-59 with the neutrons produced during fission.

$$\ce{^{59}_{27}Co + ^{1}_{0}n \longrightarrow ^{60}_{27}Co + \gamma}$$

Cobalt-59 is the only naturally occurring stable isotope of cobalt.

The neutrons generated by fission may also be used to *induce* other fission events. Over a dozen nuclei may be induced by neutrons to undergo fission, but uranium-233, uranium-235, and plutonium-239 are the only ones that can be practically used for this purpose. They are said to be **fissile nuclei**. Uranium-235 and plutonium-239 are used as the fissile material in both nuclear reactors and nuclear warheads. Plutonium-239 arguably is more desirable for use in nuclear bombs, because it can be produced from nonfissile uranium-238 and a smaller mass is needed as the nuclear fuel when compared to the mass of uranium-235.

For use in nuclear reactors and nuclear warheads, fissile nuclei are induced to undergo fission by exposing them to low-energy neutrons. Unstable uranium-236 first forms when uranium-235 captures a low-energy neutron. To achieve stability, most uranium-236 nuclei divide into two other nuclei, simultaneously releasing energy and several neutrons. One such event is represented as follows, where the asterisk denotes the unstable nucleus:

$$\ce{^{235}_{92}U + ^{1}_{0}n \longrightarrow [^{236}_{92}U]^{*} \longrightarrow ^{89}_{35}Br + ^{145}_{57}La + 2^{1}_{0}n + 192 MeV}$$

Dozens of these individual fission events occur as the fuel undergoes fission within a reactor. Several radioactive fission products are listed in Table 16.5. Their charge and mass are distributed among the elements from zinc to terbium; that is, their Z values range from 30 to 65, and their A values range from 72 to 161. Mixtures of fission products comprise one type of nuclear waste generated by a nuclear reactor. They are also the components of the **atmospheric fallout** generated when a nuclear bomb is detonated.

fissile nuclei ■ Isotopes that may be initiated to undergo fission

atmospheric fallout ■ Fission products produced during the detonation of a nuclear bomb and carried into the atmosphere where they move by air currents from place to place, and return to Earth's surface as dust or precipitation

SOLVED EXERCISE 16.4

The following equation describes a fission event that produces barium-139 and another isotope.

$$\ce{^{235}_{92}U + ^{1}_{0}n \longrightarrow ^{139}_{56}Ba + \_\_\_ + 3^{1}_{0}n}$$

Identify this second isotope.

Solution: A nuclear equation is balanced when two conditions are fulfilled: the sums of the charges are the same on each side of the arrow, and the sums of the mass numbers are the same on each side of the arrow. On the left, the sum of the charges is $92 + 0 = 92$; on the right, the sum is $56 + x + 0 = 56 + x$, where x is the charge of the second isotope.

$$56 + x = 92$$
$$x = 36$$

Referring to either Figure 4.3 or Appendix B at the end of this text, we see that the element having an atomic number of 36 is krypton.

On the left, the sum of the mass numbers is $235 + 1 = 236$; on the right, the sum is $139 + 3 + y = 142 + y$, where y is the mass number of the second isotope.

$$236 = 142 + y$$
$$y = 36$$

Hence, the fission product produced by this event is krypton-94, or $\ce{^{94}_{36}Kr}$.

TABLE 16.5	Some Radioactive Fission Products	
FISSION PRODUCT	HALF-LIFE[a]	MODE OF DECAY
Barium-140	12.5727 d	β^-
Cesium-134	2.0652 y	β^-
Cesium-137	30.08 y	β^-
Cerium-141	32.508 d	β^-
Cerium-144	284.91 d	β^-
Iodine-131	8.0252 d	β^-
Iodine-132	2.295 h	β^-
Krypton-85*m*	4.480 hr	β^-, IT
Lanthanum-140	1.67855 d	β^-
Niobium-95	34.991 d	β^-
Praseodymium-143	13.57 d	β^-
Praseodymium-144	17.28 min	β^-
Promethium-147	2.6234 y	β^-
Rhodium-106	131 min	β^-
Ruthenium-103	39.247 d	β^-
Ruthenium-106	371.8 d	β^-
Strontium-89	50.53 d	β^-
Strontium-90	28.90 y	β^-
Technetium-99*m*	6 hr	β^-, IT
Tellurium-129*m*	33.6 d	β^-, IT
Tellurium-132	78 hr	β^-
Xenon-133*m*	5.2475 d	IT
Zirconium-95	64.032 d	β^-

[a]Chart of Nuclides, National Nuclear Data Center, Brooklyn National Laboratory, Long Island, New York (2012).

Natural uranium is composed of the isotopes uranium-235 and uranium-238 in the ratio of 1:138. This means that only 0.71% by mass of naturally occurring uranium is uranium-235. Uranium-238 is the more abundant uranium isotope, but it is not fissile.

The neutron-induced fission of uranium-235 is a particularly unique phenomenon, because for each single neutron that is consumed by reaction, additional neutrons are produced. These newly formed neutrons are used to induce the fission of other uranium-235 nuclei. The process thereby initiates a self-sustaining **chain reaction** like that illustrated in Figure 16.12. The condition of sustaining the chain reaction is called **criticality.** In a nuclear reactor, the chain reaction is controlled by workers, and the energy is harnessed for the peaceful production of energy. By contrast, in the nuclear bomb or nuclear-armed missile, the chain reaction builds at an explosive rate, and the energy is released into the environment virtually at once.

16.9-A CRITICALITY

Can the nuclear reactor at a power plant detonate like a nuclear bomb or nuclear-armed missile? Answering this question requires an understanding of certain details about the fission phenomenon.

A certain minimum mass of fissile material must always be present for fission to occur; this is called the **critical mass**. When the mass of the fissile material is less than the

chain reaction ■ A self-sustaining nuclear reaction in which the neutrons produced in one step trigger the next step, which in turn produces neutrons that trigger the next step, and so on

criticality ■ The condition wherein a system is capable of sustaining a nuclear chain reaction by retaining at least as many neutrons from a fission reaction as are consumed in future fission events

critical mass ■ The smallest mass of fissionable material that supports a self-sustaining chain reaction

FIGURE 16.12 A nuclear chain reaction. The neutron (*n*) adjacent to the arrow is absorbed by a uranium-235 nucleus, which subsequently undergoes fission. The fission by-products (*f*) and additional neutrons are thus produced. These neutrons induce the fission of other uranium-235 nuclei. The chain reaction is self-perpetuating until either the fuel is consumed or the neutrons are absorbed by matter other than the fuel.

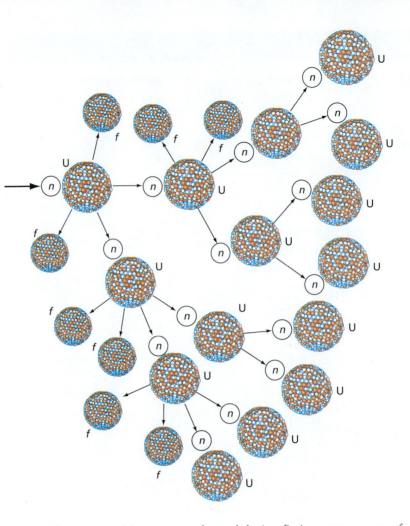

critical mass, most of the neutrons formed during fission events escape from the mass and do not induce the fission of other nuclei. The perpetuation of a chain reaction requires the system to retain at least as many neutrons as were consumed in the fission reaction from which they were produced. In this way, the neutrons may react in further fission events. Only then has the system achieved criticality.

SOLVED EXERCISE 16.5

In 2006, researchers at the University of Utah showed that young children living in Utah and Nevada during the 1950s and 1960s downwind from aboveground weapons-testing sites were 7.5 times more likely to develop thyroid neoplasms, the precursors to thyroid cancer, compared with individuals in the general population [J. L. Lyon et al., "Thyroid disease associated with exposure to the Nevada nuclear weapons test site radiation: A reevaluation based on corrected dosimetry and examination data," *Epidemiology*, Vol. 14 (2006), pp. 604–614]. What is the most likely cause of this childhood cancer?

Solution: Table 16.5 shows that iodine-131 and iodine-132 are produced during fission. When humans are exposed to airborne releases of these radioisotopes, localized damage to the thyroid occurs. Thyroid cancer is a commonly observed abnormality associated with human exposure to airborne releases of iodine radioisotopes produced during the testing of nuclear weapons. The iodine radioisotopes are discharged into the air, from which they are subsequently removed by rainwater. The rain falls on pastures where grazing dairy cattle consume the grass and produce iodine-contaminated milk. Children who drink this milk are more likely to develop thyroid cancer than individuals who do not drink iodine-contaminated milk. Thus, those children who live in areas where iodine radioisotopes tend to accumulate are more likely to develop thyroid cancer than children who live in the general population.

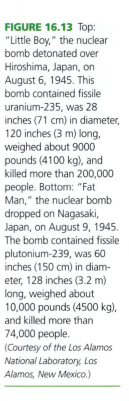

FIGURE 16.13 Top: "Little Boy," the nuclear bomb detonated over Hiroshima, Japan, on August 6, 1945. This bomb contained fissile uranium-235, was 28 inches (71 cm) in diameter, 120 inches (3 m) long, weighed about 9000 pounds (4100 kg), and killed more than 200,000 people. Bottom: "Fat Man," the nuclear bomb dropped on Nagasaki, Japan, on August 9, 1945. The bomb contained fissile plutonium-239, was 60 inches (150 cm) in diameter, 128 inches (3.2 m) long, weighed about 10,000 pounds (4500 kg), and killed more than 74,000 people. (*Courtesy of the Los Alamos National Laboratory, Los Alamos, New Mexico.*)

To end World War II, the nuclear bombs shown in Figure 16.13 were detonated 1800 feet (550 m) over the Japanese cities of Hiroshima and Nagasaki. Their fissionable material was first compressed into a relatively small volume by the detonation of a mantle of explosive charges that caused the nuclear fuel to increase in density and achieve a super-critical mass. Then, a chain reaction occurred that released a cataclysmic amount of energy into the environment. The nuclear bomb that detonated over Hiroshima released an estimated 6.3×10^{13} joules of energy to the environment. This is equivalent to the energy released during the detonation of 15,000 tons (13,600 t) of TNT. About half of it was released in the air blast, 35% was heat, and 15% was nuclear radiation. The fireball resulting from the Hiroshima explosion was 50% hotter than the surface of the sun.

The amount of nuclear fuel in the bombs detonated over Hiroshima and Nagasaki contained far less fissionable material per unit volume than the amount used in a nuclear reactor. Furthermore, the number of neutrons available for future fission events is restricted in a reactor through the use of control rods. These are long rods made from cadmium or boron steel that are inserted into or near the fuel, or withdrawn from the fuel. Both cadmium and boron have naturally occurring isotopes that readily absorb neutrons. Thus, the insertion and withdrawal of control rods effectively mediate the fission reactions by absorbing extra neutrons and controlling their propagation. With their use, a nuclear explosion could never occur. Furthermore, even if all the control rods were removed, a nuclear power plant could never explode in the same manner as a nuclear bomb. If the chain reaction in a nuclear reactor were to become uncontrollable, the energy generated by the reaction would

meltdown ■ Severe overheating of a nuclear reactor core, resulting in the melting of the nuclear fuel and its containment encasing

only cause a **meltdown** that could initiate a hydrogen or steam explosion, but not a nuclear explosion.

Notwithstanding its improbability, a disaster is likely to occur whenever individuals are exposed to fissionable material in an amount that is equal to or greater than its critical mass. Sixty incidents involving the accumulation of a critical mass of fissionable material have occurred worldwide, generally because safety practices were disregarded. In 1999, for example, a critical mass of uranium-235 unintentionally was accumulated at a uranium-reprocessing facility in Tokaimura, Japan that resulted in exposing three workers to neutrons, two of whom subsequently died from the exposure.

16.9-B TERRORISTS' POTENTIAL MISUSE OF FISSILE MATERIAL

During the past decade, Americans have been subjected to a prevailing fear that terrorists may use fissile material to construct a weapon of mass destruction. Weapons-grade uranium and plutonium are reasonably available for such use, although estimates of their amounts differ. To clandestinely build a nuclear bomb, terrorists need only 55 pounds (25 kg) of weapons-grade uranium (Section 16.9-E) or 18 pounds (8 kg) of plutonium-239.

A nuclear-armed terrorist attack could negatively impact our way of life in a manner worse than that encountered in an average war scene. This fact is emphasized by a Rand Corporation study[4] commissioned by the U.S. Department of Homeland Security that concluded a nuclear-armed terrorist attack would kill 60,000 people in San Jose, California, and cost as much as $1 trillion in damages and cleanup. It is evident that even a small blast in a heavily populated city would cause an enormous number of casualties.

16.9-C PLUTONIUM-239

Prior to and through the years of the Cold War, vast amounts of fissile plutonium-239 were produced as plutonium(IV) oxide (PuO_2). Amassing plutonium-239 served as a deterrent to stabilize worldwide political power. The production was accomplished in **breeder reactors** from natural uranium (99.3% uranium-238 and 0.71% uranium-235). The initial nuclear reaction produced uranium-239, which decayed in a stepwise process to neptunium-239 and plutonium-239, as follows:

breeder reactor ■ A device in which nonfissionable uranium-238 is converted into fissionable plutonium-239

$$^{238}_{92}U + ^{1}_{0}n \longrightarrow ^{239}_{92}U + \gamma$$

$$^{239}_{92}U \longrightarrow ^{239}_{93}Np + \beta^-$$

$$^{239}_{93}Np \longrightarrow ^{239}_{94}Pu + \beta^-$$

Even today, Pakistan and India, bitter archrivals, continue to breed plutonium-239 in reactors and stockpile it as a deterrent for calming disputes that have existed for decades. The U.S. arsenal of at least 90 tons (82 t) of plutonium-239 is stored within sealed bunkers at the Pantex DOE Plant near Armadillo, Texas.

There are at least two reasons why the world's supply of plutonium-239 may be troublesome:

■ As we noted in Section 16.9-B, there is always the psychological fear that plutonium may be acquired by rogue nations and terrorists for clandestine purposes.

■ Because the half-life of plutonium-239 is 24,110 years, it is evident that once released from its confinement, plutonium-239 poses a risk to Earth's inhabitants for tens of thousands of years.

---

[4]Charles Meade and Roger C. Molander (Rand Corporation) "Considering the Effects of a Catastrophic Terrorist Attack" (2006).

The only viable way to diminish supplies of plutonium-239 is to blend them with the oxides of natural or depleted uranium (Section 16.9-D) into the **mixed oxide fuel** called **MOX** and use it to fuel nuclear reactors.

16.9-D URANIUM-ENRICHMENT PROCESSES

Fissile uranium-235 is the primary fuel that serves as the source of the power generated at most nuclear power plants, but to effectively serve in this capacity, uranium must be isotopically enriched in uranium-235 to approximately 5% by mass. (For use as fissile warhead material, uranium-235 must be enriched to more than 80% by mass.) Although the basic nature of the **uranium-enrichment process** is not secret, some features of the engineering details are highly classified.

An essential element of the uranium-enrichment process is the production of uranium(VI) fluoride, frequently called uranium hexafluoride. Although it does not react appreciably with oxygen, nitrogen, or carbon dioxide, uranium hexafluoride reacts vigorously with water to produce uranyl fluoride and hydrogen fluoride.

$$UF_6(g) \quad + \quad 2H_2O(g) \quad \longrightarrow \quad UO_2F_2(s) \quad + \quad 4HF(g)$$

Uranium(VI) fluoride   Water   Uranyl fluoride   Hydrogen fluoride

Because the inhalation of hydrogen fluoride is highly dangerous to health (Section 8.11-B), uranium hexafluoride is considered hazardous primarily because of its corrosivity, not its radioactive nature. As first noted in Section 6.6-B, DOT acknowledges this property of uranium hexafluoride by requiring shippers and carriers to post RADIOACTIVE *and* CORROSIVE placards on the transport vehicle or freight container used to transport 1001 pounds (454 kg) or more of this substance.

The production of uranium hexafluoride occurs during the processing of a uranium-bearing ore like pitchblende. Minerals containing uranium are not rare in nature, but the uranium is economically recoverable only from deposits in which the uranium is relatively concentrated. These sources exist only in distinctive mines in Australia, Canada, and Kazakhstan.

Uranium generally exists as a mixture of uranium(V) oxide (U_2O_5) and uranium(VI) oxide (UO_3), collectively called triuranium octoxide (U_3O_8). It is recovered from the ore by sequentially implementing the series of steps briefly summarized as follows:

■ The ore is crushed and then mixed with nitric acid, resulting in the production of *uranyl nitrate*.

$$U_3O_8(s) \quad + \quad 8HNO_3(aq) \quad \longrightarrow \quad 3UO_2(NO_3)_2(aq) \quad + \quad 2NO_2(g) \quad + \quad 4H_2O(l)$$

Triuranium octoxide   Nitric acid   Uranyl nitrate   Nitrogen dioxide   Water

■ Uranyl nitrate is thermally decomposed to produce uranium(VI) oxide, commonly called **orange oxide**.

$$2UO_2(NO_3)_2(aq) \quad \longrightarrow \quad 2UO_3(s) \quad + \quad 4NO_2(g) \quad + \quad O_2(g)$$

Uranyl nitrate   Uranium(VI) oxide   Nitrogen dioxide   Oxygen

■ Uranium(VI) oxide is reduced with hydrogen, which produces uranium(IV) oxide, called **yellowcake**.

$$UO_3(s) \quad + \quad H_2(g) \quad \longrightarrow \quad UO_2(s) \quad + \quad H_2O(g)$$

Uranium(VI) oxide   Hydrogen   Uranium(IV) oxide   Water

■ Uranium(IV) oxide is reacted with anhydrous hydrogen fluoride at 932°F (500°C) forming uranium(IV) fluoride, called **green salt**.

$$UO_2(s) \quad + \quad 4HF(g) \quad \longrightarrow \quad UF_4(s) \quad + \quad 2H_2O(g)$$

Uranium(IV) oxide   Hydrogen fluoride   Uranium(IV) fluoride   Water

mixed oxide fuel (MOX) ■ A blend of plutonium oxide with the oxides of natural or depleted uranium

uranium-enrichment processes ■ Isolation techniques for concentrating fissile U-235

orange oxide ■ Uranium(VI) oxide (UO_3), an intermediate in the production of uranium hexafluoride

yellowcake ■ Uranium(IV) oxide (UO_2), an intermediate in the production of uranium hexafluoride

green salt ■ Uranium(IV) fluoride (UF_4), an intermediate in the production of uranium hexafluoride

- Uranium(IV) fluoride is further fluorinated at 482°F (250°C) to produce uranium hexafluoride, which is a gas above its sublimation temperature of 147°F (64°C).

$$UF_4(s) \quad + \quad F_2(g) \quad \longrightarrow \quad UF_6(g)$$

Uranium(IV) fluoride   Fluorine   Uranium(VI) fluoride

As it is produced, the uranium hexafluoride is an isotopic mixture of $^{235}UF_6$ and $^{238}UF_6$. Although these isotopic forms may be separated by several means, the least expensive procedure is a gas-centrifugation process that takes into account the mass difference of the two components. In essence, the mixture of $^{235}UF_6$ and $^{238}UF_6$ is spun at high speeds, thereby creating a centrifugal force. As the centrifuge operates, the heavier $^{238}UF_6$ concentrates outward. The periodic removal of $^{238}UF_6$ leaves behind the fissile $^{235}UF_6$. It is called **enriched uranium**. For use as a reactor fuel, the uranium-235-enriched hexafluoride generally is reduced to uranium(IV) oxide, which is ground into a powder and pressed into ceramic pellets approximately 1.5 centimeters in diameter and 1.5 centimeters long. The pellets are inserted into thin zirconium tubes to form **fuel rods**, often as long as 13 feet (4 m). Approximately 200 sealed rods are then assembled into bundles that comprise the core of the reactor.

When a process is conducted to reduce or eliminate the concentration of uranium-235 in uranium metal or a uranium compound, the compound that remains is called **depleted uranium**. Depleted uranium hexafluoride, or **DUF$_6$**, is a waste product generated during the uranium-enrichment process. In the United States, approximately 825,000 tons (750,000 t) of DUF$_6$ is now stored in steel cylinders at three locations: Paducah, Kentucky; Portsmouth, Ohio; and East Tennessee Technology Park, Tennessee. The majority of this material was generated during past decades in connection with the U.S. military defense program. Its management is a responsibility of DOE.

Peaceful nations are severely limited in their use of fissile uranium for military purposes. Nonetheless, in contemporary times, they are obliged to implement appropriate diplomatic steps aimed at preventing rogue nations from acquiring nuclear fuel for military purposes. The fear is that rogue nations operate nuclear power plants under the guise of energy production when their real intention is to maximize the production of fissile material for use in nuclear arms. The diplomatic efforts of peaceful nations are directed at preventing them from conducting uranium-enrichment processes or acquiring fissile material from other nations for military purposes.

Clandestine operations associated with the production of uranium hexafluoride may often be identified by inspecting the sites where the production is suspected. Such inspections are conducted by representatives of IAEA, which promotes the international development of peaceful uses of nuclear technology and discourages their military applications.

enriched uranium
■ Uranium and its compounds in which the concentration of uranium-238 has been reduced or eliminated by processing

fuel rod ■ A rod-shaped metal assembly often made of a zirconium alloy and used to contain pellets of nuclear fuel

depleted uranium (DUF$_6$) ■ Uranium and its compounds in which the concentration of uranium-235 has been reduced or eliminated by processing

16.9-E SPONTANEOUS FISSION

As first noted in Section 16.2, spontaneous fission occurs only when certain very heavy radioisotopes decay. These radioisotopes have an atomic number greater than 92. As their nuclei undergo spontaneous fission, they split into two nuclei and simultaneously generate neutrons.

An example of a radioisotope that decays by spontaneous fission is curium-250. When curium-250 undergoes spontaneous fissions, it splits into two other nuclei. One such event is denoted as follows:

$$^{250}_{96}Cm \longrightarrow {}^{137}_{52}Te + {}^{112}_{44}Ru + {}^{1}_{0}n$$

16.9-F THE GLOBAL FEAR OF USING NUCLEAR BOMBS DURING WARFARE

Throughout the history of civilization, the use of nuclear weapons was required only twice—when the United States detonated them over Japan and ended World War II. Thereafter, especially through the years of the Cold War, it was the *threat* of nuclear weapon deployment that facilitated the balance of political power. There was not a target on Earth that the United States and the former Soviet Union could not hit with long-range nuclear-armed ballistic missiles. Nine nations now possess nuclear weapons: the United States, Russia, the United Kingdom, France, China, Israel, India, Pakistan, and North Korea. Iran is probably close to possessing nuclear weapons, and Pakistan and India have increased their capabilities to construct them.

In the 1990s, the United States and the Russian Federation engaged in measures aimed at reducing their nuclear arsenals by means of bilateral agreements formally known as Measures for the Further Reduction and Limitation of Strategic Offensive Arms, more commonly called the **Strategic Reduction Arms Treaty**, or **START**. In 2011, each nation agreed to reduce their respective numbers of deployable nuclear warheads to 1550 by 2018—either as intercontinental ballistic missiles or deployed submarine-launched ballistic missiles.[5]

> **Strategic Reduction Arms Treaty (START)** ■ A bilateral agreement between the Russia Federation and the United States to achieve nuclear-arms-reduction goals by 2018

Even when START is completely implemented, however, both nations still will possess substantial numbers of nuclear warheads. Furthermore, while the United States and Russia are diminishing their arsenals, other nations are expanding them. At the start of 2013, for example, the United States and Russia had reduced their nuclear warheads from 8000 to 7700, and from 10,000 to 8000, respectively, but both China and India had ten more warheads than they did at the start of 2012.[6] In this regard, the United States considers the actions of Iran to be the most fearsome. Although Iran insists that its program is peaceful and meant only to power a future generation of reactors, the United States and other Western nations believe that Iran intends to produce nuclear weapons.

Pakistan's activities, too, are fearsome, because the country has taken steps to increase its capacity to produce plutonium. Pakistan's stockpile of over 100 deployable weapons is now estimated to surpass India's stockpile. Both nations, who have fought each other for decades, have launched initiatives to modernize their nuclear warhead and delivery systems. The all-pervading fear is that the flagrant misuse of nuclear bombs by any nation would not only destabilize the immediate locality but could potentially decimate individual civilizations and alter life on the entire planet.

To encourage international nuclear disarmament and reduce the further proliferation of nuclear weapons, the United Nations proposed a **Comprehensive Nuclear-Test-Ban Treaty**, or **CTBT**. One condition of this treaty prohibits the testing of a nuclear device of any size within any environment. As of 2011, the CTBT was signed by 182 countries, of which 153 have also ratified it. Included among the countries that have ratified the treaty are France, the Russian Federation, and the United Kingdom, but China, Egypt, India, Iran, Israel, North Korea, Pakistan, Syria, and the United States have not signed it.

> **Comprehensive Nuclear-Test-Ban Treaty (CTBT)** ■ A treaty proposed by the United Nations, not yet a component of international law, that aims to prohibit the testing of all types of nuclear devices

Inspectors for the IAEA periodically monitor nuclear facilities to ensure that nuclear material has not been diverted to military uses. Notwithstanding the responsibilities of the agency, the proliferation of fuel for nuclear weapons continues as demonstrated by the following:

- In 2006, 2007, and 2009, North Korea demonstrated that it could detonate a nuclear bomb.
- Without IAEA's knowledge, Syria secretly constructed a nuclear reactor to breed plutonium-239, but Israel destroyed the reactor in 2007 before it was activated.
- Iran has constructed a nuclear reactor, allegedly to use for the production of electrical power, but the prevailing fear is that it actually plans to create plutonium-239 for military use.
- Pakistan and India have accelerated the breeding of plutonium-239 in reactors.

The combination of these incidents signals all too clearly that mass killing and destruction from the deployment of nuclear weapons may again be realized. The energy

---

[5]Noteworthy is the fact that in 2011, the United States removed the uranium fuel from the last 9-megaton B53 bomb at the Pantex DOE Plant near Armadillo, Texas. Each B53 bomb was 600 times more powerful than the bomb dropped over Hiroshima. The nation's largest nuclear bomb is now the 1.2-megaton B83.
[6]SIPRI Yearbook, Stockholm International Peace Research (2013).

released by the detonation of a nuclear bomb or the engagement of a nuclear warhead far surpasses the energy associated with the detonation of ordinary chemical explosives. The explosion of a nuclear bomb, for instance, releases more energy than that involved in the detonation of approximately 1 million tons (~909,000 t) of TNT. The release of this amount of energy is globally fearsome for the following combination of reasons:

- This amount of energy could instantly kill tens of thousands of people within the immediate blast zone.
- The survivors of the initial blast would not only suffer from the acute effects of radiation exposure, but could acquire long-term health-related problems like cancer.
- Climatic changes would occur. Computer modeling illustrates that the detonation of 100 Hiroshima-sized bombs, for example, would dim the sunlight reaching Earth for 10 or more years, reduce rainfall, and cool the globe to temperatures resembling those experienced during the Little Ice Age (Section 10.12-B). The reduced temperatures would also cause crop failures, malnutrition, and famine throughout most of the world and cause diseases to multiply.

16.9-G NUCLEAR POWER PLANTS AND THEIR HAZARDS

Fissile nuclei are used by the nuclear power industry as fuel for the generation of electricity. Four hundred thirty-five (435) nuclear power plants produce approximately 20% of the electrical energy used worldwide. One hundred four, roughly one-fourth of the world's plants, operate within the United States, but most are approaching their projected operating lives of 40 years, seven are 40+ years old, and all but three have been operating for at least 20 years.[7] Thirty new plants are contemplated for construction during the next decade, with the first two scheduled to come online in 2014 and 2015. In 2012, the NRC issued a license for construction of the first new nuclear power plant in the United States since 1978.

If emissions from uranium mining and enrichment, fuel production, and spent-fuel production are ignored, the operation of nuclear power plants provides certain environmental benefits that compare favorably over the operation of fossil-fuel-fired power plants. In particular, operating nuclear power plants do not emit the greenhouse gases blamed for climate change. They also generate far less waste than do fossil-fuel-fired power plants.

Notwithstanding these benefits, a fear is perceived by the general public in connection with the siting and operation of nuclear power plants. A component of this fear is related with the longevity of the radiation generated during their operation. A number of radioisotopes having very long half-lives are produced during the operation of nuclear reactors. Examples include neptunium-237 and plutonium-239, whose half-lives are 2,000,000 years and 24,110 years, respectively. What can be done when radiation is released inadvertently through mismanagement, the failure to follow established safety practices, catastrophic natural disasters, or terrorist events?

NRC subjects nuclear power plants to especially rigid regulatory controls and licensing requirements. Nonetheless, nuclear power plants could release radioactivity from their confinement vessels into the environment. Such events are not merely hypothetical possibilities, as there have been well-documented instances of airborne releases of radioactivity from nuclear power plants in the United States, Russia, and Japan. Three occurrences are noted here.

---

[7]In 2012, the NRC granted a license to Southern Company to construct and operate two nuclear reactors adjacent to two operating plants at its facility near Waynesboro, Georgia. This was the first license issued by the NRC in 34 years.

Chernobyl Nuclear Power Plant

In April 1986, a series of operator errors unleashed a power surge that triggered an explosion and a partial meltdown of the fuel at the Chernobyl nuclear power plant in Ukraine, a member country of the former Soviet Union. Radioactive fuel and fission by-products were discharged into the environment, the majority of which ultimately spread over not only Ukraine, but also Russia, Belarus, and parts of western Europe. The radiation level at the reactor site ranged from 10,000 to 300,000 mSv/hr. Immediately after the disaster, approximately 24,000 individuals living within 9 miles (15 km) of the site received an average dose of 450 millisieverts before they were evacuated.

The Chernobyl disaster turned bustling villages and towns into unpopulated areas and directly threatened the health of more than 5 million people. As of 2005, however, fewer than 28 deaths had been directly attributed to radiation exposure from the disaster, almost all of which were rescue workers. Afterwards, approximately 6000 people were diagnosed with early signs of thyroid cancer caused in part by inhalation of airborne iodine radioisotopes. Many survived owing to prompt medical treatment. Even years after the disaster, thyroid cancer was identified as the most commonly identified ill effect associated with exposure to the airborne radiation from the Chernobyl reactor.[8]

Three-Mile-Island Nuclear Plant

In 1979, an accident occurred at the Three-Mile-Island plant in southeastern Pennsylvania when the coolant used to modulate the generated heat was inadvertently discharged into the atmosphere and the Susquehanna River.[9] The operators quickly shut down the power plant, and all auxiliary equipment remained operative. A partial meltdown occurred, resulting in the venting of approximately 43,000 Ci of krypton radioisotopes from the reactor building into the surrounding environment, but the confinement buildings were not destroyed. The maximum dose received by those living within 10 miles (16 km) of the facility was 0.08 millisieverts. No one died as an immediate result of the accident, but the cancer rates in the region surrounding the plant remain points of fierce debate even today.

Fukushima Dai-Ichi Nuclear Power Station

In 2011, a magnitude-9 earthquake occurred in Japan, spawning a tsunami and disabling the electrical power system at the Fukushima Dai-ichi Power Station.[10] Three of its six reactors were operating at the time of the earthquake, and three others were undergoing maintenance. Without power, it was impossible to cool the three operating reactors. Within five hours of the onslaught, triple meltdowns occurred which, along with fires and hydrogen explosions, ultimately damaged the integrity of four reactors and spread radioactivity over an area of approximately 230 square miles (596 km^2).

The hydrogen explosions caused radioactive fission products to be spewed into the atmosphere. These radioisotopes included iodine-131, iodine-132, cesium-134, cesium-137, tellurium-132, and xenon-133. Japanese water supplies, milk, vegetation, and seafood were tainted with iodine-131 and cesium-137 at unacceptable levels. The highest radiation level recorded at the plant, obtained robotically, was 57 Sv/hr. Individuals living 19 miles (31 km) from the post-disaster plant experienced an annual maximum concentration of 20 millisieverts.

In the United States, iodine-131 and other radioisotopes were detected at very low concentrations, not only in Hawaii and the western coastal states, but also in Pennsylvania and Massachusetts.

---

[8]International Atomic Energy Agency, "Chernobyl's Legacy: Health, Environmental and Socio-Economic Impacts," IAEA06-09181 (Vienna, Austria, 2006). (The findings noted in this report have been highly controversial regarding their validity. Interested parties may scan the Internet for dozens of follow-up articles.)

[9]U.S. Nuclear Regulatory Commission, *Backgrounder on the Three-Mile-Island Accident* (2009).

[10]Emilie van Deventer et al., "WHO's public health agenda in response to the Fukushima Daiichi nuclear accident," *J. Radiol. Prot.*, Vol. 32 (2012), pp. 99–122.

16.9-H FUTURE USE OF NUCLEAR POWER PLANTS

The disaster at Fukushima Dai-ichi generated an international wake-up call regarding the siting, design, and operation of nuclear power plants. At least one point became obvious: Nuclear power plants should not be sited in seismologically risky regions.

Nonetheless, even with worries about the safety of nuclear technology, unanimity is lacking among the countries of the world over the continued use of nuclear reactors. Germany has announced that it will discontinue use of its 17 nuclear reactors for the generation of electricity by 2022; Japan turned off the last of its 50 reactors in 2012; and Switzerland discontinued use of its nuclear power facilities in 2013. Also in 2013, twin reactors at California's San Onofre's nuclear power facility were shut down permanently, following the detection of a small radiation leak. On the other hand, China and India appear to have an undiminished appetite for nuclear power plants, because they are positioned to build additional ones. Indonesia, Thailand, South Korea, and Vietnam also are proceeding with construction of new plants.

In the United States, public support for nuclear reactors has plummeted, because expert designers, engineers, and regulators appear incapable of eliminating or minimizing their potential hazards. Even so, four new reactors probably will be constructed using current technology during the next decade.

The Fukushima Dai-ichi incident has also generated a rethinking about the storage requirements for spent nuclear fuel. In the future, the operators of nuclear power plants may be required to transfer their spent fuel rods into dry casks. [Spent fuel rods containing residual MOX (Section 16.9-B) were stored at the Fukushima Dai-Ichi Power Plant in water-filled pools.] A **dry cask,** one type of which is illustrated in Figure 16.14, is a device that provides stand-alone containment of spent nuclear fuel and other high-level radioactive wastes that have first "cooled" for one to five years. The waste is robotically inserted into a 10-inch-thick steel canister that is vacuum-sealed, filled with an inert gas like helium to prevent corrosion, and welded shut. Each canister is then placed inside a three-foot-thick concrete barrel that is equipped with small vents to keep its contents cool. This

dry cask ■ A means of containment for spent nuclear fuel during its storage

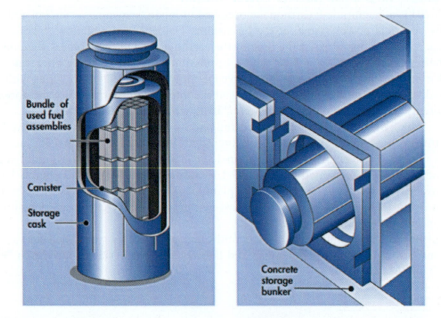

FIGURE 16.14 This steel and concrete "dry cask" resting on a concrete pad is a cylinder in which high-level radioactive waste such as spent nuclear fuel may be stored. Each dry cask is routinely entombed in a concrete storage bunker until a permanent disposal facility is available. Approximately 1300 dry casks are now in use at 55 sites at nuclear power stations and other locations nationwide. Some have also been used for containment of radioactive materials during their transportation. (*Courtesy of U.S. Nuclear Regulatory Commission, Washington, DC*)

design and construction provides shielding and performs the functions of confinement, radiological shielding, decay heat removal, and physical protection of the spent fuel during normal, abnormal, and accident conditions.

The disaster at the Fukushima Dai-ichi Nuclear Power Station also prompts lawmakers and policy planners to develop a satisfactory course of action for disposing of the nation's most fearsome nuclear wastes. The production of these wastes is ongoing at the nation's nuclear power plants, but nuclear wastes generated as early as the 1940s still remain in storage. For example, wastes formerly generated in conjunction with production of the nation's nuclear weapons are still stored at former defense plants such as the Hanford Nuclear Reservation in Hanford, Washington. This site occupies an area of 560 square miles (1400 km^2) and is located within the Columbia River Basin in south-central Washington. In 1943, when it was operational, the facility produced plutonium-239 for the U.S. nuclear weapons arsenal during World War II and through the Cold War. It now is regarded as the nation's most seriously contaminated radioactive site, because the wastes in its 177 underground tanks are leaking into the surrounding soils.

Certain other nuclear wastes are now stored at temporary storage sites in 35 states. They were generated not only within the United States but outside the country as well. The latter wastes are stored to keep nuclear fuel from reaching international terrorist organizations or rogue nations.

Acknowledging the magnitude of the nuclear waste disposal problem, Congress directed the DOE to locate and construct a national site in which high-level nuclear wastes could be stored permanently. An essential element of the site is an acceptable means for protecting the next 25,000 generations of Americans from exposure to the nuclear waste generated contemporaneously.

The site initially selected by DOE and approved by Congress as a geologic repository for spent fuel and high-level radioactive wastes was a volcanic-rock ridge located within Yucca Mountain, approximately 100 miles (161 km) northwest of Las Vegas, Nevada. However, in 2010, political pressures resulted in the issuance of a presidential order that essentially ceased the government's long-standing plan to dispose of these wastes at Yucca Mountain. After two decades of work, technical problems, legal challenges, and political opposition caused a scuttling of the project with the stroke of a pen.

In 2013, DOE set a goal of locating an underground, permanent disposal site by 2026, designing and licensing it by 2042, and constructing and accepting nuclear waste for burial by 2048.

In the United States, there are two active low-level nuclear waste storage sites:

- Waste Isolation Pilot Project, or WIPP, near Carlsbad, New Mexico, which is used today mainly for the temporary storage of plutonium wastes. WIPP is a massive geologic repository in which wastes are stored in chambers excavated in vast salt beds nearly a half mile underground.
- The Nevada National Security Site, approximately 65 miles north of Las Vegas, is used to store the government's dregs associated with the cleanup of Cold War laboratories and factories.

To date, no country has implemented a plan that addresses the long-term disposal of nuclear waste. France is the only country that treats its spent fuel, and Finland and Sweden are the only countries that have selected locations as permanent depositories for nuclear waste.

16.10 TRANSPORTING RADIOACTIVE MATERIALS

A stringent set of regulations apply to the transportation of a radioactive material. The portion of these regulations most immediately applicable to emergency responders is summarized in the subsections that follow.

TABLE 16.6	Specific Activity and Total Activity for Some Selected Radioisotopes[a]			
	SPECIFIC ACTIVITY		**TOTAL ACTIVITY**	
RADIOISOTOPE	TBq/g	Ci/g	TBq	Ci
Bismuth-207	1.0×10^1	2.7×10^{-10}	1.0×10^6	2.7×10^{-5}
Cobalt-60	1.0×10^1	2.7×10^{-10}	1.0×10^6	2.7×10^{-5}
Cesium-137	1.0×10^1	2.7×10^{-10}	1.0×10^6	2.7×10^{-5}
Molybdenum-99	1.0×10^2	2.7×10^{-9}	1.0×10^6	2.7×10^{-5}
Potassium-40	1.0×10^2	2.7×10^{-9}	1.0×10^6	2.7×10^{-5}

[a]Excerpted from 49 C.F.R. §173.436.

When shippers offer a radioactive material for transportation, it is first necessary to determine whether DOT applies an exemption that is applicable to the potential consignment. The shippers determine the activity concentration and the total activity in the potential consignment and compare these values with those published at 49 C.F.R. §173.436. The DOT regulations apply when the activity concentration and the total activity in the consignment exceed the published values. Table 16.6 lists some representative values for several selected radioisotopes.

Some radioisotopes are transported as "hazardous substances" within the meaning of CERCLA (Section 6.2-B), examples of which are provided in Table 16.7 along with their **reportable quantities (RQ)**. There are six reportable quantities for radioisotopes: 0.00037 TBq, 0.0037 TBq, 0.037 TBq, 0.37 TBq, 3.7 TBq, and 37 TBq.

reportable quantity (RQ) ■ The amount of a radioisotope listed in Appendix B at 40 C.F.R. §302.4 and 49 C.F.R. §172.101, the release of which triggers mandatory notification to the National Response Center

16.10-A SHIPPING DESCRIPTIONS

When shippers offer a radioactive material for transportation, DOT requires them to include the following information in its shipping description:

■ Proper shipping name, hazard class 7, and identification number. The proper shipping names of the hazardous materials in class 7 always begin with the words "Radioactive material." Some examples are listed in Appendix C.

TABLE 16.7	Reportable Quantities of Some Selected Radioisotopic Hazardous Substances[a]		
HAZARDOUS SUBSTANCE	**RQ [TBq]**	**HAZARDOUS SUBSTANCE**	**RQ [TBq]**
Arsenic-77	37	Plutonium-239	0.0037
Copper-64	37	Potassium-40	0.037
Iodine-131	0.00037	Promethium-147	0.37
Iridium-137	3.7	Radium-228	0.0037
Lead-212	0.37	Radon-222	0.0037
Mercury-195	3.7	Strontium-90	0.0037
Molybdenum-99	3.7	Uranium-233	0.0037
Phosphorus-32	0.0037	Uranium-235	0.0037
Platinum-189	3.7	Zinc-71	3.7

[a]The reportable quantities of radioisotopes that are hazardous substances are provided in Table 2 following Occupational Safety and Health Administration Regulation Standards, 49 C.F.R. §172.101. Occupational Safety and Health Administration.

- The capital letters *RQ*, when an activity equal to or greater than the reportable quantity is offered for transportation.
- The name of the radioisotope.
- A description of the physical and chemical form of the material.
- The activity of the radioisotope expressed in becquerels or its multiples or fractions. The activity is determined by using appropriate radiation detection equipment. The activity may also be expressed in curies or other units and entered parenthetically immediately next to the measurement in becquerels.
- Identification of the labels that DOT requires to be affixed to the relevant packaging.
- The transport index, when DOT requires either RADIOACTIVE YELLOW-II or RADIOACTIVE YELLOW-III labels to be affixed to the packaging. The **transport index** is a dimensionless number used to designate the degree of control to be exercised during transportation of a nonfissile radioactive material. It is determined by multiplying the maximum radiation level in mSv/hr at 3.3 feet (1 m) from the external surface of the package by 100.
- The criticality safety index (Section 16.10-F), when DOT requires FISSILE labels to be affixed to the packaging.
- The expression "highway route–controlled quantity," or HRCQ, when shippers offer for transportation a "highway route–controlled quantity" (Section 16.10-G) of a class 7 hazardous material [49 C.F.R. §172.203(d)(10)].

transport index ▪ For purposes of DOT regulations, a dimensionless number that shippers and carriers assign to certain radioactive materials to designate the degree of control to be exercised during their transit

16.10-B LABELING REQUIREMENTS

When shippers offer packages of radioactive materials for transportation, DOT requires them to affix on their opposite sides other than the bottom the applicable RADIOACTIVE WHITE-I, RADIOACTIVE YELLOW-II, RADIOACTIVE YELLOW-III, or FISSILE labels. RADIOACTIVE YELLOW-III labels are always affixed on the opposite sides of packages containing a "highway route–controlled quantity" of a class 7 hazardous material, and FISSILE labels are affixed adjacent to the applicable RADIOACTIVE YELLOW-II or RADIOACTIVE YELLOW-III labels. The features of these labels were previously illustrated in Figure 6.5.

DOT requires shippers to choose the applicable RADIOACTIVE label by measuring the radiation level at any point on the external surface of the packaging and determining the transport index. The applicable label is then chosen from Table 16.8. DOT

TABLE 16.8	Categories of the DOT Class 7 Labels[a]	
TRANSPORT INDEX[b]	**MAXIMUM RADIATION LEVEL AT ANY POINT ON A PACKAGE'S EXTERNAL SURFACE**	**LABEL CATEGORY**
0	Less than or equal to 0.005 mSv/hr (0.5 mrem/hr)	RADIOACTIVE WHITE-I
More than 0 but not more than 1	Greater than 0.005 mSv/hr (0.5 mrem/hr) but less than or equal to 0.5 mSv/hr (50 mrem/hr)	RADIOACTIVE YELLOW-II
More than 1 but not more than 10	Greater than 0.5 mSv/hr (50 mrem/hr) but less than or equal to 2 mSv/hr (200 mrem/hr)	RADIOACTIVE YELLOW-III
More than 10	Greater than 2 mSv/hr (200 mrem/hr) but less than or equal to 10 mSv/hr (1000 mrem/hr)[c]	RADIOACTIVE YELLOW-III

[a]49 C.F.R. §172.403.
[b]When the transport index is calculated as a number less than 0.05, the value is considered zero.
[c]DOT subjects shippers to the exclusive-use provisions set forth at 49 C.F.R. §173.441 when transporting radioactive material having this maximum radiation level.

also requires shippers to inscribe the following information within the contents-entry space:

- The name of the radioisotope to be transported and its activity on the applicable RADIOACTIVE WHITE-I, RADIOACTIVE YELLOW-II, or RADIOACTIVE YELLOW-III label
- LSA-I material, or LSA-I (Section 16.10-C), on the applicable RADIOACTIVE WHITE-I, RADIOACTIVE YELLOW-II, or RADIOACTIVE YELLOW-III label, when shippers offer LSA-I material for transportation
- The transport index on RADIOACTIVE YELLOW-II and RADIOACTIVE YELLOW-III labels
- The criticality safety index on the FISSILE label when shippers offer subcritical masses of uranium-233, uranium-235, plutonium-239 or plutonium-241 for transportation

When shippers intend to offer packages that have been emptied of their class 7 contents for transportation, DOT may require them to affix EMPTY labels on opposite sides of the packages. The EMPTY label is a white square with black lettering.

```
┌─────────────┐
│             │
│             │
│    EMPTY    │
│             │
│             │
└─────────────┘
```

DOT requires EMPTY labels to be affixed to the packages when the exterior radiation levels reveal the presence of internal contamination following the removal of their radioactive contents.

16.10-C MARKING REQUIREMENTS

At 49 C.F.R. §§173.471 - 173.473, DOT requires shippers to mark each package containing a class 7 hazardous material with the following information:

- The name and address of the shipper and receiver of the package
- The proper shipping name and UN identification number of the radioactive material
- *RQ*, if the package contains an amount equal to or greater than the reportable quantity of the radioactive material
- The gross mass when the package weighs over 110 pounds (50 kg)
- Orientation arrows, if the contents are liquid
- "Radioactive—LSA" or "Radioactive—SCO," if applicable. **Low specific activity material (LSA)** refers to radioactive material that complies with the DOT descriptions and activity limitations provided in Table 16.9. **Surface contaminated object (SCO)** refers to a solid object that is not itself radioactive but has radioactive material distributed on its surface.
- The package types TYPE A or TYPE B, as appropriate (Section 16.10-E)
- The trefoil when the class 7 hazardous material is contained within a type B package
- USA, if destined for export
- For industrial packages (Section 16.10-E), TYPE IP-1, TYPE IP-2, or TYPE IP-3, as applicable, with the international vehicle registration code of the country of origin of the design
- For type B packages (Section 16.10-E), TYPE B(U) or TYPE B(M), when applicable, and if destined for export, USA in conjunction with the specification marking or other package-certificate identification

low specific activity material (LSA) ■ For purposes of DOT regulations, certain radioactive material, including uranium and thorium ores and tritium-labeled water, that complies with the descriptions and activity limitations published at 49 C.F.R. §173.403

surface contaminated object (SCO) ■ For purposes of DOT regulations, a solid object that is not itself radioactive but has radioactive material distributed on its surface

TABLE 16.9	Low Specific Activity Materials[a]

GROUP	FEATURES
LSA-I	Uranium and thorium ores, concentrates of uranium and thorium ores, and other ores containing naturally occurring radionuclides that are intended to be processed for the use of the radionuclides, *or*
	Solid unirradiated natural uranium or thorium or depleted uranium or natural thorium or their solid or liquid compounds or mixtures, *or*
	Radioactive material other than fissile material for which the A_2 value is unlimited, *or*
	Other radioactive material, excluding fissile material in quantities not excepted under §173.453, in which the radioactive material is essentially uniformly distributed and the average specific activity does not exceed 30 times the values for activity concentration specified in §173.436, or 30 times the default values listed in Table 8 of §173.433.
LSA-II	Water with a tritium concentration of up to 0.8 TBq/L (20.0 Ci/L), *or*
	Other radioactive material in which the activity is distributed throughout and the average specific activity does not exceed $10^{-4}A_2$/g for solids and gases and $10^{-5}A_2$/g for liquids. The significance of A_2 is noted in Section 16.10-E.
LSA-III	Solids (e.g., consolidated wastes, activated materials), excluding certain powders in which: The radioactive material is distributed throughout a solid or a collection of solid objects, or is essentially uniformly distributed in a solid compact binding agent such as concrete, bitumen, or ceramic;
	The radioactive material is relatively insoluble, or it is intrinsically contained in a relatively insoluble material, so that, even under loss of packaging, the loss of radioactive material per package by leaching when placed in water for seven days would not exceed 0.1 A_2; and
	The estimated average specific activity of the solid, excluding any shielding material, does not exceed $2 \times 10^{-3}A_2$/g.

[a]Excerpted from 49 C.F.R. §173.403.

16.10-D PLACARDING REQUIREMENTS

DOT requires carriers to post the RADIOACTIVE placard shown in Figure 6.13 on each side and each end of a transport vehicle or freight container used to ship the following by highway or rail:

- A package to which DOT requires RADIOACTIVE YELLOW-III labels to be affixed
- "Exclusive-use" LSA materials or SCOs within "excepted packages" (Section 16.10-E)
- A "highway route–controlled quantity" of a class 7 hazardous material (Section 16.10-G)

DOT prohibits carriers from displaying the UN/NA identification number of a radioactive material across the center face of a RADIOACTIVE placard.

16.10-E PACKAGING TYPES

When shippers offer a radioactive material for transportation, DOT requires them to containerize it within one of the following types of packaging: excepted package, industrial package (IP), type A package, or type B package. These packages have been designed to protect workers, the public, and the environment from the inherent risks associated with transporting radioactive materials. Each is described next:

■ An **excepted package** refers to a type of packaging that DOT authorizes when LSA materials and SCOs are transported by an exclusive-use means. **Exclusive use** refers to the sole use by a single consignor of a conveyance or a large freight container for which all initial, intermediate, and final loading and unloading are carried out in accordance with the written directions of the consignor or consignee. An example of an excepted package is a cardboard box that complies with certain specified test conditions. It is "excepted" from the routine labeling and marking requirements but is subject to other requirements published at 49 C.F.R. §§173.421–173.426. Shippers must also comply with

excepted package ■ For purposes of DOT regulations, a package type in which LSA materials and SCOs are transported by exclusive-use means

exclusive use ■ For purposes of DOT and NRC regulations, the sole use by a single consignor of a conveyance or a large freight container for which all initial, intermediate, and final loading and unloading are carried out in accordance with the written directions of the consignor or consignee

FIGURE 16.15 When a radioactive material is offered for transportation as an excepted package, shippers post this marking on its packaging. The crosshatched border is typically red on a white background, and the wording is printed in black.

Radioactive Material, Excepted Package

This package contains radioactive material, excepted package and is in all respects in compliance with the applicable international and national governmental regulations.

UN_____

The information for this package need not appear on the Notification to Captain (NOTOC)

the conveyance activity limits published at 49 C.F.R. §173.427(e). An excepted package of a radioactive material is generally apparent since it is marked as shown in Figure 16.15.

■ An industrial package is another type of authorized packaging in which DOT permits the shipment of certain quantities of radioactive material, instruments, or articles manufactured from natural or depleted uranium or natural thorium. The DOT regulations require shippers to use one of three types designated as IP-1, IP-2, and IP-3. Each type must be designed and fabricated to retain the integrity of containment and shielding when it is subjected to normal conditions of transport. In addition, IP-2 and IP-3 satisfy certain prescribed DOT tests published at 49 C.F.R. §§178.603 and 178.606 relating to their ability to remain intact when dropped and stacked, respectively.

■ A **type A package** has been designed to retain the integrity of containment and shielding when subjected to normal conditions of transport. Four examples of a type A package are shown in Figure 16.16: a fiberboard box, wooden box, steel drum, and lead canister.

■ A **type B package** has been designed to retain the integrity of containment and shielding when subjected to normal conditions of transport and to retain the integrity of containment and shielding when subjected to prescribed hypothetical accident–test conditions. Three examples are shown in Figure 16.17: a steel outer drum with an inner container, a concrete and steel cask, and a massive transport container shielded with lead.

type A package ■ For purposes of DOT regulations, a package designed to transport an A_1 or A_2 quantity of radioactive material, as appropriate

type B package ■ For purposes of DOT regulations, a package designed to transport greater than an A_1 or A_2 quantity of radioactive material, as appropriate

FIGURE 16.16 These typical forms of type A packaging have been shown to withstand certain DOT test procedures that were designed to maintain the integrity of containment and shielding under the normal conditions of transport.

Package must withstand normal conditions of transport only without loss or dispersal of the radioactive control contents.

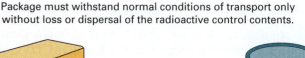

DOT Specification fiberboard box

DOT Specification steel drum

DOT Specification wooden box

DOT Specification type A package

Package must stand both normal and accident test conditions without loss of contents.

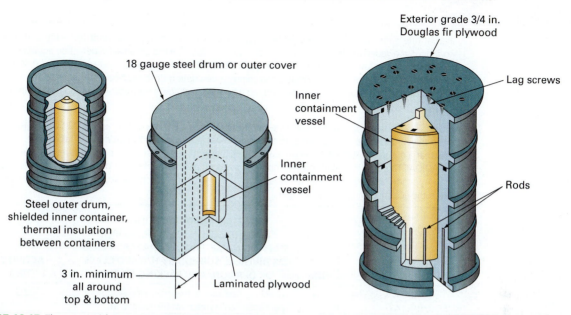

FIGURE 16.17 These typical forms of type B packaging have been shown to withstand certain DOT test procedures that were designed to maintain the integrity of containment and shielding under normal conditions of transport and the damaging conditions of a transportation accident.

When shippers offer a radioactive material for transportation, DOT requires them to select a type A or type B package based in part on the magnitude of two activity values, called A_1 and A_2.

■ A_1 is the maximum activity of a special-form class 7 material permitted in a type A package. A **special form class 7 material** refers to an indispersible solid radioactive material that is either a single solid piece or a sealed capsule, at least one dimension of which is not less than 0.2 inch (5 mm), and in the case of the capsule, contains radioactive material that may be opened only by destroying it. A radioactive material that is not a special-form class 7 material is a **normal form class 7 material**, or **nonspecial form class 7 material**.

■ A_2 is the maximum activity of a class 7 material, *other than* a special-form class 7 material, LSA, or SCO, that is permitted for transportation within a type A package.

DOT publishes the A_1 and A_2 values at 49 C.F.R. §173.435, some excerpts of which are provided for several radioisotopes in Table 16.10.

A_1 ■ For purposes of DOT regulations, the maximum activity of special form class 7 (radioactive) material permitted in a type A package

A_2 ■ For purposes of DOT regulations, the maximum activity of class 7 (radioactive) material, other than special form class 7 material, LSA, and SCO, that is permitted in a type A package

special form class 7 material ■ For purposes of DOT regulations, an indispersible, solid radioactive material that is either a single solid piece or a sealed capsule, at least one dimension of which is not less than 0.2 inch (5 mm), and in the case of the capsule, contains radioactive material that can be opened only by destroying it

TABLE 16.10	**A_1 and A_2 Values for Some Selected Radioisotopes[a]**					
	A_1		**A_2**		**SPECIFIC ACTIVITY**	
RADIOISOTOPE	**TBq**	**Ci**	**TBq**	**Ci**	**TBq/g**	**Ci/g**
Bismuth-207	0.7	19	0.7	19	1.9	52
Cobalt-60	0.4	11	0.4	11	42	1100
Cesium-137	2	54	0.6	16	3.2	87
Molybdenum-99	1.0	2.7	0.74[b]	20[b]	1.8×10^4	4.8×10^5
Potassium-40	0.9	24	0.9	24	2.4×10^{-7}	6.4×10^{-6}

[a]Excerpted from 49 C.F.R. §173.435 and 10 C.F.R. Part 71, Appendix A, Table A-1.
[b]Domestic use only.

A shipper offers for domestic transportation a type A package weighing 65 pounds (29.5 kg) and containing $^{99}_{42}$Mo-tagged molybdenum dioxide having an activity of 2.9 TBq. The use of radiation detection equipment reveals a maximum radiation level of 0.005 mSv/hr on the surface of the package and a radiation level of 0.0002 mSv/hr at 3.3 feet (1 m) from the surface. What is the shipping description of this radioactive material, and which DOT labels, markings, and placards, if any, are required when the package is transported by means of a motor vehicle on public highways?

Solution: It is first appropriate to determine the transport index. We calculate it as 0.0002 Sv/hr × 100, or 0.02 (dimensionless). Footnote "b" to Table 16.8 indicates that this calculated value may be taken as 0. Table 16.8 reveals that a maximum surface radiation level of 0.005 mSv/hr and a transport index of 0 require the shipper to affix RADIOACTIVE WHITE-I labels on two opposite sides other than the bottom of the package. Then, by reference to Appendix C, Table 16.6, and Section 16.10-A, the shipper prepares the following shipping description to enter on the accompanying shipping paper:

UNITS	HM	SHIPPING DESCRIPTION (IDENTIFICATION NUMBER, PROPER SHIPPING NAME, PRIMARY HAZARD CLASS OR DIVISION, SUBSIDIARY HAZARD CLASS OR DIVISION, AND PACKING GROUP)	ACTIVITY (TBq)
1 steel drum weighing 65 lb (UN1A2)	X	UN2915, Radioactive material, Type A package, 7 (Molybdenum-99, tagged in solid molybdenum dioxide) (RADIOACTIVE WHITE-I label)	2.9

From Table 16.7, we determine that the shipper is not required to enter the letters *RQ* in the shipping description, since the reportable quantity for molybdenum-99 is 3.7 terabecquerels. Then, following information in Section 16.10-C, the shipper marks the package with the name and address of the shipper and receiver, the designation "Radioactive material, n.o.s.," Type A package, and the identification number UN2915. Because the package weighs less than 110 pounds (50 kg), DOT does not require the shipper to mark the weight on the package. When DOT requires RADIOACTIVE WHITE-I labels to be affixed to the packaging, the carrier is not obligated to placard the motor vehicle used to transport the package.

---

normal form class 7 material (nonspecial form class 7 material)
■ For purposes of DOT regulations, radioactive material that does not satisfy the definition of a special-form class 7 material

DOT uses A_1 and A_2 values to define type A and type B packages. These definitions are paraphrased as follows:

■ A "type A package" refers to a package designed to transport an A_1 or A_2 quantity of radioactive material, as appropriate. The purpose of the design is to retain the integrity of containment and shielding when subjected to normal conditions of transport.

■ A "type B package" refers to a package designed to transport greater than an A_1 or A_2 quantity of radioactive material, as appropriate. The purpose of the design is to retain the integrity of containment and shielding when subjected to normal conditions of transport *and* DOE-prescribed hypothetical accident–test conditions.

■ A "type B(U) package" refers to a type B package that together with its radioactive contents for international shipment requires unilateral approval only of the package design and of any stowage provisions that may be necessary for heat dissipation.

■ A "type B(M)" package refers to a type B package that together with its radioactive contents for international shipment requires unilateral approval of the package design and may require approval of the conditions of shipment. A type B(M) package has a maximum normal operating pressure of more than 100 psig (700 kPa/cm²) or a relief device that allows the release of class 7 radioactive material into the environment under accident conditions.

When exporting fissile material, shippers are required to use the package types designated as AF, B(U)F, and B(M)F. These package types have been designed to meet especially stringent DOE criteria for shipping fissile material.

16.10-F CRITICALITY SAFETY INDEX

Although DOT and NRC require the use of the criticality safety index for several purposes, the primary concern here is its use in connection with the transportation of fissile material. Although NRC bears the responsibility of issuing general licenses to those who transport fissile material, DOT promulgates the regulations to which the licensees must adhere when transporting it.

As noted earlier, one DOT regulation is the requirement to inscribe the criticality safety index on the FISSILE labels that are affixed to opposite sides of packages containing fissile material. The **criticality safety index**, or **CSI**, is a dimensionless number used to designate the degree of control over the accumulation of packages, overpacks, or freight containers holding fissile material during their transportation. Shippers calculate the CSI using formulas and tables published at 10 C.F.R. §§71.22, 71.23 and 71.59.

DOT and NRC regulations at 10 C.F.R. §71.59 require a shipper or carrier to control a package containing fissile material to assure that an array of such packages remains subcritical. To enable this control, the designer of the package derives a number "N" based on the satisfaction of each of the following conditions, assuming that the packages are stacked together in any arrangement and with close full reflection on all sides of the stack by water:

- Five times "N" undamaged packages with nothing between them is subcritical
- Two times "N" damaged packages, if each package is subjected to hypothetical accident conditions specified at 10 C.F.R. §71.73
- The value of "N" cannot be less than 0.5

The CSI is determined by dividing the number 50 by the value of "N" that has been derived using these conditions. The value of the CSI may be zero provided that an unlimited number of packages are subcritical, such that the value of "N" is effectively equal to infinity.

When a package containing fissile material is assigned a CSI value less than or equal to 50, the package may be shipped in a non–exclusive-use conveyance, provided that the sum of the CSIs is limited to less than or equal to 50; or the package may be shipped in an exclusive-use conveyance provided that the sum of the CSIs is equal to or less than 100. When the package containing fissile material is assigned a CSI value greater than 50, the package must be shipped by a carrier in an exclusive-use conveyance, provided that the sum of the CSIs is equal to or less than 100.

> **criticality safety index (CSI)** ▪ For purposes of DOT and NRC regulations, a dimensionless number used to provide control over the accumulation of packages, overpacks, or freight containers holding fissile material during their transportation

16.10-G HIGHWAY ROUTE–CONTROLLED QUANTITY

Under certain unique circumstances, DOT allows carriers to transport a specified amount of a radioactive material by a special route. It is called a **highway route–controlled quantity**, or **HRCQ**. DOT defines HRCQ as 3000 times the A_1 value for special-form class 7 material, 3000 times the A_2 value for normal form class 7 radioactive material, or 1000 TBq (27,000 Ci), whichever is less.

When carriers intend to transport a highway route–controlled quantity of a radioactive material, DOT requires them to ensure that the public's risk of radiation exposure is eliminated or minimized, by considering accident rates, the transit time, population density, anticipated activities, and the time of day and week during which the transportation occurs. DOT then directs the carrier to transport the radioactive material on the preferred route, or preferred highway (Section 15.4-F). DOT permits the carrier to deviate from using the preferred route or preferred highway only under emergency conditions that would make continued use of the preferred route or preferred highway unsafe or, when necessary, to stop for rest, fuel, or vehicle repairs.

When carriers are directed to use a preferred route or preferred highway, DOT requires them to post RADIOACTIVE placards within squares having a white background and

> **highway route–controlled quantity (HRCQ)** ▪ For purposes of DOT regulations, the least of the following amounts: 3000 times the A_1 value for a special-form class 7 (radioactive) material; 3000 times the A_2 value for a normal form class 7 material; or 1000 TBq (27,000 Ci)

a black border on each side and each end of the motor vehicle. The posting of these placards within the white squares serves to designate that DOT has authorized the shipment by a preferred route.

When carriers transport a highway route–controlled quantity of a radioactive material in a motor vehicle, freight container, or railcar, DOT also requires them to prepare and implement a security plan whose components comply with the requirements of 49 C.F.R. §172.802. Motor carriers are also required to obtain a hazardous materials safety permit (Section 6.10) before transporting a highway route–controlled quantity of a radioactive material. Issuance of the safety permit requires motor carriers to prepare a written route plan that complies with the requirements of 49 C.F.R. §397.101. Motor carriers may also be subject to other special controls.

16.11 RESPONDING TO INCIDENTS INVOLVING A RELEASE OF RADIOACTIVE MATERIALS

When radiation sources are located nearby, government regulations require the authorized regulatory body to warn individuals of their presence. For example, NRC and OSHA require the posting of the signs previously noted in Figure 16.9 in the workplace to warn employees that exposure to radiation is likely. Emergency responders are also warned of the presence of radiation sources when they encounter these signs.

A radioactive material may also be encountered at the scenes of transportation mishaps. Its presence is rapidly verified by observing the following:

- The number 7 as a component of a shipping description of a hazardous material on a shipping paper
- The word *RADIOACTIVE* and the number 7 on yellow-and-white labels or the word *FISSILE* and the number 7 on white-and-black labels affixed to packages
- The word *RADIOACTIVE* and the number 7 on yellow-and-white placards posted on the relevant transport vehicle

Once the presence of a radioactive material has been verified, how may responders protect themselves from undue exposure to ionizing radiation? The answer to this question requires the implementation of three basic principles: shielding, time, and distance.

Because moving behind a barrier or other shield may be impractical during an emergency response action, the best means of personal protection for responders is to limit the time of exposure and to maintain a position as far removed from the source as practical so as to receive the minimum radiation dose.

Emergency responders may appreciate the importance of separation from a radioactive source by noting the **inverse square law of radiation**. This law is arithmetically expressed as follows:

inverse square law of radiation ■ The observation that the intensity of radiation decreases as the inverse square of the distance from its source increases

$$I = \frac{I_o}{r^2}$$

In this equation, I_o is the original intensity of a "small" radiation source, and I is the intensity at a distance r. It summarizes the observation that the intensity of ionizing radiation decreases as the inverse square of the distance from the source increases. For example, if a radioactive material registers 1000 counts/min on a Geiger counter held 1 foot (0.3 m) from a source, the counter will register only 250 counts/min when it is held 2 feet (0.6 m) from the source. Hence, doubling the distance between an individual and the source of radiation reduces the radiation exposure to one-fourth the original value.

Use of the inverse square law of radiation is limited to radiation measurements from a small source. The closest measurement must be a distance equal to or greater than three times the maximum dimension of the source. The law cannot be used when the radiation is spread over a relatively large area.

When a radioactive material has spilled, or when the package used for containment of a radioactive material is leaking or damaged, emergency responders should consider procedures to protect public health and the environment. Because special training in the handling of radioactive materials is essential, the regional offices of DOE's National Nuclear Security Administration (NNSA) or the appropriate state or local radiological authorities should be notified when there is a need for an emergency response effort involving a radioactive material. NNSA is prepared to respond to any type of radiological accident or terrorist event occurring anywhere in the world.

When firefighters and other emergency responders are called to such incidents, care should be exercised to avoid possible inhalation, ingestion, or contact with radioactive materials. Loose radioactive material and its associated package should be segregated in an area pending disposal instructions from EPA or responsible radiological authorities. If a fire is ongoing, it should be extinguished only from a distance that is considered suitable for protection of personnel. All appropriate measures should be exercised to prevent the spread of radioactivity by minimizing the quantity of runoff water.

16.12 RESIDENTIAL RADON

Radioisotopes of radium continuously form during the decay of naturally occurring uranium and thorium radioisotopes. They form in geologic areas where uranium and thorium are components of metal-bearing ores, granite, or black shale. For example, radium-220 forms when the naturally occurring thorium-232 in a metal-bearing ore decays by the following series of steps:

$$^{232}_{90}\text{Th} \longrightarrow {}^{228}_{88}\text{Ra} \longrightarrow {}^{228}_{89}\text{Ac} \longrightarrow {}^{228}_{90}\text{Th} \longrightarrow {}^{224}_{88}\text{Ra} \longrightarrow {}^{220}_{86}\text{Rn}$$

Although geologic formations containing uranium and thorium are relatively uncommon, they do exist below Earth's upper stratum in specific areas of the world. In the United States, the highest radon levels appear to occur along a geologic formation called the Reading Prong,[11] which exists across southeastern Pennsylvania into northern New Jersey and New York.

Buried deep beneath Earth's surface, the presence of radium radioisotopes presents little problem. However, the decay of some radium radioisotopes produces the colorless, odorless noble gas radon, as illustrated by the decay of radium-226.

$$^{226}_{88}\text{Ra} \longrightarrow {}^{222}_{86}\text{Rn} + \alpha$$

Thus formed, radon slowly percolates upward through the rock and soil and enters the atmosphere, where it dissipates. The average atmospheric concentration of the radon radioisotopes is 0.4 pCi/L.

In the basement of homes, radon may find its way through mortar joints, the cracks in foundation floors, gaps between wall-to-wall joints, and pores in concrete blocks and other building materials. Radon also dissolves within groundwater aquifers from which it may enter a home via the water supply and open sump pumps.

The radon that enters a dwelling is called **residential radon**. It consists primarily of radon-220 and radon-222, which have half-lives of 54.5 seconds and 3.82 days, respectively.

residential radon ■ The radon that accumulates within a dwelling from the decay of naturally occurring uranium and thorium radioisotopes

---

[11]Thomas M. Gerusky, "The Pennsylvania Radon Story" (Pennsylvania Department of Environmental Protection, 2012).

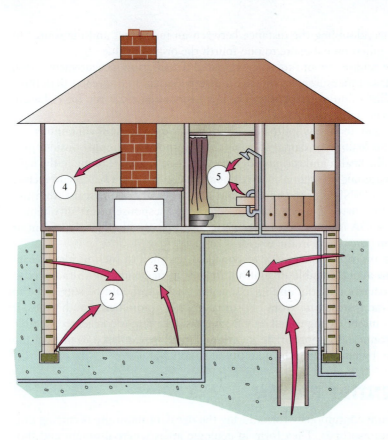

FIGURE 16.18 The six most common routes by which radon enters a home: (1) an open sump pump; (2) the gaps between basement wall-to-wall joints; (3) the foundation cracks in the basement floor; (4) the pores in concrete blocks and mortar joints; and (5) the residential water supply.

Each undergoes alpha decay. The products resulting from the decay are polonium radioisotopes, which also undergo alpha decay. The average year-round residential-radon concentration is approximately 1.3 pCi/L.

Within the building shown in Figure 16.18, there is little exchange between the inside and outside air. Under this condition, the residential-radon concentration may become so high that its presence poses an inhalation health hazard. This is an especially serious situation, because there is compelling evidence, based primarily on studies of uranium miners, showing that exposure to radon causes lung cancer by mutating the genes of normal cells.

The risk of contracting cancer from exposure to radon is proportional to the amount that enters a home and the length of time it remains within living areas. The majority of radon that is inhaled is exhaled back to the atmosphere immediately, but because the radon radioisotopes have short half-lives, some radon decays *before* it is exhaled. The polonium by-products then lodge within lung tissue and radiate the nearby cells with alpha particles. The radon radioisotopes may also decay in the air, whereupon the polonium radioisotopes adhere to dust particles. Individuals are exposed to them when the dust is inhaled.

Several types of radiation detectors are designed to determine the concentration of residential radon. Some are available for sale in hardware stores. The types used in dwellings and occupational settings collect radon decay products on an air filter or charcoal canister for a pre-established time. Then, the filter or canister is mailed to a radiation laboratory where it is analyzed for its radiation content.

Because the inhalation of radon has been shown to cause lung cancer in humans, radon is denoted as a known human carcinogen. The specific radioisotopes, radon-220 and radon-222, are also denoted as human carcinogens. Radium-224, -226, and -228 are also human carcinogens.

The risk of developing lung cancer from exposure to a radon-enriched atmosphere is exacerbated for individuals who are smokers; that is, in combination, radon and the carcinogenic components of tobacco smoke act synergistically. Smoking increases an individual's risk of acquiring cancer from radon exposure by as much as 15 times. Exposure to radon is the leading cause of lung cancer in nonsmokers and the second leading cause overall. Smoking tobacco products is the leading cause of lung cancer.

It is apparent that to reduce the number of radon-induced deaths due to lung cancer, the levels of residential radon must be minimized or eliminated within dwellings. EPA recommends that home occupants use a concentration of 4 pCi per liter of air (4 pCi/L) as an action guideline. The World Health Organization's action guideline is set at 2.7 pCi/L. Radon levels may be reduced in homes by sealing basement floors, increasing the air flowing throughout a residence by opening windows and using fans, and when the dwellings have a crawl space, keeping the vents in this area open throughout the year. The implementation of these measures assures the occupants that the radon concentration inside their homes is no higher than that of the ambient air outside.

In the late 1990s, scientific researchers[12] estimated that residential-radon exposure causes 15,400 to 21,800 cases of lung cancer each year. They also estimated that 260 individuals die each year from the exposure to household water containing dissolved radon. Approximately one-third of these incidents could have been prevented if the residential-radon concentration had been reduced to a level below EPA's action guideline.

16.13 RADIOLOGICAL DISPERSAL DEVICE

A **dirty bomb** may be regarded as any conventional explosive device charged with a hazardous material that disperses into the environment as the explosive is detonated. In theory, the hazardous material could be a poisonous gas, flammable material, corrosive material, biological or chemical warfare agent, or radioactive material. However, it is by the widespread dissemination of a radioactive material that terrorists instill the highest degree of fear, massive panic, and chaos in the affected population because of misinformation regarding the basic principles of radiation exposure. This device is characteristic of the radiological dirty bomb. The threat of involvement in a radiological-dirty-bomb incident is highest in densely populated areas, such as during a sports event at which large numbers of people congregate.

When a radiological dirty bomb is initially activated, the immediate area of the incident becomes contaminated with one or more radioisotopes. This area itself could constitute a sizable swath of land that would remain uninhabitable for years. Following the dispersal of radioactive material by explosive action, the radioisotopes then could diffuse throughout the air, where currents could carry them to far-removed locations including street canyons, subway tunnels, and inside buildings. In time, the radioactive material could even disperse worldwide. It is for this reason that a radiological dirty bomb is also called a **radiological dispersal device**.

The detonation of a radiological dispersal device could never produce the immediate mass casualties or devastation that we associate with the detonation of nuclear weapons. Nonetheless, the level of radiation could induce regulatory agencies to cordon off sections of entire cities for long periods.

Experts fear that terrorists are most likely to load a radioisotope listed in Table 16.11 into a radiological dispersal device. These choices represent the most common radioisotopes used in industrial and medical settings. Also listed in Table 16.11 is a "quantity of concern" for each radioisotope. It represents the activity in curies that is worrisome when the radioisotope is encountered because the activity poses a health risk. The third column in Table 16.11 provides the threshold concentration needed to contaminate 0.4 square miles (1 km^2). It is the activity of the radioisotope in curies that contaminates the area to

dirty bomb ■ A clandestine explosive device charged with a radioactive or other hazardous material that is disseminated into the environment as the bomb detonates

radiological dispersal device ■ Any unconventional weapon used to deliberately disperse radioactive material in the environment to create terror or inflict harm

[12]National Academy of Sciences, *Biological Effects of Ionizing Radiation VI (BEIR VI) Report* (Washington, DC: National Academies Press, 1998).

TABLE 16.11	Radioisotopes Most Likely to Be Used by Terrorists in a Radiological Dispersal Device[a]	
RADIOISOTOPE	**QUANTITY OF CONCERN (Ci)**	**THRESHOLD (Ci) TO CONTAMINATE 0.4 mi^2 (1 km^2)**
Americium-241	16.22	78
Californium-252	5.41	49
Cesium-137	27.03	42
Cobalt-60	8.11	11
Curium-244	13.51	130
Gadolinium-153	270.17	390
Iridium-192	21.62	100
Plutonium-238	16.22	220
Promethium-147	10,810.81	410
Radium-226	10.81	13
Selenium-75	54.05	150
Strontium-90	270.27	200
Thulium-170	5405.41	2000
Ytterbium-169	81.08	600

[a]Adapted from Jonathan Medalia, "'Dirty bombs': Technical background, attack prevention, and response, Issues for Congress," Congressional Research Service, DRS Report R41890 (June 24, 2011).

a level that a person living within it for a year would receive a dose of 2 rem during the first year after the attack. EPA and FEMA use this threshold as a protective action guide for relocation because personal damage from the radiation exposure is likely.

For illustrative purposes, suppose terrorists load cesium chloride powder tagged with the radioisotope cesium-137 into a device. Because the powder is used as a sealed source to irradiate food, it is reasonably accessible to anyone. If terrorists load the cesium chloride into a bomb that is subsequently detonated, the cesium-137 would readily disperse into the immediate environment as a powder. Any cesium-137 activity equal to or greater than 27.03 Ci is considered a quantity of concern. The threshold concentration that can potentially contaminate an area of 0.4 square miles (1 km^2) is 42 Ci. Any cesium-137 activity equal to or greater than this concentration causes EPA and FEMA to require the residents living in the contaminated area to relocate elsewhere.

What may an emergency response team do to best serve the public when terrorists have detonated a radiological dispersal device? The three basic tenets associated with radiation protection—shielding, time, and distance—come into play when answering this question. Translated into action, this means that the affected population should be removed as quickly as feasible from the areas where the concentrations of radioactive material are highest. These individuals may then seek out official protective information from local radio and television broadcasts.

EPA has issued protective-action guidelines for the personnel who respond to emergencies involving the activation of a radiological dispersal device.[13] Not only must responders decide how to accomplish an action, they must also acknowledge that implementing a response action involving radiation exposure could jeopardize their own health.

---

[13]Manual of Protective-Action Guides and Protective Actions for Nuclear Incidents (Washington, DC: U.S. Environmental Protection Agency, May 1992), EPA-400-R-92-001; *Federal Register* 71 (January 3, 2006): 173–96.

The adverse health effects that potentially result from radiation exposure were previously cited in Tables 16.3 and 16.4.

The protective-action guidelines are based on the nature of activities to be conducted at a site that is contaminated with various levels of radiation. They may be paraphrased as follows:

- When the prevailing radiation level is equal to or less than 5 rem, emergency responders may initiate operations in a normal fashion.
- When the radiation level ranges from 5 to 10 rem, careful consideration must be given to the relative importance of implementing response actions at once or waiting until the radiation level has subsided. Will multiple lives be saved, or will large populations be protected only if the actions are conducted at once?
- When the radiation level ranges from 10 to 25 rem, emergency responders are at risk of severe personal damage from radiation exposure, especially when executing the response action requires exposure over a significant time. Once again, the unique conditions of the incident at hand must be evaluated to decide whether the responders are likely to immediately save multiple lives or protect large populations.
- When the level exceeds 25 rem, EPA recommends the use of robotic equipment to the maximum extent feasible. The use of robotic equipment minimizes the radiation-exposure period that ordinarily would be experienced by the emergency responders. This could prevent the onset of radiation sickness.

Features of Atomic Nuclei

1. Identify the number of protons and neutrons contained in each of the following radioisotopes:
   (a) tritium
   (b) fluorine-18
2. What fraction of a radioisotope remains after the passage of five half-lives?
3. A radiologist determines that exposure to a particular radioisotope, X, is considered safe at concentrations of less than 1 mg/mi^2. The concentration of X at a nuclear explosion testing site is measured to be 10 mg/mi^2 immediately following the explosion of a nuclear bomb. If the half-life of X is 24 hours, will the site be safe to enter in 4 days? If not, how many days must lapse before the concentration of X has been reduced to a level that entry to the site is considered safe?

Modes of Radioisotopic Decay

4. Identify the nucleus that is produced when each of the following nuclear transformations occurs:

   (a) $^{18}_{9}F \longrightarrow$ ____ $+ \beta^+$

   (b) $^{28}_{13}Al \longrightarrow {}^{28}_{14}Si +$ ___

   (c) $^{231}_{91}Pa \longrightarrow {}^{227}_{89}Ac +$ ___

   (d) $^{37}_{18}Ar + {}^{0}_{-1}e \longrightarrow$ ___

   (e) $^{149}_{60}Nd \longrightarrow {}^{149}_{61}Pm +$ ___

   $\downarrow$

   ___ $+ \beta^-$

   (f) $^{81m}_{34}Se \longrightarrow {}^{81}_{34}Se +$ ___

   $\downarrow$

   $^{81}_{35}Br +$ ___

Measurement of Radioactivity

5. The average annual dose from medical and dental X rays is approximately 1 rad. When expressed in millisieverts (mSv), what is the average dose that a person receives annually during medical and dental examinations?
6. What is the specific activity in becquerels per gram of 10 grams of 1000 pCi of $^{14}_{6}C$-tagged sodium bicarbonate?
7. To protect public health, EPA proposes 4 mrem/y as the maximum radiation limit an individual should receive from consuming radiation-contaminated groundwater. If an individual receives a dose of 5 μSv/mo through the consumption of drinking water, has the person been exposed to radiation in excess of EPA's recommended limit?
8. One type of smoke detector is an ionization device containing americium-241. This radioisotope ionizes the air molecules between a pair of electrodes, thereby permitting

the passage of a small electrical current. When smoke particles enter this space, they adhere to the ionized molecules, thereby reducing the flow of current. This reduction activates an alarm. Ion detectors are especially useful for recognizing fast-moving fires, such as wastepaper basket and grease fires. If the specific activity of a source of americium-241 for use in manufacturing smoke detectors is measured as 3.4 Ci/g, what is its specific activity in terabecquerels per gram?

III Effects Resulting from Exposure to Radiation

9. The OSHA regulation at 29 C.F.R. §1910.96(b)(3) prohibits an employer from exposing individuals under 18 years of age to a dose of ionizing radiation greater than 750 mrem per calendar quarter. What is the most likely reason OSHA selected 18 as the maximum age to which radiation exposure should be strictly limited?
10. What biological effects are likely to be experienced by emergency responders who have received a short-term radiation dose of 400 millisieverts from exposure to cesium-137?
11. To reduce or eliminate the likelihood of metastasis, oncologists often use iodine-131 to treat patients with thyroid cancer following the surgical removal of their thyroid glands. A typical dosage is 30 mCi of ^{131}I-labeled sodium iodide. Why is the selection of iodine-131 for this treatment process likely to be more uniquely effective when compared to the use of other chemotherapeutic drugs?

Nuclear Fission

12. Identify the number of neutrons produced during the fission event represented by the following equation:

$$^{239}_{94}\text{Pu} + ^{1}_{0}n \longrightarrow ^{145}_{57}\text{La} + ^{92}_{37}\text{Rb} + \underline{\hspace{1cm}} + ^{1}_{0}n$$

13. What is the most likely reason the United States contributes low-level enriched uranium to a multinational fuel bank, which provides the fuel to countries that refrain from enriching uranium?

Transporting Radioactive Materials

14. The DOT regulation at 49 C.F.R. §173.417 requires carriers to limit packages of fissile radioactive material to certain prescribed amounts per transport vehicle intended for domestic shipment. What is the most likely reason that DOT limits the amount of fissile material contained within a transport vehicle?
15. The DOT regulation at 49 C.F.R. §177.842 requires shippers and carriers to avoid placing packages of radioactive materials bearing the RADIOACTIVE YELLOW-II and RADIOACTIVE YELLOW-III labels within a transport vehicle, storage location, or any other place closer than certain prescribed distances to any area that may be occupied by a passenger, employee, or animal. What is the most likely reason that DOT requires these packages to be placed in this fashion?
16. What shipping description does DOT require shippers to enter on the accompanying shipping paper when offering each of the following radioisotopes for domestic transportation by cargo air within the same aircraft:
    (a) Arsenic trichloride oil containing 30 TBq of arsenic-77 within a glass tube packaged in a wooden box and having a transport index of 0 and a maximum surface radiation level of 0.004 mSv/hr.

(b) Solid cesium chloride containing 1.2 Ci of cesium-137 within a glass tube packaged in an aluminum drum and having a transport index of 0.7 and a maximum surface radiation level of 0.4 mrem/h.

Responding to Emergencies Involving a Release of Radioactive Materials

17. When first-on-the-scene responders arrive at the scene of a highway traffic accident, they note that yellow-and-white RADIOACTIVE placards are posted on each side and each end of an overturned motor van. A representative of the carrier provides the accompanying shipping paper, which reads as follows:

UNITS	HM	SHIPPING DESCRIPTION (IDENTIFICATION NUMBER, PROPER SHIPPING NAME, PRIMARY HAZARD CLASS OR DIVISION, SUBSIDIARY HAZARD CLASS OR DIVISION, AND PACKING GROUP)	ACTIVITY (TBq)
1 reinforced steel drum with inner containment vessel (UN1A2)	X	RQ, UN2917, Radioactive material Type B(M) package, 7 (Cobalt-60 as metallic cobalt) (RADIOACTIVE YELLOW-III label) (Transport Index = 7.8)	37.5

What immediate actions should the responders execute to protect public health, safety, and the environment?

Residential Radon

18. In most states, why is it mandatory for homeowners to retain the services of a professional to determine the residential radon concentration and disclose this level to potential homebuyers?

Safety Data Sheet for Hydrogen Peroxide

Section 1. Hazardous Chemical Identification

Synonyms: Peroxide; Hydrogen peroxide solution
CAS No.: 7722-84-1
Molecular Weight: 34.01
Chemical Formula: H_2O_2
Supplier: My Company
My Street
My Town
My State 00000
Telephone (000) 000-0000

Section 2. Hazards Identification

Emergency Overview

OSHA Hazards: Oxidizer; Target Organ Effect; Toxic by Ingestion; Corrosive; Carcinogen

Target Organs: Eyes; Skin; Respiratory system

GHS Classification: Oxidizing liquid (Category 2); Acute Toxicity, oral (Category 4);
Acute toxicity, inhalation (Category 5); Skin corrosion (Category 1A);
Serious eye damage (Category 1); Acute aquatic toxicity (Category 3)

GHS Label Elements, including Hazard and Precautionary Statements:

Pictograms:

Signal Word	DANGER	
Hazard statements:	H272	May intensify fire; oxidizer
	H302	Harmful if swallowed
	H314	Causes severe skin burns and eye damage
	H333	May be harmful if inhaled
	H402	Harmful to aquatic life
Precautionary statements:	P220	Keep/Store away from clothing/combustible materials
	P280	Wear protective gloves/protective clothing/eye protection/face protection
		IF IN EYES:
	P305	Rinse cautiously with water for several minutes.
	P351	Remove contact lenses, if present and easy to do.
	P338	Continue rinsing.
	P310	Immediately call a POISON CENTER or doctor/physician

Section 3. Chemical Composition/Information on Ingredients

Ingredient	CAS No.	Percent
Hydrogen Peroxide	7722-84-1	35%
Water	7732-18-5	65%

Section 4. First-Aid Measures

Eye contact: Immediately flush with plenty of water for at least 15 minutes. If easy to do, remove contact lenses. Call a physician or poison control center immediately. In case of irritation from airborne exposure, move to fresh air. Get medical attention immediately.

Skin contact: Immediately flush with plenty of water for at least 15 minutes while removing contaminated clothing and shoes. Call a physician or poison control center immediately. Wash clothing separately before use. Destroy or thoroughly clean contaminated shoes.

Inhalation: Move to fresh air. If breathing stops, provide artificial respiration. If breathing is difficult, give oxygen. Call a physician or poison control center immediately.

Ingestion: Call a physician or poison control center immediately. Do not induce vomiting. If vomiting occurs, the head should be kept low so that stomach vomit doesn't enter the lungs.

Notes to Physician: Keep victim under observation. Treat symptomatically.

General advice: In the case of accident, or if you feel unwell, seek medical advice immediately (show the label where possible). Show this safety data sheet to the doctor in attendance. Ensure that medical personnel are aware of the material involved, and take precautions to protect themselves.

Section 5. Firefighting Measures

Extinguishing Media: Flood with water.

Fire/Explosion Hazards: Product does not burn. On decomposition, it releases oxygen, which may intensify fire.

Firefighting Procedures: Any tank or container surrounded by fire should be flooded with water for cooling. Wear full protective clothing and use self-contained breathing apparatus.

Flammable Limits: Product does not burn.

Sensitivity to Impact: No data available.

Sensitivity to Static Discharge: No data available.

Section 6. Accidental Release Measures

Personal Precautions: Eliminate all sources of ignition. Wear appropriate protective equipment and clothing during clean-up. Keep upwind. Keep out of low areas. Ventilate closed spaces before entering them. Local authorities should be advised if significant spillages cannot be contained.

Environmental Precautions: Prevent further leakage or spillage if safe to do so. Avoid discharge into drains, water courses, or on the ground.

Methods for Containment: Stop the flow of material, if this is without risk. Keep combustible material (wood, paper, oil, etc.) away from spilled material. Dike the spilled material, where this is possible. Prevent entry into waterways, sewers, basements, or confined areas.

Section 7. Handling and Storage Information

Handling: Wear chemical splash-type goggles and full-face shield, impervious clothing (rubber, PVC, etc.) and rubber or neoprene gloves and shoes. Avoid cotton, wool, and leather. Avoid excessive heat and contamination since they may cause decomposition and generation of oxygen gas that could produce high-pressure environment and result in container rupture.

Storage: Store in a well-ventilated, cool (<35°C), and dark area separated from combustible materials, reducing agents, strong bases, and organic compounds. Do not store on wooden shelves or floors. Vented containers should be periodically checked for bulging.

Section 8. Physical and Chemical Properties

Appearance:	Clear liquid
Odor:	Slight acrid odor
Solubility:	Infinitely soluble
Density:	1.11
pH:	No information located
% Volatiles by volume @70°F (21°C):	No information located
Boiling Point:	226°F (108°C)
Melting Point:	−27°F (−33°C)
Vapor Density (Air = 1):	No information located
Vapor pressure (mmHg):	No information located
Evaporation Rate (BuAc = 1):	No information located

Section 9. Chemical Stability and Reactivity

Stability:	Normally stable if uncontaminated, but slowly decomposes to release oxygen. Unstable with heat, may result in dangerous pressures. A strong oxidizer, reacts violently upon contact with many organic substances, particularly textile and paper. Avoid light and keep in a closed but vented container to prevent evaporation (concentration and contamination).
Hazardous Decomposition Products:	Decomposes to water and oxygen with rapid release of heat. Use vented containers. The solution may decompose violently upon heating.
Hazardous Polymerization:	Will not occur.
Incompatibilities:	Heat, reducing agents, organic materials, dirt, alkalis, rust, and many metals. Spontaneous combustion may occur upon standing in contact with readily flammable materials.
Conditions to Avoid:	Avoid excess heat and light. Avoid contact with combustible or organic materials.

Section 10. Toxicological Properties

Inhalation (rat) LC_{50}:	2000 mg/m^3
Oral (rat) LD_{50}:	1232 mg/kg
Skin (rabbit) LD_{50}:	>2000 mg/kg
Cancer Potential	IRAC has concluded that there is inadequate evidence for human carcinogenicity from exposure to hydrogen peroxide, but limited evidence for animal carcinogenicity. ACGIH has concluded that hydrogen peroxide is a confirmed animal carcinogen with unknown relevance to humans.

Section 11. Ecological Information

Channel catfish, 96-hour LC_{50}:	37.4 mg/L
Fathead minnow, 96-hour LC_{50}:	16.4 mg/L
Daphnia magna, 24-hour EC_{50}:	7.7 mg/L

Daphnia pulex, 48-hour LC$_{50}$:   2.4 mg/L

Freshwater snail, 96-hour LC$_{50}$:   17.7 mg/L

Environmental Fate: Hydrogen peroxide in the aquatic environment is subject to various oxidation or reduction processes and decomposes into water and oxygen. The hydrogen peroxide half-life in freshwater ranges from 8 hours to 20 days, in air from 10 to 20 hours, and in soils from minutes to hours depending upon microbiological activity and metal contaminants.

Section 12. Disposal Considerations

An acceptable method of disposal is to dilute with a large volume of water and allow the hydrogen peroxide to decompose, followed by discharge into a suitable water treatment system in accordance with all applicable environmental regulations. The appropriate regulatory agencies should be contacted prior to disposal.

Section 13. Transport Information

Proper Shipping Name: Hydrogen peroxide, aqueous solution (stabilized) (contains 20%–40% hydrogen peroxide)

Hazard Classes: 5.1, 8

UN/NA Number: UN2014

Packing Group: II

DOT Labels:

DOT Placard:

Section 14. Regulatory Information

SARA, 302/304/311/312:	Extremely hazardous substance
CERCLA:	Hazardous substance
TSCA:	8(b) Inventory
OSHA:	Hazardous chemical
New York:	Acutely hazardous substance
Rhode Island:	Right-To-Know hazardous substance
Pennsylvania:	Right-To-Know hazardous substance
Florida:	Hazardous substance
Minnesota:	Hazardous substance
Massachusetts:	Right-To-Know hazardous substance
New Jersey:	Hazardous substance

Section 15. Other Information

NFPA Ratings:
Health, 3; Flammability, 0;
Instability, 1; Other, OXY

WHMIS Symbols:

Table of Elements and Their Atomic Weights[a]

	SYMBOL	ATOMIC NUMBER	ATOMIC WEIGHT		SYMBOL	ATOMIC NUMBER	ATOMIC WEIGHT
Actinium*	Ac	89		Fermium*	Fm	100	
Aluminum	Al	13	26.98154	Fluorine	F	9	18.9984
Americium*	Am	95		Francium*	Fr	87	
Antimony	Sb	51	121.760	Gadolinium	Gd	64	157.25
Argon	Ar	18	39.948	Gallium	Ga	31	69.723
Arsenic	As	33	74.92160	Germanium	Ge	32	72.63
Astatine*	At	85		Gold	Au	79	196.9666
Barium	Ba	56	137.327	Hafnium	Hf	72	178.49
Berkelium*	Bk	97		Hassium*	Hs	108	
Beryllium	Be	4	9.01218	Helium	He	2	4.00260
Bismuth	Bi	83	208.98040	Holmium	Ho	67	164.93032
Bohrium*	Bh	107		Hydrogen	H	1	[1.00784; 1.00811]
Boron	B	5	[10.806; 10.821]	Indium	In	49	114.818
Bromine	Br	35	79.904	Iodine	I	53	126.9045
Cadmium	Cd	48	112.411	Iridium	Ir	77	192.217
Calcium	Ca	20	40.078	Iron	Fe	26	55.845
Californium*	Cf	98		Krypton	Kr	36	83.798
Carbon	C	6	[12.0096; 12.0116]	Lanthanum	La	57	138.90547
Cerium	Ce	58	140.116	Lawrencium*	Lw	103	
Cesium	Cs	55	132.90545	Lead	Pb	82	207.2
Chlorine	Cl	17	[35.446; 35.457]	Lithium	Li	3	[6.938; 6.997]
Chromium	Cr	24	51.9961	Lutetium	Lu	71	174.9668
Cobalt	Co	27	58.93320	Magnesium	Mg	12	24.3050
Copernicium*	Cn	112		Manganese	Mn	25	54.93805
Copper	Cu	29	63.546	Meitnerium*	Mt	109	
Curium	Cm	96		Mendelevium	Md	101	
Darmstadtium*	Ds	110		Mercury	Hg	80	200.59
Dubnium*	Db	105		Molybdenum	Mo	42	95.96
Dysprosium	Dy	66	162.500	Neodymium	Nd	60	144.242
Einsteinium*	Es	99		Neon	Ne	10	20.1797
Erbium	Er	68	167.259	Neptunium	Np	93	
Europium	Eu	63	151.964	Nickel	Ni	28	58.6934

	SYMBOL	ATOMIC NUMBER	ATOMIC WEIGHT		SYMBOL	ATOMIC NUMBER	ATOMIC WEIGHT
Niobium	Nb	41	92.90638	Silver	Ag	47	107.8682
Nitrogen	N	7	[14.00643; 14.00728]	Sodium	Na	11	22.98977
Nobelium*	No	102		Strontium	Sr	38	87.62
Osmium	Os	76	190.23	Sulfur	S	16	[32.059; 32.076]
Oxygen	O	8	[15.99903; 15.99977]	Tantalum	Ta	73	180.94788
Palladium	Pd	46	106.42	Technetium*	Tc	43	
Phosphorus	P	15	30.97376	Tellurium	Te	52	127.60
Platinum	Pt	78	195.084	Terbium	Tb	65	158.92535
Plutonium*	Pu	94		Thallium	Tl	81	[204.382; 204.385]
Polonium*	Po	84		Thorium	Th	90	232.03806
Potassium	K	19	39.0983	Thulium	Tm	69	168.93421
Praseodymium	Pr	59	140.90765	Tin	Sn	50	118.710
Promethium	Pm	61		Titanium	Ti	22	47.867
Protactinium*	Pa	91	231.03588	Tungsten	W	74	183.84
Radium*	Ra	88		Ununhexium*	Uuh	116	
Radon*	Rn	86		Ununoctium*	Uuo	118	
Rhenium	Re	75	186.207	Ununpentium*	Uup	115	
Rhodium	Rh	45	102.90550	Ununquadium*	Uuq	114	
Roentgenium*	Rg	111		Ununseptium*,b	Uns	117	
Rubidium	Rb	37	85.4678	Ununtrium*	Uut	113	
Ruthenium	Ru	44	101.07	Uranium*	U	92	238.02891
Rutherfordium*	Rf	104		Vanadium	V	23	50.9415
Samarium	Sm	62	150.36	Xenon	Xe	54	131.293
Scandium	Sc	21	44.95591	Ytterbium	Yb	70	173.054
Seaborgium*	Sg	106		Yttrium	Y	39	88.90585
Selenium	Se	34	78.96	Zinc	Zn	30	65.38
Silicon	Si	14	[28.084; 28.086]	Zirconium	Zr	40	91.224

[a]Michael E. Wieser and Tyler B. Coplen, "Atomic weights of the elements (2009) (IUPAC Technical Report)," *Pure Appl. Chem.*, Vol. 83 (2011), pp. 359–396. When known, the atomic weight is provided with the first five digits after the decimal point. The use of an asterisk with the name of an element indicates that the element does not have any stable isotopes. An atomic weight can be calculated for them only when one of their radioisotopes has a half-life greater than 1×10^{10} years.
[b]For completeness, ununseptium is listed here despite the fact that it was not discovered until 2010.

APPENDIX C

Hazardous Materials Table[a]

SYMBOLS	HAZARDOUS MATERIALS DESCRIPTIONS AND PROPER SHIPPING NAMES	HAZARD CLASS OR DIVISION	IDENTIFICATION NUMBER	PACKING GROUP	LABEL CODES	SPECIAL PROVISIONS
(1)	(2)	(3)	(4)	(5)	(6)	(7)
	Acetone	3	UN1090	II	3	
	Acetonitrile	3	UN1648	II	3	
	Acetyl benzoyl peroxide, solid, or with more than 40 percent in solution	Forbidden				
	Acrolein, stabilized	6.1	UN1092	I	6.1, 3	1
	Alcoholic beverages	3	UN3065	II	3	
		3	UN3065	III	3	
	Aluminum alkyls	4.2	UN3051	I	4.2, 4.3	
D	Ammonia, anhydrous	2.2	UN1005		2.2	
I	Ammonia, anhydrous	2.3	UN1005		2.3, 8	4
D	Ammonia solution, *relative density less than 0.880 at 15°C in water, with more than 50% ammonia*	2.2	UN3318		2.2	
I	Ammonia solution, *relative density less than 0.880 at 15°C in water, with more than 50% ammonia*	2.3	UN3318		2.3, 8	4
	Ammonia solution, *relative density less than 0.880 at 15°C in water, with more than 35% but not more than 50% ammonia*	2.2	UN2073		2.2	
	Ammonia solution, *relative density between 0.880 and 0.957 at 15°C in water, with more than 10% but not more than 35% ammonia*	8	UN2672	III	8	
	Ammunition, *incendiary liquid or gel, with burster, expelling charge, or propelling charge*	1.3J	UN0247	II	1.3J	
D	Asbestos	9	NA2212	III	9	

SYMBOLS	HAZARDOUS MATERIALS DESCRIPTIONS AND PROPER SHIPPING NAMES	HAZARD CLASS OR DIVISION	IDENTIFICATION NUMBER	PACKING GROUP	LABEL CODES	SPECIAL PROVISIONS
(1)	(2)	(3)	(4)	(5)	(6)	(7)
	Barium cyanide	6.1	UN1565	I	6.1	
	Battery fluid, acid	8	UN2796	II	8	
A	Calcium oxide	8	UN1910	III	8	
	Calcium, pyrophoric *or* Calcium alloys, pyrophoric	4.2	UN1855	I	4.2	
I	Carbon, activated	4.2	UN1362	III	4.2	
I	Carbon, *animal or vegetable origin*	4.2	UN1361	II	4.2	
		4.2	UN1361	III	4.2	
	Caustic soda (*etc.*), *see* sodium hydroxide (*etc.*)					
		8	UN1824	III	8	
D	Charcoal *briquettes, shell screenings, wood, etc.*	4.2	NA1361	III	4.2	
	Chlorine	2.3	UN1017		2.3, 5.1, 8	2
	Chlorine trifluoride	2.3	UN1749		2.3, 5.1, 8	2
	Chlorobenzene	3	UN1134	III	3	
	Coal tar distillates, flammable	3	UN1136	II	3	
		3	UN1136	III	3	
D	Consumer commodity	ORM-D			None	
	Cyclohexane	3	UN1145	II	3	
	Dichloromethane	6.1	UN1593	III	6.1	
D	Diesel fuel	3	NA1993	III	3	
I	Diesel fuel	3	UN1202	III	3	
	Dinitrogen tetroxide	2.3	UN1067		2.3, 5.1, 8	1
G	Elevated-temperature liquid, flammable, n.o.s., *with a flashpoint above 37.8°C, at or above its flashpoint*	3	UN3256		3	
	Environmentally hazardous substances, liquid, n.o.s.	9	UN3082	III	9	
	Environmentally hazardous substances, solid, n.o.s.	9	UN3077	III	9	
+	Epichlorohydrin	6.1	UN2023	II	6.1, 3	
	Ethanol and gasoline mixture *or* Ethanol and motor spirit mixture *or* Ethanol and petrol mixture, *with more than 10% ethanol*	3	UN3475	II	3	
	Ethyl acetate	3	UN1173	II	3	

SYMBOLS	HAZARDOUS MATERIALS DESCRIPTIONS AND PROPER SHIPPING NAMES	HAZARD CLASS OR DIVISION	IDENTIFICATION NUMBER	PACKING GROUP	LABEL CODES	SPECIAL PROVISIONS
(1)	(2)	(3)	(4)	(5)	(6)	(7)
	Ethyl mercaptan	3	UN2363	I	3	
	Ethyl methyl ketone *or* Methyl ethyl ketone	3	UN1193	II	3	
	Ethylbenzene	3	UN1175	II	3	
	Ethylene	2.1	UN1962		2.1	
	Explosive, blasting, type A	1.1D	UN0081	II	1.1D	
	Fireworks	1.1G	UN0333	II	1.1G	
	Fireworks	1.2G	UN0334	II	1.2G	
	Fireworks	1.3G	UN0335	II	1.3G	
	Fireworks	1.4G	UN0336	II	1.4G	
	Fireworks	1.4S	UN0337	II	1.4S	
G	Flammable liquids, n.o.s.	3	UN1993	I	3	
G	Flammable liquids, toxic, n.o.s.	3	UN1992	I	3, 6.1	
G	Flammable solids, toxic, organic, n.o.s.	4.1	UN2926	II	4.1, 6.1	
		4.1	UN2926	III	4.1, 6.1	
	Fluorine, compressed	2.3	UN1045		2.3, 5.1, 8	1
	Formic acid *with not less than 10% but not more than 85% acid by mass*	8	UN3412	II	8	
	Fuel, aviation, turbine engine	3	UN1863	I	3	
		3	UN1863	II	3	
		3	UN1863	III	3	
D	Fuel oil *(No. 1, 2, 4, 5 or 6)*	3	NA1993	III	3	
	Gas oil	3	UN1202	III	3	
	Gasoline *includes gasoline mixed with ethyl alcohol, with not more than 10% alcohol*	3	UN1203	II	3	
D, G	Hazardous waste, liquid, n.o.s.	9	NA3082	III	9	
D, G	Hazardous waste, solid, n.o.s.	9	NA3077	III	9	
	Hydrazine, anhydrous	8	UN2029	I	8, 3, 6.1	
	Hydrofluoric acid, anhydrous, see Hydrogen fluoride, anhydrous					
	Hydrogen, compressed	2.1	UN1049		2.1	
	Hydrogen fluoride, anhydrous	8	UN1052	I	8, 6.1	3

SYMBOLS	HAZARDOUS MATERIALS DESCRIPTIONS AND PROPER SHIPPING NAMES	HAZARD CLASS OR DIVISION	IDENTIFICATION NUMBER	PACKING GROUP	LABEL CODES	SPECIAL PROVISIONS
(1)	(2)	(3)	(4)	(5)	(6)	(7)
	Hydrogen, refrigerated liquid (cryogenic liquid)	2.1	UN1966		2.1	
	Hydrogen sulfide	2.3	UN1053		2.3, 2.1	2
G	Infectious substances, affecting humans	6.2	UN2814		6.2	
	Lime, unslaked, see Calcium oxide					
	Liquefied petroleum gas, see Petroleum gases, liquefied					
	Lithium aluminum hydride	4.3	UN1410	I	4.3	
	Lithium hydride	4.3	UN1414	I	4.3	
	London purple	6.1	UN1621	II	6.1	
	Lye, see Sodium hydroxide, solutions					
	Mercury oxide	6.1	UN1641	II	6.1	
	Methyl bromide	2.3	UN1062		2.3	3
	Methyl ethyl ketone, see Ethyl methyl ketone					
	Methyl isobutyl ketone	3	UN1245	II	3	
	Methyl isocyanate	6.1	UN2480	I	6.1, 3	1
	Methylene chloride, see Dichloromethane					
	Nickel carbonyl	6.1	UN1259	I	6.1, 3	1
+	Nitric acid, red fuming	8	UN2032	I	8, 5.1, 6.1	2
	Nitrogen, compressed	2.2	UN1066		2.2	
	Nitrogen dioxide, see Dinitrogen tetroxide					
	Nitrous oxide	2.2	UN1070		2.2, 5.1	
	Organic peroxide type A, liquid or solid	Forbidden				
G	Organic peroxide type B, solid	5.2	UN3102	II	5.2, 1	
G	Organic peroxide type B, solid, temperature-controlled	5.2	UN3112	II	5.2, 1	
G	Organic peroxide type D, liquid	5.2	UN3105	II	5.2	
G	Organic peroxide type D, liquid, temperature-controlled	5.2	UN3115	II	5.2	
	Organophosphorus pesticides, solid, toxic	6.1	UN2783	I	6.1	

SYMBOLS	HAZARDOUS MATERIALS DESCRIPTIONS AND PROPER SHIPPING NAMES	HAZARD CLASS OR DIVISION	IDENTIFICATION NUMBER	PACKING GROUP	LABEL CODES	SPECIAL PROVISIONS
(1)	(2)	(3)	(4)	(5)	(6)	(7)
		6.1	UN2783	II	6.1	
		6.1	UN2783	III	6.1	
D, G	Other regulated substance, liquid, n.o.s.	9	NA3082	III	9	
	Oxygen, compressed	2.2	UN1072		2.2, 5.1	
+	Oxygen generator, chemical spent	9	NA3356	III	9	
	Oxygen, refrigerated liquid *(cryogenic liquid)*	2.2	UN1073		2.2, 5.1	
	Paint-related material *including paint thinning, drying, removing, or reducing compound*	3	UN1263	I	3	
	Petroleum gases, liquefied *or Liquefied petroleum gas*	2.1	UN1075		2.1	
	Phosgene	2.3	UN1076		2.3, 8	1
	Phosphoric acid, solution	8	UN1805	III	8	
	Phosphorus, amorphous	4.1	UN1338	III	4.1	
	Phosphorus, white dry *or* Phosphorus, white, under water *or* Phosphorus, white, in solution, *or* Phosphorus, yellow dry *or* Phosphorus, yellow, under water *or* Phosphorus, yellow in solution	4.2	UN1381	I	4.2, 6.1	
	Phosphorus, white, molten	4.2	UN2447	I	4.2, 6.1	
	Phthalic anhydride *with more than 0.05% maleic anhydride*	8	UN2214	III	8	
I	Radioactive material, excepted package—empty packaging	7	UN2908		Empty	
I	Radioactive material, low specific activity (LSA-III), *nonfissile or fissile excepted*	7	UN3322		7	
I	Radioactive material, Type A package, *non-special form, nonfissile or fissile excepted*	7	UN2915		7	
	Radioactive material, uranium hexafluoride, fissile	7	UN2977		7, 8	
	Selenium disulfide	6.1	UN2657	II	6.1	
	Sodium	4.3	UN1428	I	4.3	

SYMBOLS	HAZARDOUS MATERIALS DESCRIPTIONS AND PROPER SHIPPING NAMES	HAZARD CLASS OR DIVISION	IDENTIFICATION NUMBER	PACKING GROUP	LABEL CODES	SPECIAL PROVISIONS
(1)	(2)	(3)	(4)	(5)	(6)	(7)
	Sodium hydroxide, solid	8	UN1823	II	8	
	Sodium hydroxide solution	8	UN1824	II	8	
		8	UN1824	III	8	
	Sodium peroxide	5.1	UN1504	I	5.1	
	Stannic chloride, anhydrous	8	UN1827	II	8	
D	Sulfur	9	NA1350	III	9	
I	Sulfur	4.1	UN1350	III	4.1	
D	Sulfur, molten	9	NA2448	III	9	
I	Sulfur, molten	4.1	UN2448	III	4.1	
	Sulfuryl fluoride	2.3	UN2191		2.3	4
G	Tear gas substances, liquid, n.o.s.	6.1	UN1693	I	6.1	
		6.1	UN1693	II	6.1	
	1,1,2,2-Tetrachloroethane	6.1	UN1702	II	6.1	
	Toluene	3	UN1294	II	3	
	1,1,1-Trichloroethane	6.1	UN2831	III	6.1	
	Trinitrotoluene and Trinitrobenzene mixtures *or* TNT and trinitrobenzene mixtures *or* TNT and hexanitrostilbene mixtures *or* Trinitrotoluene and hexanitrostilbene mixtures	1.1D	UN0388	II	1.1D	
	Trinitrotoluene, *or* TNT, *dry or wetted with less than 30% water by mass*	1.1D	UN0209	II	1.1D	
	Trinitrotoluene, *wetted with not less than 30% water by mass*	4.1	UN1356	I	4.1	
	Trinitrotoluene mixtures containing Trinitrobenzene and Hexanitrostilbene *or* TNT mixtures containing trinitrobenzene and hexanitrostilbene mixtures	1.1D	UN0389	II	1.1D	
	Xylenes	3	UN1307	II	3	
		3	UN1307	III	3	

[a]Excerpted from the DOT Hazardous Materials Table, 49 C.F.R. §172.101. In this excerpt, column 7 contains only the four codes immediately relevant to emergency response personnel. Columns 8 through 10 in the complete table have been redacted here in their entirety.

The excerpt is provided for illustrative purposes only and is current as of April 23, 2013. However, DOT periodically deletes entries, alters proper shipping names, and adds new hazardous materials to the listing. Consequently, when intended for use, shippers and carriers should consult the complete table to assure they are using the information accurately.

INDEX

Vesicants, 597–598
Vinyl chloride
  carcinogen, 401
  ill effects from inhalation, 629
Vinyl chloride-vinylidene chloride copolymer,
    626–627
Vinyl group, 477
Vinyl polymers
  description of, 626
  polyacrylamide, 631–632
  polyacrylonitrile, 630–631
  polyethylene, 627
  poly(methyl methacrylate), 631
  polypropylene, 628
  poly(vinyl chloride), 628–630
  production of, 626
Volatile liquids, 53
Volatile organic compounds (VOCs),
    232–233
Volume
  defined, 36
  units of measurement, 39–40
Volumetric expansion, 59–61
Volumetric thermal expansion
    coefficients, 60
Vulcanization, 638
Vulcanized rubber, 638
VX nerve agent, 596

emergency response actions, 338–342
  producing acetic acid vapor, 336–338
  producing hydrogen chloride, 332–336
  transportation of, 310
Weak acids, 273
Weak bases, 273
Weapons of mass destruction, 647
Weight, of matter, 36
Wet-chemical fire extinguishing agents,
    169–170
White phosphorus
  combustion, 253, 254
  production of, 252–253
  properties of, 252–253
  shipping descriptions, 254
WHMIS (Workplace Hazardous Materials
    Information System), 23–25
Wood alcohol. See Methanol
Wool, 625–626
Workplace Hazardous Materials Information
    System (WHMIS), 23–25
Workplace regulations (OSHA)
  acetic acid, 298
  acetic anhydride, 337
  alcohols, 550–551
  amines, 583
  ammonia, 389
  ammonium nitrate, 448–451
  aromatic hydrocarbons, 498–499
  asbestos, 406
  benzene, 498–499
  bulk oxygen systems, 227–228
  carbon dioxide fire extinguishers, 164
  carbon disulfide, 595
  carbon monoxide, 368
  carcinogens, 401
  chlorine, 248
  container hazard warnings, 280
  corrosive materials, 300–301
  emergency action plans, 14
  employee right-to-know law, 12
  explosives, 647
  flammable liquid storage, 84–90
  formaldehyde, 568–569
  gas cylinders, 78
  hydrogen, 244
  hydrogen cyanide, 374

hydrogen fluoride, 294
hydrogen peroxide, 430
hydrogen sulfide, 381
lead, 403
nitric acid, 286
nitrogen oxides, 386
nitroglycerin, 662
oxidizing chromium compounds, 455
ozone, 233
phosphine, 331
polynuclear aromatic hydrocarbons
    (PAHs), 503
radiation exposure, 701–703
radioactive materials, 686
safety data sheets (SDS), 13
sulfur dioxide, 378
sulfuric acid, 282
tetryl, 672
TNT, 669
toluene, 498–499
toxic substances, 347
warning labels, 12–13
xylene(s), 498–499

X

Xylene(s)
  ill effects from inhalation, 498
  physical properties, 496
  production of, 497
  transportation of, 499
  workplace regulations, 498–499

Y

Yellow phosphorus. See White phosphorus
Yellow section, ERG, 212
Yellowcake, 711

Z

Ziegler-Natta catalyst, 325
Zinc, metallic, 323
Zirconium, metallic, 319–320
Zones, hazard, 179, 361
Zyklon B, 373